Study Guide and Solutions Manual for
Organic Chemistry

Fourth Edition

Susan McMurry
Cornell University

Brooks/Cole Publishing Company
I(T)P™ An International Thomson Publishing Company

Pacific Grove ▪ Albany ▪ Bonn ▪ Boston ▪ Cincinnati ▪ Detroit ▪ London ▪ Madrid ▪ Melbourne
Mexico City ▪ New York ▪ Paris ▪ San Francisco ▪ Singapore ▪ Tokyo ▪ Toronto ▪ Washington

Sponsoring Editor: *Elizabeth Barelli Rammel*
Editorial Associate: *Beth Wilbur*
Production Coordinator: *Dorothy Bell*

Cover Design: *Vernon T. Boes*
Cover Illustration: *Ken Eward, BioGrafx*
Printing and Binding: *Quebecor Printing, Fairfield*

COPYRIGHT © 1996 by Brooks/Cole Publishing Company
A division of International Thomson Publishing Inc.
I(T)P™ The ITP logo is a trademark under license.

For more information, contact:

BROOKS/COLE PUBLISHING COMPANY
511 Forest Lodge Rd.
Pacific Grove, CA 93950
USA

International Thomson Editores
Campos Eliseos 385, Piso 7
Col. Polanco
11560 México D. F. México

International Thomson Publishing Europe
Berkshire House 168-173
High Holborn
London, WC1V 7AA
England

International Thomson Publishing GmbH
Königswinterer Strasse 418
53227 Bonn
Germany

Thomas Nelson Australia
102 Dodds Street
South Melbourne, 3205
Victoria, Australia

International Thomson Publishing Asia
221 Henderson Road
#05-10 Henderson Building
Singapore 0315

Nelson Canada
1120 Birchmount Road
Scarborough, Ontario
Canada M1K 5G4

International Thomson Publishing Japan
Hirakawacho Kyowa Building, 3F
2-2-1 Hirakawacho
Chiyoda-ku, Tokyo 102
Japan

Printed in the United States of America

10 9 8 7 6 5 4

ISBN 0-534-23833-5

Preface

What enters your mind when you hear the words "organic chemistry?" Some of you may think, "the chemistry of life," or "the chemistry of carbon." Other responses might include "pre-med, pressure," "difficult," or "memorization." Although formally the study of the compounds of carbon, organic chemistry encompasses many skills that are common to other areas of study. Organic chemistry is as much a liberal art as a science, and mastery of the concepts and techniques of organic chemistry can lead to an enhanced competence in other fields.

As you proceed to solve the problems that accompany the text, you will bring to the task many problem-solving techniques. For example, planning an organic synthesis requires the skills of a chess player; you must plan your moves while looking several steps ahead, and you must keep your plan flexible. Structure-determination problems are like detective problems, in which many clues must be assembled to yield the most likely solution. Naming organic compounds is similar to the systematic naming of biological specimens; in both cases, a set of rules must be learned and then applied to the specimen or compound under study.

The problems in the text fall into two categories: drill and complex. Drill problems, which appear throughout the text and at the end of each chapter, test your knowledge of one fact or technique at a time. You may need to rely on memorization to solve these problems, which you should work on first. More complicated problems require you to recall facts from several parts of the text and then use one or more of the problem-solving techniques mentioned above. As each major type of problem—synthesis, nomenclature, or structure determination—is introduced in the text, a solution is extensively worked out in this *Solutions Manual*.

Here are several suggestions that may help you with problem solving:

1. The text is organized into chapters that describe individual functional groups. As you study each functional group, *make sure that you understand the structure and reactivity of that group*. In case your memory of a specific reaction fails you, you can rely on your general knowledge of functional groups for help.

2. *Use molecular models.* It is difficult to visualize the three-dimensional structure of an organic molecule when looking at a two-dimensional drawing. Models will help you to appreciate the structural aspects of organic chemistry and are indispensable tools for understanding stereochemistry.

3. Every effort has been made to make this *Solutions Manual* as clear, attractive, and error-free as possible. Nevertheless, you should *use the Solutions Manual in moderation*. The principal use of this book should be to check answers to problems you have already worked out. The *Solutions Manual* should not be used as a substitute for effort; at times, struggling with a problem is the only way to teach yourself.

4. *Look through the appendices at the end of the Solutions Manual.* Some of these appendices contain tables that may help you in working problems; others present information related to the history of organic chemistry.

Acknowledgments I would like to thank my husband, John McMurry, for offering me the opportunity to write this book many years ago and for supporting my efforts while this edition was being prepared. My appreciation goes to Virginia Severn Goodman, Sonja Erion, Melba Wallace and Sherrie Yourstone, all of whom were involved in producing previous editions of this book. Many people at Brooks/Cole Publishing company have given me encouragement during this project; special thanks are due to Harvey Pantzis, Connie Jirovsky, Elizabeth Rammel and Joan Marsh. I am grateful to Elmer Ewing, my supervisor at Cornell University, for allowing me the flexible work schedule that I needed in order to finish this book. For this edition, all manuscript preparation was done by the author, using a Macintosh computer and the programs WordPerfect and ChemIntosh,which was easy to learn and fun to use. Finally, I would like to thank our seven-year-old son Paul McMurry, who patiently watched me work on this book, while wishing that he could be playing games on the computer instead.

Contents

Chapter 1 – Structure and Bonding

1.1 The elements of the periodic table are organized into groups that are based on the number of outer-shell electrons each element has. For example, an element in group 1A has one outer-shell electron, and an element in group 5A has five outer-shell electrons. To find the number of outer-shell electrons for a given element, use the periodic table to locate its group.

(a) Potassium (group 1A) has one electron in its outermost shell.
(b) Aluminum (group 3A) has three outer-shell electrons.
(c) Krypton is a noble gas and has eight electrons in its outermost shell.

1.2 a) To find the ground-state electron configuration of an element, first locate its atomic number. For boron, the atomic number is 5; boron thus has 5 protons and 5 electrons. Next, assign the electrons to the proper energy levels, starting with the lowest level:

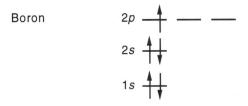

Remember that only two electrons can occupy the same orbital, and that they must be of opposite spin.

A different way to represent the ground-state electron configuration is to simply write down the occupied orbitals and to indicate the number of electrons in each orbital. For example, the electron configuration for boron is $1s^2 2s^2 2p$.

Often, we are interested only in the electrons in the outermost shell. We can then represent all filled levels by the symbol for the noble gas having the same levels filled. In the case of boron, the filled $1s$ energy level is represented by [He], and the valence shell configuration is symbolized by [He] $2s^2 2p$.

b) Let's consider an element with many electrons. Phosphorus, with an atomic number of 15, has 15 electrons. Assigning these to energy levels:

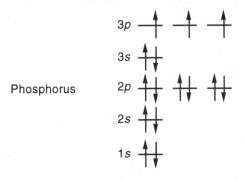

Notice that the 3p electrons are all in different orbitals. According to *Hund's rule*, we must place one electron into each orbital of the same energy level until all orbitals are half-filled.

The more concise way to represent ground-state electron configuration for phosphorus: $1s^2\,2s^2\,2p^6\,3s^2\,3p^3$ or [Ne] $3s^2\,3p^3$

c) Oxygen (atomic number 8)   d) Chlorine (atomic number 17)

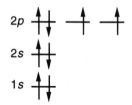

$1s^2\,2s^2\,2p^4$

[He] $2s^2\,2p^4$

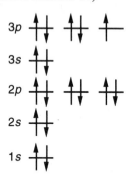

$1s^2\,2s^2\,2p^6\,3s^2\,3p^5$

[Ne] $3s^2\,3p^5$

1.3

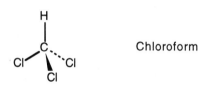

Chloroform

1.4   a) Carbon (group 4A) has four electrons in its valence shell and forms four bonds to achieve the noble-gas configuration of neon. A likely formula is CCl_4.

Element   Group Likely Formula

b) Al   3A AlH_3
c) C   4A CH_2Cl_2
d) Si   4A SiF_4
e) N   5A CH_3NH_2

1.5   Follow these three steps for drawing the Lewis structure of a molecule.

(1) Determine the number of valence, or outer-shell electrons for each atom in the molecule. For chloroform, we know that carbon has four valence electrons, hydrogen has one, and each chlorine has seven.

$$\cdot \overset{\displaystyle \cdot}{C} \cdot \qquad 4 \times 1 = 4$$

$$H\cdot \qquad 1 \times 1 = 1$$

$$:\overset{\displaystyle \cdot\cdot}{\underset{\displaystyle \cdot\cdot}{Cl}}\cdot \qquad \underline{7 \times 3 = 21}$$

$$26 \quad \text{total valence electrons}$$

(2) Next, use two electrons for each single bond.

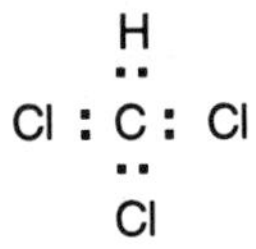

$$\text{Cl} : \overset{\displaystyle H}{\underset{\displaystyle Cl}{\overset{\cdot\cdot}{C}}} : \text{Cl}$$

(3) Finally, use the remaining electrons to achieve an noble gas configuration for all atoms.

Molecule	Lewis structure	Line-bond structure

a) $CHCl_3$

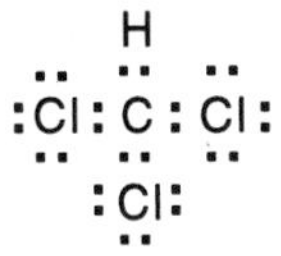

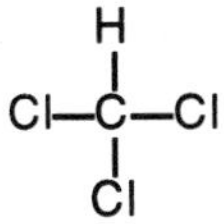

b) H_2S
    8 valence electrons

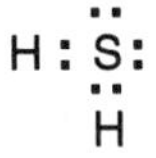

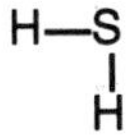

c) CH_3NH_2
    14 valence electrons

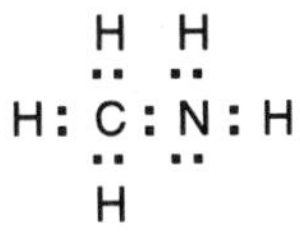

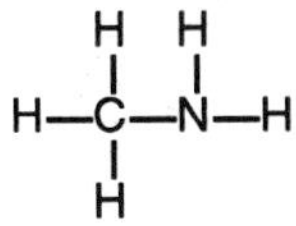

d) BH_3
    6 valence electrons

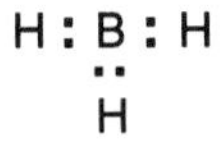

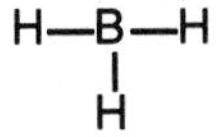

Borane can't achieve a noble-gas configuration because it has only six valence electrons.

e) NaH
    2 valence electrons

f) CH_3Li
    8 valence electrons

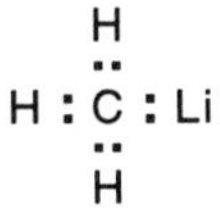

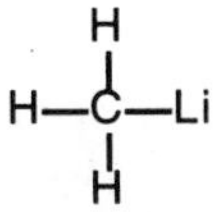

1.6 Bonds formed between an element on the right side of the periodic table and an element on the left side are ionic. Bonds formed between an element in the middle of the periodic table and another element are most often covalent.

    Ionic bonds: LiI, KBr, $MgCl_2$
    Covalent bonds: CH_4, CH_2Cl_2, Cl_2

1.7

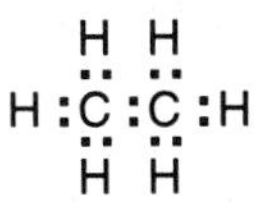

Ethane

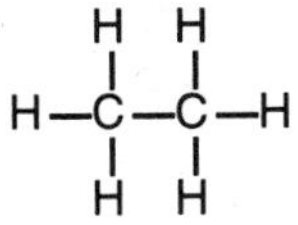

1.8

Propane

All carbon atoms are tetrahedral, and all bond angles are approximately 109.5°.

1.9   The two carbons bond to each other by overlap of two sp^3 hybrid orbitals. Six sp^3 hybrid orbitals (three from each carbon) are left over, and they can bond with a maximum of six hydrogens. Thus, a formula such as C_2H_7 is not possible.

1.10

Propene

The C3–H bonds are σ bonds formed by overlap of an sp^3 orbital of carbon 3 with an s orbital of hydrogen.

The C2–H and C1–H bonds are σ bonds formed by overlap of an sp^2 orbital of carbon with an s orbital of hydrogen.

The C2–C3 bond is a σ bond formed by overlap of an sp^3 orbital of carbon 3 with an sp^2 orbital of carbon 2.

There are two C1–C2 bonds. One is a σ bond formed by overlap of an sp^2 orbital of carbon 1 with an sp^2 orbital of carbon 2. The other is a π bond formed by overlap of a p orbital of carbon 1 with a p orbital of carbon 2. All four atoms connected to the carbon-carbon double bond lie in the same plane, and all bond angles between these atoms are 120°.

1.11

All atoms lie in the same plane, and all bond angles are approximately 120°

1.12

Acetaldehyde

1.13

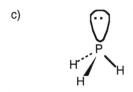

Propyne

The C3-H bonds are σ bonds formed by overlap of an sp^3 orbital of carbon 3 with an s orbital of hydrogen.

The C1-H bond is a σ bond formed by overlap of an sp orbital or carbon 1 with an s orbital of hydrogen.

The C2-C3 bond is a σ bond formed by overlap of an sp orbital of carbon 2 with an sp^3 orbital of carbon 3.

There are three C1-C2 bonds. One is a σ bond formed by overlap of an sp orbital of carbon 1 with an sp orbital of carbon 2. The other two bonds are π bonds formed by overlap of two p orbitals of carbon 1 with two p orbitals of carbon 2.

The three carbon atoms of propyne lie on a straight line; the bond angle is 180°.

1.14

H : C :: N : H   Formaldimine

Four electrons are shared in the carbon-nitrogen double bond. The nitrogen atom is sp^2 hybridized.

1.15

a)

The sp^3 -hybridized oxygen atom has tetrahedral geometry.

b)

Tetrahedral geometry

c)

Like nitrogen, phosphorus has five outer-shell electrons. PH_3 has tetrahedral geometry.

1.16

Element	Atomic Number	Number of valence electrons
a) Magnesium	12	2
b) Sulfur	16	6
c) Bromine	35	7

1.17

Element	Atomic Number	Ground-state Electron configuration
a) Sodium	11	$1s^2\,2s^2\,2p^6\,3s$
b) Aluminum	13	$1s^2\,2s^2\,2p^6\,3s^2\,3p$
c) Silicon	14	$1s^2\,2s^2\,2p^6\,3s^2\,3p^2$
d) Calcium	20	$1s^2\,2s^2\,2p^6\,3s^2\,3p^6\,4s^2$

1.18   (a) $AlCl_3$ (b) CF_2Cl_2 (c) NI_3

1.19

a)   H : C ::: C : H   10 valence electrons

b)   H : Al : H   6 valence electrons

c)   H : C : S : C : H   20 valence electrons

d)   : Cl : C : Cl :   24 valence electrons

e)   H : C :: C : C :: C : H   22 valence electrons

f)   H : C : C : O : H   24 valence electrons

1.20

H : C : C ::: N :   Acetonitrile

Nitrogen has five electrons in its outer electron shell. Three are used in the carbon-nitrogen triple bond, and two are a nonbonding electron pair.

1.21   The H_3C- carbon is sp^3 hybridized, and the $-CN$ carbon is sp hybridized.

1.22

1.23

a) 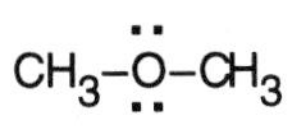 $CH_3-\overset{..}{\underset{..}{O}}-CH_3$

b) 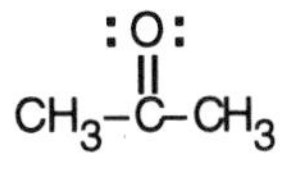 $CH_3-\overset{\overset{:O:}{\|}}{C}-CH_3$

c) $CH_3-\overset{\overset{:O:}{\|}}{C}-\overset{..}{N}H_2$

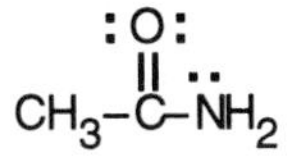

d) 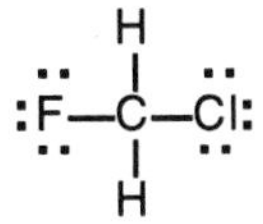 $:\overset{..}{\underset{..}{F}}-\overset{\overset{H}{|}}{\underset{\underset{H}{|}}{C}}-\overset{..}{\underset{..}{C}l}:$

1.24   In molecular formulas of organic molecules, carbon is listed first, followed by hydrogen. All other elements are listed in alphabetical order.

Compound	*Molecular Formula*
a) Phenol	C_6H_6O
b) Aspirin	$C_9H_8O_4$
c) Vitamin C	$C_6H_8O_6$
d) Nicotine	$C_{10}H_{14}N_2$
e) Novocain	$C_{13}H_{21}ClN_2O_2$
f) Glucose	$C_6H_{12}O_6$

1.25   To work a problem of this sort, you must examine all possible structures consistent with the rules of valence. You must systematically consider all possible attachments, including those that have branches, rings and multiple bonds.

a)

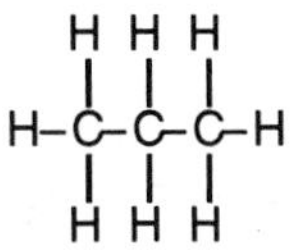

b)

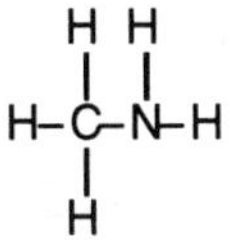

c) 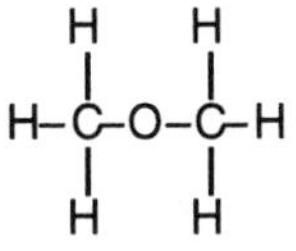   and

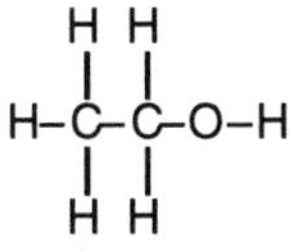

d) 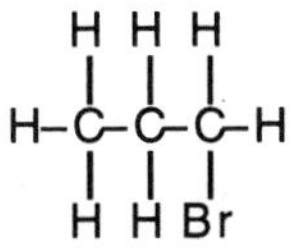   and

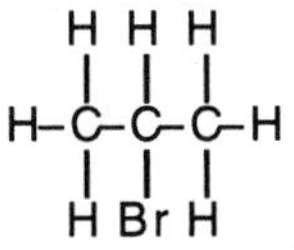

e)

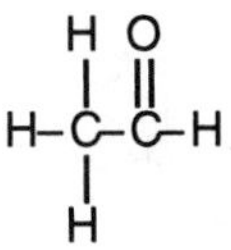

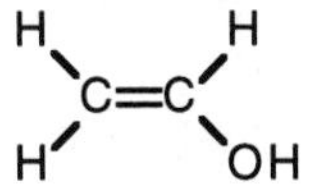

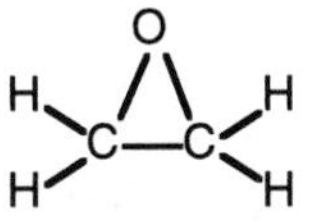

f)

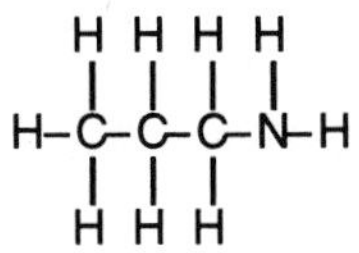

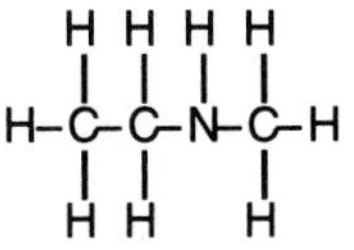

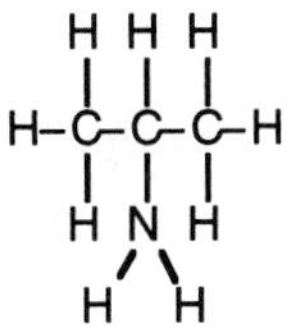

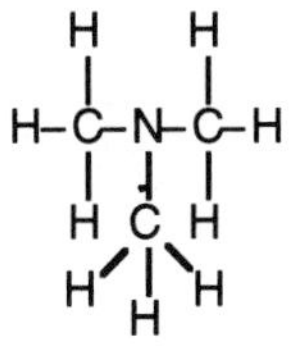

1.26

a)

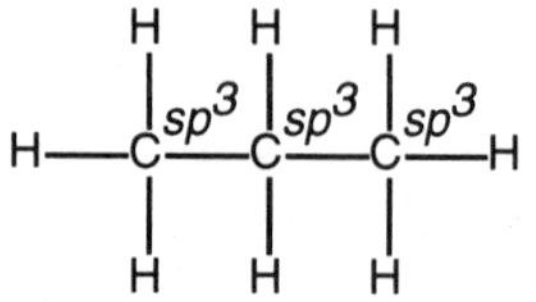

b)

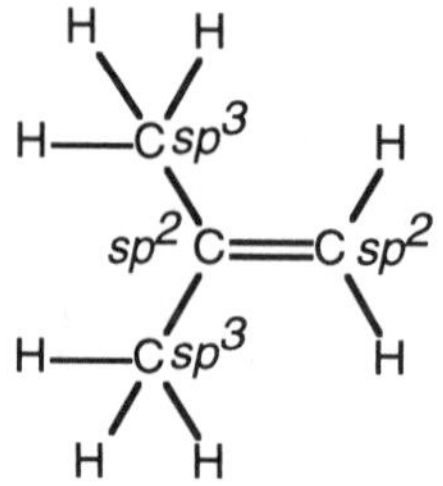

c)

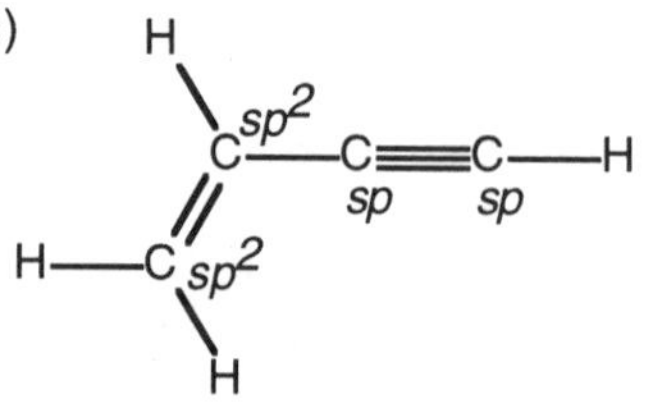

d)

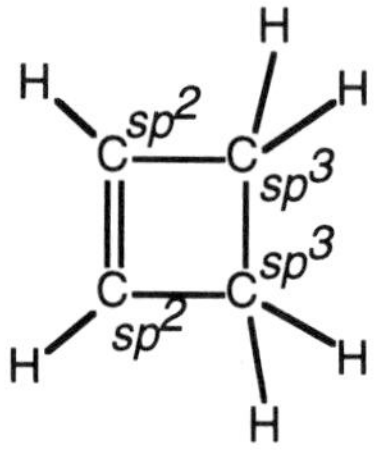

e)

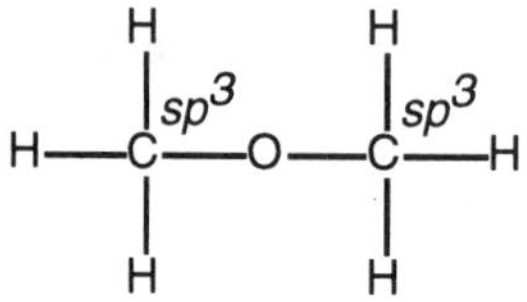

1.27

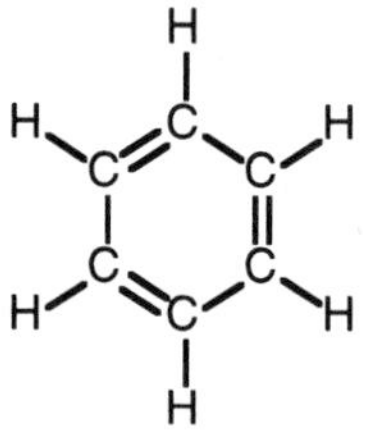

Benzene

All carbon atoms of benzene are sp^2 hybridized, and all bond angles of benzene are 120°. Benzene is a planar molecule.

1.28   All angles are approximate.
a) 109°   b) 109°   c) 109° d) 120°

1.29   a) sp^3 b) sp^3 c) sp^2

1.30

a)

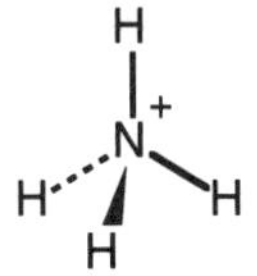

The ammonium ion is tetrahedral because nitrogen is sp^3 hybridized

b)

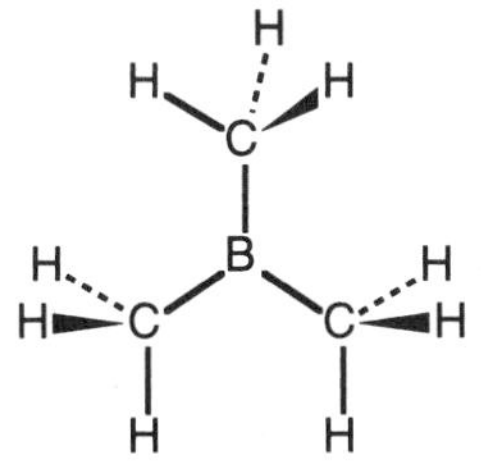

The boron-carbon portion of the molecule is planar because of sp^2 hybridization at boron. The $-CH_3$ portions are tetrahedral.

c)

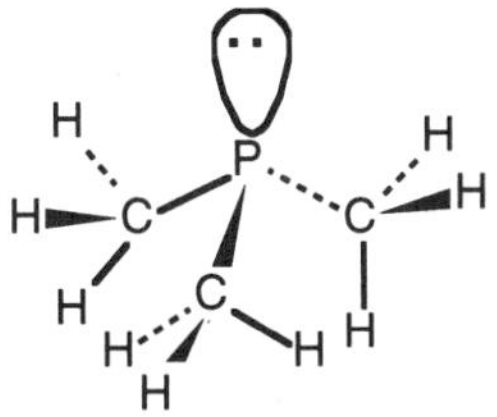

Trimethylphosphine is pyramidal. The $-CH_3$ portions are tetrahedral.

d)

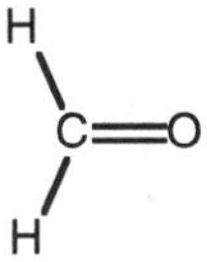

Formaldehyde is planar because carbon is sp^2 hybridized.

1.31

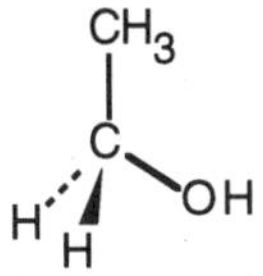

Ethanol

1.32 a) SO_2 has eighteen valence electrons (six from sulfur and six from each oxygen). The oxygen-sulfur-oxygen bond angle is approximately 120°.

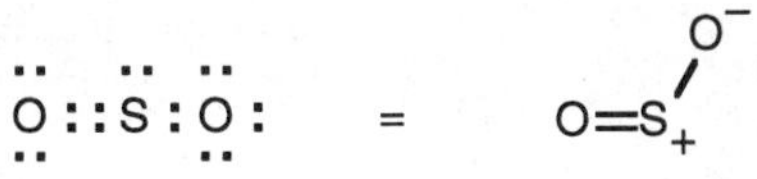

b) SO_3 has 24 valence electrons and is a planar molecule.

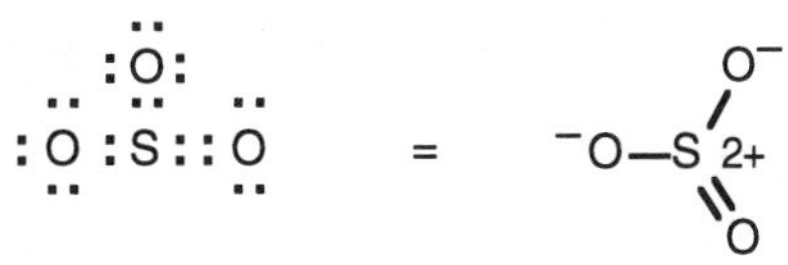

c) Four oxygens and one sulfur contribute 30 valence electrons. In addition, there are two electrons that give SO_4^{2-} its negative charge. The total number of electrons used to draw the Lewis structure is 32. SO_4^{2-} is a tetrahedral anion.

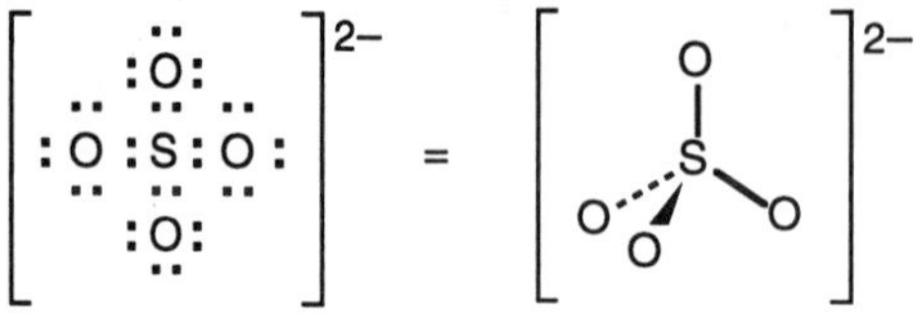

1.33

a)

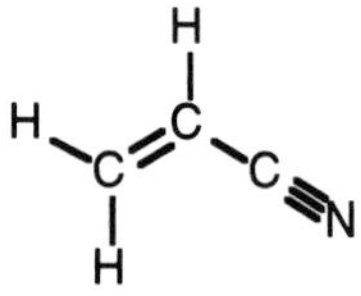

Acrylonitrile

b)

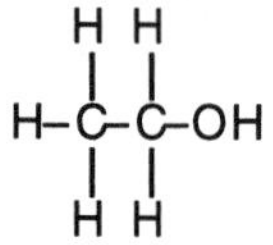

Ethanol

c)

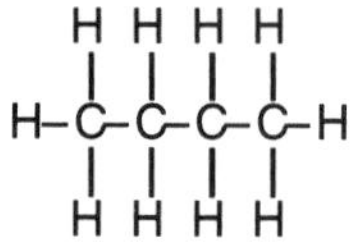

Butane

1.34

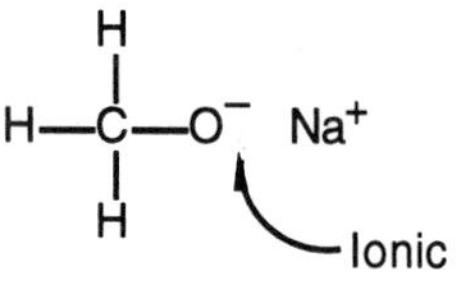

All other bonds are covalent.

1.35

a)

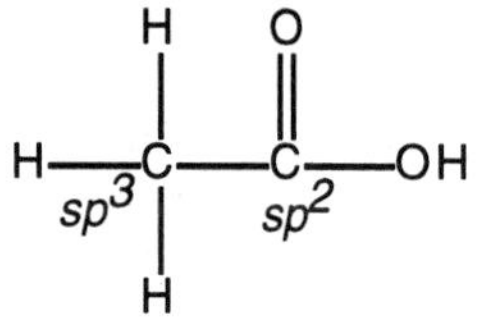

b)

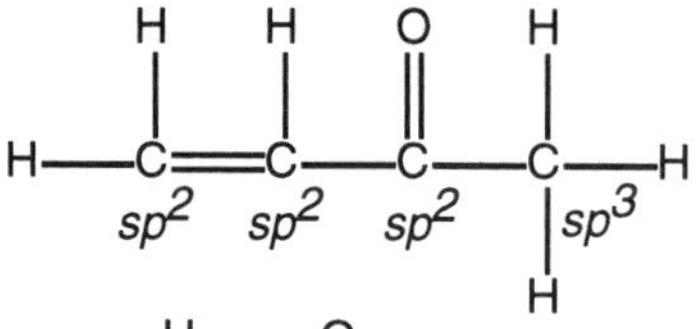

c)

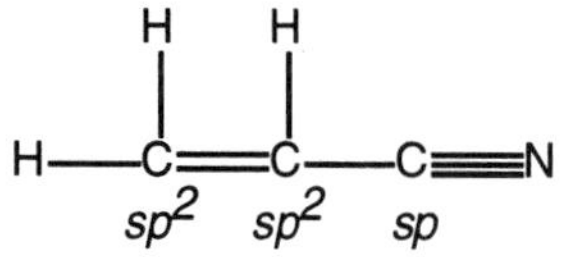

d)

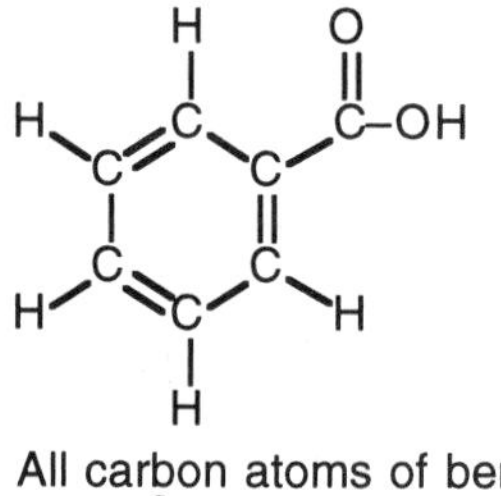

All carbon atoms of benzoic acid are sp^2 hybridized.

1.36 All angles are approximate.
a) 109°   b) 120°   c) 180°   d) 109°

1.37 a) sp^3 b) sp^2 c) sp d) sp^3

1.38 Ionic: NaCl
Covalent: CH_3Cl, Cl_2, HOCl

1.39

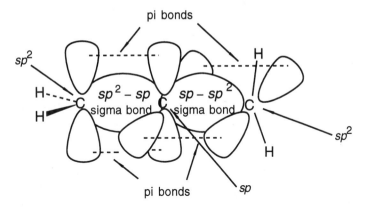

The central carbon of allene forms two σ bonds and two π bonds. The central carbon is *sp*–hybridized, and the two terminal carbons are *sp²*–hybridized. The carbon-carbon bond angle is 180°, indicating linear geometry for the carbons.

1.40

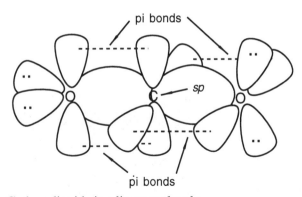

Carbon dioxide is a linear molecule.

1.41

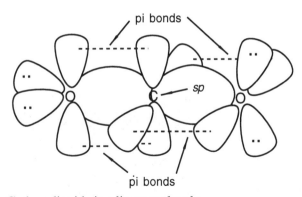

All the indicated atoms are *sp²* hybridized.

1.42 a) A carbocation is isoelectronic with (has the same number of electrons as) a trivalent boron compound.
   b) The positively charged carbon atom has six valence shell electrons.
   c) A carbocation is *sp²*–hybridized.
   d) A carbocation is planar.

1.43

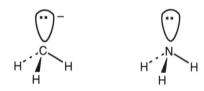

The negatively charged carbanion carbon has eight valence electrons and is sp^3–hybridized. A carbanion is tetrahedral and is isoelectronic with a trivalent nitrogen compound.

1.44   According to the Pauli Exclusion Principle, two electrons in the same orbital must have opposite spins. Thus, the two electrons of triplet (spin-unpaired) methylene must occupy different orbitals. In triplet methylene, sp–hybridized carbon forms one bond to each of two hydrogens. Each of the two unpaired electrons occupies a p orbital. In singlet (spin-paired) methylene the two electrons can occupy the same orbital because they have opposite spins. Including the two C–H bonds, there are a total of three occupied orbitals. We predict sp^2 hybridization and planar geometry for singlet methylene.

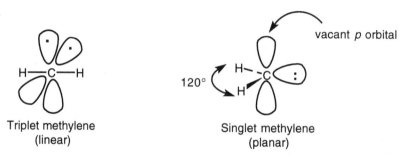

Triplet methylene
(linear)

Singlet methylene
(planar)

1.45   a) $CH_3CH_2CH=CH_2$ b) $CH_2=CHCH=CH_2$ c) $CH_2=CHC\equiv CH$

1.46

$CH_3CH_2CH_2CH_3$ $CH_3\overset{\overset{\displaystyle CH_3}{|}}{C}HCH_3$

The two compounds differ in the way that the carbon atoms are connected.

1.47

$CH_3CH=CH_2$ $H_2C\overset{\overset{\displaystyle CH_2}{\diagup\ \diagdown}}{—}CH_2$

One compound has a double bond, and one has a ring.

1.48

CH_3CH_2OH CH_3OCH_3

The two compounds differ in the location of the oxygen atom.

Study Guide for Chapter 1

After studying this chapter, you should be able to:

(1) Predict the ground-state electron configuration of atoms (Problems 1.1, 1.2, 1.17, 1.18).

(2) Draw Lewis electron-dot structures of simple compounds (Problems 1.5, 1.7, 1.12, 1.14, 1.19, 1.20, 1.22, 1.23, 1.32, 1.41).

(3) Draw simple organic compounds with the correct three-dimensional geometry (1.3, 1.31).

(4) Identify bonds as ionic or covalent (1.6, 1.34, 1.38).

(5) Predict and describe the hybridization of bonds in simple compounds (1.8, 1.10, 1.11, 1.13, 1.21, 1.26, 1.27, 1.28, 1.35, 1.36, 1.42, 1.43, 1.44, 1.45).

(6) Predict bond angles and shapes of molecules (1.15, 1.29, 1.30, 1.37, 1.39, 1.40).

(7) Convert Kekulé structures into molecular formulas, and vice versa (1.24, 1.25, 1.33).

Chapter 2 – Bonding and Molecular Properties

2.1 Use Figure 2.2 to answer this problem. The larger the number, the more electronegative the element.

More electronegative	Less electronegative
a) H (2.1)	Li (1.0)
b) Br (2.8)	Be (1.6)
c) Cl (3.0)	I (2.5)
d) C (2.5)	H (2.1)

Carbon is slightly more electronegative than hydrogen.

2.2 As in Problem 2.1, use Figure 2.2.

a) H_3C—Br
  δ^+ δ^-

b) H_3C—NH_2
  δ^+ δ^-

c) H_3C—Li
  δ^- δ^+

d) H_2N—H
  δ^- δ^+

e) H_3C—OH
  δ^+ δ^-

f) H_3C—MgBr
  δ^- δ^+

g) H_3C—F
  δ^+ δ^-

2.3

a) Carbon: EN = 2.5
   Lithium: EN = 1.0
   _____
   $\Delta EN = 1.5$

b) Carbon: EN = 2.5
   Potassium: EN = 0.8
   _____
   $\Delta EN = 1.7$

c) Fluorine: EN = 4.0
   Carbon: EN = 2.5
   _____
   $\Delta EN = 1.5$

d) Carbon: EN = 2.5
   Magnesium: EN = 1.2
   _____
   $\Delta EN = 1.3$

e) Oxygen   EN = 3.5
   Carbon: EN = 2.5
   _____
   $\Delta EN = 1.0$

The most polar bond has the largest ΔEN. Thus, in order of increasing bond polarity:

$$H_3C—OH < H_3C—MgBr < H_3C—Li < H_3C—F < H_3C—K$$

2.4 We must look at the polarization of the individual bonds of a molecule to account for an observed dipole moment. In the case of methanol, CH_3OH, the individual bond polarities can be estimated from Figure 2.2. A bond is polarized in the direction of the more electronegative element (the larger numbers in Figure 2.2). In addition, we must take into account the contribution of the two lone pairs of oxygen. Indicating the individual bond polarities by arrows, we can predict the direction of the dipole moment.

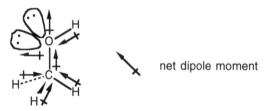

net dipole moment

It would be difficult to calculate the dipole moment from the individual bond moments. It is often possible, however, to estimate qualitatively the direction and relative magnitude of the dipole moment by estimating the net direction of the bond polarities.

2.5

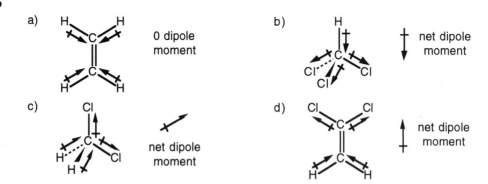

The dipole moment of CO_2 is zero because the bond polarities of the two carbon-oxygen bonds cancel.

2.6

a)   0 dipole moment

b)   net dipole moment

c)   net dipole moment

d)   net dipole moment

2.7

$$\text{Formal charge (FC)} = \begin{bmatrix} \text{\# of valence} \\ \text{electrons} \end{bmatrix} - \begin{bmatrix} \dfrac{\text{\# of bonding electrons}}{2} \end{bmatrix} - \begin{bmatrix} \text{\# non-bonding} \\ \text{electrons} \end{bmatrix}$$

For sulfur, FC   = $6 - \dfrac{6}{2} - 2 = +1$

For oxygen, FC   = $6 - \dfrac{2}{2} - 6 = -1$

2.8

$$\text{Formal charge (FC)} = \left[\begin{array}{c}\text{\# of valence}\\ \text{electrons}\end{array}\right] - \left[\frac{\text{\# of bonding electrons}}{2}\right] - \left[\begin{array}{c}\text{\# non-bonding}\\ \text{electrons}\end{array}\right]$$

a) $H_2C{=}N{=}\overset{..}{N}:$ The Lewis dot structure is $H:\overset{\overset{H}{\underset{..}{}}}{C}::\overset{1}{N}::\overset{2}{\underset{..}{N}}:$

     Hydrogen   FC $=$ $1 - \dfrac{2}{2} - 0 = 0$

     Carbon   FC $=$ $4 - \dfrac{8}{2} - 0 = 0$

     Nitrogen 1   FC $=$ $5 - \dfrac{8}{2} - 0 = +1$

     Nitrogen 2   FC $=$ $5 - \dfrac{4}{2} - 4 = -1$

Remember: <u>Valence electrons</u> are the electrons particular to a specific element. <u>Bonding electrons</u> are those electrons involved in bonding to other atoms. <u>Nonbonding electrons</u> are those electrons in lone pairs.

b) $CH_3{-}C{\equiv}N{-}O \;\;\equiv\;\;$ $H:\overset{\overset{1}{\underset{..}{H}}}{\underset{\underset{H}{}}{C}}:\overset{2}{C}:::N:\overset{..}{\underset{..}{O}}:$

     Hydrogen   FC $=$ $1 - \dfrac{2}{2} - 0 = 0$

     Carbon 1   FC $=$ $4 - \dfrac{8}{2} - 0 = 0$

     Carbon 2   FC $=$ $4 - \dfrac{8}{2} - 0 = 0$

     Nitrogen   FC $=$ $5 - \dfrac{8}{2} - 0 = +1$

     Oxygen   FC $=$ $6 - \dfrac{2}{2} - 6 = -1$

c) $H:\overset{\overset{1}{\underset{..}{H}}}{\underset{\underset{H}{}}{C}}:N:::\overset{2}{C}:$

     Hydrogen   FC $=$ $1 - \dfrac{2}{2} - 0 = 0$

     Carbon 1   FC $=$ $4 - \dfrac{8}{2} - 0 = 0$

     Carbon 2   FC $=$ $4 - \dfrac{6}{2} - 2 = -1$

     Nitrogen   FC $=$ $5 - \dfrac{8}{2} - 0 = +1$

2.9

a)

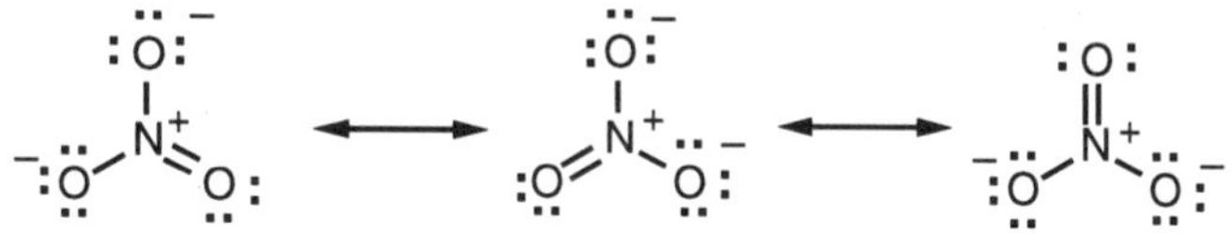

b)

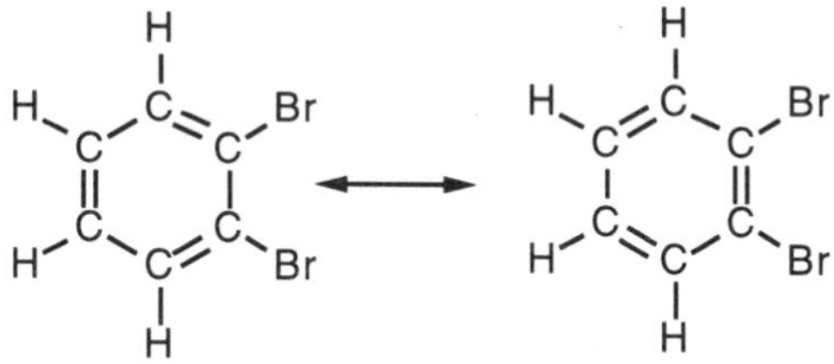

c) $H_2C=CH-CH_2^+$ ⟷ $H_2\overset{+}{C}-CH=CH_2$

d) :N≡N—Ṅ̈—H ⟷ :N̈=N=N̈—H

2.10 Recall from Section 2.6 that a strong acid has a low pK_a and a weak acid has a high pK_a. Accordingly, picric acid ($pK_a = 0.38$) is a stronger acid than formic acid ($pK_a = 3.75$).

2.11 HO–H is a stronger acid than H_2N–H. Since H_2N^- is a stronger base than HO^-, the conjugate acid of H_2N^- (H_2N–H) is a weaker acid than the conjugate acid of HO^- (HO–H).

2.12

a)   H–CN + $CH_3COO^-Na^+$ $\xrightarrow{\text{?}}$ $Na^{+-}CN$ + CH_3COOH

$pK_a = 9.3$
Weaker acid

$pK_a = 4.7$
Stronger acid

Remember that the lower the pK_a, the stronger the acid. Thus CH_3COOH, not HCN, is the stronger acid, and the above reaction will not take place in the direction written.

b) CH_3CH_2O–H + $Na^+\ ^-CN$ $\xrightarrow{\text{?}}$ $CH_3CH_2O^-\ Na^+$ + H–CN

$pK_a = 16$
Weaker acid

$pK_a = 9.3$
Stronger acid

Using the same reasoning as in part a), we can see that the above reaction will not occur.

2.13

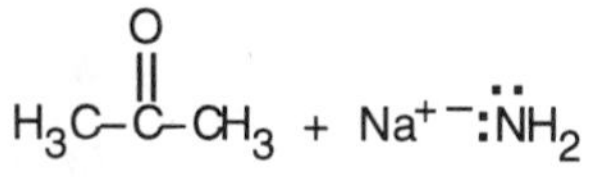

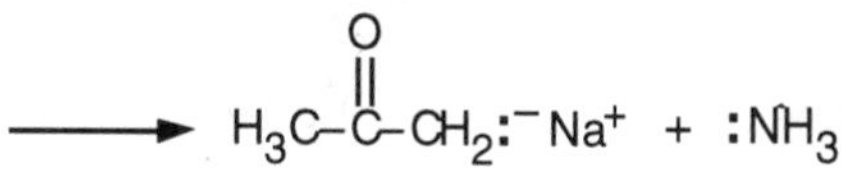

$pK_a = 19$
Stronger acid

$pK_a = 36$
Weaker acid

The above reaction will take place as written.

2.14 Enter –9.31 into the calculator and use the INV LOG function to arrive at the answer $K_a = 4.9 \times 10^{-10}$.

2.15

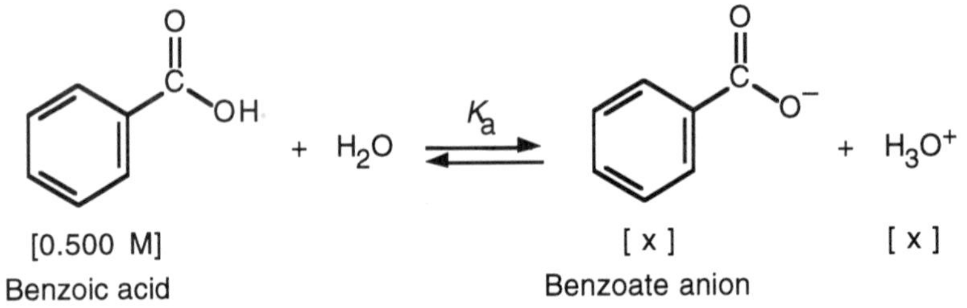

| [0.500 M] | [x] | [x] |
| Benzoic acid | Benzoate anion | |

$$K_a = \frac{[\text{Benzoate}][\text{H}_3\text{O}^+]}{[\text{Benzoic acid}]} = \frac{x^2}{0.500 - x}$$

Because x is small, $0.500 - x$ is approximately equal to 0.500.
Since $pK_a = 4.19$, $K_a = 6.46 \times 10^{-5}$.

$$6.46 \times 10^{-5} = \frac{x^2}{0.500}$$

$$32.3 \times 10^{-6} = x^2$$

$$5.68 \times 10^{-3} = [\text{H}_3\text{O}^+]; \text{pH} = 2.25$$

2.16

a)

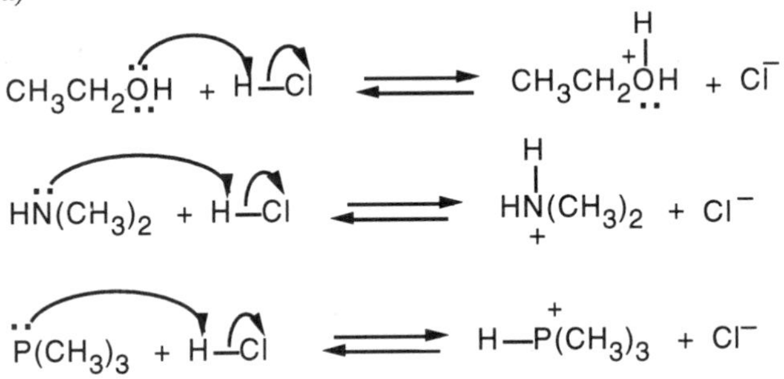

b)

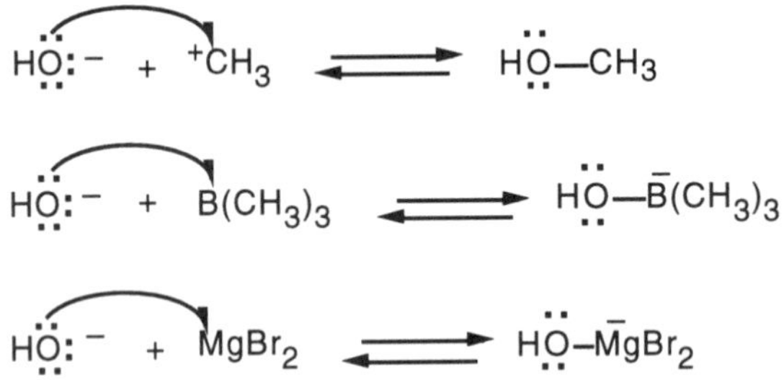

2.17

a)

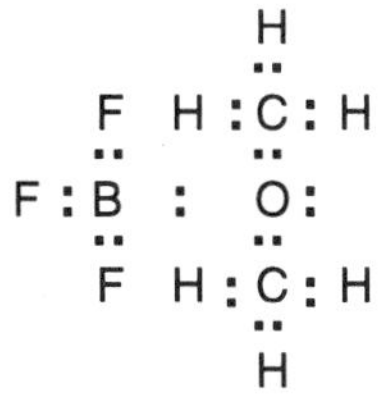

$$
\begin{array}{ccc}
 & & H \\
 & & \ddot{} \\
F & H:C:H \\
\ddot{} & & \ddot{} \\
F:B & : & O: \\
\ddot{} & & \ddot{} \\
F & H:C:H \\
 & & \ddot{} \\
 & & H
\end{array}
$$

boron: $FC = 3 - \dfrac{8}{2} - 0 = -1$

oxygen: $FC = 6 - \dfrac{6}{2} - 2 = +1$

The formal charge of –1 for boron indicates that boron has a net negative charge; oxygen has a net positive charge.

b)

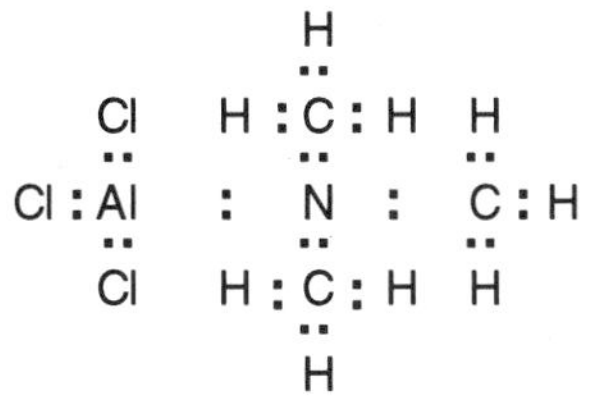

$$
\begin{array}{ccccc}
 & & H \\
 & & \ddot{} \\
Cl & H:C:H & H \\
\ddot{} & \ddot{} & \ddot{} \\
Cl:Al & : & N & : & C:H \\
\ddot{} & \ddot{} & \ddot{} \\
Cl & H:C:H & H \\
 & & \ddot{} \\
 & & H
\end{array}
$$

aluminum: $FC = 3 - \dfrac{8}{2} - 0 = -1$

nitrogen: $FC = 5 - \dfrac{8}{2} - 0 = +1$

The formal charge of –1 for aluminum indicates that it has a net negative charge; the formal charge +1 for nitrogen indicates that it has a net positive charge.
For clarity, electron dots have been left off F and Cl in the above structures.

2.18

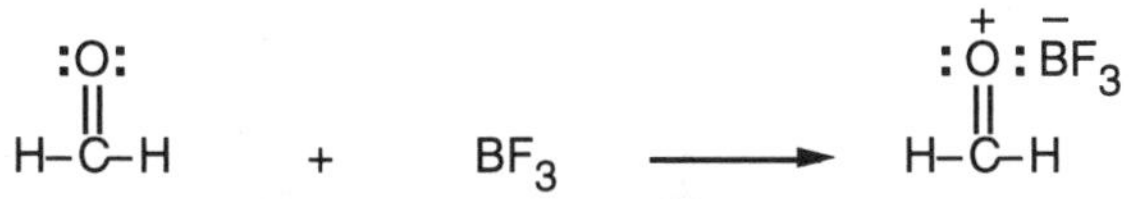

:O:
‖
H–C–H   + BF_3 $\longrightarrow$:Ö:$\overset{+}{}\overset{-}{}BF_3$
                                                         ‖
Lewis base   Lewis acid   H–C–H

2.19 a) This problem will be worked out in detail, using the rules in the text.

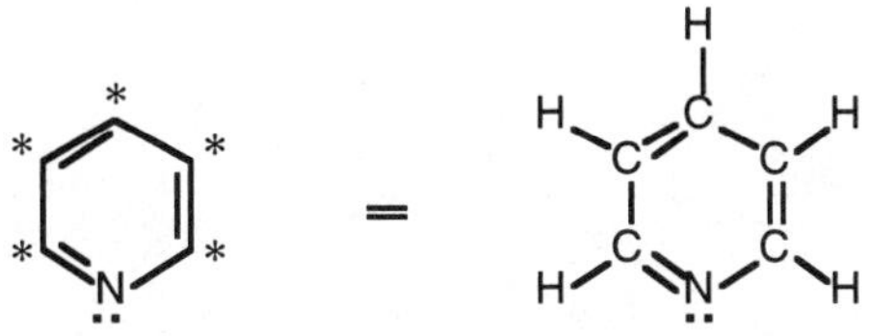

Remember:
(1) carbon atoms occur at the intersection of two line-bonds (indicated by * above);
(2) to satisfy valency, a hydrogen is understood to be bonded to each of the above carbons. The molecular formula of pyridine is C_5H_5N.

b) Cyclohexanone: $C_6H_{10}O$
c) Indole: C_8H_7N

2.20 Several possible shorthand structures can satisfy each molecular formula.

a) C_5H_{12}

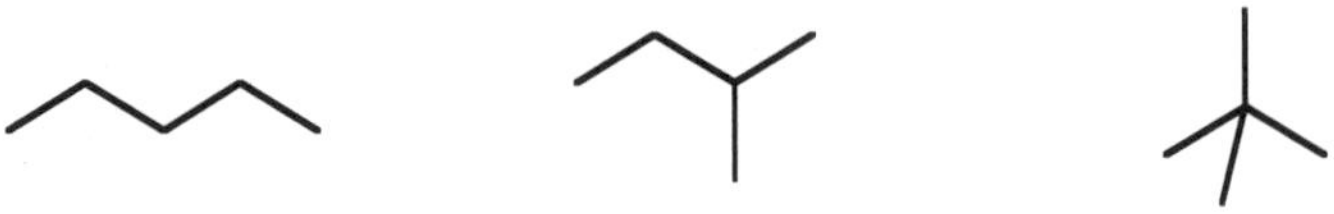

b) C_2H_7N

c) C_3H_6O

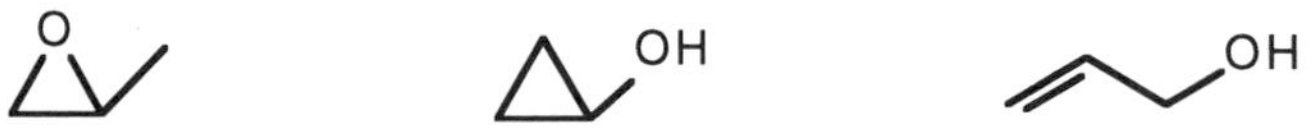

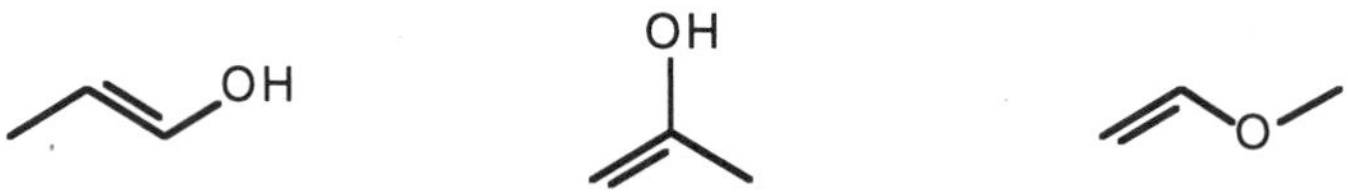

d) C_4H_9Cl

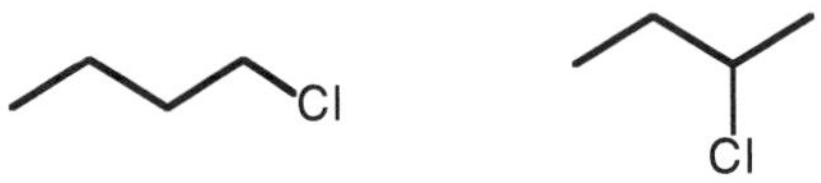

2.22

a)

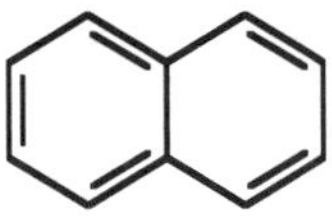

b)

c)

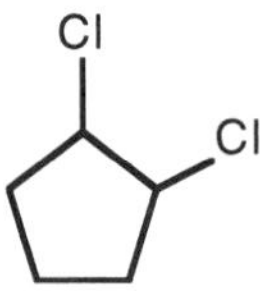

d)

2.23

a)

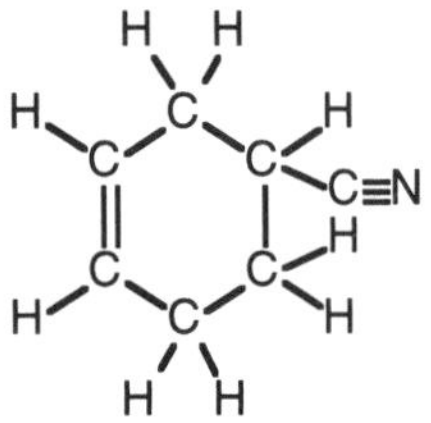

b)

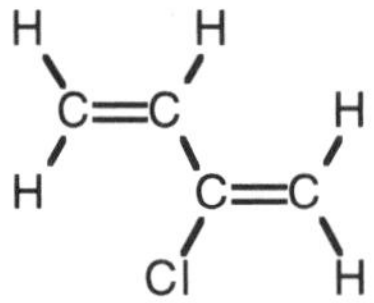

c)

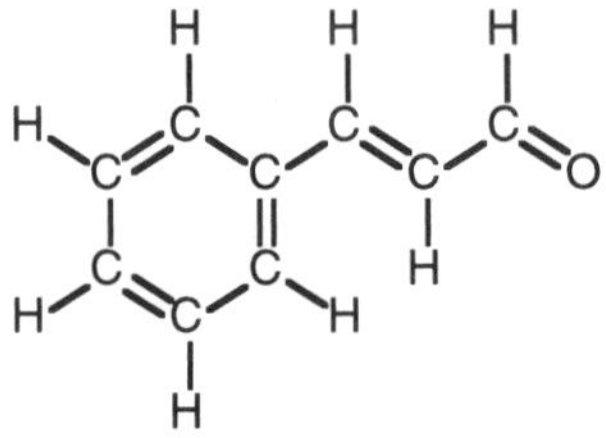

d)

2.24

a) $(CH_3)_2\overset{..}{O}BF_3$

oxygen	FC	$= 6 - \dfrac{6}{2} - 2 = +1$
boron	FC	$= 3 - \dfrac{8}{2} - 0 = -1$

b) $H_2\overset{..}{C}-\overset{1}{N}\equiv\overset{2}{N}:$

carbon	FC	$= 4 - \dfrac{6}{2} - 2 = -1$
nitrogen 1	FC	$= 5 - \dfrac{8}{2} - 0 = +1$
nitrogen 2	FC	$= 5 - \dfrac{6}{2} - 2 = 0$

c) $H_2C=\overset{1}{N}=\overset{2}{\underset{..}{N}}:$

carbon	FC	$= 4 - \dfrac{8}{2} - 0 = 0$
nitrogen 1	FC	$= 5 - \dfrac{8}{2} - 0 = +1$
nitrogen 2	FC	$= 5 - \dfrac{4}{2} - 4 = -1$

d) $:\overset{1}{\underset{..}{O}}=\overset{2}{\underset{..}{O}}-\overset{3}{\underset{..}{O}}:$

oxygen 1	FC	$= 6 - \dfrac{4}{2} - 4 = 0$
oxygen 2	FC	$= 6 - \dfrac{6}{2} - 2 = +1$
oxygen 3	FC	$= 6 - \dfrac{2}{2} - 6 = -1$

e) $H_2\overset{..}{C}-\overset{\overset{\displaystyle CH_3}{|}}{\underset{\underset{\displaystyle CH_3}{|}}{P}}-CH_3$

carbon	FC	$= 4 - \dfrac{6}{2} - 2 = -1$
phosphorus	FC	$= 5 - \dfrac{8}{2} - 0 = +1$

f) (pyridine)$N-\overset{..}{\underset{..}{O}}:$

nitrogen	FC	$= 5 - \dfrac{8}{2} - 0 = +1$
oxygen	FC	$= 6 - \dfrac{2}{2} - 6 = -1$

2.25–2.26

More polar	Less polar

a) $H_3C \overset{+\!\!\longrightarrow}{\underset{\delta^+ \quad \delta^-}{\rule{2.5em}{0pt}}} Cl$ Cl–Cl

b) $H \overset{+\!\!\longrightarrow}{\underset{\delta^+ \quad \delta^-}{\rule{2em}{0pt}}} Cl$ $H_3C \overset{\longleftarrow\!\!+}{\underset{\delta^- \quad \delta^+}{\rule{2em}{0pt}}} H$

c) $HO \overset{\longleftarrow\!\!+}{\underset{\delta^- \quad \delta^+}{\rule{2.5em}{0pt}}} CH_3$ $(CH_3)_3Si \overset{+\!\!\longrightarrow}{\underset{\delta^+ \quad \delta^-}{\rule{2.5em}{0pt}}} CH_3$

d) $Li \overset{+\!\!\longrightarrow}{\underset{\delta^+ \quad \delta^-}{\rule{2.5em}{0pt}}} OH$ $H_3C \overset{\longleftarrow\!\!+}{\underset{\delta^- \quad \delta^+}{\rule{2.5em}{0pt}}} Li$

2.27

a)

b) no dipole moment

c) $\overset{+\!\!\longrightarrow}{Li-H}$

d) $F_3\overset{-}{B} \overset{\longleftarrow\!\!+}{\rule{2.5em}{0pt}} \overset{+}{N}(CH_3)_3$

e)

f)

2.28   In phosgene, the individual bond polarities tend to cancel, but in formaldehyde, the bond polarities reinforce one another. Thus, phosgene has a smaller dipole moment.

Phosgene   Formaldehyde

2.29 $\mu = Q \times r$. For a proton and an electron separated by 1.00 Å, $\mu = 4.8$ D. If the two charges are separated by 1.36 Å, $\mu = 6.53$ D. Since the observed dipole moment is 1.08 D, the H—Cl bond has (1.08 D / 6.53 D x 100 %) = 16.5 % ionic character.

2.30   The magnitude of a dipole moment depends on both charge and distance between atoms. Fluorine is more electronegative than chlorine, but because a C–F bond is shorter than a C–Cl bond, the dipole moment of CH_3F is smaller than that of CH_3Cl.

2.31 Use Figure 2.2 for electronegativities. The most electronegative atom is starred.

a) CH₂FĊl

b) ḞCH₂CH₂CH₂Br

c) HŎCH₂CH₂NH₂

d) CH₃ŎCH₂Li

2.32 Resonance forms do not differ in the position of nuclei. The two structures in (a) are not resonance forms because the carbon and hydrogen atoms outside the ring occupy different positions in each structure.

not resonance structures

The pairs of structures in parts (b), (c), and (d) are resonance forms.

2.33

a)

b)

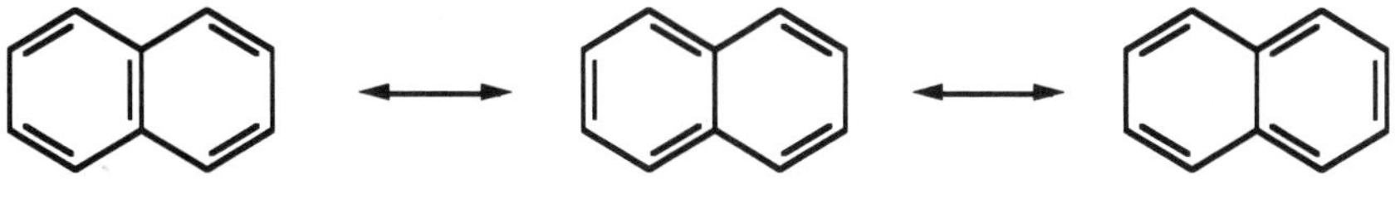

c)

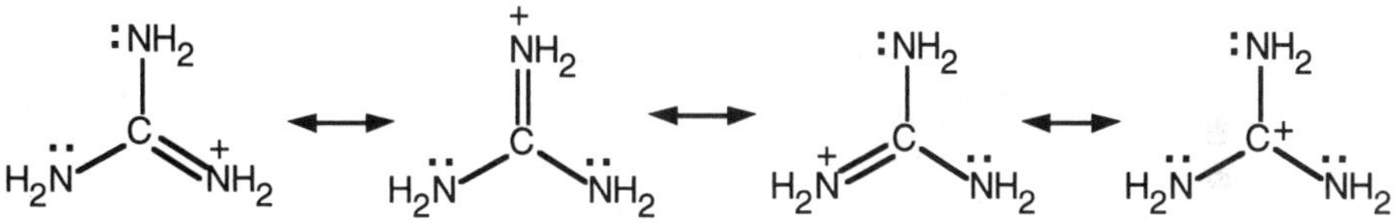

d) H₃C–S̈–ĊH₂ ⟷ H₃C–S̈=CH₂

e) H₂C=CH–ĊH₂ ⟷ H₂Ċ–CH=CH₂

f) H₂C=CH–CH=CH–ĊH–CH₃ ⟷ H₂C=CH–ĊH–CH=CH–CH₃

H₂Ċ–CH=CH–CH=CH–CH₃

2.34 The two structures are not resonance forms because the position of the carbon atoms is different in each form.

2.35

CH₃ŎH + HCl ⇌ CH₃ŎH₂⁺ + Cl⁻

CH₃ŎH + Na⁺⁻ N̈H₂ ⇌ CH₃Ö:⁻Na⁺ + N̈H₃

2.36

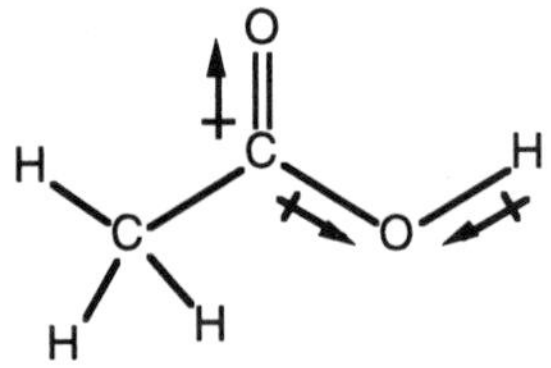

The O–H hydrogen of acetic acid is more acidic than the C–H hydrogens. The –OH oxygen is quite electronegative, and, consequently, the –O–H bond is more highly polarized than the –C–H bonds.

2.37

Lewis acids: $AlBr_3$, BH_3, HF, $TiCl_4$

Lewis bases: $CH_3CH_2\overset{..}{N}H_2$, $CH_3–\overset{..}{\underset{..}{S}}–CH_3$

2.38

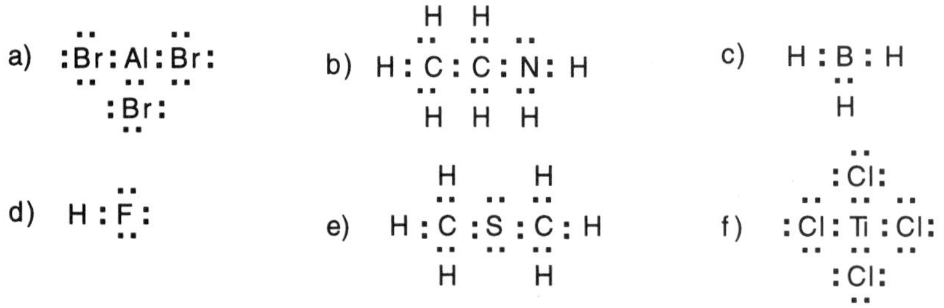

a) :Br:Al:Br:
          :Br:

b) H H
    H:C:C:N:H
    H H H

c) H:B:H
        H

d) H:F:

e) H   H
    H:C:S:C:H
    H   H

f) :Cl:
    :Cl:Ti:Cl:
    :Cl:

2.39

a) CH_3OH + H_2SO_4 ⇌ $CH_3OH_2^+$ + HSO_4^-
   stronger   stronger   weaker   weaker
    base   acid   acid   base

b) CH_3OH + $NaNH_2$ ⇌ $CH_3O^-Na^+$ + NH_3
   stronger   stronger   weaker   weaker
    acid   base   base   acid

c) $CH_3NH_3{}^+Cl^-$ + NaOH ⇌ CH_3NH_2 + H_2O + NaCl
   stronger   stronger   weaker   weaker
    acid   base   base   acid

2.40

a) $H_3C–\overset{CH_3}{\underset{CH_3}{\overset{+}{N}}}–\overset{..}{\underset{..}{O}}:^-$

b) $H_3C–\overset{\overset{-}{..}}{N}–\overset{+}{N}≡N:$

c) $H_3C–\overset{..}{N}=\overset{+}{N}=\overset{..}{\underset{..}{N}}:^-$

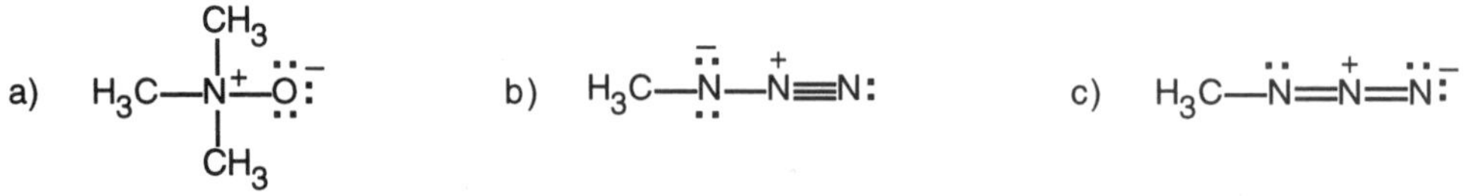

2.41

Least acidic ———————————————————————→ Most acidic

$$CH_3\overset{\displaystyle O}{\overset{\|}{C}}CH_3 \quad < \quad \langle \rangle\!\!-\!\!OH \quad < \quad CH_3\overset{\displaystyle O}{\overset{\|}{C}}CH_2\overset{\displaystyle O}{\overset{\|}{C}}CH_3 \quad < \quad CH_3\overset{\displaystyle O}{\overset{\|}{C}}OH$$

$pK_a = 19$ $\qquad\qquad pK_a = 9.9$ $\qquad\qquad pK_a = 9$ $\qquad\qquad pK_a = 4.7$

2.42 To react completely with NaOH, an acid must have a pK_a somewhat lower than the pK_a of H_2O. Thus, all substances in the previous problem except acetone will react completely with NaOH.

2.43 The stronger the acid (lower pK_a), the weaker its conjugate base. Since NH_4^+ is a stronger acid than $CH_3NH_3^+$, NH_3 is a weaker base than CH_3NH_2.

2.44

$$H_3C\!-\!\overset{\displaystyle CH_3}{\underset{\displaystyle CH_3}{\overset{|}{\underset{|}{C}}}}\!-\!O^-K^+ \; + \; H_2O \longrightarrow H_3C\!-\!\overset{\displaystyle CH_3}{\underset{\displaystyle CH_3}{\overset{|}{\underset{|}{C}}}}\!-\!OH \; + \; K^+\; {}^-OH$$

$pK_a = 15.7$ $\qquad\qquad\qquad\qquad pK_a \approx 18$
stronger acid $\qquad\qquad\qquad\qquad$ weaker acid

The reaction will take place as written because water is a stronger acid than *tert*–butyl alcohol. Thus, a solution of potassium *tert*-butoxide in water can't be prepared.

2.45 a) Acetone: $K_a = 5 \times 10^{-20}$ $\qquad\qquad$ b) Formic acid: $K_a = 1.8 \times 10^{-4}$

2.46 a) Nitromethane: $pK_a = 10.30$ $\qquad\qquad$ b) Acrylic acid: $pK_a = 4.25$

2.47

$$\text{Formic acid} \; + \; H_2O \;\underset{\longleftarrow}{\overset{K_a}{\rightleftharpoons}}\; \text{Formate}^- \; + \; H_3O^+$$
[0.050 M] $\qquad\qquad\qquad\qquad$ [x] $\qquad$ [x]

$$K_a = 1.8 \times 10^{-4} = \frac{x^2}{0.050 - x}$$

If you let $0.050 - x = 0.050$, then $x = 3.0 \times 10^{-3}$ and pH = 2.52. If you calculate x exactly, then $x = 2.9 \times 10^{-3}$ and pH = 2.54.

2.48 Only acetic acid will react with sodium bicarbonate.

2.49 Sodium bicarbonate reacts with acetic acid to produce carbonic acid, which breaks down to form CO_2. The resulting CO_2 bubbles indicate the presence of acetic acid. Phenol does not react with sodium bicarbonate.

2.50

a) $CH_3\ddot{O}H$ + H^+ $\longrightarrow$ $CH_3\overset{+}{\underset{\cdot\cdot}{O}}H_2$

  base   acid

b) $CH_3\ddot{O}H$ + $:\bar{N}H_2$ $\longrightarrow$ $CH_3\ddot{\underset{\cdot\cdot}{O}}:^-$ + $:NH_3$

  acid   base

c)

  base   acid

d)

  acid   base

e)

  base   acid

f) $(CH_3)_3\overset{+}{O}\overset{-}{BF_4}$ +

  acid   base

2.51 Pairs (a) and (d) represent resonance structures; pairs (b) and (c) do not. For (b) and (c), a proton differs in position in each structure of the pair.

2.52

a) $H_3C-\overset{+}{N}$ ... $\longleftrightarrow$ ... $H_3C-\overset{+}{N}$ ...

b) $:\overset{+}{O}=\overset{\cdot\cdot}{\underset{\cdot\cdot}{O}}-\overset{\cdot\cdot}{\underset{\cdot\cdot}{\bar{O}}}:$ $\longleftrightarrow$ $:\overset{\cdot\cdot}{\underset{\cdot\cdot}{\bar{O}}}-\overset{+}{O}=\overset{\cdot\cdot}{O}:$

c) $H_2C=\overset{+}{N}=\overset{-}{\underset{\cdot\cdot}{N}}:$ $\longleftrightarrow$ $H_2\overset{-}{C}-\overset{+}{N}\equiv N:$ $\longleftrightarrow$ $H_2\overset{\cdot\cdot}{\underset{\cdot\cdot}{C}}-\overset{+}{N}\equiv N:$

2.53

$$:\overset{..}{\underset{..}{O}}:^-$$
$$H_3C{-}\overset{\displaystyle |}{\underset{\displaystyle |}{S}}{}^{2+}{-}CH_3$$
$$:\overset{..}{\underset{..}{O}}:^-$$

FC of each oxygen $= 6 - \dfrac{2}{2} - 6 = -1$

FC of sulfur $= 6 - \dfrac{8}{2} - 0 = +2$

The presence of formal charges indicates that electrons are not shared equally between S and O; they are strongly attracted to oxygen. The S–O bonds, therefore, are strongly polar. If the geometry of the molecule were planar (as drawn above), the dipole moments of the individual S–O bonds would cancel, resulting in a net dipole moment of zero. Tetrahedral geometry, however, predicts a large dipole moment.

2.54

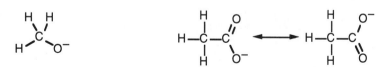

a)   b)   c)   d)

2.55

Acetic acid is much more acidic than methanol because the resulting acetate anion is stabilized by resonance.

Study Guide for Chapter 2

After studying this chapter, you should be able to:

(1)   Predict which of a pair of elements is more electronegative (2.1, 2.2, 2.3, 2.31).

(2)   Predict the direction of polarity of a chemical bond, and predict the dipole moment of a compound (2.4, 2.5, 2.6, 2.26, 2.27, 2.28, 2.29, 2.30, 2.31, 2.36).

(3)   Calculate the formal charge for atoms in molecules (2.7, 2.8, 2.17, 2.24, 2.40, 2.53).

(4)   Draw resonance forms of molecules (2.9, 2.32, 2.33, 2.34, 2.51, 2.52, 2.55).

(5)   Predict the relative acid/base strength of Brønsted acids and bases (2.10, 2.11, 2.41, 2.42, 2.43).

(6)   Predict the direction of Brønsted acid-base reactions (2.12, 2.13, 2.35, 2.39, 2.44, 2.48, 2.49).

(7)   Calculate:
(a) pK_a from K_a, and vice-versa (2.14, 2.45, 2.46).
(b) pH of a solution of a weak acid (2.15, 2.47).

(8)   Identify Lewis acids and bases (2.16, 2.18, 2.37, 2.38, 2.50).

(9)   Draw chemical structures from molecular formulas, and vice-versa (2.19, 2.20, 2.22, 2.23).

3.1

a)

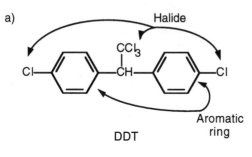

Halide

CCl_3

Cl— —CH— —Cl

Aromatic ring

DDT

b)

Amine

NH_2

$CH_2CH-COOH$

Carboxylic acid

Aromatic ring

Phenylalanine

c)

Aldehyde

CHO

Double bond

Acrolein

d)

Double bond

Aromatic ring

Styrene

3.2

a) CH_3OH

Methanol

b)

CH_3

Toluene

c) CH_3COOH

Acetic acid

d) CH_3NH_2

Methylamine

e)

$$O$$
$$\|$$
$CH_3CCH_2NH_2$

Aminoacetone

f)

1,3–Butadiene

3.3 We know that carbon forms four bonds and hydrogen forms one. Thus, if you draw all possible six carbon skeletons and add hydrogens so that all carbons have four bonds, you will arrive at the following structures:

$CH_3CH_2CH_2CH_2CH_2CH_3$

CH_3
$|$
$CH_3CH_2CH_2CHCH_3$

CH_3
$|$
$CH_3CH_2CHCH_2CH_3$

CH_3
$|$
$CH_3CH_2CCH_3$
$|$
CH_3

CH_3
$|$
$CH_3CHCHCH_3$
$|$
CH_3

3.4   This problem becomes easier when you realize that the isomers can be alcohols and ethers. A systematic approach to this type of problem is helpful. Let's start with the alcohol isomers.

1) Draw the simplest long-chain parent alkane. Here, the alkane is butane, $CH_3CH_2CH_2CH_3$.
2) Find the number of different sites to which a functional group may be attached. For butane, two different sites are possible ($-CH_3$ and $-CH_2-$).
3) At each different site, replace an –H by an –OH and draw the isomer.

$$CH_3CH_2CH_2CH_2-OH \quad \text{and} \quad CH_3CH_2\overset{\overset{\displaystyle OH}{|}}{C}HCH_3$$

4) Draw the simplest branched C_4H_{10} alkane.

$$CH_3\overset{\overset{\displaystyle CH_3}{|}}{C}HCH_3$$

5) Find the number of different sites. (There are two for the above alkane.)
6) For each site, replace an –H with an –OH and draw the isomer.

$$CH_3\overset{\overset{\displaystyle CH_3}{|}}{C}HCH_2-OH \quad \text{and} \quad CH_3\overset{\overset{\displaystyle CH_3}{|}}{\underset{\underset{\displaystyle OH}{|}}{C}}CH_3$$

7) Proceed with the next simplest branched C_4H_{10} alkane. In this problem, we have already drawn all alcohol isomers.

For the ethers, start with the $-OCH_3$ isomers. There are two possible sites for attachment of an $-OCH_3$ group to propane ($CH_3CH_2CH_3$) and thus there are two $-OCH_3$ isomers.

$$CH_3CH_2CH_2-OCH_3 \quad \text{and} \quad CH_3\overset{\overset{\displaystyle OCH_3}{|}}{C}HCH_3$$

Finally, there is one $-OCH_2CH_3$ ether, diethyl ether, $CH_3CH_2OCH_2CH_3$.

3.5   a) Nine isomeric esters of formula $C_5H_{10}O_2$ can be drawn.

$$CH_3CH_2CH_2\overset{\overset{\displaystyle O}{||}}{C}OCH_3 \qquad CH_3\overset{\overset{\displaystyle O}{||}}{\underset{\underset{\displaystyle CH_3}{|}}{C}H}COCH_3 \qquad CH_3CH_2\overset{\overset{\displaystyle O}{||}}{C}OCH_2CH_3$$

$$CH_3\overset{\overset{\displaystyle O}{||}}{C}OCH_2CH_2CH_3 \qquad CH_3\overset{\overset{\displaystyle O}{||}}{C}O\overset{\overset{\displaystyle CH_3}{|}}{C}HCH_3 \qquad H\overset{\overset{\displaystyle O}{||}}{C}OCH_2CH_2CH_2CH_3$$

$$\underset{\text{HCOCHCH}_2\text{CH}_3}{\overset{\text{O CH}_3}{\parallel \ \ \ |}}$$

$$\underset{\text{HCOCH}_2\text{CHCH}_3}{\overset{\text{O} \ \ \ \ \text{CH}_3}{\parallel \ \ \ \ \ \ |}}$$

$$\underset{\underset{\text{CH}_3}{|}}{\overset{\text{O CH}_3}{\parallel \ \ |}}{\text{HCOCCH}_3}$$

b) $CH_3CH_2CH_2C{\equiv}N$

$$\underset{\text{CH}_3\text{CHC}{\equiv}\text{N}}{\overset{\overset{\text{CH}_3}{|}}{}}$$

3.6   a) Two alcohols have the formula C_3H_8O.

$CH_3CH_2CH_2OH$

$$\underset{\text{CH}_3\text{CHCH}_3}{\overset{\overset{\text{OH}}{|}}{}}$$

b) Four bromoalkanes have the formula C_4H_9Br. Refer to the solution to Problem 3.4 if you need help.

$CH_3CH_2CH_2CH_2Br$

$$\underset{\text{CH}_3\text{CH}_2\text{CHCH}_3}{\overset{\overset{\text{Br}}{|}}{}}$$

$$\underset{\text{CH}_3\text{CHCH}_2\text{Br}}{\overset{\overset{\text{CH}_3}{|}}{}}$$

$$\underset{\underset{\text{Br}}{|}}{\overset{\overset{\text{CH}_3}{|}}{\text{CH}_3\text{CCH}_3}}$$

3.7

$CH_3CH_2CH_2CH_2CH_2{-}\xi$

$$\underset{\underset{\text{CH}_3}{|}}{\text{CH}_3\text{CH}_2\text{CH}_2\text{CH}{-}\xi}$$

$$\underset{\underset{\text{CH}_2\text{CH}_3}{|}}{\text{CH}_3\text{CH}_2\text{CH}{-}\xi}$$

$$\underset{\underset{\text{CH}_3}{|}}{\text{CH}_3\text{CH}_2\text{CHCH}_2{-}\xi}$$

$$\underset{\underset{\text{CH}_3}{|}}{\text{CH}_3\text{CHCH}_2\text{CH}_2{-}\xi}$$

$$\underset{\underset{\text{CH}_3}{|}}{\overset{\overset{\text{CH}_3}{|}}{\text{CH}_3\text{CH}_2\text{C}{-}\xi}}$$

$$\underset{\underset{\text{CH}_3}{|}}{\overset{\overset{\text{CH}_3}{|}}{\text{CH}_3\text{CHCH}{-}\xi}}$$

$$\underset{\underset{\text{CH}_3}{|}}{\overset{\overset{\text{CH}_3}{|}}{\text{CH}_3\text{CCH}_2{-}\xi}}$$

3.8

a) $$\underset{\underset{t}{\uparrow \ \underset{\text{CH}_3}{|}}}{\overset{\overset{\text{CH}_3}{|} \ \nearrow t}{\text{CH}_3\text{CHCHCH}_3}}$$

b) $\boxed{CH_3CHCH_3}$
      $CH_3CH_2CHCH_2CH_3$

c) $$\underset{\underset{s}{\uparrow \ \underset{\text{CH}_3}{|}}}{\overset{q \searrow \ \text{CH}_3}{\text{CH}_3\text{CH}_2\text{CCH}_3}}$$

3.9

a) $$\overset{p}{\underset{\underset{p \ \ t \ \ s \ \ s \ \ p}{\text{CH}_3\text{CHCH}_2\text{CH}_2\text{CH}_3}}{\overset{\overset{\text{CH}_3}{|}}{}}}$$

b) $$\overset{p \ \ t \ \ p}{\text{CH}_3\text{CHCH}_3} \\ \underset{p \ \ s \ \ t \ \ s \ \ p}{\text{CH}_3\text{CH}_2\text{CHCH}_2\text{CH}_3}$$

c) $$\underset{\underset{p}{\text{CH}_3}}{\overset{p}{\text{CH}_3}} \ \ \ \underset{\underset{p}{\text{CH}_3}}{\overset{p}{\text{CH}_3}} \\ \underset{p \ \ t \ \ s}{\text{CH}_3\text{CHCH}_2}{-}\underset{q \ p}{\overset{}{\text{C}}}{-}\text{CH}_3$$

p = primary; s = secondary; t = tertiary; q = quaternary

3.10

a) $$\overset{p}{\underset{\underset{p \ \ t \ \ s \ \ s \ \ p}{\text{CH}_3\text{CHCH}_2\text{CH}_2\text{CH}_3}}{\overset{\overset{\text{CH}_3}{|}}{}}}$$

b) $$\overset{p \ \ t \ \ p}{\text{CH}_3\text{CHCH}_3} \\ \underset{p \ \ s \ \ t \ \ s \ \ p}{\text{CH}_3\text{CH}_2\text{CHCH}_2\text{CH}_3}$$

c) $$\underset{\underset{p}{\text{CH}_3}}{\overset{p}{\text{CH}_3}} \ \ \ \underset{\underset{p}{\text{CH}_3}}{\overset{p}{\text{CH}_3}} \\ \underset{p \ \ t \ \ s}{\text{CH}_3\text{CHCH}_2}{-}\underset{q \ p}{\overset{}{\text{C}}}{-}\text{CH}_3 \\ \underset{p \ \text{CH}_3}{}$$

3.11

a) $CH_3CH_2CH_2CH_2CH_3$ $CH_3CH_2\overset{\overset{\displaystyle CH_3}{|}}{C}HCH_3$ $CH_3\overset{\overset{\displaystyle CH_3}{|}}{\underset{\underset{\displaystyle CH_3}{|}}{C}}CH_3$

Pentane   2–Methylbutane   2,2–Dimethylpropane

b)

The longest chain is a *hexane*.

The substituents are: 3–methyl, 4–methyl.

The IUPAC name is 3,4–dimethylhexane.

c) $(CH_3)_2CHCH_2\overset{\overset{\displaystyle CH_3}{|}}{C}HCH_3$ d) $(CH_3)_3CCH_2CH_2\overset{\overset{\displaystyle CH_3}{|}}{\underset{\underset{\displaystyle CH_2CH_3}{|}}{C}}H$

2,4–Dimethylpentane

2,2,5–Trimethylheptane

3.12

a)   3,4–Dimethylnonane $CH_3CH_2CH_2CH_2CH_2\overset{\overset{\displaystyle CH_3}{|}}{\underset{\underset{\displaystyle CH_3}{|}}{C}}HCHCH_2CH_3$

b)   3–Ethyl–4,4–dimethylheptane $CH_3CH_2CH_2\overset{\overset{\displaystyle CH_3}{|}}{\underset{\underset{\displaystyle CH_3}{|}}{C}}\text{———}CHCH_2CH_3 \atop \quad\quad\quad\quad\quad CH_2CH_3$

c)   2,2–Dimethyl–4–propyloctane $CH_3CH_2CH_2CH_2\overset{}{\underset{\underset{\displaystyle CH_2CH_2CH_3}{|}}{C}}HCH_2C(CH_3)_3$

d)   2,2,4–Trimethylpentane $CH_3\overset{\overset{\displaystyle CH_3}{|}}{C}HCH_2\overset{\overset{\displaystyle CH_3}{|}}{\underset{\underset{\displaystyle CH_3}{|}}{C}}CH_3$

3.13

a)

The longest chain is a *hexane*.
The correct name is 2–methylhexane.

b)

The longest chain is an *octane*. The
correct name is 4,5–dimethyloctane.

c)

$$CH_3\overset{\overset{\displaystyle CH_3}{|}}{\underset{\underset{\displaystyle CH_3}{|}}{C}}\!\!-\!\!\overset{}{\underset{\underset{\displaystyle CH_2CH_3}{|}}{C}}HCH_2CH_3$$

The numbering of substituents should start from the other end of the chain. Also, substituents should be cited in alphabetical order. The correct name is 3–ethyl–2,2–dimethylpentane.

d)

$$CH_3\overset{\overset{\displaystyle CH_3}{|}}{\underset{\underset{\displaystyle CH_2CH_3}{|}}{CH}}CHCH_2CH_2CH_3$$

The longest chain is a *heptane;* the numbering of substituents should start from the other end. The correct name is 3,4–dimethylheptane.

e)

$$CH_3CH_2CH_2\overset{\overset{\displaystyle CH_3}{|}}{\underset{\underset{\displaystyle CH_3}{|}}{CH}}CHCH_3$$

The name must include the prefix "di-" when two substituents are the same. The correct name is 2,3–dimethylhexane.

f)

$$CH_3CH_2\overset{\overset{\displaystyle CH_3}{|}}{\underset{\underset{\displaystyle CH_3}{|}}{C}}CH_2CH_3$$

The number "3" must be repeated in the name. The correct name is 3,3–dimethylpentane.

3.14

$$CH_3CH_2CH_2CH_2CH_2\text{—}$$

Pentyl

$$CH_3CH_2CH_2\overset{}{\underset{\underset{\displaystyle CH_3}{|}}{CH}}\text{—}$$

1–Methylbutyl

$$CH_3CH_2\overset{}{\underset{\underset{\displaystyle CH_2CH_3}{|}}{CH}}\text{—}$$

1–Ethylpropyl

$$CH_3CH_2\overset{}{\underset{\underset{\displaystyle CH_3}{|}}{CH}}CH_2\text{—}$$

2–Methylbutyl

$$CH_3\overset{}{\underset{\underset{\displaystyle CH_3}{|}}{CH}}CH_2CH_2\text{—}$$

3–Methylbutyl

$$CH_3CH_2\overset{\overset{\displaystyle CH_3}{|}}{\underset{\underset{\displaystyle CH_3}{|}}{C}}\text{—}$$

1,1–Dimethylpropyl

$$CH_3\overset{\overset{\displaystyle CH_3}{|}}{\underset{\underset{\displaystyle CH_3}{|}}{CH}}CH\text{—}$$

1,2–Dimethylpropyl

$$CH_3\overset{\overset{\displaystyle CH_3}{|}}{\underset{\underset{\displaystyle CH_3}{|}}{C}}CH_2\text{—}$$

2,2–Dimethylpropyl

3.15

a)

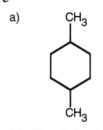

1,4–Dimethylcyclohexane

b)

1–Methyl–3–propylcyclopentane

c)

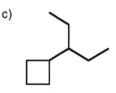

3–Cyclobutylpentane

d)

CH₂CH₃

Br

1–Bromo–4–ethylcyclodecane

e)

CH₃

CH(CH₃)₂

1–Isopropyl–2–methylcyclohexane

f)

Br

CH₃

C(CH₃)₃

4–Bromo–1–*tert*–butyl–2–
methylcycloheptane

3.16

a)

CH₃
CH₃

1,1–Dimethylcyclooctane

b)

3–Cyclobutylhexane

c)

Cl

Cl

1,2–Dichlorocyclopentane

d)

CH₃

Br   Br

1,3–Dibromo–5–methylcyclohexane

3.17

a)

CH₃

H

H

Br

trans–1–Bromo–3–methylcyclohexane

b)

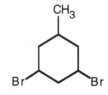

CH₃

CH₃

H

H

cis–1,2–Dimethylcyclopentane

c)

CH₂CH₃

H

H

C(CH₃)₃

trans–1–*tert*-Butyl-2-ethylcyclohexane

3.18

a)

OH

hydroxyl

aromatic
ring

Phenol

b)

O

ketone

double
bond

2–Cyclohexenone

c)

NH₂   amine

CH₃CHCOOH

carboxylic
acid

Alanine

d)

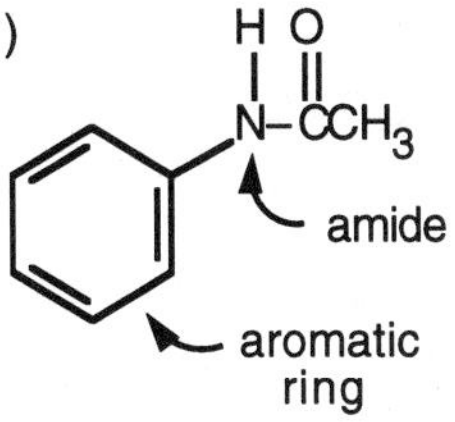

Acetanilide

e)

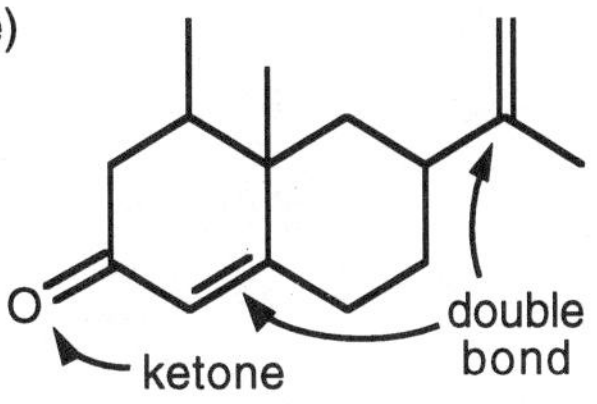

Nootkatone

f)

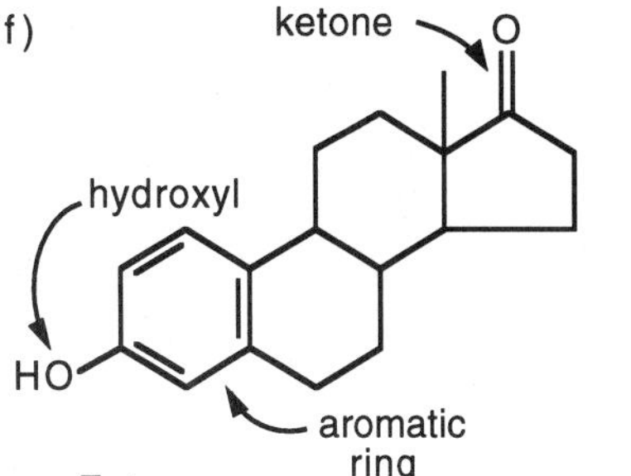

Estrone

g)

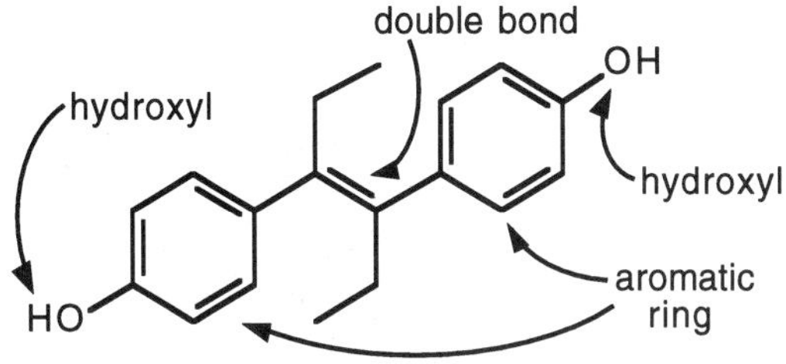

Diethylstilbestrol

h)

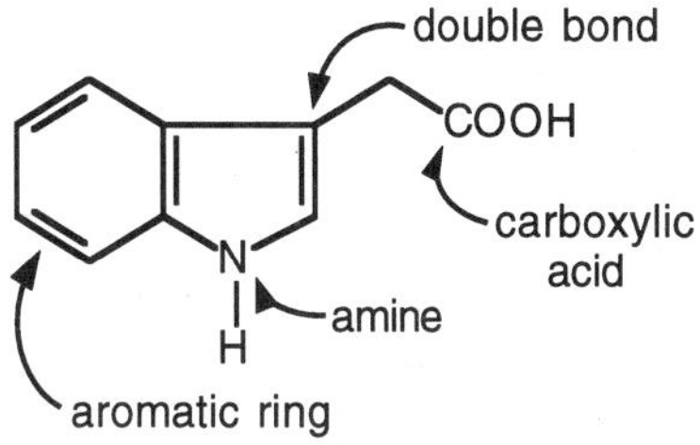

3–Indoleacetic acid

3.19 a) Eighteen isomers have the formula C_8H_{18}. Three are pictured.

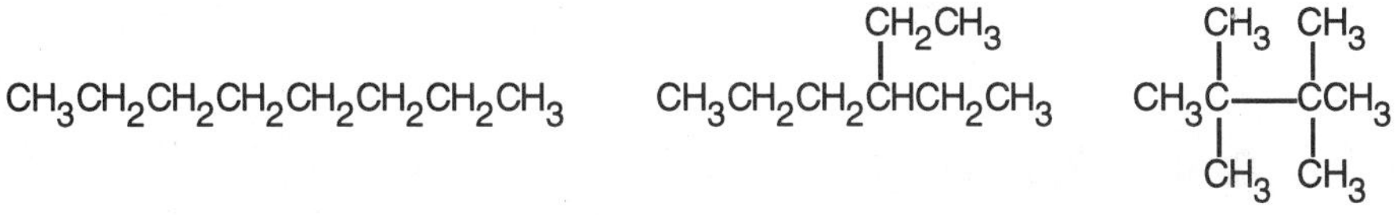

b) Structures with the formula $C_4H_8O_2$ may represent esters, carboxylic acids or many other complicated molecules. Two possibilities:

3.20

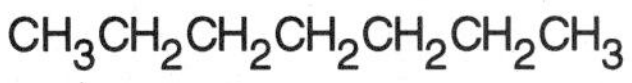

Heptane

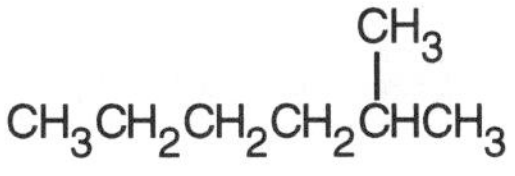

2–Methylhexane

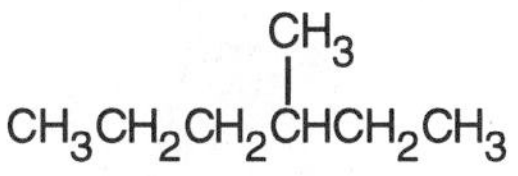

3–Methylhexane

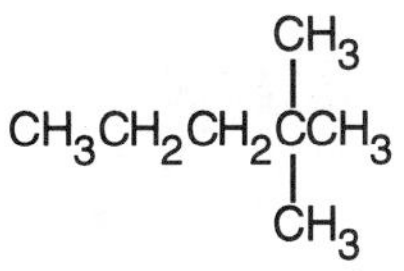

2,2–Dimethylpentane

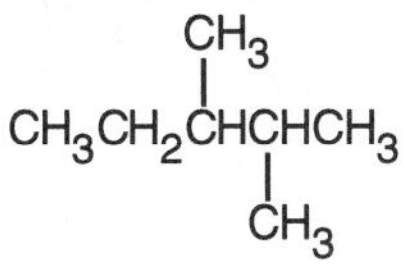

2,3–Dimethylpentane

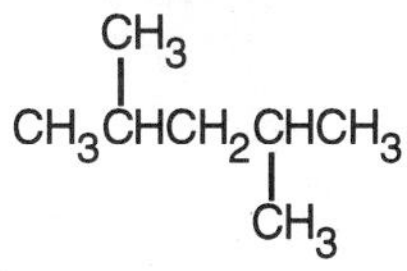

2,4–Dimethylpentane

$$CH_3CH_2\overset{\overset{\displaystyle CH_3}{|}}{\underset{\underset{\displaystyle CH_3}{|}}{C}}CH_2CH_3$$

3,3–Dimethylpentane

$$CH_3CH_2\overset{}{\underset{\underset{\displaystyle CH_2CH_3}{|}}{CH}}CH_2CH_3$$

3–Ethylpentane

$$CH_3\overset{\overset{\displaystyle CH_3}{|}}{CH}{-}\overset{\overset{\displaystyle CH_3}{|}}{\underset{\underset{\displaystyle CH_3}{|}}{C}}CH_3$$

2,2,3–Trimethylbutane

3.21   Other answers to this problem and to Problem 3.22 are acceptable.

a) $CH_3CH_2\overset{\overset{\displaystyle O}{\|}}{C}CH_2CH_3$

b) $CH_3\overset{\overset{\displaystyle O}{\|}}{C}NHCH_2CH_3$

c) $CH_3\overset{\overset{\displaystyle O}{\|}}{C}OCH_2CH_2CH_3$

d)

e) $CH_3\overset{\overset{\displaystyle O}{\|}}{C}CH_2\overset{\overset{\displaystyle O}{\|}}{C}OCH_2CH_3$

f) $H_2NCH_2CH_2OH$

3.22

a) C_4H_8O : $CH_3\overset{\overset{\displaystyle O}{\|}}{C}CH_2CH_3$

b) C_5H_9N : $CH_3CH_2CH_2CH_2C{\equiv}N$

c) $C_4H_6O_2$: $H\overset{\overset{\displaystyle O}{\|}}{C}CH_2CH_2\overset{\overset{\displaystyle O}{\|}}{C}H$

d) $C_6H_{11}Br$: $CH_3CH_2CH{=}CHCH_2CH_2Br$

e) C_6H_{14} : $CH_3CH_2CH_2CH_2CH_2CH_3$

f) C_6H_{12} :

g) C_5H_8 : $CH_3CH{=}CHCH{=}CH_2$

h) C_5H_8O : $H_2C{=}CH\overset{\overset{\displaystyle O}{\|}}{C}CH_2CH_3$

3.23   First, draw all straight-chain isomers. Then proceed to the simplest branched structure.

a) $CH_3CH_2CH_2CH_2OH$ $CH_3CH_2\overset{\overset{\displaystyle OH}{|}}{CH}CH_3$ $CH_3\overset{\overset{\displaystyle CH_3}{|}}{CH}CH_2OH$ $CH_3\overset{\overset{\displaystyle OH}{|}}{\underset{\underset{\displaystyle CH_3}{|}}{C}}CH_3$

There are 4 alcohol isomers of $C_4H_{10}O$.

b) $CH_3CH_2CH_2CH_2CH_2NH_2$ $CH_3CH_2CH_2\overset{\overset{\displaystyle NH_2}{|}}{CH}CH_3$ $CH_3CH_2\overset{\overset{\displaystyle NH_2}{|}}{CH}CH_2CH_3$

$CH_3CH_2\overset{}{\underset{\underset{\displaystyle CH_3}{|}}{CH}}CH_2NH_2$ $CH_3CH_2\overset{\overset{\displaystyle NH_2}{|}}{\underset{\underset{\displaystyle CH_3}{|}}{C}}CH_3$ $CH_3\overset{\overset{\displaystyle NH_2}{|}}{CH}\overset{}{\underset{\underset{\displaystyle CH_3}{|}}{CH}}CH_3$ $H_2NCH_2CH_2\overset{}{\underset{\underset{\displaystyle CH_3}{|}}{CH}}CH_3$

$$\underset{\underset{CH_3}{|}}{\overset{\overset{CH_3}{|}}{CH_3CCH_2NH_2}}$$

$$CH_3CH_2CH_2CH_2NHCH_3$$

$$CH_3CH_2CH_2NHCH_2CH_3$$

$$\overset{\overset{CH_3}{|}}{CH_3CH_2CHNHCH_3}$$

$$\overset{\overset{CH_3}{|}}{CH_3CHCH_2NHCH_3}$$

$$\underset{\underset{CH_3}{|}}{\overset{\overset{CH_3}{|}}{CH_3CNHCH_3}}$$

$$\overset{\overset{CH_3}{|}}{CH_3CH_2NHCHCH_3}$$

$$\overset{\overset{CH_3}{|}}{CH_3CH_2CH_2NCH_3}$$

$$\overset{\overset{CH_3}{|}}{CH_3CH_2NCH_2CH_3}$$

$$\underset{\underset{CH_3}{|}}{\overset{\overset{CH_3}{|}}{CH_3CHNCH_3}}$$

There are 17 isomers of $C_5H_{13}N$. Nitrogen can be bonded to one, two or three alkyl groups.

c) $\quad CH_3CH_2CH_2\overset{\overset{O}{||}}{C}CH_3 \qquad CH_3CH_2\overset{\overset{O}{||}}{C}CH_2CH_3 \qquad \underset{\underset{CH_3}{|}}{CH_3CH\overset{\overset{O}{||}}{C}CH_3}$

There are 3 ketone isomers with the formula $C_5H_{10}O$.

d) $\quad CH_3CH_2CH_2CH_2\overset{\overset{O}{||}}{CH} \qquad \underset{\underset{CH_3}{|}}{CH_3CHCH_2\overset{\overset{O}{||}}{CH}} \qquad \underset{\underset{CH_3}{|}}{CH_3CH_2CH\overset{\overset{O}{||}}{CH}} \qquad \underset{\underset{CH_3}{|}}{\overset{\overset{CH_3}{|}}{CH_3C}}\!\!-\!\!\overset{\overset{O}{||}}{CH}$

There are 4 isomeric aldehydes with the formula $C_5H_{10}O$. Remember that the aldehyde functional group can occur only at the end of a chain.

e) $\quad CH_3CH_2\overset{\overset{O}{||}}{C}OCH_3 \qquad CH_3\overset{\overset{O}{||}}{C}OCH_2CH_3 \qquad H\overset{\overset{O}{||}}{C}OCH_2CH_2CH_3 \qquad H\overset{\overset{O}{||}}{C}O\underset{\underset{CH_3}{|}}{CH}CH_3$

There are 4 esters with the formula $C_4H_8O_2$.

f) $\quad CH_3CH_2OCH_2CH_3 \qquad CH_3OCH_2CH_2CH_3 \qquad \underset{\underset{CH_3}{|}}{CH_3OCHCH_3}$

There are 3 ethers with the formula $C_4H_{10}O_2$.

3.24

a) CH_3CH_2OH

b) $CH_3\underset{\underset{\displaystyle CH_3}{|}}{\overset{\overset{\displaystyle CH_3}{|}}{C}}C\equiv N$

c) $CH_3\underset{\overset{|}{Br}}{CH}CH_3$

d) $CH_3\underset{\overset{|}{OH}}{CH}CH_2OH$

e) $CH_3\underset{\overset{|}{CH_3}}{CH}OCH_3$

f) $CH_3\underset{\underset{\displaystyle CH_3}{|}}{\overset{\overset{\displaystyle CH_3}{|}}{C}}CH_3$

3.25

$CH_3CH_2CH_2CH_2CH_2Br$
1–Bromopentane

$CH_3CH_2CH_2\underset{\overset{|}{Br}}{CH}CH_3$
2–Bromopentane

$CH_3CH_2\underset{\overset{|}{Br}}{CH}CH_2CH_3$
3–Bromopentane

3.26

$CH_3\underset{\overset{|}{CH_3}}{CH}CH_2CH_2\underset{\overset{|}{CH_3}}{CH}CH_2Cl$

$CH_3\underset{\overset{|}{CH_3}}{CH}CH_2CH_2\underset{\underset{\displaystyle Cl}{|}}{\overset{\overset{\displaystyle CH_3}{|}}{C}}CH_3$

$CH_3\underset{\overset{|}{CH_3}}{CH}CH_2\underset{\underset{\displaystyle Cl}{|}}{\overset{\overset{\displaystyle CH_3}{|}}{CH}}CH CH_3$

1–Chloro–2,5–dimethylhexane   2–Chloro–2,5–dimethylhexane   3–Chloro–2,5–dimethylhexane

3.27

a) sp^2

b) $—C\equiv N$ sp

c) sp^2

d) sp^3

3.28

a) $CH_3CH_2CH_2CH_2CH_2CHCH_3$
with CH_3 on the CH

2–Methylheptane

b)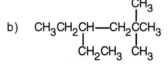

4–Ethyl–2,2–dimethylhexane

c)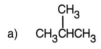

4–Ethyl–3,4–dimethyloctane

d) $CH_3CH_2CH_2CCH_2CHCH_3$
with two CH_3 groups on the C and CH_3 below

2,4,4–Trimethylheptane

e)

3,3–Diethyl–2,5–dimethylnonane

f) $CH_3CH_2CH_2CHCHCH_2CH_3$
with CH_3CHCH_3 above and CH_3 below

4–Isopropyl–3–methylheptane

3.29

a) CH_3CHCH_3
with CH_3 above

b)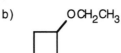

c) $CH_3CH_2CH_2CH_2CH_2CH_3$

3.30

a)   □

b) CH_3CH—$CHCH_3$
with CH_3 and CH_3 above

3.31

a) $CH_3CH_2CH_2CH_2CH_2Br$

b)

c) $CH_3CHC{\equiv}N$
with CH_3 above

d)   cyclopentane with CH_2OH

e)   There are no aldehyde isomers. However,

$$CH_3\overset{O}{\overset{\|}{C}}CH_3$$

is a ketone isomer.

f)   benzene ring with $COOH$ and CH_3

3.32   The purpose of this problem is to teach you to recognize identical structures when they are drawn slightly differently.

a) 1 and 2 are the same.   b) All structures are the same.
c) 1 and 2 are the same.   d) 1 and 3 are the same.
e) 1 and 3 are the same.

3.33

a)
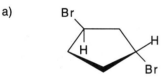

trans−1,3−Dibromocyclopentane

b)
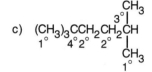

CH₃CH₂ ... CH₂CH₃

cis−1,4−Diethylcyclohexane

c)
CH₃

H

H

CH(CH₃)₂

trans−1−Isopropyl−3−methylcycloheptane

d)
⬡—CH₂—⬡

Dicyclohexylmethane

3.34

a) CH₃CHCH₂CH₃
   1° 3° 2° 1°
with CH₃ 1° above

b) (CH₃)₂CHCH(CH₂CH₃)₂
   1° 3° 3° 2° 1°

c) (CH₃)₃CCH₂CH₂CH
   1°   4°2° 2°
with 1° CH₃, 3° CH, 1° CH₃

d)
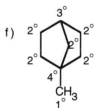

e)

f)

3.35   a) 2−Methylpentane
        c) 2,3,3−Trimethylhexane
        e) 3,3,5−Trimethyloctane
        g) 5−Ethyl−3,5−dimethyloctane

b) 2,2−Dimethylbutane
d) 5−Ethyl−2−methylheptane
f) 2,2,3,3−Tetramethylhexane

3.36

CH₃CH₂CH₂CH₂CH₂CH₃

Hexane

             CH₃
CH₃CH₂CH₂CHCH₃

2−Methylpentane

         CH₃
CH₃CH₂CHCH₂CH₃

3−Methylpentane

    CH₃
CH₃CH₂CCH₃
    CH₃

2,2−Dimethylbutane

    CH₃
CH₃CHCHCH₃
    CH₃

2,3−Dimethylbutane

3.37

a) $CH_3\overset{\displaystyle CH_2CH_3}{\underset{}{C}}HCH_2CH_2CH_2\overset{\displaystyle CH_3}{\underset{\displaystyle CH_3}{C}}CH_3$

The longest chain is an octane.

correct name: 2,2,6–Trimethyloctane

b) $CH_3\overset{\displaystyle CH_3}{\underset{\displaystyle CH_2CH_3}{C}}HCHCH_2CH_2CH_3$

The longest chain is a hexane;
numbering should start from
the other end.

correct name: 3–Ethyl–2–methylhexane

c) $CH_3CH_2\overset{\displaystyle CH_3}{\underset{\displaystyle CH_3}{C}}\text{———}\overset{}{\underset{\displaystyle CH_2CH_3}{C}}HCH_2CH_3$

Numbering should start from
the other end.

correct name: 4–Ethyl–3,3–dimethylhexane

d) $CH_3CH_2\overset{\displaystyle CH_3}{\underset{}{C}}H\text{———}\overset{\displaystyle CH_3}{\underset{\displaystyle CH_3}{C}}CH_2CH_2CH_2CH_3$

Numbering should start from
the other end.

correct name: 3,4,4–Trimethyloctane

e) $CH_3CH_2CH_2\overset{\displaystyle CH_3}{\underset{\displaystyle CH_3\overset{}{C}HCH_3}{C}}HCH_2CHCH_3$

The longest chain is an octane.

correct name: 2,3,5–Trimethyloctane

f)

The substituents should have the
lowest possible numbers.

correct name: *cis*–1,3–Dimethylcyclohexane

3.38

a)

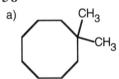

CH$_3$

CH$_3$

1,1–Dimethylcyclooctane

b)

CH$_3$ CH$_2$CH$_3$

CH$_3$CCH$_2$CCH$_2$CH$_3$

CH$_3$ CH$_2$CH$_3$

4,4–Diethyl–2,2–dimethylhexane

c)

CH$_3$

CH$_3$

CH$_3$

1,1,2–Trimethylcyclohexane

d)

CH$_3$

CH$_2$CH$_2$CHCH$_3$

CH$_3$CH$_2$CH$_2$CH$_2$CH$_2$CHCH$_2$CH$_2$CH$_2$CH$_2$CH$_3$

6–(3–Methylbutyl)–undecane

Remember that you must
choose an alkane whose principal chain is long enough so
that the substituent does not
become part of the principal
chain.

3.39

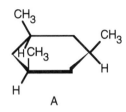

CH$_3$

CH$_3$

H CH$_3$

H

H

A

CH$_3$

CH$_3$

H CH$_3$

H

CH$_3$

H

B

Two cis–trans isomers of 1,3,5–trimethylcyclohexane are possible. In one isomer (A), all
methyl groups are cis; in B, one methyl group is trans to the other two.

3.40

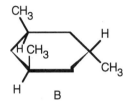

a)

Br

Br

H

H

cis–1,3–Dibromocyclohexane

H

Br

Br

H

trans–1,4–Dibromocyclohexane

⎤
⎥ constitutional
⎥ isomers
⎦

b)

CH$_3$

CH$_3$CH$_2$CH$_2$CHCHCH$_3$

CH$_3$

2,3–Dimethylhexane

CH$_3$ CH$_3$

CH$_3$CHCH$_2$CH$_2$CHCH$_3$

2,5,5–Trimethylpentane
(correct name: 2,5–Dimethylhexane)

⎤
⎥ constitutional
⎥ isomers
⎦

c)

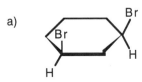

Cl   Cl

Cl

Cl

⎤
⎥ identical
⎥
⎦

3.41

trans–1,3–Dibromocyclopentane   cis–1,3–Dibromocyclopentane

3.42

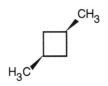

cis–1,3–Dimethylcyclobutane

3.43 Because malic acid has 2–COOH groups, the formula for the rest of the molecule is C_2H_4O. Possible structures for malic acid are:

primary alcohol

secondary alcohol

ether

tertiary alcohol

ester

ester

Because only one of these compounds (the second one) is also a secondary alcohol, it must be malic acid.

3.44

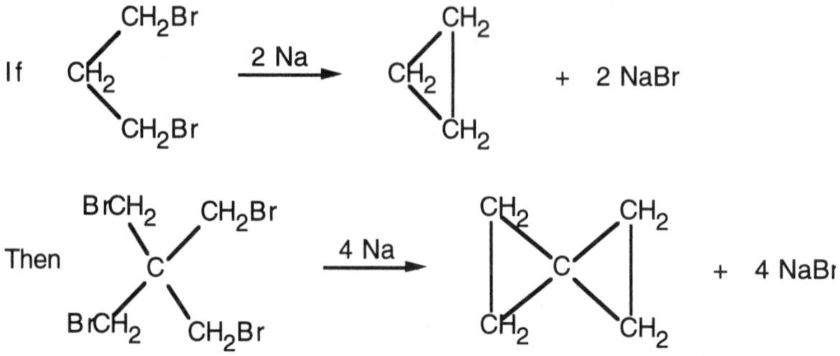

If

$$\underset{CH_2Br}{\overset{CH_2Br}{CH_2}} \quad \xrightarrow{\text{2 Na}} \quad \underset{CH_2}{\overset{CH_2}{CH_2}} \quad + \quad 2 \text{ NaBr}$$

Then

$$\underset{BrCH_2 \quad CH_2Br}{\overset{BrCH_2 \quad CH_2Br}{C}} \quad \xrightarrow{\text{4 Na}} \quad \underset{CH_2 \quad CH_2}{\overset{CH_2 \quad CH_2}{C}} \quad + \quad 4 \text{ NaBr}$$

The two rings are perpendicular in order to keep the geometry of the central carbon as close to tetrahedral as possible.

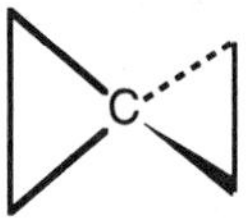

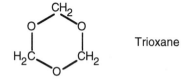

3.45 Many students do not know where to begin when they are assigned this type of problem. To start, read the problem carefully, word for word. Then try to interpret parts of the problem. For example:

1) Formaldehyde is an aldehyde, $\overset{\overset{\displaystyle O}{\parallel}}{H-C-H}$.
2) It trimerizes — that is, 3 formaldehydes come together to form a compound $C_3H_6O_3$. No atoms are eliminated, so all of the original atoms are still present.
3) There are no carbonyls. This means that trioxane cannot contain any –C=O functional groups. If you look back to Table 3.1, you can see that the only oxygen functional groups that can be present are either ethers or alcohols.
4) A monobromo derivative is a compound in which one of the H's has been replaced by a Br. Because only one monobromo derivative is possible, we know that there can only be one type of hydrogen in trioxane. The only possibility for trioxane is:

Trioxane

3.46

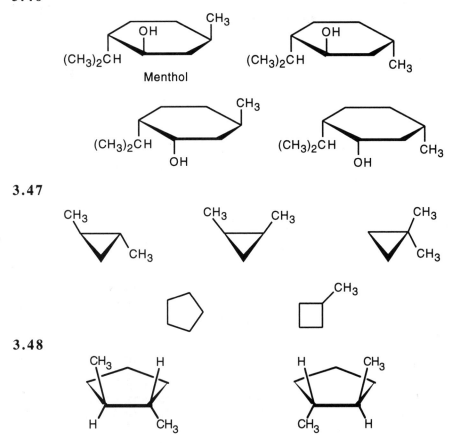

Menthol

3.47

3.48

The two *trans*–1,2–dimethylcyclopentanes are mirror images.

3.49 A puckered ring allows all the bonds in the ring to have a nearly tetrahedral bond angle.

Study Guide for Chapter 3

After studying this chapter, you should be able to:

(1) Identify functional groups in molecules, and draw molecules containing a given functional group (3.1, 3.2, 3.18, 3.21, 3.22, 3.23, 3.24, 3.31, 3.45).

(2) Systematically draw all possible isomers of a given molecular formula (3.3, 3.4, 3.5, 3.6, 3.19, 3.20, 3.23, 3.25, 3.26, 3.43, 3.47).

(3) Name and draw alkanes and alkyl groups (3.7, 3.11, 3.12, 3.13, 3.14, 3.28, 3.35, 3.36, 3.37, 3.38).

(4) Identify carbon and hydrogen as being primary, secondary or tertiary (3.8, 3.9, 3.10, 3.29, 3.30, 3.34).

(5) Name and draw cycloalkanes, indicating cis-trans geometry if required (3.15, 3.16, 3.17, 3.33, 3.39, 3.40, 3.41, 3.42, 3.46, 3.48).

4.2

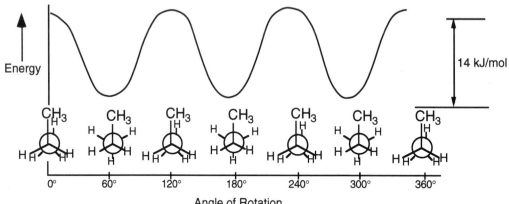

Angle of Rotation

4.3

a)

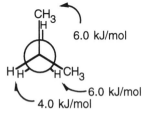

This conformer of 2,3–dimethylbutane is the most stable since it is staggered and has the fewest $CH_3 - CH_3$ *gauche* interactions.

4.4

a)

The *most stable* conformer occurs at 60° 180°, and 300°.

b)

6.0 kJ/mol

6.0 kJ/mol

4.0 kJ/mol

The *least stable* conformer occurs at 0°, 120°, 240°, and 360°.

c,d)

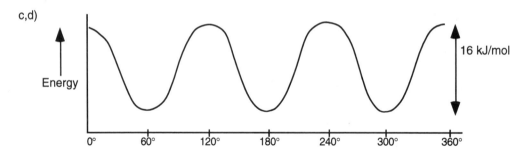

4.5 Cyclopropane is a more efficient fuel. Combustion of cyclopropane releases more heat per gram than combustion of cyclohexane because of the high strain energy of the cyclopropane ring.

4.6

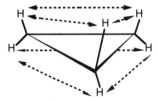

All hydrogen atoms on the same side of the cyclopropane ring are eclipsed by each other. If we draw each hydrogen-hydrogen interaction, we count six eclipsing interactions. Since each of these interactions "costs" 4.0 kJ/mol, all six cost 24.0 kJ/mol. (24 kJ/mol ÷115 kJ/mol) = 0.21; thus, 21% of the total strain energy of cyclopropane is due to eclipsing strain.

4.7

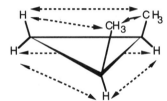 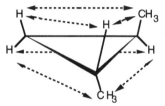

		cis isomer		trans isomer	
eclipsing interaction	energy cost (kJ/mol)	# of interactions	total energy cost (kJ/mol)	# of interactions	total energy cost (kJ/mol)
H–H	4.0	3	12.0	2	8.0
H–CH$_3$	6.0	2	12.0	4	24.0
CH$_3$–CH$_3$	11	1	11	0	0
			35		32

The added energy cost of eclipsing interactions causes *cis*–1,2–dimethylcyclopropane to be of higher energy, and to be less stable than the trans isomer. Since the cis isomer is of higher energy, its heat of combustion is also greater.

4.8

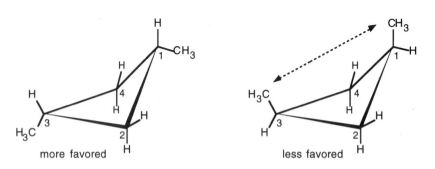

Two types of interaction are present in *cis*–1,2–dimethylcyclobutane. One interaction occurs between the two methyl groups, which are almost eclipsed. The other is an across-the-ring interaction between methyl group at position 1 of the ring and a hydrogen at position 3. Because neither of these interactions are present in trans isomer, it is more stable than the cis isomer.

In *trans*–1,3–dimethylcyclobutane an across-the-ring interaction occurs between the methyl group at position 1 of the ring and a hydrogen at position 3. Because no interactions are present in the cis isomer, it is more stable than the trans isomer.

4.9

4.10

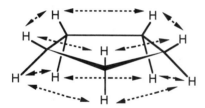

If cyclopentane were planar, it would have ten hydrogen–hydrogen interactions with a total energy cost of 40 kJ/mol. The measured total strain energy of 26 kJ/mol indicates that 14 kJ/mol of eclipsing strain in cyclopentane has been relieved by puckering.

4.11

The conformation with bromine in the equatorial position is more stable.

4.12 Make a model of *cis*–1,2–dichlorocyclohexane. Take note of the fact that all cis substituents are on the same side of the ring and that two adjacent cis substituents have an axial-equatorial relationship. Now, perform a ring-flip on the cyclohexane.

After the ring-flip, the relationship of the two substituents is still axial-equatorial. No two adjacent cis substituents can be converted to being both axial or both equatorial without breaking bonds.

4.13 For a *trans*–1,2–disubstituted cyclohexane, two adjacent substituents must be either both axial or both equatorial.

A ring flip converts two adjacent axial substituents to equatorial substituents, and vice versa. As in Problem 4.12, no two adjacent trans substituents can be converted to an axial-equatorial relationship without bond breaking.

4.14

trans–1,4–Dimethylcyclohexane

4.15

The most stable conformation of axial *tert*–butylcyclohexane is pictured. One methyl group is positioned above the ring and competes for space with two axial ring protons. In the other axial alkylcyclohexanes, this methyl group is replaced by hydrogen, which has a much smaller space requirement. The steric strain caused by an axial *tert*–butyl group is therefore higher than the strain caused by axial methyl, ethyl or isopropyl groups.

4.16 The energy difference between an axial and an equatorial cyano group is very small because there are no 1–3 diaxial interactions for a cyano group.

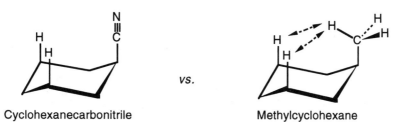

vs.

Cyclohexanecarbonitrile   Methylcyclohexane

4.17   Table 4.2 shows that an axial bromine causes 2 x 1.0 kJ/mol of steric strain. Thus, the energy difference between axial and equatorial bromocyclohexane is 2.0 kJ/mol. According to Figure 4.20, this energy difference corresponds to a 70:30 ratio of more stable : less stable conformer. Thus, 70% of bromocyclohexane molecules are in the equatorial conformation, and 30% are in the axial conformation at any given moment.

4.18

a)

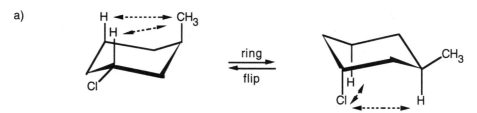

trans–1–Chloro–3–methylcyclohexane

2 (H–CH₃) = 7.6 kJ/mol   2 (H–Cl) = 2.0 kJ/mol

The second conformation is more stable than the first.

b)

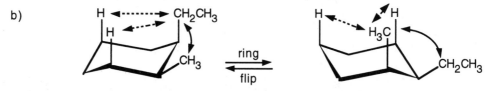

cis–1–Ethyl–2–methylcyclohexane

one CH₃–CH₂CH₃ gauche   one CH₃–CH₂CH₃ gauche
interaction = 3.8 kJ/mol   interaction = 3.8 kJ/mol
2 (H–CH₂CH₃) = 8.0 kJ/mol   2 (H–CH₃) = 7.6 kJ/mol
_____   _____
Total = 11.8 kJ/mol   Total = 11.4 kJ/mol

The second conformation is more stable than the first.

c)

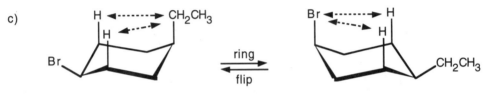

cis–1–Bromo–4–ethylcyclohexane

2 (H–CH₂CH₃) = 8.0 kJ/mol   2 (H–Br) = 2.0 kJ/mol

The second conformation is more stable than the first.

d)

cis–1–*tert*–Butyl–4–ethylcyclohexane

2 [H–C(CH₃)₃] = 22.8 kJ/mol   2 (H–CH₂CH₃) = 8.0 kJ/mol

The second conformation is more stable than the first.

4.19

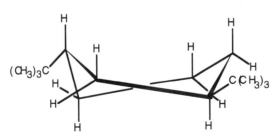

trans–1,3–Di–*tert*–butylcyclohexane

In the chair conformation of *trans*–1,3–di–*tert*–butylcyclohexane, one *tert*–butyl group is axial and one is equatorial. The 1,3–diaxial interactions of the axial *tert*–butyl group make the chair form of *trans*–1,3–di–*tert*–butylcyclohexane 22.8 kJ/mol (5.4 kcal/mol) less stable than a cyclohexane with no axial substituents. Since a twist boat conformation is 23 kJ/mol (5.5 kcal/mol) less stable than a chair, it is almost as likely that the compound will assume a twist-boat conformation as a chair conformation. The twist-boat removes the 1,3–diaxial interaction present in the chair form of *trans*–1,3–di–*tert*–butylcyclohexane.

4.20

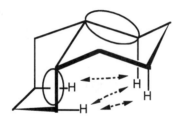

Trans–decalin is more stable than *cis*–decalin. Three 1,3–diaxial interactions cause *cis*–decalin to be of higher energy than *trans*–decalin. You may be able to visualize these interactions by thinking of the circled parts of *cis*–decalin as similar to axial methyl groups. The gauche interactions that occur with axial methyl groups also occur in *cis*–decalin.

4.21 a) *Angle strain* is the strain caused by the deformation of a bond angle from its normal value.

b) *Steric strain* is the repulsive interaction caused by atoms attempting to occupy the same space.

c) *Torsional strain* is the repulsive interaction between two bonds as they rotate past each other. Torsional strain is responsible for the barrier to rotation in ethane and makes the eclipsed form higher in energy than the staggered form.

d) The *heat of combustion* of an organic compound is the heat liberated when the compound burns completely in oxygen to form CO_2 and H_2O. The heat of combustion for two compounds having the same formula is a larger negative number for a strained compound, indicating that the strained compound is of higher energy.

e) A *conformation* is one of the many possible arrangements of atoms caused by rotation about a single bond.

f) A *staggered* conformation is the conformation in which all groups on two adjacent carbons are as far from each other as possible.

g) An *eclipsed* conformation is the conformation in which all groups on two adjacent carbons are as close to each other as possible.

h) *Gauche butane* is the conformation of butane in which the C1 and C4 methyl groups are 60° apart. Gauche butane is of higher energy than anti butane because of steric strain.

4.22

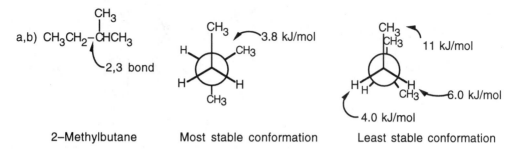

The energy difference between the two conformations is $(11 + 6.0 + 4.0) - 3.8 = 17$ kJ/mol.

c) Consider the least stable conformation to be at zero degrees. Keeping the "front" of the projection unchanged, rotate the "back" by 60° to obtain each conformation.

at 60°: energy = 3.8 kJ/mol

at 120°: energy = 18.0 kJ/mol

at 180°: energy = 3.8 kJ/mol

at 240°: energy = 21 kJ/mol

at 300°: energy = 7.6 kJ/mol

Use the lowest energy conformation as the energy minimum. The highest energy conformation is 17 kJ/mol higher in energy than the lowest energy conformation.

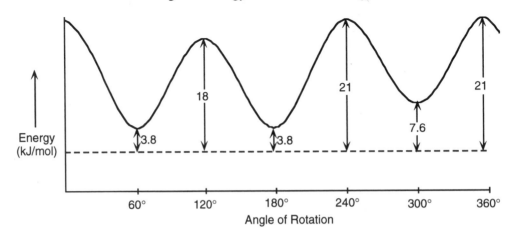

4.23

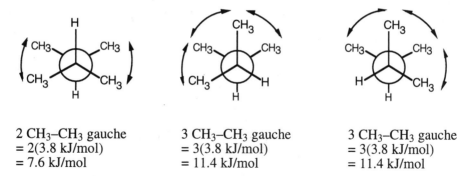

2 CH₃–CH₃ gauche
= 2(3.8 kJ/mol)
= 7.6 kJ/mol

3 CH₃–CH₃ gauche
= 3(3.8 kJ/mol)
= 11.4 kJ/mol

3 CH₃–CH₃ gauche
= 3(3.8 kJ/mol)
= 11.4 kJ/mol

4.24 Since we are not told the values of the interactions for 1,2–dibromoethane, the diagram can only be qualitative.

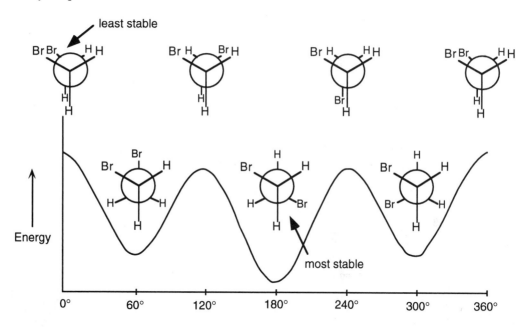

The anti conformer is at 180°.
The gauche conformers are at 60°, 300°.

4.25 The anti conformer has no net dipole moment because the bond polarities of the individual bonds cancel. The gauche conformer, however, has a dipole moment. Because the observed dipole moment is 1.0 D at room temperature, a mixture of conformers must be present.

4.26 The highest energy conformation of bromoethane is 15 kJ/mol (3.6 kcal/mol). Because this includes two H–H eclipsing interactions of 4.0 kJ/mol (1.0 kcal/mol) each, the value of an H–Br eclipsing interaction is 15-2(4.0) = 7 kJ/mol (3.6 – 2(1.0) = 1.6 kcal/mol).

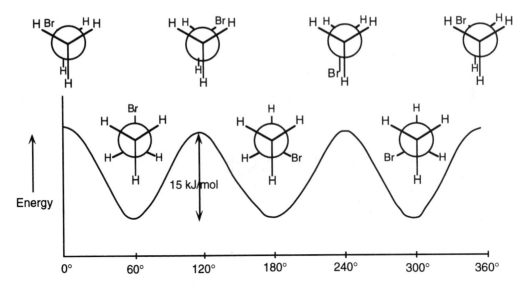

4.27 a) An *axial bond* of a cyclohexane ring is a bond that is perpendicular to the plane of the ring.

b) An *equatorial bond* of a cyclohexane ring lies more or less in the plane of the ring.

c) The *chair conformation* of a cyclohexane ring is the "puckered" conformation that allows all carbon–carbon bond angles to have a value very close to the ideal tetrahedral angle.

d) A *1,3–diaxial interaction* is a type of steric strain between an axial functional group on C1 of a cyclohexane ring and an axial group on C3 (or C5).

4.28

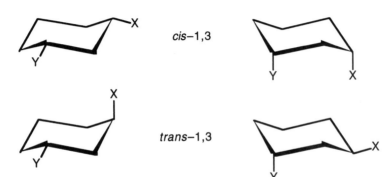

4.29

![cyclohexane chair conformations labeled cis-1,3 and trans-1,3 with X and Y substituents]

A cis–1,3–disubstituted isomer exists almost exclusively in the diequatorial conformation, which has no 1,3–diaxial interactions. The trans isomer must have one group axial, leading to 1,3–diaxial interactions. Thus, the trans isomer is less stable than the cis isomer.

4.30

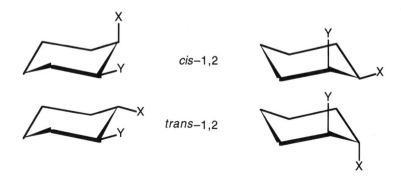

cis–1,2

trans–1,2

Reasoning similar to that in the previous problem can be used to show that the trans–1,2 disubstituted isomer, which exists principally in the diequatorial conformation, is more stable than the cis isomer.

4.31

trans

cis

The *trans*–1,4–isomer is more stable.

4.32 Since the methyl group of *N*–methylpiperidine prefers an equatorial conformation, the steric requirements of a methyl group must be greater than those of an electron lone pair.

4.33

Use Table 4.2 to find the values of 1,3–diaxial interactions. For the first conformation, the steric strain is 2 x 1.0 kJ/mol = 2.0 kJ/mol (2 x 0.25 kcal/mol = 0.5 kcal/mol). The steric strain in the second conformation is 2 x 3.8 kJ/mol, or 7.6 kJ/mol (2 x 0.9 kcal/mol, or 1.8 kcal/mol). The first conformation is more stable than the second conformation by 5.6 kJ/mol (1.3 kcal/mol).

4.34

 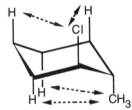

no 1,3–diaxial
interactions

2 x 3.8 kJ/mol = 7.6 kJ/mol
2 x 1.0 kJ/mol = 2.0 kJ/mol
_____
9.6 kJ/mol

The first conformation is more stable than the second conformation by 9.6 kJ/mol (2.3 kcal/mol).

4.35

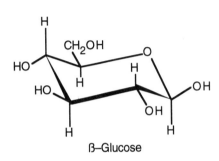

β–Glucose

4.36 To solve this problem: (1) Find the energy cost of a 1–3 diaxial interaction by using Table 4.2. (2) Convert this energy difference into a percent by using Figure 4.20.

a)

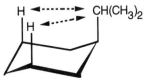

$2(H-CH(CH_3)_2)$ = 9.2 kJ/mol

% equatorial = 97.5

% axial = 2.5

b)

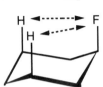

$2(H-F)$ = 1.0 kJ/mol

% equatorial = 60

% axial = 40

c)

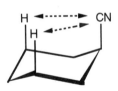

$2(H-CN)$ = 0.8 kJ/mol

% equatorial = 58

% axial = 42

d)

$$2(H{-}OH) = 4.2 \text{ kJ/mol}$$
$$\% \text{ equatorial} = 85$$
$$\% \text{ axial} = 15$$

4.37   Be aware of the distinction between axial–equatorial and cis–trans. Axial substituents are parallel to the axis of the ring; equatorial substituents lie around the "equator" of the ring. Cis substituents are on the same side of the ring; trans substituents are on opposite side of the ring.

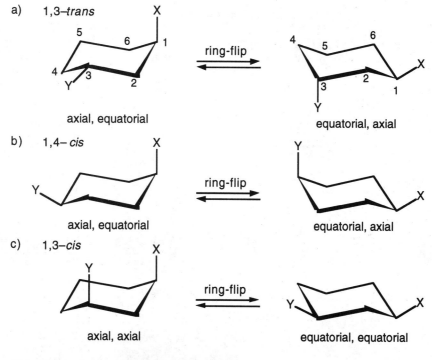

a)   1,3–*trans*

axial, equatorial   equatorial, axial

b)   1,4–*cis*

axial, equatorial   equatorial, axial

c)   1,3–*cis*

axial, axial   equatorial, equatorial

d) 1,5–trans is the same as 1,3–trans

e) 1,5–cis is the same as 1,3–cis

f)   1,6–*trans*

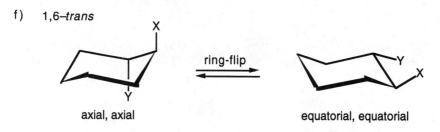

axial, axial   equatorial, equatorial

4.38

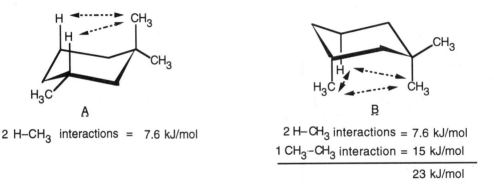

diequatorial   diaxial

The large energy difference between conformations is due to the severe 1,3–diaxial interaction between the two methyl groups.

4.39 Diaxial cis–1,3–dimethylcyclohexane contains three 1,3–diaxial interactions -- two H–CH$_3$ interactions of 3.8 kJ/mol (0.9 kcal/mol) each, and one CH$_3$–CH$_3$ interaction. If the diaxial conformation is 23 kJ/mol (5.4 kcal/mol) less stable than the diequatorial, 23 – 2(3.8) = 15 kJ/mol (3.6 kcal/mol) of this strain energy must be due to the CH$_3$–CH$_3$ interaction.

4.40

A

2 H–CH$_3$ interactions = 7.6 kJ/mol

B

2 H–CH$_3$ interactions = 7.6 kJ/mol
1 CH$_3$–CH$_3$ interaction = 15 kJ/mol
_____
23 kJ/mol

Conformation **A** is favored because it is 15 kJ/mol lower in energy than conformation **B**.

4.41

There are two cis-trans stereoisomers of 1,3,5-trimethylcyclohexane. In one isomer, all methyl groups are cis to each other; in the other isomer, one methyl group is trans to the other two.

4.42 The isomer with all substituents cis to each other is more stable because it has no 1,3-diaxial interactions.

4.43 Note: In working with decalins, it is essential to use models. Many structural features of decalins that are obvious with models are not easily visualized with drawings.

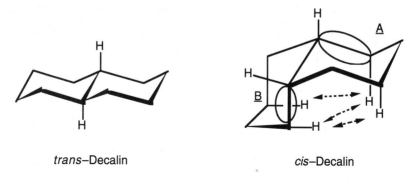

trans–Decalin   *cis*–Decalin

No 1,3–diaxial interactions are present in *trans*–decalin.

At the ring junction of *cis*–decalin, one ring acts as an axial substituent of the other (see circled bonds). The circled part of ring <u>B</u> has two 1,3–diaxial interactions with ring <u>A</u> (indicated by arrows). Similarly, the circled part of ring <u>A</u> has two 1,3–diaxial interactions with ring <u>B</u>; one of these interactions is the same as an interaction of part of the <u>B</u> ring with ring <u>A</u>. These three 1,3–diaxial interactions have a total energy cost of 11.4 kJ/mol (2.7 kcal/mol). *Cis*–decalin is therefore less stable than *trans*–decalin by 11.4 kJ/mol (2.7 kcal/mol).

4.44

A ring-flip converts an axial substituent into an equatorial substituent and vice versa. At the ring junction of *trans*–decalin, each ring is a trans–trans diequatorial substituent of the other. If a ring-flip were to occur, the two rings would become axial substituents of each other. You can see with models that a diaxial ring junction is impossibly strained. Consequently, *trans*–decalin does not ring-flip.

The rings of *cis*–decalin are joined by an axial bond and an equatorial bond. After a ring-flip, the rings are still linked by an equatorial and an axial bond. No additional strain or interaction is introduced by a ring-flip of *cis*–decalin.

4.45

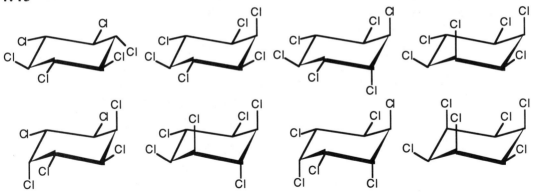

The first isomer is the most stable because all chlorine atoms can assume an equatorial configuration.

4.46

	Most stable	Strain energy	Least stable	Strain energy
a)	CH₃ structure	3.8 kJ/mol (0.9 kcal/mol)	CH₃ structure	21 kJ/mol (5.0 kcal/mol)
b)	CH₃ structure	7.6 kJ/mol (1.8 kcal/mol)	CH₃ structure	23 kJ/mol (5.4 kcal/mol)
c)	CH₃ structure	7.6 kJ/mol (1.8 kcal/mol)	CH₃ structure	26 kJ/mol (6.2 kcal/mol)
d)	CH₃ structure	15.2 kJ/mol (3.6 kcal/mol)	CH₃ structure	28 kJ/mol (6.6 kcal/mol)

* = most stable overall
** = least stable overall

4.47

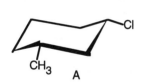

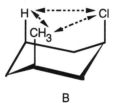

A

B

Conformation A of *cis*–1–chloro–3–methylcyclohexane has no 1,3–diaxial interactions and is the more stable conformation. Steric strain in B is due to one CH_3–H interaction (3.8 kJ/mol), one Cl–H interaction (1.0 kJ/mol) and one CH_3–Cl interaction. Since the total-strain energy of B is 15.5 kJ/mol, 15.5 - 3.8 - 1.0 = 10.7 kJ/mol (2.55 kcal/mol) of strain is caused by a CH_3–Cl interaction.

4.48

If you build a model of 1–norbornene, you will find that it is almost impossible to form the bridgehead double bond. sp^2–Hybridization at the double bond requires all carbons bonded to the starred carbon to lie in a plane, yet the angle strain caused by sp^2 hybridization is too severe to allow a bridgehead double bond to exist.

4.49 A steroid ring system is fused, and ring-flips don't occur. Thus, substituents such as the methyl groups shown remain axial. Substituents on the same side as the methyl groups are in alternating axial and equatorial positions. Thus, an "up" substituent at C3 (a) is equatorial.

Substituents on the bottom side of the ring system also alternate axial and equatorial positions. A substituent at C7 (b) is axial, and one at C11 (c) is equatorial

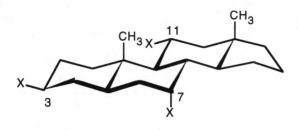

4.50

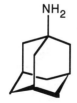

4.51

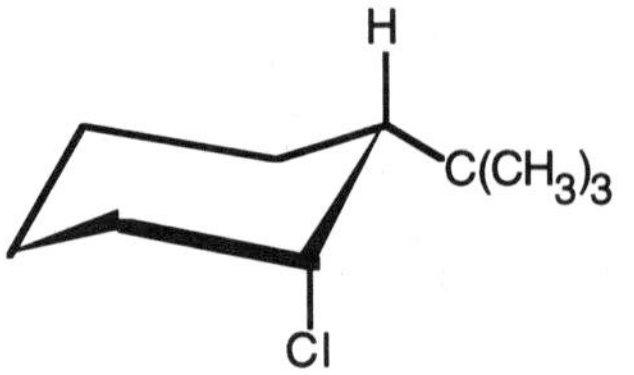

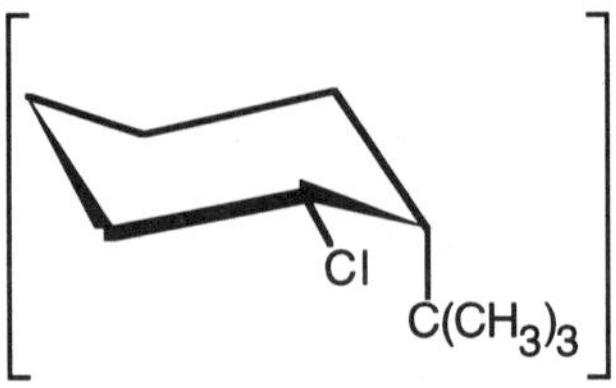

cis

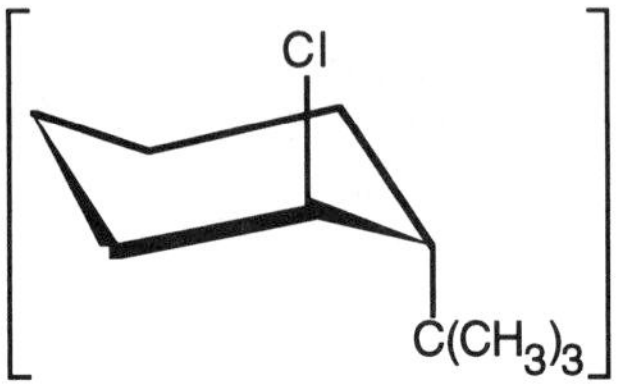

trans

All four conformations of the two isomers are illustrated. The second conformation of each pair has a high degree of steric strain, and thus each isomer adopts the first conformation. Since only the cis isomer has chlorine in the necessary axial position, it is more reactive than the trans isomer.

4.52   Draw the four possible isomers of 4–*tert*–butylcyclohexane–1,3–diol. Make models of these isomers also.

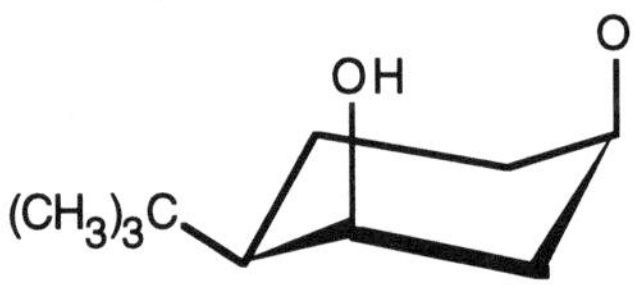

1

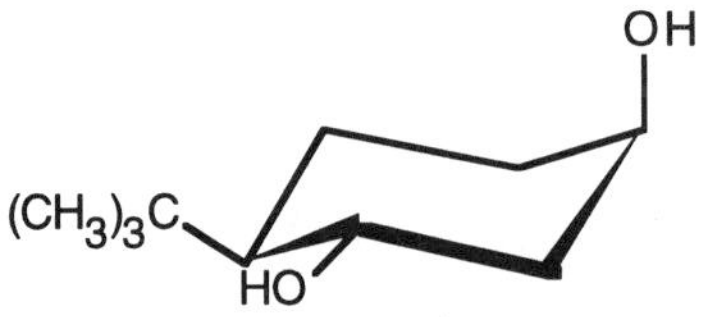

2

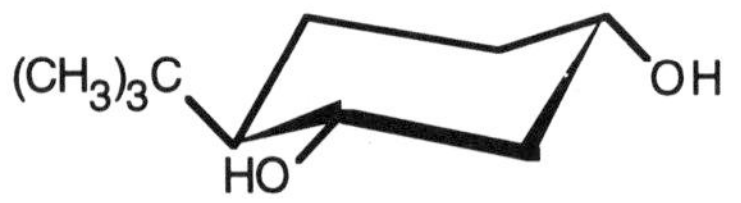

3

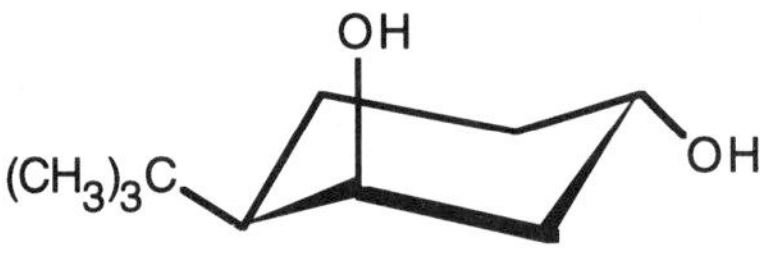

4

Only when the two hydroxyl groups are cis–diaxial (structure 1) can the acetal ring form. In any other conformation, the oxygen atoms are too far apart to be incorporated into a six-membered ring.

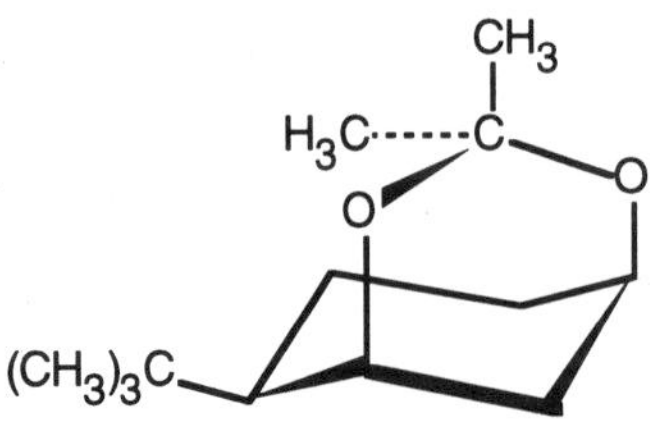

Study Guide for Chapter 4

After studying this chapter, you should be able to:

(1)   Draw energy vs. angle of rotation graphs for single bond conformations (4.2, 4.24, 4.26).

(2)   Draw Newman projections of bond conformations and predict their relative stability (4.3, 4.4, 4.22, 4.23, 4.46).

(3)   Understand the geometry of, and predict the stability of cycloalkanes having fewer than six carbons (4.5, 4.6, 4.7, 4.8, 4.9, 4.10).

(4)   Draw and name substituted cyclohexanes (4.11, 4.12, 4.13, 4.14, 4.28, 4.35, 4.37, 4.41 4.51).

(5)   Predict the stability of substituted cyclohexanes by estimating steric interactions (4.15, 4.16, 4.17, 4.18, 4.19, 4.20, 4.29, 4.30, 4.31, 4.32, 4.33, 4.34, 4.36, 4.38, 4.39, 4.40, 4.42, 4.43, 4.44, 4.45, 4.47, 4.48).

(6)   Define important terms relating to alkane and cycloalkane stereochemistry (4.21, 4.27).

5.1

a) $CH_3Br + KOH \longrightarrow CH_3OH + KBr$ substitution

b) $CH_3CH_2OH \longrightarrow H_2C=CH_2 + H_2O$ elimination

c) $H_2C=CH_2 + H_2 \longrightarrow CH_3CH_3$ addition

5.2

$$CH_3CH_2CH_2\overset{\overset{\displaystyle CH_3}{|}}{C}HCH_2Cl \quad + \quad CH_3CH_2CH_2\overset{\overset{\displaystyle CH_3}{|}}{\underset{\underset{\displaystyle Cl}{|}}{C}}CH_3 \quad +$$

1–Chloro–2–methylpentane   2–Chloro–2–methylpentane

$$CH_3CH_2CH_2\overset{\overset{\displaystyle CH_3}{|}}{C}HCH_3 \quad \xrightarrow[hv]{Cl_2} \quad CH_3CH_2\overset{\overset{\displaystyle CH_3}{|}}{C}H\overset{\underset{\underset{\displaystyle Cl}{|}}{}}{C}HCH_3 \quad + \quad CH_3\overset{\underset{\underset{\displaystyle Cl}{|}}{}}{C}HCH_2\overset{\overset{\displaystyle CH_3}{|}}{C}HCH_3 \quad +$$

2–Methylpentane   3–Chloro–2–methylpentane   2–Chloro–4–methylpentane

$$ClCH_2CH_2CH_2\overset{\overset{\displaystyle CH_3}{|}}{C}HCH_3$$

1–Chloro–4–methylpentane

5.3   Pentane has three types of hydrogen atoms, $\overset{\text{a b c b a}}{CH_3CH_2CH_2CH_2CH_3}$. Although monochlorination produces $CH_3CH_2CH_2CH_2CH_2Cl$, it is not possible to avoid producing $CH_3CH_2CH_2CHClCH_3$ and $CH_3CH_2CHClCH_2CH_3$ as well. Since neopentane has only one type of hydrogen, monochlorination yields a single product.

5.4

a)
$$CH_3\overset{\overset{\displaystyle O^{\delta-}}{\|}}{\underset{\delta+}{C}}CH_3$$

ketone

b)
$$H_2C=CH\overset{\overset{\displaystyle \delta^-O}{\|}}{C}O\,CH_2CH_3$$
$$\underset{\delta+ \ \ \delta-}{}$$

— ester
— double bond

c)

d)

organometallic

5.5 To identify electrophiles, look for Lewis acids, positively charged species, or positively polarized atoms. To identify nucleophiles, look for Lewis bases or negatively charged species.

Electrophiles: H^+, Mg^{2+} Nucleophiles: $H\overset{..}{\underset{..}{O}}:^-$, $:NH_3$

5.6

5.7 According to Table 5.3, a negative $\Delta G°$ indicates that a reaction is favorable. Thus, a reaction with $\Delta G° = -44$ kJ/mol (-11 kcal/mol) is more favorable than a reaction with $\Delta G° = +44$ kJ/mol ($+11$ kcal/mol).

5.8 From the expression $\Delta G° = -RT\ln K_{eq}$, we can see that a large K_{eq} is related to a large negative $\Delta G°$ and thus a large negative $\Delta H°$ (if $\Delta S°$ is small). Consequently, a reaction with $K_{eq} = 1000$ is probably more exothermic than a reaction with $K_{eq} = 0.001$.

5.9 $\Delta G° = -RT\ln K_{eq}$; R = 0.00831 kJ/(K·mol); T = 298 K
     = $-[0.00831$ kJ/(K·mol)$]$ (298 K) $(\ln K_{eq})$ = $(-2.476$ kJ/mol$)$ $(\ln K_{eq})$

If $K_{eq} = 1000$, then $\ln K_{eq} = 6.91$ and $\Delta G° = (-2.476$ kJ/mol$)$ $(6.91) = -17.1$ kJ/mol

If $K_{eq} = 1$, then $\ln K_{eq} = 0$ and $\Delta G° = 0$

If $K_{eq} = 0.001$, then $\Delta G° = +17.1$ kJ/mol

$\Delta G° = -RT\ln K_{eq}$; $\ln K_{eq} = -\Delta G°/RT$; T = 298 K
$\ln K_{eq} = (-\Delta G°) / (2.476$ kJ/mol$)$

If $\Delta G° = -40$ kJ/mol, then $\ln K_{eq} = -(-40$ kJ/mol $) / (2.476$ kJ/mol$) = 16.15$;
     $K_{eq} = 1.0 \times 10^7$

If $\Delta G° = 0$, then $K_{eq} = 1$

If $\Delta G° = +40$ kJ/mol, then $\ln K_{eq} = (-40$ kJ/mol$) / (2.476$ kJ/mol$) = -16.15$;
     $K_{eq} = 1.0 \times 10^{-7}$

5.10

a) $Cl_2 \longrightarrow 2Cl\cdot$

Bond broken	D
Cl—Cl	243 kJ/mol

$\Delta H_a^{\circ} = +243$ kJ/mol

b) $CH_4 + Cl\cdot \longrightarrow \cdot CH_3 + HCl$

Bond broken	D	Bond formed	D
CH$_3$—H	+438 kJ/mol	H—Cl	+432 kJ/mol

$\Delta H_b^{\circ} = \underset{\substack{\text{bond} \\ \text{broken}}}{D} - \underset{\substack{\text{bond} \\ \text{formed}}}{D} = 438$ kJ/mol $- 432$ kJ/mol $= +6$ kJ/mol

c) $\cdot CH_3 + Cl_2 \longrightarrow CH_3Cl + \cdot Cl$

Bond broken	D	Bond formed	D
Cl—Cl	+243 kJ/mol	CH$_3$—Cl	+351 kJ/mol

$\Delta H_c^{\circ} = \underset{\substack{\text{bond} \\ \text{broken}}}{D} - \underset{\substack{\text{bond} \\ \text{formed}}}{D} = 243$ kJ/mol $- 351$ kJ/mol $= -108$ kJ/mol

d) $\Delta H_{\text{overall}}^{\circ} = \Delta H_b^{\circ} + \Delta H_c^{\circ} = -108$ kJ/mol $+ 6$ kJ/mol $= -102$ kJ/mol

The overall reaction between chlorine and methane is exothermic.

5.11

a) $CH_3CH_2OCH_3 + HI \longrightarrow CH_3CH_2OH + CH_3I$

Bonds broken	D	Bonds formed	D
CH$_3$CH$_2$O–CH$_3$	339 kJ/mol	CH$_3$CH$_2$O–H	436 kJ/mol
H–I	298 kJ/mol	CH$_3$–I	234 kJ/mol
$\underset{\substack{\text{bonds} \\ \text{broken}}}{D} = 637$ kJ/mol		$\underset{\substack{\text{bonds} \\ \text{formed}}}{D} = 670$ kJ/mol	

$\Delta H_{\text{overall}}^{\circ} = \underset{\substack{\text{bonds} \\ \text{broken}}}{D} - \underset{\substack{\text{bonds} \\ \text{formed}}}{D} = 637$ kJ/mol $- 670$ kJ/mol $= -33$ kJ/mol

b) $CH_3Cl + NH_3 \longrightarrow CH_3NH_2 + HCl$

Bonds broken	D	Bonds formed	D
CH_3-Cl	351 kJ/mol	CH_3-NH_2	335 kJ/mol
NH_2-H	449 kJ/mol	$H-Cl$	432 kJ/mol
$D_{\substack{bonds \\ broken}} =$	800 kJ/mol	$D_{\substack{bonds \\ formed}} =$	767 kJ/mol

$\Delta H°_{overall} = D_{\substack{bonds \\ broken}} - D_{\substack{bonds \\ formed}} =$ 800 kJ/mol − 767 kJ/mol = + 33 kJ/mol

5.12   A reaction with $\Delta G^{\ddagger}$ = 45 kJ/mol is faster than a reaction with $\Delta G^{\ddagger}$ = 70 kJ/mol. It is not possible to measure the size of K_{eq} from $\Delta G^{\ddagger}$ because $\Delta G^{\ddagger}$ measures the energy difference between reactant and *transition state* rather than between reactant and *product*. The energy difference between reactant and product is described by $\Delta G°$, and thus also by K_{eq}.

5.13

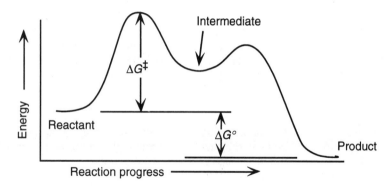

5.14

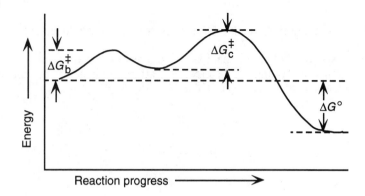

Refer to problem 5.10 for a description of steps b and c of the chlorination reaction. Since $\Delta H°_{overall}$ is negative, $\Delta G°$ is also negative.

5.15 – 5.16

a) $\overset{\delta+}{C}H_3\overset{}{C}H_2\overset{\delta+\ \delta-}{C}\equiv N$

nitrile

b)

ether

c)

ketone   ester

d) ketone

double bonds

e) double bond

amide

f)

aldehyde

aromatic ring

5.17   a) substitution   b) elimination   c) addition   d) substitution

5.18   An *addition reaction* takes place when two reactants form a single product.
An *elimination reaction* takes place when one reactant splits apart to give two products.
A *substitution reaction* occurs when two reactants exchange parts to yield two different products.
A *rearrangement reaction* occurs when a reactant undergoes a reorganization of bonds to give a different product.

5.19   a) A *polar reaction* is a process that involves unsymmetrical bond making and breaking. In a polar reaction, electron-rich sites in the functional groups of one molecule react with electron-poor sites in the functional groups of another molecule.
b) *Heterolytic bond breakage* occurs when both bonding electrons leave with one fragment.
c) *Homolytic bond breakage* occurs when one bonding electron leaves with each fragment.
d) A *radical reaction* is a reaction in which odd-electron species are produced or consumed.
e) A *functional group* is a group of atoms that has a characteristic reactivity.
f) *Polarization* is the temporary change in the electron distribution in atoms or functional groups due to interactions with reagents or solvent.

5.20

a) $H\ddot{O}:^-$

b) H^+

c), d) $CH_3Br \ + \ H\ddot{O}:^- \ \longrightarrow \ CH_3OH \ + :\ddot{B}r:^-$

e) $H\ddot{O}-H \ \longrightarrow \ H\ddot{O}:^- \ + \ H^+$

f) $Cl-Cl \ \longrightarrow \ 2 \ :\ddot{C}l\cdot$

5.21

Nucleophiles: $:\ddot{\underset{..}{C}}\overset{..}{l}:^-$, $CH_3\overset{..}{N}H_2$

Electrophile: BF_3

5.22

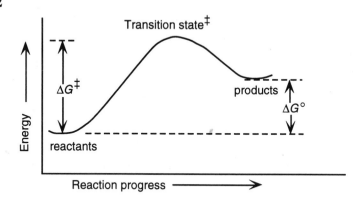

$\Delta G°$ is positive.

5.23

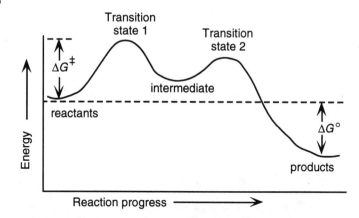

$\Delta G°$ is negative.

5.24   A transition state represents a structure occurring at an energy maximum. An intermediate occurs at an energy minimum between two transition states. Even though an intermediate may be of such high energy that it cannot be isolated, it is still of lower energy than a transition state.

5.25   Problem 5.23 shows a reaction energy diagram of a two-step exothermic reaction. Step 2 is faster than step 1 because $\Delta G^{\ddagger}_2 < \Delta G^{\ddagger}_1$.

5.26

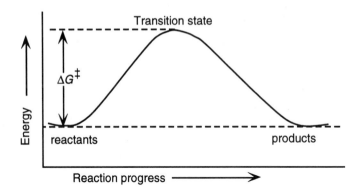

A reaction with $K_{eq} = 1$ has $\Delta G° = 0$.

5.27

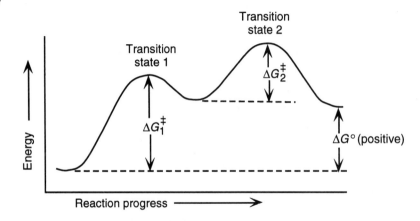

(a) $\Delta G°$ is positive.
(b)   There are two steps in the reaction.
(c) Step 2 is faster because $\Delta G^{\ddagger}_2 < \Delta G^{\ddagger}_1$
(d) There are two transition states, as indicated on the diagram.

5.28

a) $CH_3—OH$ + $H—Br$ $\longrightarrow$ $CH_3—Br$ + $H—OH$

    380 kJ/mol   366 kJ/mol   293 kJ/mol   498 kJ/mol

    $\Delta H°$ = −45 kJ/mol

b) $CH_3CH_2O–H$ + $CH_3–Cl$ $\longrightarrow$ $CH_3CH_2O–CH_3$ + $H–Cl$

    436 kJ/mol   351 kJ/mol   339 kJ/mol   432 kJ/mol

    $\Delta H°$ = + 16 kJ/mol

5.29

a) $CH_3CH_2–H$ + Cl_2 $\longrightarrow$ $CH_3CH_2–Cl$ + $H–Cl$

    420 kJ/mol   243 kJ/mol   338 kJ/mol   432 kJ/mol

    $\Delta H°$ = −107 kJ/mol

b) CH_3CH_2-H + Br_2 $\longrightarrow$ CH_3CH_2-Br + $H-Br$

420 kJ/mol   193 kJ/mol   285 kJ/mol   366 kJ/mol

$\Delta H° = -38$ kJ/mol

c) CH_3CH_2-H + I_2 $\longrightarrow$ CH_3CH_2-I + $H-I$

420 kJ/mol   151 kJ/mol   222 kJ/mol   298 kJ/mol

$\Delta H° = +51$ kJ/mol

Of the three halogenation reactions, chlorination is energetically the most favorable.

5.30

Bonds broken	_D_	Bonds formed	_D_
CH_3-CH_3	376 kJ/mol	CH_3-Br	293 kJ/mol
Br_2	193 kJ/mol	CH_3-Br	293 kJ/mol
	_D_ = 569 kJ/mol		_D_ = 586 kJ/mol

$\Delta H°_{overall} = \underset{\text{bonds broken}}{D} - \underset{\text{bonds formed}}{D} = 569$ kJ/mol $- 586$ kJ/mol $= -17$ kJ/mol

$\Delta H°$ for bromoethane formation is -38 kJ/mol; $\Delta H°$ for bromomethane formation is -17 kJ/mol. Although both of these reactions have negative $\Delta H°$, the reaction that forms bromoethane is more favorable.

5.31   Irradiation initiates the chlorination reaction by producing chlorine radicals. Although these radicals are consumed in the propagation steps, new Cl· radicals are formed to carry on the reaction. After irradiation stops, chlorine radicals are still present to carry on the propagation steps, but, as time goes on, radicals combine with each other in termination reactions that remove radicals from the reaction mixture. Because the number of radicals decreases, fewer propagation cycles occur, and the reaction gradually slows down and stops.

5.32, 5.33

	Product			$\Delta H°$

$$CH_3CH_2\overset{\underset{\displaystyle |}{CH_3}}{C}HCH_3 + Cl_2 \xrightarrow{\text{light}}$$

$$\overset{\underset{\displaystyle |}{CH_3}}{CH_3CH_2CHCH_2Cl} \qquad + \quad HCl \qquad -107 \text{ kJ/mol}$$
1–Chloro–2–methylbutane

$$\overset{\underset{\displaystyle |}{CH_3}}{ClCH_2CH_2CHCH_3} \qquad + \quad HCl \qquad -107 \text{ kJ/mol}$$
1–Chloro–3–methylbutane

$$\overset{\underset{\displaystyle |}{CH_3}}{CH_3CHClCHCH_3} \qquad + \quad HCl \qquad -127 \text{ kJ/mol}$$
2–Chloro–3–methylbutane

$$\overset{\underset{\displaystyle |}{CH_3}}{CH_3CH_2CClCH_3} \qquad + \quad HCl \qquad -129 \text{ kJ/mol}$$
2–Chloro–2–methylbutane

Formation of the tertiary chloride is favored, but a mixture of products is expected since all three $\Delta H°$ values are quite close to one another.

5.34 The following compounds yield single monohalogenation products because each has only one kind of hydrogen atom.

$$C_2H_6, \qquad \bigcirc, \qquad CH_3C{\equiv}CCH_3$$

5.35 For the first series of steps:
a) $\Delta H° = +243$ kJ/mol
b) $\Delta H° = +6$ kJ/mol $\quad | \quad \Delta H°_{\text{overall}} = -102$ kJ/mol
c) $\Delta H° = -108$ kJ/mol

For the alternate series:
a) $\Delta H° = +243$ kJ/mol
b) $\Delta H° = +87$ kJ/mol $\quad | \quad \Delta H°_{\text{overall}} = -102$ kJ/mol
c) $\Delta H° = -189$ kJ/mol

For both series of reactions, the radical-producing initiation step has $\Delta H° = +243$ kJ/mol and the propagation steps b + c have $\Delta H° = -102$ kJ/mol. In series 2, however, one step has $\Delta H° = +87$ kJ/mol; this step is energetically much less favorable than any step in series 1 and disfavors the second series as a whole. The first route for chlorination of methane is thus more likely to occur.

5.36 a) The reaction is a polar rearrangement.

b)

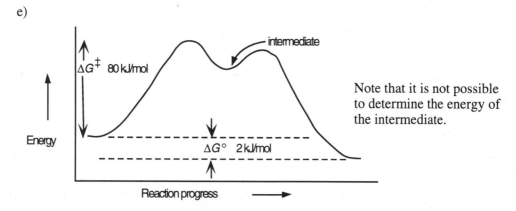

c) $K_{eq} = \dfrac{[Products]}{[Reactants]} = \dfrac{.70}{.30} = 2.3$

d) Section 5.8 states that reactions that occur spontaneously have $\Delta G^{\ddagger}$ of less than 80 kJ/mol at room temperature. Since this reaction proceeds slowly at room temperature, $\Delta G^{\ddagger}$ is probably close to 80 kJ/mol.

e)

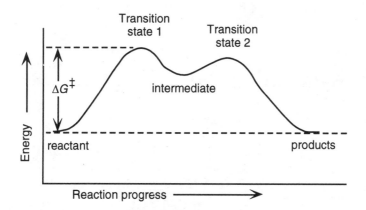

Note that it is not possible to determine the energy of the intermediate.

5.37

$\Delta G^{\ddagger}$ is approximately 80 kJ/mol. ΔG° is approximately zero.

5.38 $\Delta G° = \Delta H° - T\Delta S°$
$= -75 \text{ kJ/mol} - (298 \text{ K}) [0.054 \text{ kJ/(K·mol)}]$
$= -75 \text{ kJ/mol} - 16 \text{ kJ/mol}$
$= -91 \text{ kJ/mol}$

The reaction is exothermic.

5.39 $\Delta G° = -RT\ln K_{eq}$
$= [-0.008315 \text{ kJ/(K·mol)}] (298 \text{ K}) (\ln K_{eq}) = 91 \text{ kJ/mol}$

$\ln K_{eq} = (-91 \text{ kJ/mol}) / (-2.48 \text{ kJ/mol}) = 36.7$

$K_{eq} = 8.7 \text{ x } 10^{15}$

5.40

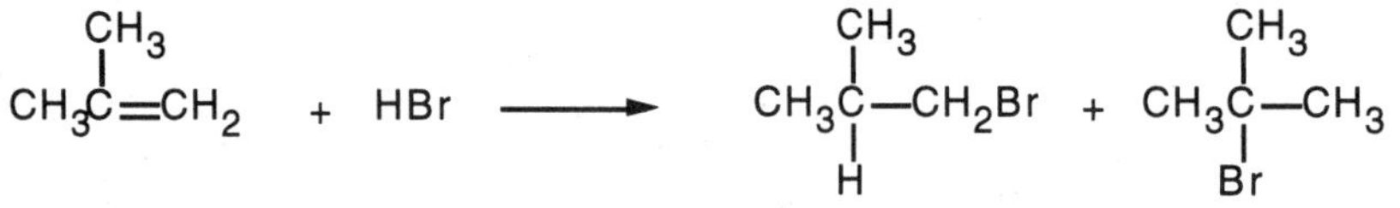

2-Methylpropene   1-Bromo-2-methyl-   2-Bromo-2-methyl-
                              propane   propane

5.41

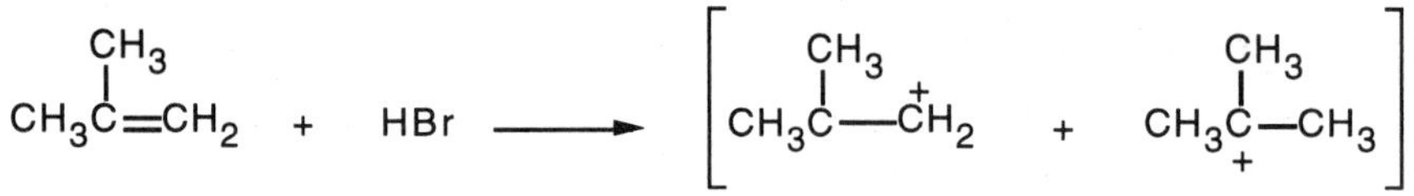

The second carbocation is more stable because more alkyl substituents are bonded to it.

5.42

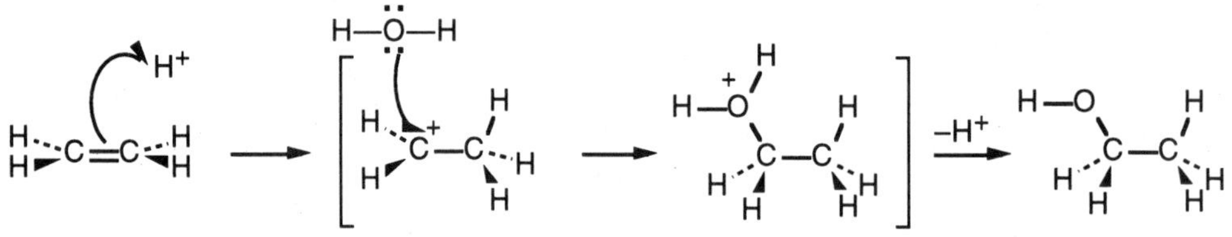

Study Guide for Chapter 5

After studying this chapter, you should be able to:

(1)   Identify reactions as polar, radical, substitution, elimination, addition or rearrangement reactions, and define these terms (5.1, 5.17, 5.18, 5.19).

(2)   Understand the mechanism of radical reactions (5.3, 5.4, 5.31, 5.34).

(3)   Predict the direction of bond polarity for a given functional group (5.4, 5.16).

(4)   Identify reagents as electrophiles or nucleophiles (5.5, 5.20, 5.21).

(5)   Understand the concepts of equilibrium and rate (5.7, 5.8, 5.12, 5.24).

(6)   Calculate $\Delta G°$ and K_{eq} of reactions (5.9, 5.36, 5.38, 5.39).

(7)   Use bond dissociation energies to calculate $\Delta H°$ of simple reactions (5.10, 5.11, 5.28, 5.29, 5.30, 5.32, 5.35).

(8)   Draw reaction energy diagrams and label them properly (5.13, 5.14, 5.22, 5.23, 5.25, 5.26, 5.27, 5.36, 5.37).

6.1 "Degree of unsaturation" refers to the number of rings, double bonds, or triple bonds that a compound may contain. As an example, consider the hydrocarbon C_8H_{14} in part (a). The formula of a C_8 alkane is C_8H_{18}. C_8H_{14}, which contains four fewer (or two pairs fewer) hydrogens than C_8H_{18}, may have two double bonds, or two rings, or one of each, or one triple bond. C_8H_{14} thus has a degree of unsaturation of 2.

Compound	Degree of Unsaturation
a) C_8H_{14}	2
b) C_5H_6	3
c) $C_{12}H_{20}$	3
d) $C_{20}H_{32}$	5
e) $C_{40}H_{56}$	13

6.2

Compound	Degree of Unsaturation	Structures
a) C_4H_8	1	$CH_3CH_2CH=CH_2$, $CH_3CH=CHCH_3$, $(CH_3)_2C=CH_2$
b) C_4H_6	2	$CH_2=CHCH=CH_2$, $CH_3CH=C=CH_2$, $CH_3C\equiv CCH_3$
c) C_3H_4	2	$CH_2=C=CH_2$, $CH_3C\equiv CH$,

6.3

a) Subtract one hydrogen for each nitrogen present to find the equivalent hydrocarbon formula C_6H_4. Compared to the alkane C_6H_{14}, the compound of formula C_6H_4 has ten fewer hydrogens, or five fewer hydrogen pairs, and contains five degrees of unsaturation.

b) $C_6H_5NO_2$ also contains five degrees of unsaturation because oxygen does not affect the equivalent hydrocarbon formula of a compound.

c) A halogen atom is equivalent to a hydrogen atom in the calculation of the equivalent hydrocarbon formula. Here, the formula is C_8H_{12}. $C_8H_9Cl_3$ has three degrees of unsaturation.

d) $C_9H_{16}Br_2$ – one degree of unsaturation.

e) $C_{10}H_{12}N_2O_3$ – six degrees of unsaturation.

f) $C_{20}H_{32}O_2$ – five degrees of unsaturation.

6.4

a)

$$\underset{H_2C=CHCH-CCH_3}{\overset{\overset{\displaystyle H_3C \quad CH_3}{|\quad\ |}}{\underset{|}{\overset{1\ \ 2\ \ |3\ \ 4|5}{}}}}$$
$$CH_3$$

1) Find the longest carbon chain containing the double bond, and name the parent compound. Here the longest chain contains five carbons, and the compound is a *pentene*.
2) Number the carbon atoms, giving to the double bond the lowest possible number.
3) Name the compound: 3,4,4–trimethyl–1–pentene.

b)
$$\overset{\overset{\displaystyle CH_3}{|}}{CH_3CH_2CH=CCH_2CH_3}$$
3–Methyl–3–hexene

c)
$$\overset{\overset{\displaystyle CH_3 \qquad\ CH_3}{|\qquad\quad\ |}}{CH_3CH=CHCHCH=CHCHCH_3}$$
4,7–Dimethyl–2,5–octadiene

6.5

a)
$$\overset{\overset{\displaystyle CH_3}{|}}{CH_2=CHCH_2CH_2C=CH_2}$$
2–Methyl–1,5–hexadiene

b)
$$\overset{\overset{\displaystyle CH_2CH_3}{|}}{CH_3CH_2CH_2CH=CC(CH_3)_3}$$
3–Ethyl–2,2–dimethyl–3–heptene

c)
$$\overset{\overset{\displaystyle CH_3}{|}}{CH_3CH=CHCH=CHC-\!\!\underset{\underset{\displaystyle CH_3\ CH_3}{|\quad\ |}}{}\!\!C=CH_2}$$
2,3,3–Trimethyl–1,4,6–octatriene

d)

$$\begin{array}{cc} CH_3 & CH_3 \\ | & | \\ CH_3CH & CHCH_3 \\ \diagdown & \diagup \\ C=C \\ \diagup & \diagdown \\ CH_3CH & CHCH_3 \\ | & | \\ CH_3 & CH_3 \end{array}$$

3,4–Diisopropyl–2,5–dimethyl–3–hexene

e)
$$\overset{\overset{\displaystyle C(CH_3)_3}{|}}{CH_3CH_2CH_2CHCH_2CHCH_3}$$
$$CH_3$$
4–*tert*–Butyl–2–methylheptane

6.6

a)

1,2–Dimethylcyclohexene

b)

4,4–Dimethylcycloheptene

c)

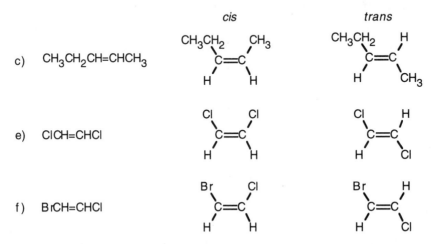

3-Isopropylcyclopentene

6.7   Compounds (c), (e), and (f) can exist as pairs of cis–trans isomers.

cis   *trans*

c) $CH_3CH_2CH=CHCH_3$

e) $ClCH=CHCl$

f) $BrCH=CHCl$

6.8   Models are essential here! A model of cyclohexene shows that a six-membered ring is too small to contain a trans double bond without causing severe strain to the ring. A ten-membered ring is flexible enough to accommodate either a cis or a trans double bond, although the cis isomer has less ring strain than the trans isomer.

6.9   Review the sequence rules of Section 6.6. In summary:

Rule 1: A high-atomic-weight atom has priority over a low-atomic weight atom.

Rule 2: If a decision can't be reached by Rule 1, look at the second, third or fourth atom out until a decision can be made.

Rule 3: Multiple-bonded atoms are considered to be equivalent to the same number of single bonded atoms.

	High	Low	Rule
a)	–Br	–H	1
b)	–Br	–Cl	1
c)	–CH$_2$CH$_3$	–CH$_3$	2
d)	–OH	–NH$_2$	1
e)	–CH$_2$OH	–CH$_3$	2
f)	–CH=O	–CH$_2$OH	3

6.10

Highest priority ⟶ Lowest priority

a)   –Cl, –OH, –CH$_3$, –H

b)   –CH$_2$OH, –CH=CH$_2$, –CH$_2$CH$_3$, –CH$_3$

c)   –COOH, –CH$_2$OH, –CN, –CH$_2$NH$_2$

d)   –CH$_2$OCH$_3$, –CN, –C≡CH, –CH$_2$CH$_3$

6.11

a)

(L)   H$_3$C   CH$_2$OH   (L)

C=C

(H) CH$_3$CH$_2$ Cl   (H)

Z

First, consider substituents on the right side of the double bond. –Cl ranks higher than –CH$_2$OH by Cahn-Ingold-Prelog rules. On the left side, –CH$_2$CH$_3$ ranks higher than –CH$_3$. The isomer has Z configuration because the higher priority groups are on the same side of the double bond.

b)

(H) Cl   CH$_2$CH$_3$ (L)

C=C

(L) CH$_3$O   CH$_2$CH$_2$CH$_3$ (H)

E

c)

(H) CH$_3$

COOH   (H)

C=C

(L)   CH$_2$OH   (L)

Z

Notice that in ranking substituents on the left side of the bond, the upper substituent is of higher priority because of the methyl group attached to the ring.

d)

(L) H   CN   (H)

C=C

(H)   CH$_3$ CH$_2$NH$_2$ (L)

E

6.12

	More stable	Less stable

a)

CH$_3$ H

C=C

CH$_3$ H

disubstituted double bond

CH$_3$CH$_2$ H

C=C

H   H

monosubstituted double bond

	More stable	Less stable

b)

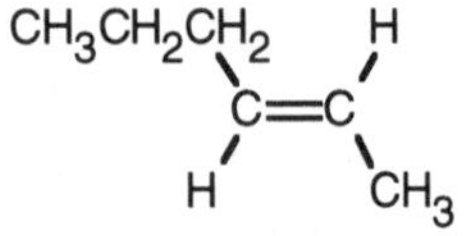

no steric strain

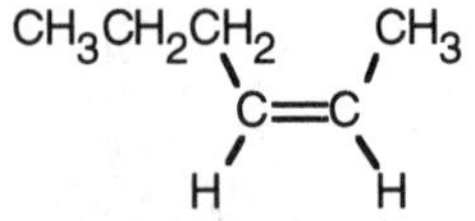

E

steric strain of groups on
same side of double bond

Z

c)

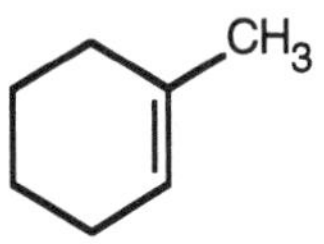

trisubstituted double bond

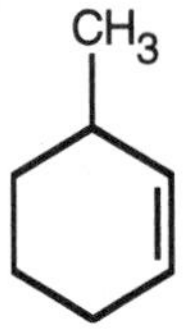

disubstituted double bond

6.13

a)

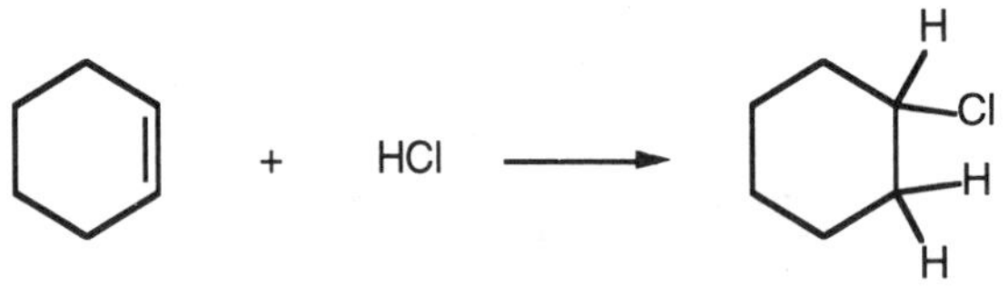

+   HCl   →   Chlorocyclohexane

Since the starting material is symmetrical, only one product is possible.

b) $(CH_3)_2C=CHCH_2CH_3$ $\xrightarrow{\text{HBr}}$ $(CH_3)_2\overset{\overset{\text{Br}}{|}}{C}CH_2CH_2CH_3$

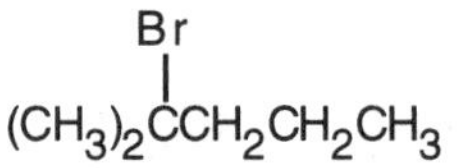

2–Bromo–2–methylpentane

c) $CH_3CH_2CH_2CH=CH_2$ $\xrightarrow{H_3PO_4,\ KI}$ $CH_3CH_2CH_2\overset{\overset{\text{I}}{|}}{C}HCH_3$

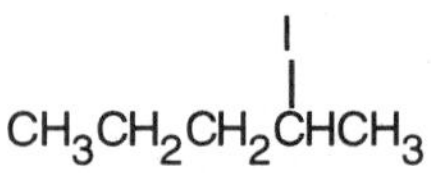

2–Iodopentane

d)

+   HBr   →

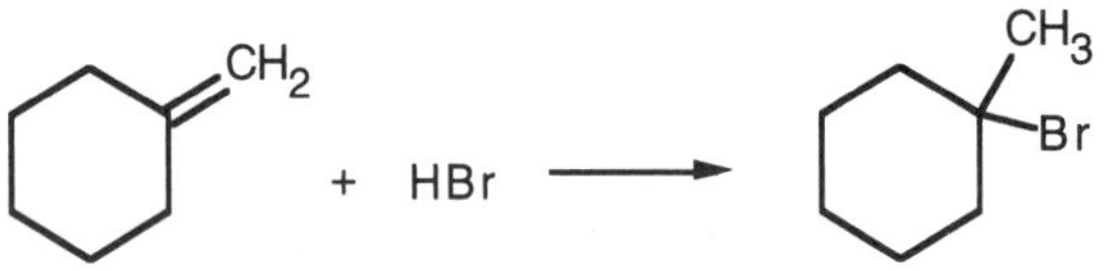

1–Bromo–1–methylcyclohexane

6.14

a)

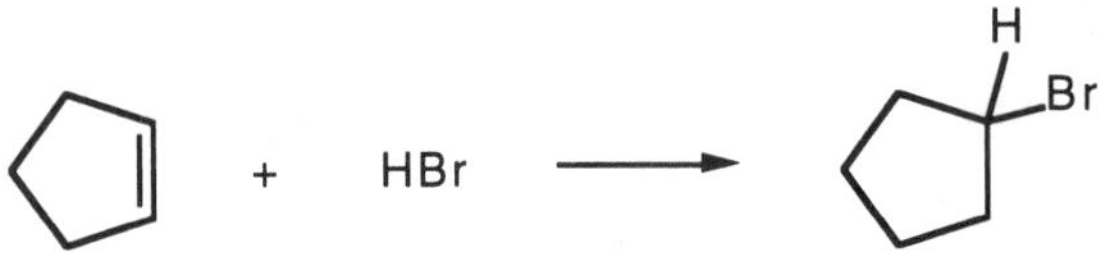

+   HBr   →

Cyclopentene

b)   CH₃CH₂CH=CHCH₂CH₃ + HBr ⟶ CH₃CH₂CHBrCH₂CH₂CH₃

3–Hexene

c)

or

KI
⟶
H₃PO₄

d)

+   HCl   ⟶

6.15

a)   CH₃CH₂C=CHCHCH₃ + HBr ⟶ [CH₃CH₂C—CH₂CHCH₃] ⟶ CH₃CH₂C—CH₂CHCH₃

carbocation intermediate

b)

=CHCH₃ + HI ⟶ [—CH₂CH₃] ⟶

carbocation intermediate

6.16 The second step in the electrophilic addition of HCl to alkenes is exothermic. According to the Hammond postulate, the transition state should resemble the carbocation intermediate.

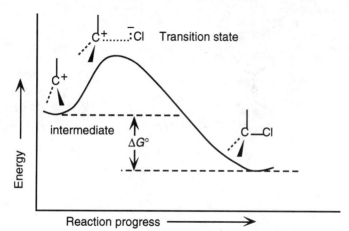

6.17

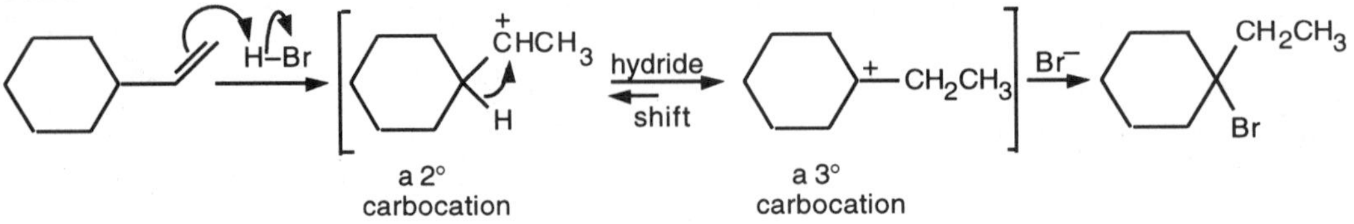

a 2° carbocation	a 3° carbocation

6.18   *Formula*   *Degree of Unsaturation*

a) C_6H_6 4
b) C_6H_{10} 2
c) $C_{10}H_{16}$ 3
d) $C_6H_6Cl_6$ 1
e) C_5H_5N 4
f) $C_{10}H_{10}O_2$ 6

6.19   The purpose of this problem is to give you experience in calculating the number of double bonds and/or rings in a formula. Additionally, you will gain practice in writing structures containing various functional groups. Remember that any formulas that satisfy the rules of valency are acceptable. Try to identify functional groups in the formulas you draw.

a) $C_{10}H_{16}$ — three degrees of unsaturation. Examples:

$$CH_3CH_2CH=CHCH=CHCH=CHCH_2CH_3$$

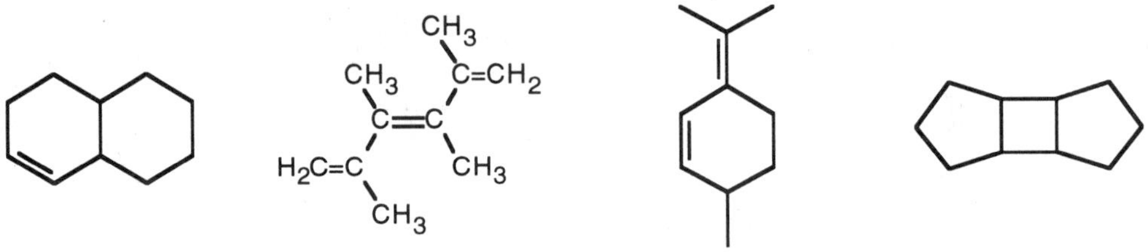

b) C_8H_8O. The equivalent hydrocarbon is C_8H_8; each structure contains five degrees of unsaturation. Examples:

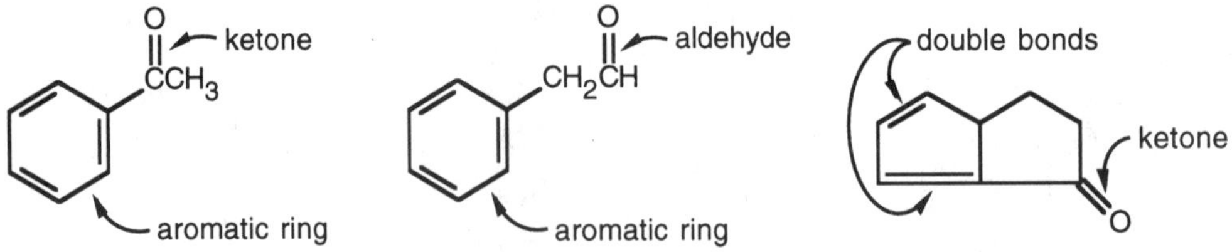

c) This compound has C_7H_{12} as its equivalent hydrocarbon formula. $C_7H_{10}Cl_2$ has two degrees of unsaturation. Examples:

d) $C_{10}H_{16}O_2$ -- three degrees of unsaturation.

e) $C_5H_9NO_2$ -- two degrees of unsaturation.

ketone
amide

$CH_3CH_2CCH_2CNH_2$

double bond
nitro group

$CH_2=CHCH_2CH_2CH_2N^+O^-$

carboxylic acid

$CH_2=CHCH_2NHCH_2COH$

amine
double bond

alcohol
OH
amide

ketone
NH_2
amine
ether

f) $C_8H_{10}ClNO$ -- four degrees of unsaturation.

halide
amide

$ClCH=CHCH=CHCH=CHCH_2CNH_2$

double bonds

double bonds
ether
halide
Cl
amine

alcohol CH_3 aromatic ring
amine
HO CH_2NH_2
Cl
halide

halide
Cl
double bonds
ketone
H_2N O
amine

ketone
H N   amine
O
halide
Cl

6.20 Interpreting problems of this sort is often difficult. To start, you should train yourself to *read every word* of the problem. Then you should try to solve the problem phrase by phrase. For example, "A compound of formula $C_{10}H_{14}$" describes a compound having four degrees of unsaturation ($C_{10}H_{14}$ has four fewer hydrogen pairs than a C_{10} acyclic alkane). The phrase "undergoes catalytic hydrogenation" means that H_2 is added to the double bonds. "Absorbs only two equivalents of H_2" means that only two of the degrees of unsaturation are double bonds (or a triple bond). The other two must be rings.

6.21 A compound of formula $C_{12}H_{13}N$ has as its equivalent hydrocarbon $C_{12}H_{12}$, which has seven degrees of unsaturation. Since two of the unsaturations are due to rings, the other five are due to double or triple bonds. Thus, $C_{12}H_{13}N$ absorbs five equivalents of hydrogen.

6.22   a) 4–Methyl–2–hexene   b) 4–Butyl–7–methyl–2–octene
         c) 2–Ethyl–1–butene   d) 3,4–Dimethyl–1,5–heptadiene
         e) 4–Methyl–1,3–hexadiene   f) 1,2–Butadiene
         g) 3,3–Dimethyl–1–butene   h) 2,2,5,5–Tetramethyl–3–hexene

6.23   Because the longest carbon chain contains 8 carbons, and because there are three double bonds present, ocimene is an *octatriene*. Start numbering at the end that will give the lower number to the first double bond (1,3,6 is lower than 2,5,7). Number the methyl substituents and, finally, name the compound.

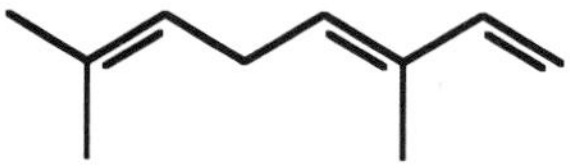

(3*E*)–3,7–Dimethyl–1,3,6–octatriene

6.24, 6.25

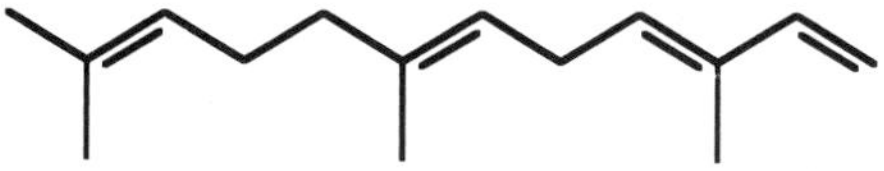

(3*E*, 6*E*)–3,7,11–Trimethyl–1,3,6,10–dodecatetraene

6.26

a)

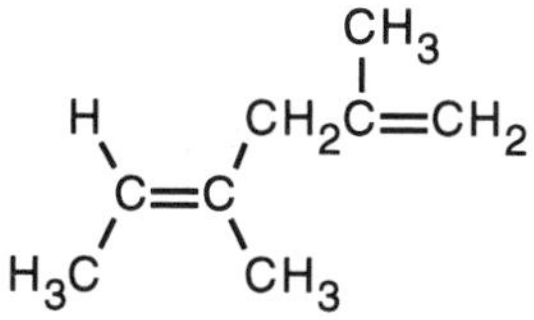

b)

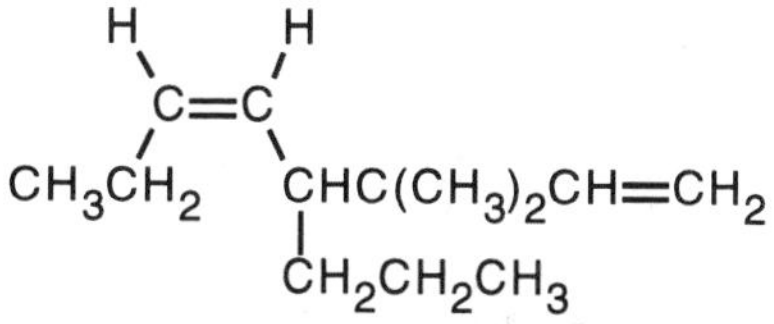

c)

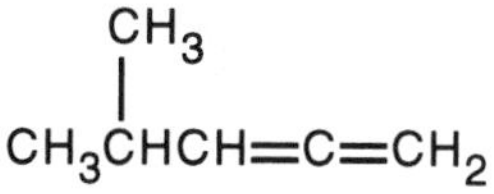

d)

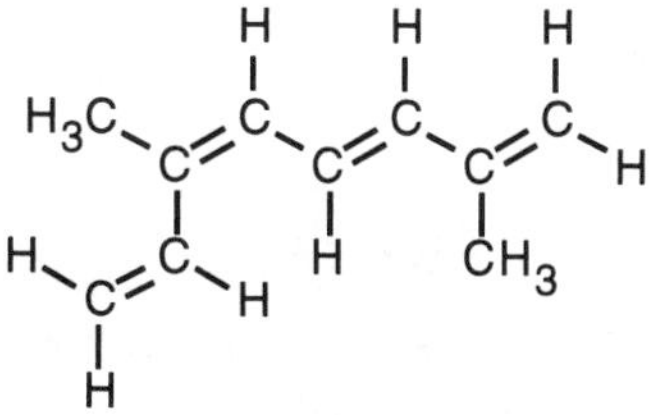

e)

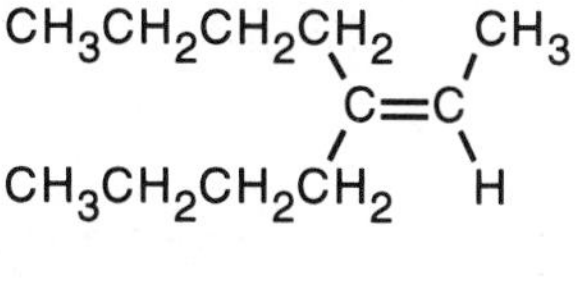

f)

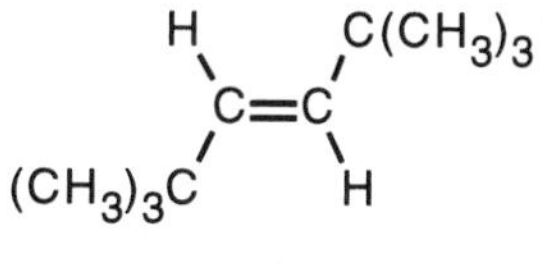

6.27

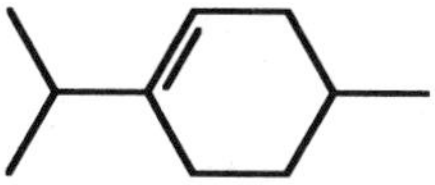

Menthene

6.28

a) H_2C=$CHCH$=$C(CH_3)_2$

Correct name: 4-Methyl-1,3-pentadiene.
Numbering must start at the other end.

b)
$$CH_3CH_2\overset{\overset{\displaystyle CH_2}{\|}}{C}CH=CH_2$$

Correct name: 2-Ethyl 1,3-butadiene.
The parent chain must contain both
double bonds.

c)

Correct name: (2Z,5E)-2,5-Octadiene.
Numbering must start at the other end.

d)

Correct name: Z-4-Ethyl-4-octene.
Numbering must start at the other end.

e)

Correct name: E-4-Ethyl-4-octene.
The longest chain containing the double
bond is an octene.

f)

H_2C=$CHCH_2CH$=CH_2

Correct name: 1,4-Pentadiene.
The parent chain must contain both
double bonds.

6.29

$CH_3CH_2CH_2CH$=CH_2

1–Pentene

CH_3CH_2CH=$CHCH_3$

2–Pentene

$$CH_3CH_2\overset{\overset{\displaystyle CH_3}{|}}{C}=CH_2$$

2–Methyl–1–butene

$$CH_3\overset{\overset{\displaystyle CH_3}{|}}{C}HCH=CH_2$$

3–Methyl–1–butene

$$CH_3CH=\overset{\overset{\displaystyle CH_3}{|}}{C}CH_3$$

2–Methyl–2–butene

6.30

$CH_3CH_2CH_2CH_2CH$=CH_2

1–Hexene

$CH_3CH_2CH_2CH$=$CHCH_3$

2–Hexene

CH_3CH_2CH=$CHCH_2CH_3$

3–Hexene

$$CH_3CH_2CH_2\overset{\overset{\displaystyle CH_3}{|}}{C}=CH_2$$

2–Methyl–1–pentene

$$CH_3CH_2\overset{\overset{\displaystyle CH_3}{|}}{C}HCH=CH_2$$

3–Methyl–1–pentene

$$CH_3\overset{\overset{\displaystyle CH_3}{|}}{C}HCH_2CH=CH_2$$

4–Methyl–1–pentene

$$CH_3CH_2CH=\overset{\overset{\displaystyle CH_3}{|}}{C}CH_3$$

2–Methyl–2–pentene

$$CH_3CH_2\overset{\overset{\displaystyle CH_3}{|}}{C}=CHCH_3$$

3–Methyl–2–pentene

$$CH_3\overset{\overset{\displaystyle CH_3}{|}}{C}HCH=CHCH_3$$

4–Methyl–2–pentene

$$\underset{\overset{|}{CH_3}}{CH_3CHC=CH_2}$$

CH₃
|
CH₃CHC=CH₂
|
CH₃

2,3–Dimethyl–1–butene

CH₃
|
CH₃CCH=CH₂
|
CH₃

3,3–Dimethyl–1–butene

CH₃   CH₃
\   /
C=C
/   \
CH₃   CH₃

2,3–Dimethyl–2–butene

CH₂CH₃
|
CH₃CH₂C=CH₂

2–Ethyl–1–butene

6.31   In Problem 6.29, only 2–pentene shows cis–trans isomerism. In Problem 6.30, 2–hexene, 3–hexene, 3–methyl–2–pentene, and 4–methyl–2–pentene show cis–trans isomerism.

6.32   As expected, the two trans compounds are more stable than their cis counterparts. The cis–trans difference is much more extreme for the tetramethyl compound, however. Build a model of *cis*–2,2,5,5–tetramethyl–3–hexene and notice the extreme crowding of the methyl groups. Steric interference makes the cis isomer much less stable than the trans isomer and causes *cis* ΔH_{hydrog} to have a much larger negative value than *trans* ΔH_{hydrog} for the hexane isomers.

6.33   In all these examples, ΔH_{hydrog} for the trans cycloalkenes is a larger negative number than ΔH_{hydrog} for the cis compounds, indicating that the trans cycloalkenes are less stable than cis cycloalkenes. Build models of the two cyclooctenes and notice the large amount of strain in *trans*–cyclooctene, relative to *cis*–cyclooctene. This strain causes the trans isomer to be of higher energy and to have a ΔH_{hydrog} larger than the *cis* isomer. Use models to construct the other four cycloalkenes. As ring size increases, the problem of strain for the *trans* rings becomes less severe, and ΔH_{hydrog} becomes a smaller negative number.

6.34   The central carbon of allene forms two sigma bonds and two pi bonds. The central carbon is *sp* hybridized, and the carbon-carbon bond angle is 180°, indicating linear geometry for the carbons of allene.

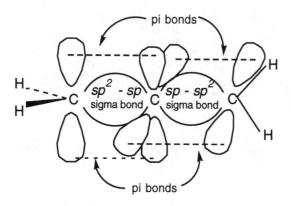

6.35   Because its heat of hydrogenation is so much larger, 1,2–pentadiene is less stable than 1,4–pentadiene. This instability may be due to strain encountered when one carbon must form a double bond to each of two different carbons.

6.36

a) 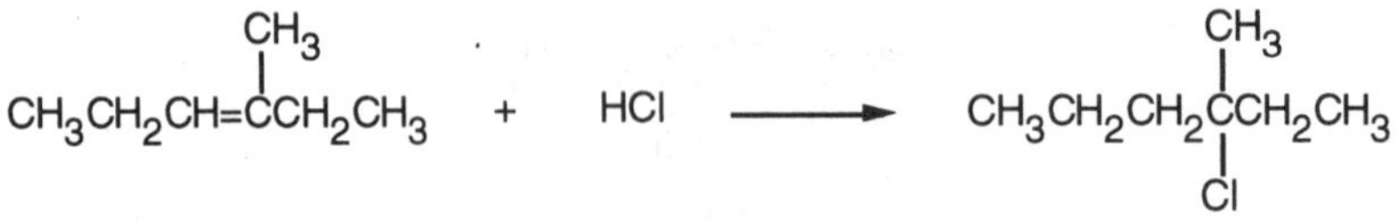 $CH_3CH_2CH=\overset{\underset{|}{CH_3}}{C}CH_2CH_3$ + HCl ⟶ $CH_3CH_2CH_2\overset{\underset{|}{CH_3}}{\underset{|}{\underset{Cl}{C}}}CH_2CH_3$

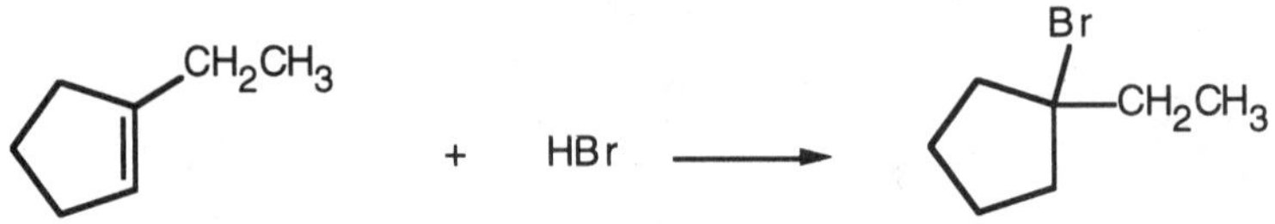

b)   + HBr ⟶

c) $CH_3CH_2\overset{\underset{|}{CH_3}}{C}=CHC(CH_3)_3$ + HI ⟶ $CH_3CH_2\overset{\underset{|}{CH_3}}{\underset{|}{\underset{I}{C}}}CH_2C(CH_3)_3$

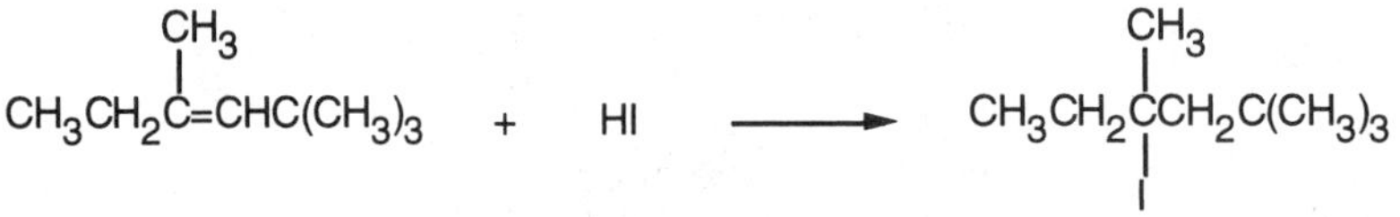

d) 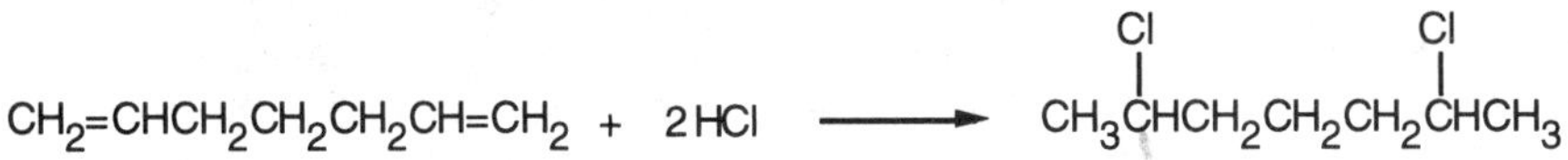 $CH_2=CHCH_2CH_2CH_2CH=CH_2$ + 2HCl ⟶ $CH_3\overset{\underset{|}{Cl}}{C}HCH_2CH_2CH_2\overset{\underset{|}{Cl}}{C}HCH_3$

e) 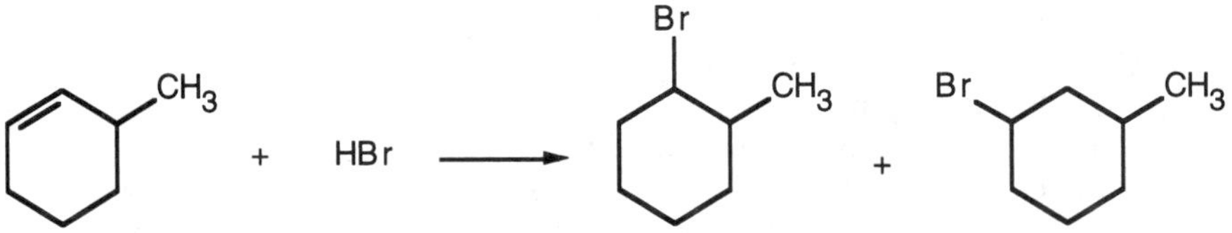   + HBr ⟶ +

Both products are likely to
be formed.

6.37

a) 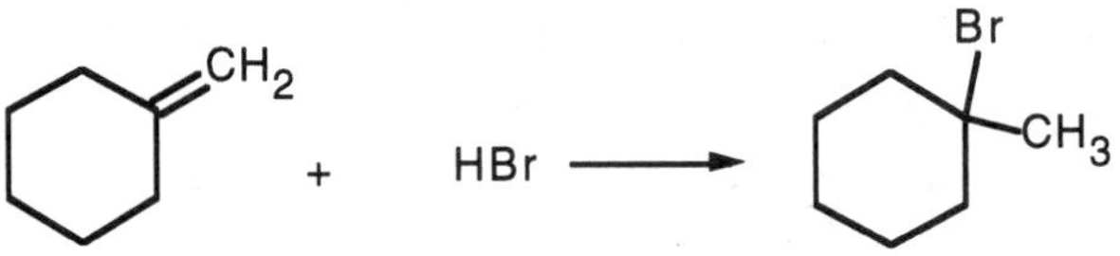   + HBr ⟶

b) 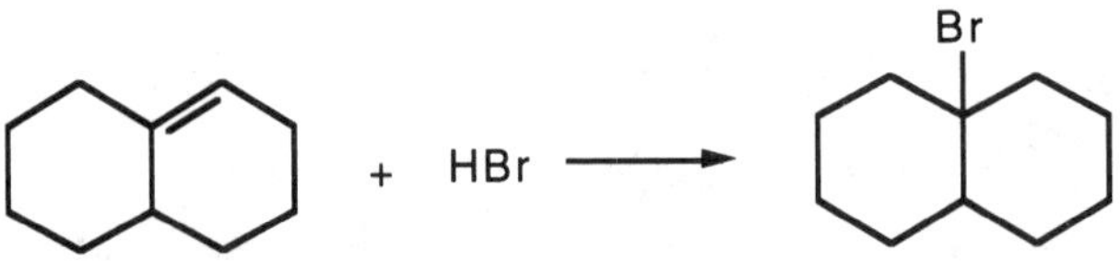   + HBr ⟶

c) 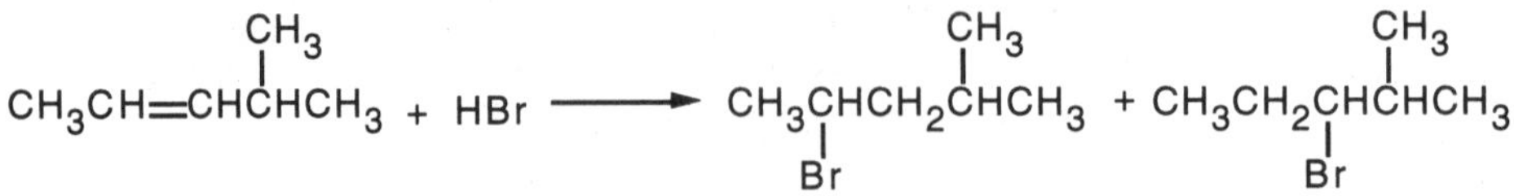 $CH_3CH=CH\overset{\underset{|}{CH_3}}{C}HCH_3$ + HBr ⟶ $CH_3\overset{\underset{|}{\underset{Br}{C}}}{C}HCH_2\overset{\underset{|}{CH_3}}{C}HCH_3$ + $CH_3CH_2\overset{\underset{|}{CH_3}}{\underset{|}{\underset{Br}{C}}}HCH_3$

6.38

Highest priority ———➤ Lowest priority

a) –I, –Br, –CH$_3$, –H

b) –OCH$_3$, –OH, –COOH, –H

c) –COOCH$_3$, –COOH, –CH$_2$OH, –CH$_3$

d) –COCH$_3$, –CH$_2$CH$_2$OH, –CH$_2$CH$_3$, –CH$_3$

e) –CH$_2$Br, –C≡N, –CH$_2$NH$_2$, –CH=CH$_2$

f) –CH$_2$OCH$_3$, –CH$_2$OH, –CH=CH$_2$, –CH$_2$CH$_3$

6.39

a) (H) HOCH$_2$ CH$_3$ (H)
             C=C   **Z**
(L)   CH$_3$ H (L)

b) (L) HOOC   H (L)
           C=C   **Z**
(H) Cl   OCH$_3$ (H)

c) (H) NC   CH$_3$ (L)
          C=C   **E**
(L) CH$_3$CH$_2$ CH$_2$OH (H)

d) (H) CH$_3$O$_2$C   CH=CH$_2$ (H)
          C=C   **Z**
(L) HO$_2$C   CH$_2$CH$_3$ (L)

6.40 a) 3–Methylcyclohexene   b) 1,5–Dimethylcyclopentene
      c) Ethylcyclobutadiene   d) 1,2–Dimethyl–1,4–cyclohexadiene
      e) 5–Methyl–1,3–cyclohexadiene   f) 1,5–Cyclooctadiene

6.41

a) (H) CH$_3$
                COOH (H)
        C=C   **Z**   (correct)
(L)   H (L)

b) (L) H   CH$_2$CH=CH$_2$ (H)
        C=C   **E**   (correct)
(H) CH$_3$ CH$_2$CH(CH$_3$)$_2$ (L)

c) (H) Br   CH$_2$NH$_2$ (L)
        C=C   **E**   (incorrect)
(L) H   CH$_2$NHCH$_3$ (H)

d) (H)NC   CH$_3$ (L)
        C=C   **E**   (correct)
(L) (CH$_3$)$_2$NCH$_2$ CH$_2$CH$_3$ (H)

e)

This compound does not exhibit E–Z isomerism.

f)   (L) HOCH$_2$ COOH (H)

E (correct)

(H) CH$_3$OCH$_2$ COCH$_3$ (L)

6.42

a)   H$_2$C=CH$_2$ +   H—Cl $\longrightarrow$ H—CH$_2$CH$_2$—Cl $\Delta H° = $ -91 kJ/mol
     235 kJ/mol   432 kJ/mol   420 kJ/mol   338 kJ/mol

b)   H$_2$C=CH$_2$ +   H—Br $\longrightarrow$ H—CH$_2$CH$_2$—Br $\Delta H° = $ -104 kJ/mol
     235 kJ/mol   366 kJ/mol   420 kJ/mol   285 kJ/mol

c)   H$_2$C=CH$_2$ +   H—I $\longrightarrow$ H—CH$_2$CH$_2$—I $\Delta H° = $ -109 kJ/mol
     235 kJ/mol   298 kJ/mol   420 kJ/mol   222 kJ/mol

The reaction of ethylene with HI is the most favorable because its $\Delta H°$ is the largest negative number.

6.43

3°
carbocation

3°
carbocation

6.44

6.45   a) C$_{27}$H$_{46}$O   five degrees of unsaturation
          b) C$_{14}$H$_9$Cl$_5$ eight degrees of unsaturation
          c) C$_{20}$H$_{34}$O$_5$ four degrees of unsaturation
          d) C$_8$H$_{10}$N$_4$O$_2$ six degrees of unsaturation
          e) C$_{21}$H$_{28}$O$_5$ eight degrees of unsaturation
          f) C$_{17}$H$_{23}$NO$_3$ seven degrees of unsaturation

6.46

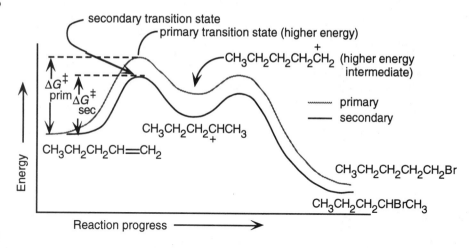

6.47

	Transition State #1	Transition State #2
2–bromopentane path		
1–bromopentane path		

The first step (protonation) for both reaction paths is endothermic, and both transition states resemble the carbocation intermediate. Transition states for the exothermic second step also resemble the carbocation intermediate. Transition state #1 for 1–bromopentane is more like the carbocation intermediate than is transition state #1 for 2–bromopentane.

6.48

$$CH_3CH_2\underset{\overset{|}{CH_3}}{C}=CHCH_3 \;+\; H_2O \quad\xrightarrow[\text{catalyst}]{\text{acid}}\quad CH_3CH_2\underset{\overset{|}{OH}}{\overset{\overset{CH_3}{|}}{C}}-CH_2CH_3$$

6.49

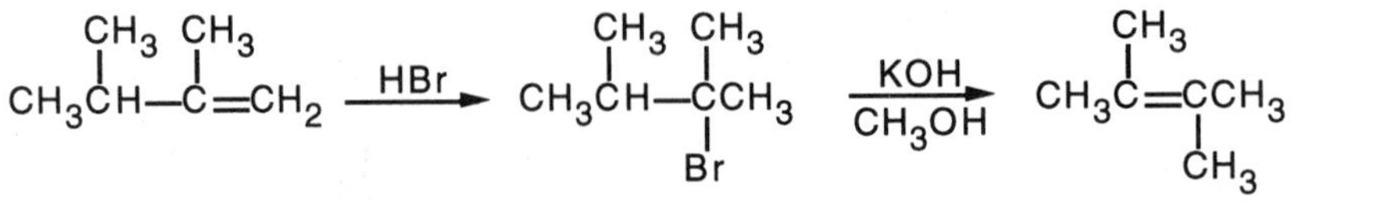

CH₃CH—C=CH₂ (2,3-Dimethyl-1-butene) → HBr → 2-Bromo-2,3-dimethylbutane → KOH/CH₃OH → 2,3-Dimethyl-2-butene

The product, 2,3-dimethyl-2-butene, is formed by removal of HBr from 2-bromo-2,3-dimethylbutane. Note that the product has the more substituted double bond.

Study Guide for Chapter 6

After studying this chapter, you should be able to:

(1)   Calculate the degree of unsaturation of any compound, including those containing N, O and halogen (6.1, 6.2, 6.3, 6.18, 6.19, 6.20, 6.21, 6.45).

(2)   Name cyclic and acyclic alkenes, and draw structures corresponding to given names (6.4, 6.5, 6.6, 6.22, 6.23, 6.24, 6.26, 6.27, 6.28, 6.29, 6.30, 6.40).

(3)   Draw cis–trans isomers of alkenes (6.7, 6.8, 6.32).

(4)   Assign priorities to double-bond substituents according to sequence rules (6.9, 6.10, 6.38).

(5)   Assign E, Z configurations to double bonds (6.11, 6.25, 6.26, 6.39, 6.41).

(6)   Predict the relative stability of alkene double bonds (6.12, 6.33, 6.35).

(7)   Understand the mechanism of electrophilic addition reactions (6.15, 6.17, 6.43, 6.44, 6.46, 6.47,).

(8)   Understand the Hammond postulate (6.16, 6.46, 6.47).

(9)   Predict the products of simple reactions of alkenes (6.13, 6.14, 6.36, 6.37, 6.48, 6.49).

Chapter 7 – Alkenes: Reactions and Synthesis

7.1

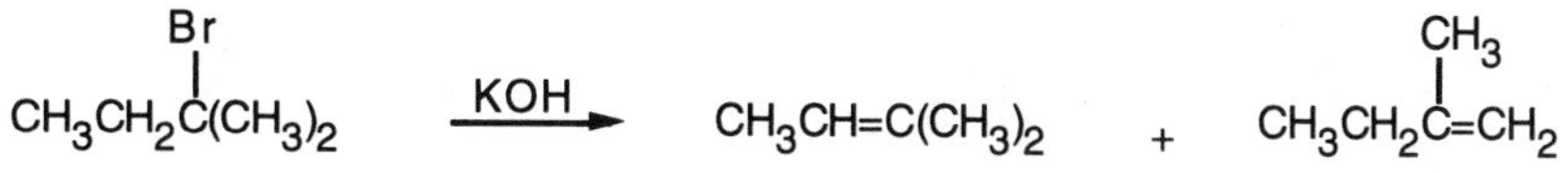
Br
CH3CH2C(CH3)2
KOH
CH3CH=C(CH3)2
+
CH3
CH3CH2C=CH2

Dehydrobromination can occur in two directions to yield a mixture of products.

7.2

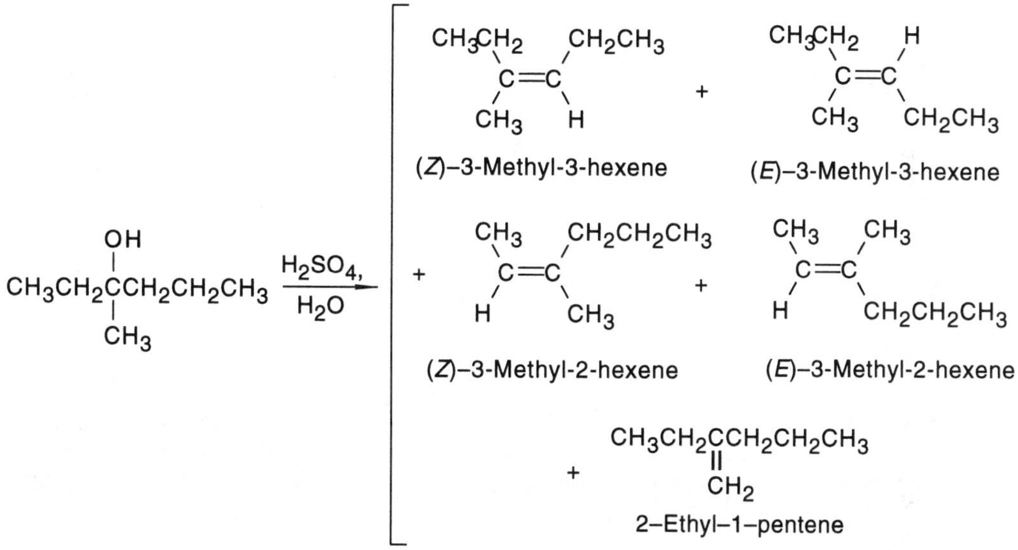
OH
CH3CH2CCH2CH2CH3
CH3
H2SO4,
H2O
(Z)–3-Methyl-3-hexene
(E)–3-Methyl-3-hexene
(Z)–3-Methyl-2-hexene
(E)–3-Methyl-2-hexene
CH3CH2CCH2CH2CH3
CH2
2–Ethyl–1–pentene

Five alkene products, including *E*, *Z* isomers, might be obtained by dehydration of 3–methyl–3–hexanol.

7.3

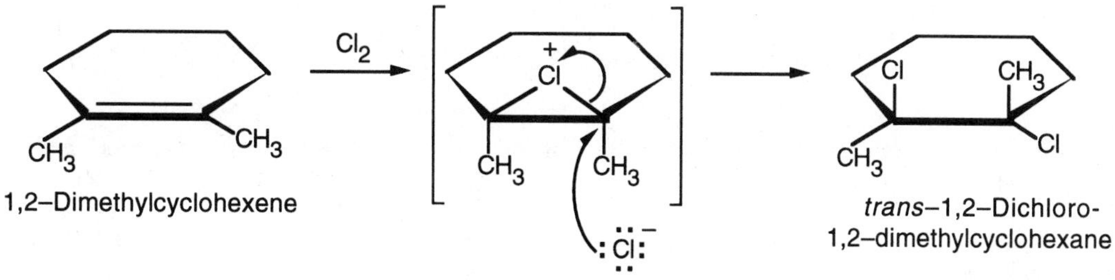
Cl2
1,2–Dimethylcyclohexene
trans–1,2–Dichloro-
1,2–dimethylcyclohexane

The chlorines are *trans* to one another in the product, as are the methyl groups.

7.4

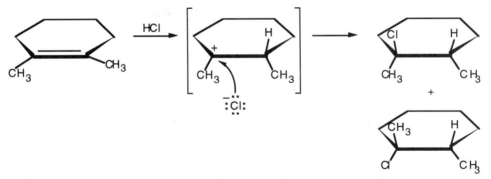

Addition of hydrogen halides involves formation of an open carbocation, not a cyclic halonium ion intermediate. The carbocation, which is sp^2 hybridized and planar, can be attacked by chloride from either top or bottom, yielding products in which the two methyl groups can be either *cis* or *trans* to each other.

7.5

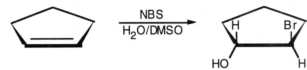

Br and OH are trans in the product.

7.6   NBS is the source of the electrophilic Br^+ ion in bromohydrin formation. Attack of the alkene pi electrons on Br^+ forms a cyclic bromonium ion. When this bromonium ion is opened by water, a partial positive charge develops at the carbon whose bond to bromine is being cleaved.

Since a secondary carbon can stabilize this charge better than a primary carbon, opening of the bromonium ion occurs at the secondary carbon to yield Markovnikov product.

7.7   Keep in mind that oxymercuration corresponds to Markovnikov addition of H_2O to an alkene.

a) $CH_3CH_2CH_2CH=CH_2$ $\xrightarrow[\text{2. NaBH}_4]{\text{1. Hg (OAc)}_2, \text{H}_2\text{O}}$ $CH_3CH_2CH_2\overset{\overset{\displaystyle OH}{|}}{C}HCH_3$

b) $CH_3CH_2CH=\overset{\overset{\displaystyle CH_3}{|}}{C}CH_3$ $\xrightarrow[\text{2. NaBH}_4]{\text{1. Hg (OAc)}_2, \text{H}_2\text{O}}$ $CH_3CH_2CH_2\overset{\overset{\displaystyle CH_3}{|}}{\underset{\underset{\displaystyle OH}{|}}{C}}CH_3$

7.8

a) $CH_3C=CHCH_2CH_2CH_3$
           |
           CH_3

            or

   $CH_3CCH_2CH_2CH_2CH_3$
        ‖
        CH_2

       1. $Hg(OAc)_2$, H_2O
       ⟶
       2. $NaBH_4$

            OH
            |
   $CH_3CCH_2CH_2CH_2CH_3$
            |
           CH_3

b)

       1. $Hg(OAc)_2$, H_2O
       ⟶
       2. $NaBH_4$

Oxymercuration occurs with Markovnikov orientation.

7.9   Recall the mechanism of hydroboration and note that the hydrogen added to the double bond comes from borane. The product of hydroboration with BD_3 has deuterium bonded to the more substituted carbon; –D and –OH are cis to one another.

7.10

a) $(CH_3)_2CHCH=CH_2$ 1. BH_3, THF   ⟶   2. H_2O_2, ^-OH $(CH_3)_2CHCH_2CH_2OH$

b) $(CH_3)_2C=CHCH_3$ 1. BH_3, THF   ⟶   2. H_2O_2, ^-OH $(CH_3)_2CHCHCH_3$
                                                          |
                                                         OH

c)

       1. BH_3, THF
       ⟶
       2. H_2O_2, ^-OH

7.11

a)

 + $CHCl_3$ $\xrightarrow{KOH}$

b) $(CH_3)_2CHCH_2CH=CHCH_3$ + CH_2I_2 $\xrightarrow{Zn/Cu}$ $(CH_3)_2CHCH_2CH-CHCH_3$

7.12

A   B

Focus on the stereochemistry of the three-membered ring. Simmons-Smith reaction of
1,1–diiodoethane with the double bond occurs with syn stereochemistry and can produce
two isomers. In one of these isomers (A), the methyl group is on the same side of the
three-membered ring as the cyclohexane ring carbons. In B, the methyl group is on the
side of the three-membered ring opposite to the cyclohexane ring carbons.

7.13

a) $(CH_3)_2C=CHCH_2CH_3$ $\xrightarrow[\text{Ethanol}]{\text{H}_2,\,\text{Pd/C}}$ $(CH_3)_2CHCH_2CH_2CH_3$

       2–Methyl–2–pentene   2–Methylpentane

b)

       3,3–Dimethylcyclopentene   1,1-Dimethylcyclopentane

7.14

a)

       1–Methylcyclohexene

b) $CH_3CH_2CH=C(CH_3)_2$ $\xrightarrow{\begin{array}{c}1.\ \text{OsO}_4,\ \text{pyridine}\\ 2.\ \text{NaHSO}_3,\ \text{H}_2\text{O}\end{array}}$

       2–Methyl–2–pentene

c) $CH_2=CHCH=CH_2$ $\xrightarrow{\begin{array}{c}1.\ 2\ \text{OsO}_4,\ \text{pyridine}\\ 2.\ \text{NaHSO}_3,\ \text{H}_2\text{O}\end{array}}$ $HOCH_2\overset{\text{HO}}{\underset{}{C}}H\overset{\text{OH}}{\underset{}{C}}HCH_2OH$

       1,3–Butadiene

7.15

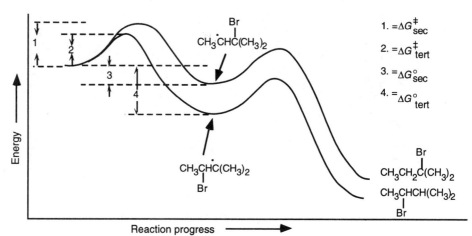

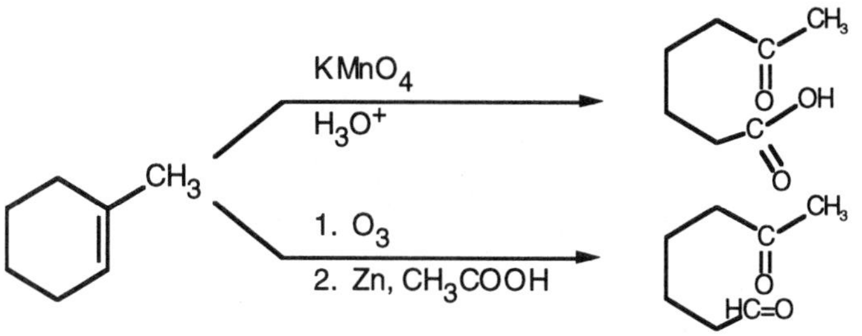

7.16

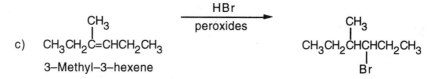

a) $(CH_3)_2C=CH_2$ $\xrightarrow[\text{2. Zn, H}_3\text{O}^+]{\text{1. O}_3}$ $(CH_3)_2C=O$ + $O=CH_2$

b) $CH_3CH_2CH=CHCH_2CH_3$ $\xrightarrow[\text{2. Zn, H}_3\text{O}^+]{\text{1. O}_3}$ $CH_3CH_2CH=O$ + $O=CHCH_2CH_3$

7.17

a) $CH_3CH_2\overset{\overset{\displaystyle CH_3}{|}}{C}=CH_2$ $\xrightarrow{\text{HBr}}$ $CH_3CH_2\overset{\overset{\displaystyle Br}{|}}{C}(CH_3)_2$

     2–Methyl–1–butene

b) $CH_3CH_2CH=CH_2$ $\xrightarrow[\text{peroxides}]{\text{HBr}}$ $CH_3CH_2CH_2CH_2Br$

     1–Butene

c) $CH_3CH_2\overset{\overset{\displaystyle CH_3}{|}}{C}=CHCH_2CH_3$ $\xrightarrow[\text{peroxides}]{\text{HBr}}$ $CH_3CH_2\overset{\overset{\displaystyle CH_3}{|}}{\underset{\underset{\displaystyle Br}{|}}{CH}}CHCH_2CH_3$

     3–Methyl–3–hexene

7.18

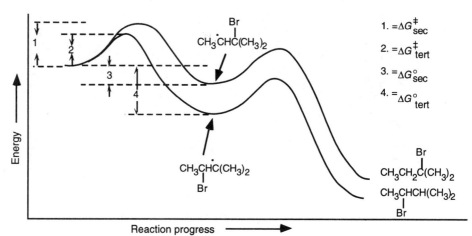

1. $= \Delta G^{\ddagger}_{sec}$
2. $= \Delta G^{\ddagger}_{tert}$
3. $= \Delta G^{\circ}_{sec}$
4. $= \Delta G^{\circ}_{tert}$

The reaction producing the more stable, tertiary radical has a lower $\Delta G°$ for the first step. (Remember that a larger negative $\Delta G°$ indicates that a reaction is more favorable.) According to the Hammond Postulate, the more stable radical also forms faster, and $\Delta G^{\ddagger}$ for its formation is lower.

7.19

Monomer   Polymer

a) $H_2C=CHOCH_3$

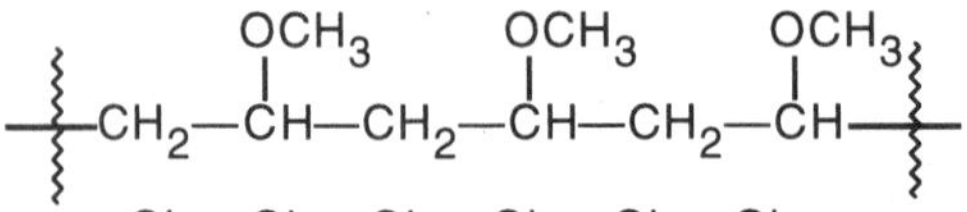

b) $ClHC=CHCl$

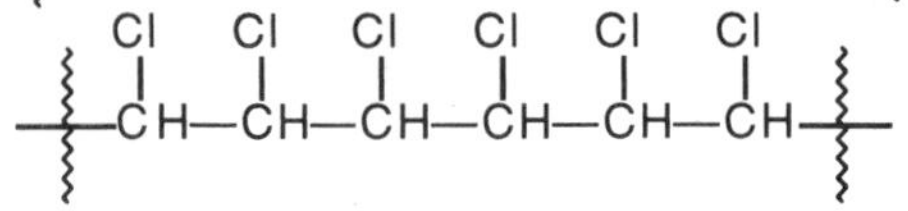

7.20 Addition of IN_3 to the alkene yields a product in which $\underline{I}$ is bonded to the primary carbon and N_3 is bonded to the secondary carbon. If addition occurs with Markovnikov orientation, I^+ must be the electrophile, and the reaction must proceed through an iodonium ion intermediate. Opening of the iodonium ion gives Markovnikov product for the reasons discussed in Problem 7.6. The bond polarity of iodine azide is:

$$I \overset{+}{\underset{-}{\rightleftharpoons}} N_3$$

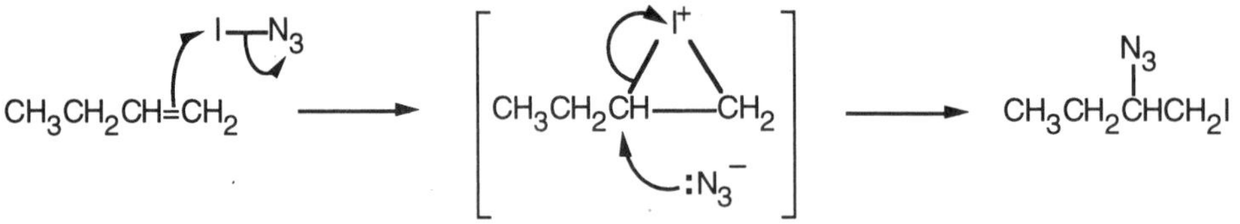

7.21

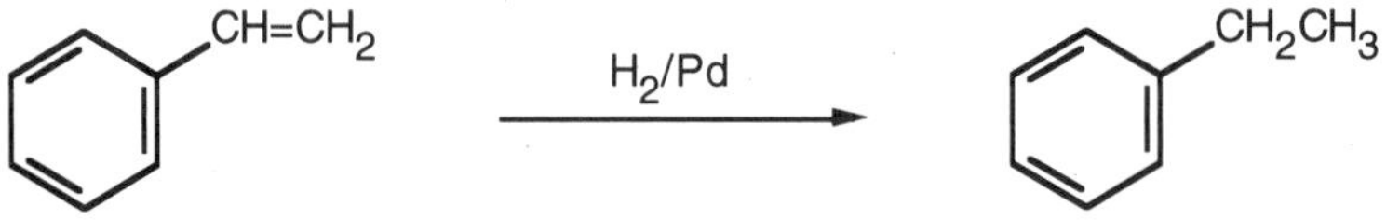

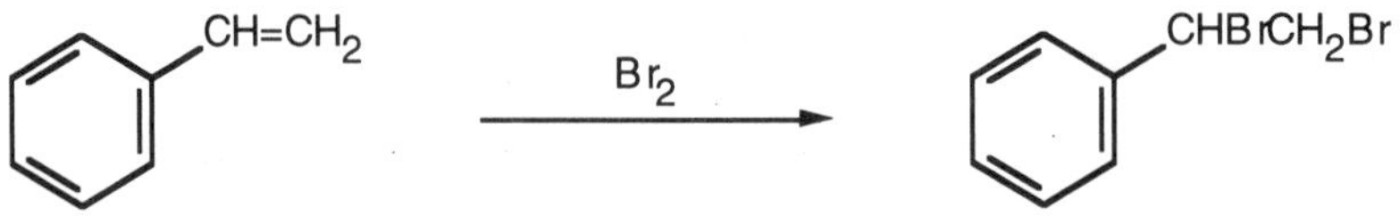

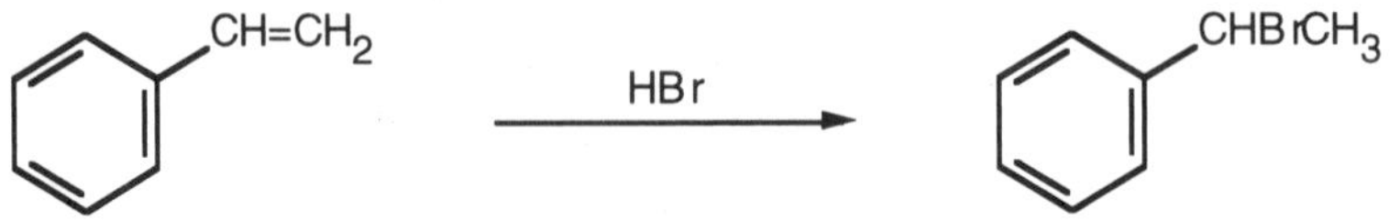

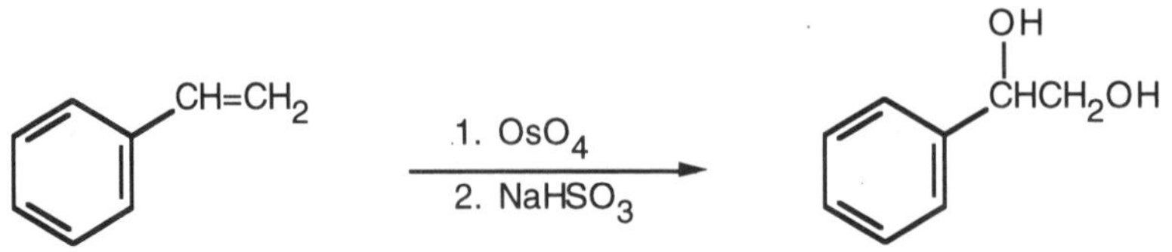

e)

$$\xrightarrow{D_2/Pd}$$

7.22

a)

$$CH_3CH_2CH_2CH_2\overset{\overset{\displaystyle CH_3}{|}}{C}=CH_2$$ 2–Methyl–1–hexene

$$CH_3CH_2CH_2CH=C(CH_3)_2$$ 2–Methyl–2–hexene

$$CH_3CH_2CH=CHCH(CH_3)_2$$ 2–Methyl–3–hexene

$$CH_3CH=CHCH_2CH(CH_3)_2$$ 5–Methyl–2–hexene

$$CH_2=CHCH_2CH_2CH(CH_3)_2$$ 5–Methyl–1–hexene

$$\xrightarrow{H_2/Pd}$$ $CH_3CH_2CH_2CH_2CH(CH_3)_2$

2–Methylhexane

b)

3,3–Dimethylcyclohexene

4,4–Dimethylcyclohexene

$$\xrightarrow{H_2/Pd}$$

1,1–Dimethylcyclohexane

c) $CH_3CH=CHCH_2CH(CH_3)_2$ $\xrightarrow{Br_2/CCl_4}$ $CH_3CHBrCHBrCH_2CH(CH_3)_2$

5–Methyl–2–hexene   2,3–Dibromo–5–methylhexane

d) $CH_3CH_2CH_2CH=CH_2$ $\xrightarrow[\text{2. NaBH}_4]{\text{1. Hg(OAc)}_2,\ H_2O}$ $CH_3CH_2CH_2CH(OH)CH_3$

1-Pentene   2-Pentanol

e) $CH_3CH_2CH_2CH_2\overset{\overset{\displaystyle}{|}}{C}=CHCH_3$ $\xrightarrow[\text{peroxides}]{HBr}$ $CH_3CH_2CH_2CH_2\overset{\overset{\displaystyle Br}{|}}{C}HCHCH_3$

$\quad\quad\quad\quad\quad CH_3$ $\quad\quad\quad\quad\quad\quad\quad\quad CH_3$

3–Methyl–2–heptene   2–Bromo–3–methylheptane

f) $CH_3CH_2CH_2CH_2CHCH=CH_2$ $\xrightarrow[\text{ether}]{HCl}$ $CH_3CH_2CH_2CH_2\overset{\overset{\displaystyle Cl}{|}}{C}HCHCH_3$

$\quad\quad\quad\quad\quad CH_3$ $\quad\quad\quad\quad\quad\quad\quad\quad CH_3$

3–Methyl–1–heptene   2–Chloro–3–methylheptane

7.23

a)

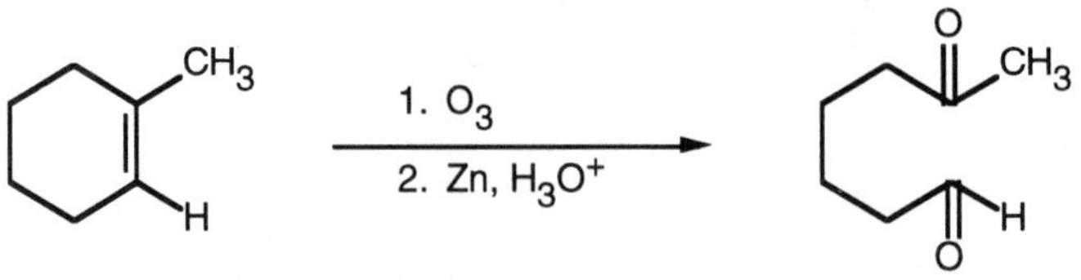

$$\xrightarrow[\text{2. Zn, H}_3\text{O}^+]{\text{1. O}_3}$$

b)

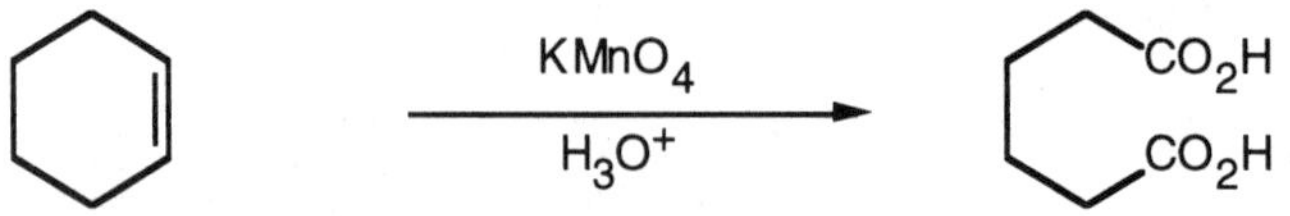

$$\xrightarrow[\text{H}_3\text{O}^+]{\text{KMnO}_4}$$

c)

$$\xrightarrow[\text{2. H}_2\text{O}_2, \text{ }^-\text{OH}]{\text{1. BH}_3}$$

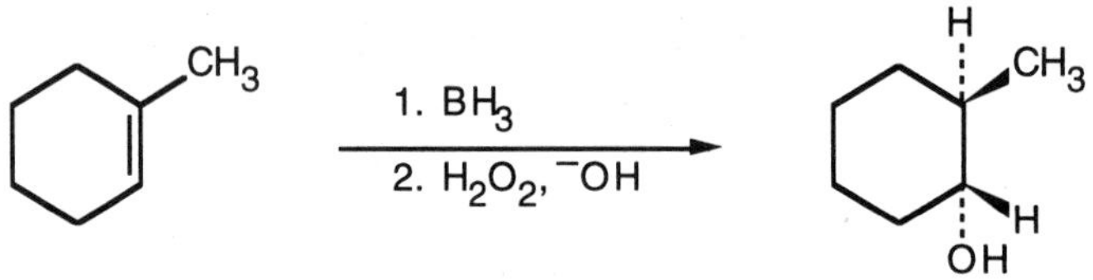

Remember that –H and –OH add *syn* across the double bond.

d)

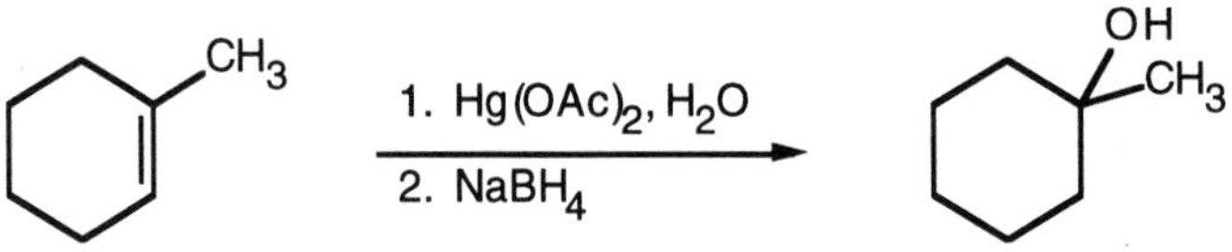

$$\xrightarrow[\text{2. NaBH}_4]{\text{1. Hg(OAc)}_2, \text{ H}_2\text{O}}$$

7.24

a)

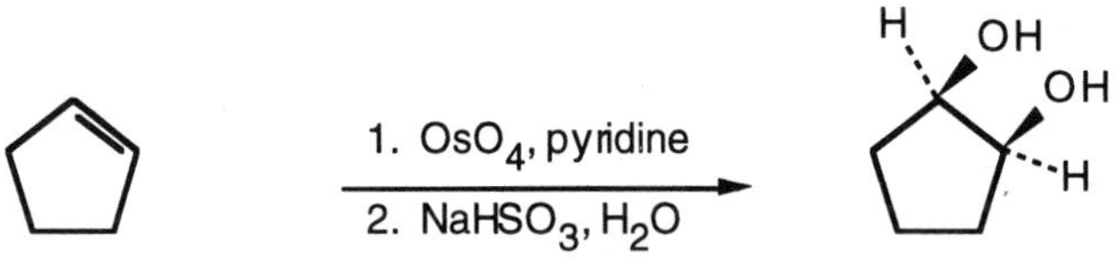

$$\xrightarrow[\text{2. NaHSO}_3, \text{ H}_2\text{O}]{\text{1. OsO}_4, \text{ pyridine}}$$

b)

$$\xrightarrow[\text{2. NaBH}_4]{\text{1. Hg(OAc)}_2, \text{ H}_2\text{O}}$$

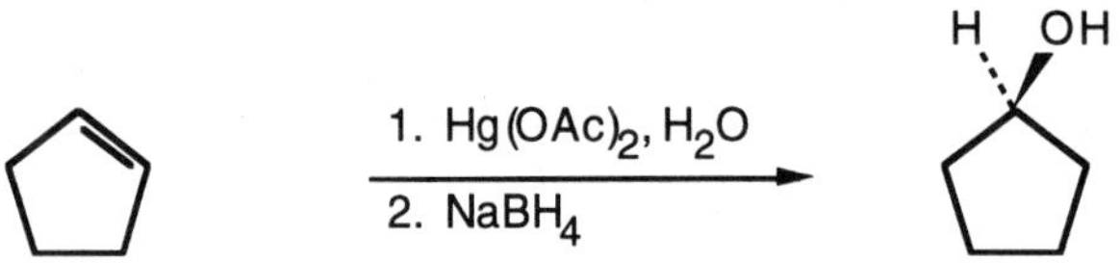

Hydroboration/oxidation is another route to this product.

c)

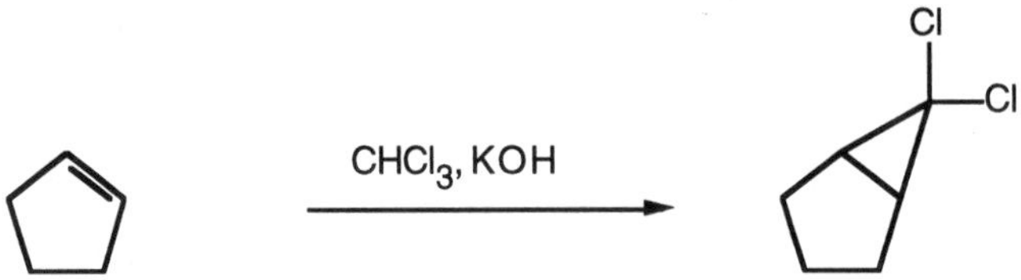

$$\xrightarrow{\text{CHCl}_3, \text{ KOH}}$$

d) $CH_3CH=CHCH(CH_3)_2$

$$\xrightarrow[\text{2. Zn, H}_3\text{O}^+]{\text{1. O}_3}$$

CH_3CHO + $(CH_3)_2CHCHO$

e) $(CH_3)_2C=CH_2$

$$\xrightarrow[\text{2. H}_2\text{O}_2, \text{ }^-\text{OH}]{\text{1. BH}_3}$$

$(CH_3)_2CHCH_2OH$

f)

$$H_2SO_4, H_2O \xrightarrow{\quad} heat$$

7.25 Because ozonolysis gives only one product, we can assume that the alkene is symmetrical.

$$\xleftarrow[\text{2. Zn, } H_3O^+]{\text{1. } O_3}$$

2,3–Dimethyl–2–butene

7.26 a) If the hydrocarbon reacts with only one equivalent of hydrogen, it has only one double bond.

b) If only one type of aldehyde is produced on ozonolysis, the alkene must be symmetrical.

Putting these two facts together allows us to deduce that the unknown hydrocarbon is $CH_3CH_2CH_2CH_2CH=CHCH_2CH_2CH_2CH_3$.

$$CH_3CH_2CH_2CH_2CH=CHCH_2CH_2CH_2CH_3 \xrightarrow[\text{2. Zn, } H_3O^+]{\text{1. } O_3} 2\ CH_3CH_2CH_2CH_2\overset{\displaystyle O}{\overset{\displaystyle \|}{C}}H$$

5-Decene   Pentanal

$$\Big\downarrow H_2/Pd$$

$$CH_3CH_2CH_2CH_2CH_2CH_2CH_2CH_2CH_2CH_3$$

Decane

7.27 Remember that alkenes can give ketones, carboxylic acids, and CO_2 on oxidative cleavage with $KMnO_4$ in acidic solution.

a) $CH_3CH_2CH=CH_2 \xrightarrow[H_3O^+]{KMnO_4} CH_3CH_2COOH + CO_2$

b) $CH_3CH_2CH_2CH=C(CH_3)_2 \xrightarrow[H_3O^+]{KMnO_4} CH_3CH_2CH_2COOH + (CH_3)_2C=O$

c)

$$\xrightarrow[H_3O^+]{KMnO_4}$$

$+ (CH_3)_2C=O$

7.28 Compound $\underline{A}$ has three degrees of unsaturation. Because compound $\underline{A}$ contains only one double bond, the other two degrees of unsaturation must be rings.

$$\underset{B}{} \xleftarrow[\text{2. Zn, H}_3\text{O}^+]{\text{1. O}_3} \underset{A}{} \xrightarrow{\text{H}_2/\text{Pd}}$$

Other structures having two fused rings are possible.

7.29 Don't get discouraged by the amount of information in this problem. Read slowly and interpret piece by piece. We know the following.

1) Hydrocarbon $\underline{A}$ (C_6H_{12}) has one double bond or ring.
2) Because $\underline{A}$ reacts with one equivalent of H_2, it has one double bond and no ring.
3) Compound $\underline{A}$ forms a diol when reacted with OsO_4.
4) When alkenes are oxidized with $KMnO_4$ they give either carboxylic acids or ketones, depending on the substitution pattern of the double bond.
   a) A ketone is produced from what was originally a disubstituted carbon in the double bond.
   b) A carboxylic acid is produced from what was originally a monosubstituted carbon in the double bond.
5) One fragment from $KMnO_4$ oxidation is a carboxylic acid, CH_3CH_2COOH.
   a) This fragment was $CH_3CH_2CH=$ (a monosubstituted double bond) in compound $\underline{A}$.
   b) It contains three of the six carbons of compound $\underline{A}$.
6) a) The other fragment contains three carbons.
   b) It is a ketone.
   c) The only three carbon ketone is acetone, $O=C(CH_3)_2$.
   d) This fragment was $=C(CH_3)_2$ in compound $\underline{A}$.
7) If we join the fragment in 5a with the one in 6d, we get:

$$\underset{A}{CH_3CH_2CH=C(CH_3)_2} \qquad C_6H_{12}$$

The complete scheme:

7.30 The oxidative cleavage reaction of alkenes with O_3, followed by Zn in acid, produces aldehyde and ketone functional groups at sites where double bonds used to be. On ozonolysis, these two dienes yield only aldehydes because all double bonds are monosubstituted.

Because the other diene is symmetrical, only one dialdehyde, $OCHCH_2CHO$, is produced.

7.31 Try to solve this problem phrase by phrase.
1) $C_{10}H_{18}O$ has two double bonds and/or rings.
2) $C_{10}H_{18}O$ must be an alcohol because it undergoes reaction with H_2SO_4.
3) When $C_{10}H_{18}O$ is treated with dilute H_2SO_4, a mixture of alkenes of the formula $C_{10}H_{16}$ is produced.
4) Since the major alkene product **B** yields only cyclopentanone, C_5H_8O, on ozonolysis, **B** and **A** contain two rings. **A** therefore has no double bonds.

7.32

Cyclohexene   1–Methylcyclohexene

Addition to 1–methylcyclohexene occurs at a faster rate. Positive charge generated during the reaction is better stabilized by a tertiary carbocation intermediate than by a secondary carbocation intermediate.

7.33

a) $CH_3CH=CHCH_3$ $\xrightarrow{HBr}$ $CH_3CH_2\overset{\overset{\displaystyle Br}{|}}{C}HCH_3$

2–Butene

b)   3 CH₃CH=CHCH₃ $\xrightarrow[\text{THF}]{\text{BH}_3}$ (CH₃CH₂CH)₃B $\xrightarrow[^-\text{OH}]{\text{H}_2\text{O}_2}$ 3 CH₃CH₂CHCH₃

 2–Butene

c)   (CH₃)₂C=CH₂ $\xrightarrow[\text{peroxides}]{\text{HBr}}$ (CH₃)₂CHCH₂Br

 2–Methylpropene

d)   CH₃CH=C(CH₃)₂ $\xrightarrow[\text{peroxides}]{\text{HI}}$ CH₃CH₂C(CH₃)₂

 2–Methyl–2–butene

Addition of HI always gives Markovnikov products.

7.34

Cyclooctane $\xleftarrow{2\text{H}_2/\text{Pd}}$ 1,5–Cyclooctadiene $\xrightarrow[\text{2. Zn, H}_3\text{O}^+]{\text{1. O}_3}$ 2 HCCH₂CH₂CH

7.35 a) Non-Markovnikov addition using peroxides succeeds only with HBr. Other hydrogen halides yield products of Markovnikov addition.
 b) Hydroxylation of double bonds produces cis, not trans diols.
 c) Ozone reacts with both double bonds of 1,4–cyclohexadiene.
 d) Because hydroboration is a syn addition, the –H and the –OH added to the double bond must be cis to each other.

7.36 a) This alcohol can't be synthesized selectively by hydroboration/oxidation. Consider the two possible starting materials.

 1.   CH₃CH₂CH₂CH=CH₂ $\xrightarrow[\text{2. H}_2\text{O}_2, \,^-\text{OH}]{\text{1. BH}_3}$ CH₃CH₂CH₂CH₂CH₂OH

 1–Pentene yields only the primary alcohol.

 2.   CH₃CH₂CH=CHCH₃ $\xrightarrow[\text{2. H}_2\text{O}_2, \,^-\text{OH}]{\text{1. BH}_3}$ CH₃CH₂CHCH₂CH₃ + CH₃CH₂CH₂CHCH₃

 2–Pentene yields a mixture of alcohols.

 b)   (CH₃)₂C=C(CH₃)₂ $\xrightarrow[\text{2. H}_2\text{O}_2, \,^-\text{OH}]{\text{1. BH}_3}$ (CH₃)₂CHCH(CH₃)₂

 2,3–Dimethyl–2–butene yields the desired alcohol exclusively.

 c) This alcohol can't be formed cleanly by a hydroboration reaction. The –H and –OH added to a double bond must be cis to each other.

d) The product shown is not a hydroboration product; hydroboration yields an alcohol in which ⁻OH is bonded to the less substituted carbon.

7.37

a) 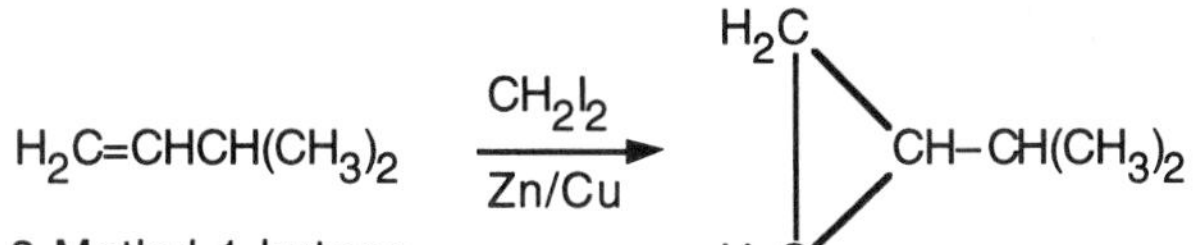 $H_2C=CHCH(CH_3)_2$ $\xrightarrow[\text{Zn/Cu}]{CH_2I_2}$

3-Methyl-1-butene

b) 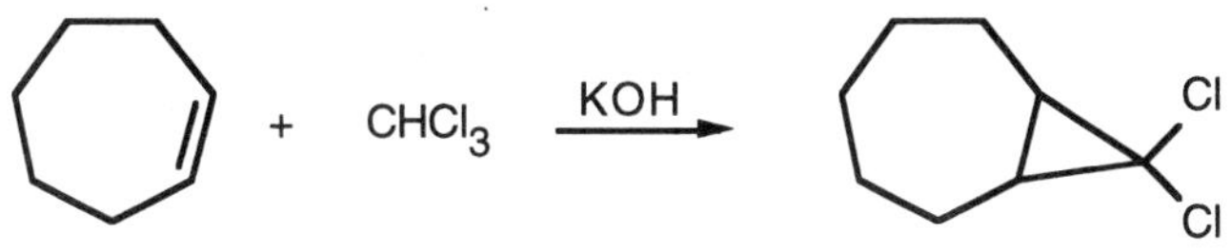   + $CHCl_3$ $\xrightarrow{KOH}$

Cycloheptene

7.38

Cholesterol

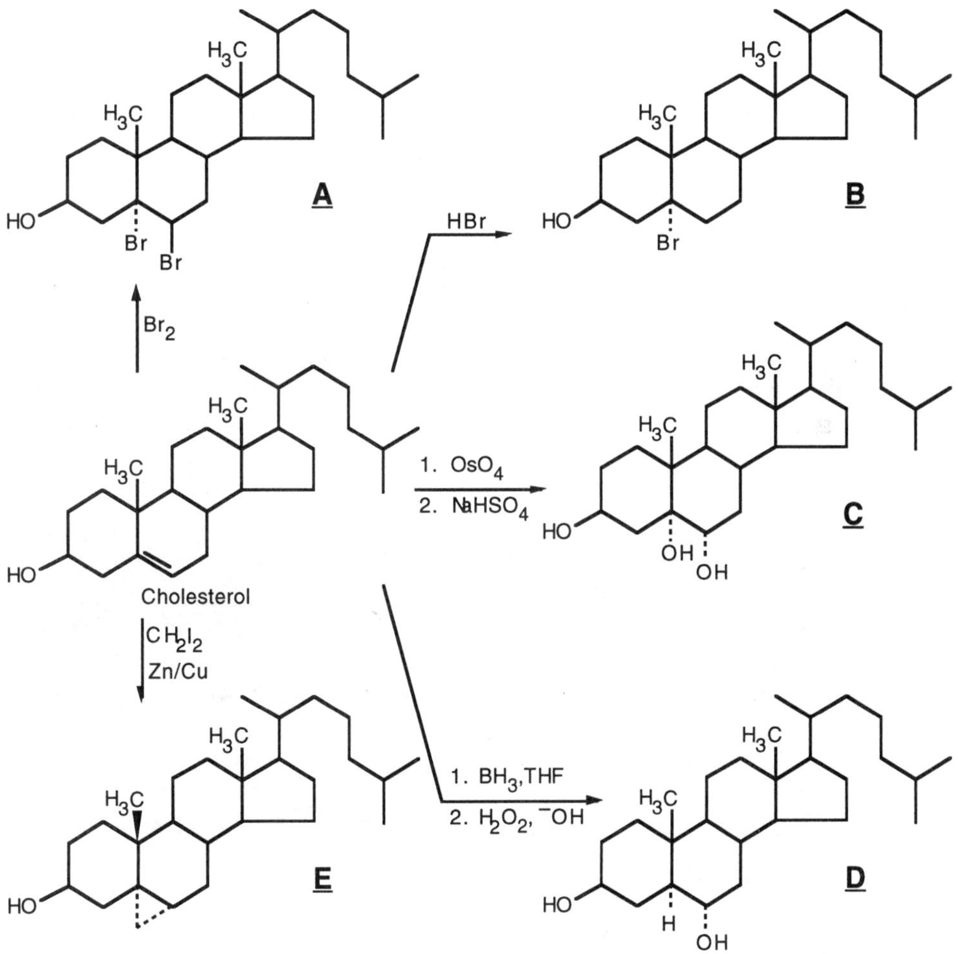

7.39

$CH_3(CH_2)_{12}CH=CH(CH_2)_7CH_3$ $\xrightarrow[H_3O^+]{KMnO_4}$ $CH_3(CH_2)_{12}COOH$ + $CH_3(CH_2)_7COOH$

7.40 C_8H_8 has five double bonds and/or rings. One of these double bonds reacts with H_2/Pd. Stronger conditions cause the uptake of four equivalents of H_2. C_8H_8 thus contains four double bonds, three of which are in an aromatic ring, and one C=C double bond. A good guess for C_8H_8 at this point is:

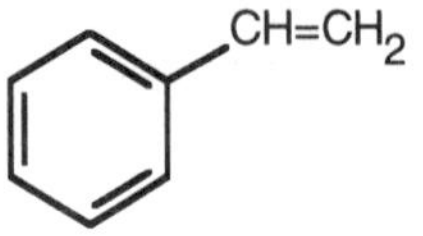

Reaction of a double bond with $KMnO_4$ yields cleavage products of the highest possible degree of oxidation. In this case, the products are $CO_2 + C_6H_5CO_2H$.

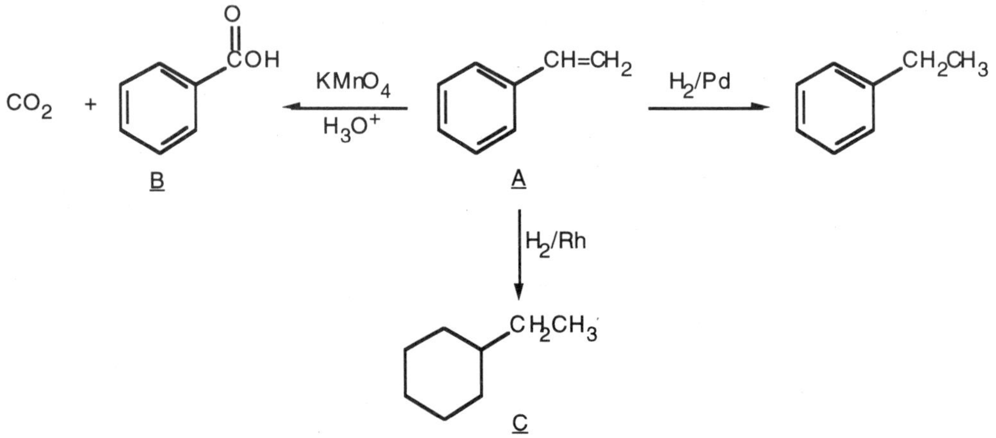

7.41

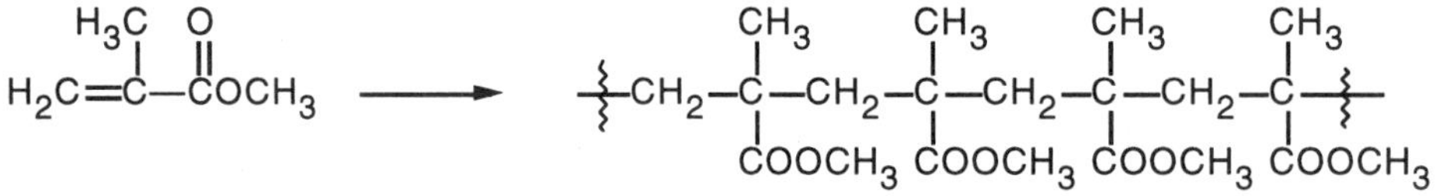

7.42

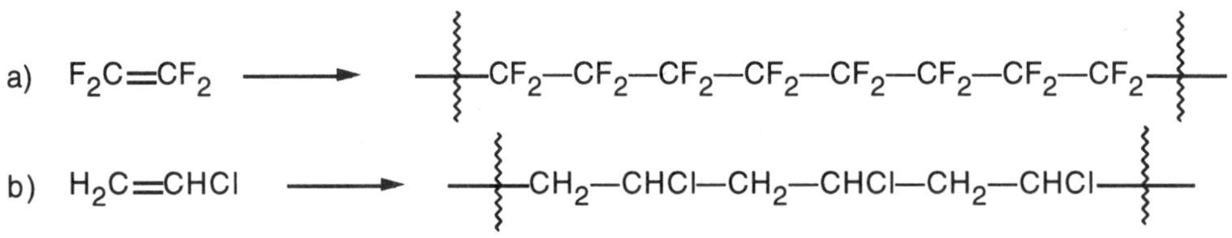

a)

b)

7.43

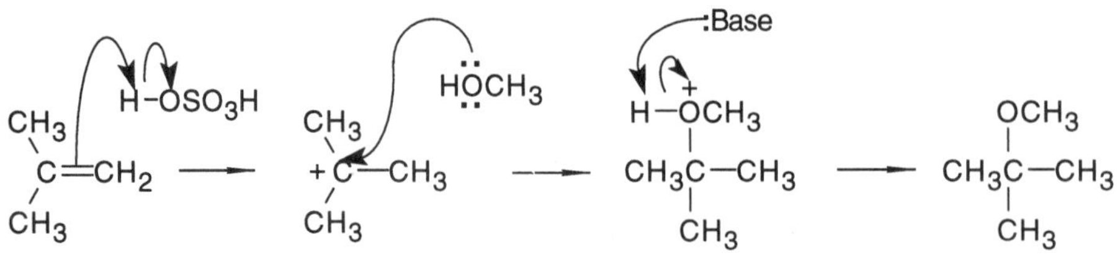

The above mechanism is the same as the mechanism shown in Section 7.4 with one exception: In this problem, methanol, rather than water, is the nucleophile, and an ether, rather than an alcohol, is the observed product.

7.44

Reaction of the double bond with Br_2 forms a cyclic bromonium ion.

The bromonium ion can be attacked by an electron pair from the nucleophilic –OH group to form the cyclic bromo ether

2–(Bromomethyl)tetrahydrofuran

The above mechanism is the same as that for halohydrin formation, shown in Section 7.3. In this case, the nucleophile is the hydroxyl group of 4–penten–1–ol.

7.45 a) Bromine dissolved in CCl_4 has a reddish-brown color. When an alkene is dissolved in Br_2/CCl_4, the double bond reacts with bromine, and the color disappears. This test distinguishes cyclopentene from cyclopentane, which does not react with Br_2.

  b) An aromatic compound such as benzene is unreactive to the Br_2/CCl_4 reagent and can be distinguished from 2–hexene, which decolorizes Br_2/CCl_4.

7.46

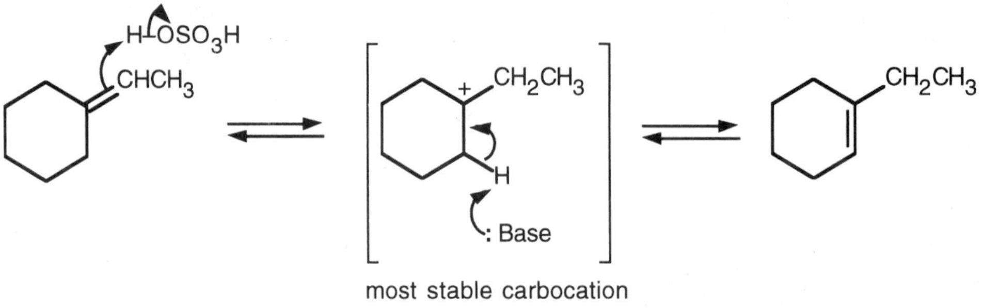

most stable carbocation

Protonation occurs to produce the most stable cation, which can then lose a proton to form either of two alkenes. Because 1–ethylcyclohexene is the major product of this equilibrium, it must be the more stable product.

7.47

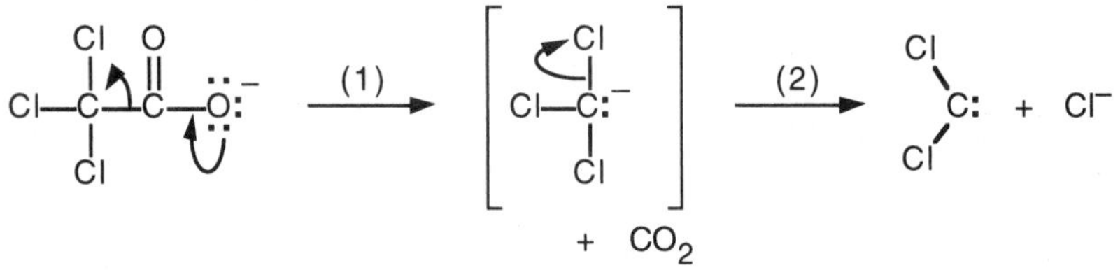

In step 1, carbon dioxide is lost from the trichloroacetate anion. In step 2, elimination of chloride anion produces dichlorocarbene. Step 2 is the same for both the above reaction and the base-induced elimination of HCl from chloroform.

7.48   α–Terpinene, $C_{10}H_{16}$, has three degrees of unsaturation – two double bonds and one ring.

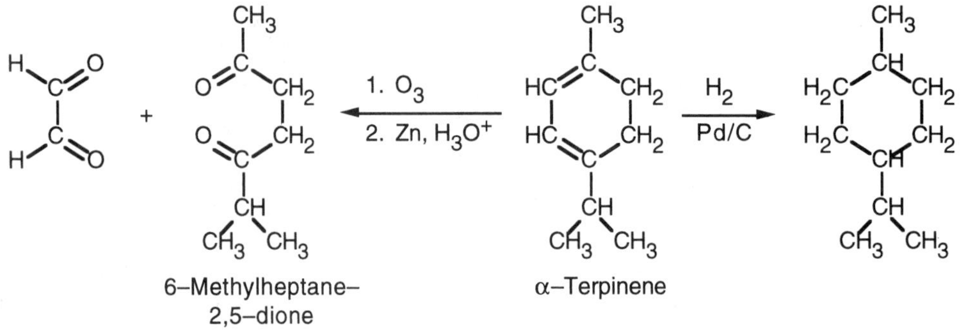

6–Methylheptane–
2,5–dione

α–Terpinene

7.49   Make models of the cis and trans diols. Notice that it is much easier to form a five-membered cyclic periodate from the cis diol than from the trans diol. We therefore predict that the cis periodate intermediate will be of lower energy than the trans periodate intermediate because of the lack of strain in the cis periodate ring.

Because any factor that lowers the energy of a transition state or intermediate also lowers $\Delta G^{\ddagger}$ and increases the rate of reaction, we predict that diol cleavage should proceed more rapidly for cis diols than for trans diols.

7.50

The most stable cation intermediate from protonation of 3-bromocyclohexene is a cyclic bromonium ion, which can be attacked from the opposite side to yield anti product.

In 3-methylcyclohexene, there are two intermediate carbocations of approximately equal stability that are attacked by bromide ion to give four different products.

7.51

7.52   Hydroboration of 2-methyl-2-pentene at 160°C is reversible. The initial organoborane intermediate can eliminate BH_3 in either of two ways, yielding either 2-methyl-2-pentene or 4-methyl-2-pentene, which in turn can undergo reversible hydroboration to yield either 4-methyl-2-pentene or 4-methyl-1-pentene. The effect of these reversible reactions is to migrate the double bond along the carbon chain. A final hydroboration then yields the most stable (primary) organoborane, which is oxidized to form 4-methyl-1-pentanol.

7.53

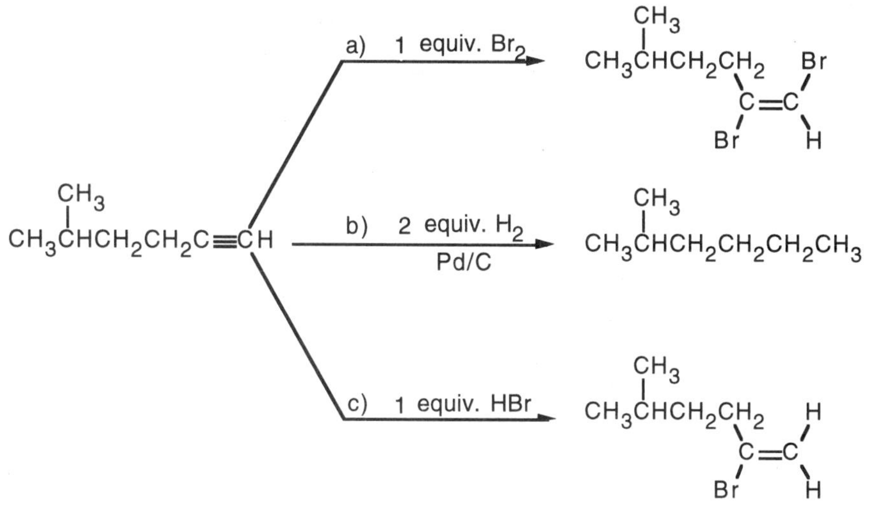

a) 1 equiv. Br₂

$$CH_3CHCH_2CH_2 \quad Br$$
with CH₃ on the CH, and C=C with Br and H

b) 2 equiv. H₂ / Pd/C

$$CH_3CHCH_2CH_2CH_2CH_3$$ with CH₃ substituent

c) 1 equiv. HBr

$$CH_3CHCH_2CH_2 \quad H$$ with CH₃ on CH, and C=C with Br and H

Starting material: $CH_3CHCH_2CH_2C\equiv CH$ with CH₃ substituent

Addition of one equivalent of HX or X₂ to a triple bond occurs with Markovnikov regiochemistry to yield a product in which the two added atoms usually have a trans-relationship across the double bond.

7.54

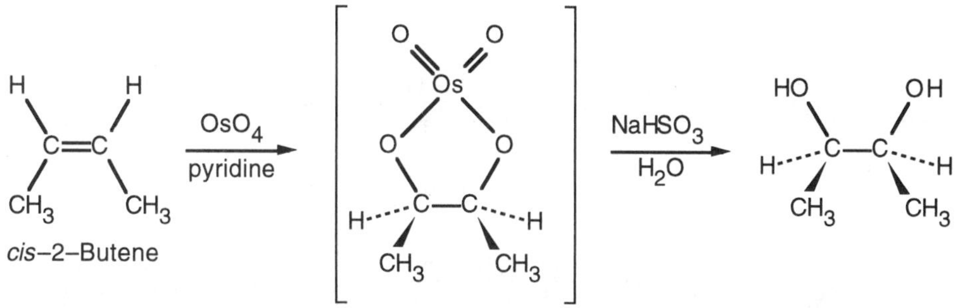

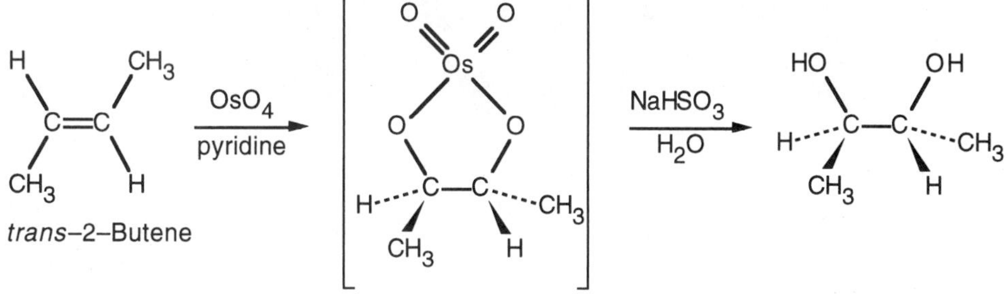

Formation of the cyclic osmate, which occurs with syn stereochemistry, retains the cis-trans stereochemistry of the double bond because osmate formation is a single-step reaction. Treatment of the osmate ester with NaHSO₃ does not affect the stereochemistry of the carbon-oxygen bond. The diol produced from *cis*–2–butene is isomeric with the diol produced from *trans*–2–butene.

7.55

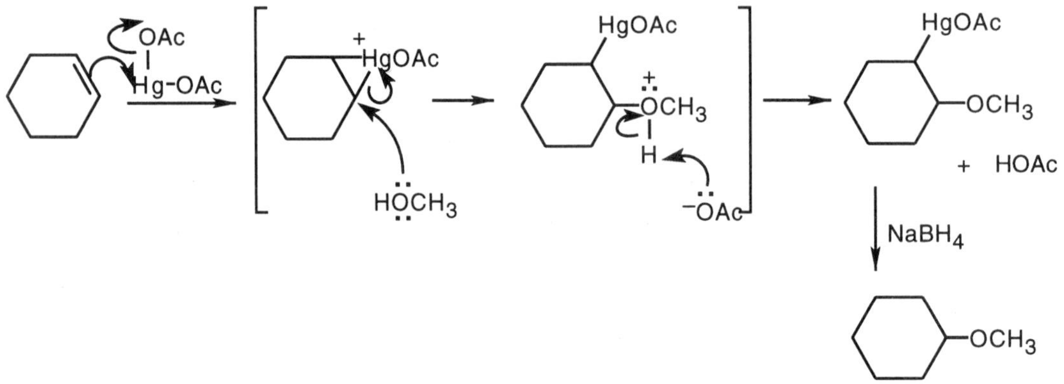

Cyclohexyl
methyl ether

The above mechanism is the same as the mechanism shown in Section 7.4 with one exception: In this problem, methanol, rather than water, is the nucleophile, and an ether, rather than an alcohol, is the observed product.

Study Guide for Chapter 7

After studying this chapter, you should be able to:

(1)   Predict the products of reactions of alkenes (7.3, 7.4, 7.5, 7.6, 7.7, 7.9, 7.11, 7.12, 7.13, 7.15, 7.16, 7.17, 7.19, 7.20, 7.21, 7.23, 7.24, 7.33, 7.35, 7.38, 7.41, 7.42, 7.53).

(2)   Choose the correct alkene starting material to yield a given product (7.8, 7.10, 7.14, 7.16, 7.22, 7.36, 7.37).

(3)   Deduce the structure of an alkene from its molecular formula and products of cleavage (7.25, 7.26, 7.27, 7.28, 7.29, 7.30, 7.31, 7.34, 7.39, 7.40, 7.48).

(4)   Formulate mechanisms for reactions of alkenes (7.20, 7.32, 7.43, 7.44, 7.46, 7.47, 7.49, 7.50, 7.51, 7.52, 7,54, 7.55).

8.1

a) 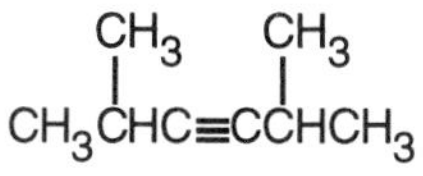 $CH_3CHC\equiv CCHCH_3$ with CH_3, CH_3

2,5–Dimethyl–3–hexyne

b) 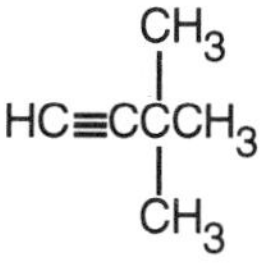 $HC\equiv CCCH_3$ with CH_3, CH_3

3,3–Dimethyl–1–butyne

c) $CH_3CH=CHCH=CHC\equiv CCH_3$

2,4–Octadiene–6–yne
(not 4,6–Octadien–2–yne)

d) 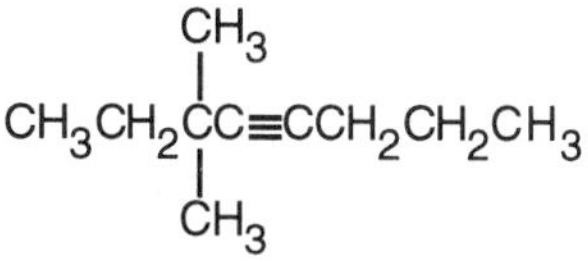 $CH_3CH_2CC\equiv CCH_2CH_2CH_3$ with CH_3, CH_3

3,3–Dimethyl–4–octyne

e) 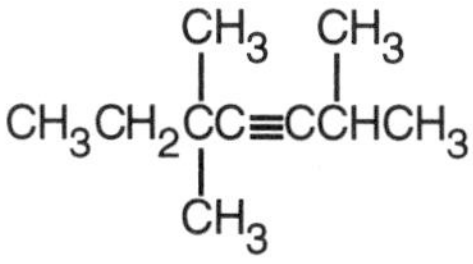 $CH_3CH_2CC\equiv CCHCH_3$ with CH_3, CH_3, CH_3

2,5,5–Trimethyl–3–heptyne

f)

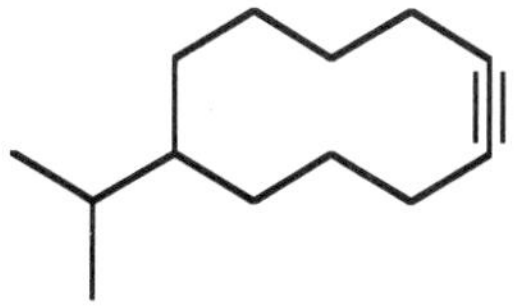

6–Isopropylcyclodecyne

8.2

$CH_3CH_2CH_2CH_2C\equiv CH$

1–Hexyne

$CH_3CH_2CH_2C\equiv CCH_3$

2–Hexyne

$CH_3CH_2C\equiv CCH_2CH_3$

3–Hexyne

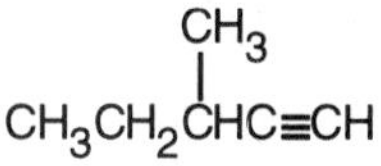

$CH_3CH_2CHC\equiv CH$ with CH_3

3–Methyl–1–pentyne

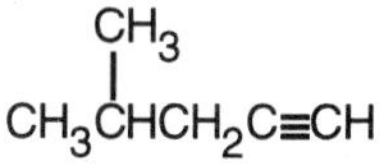

$CH_3CHCH_2C\equiv CH$ with CH_3

4–Methyl–1–pentyne

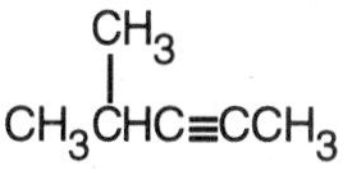

$CH_3CHC\equiv CCH_3$ with CH_3

4–Methyl–2–pentyne

$CH_3CC\equiv CH$ with CH_3, CH_3

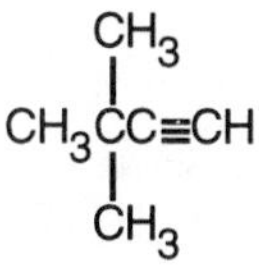

3,3–Dimethyl–1–butyne

8.3

a) $CH_3CH_2CH_2C\equiv CH$ + $2Cl_2$ $\longrightarrow$ $CH_3CH_2CH_2CCl_2CHCl_2$

b) 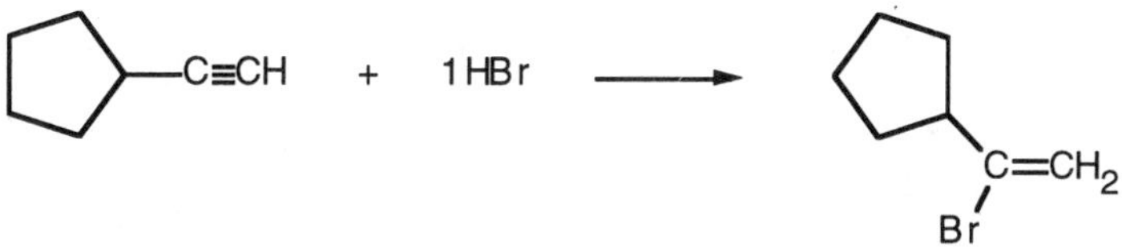   ⬡—$C\equiv CH$ + $1\,HBr$ $\longrightarrow$ ⬡—$C=CH_2$ with Br

c) $CH_3CH_2CH_2CH_2C\equiv CCH_3$ +   1 HBr $\longrightarrow$

$CH_3CH_2CH_2CH_2$, H
$C=C$
Br, CH_3

+

$CH_3CH_2CH_2CH_2$, Br
$C=C$
H, CH_3

8.4

$CH_3CH_2CH_2C\equiv CCH_2CH_2CH_3$ $\xrightarrow[HgSO_4]{H_3O^+}$ $CH_3CH_2CH_2CH_2\overset{O}{\overset{||}{C}}CH_2CH_2CH_3$

This symmetrical alkyne yields only one product.

$CH_3CH_2CH_2C\equiv CCH_2\overset{CH_3}{\overset{|}{C}}HCH_3$ $\xrightarrow[HgSO_4]{H_3O^+}$ $CH_3CH_2CH_2CH_2\overset{O}{\overset{||}{C}}CH_2\overset{CH_3}{\overset{|}{C}}HCH_3$

+

$CH_3CH_2CH_2\overset{O}{\overset{||}{C}}CH_2CH_2\overset{CH_3}{\overset{|}{C}}HCH_3$

Two ketone products result from hydration of 2–methyl–4–octyne.

8.5

a)
$CH_3CH_2CH_2C\equiv CH$ $\xrightarrow[HgSO_4]{H_3O^+}$ $\left[CH_3CH_2CH_2\overset{OH}{\overset{|}{C}}=CH_2 \right]$ $\longrightarrow$ $CH_3CH_2CH_2\overset{O}{\overset{||}{C}}CH_3$

b)
$CH_3CH_2C\equiv CCH_3$ $\xrightarrow[HgSO_4]{H_3O^+}$ $CH_3CH_2\overset{O}{\overset{||}{C}}CH_2CH_3$ + $CH_3CH_2CH_2\overset{O}{\overset{||}{C}}CH_3$

8.6

a)   ⬡—$C\equiv CH$ $\xrightarrow[2.\,H_2O_2]{1.\,BH_3,\;THF}$ ⬡—CH_2CHO

b) $(CH_3)_2CHC\equiv CCH(CH_3)_2$ $\xrightarrow[2.H_2O_2,\;^-OH]{1.BH_3,\;THF}$ $(CH_3)_2CHCH_2\overset{O}{\overset{||}{C}}CH(CH_3)_2$

8.7

a) $CH_3CH_2CH_2CH_2CH_2C\equiv CCH_3$ $\xrightarrow{Li/NH_3}$ $CH_3CH_2CH_2CH_2CH_2$, H
$C=C$
H, CH_3

2–Octyne

trans–2–Octene

b) $CH_3CH_2CH_2C\equiv CCH_2CH_3$ $\xrightarrow[\text{Lindlar}]{H_2}$

3–Heptyne

cis–3–Heptene

c) $CH_3CH_2\overset{\underset{\displaystyle CH_3}{|}}{C}HC\equiv CH$ $\xrightarrow{\text{Li/NH}_3}$ *or* $\xrightarrow[\text{Lindlar}]{H_2}$ $CH_3CH_2\overset{\underset{\displaystyle CH_3}{|}}{C}HCH=CH_2$

3–Methyl–1–pentyne   3–Methyl–1–pentene

8.8

a)

$\xrightarrow[\text{H}_3\text{O}^+]{\text{KMnO}_4}$ + CO_2

b) $CH_3(CH_2)_7C\equiv C(CH_2)_7C\equiv C(CH_2)_7CH_3$

$\downarrow \begin{matrix}\text{KMnO}_4\\ \text{H}_3\text{O}^+\end{matrix}$

$2\ CH_3(CH_2)_7COOH$ + $HOOC(CH_2)_7COOH$

8.9   A base that is strong enough to deprotonate acetone must be the conjugate base of an acid weaker than acetone. In this problem, only $Na^+\ ^-C\equiv CH$ is a base strong enough to deprotonate acetone.

8.10

Alkyne	R'X (X=Br or I)	Product
a) $CH_3CH_2CH_2C\equiv CH$	CH_3X	$CH_3CH_2CH_2C\equiv CCH_3$
or		2–Hexyne
$HC\equiv CCH_3$	$CH_3CH_2CH_2X$	
b) $(CH_3)_2CHC\equiv CH$	CH_3CH_2X	$(CH_3)_2CHC\equiv CCH_2CH_3$
		2–Methyl–3–hexyne
c)	CH_3X	

(alkyne) CH_3X (product)

d) $(CH_3)_2CHCH_2C\equiv CH$	CH_3X	$(CH_3)_2CHCH_2C\equiv CCH_3$
or		5–Methyl–2–hexyne
$HC\equiv CCH_3$	$(CH_3)_2CHCH_2X$	

e) $HC\equiv CC(CH_3)_3$ CH_3CH_2X $CH_3CH_2C\equiv CC(CH_3)_3$

                                                     2,2–Dimethyl–3–hexyne

Products (b), (c), and (e) can be synthesized by only one route because only primary halides can be used for acetylide alkylations.

8.11

$CH_3C\equiv CH$ $\xrightarrow[\text{2. }CH_3Br,\text{ THF}]{\text{1. }NaNH_2,\text{ }NH_3}$ $CH_3C\equiv CCH_3$ $\xrightarrow[\substack{\text{Lindlar}\\\text{catalyst}}]{H_2}$

cis–2–Butene

8.12   The starting material is $CH_3CH_2CH_2C\equiv CCH_2CH_2CH_3$.

a) $KMnO_4$ cleaves 4–octyne into two four–carbon fragments.

$CH_3CH_2CH_2C\equiv CCH_2CH_2CH_3$ $\xrightarrow[H_3O^+]{KMnO_4}$ 2 $CH_3CH_2CH_2COOH$

                                                     Butanoic acid

b) To reduce a triple bond to a double bond with *cis* stereochemistry use H_2 with Lindlar catalyst.

$CH_3CH_2CH_2C\equiv CCH_2CH_2CH_3$ $\xrightarrow[\text{Lindlar}]{H_2}$

cis–4–Octene

c) Addition of HBr to *cis*–4–octene (part b) yields 4–bromooctane.

$\xrightarrow{HBr}$ $CH_3CH_2CH_2CHBrCH_2CH_2CH_2CH_3$

                                                     4–Bromooctane

Alternatively, lithium/ammonia reduction of 4–octyne, followed by addition of HBr, gives 4–bromooctane.

d) Hydration of *cis*–4–octene (part b) yields 4–hydroxyoctane (4–octanol).

$\xrightarrow[\text{2. }NaBH_4]{\text{1. }Hg(OAc)_2,\text{ }H_2O}$ $CH_3CH_2CH_2\overset{\overset{\displaystyle OH}{|}}{C}HCH_2CH_2CH_2CH_3$

                                                     4–Hydroxyoctane

e) Addition of Cl_2 to 4–octene (part b) yields 4,5–dichlorooctane.

$$CH_3CH_2CH_2CHClCHClCH_2CH_2CH_3$$
4,5–Dichlorooctane

8.13   The following syntheses are explained in detail in order to illustrate "retrosynthetic" logic --
the system of planning syntheses by working backwards.

a) 1. $CH_3CH_2CH_2CH_2CH_2CH_2CH_2CH_2CH_2CH_3$. An immediate precursor might be
an alkene or alkyne. Try $n–C_8H_{17}C{\equiv}CH$, which can be reduced to decane by
H_2/Pd.

2. The alkyne $n–C_8H_{17}C{\equiv}CH$ can be formed by alkylation of $HC{\equiv}C{:}^-Na^+$ by
$n–C_8H_{17}Br$.

3. $HC{\equiv}C{:}^-Na^+$ can be formed by treatment of $HC{\equiv}CH$ with $NaNH_2$, NH_3.

The complete sequence:

$$HC{\equiv}CH \xrightarrow[NH_3]{NaNH_2} HC{\equiv}C{:}^-Na^+ \xrightarrow[THF]{n–C_8H_{17}Br} n–C_8H_{17}C{\equiv}CH \xrightarrow{H_2/Pd} n–C_{10}H_{22}$$

b) 1. An immediate precursor to $CH_3CH_2CH_2CH_2C(CH_3)_3$ might be
$HC{\equiv}CCH_2CH_2C(CH_3)_3$, which, when hydrogenated, yields 2,2–dimethylhexane.

2. $HC{\equiv}CCH_2CH_2C(CH_3)_3$ can be formed by alkylation of $HC{\equiv}C{:}^-Na^+$ with
$BrCH_2CH_2C(CH_3)_3$.

The complete sequence:

$$HC{\equiv}CH \xrightarrow[NH_3]{NaNH_2} HC{\equiv}C{:}^-Na^+$$

$$HC{\equiv}C{:}^-Na^+ + BrCH_2CH_2C(CH_3)_3 \longrightarrow HC{\equiv}CCH_2CH_2C(CH_3)_3 \xrightarrow[Pd]{2\ H_2} CH_3CH_2CH_2CH_2C(CH_3)_3$$

c) 1. $CH_3CH_2CH_2CH_2CH_2CHO$ can be made by treating $CH_3CH_2CH_2CH_2C{\equiv}CH$ with
borane followed by H_2O_2.

2. $CH_3CH_2CH_2CH_2C{\equiv}CH$ can be synthesized from $CH_3CH_2CH_2CH_2Br$ and
$HC{\equiv}C{:}^-Na^+$.

The complete sequence:

$$HC\equiv CH \xrightarrow[NH_3]{NaNH_2} HC\equiv C:^-Na^+ \xrightarrow[THF]{CH_3CH_2CH_2CH_2Br} CH_3CH_2CH_2CH_2C\equiv CH$$

$$\downarrow \begin{array}{l} 1.\ BH_3,\ THF \\ 2.\ H_2O_2 \end{array}$$

$$CH_3CH_2CH_2CH_2CH_2\overset{\displaystyle O}{\overset{\displaystyle \|}{C}}H$$

d) 1. $CH_3CH_2CH_2CH_2CH_2\overset{\displaystyle O}{\overset{\displaystyle \|}{C}}CH_3$. This ketone is the product of hydration of

$CH_3CH_2CH_2CH_2CH_2C\equiv CH$ with $H_2SO_4, H_2O, HgSO_4$.

2. $CH_3CH_2CH_2CH_2CH_2Br + HC\equiv C:^-Na^+ \xrightarrow{THF} CH_3CH_2CH_2CH_2CH_2C\equiv CH$

The complete sequence:

$$HC\equiv CH \xrightarrow[NH_3]{NaNH_2} HC\equiv C:^-Na^+ \xrightarrow[THF]{CH_3CH_2CH_2CH_2CH_2Br} CH_3CH_2CH_2CH_2CH_2C\equiv CH$$

$$\downarrow \begin{array}{l} H_2SO_4,\ H_2O \\ Hg^{2+} \end{array}$$

$$CH_3CH_2CH_2CH_2CH_2\overset{\displaystyle O}{\overset{\displaystyle \|}{C}}CH_3$$

8.14 a) 2,2–Dimethyl–3–hexyne   b) 2,5–Octadiyne
c) 3,6–Dimethyl–2–hepten–4–yne   d) 3,3–Dimethyl–1,5–hexadiyne
e) 1,3–Hexadien–5–yne   f) 3,6–Diethyl–2–methyl–4–octyne

8.15

a) $CH_3CH_2CH_2C\equiv C\overset{\displaystyle CH_3}{\underset{\displaystyle CH_3}{\overset{\displaystyle |}{\underset{\displaystyle |}{C}}}}CH_2CH_3$

b) $CH_3C\equiv CC\equiv C\overset{\displaystyle CH_3}{\overset{\displaystyle |}{C}}HCH_2\underset{\displaystyle CH_2CH_3}{\underset{\displaystyle |}{C}}HC\equiv CH$

c) $(CH_3)_3CC\equiv CC(CH_3)_3$

d) (cyclic structure)
$CH_2C\equiv C\overset{\displaystyle CH_3}{\overset{\displaystyle |}{C}}H$
$CH_2 \qquad CHCH_3$
$CH_2CH_2CH_2CH_2$

e) $HC\equiv CCH=CHCH=CHCH_3$

f) $CH_3CH_2C\equiv CCH_2C(CH_3)_2CHClCH=CH_2$

g) $CH_3CH_2CH_2CH_2\overset{\displaystyle CH_3CHCH_2CH_3}{\overset{\displaystyle |}{C}}HC\equiv CH$

h) $CH_3CH_2CH_2\overset{\displaystyle C(CH_3)_3}{\overset{\displaystyle |}{C}}HC\equiv CCH(CH_3)_2$

8.16 a) $(CH_3)_3CCH_2CH_2C \equiv CCH_2CH_3$. <u>correct name</u>; 7,7–Dimethyl–3–octyne
   This compound is an *octyne*.

   b) $HC \equiv CC(CH_3)_2CH_2CH_2CH(CH_3)_2$. <u>correct name</u>; 3,3,6–Trimethyl–1–heptyne
   Start numbering from the opposite end.

   c) $CH_3CH=CHCH_2CH(CH_3)C \equiv CH$. <u>correct name</u>; 3–Methyl–5–hepten–1–yne
   Try not to break up the name of the compound more than necessary.

   d) $HC \equiv CCH_2CHCH_2CH_2CHCH_3$.   <u>correct name</u>; 4,7,8–Trimethyl–1–nonyne
   Choose the longest chain, and number from the opposite end.

   e) $CH_3CH_2CH=CHC \equiv CH$.   <u>correct name</u>; 3–Hexen–1–yne.
   Start numbering from the opposite end.

   f)

   <u>correct name</u>; 1–Ethynyl–3–methylcyclohexane
   The ring positions are numbered incorrectly.

8.17 a) $CH_3CH=CHC \equiv CC \equiv CCH=CHCH=CHCH=CH_2$.
   1,3,5,11–Tridecatetraen–7,9–diyne

   Using *E–Z* notation: (3*E*,5*E*,11*E*)–1,3,5,11–Tridecatetraen–7,9–diyne
   The parent alkane of this hydrocarbon is tridecane.

   b) $CH_3C \equiv CC \equiv CC \equiv CC \equiv CCH=CH_2$. 1–Tridecen–3,5,7,9,11–pentayne
   This hydrocarbon is also of the tridecane family.

8.18

8.19 a) An acyclic alkane with eight carbons has the formula C_8H_{18}. C_8H_{10} has eight fewer
   hydrogens, or four fewer pairs of hydrogens, than C_8H_{18}. Thus, C_8H_{10} contains four
   degrees of unsaturation.

   b) Because only one equivalent of H_2 is absorbed over the Lindlar catalyst, *one* triple
   bond is present.

   c) Three equivalents of H_2 are absorbed when reduction is done over a palladium catalyst;
   two of them hydrogenate the triple bond already found to be present. Therefore, one
   double bond must also be present.

   d) C_8H_{10} must contain one ring. Many structures are possible.

8.20

8.21

8.22

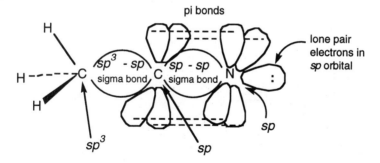

CH₃CH₂CH₂C≡CCH₃

a) $\xrightarrow{\text{2 equiv Br}_2}$ CH₃CH₂CH₂CBr₂CBr₂CH₃

b) $\xrightarrow{\text{1 equiv HBr}}$
$$\underset{\underset{\text{Br}}{|}}{\overset{\overset{\text{CH}_3\text{CH}_2\text{CH}_2}{|}}{C}} = \underset{\underset{\text{CH}_3}{|}}{\overset{\overset{\text{H}}{|}}{C}} \quad + \quad \underset{\underset{\text{H}}{|}}{\overset{\overset{\text{CH}_3\text{CH}_2\text{CH}_2}{|}}{C}} = \underset{\underset{\text{CH}_3}{|}}{\overset{\overset{\text{Br}}{|}}{C}}$$

c) $\xrightarrow{\text{xs HBr}}$ CH₃CH₂CH₂CBr₂CH₂CH₃ + CH₃CH₂CH₂CH₂CBr₂CH₃

d) $\xrightarrow{\text{Li, NH}_3}$
$$\underset{\underset{\text{H}}{|}}{\overset{\overset{\text{CH}_3\text{CH}_2\text{CH}_2}{|}}{C}} = \underset{\underset{\text{CH}_3}{|}}{\overset{\overset{\text{H}}{|}}{C}}$$

e) $\xrightarrow[\text{HgSO}_4]{\text{H}_2\text{O, H}_2\text{SO}_4}$ CH₃CH₂CH₂C(=O)CH₂CH₃ + CH₃CH₂CH₂CH₂C(=O)CH₃

8.23

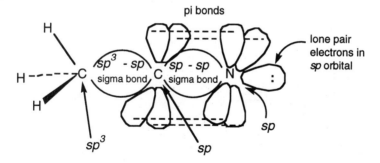

Each of the two pi bonds between carbon and nitrogen is formed by overlap of one *p* orbital of carbon with one *p* orbital of nitrogen.

8.24

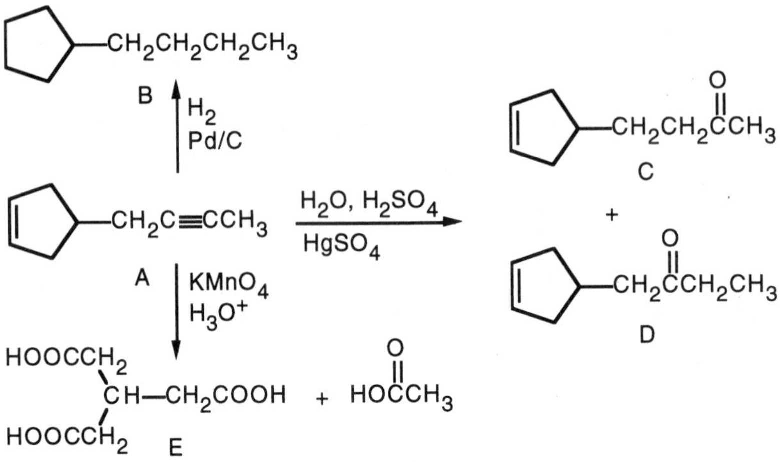

8.25

a) $CH_3CH_2C\equiv CH$ $\xrightarrow[\text{HgSO}_4]{\text{H}_2\text{O, H}_2\text{SO}_4}$ $CH_3CH_2\overset{\overset{\displaystyle O}{\|}}{C}CH_3$

b) $CH_3CH_2C\equiv CH$ $\xrightarrow[\text{2. H}_2\text{O}_2]{\text{1. BH}_3\text{, THF}}$ $CH_3CH_2CH_2CHO$

c) 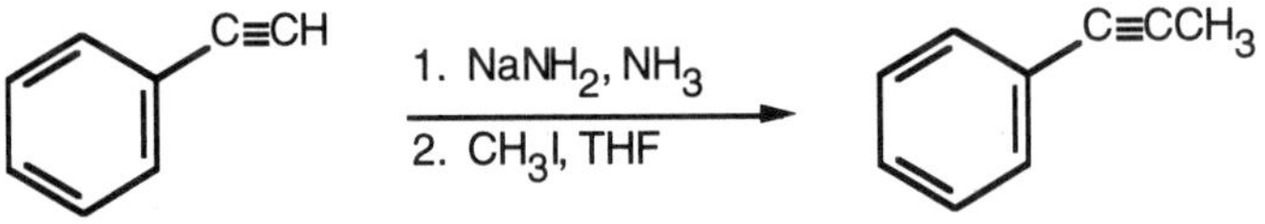 $\xrightarrow[\text{2. CH}_3\text{I, THF}]{\text{1. NaNH}_2\text{, NH}_3}$

d) 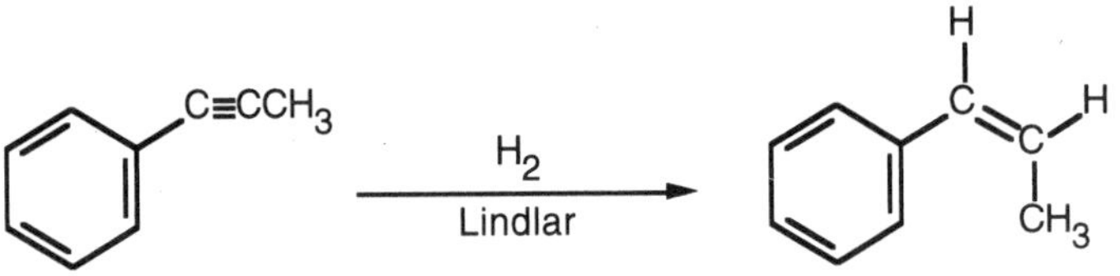 $\xrightarrow[\text{Lindlar}]{\text{H}_2}$

e) $CH_3CH_2C\equiv CH$ $\xrightarrow[\text{H}_3\text{O}^+]{\text{KMnO}_4}$ CH_3CH_2COOH + CO_2

f) $CH_3CH_2CH_2CH_2CH=CH_2$ $\xrightarrow[\text{CCl}_4]{\text{Br}_2}$ $CH_3CH_2CH_2CH_2CHBrCH_2Br$

$\xrightarrow[\text{2. H}_3\text{O}^+]{\text{1. 2 NaNH}_2}$

$CH_3CH_2CH_2CH_2C\equiv CH$

8.26

a)

trans-5-Decene

$\xrightarrow[CCl_4]{Br_2}$

$\xrightarrow{2\ NaNH_2}$

$\downarrow H_2$ Lindlar

cis-5-Decene

b)

cis-5-Decene

$\xrightarrow[CCl_4]{Br_2}$

$\xrightarrow{2\ NaNH_2}$

$\downarrow Li/NH_3$

trans-5-Decene

8.27 Both $KMnO_4$ and O_3 oxidation of alkynes yield carboxylic acids; terminal alkynes give CO_2 also.

a) $CH_3(CH_2)_5C\equiv CH$ $\xrightarrow[H_3O^+]{KMnO_4}$ $CH_3(CH_2)_5COOH$ + CO_2

b)

$C\equiv CCH_3$ $\xrightarrow[H_3O^+]{KMnO_4}$ COOH   + CH_3COOH

c) Since only one cleavage product is formed, the parent hydrocarbon must have had a triple bond as part of a ring.

$\xrightarrow[H_3O^+]{KMnO_4}$ $HOOC(CH_2)_8COOH$

d) Notice that the products of this ozonolysis contain aldehyde and ketone functional groups, as well as a carboxylic acid and CO_2. The parent hydrocarbon must thus contain a double and a triple bond.

$\underset{\displaystyle CH_3CH=CCH_2CH_2C\equiv CH}{\overset{\displaystyle CH_3}{|}}$ $\xrightarrow[\text{2. Zn, }H_3O^+]{\text{1. }O_3}$ CH_3CHO + $CH_3\overset{O}{\overset{||}{C}}CH_2CH_2COOH$ + CO_2

e)

$C\equiv CH$ $\xrightarrow[\text{2. Zn, }H_3O^+]{\text{1. }O_3}$ $OHCCH_2CH_2CH_2CH_2\overset{O}{\overset{||}{C}}COOH$ + CO_2

8.28

a) 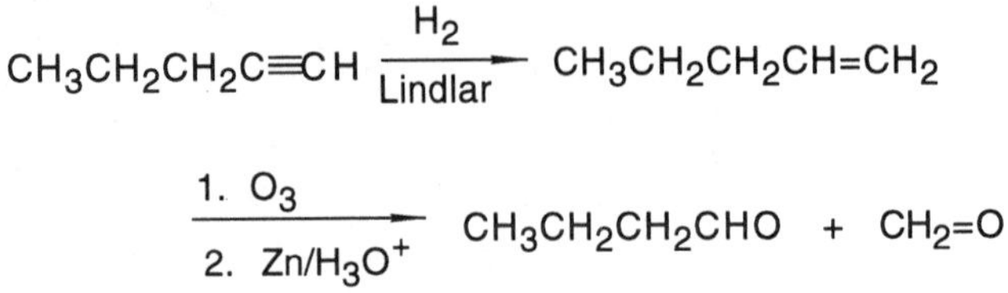 $CH_3CH_2CH_2C{\equiv}CH$ $\xrightarrow[\text{Lindlar}]{H_2}$ $CH_3CH_2CH_2CH{=}CH_2$

$\xrightarrow[\text{2. Zn/H}_3\text{O}^+]{\text{1. O}_3}$ $CH_3CH_2CH_2CHO$ + $CH_2{=}O$

b) 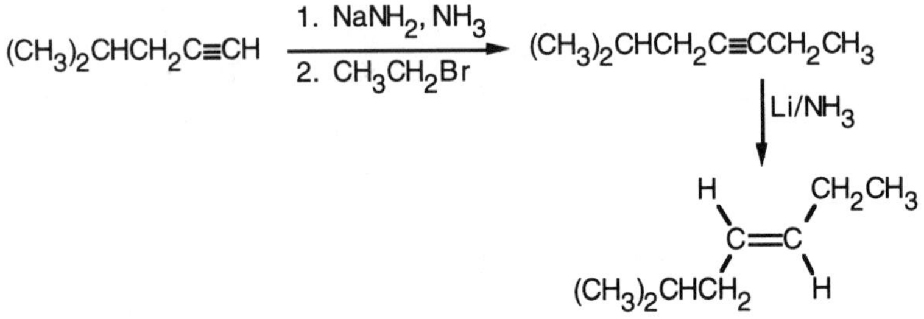 $(CH_3)_2CHCH_2C{\equiv}CH$ $\xrightarrow[\text{2. CH}_3\text{CH}_2\text{Br}]{\text{1. NaNH}_2,\ \text{NH}_3}$ $(CH_3)_2CHCH_2C{\equiv}CCH_2CH_3$

$\downarrow$ Li/NH$_3$

8.29

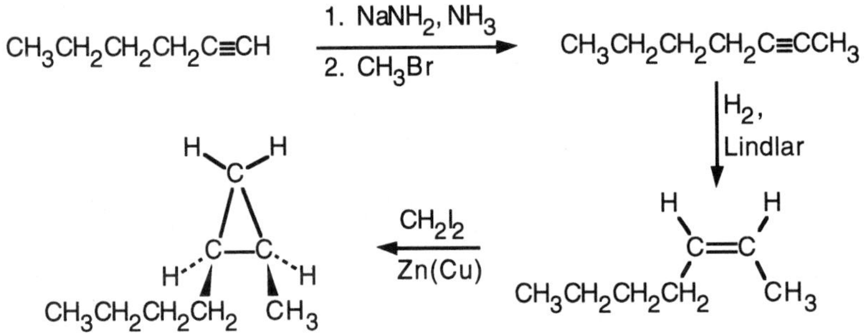

$CH_3CH_2CH_2CH_2C{\equiv}CH$ $\xrightarrow[\text{2. CH}_3\text{Br}]{\text{1. NaNH}_2,\ \text{NH}_3}$ $CH_3CH_2CH_2CH_2C{\equiv}CCH_3$

$\downarrow$ H$_2$, Lindlar

$\xleftarrow[\text{Zn(Cu)}]{\text{CH}_2\text{I}_2}$

8.30

$\xrightarrow[\text{CCl}_4]{\text{Br}_2}$

$\xrightarrow{\text{2 NaNH}_2}$

A

$\xrightarrow[\text{peroxides}]{\text{HBr}}$

B

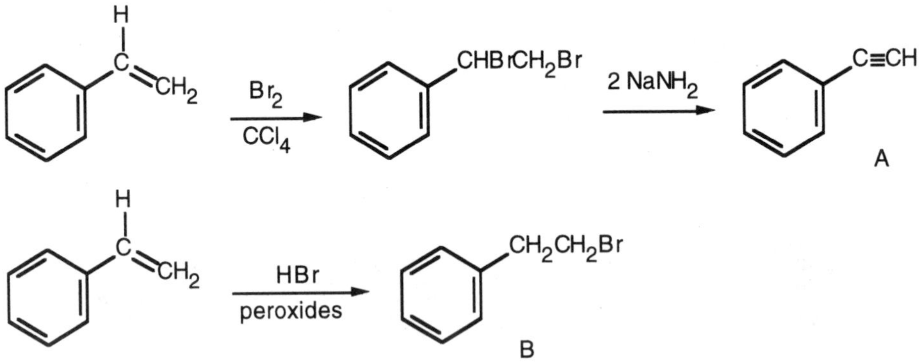

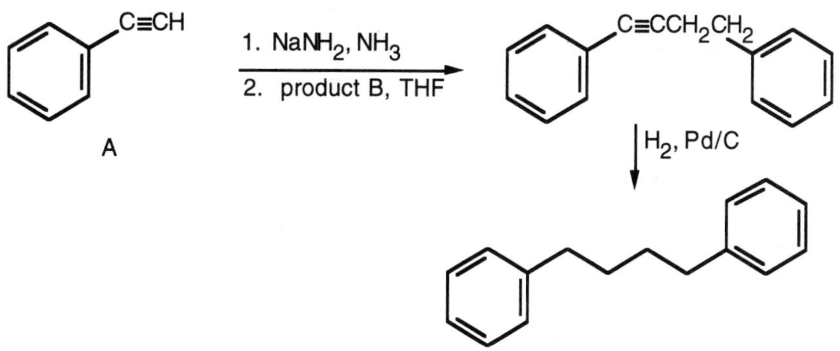

8.31

a) $CH_3(CH_2)_4C{\equiv}CH$ $\xrightarrow[\text{2. } H_2O_2]{\text{1. } BH_3, \text{ THF}}$ $CH_3(CH_2)_5CHO$

b)

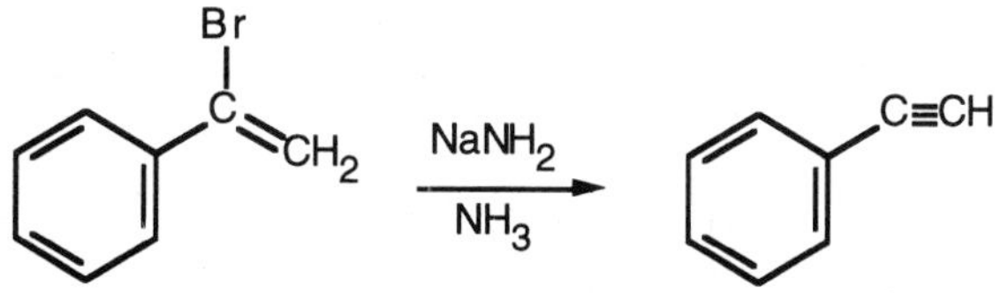

c)

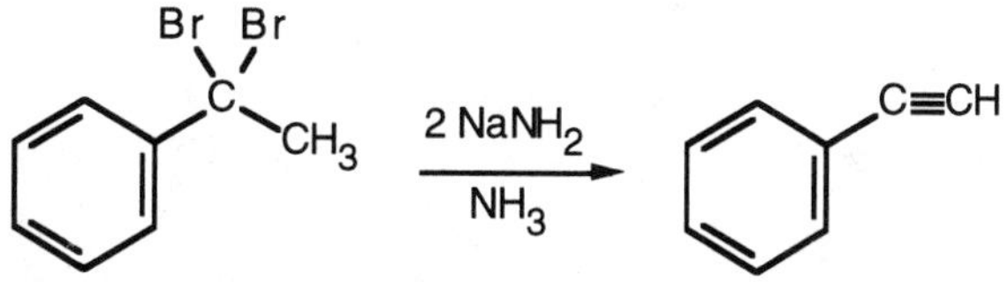

8.32

a) $CH_3CH_2C{\equiv}CH$ $\xrightarrow{2\ Cl_2}$ $CH_3CH_2CCl_2CHCl_2$

1,1,2,2–Tetrachlorobutane

b) $CH_3CH_2C{\equiv}CH$ $\xrightarrow[\text{Lindlar}]{H_2}$ $CH_3CH_2CH{=}CH_2$ $\xrightarrow[\text{peroxides}]{HBr}$ $CH_3CH_2CH_2CH_2Br$

$CH_3CH_2C{\equiv}CH$ $\xrightarrow[\text{2. } CH_3CH_2CH_2CH_2Br, \text{ THF}]{\text{1. } NaNH_2, NH_3}$ $CH_3CH_2C{\equiv}CCH_2CH_2CH_2CH_3$

$\xrightarrow[\text{Pd/C}]{H_2}$ Octane

c) $CH_3CH_2C{\equiv}CH$ $\xrightarrow[\text{2. } H_2O_2]{\text{1. } BH_3, \text{ THF}}$ $CH_3CH_2CH_2CHO$

Butanal

8.33

a) $HC{\equiv}CH$ $\xrightarrow[\text{NH}_3]{NaNH_2}$ $HC{\equiv}C{:}^-Na^+$ $\xrightarrow[\text{THF}]{CH_3CH_2CH_2Br}$ $CH_3CH_2CH_2C{\equiv}CH$

b) $HC \equiv CH$ $\xrightarrow[\text{NH}_3]{\text{NaNH}_2}$ $HC \equiv C:^- Na^+$ $\xrightarrow[\text{THF}]{\text{CH}_3\text{CH}_2\text{Br}}$ $CH_3CH_2C \equiv CH$

$\downarrow$ $\begin{array}{l}\text{NaNH}_2\\ \text{NH}_3\end{array}$

$CH_3CH_2C \equiv CCH_2CH_3$ $\xleftarrow[\text{THF}]{\text{CH}_3\text{CH}_2\text{Br}}$ $[CH_3CH_2C \equiv C:^- Na^+]$

c) $HC \equiv CH$ $\xrightarrow[\text{NH}_3]{\text{NaNH}_2}$ $HC \equiv C:^- Na^+$ $\xrightarrow{(CH_3)_2CHCH_2Br}$ $(CH_3)_2CHCH_2C \equiv CH$

$\downarrow$ $\begin{array}{l}\text{H}_2\\ \text{Lindlar}\end{array}$ or Li/NH$_3$

$(CH_3)_2CHCH_2CH=CH_2$

d)   Product from a) $\xrightarrow[\text{NH}_3]{\text{NaNH}_2}$ $[CH_3CH_2CH_2C \equiv C:^- Na^+]$ $\xrightarrow[\text{THF}]{\text{CH}_3\text{CH}_2\text{CH}_2\text{Br}}$

$CH_3CH_2CH_2C \equiv CCH_2CH_2CH_3$ $\xrightarrow[\text{Hg}^{2+}]{\text{H}_2\text{SO}_4, \text{H}_2\text{O}}$ $CH_3CH_2CH_2\overset{\overset{\displaystyle O}{\|}}{C}CH_2CH_2CH_2CH_3$

e) $HC \equiv CH$ $\xrightarrow[\text{NH}_3]{\text{NaNH}_2}$ $HC \equiv C:^- Na^+$ $\xrightarrow[\text{THF}]{\text{CH}_3\text{CH}_2\text{CH}_2\text{CH}_2\text{Br}}$ $CH_3CH_2CH_2CH_2C \equiv CH$

$\downarrow$ $\begin{array}{l}\text{1. BH}_3\text{, THF}\\ \text{2. H}_2\text{O}_2\end{array}$

$CH_3CH_2CH_2CH_2CH_2CHO$

8.34

a) $CH_3CH_2C \equiv CCH_2CH_3$ $\xrightarrow[\text{Lindlar}]{\text{D}_2}$

b) $CH_3CH_2C \equiv CCH_2CH_3$ $\xrightarrow{\text{Li, ND}_3}$

c) $CH_3CH_2CH_2C \equiv CH$ $\xrightarrow[\text{NH}_3]{\text{NaNH}_2}$ $[CH_3CH_2CH_2C \equiv C:^- Na^+]$ $\xrightarrow{\text{D}_3\text{O}^+}$ $CH_3CH_2CH_2C \equiv CD$

d)

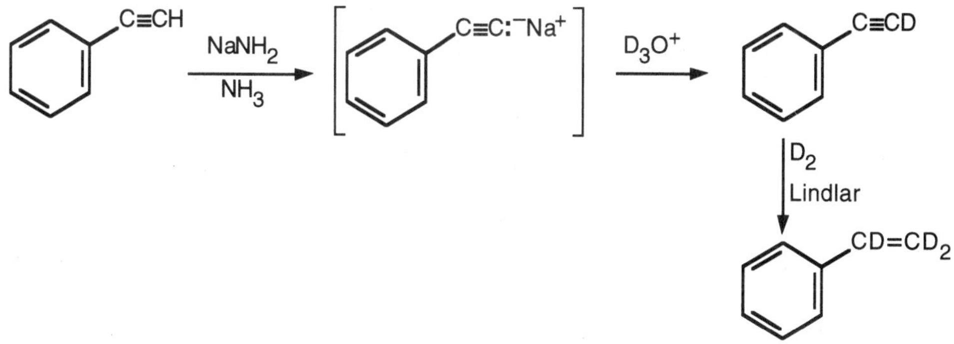

8.35

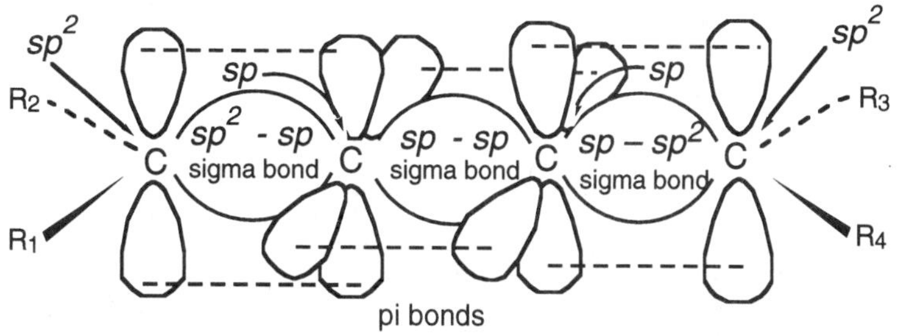

pi bonds

This simplest cumulene is pictured above. The carbons at the end of the cumulated double bonds are sp^2 hybridized and form one pi bond to the "interior" carbons. The interior carbons are sp hybridized; each carbon forms two pi bonds – one to an "exterior" carbon and one to the other interior carbon. If you build a model of this cumulene, you can see that the substituents all lie in the same plane. This cumulene can thus exhibit cis–trans isomerism, just as simple alkenes can.

In general, the substituents of any compound with an odd number of adjacent double bonds lie in a plane; these compounds can exhibit cis–trans isomerism. The relationship of substituents at the ends of any compound with an even number of adjacent double bonds will be explained later.

8.36

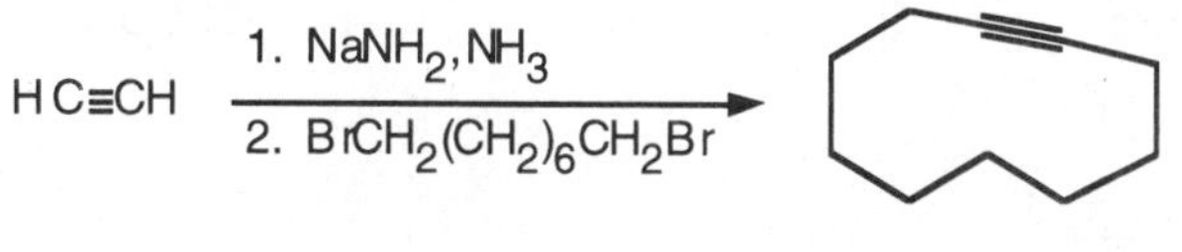

8.37 Muscalure is a C_{23} alkene. The only functional group present is the double bond between C_9 and C_{10}. Since our synthesis begins with acetylene, we can assume that the double bond can be produced by hydrogenation of a triple bond.

$HC≡CH$ $\xrightarrow[\text{NH}_3]{\text{NaNH}_2}$ $HC≡C:^-Na^+$ $\xrightarrow[\text{THF}]{\text{CH}_3(\text{CH}_2)_6\text{CH}_2\text{Br}}$ $CH_3(CH_2)_7C≡CH$

$\downarrow$ NaNH₂, NH₃

$CH_3(CH_2)_7C≡C(CH_2)_{12}CH_3$ $\xleftarrow[\text{THF}]{\text{CH}_3(\text{CH}_2)_{11}\text{CH}_2\text{Br}}$ $[CH_3(CH_2)_7C≡C:^-Na^+]$

$\downarrow$ H₂, Lindlar

$CH_3(CH_2)_6CH_2$ $\quad$ $CH_2(CH_2)_{11}CH_3$
$$\underset{H}{\overset{}{C}}=\underset{H}{\overset{}{C}}$$

(Z)–9–Tricosene

8.38

3 H₂/Pd → B (propylcyclohexane)

A $\xrightarrow{\text{1. O}_3 \quad \text{2. Zn, H}_3\text{O}^+}$ cyclohexanone (O) plus other fragments

$\xrightarrow{\text{1. NaNH}_2, \text{NH}_3 \quad \text{2. CH}_3\text{I}}$ C

8.39

$\xleftarrow[\text{Pd}]{\text{H}_2}$ $\begin{array}{l} CH_2-C≡C-C≡C-CH_2 \\ | \qquad\qquad\qquad\quad | \\ CH_2-C≡C-C≡C-CH_2 \end{array}$ $\xrightarrow[\text{2. Zn, H}_3\text{O}^+]{\text{1. O}_3}$ $\begin{array}{l} \text{2 HOOCCH}_2\text{CH}_2\text{COOH} \\ + \\ \text{2 HOOC-COOH} \end{array}$

A

8.40

$CH_3\overset{O}{\overset{||}{C}}CH_3$ $\xrightarrow[\text{2. H}_3\text{O}^+]{\text{1. HC≡C:}^-\text{Na}^+}$ $CH_3\overset{OH}{\underset{C≡CH}{\overset{|}{\underset{|}{C}}}}CH_3$ $\xrightarrow{\text{H}_3\text{O}^+}$ $CH_2=\overset{CH_3}{\overset{|}{C}}C≡CH$ $\xrightarrow[\text{Lindlar}]{\text{H}_2}$ $CH_2=\overset{CH_3}{\overset{|}{C}}CH=CH_2$

2–Methyl–
1,3–butadiene

8.41 1) Erythrogenic acid contains six degrees of unsaturation (see Sec. 6.2 for the method of calculating unsaturation equivalents for compounds containing elements other than C and H).

2) One of these double bonds is contained in the carboxylic acid functional group –COOH; thus, five other degrees of unsaturation are present.

3) Because five equivalents of H_2 are absorbed on catalytic hydrogenation, erythrogenic acid contains no rings.

4) The presence of both aldehyde and carboxylic acid products of ozonolysis indicates that both double and triple bonds are present in erythrogenic acid.

5) Only two ozonolysis products contain aldehyde functional groups; these fragments must have been double-bonded to each other in erythrogenic acid.

$H_2C=CH(CH_2)_4C\equiv$

6) The other ozonolysis products result from cleavage of triple bonds. However, not enough information is available to tell in which order the fragments were attached. The two possible structures:

A $H_2C=CH(CH_2)_4C\equiv C-C\equiv C(CH_2)_7COOH$

B $H_2C=CH(CH_2)_4C\equiv C(CH_2)_7C\equiv CCOOH$

One method of distinguishing between the two possible structures is to treat erythrogenic acid with two equivalents of H_2, using Lindlar catalyst. The resulting trialkene can then be ozonized. The fragment that originally contained the carboxylic acid can then be identified.

8.42 This reaction mechanism is similar to the mechanism of halohydrin formation.

8.43

Repeating this process several times replaces all hydrogen atoms with deuterium atoms.

Study Guide for Chapter 8

After studying this chapter, you should be able to:

(1)   Name and draw alkynes, according to IUPAC conventions (8.1, 8.2, 8.14, 8.15, 8.16, 8.17).

(2)   Predict the products of reactions of alkynes (8.3, 8.4, 8.6, 8.7, 8.11, 8.18, 8.20, 8.21, 8.22, 8.25, 8.28, 8.29, 8.31).

(3)   Choose the correct alkyne starting material to yield a given product (8.5, 8.10).

(4)   Use cleavage and/or hydrogenation products to determine the structure of an unknown alkyne (8.8, 8.19, 8.24, 8.27, 8.38, 8.39, 8.41).

(5)   Explain the hybridization of the triple bond, and account for the weak acidity of alkynes (8.9, 8.23, 8.35).

(6)   Carry out transformation and syntheses involving alkynes (8.12, 8.13, 8.26, 8.30, 8.32, 8.33, 8.34, 8.36, 8.37, 8.40).

Chapter 9 – Stereochemistry

9.1 Chiral: screw, beanstalk, shoe.
Achiral: screwdriver, hammer.

9.2 Use the following rules to locate centers that are *not* stereogenic.

1. All $-CH_3$ and $-CX_3$ carbons are nonstereogenic.
2. All $-CH_2$ and $-CX_2-$ carbons are nonstereogenic.
3. All $-\overset{|}{C}=\overset{|}{C}-$ and $-C\equiv C-$ carbons are nonstereogenic.

By rule 3, all benzene-ring carbons are nonstereogenic

a) 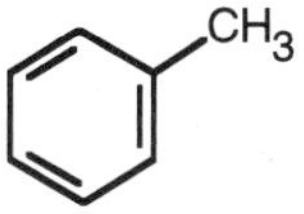

Toluene
achiral

b) 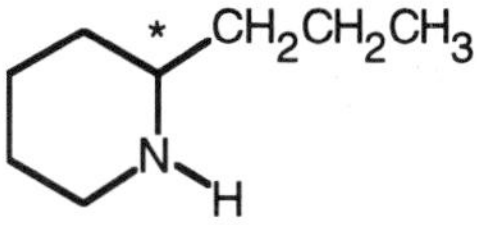

Coniine
chiral

c) 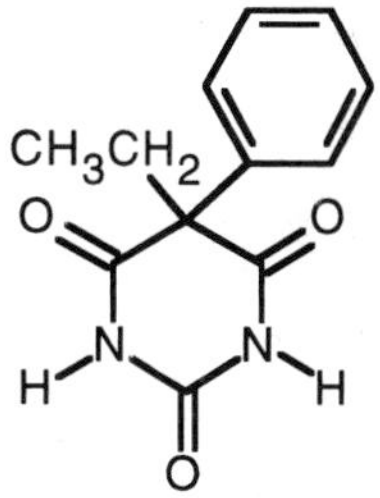

Phenobarbital
achiral

9.3

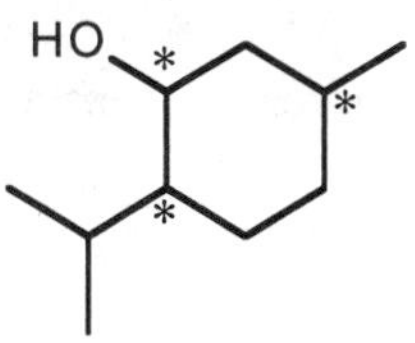

Menthol

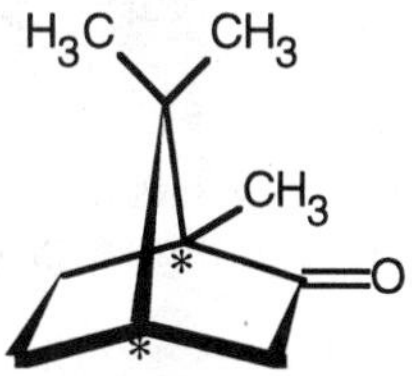

Camphor

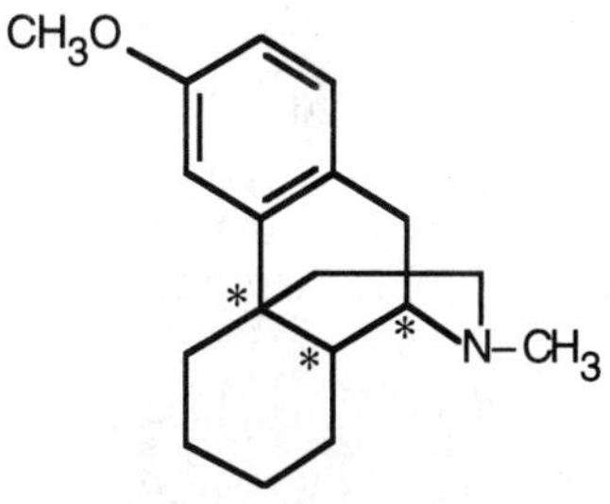

Dextromethorphan

9.4

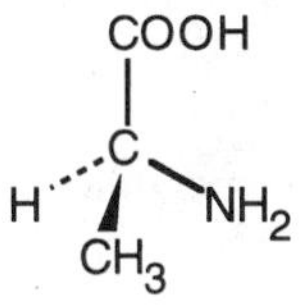

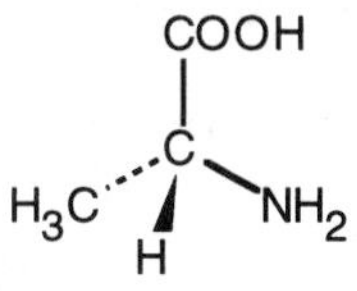

Alanine

9.5

Use the formula $[\alpha]_D = \dfrac{\alpha}{l \times C}$ where

$\quad [\alpha]_D$ = specific rotation

$\quad \alpha$ = observed rotation

$\quad l$ = path length of cell (in dm)

$\quad C$ = concentration (in g/mL)

In this problem:

$\quad \alpha = +1.21°$

$\quad l = 5.00 \text{ cm} = 0.500 \text{ dm}$

$\quad C = 1.50 \text{ g}/10.0 \text{ mL} = 0.150 \text{ g/mL}$

$$[\alpha]_D = \frac{+1.21°}{0.500 \text{ dm} \times 0.150 \text{ g/mL}} = +16.1°$$

9.6 Use the sequence rules in Section 9.6.

a) By rule 1, –H is of lowest priority, and –Br is of highest priority. By rule 2, –CH$_2$CH$_2$OH is of higher priority than –CH$_2$CH$_3$.

Highest ⟶ Lowest

–Br, –CH$_2$CH$_2$OH, –CH$_2$CH$_3$, –H

b) By rule 3, –COOH can be considered as $\overset{\displaystyle -O \quad O-}{\underset{}{-C-OH}}$. Because three oxygens are attached

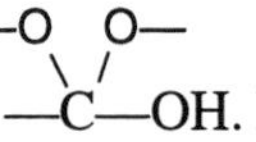

to a –COOH carbon and only one oxygen is attached to –CH$_2$OH, –COOH is of higher priority than –CH$_2$OH. –CO$_2$CH$_3$ is of higher priority than –COOH by rule 2, and –OH is of highest priority by rule 1.

Highest ⟶ Lowest

–OH, –CO$_2$CH$_3$, –COOH, –CH$_2$OH

c)   –NH$_2$, –CN, –CH$_2$NHCH$_3$, –CH$_2$NH$_2$

d)   –Br, –Cl, –CH$_2$Br, –CH$_2$Cl

9.7 The following scheme may be used to assign *R,S* configurations to stereogenic centers:

Step 1. For each stereogenic center, rank substituents by the Cahn–Ingold–Prelog system; give the number 4 to the lowest priority substituent. For part (a):

Substituent	Priority
–Br	1
–COOH	2
–CH$_3$	3
–H	4

Step 2. With practice, you can learn to mentally point your thumb from the stereogenic carbon to the fourth priority group (–H) and determine which hand has fingers that curl from group 1 to group 2 to group 3. Alternatively, you might find it easiest to redraw the molecule so that the lowest priority group is at the top. To avoid errors, use a molecular model.

Step 3. With your thumb pointing from the stereogenic carbon to the fourth priority group, determine which hand has fingers that curl from group 1 to group 2 to group 3. This compound is assigned the *S* configuration because the fingers of the left hand curl in the correct direction.

b)

c)

9.8 As in the previous problem, assign priorities to the substituents, giving the number 4 to the lowest priority substituent. Draw a tetrahedral carbon, and place the lowest priority group at the top. Since we are drawing an *S* stereoisomer, arrange the other groups so that an arrow going from group 1 to 2 to 3 curls the same way as the fingers of the left hand when the left thumb points from carbon to the lowest priority group.

$$CH_3CH_2CH_2\overset{\overset{\displaystyle OH}{|}}{C}HCH_3 \qquad \text{2–Pentanol}$$

Substituent	Priority
—OH	1
—CH$_2$CH$_2$CH$_3$	2
—CH$_3$	3
—H	4

9.9   *R,S* assignments for more complicated molecules can be made using the same method as in Problem 9.7, taking one carbon atom at a time. It's especially important to use molecular models when a compound has more than one stereogenic center. For part (a):

Step 1. Assign priorities to groups of the top stereogenic center.

Substituent	Priority
–Br	1
–CH(OH)CH$_3$	2
–CH$_3$	3
–H	4

Step 2. Orient the model so that the lowest priority group of the first stereogenic center points up.

Step 3. Pointing a thumb from –C to –H, observe the curl of the fingers. For this stereogenic center, the curl of the fingers of the right hand corresponds to the direction of the arrows traveling from group 1 to group 2 to group 3, and the configuration is *R*.

Step 4. Repeat steps 1-3 for the next stereogenic center.

b) *S,R*
c) *R,S*
d) *S,S*

a, d are enantiomers and are diastereomeric with b, c.
b, c are enantiomers and are diastereomeric with a, d.

9.10

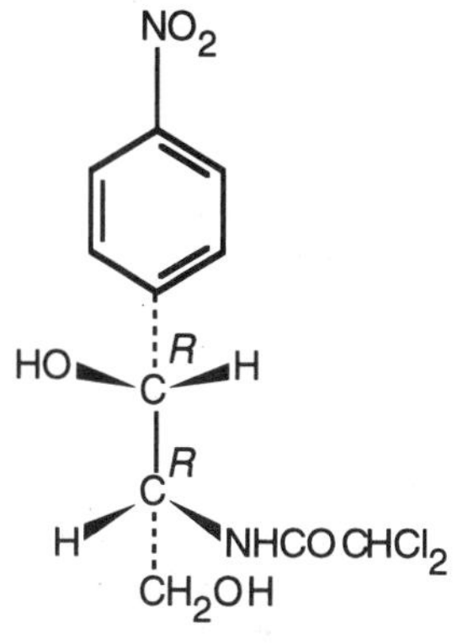

Chloramphenicol

9.11 To decide if a structure represents a *meso* compound, try to locate a plane of symmetry that divides the molecule into two halves that are mirror images. Molecular models may be helpful.

a)

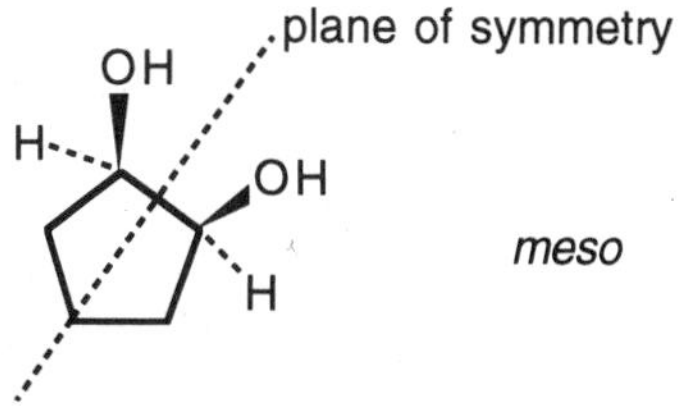

meso

b) and c) are not *meso* structures.

d)

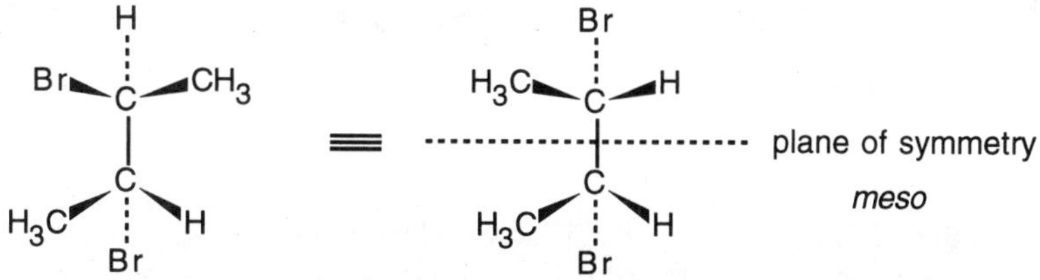

plane of symmetry

meso

9.12 For a molecule to exist as a meso form, it must possess a plane of symmetry. 2,3–Dibromobutane can exist as a pair of enantiomers *or* as a meso compound, depending on the configurations at carbons 2 and 3.

a)

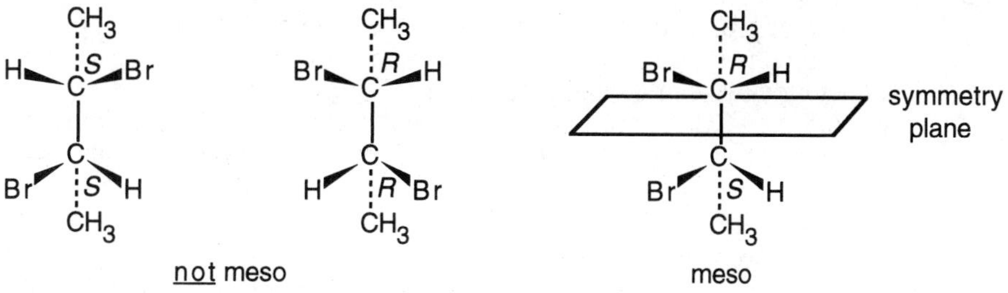

<u>not</u> meso   meso

b) 2,3–Dibromopentane has no symmetry plane and thus can't exist in a meso form.

c) 2,4–Dibromopentane can exist in a meso form.

symmetry
plane

2,4–Dibromopentane can also exist as a pair of enantiomers (2R,4R) and (2S,4S) that are not meso compounds.

9.13

Morphine has five stereogenic centers and, in principle, can have $2^5 = 32$ stereoisomers. Many of these stereoisomers are too strained to exist.

9.14   a)

S -5-Chloro-2-hexene   Chlorocyclohexane
These two compounds are constitutional isomers.

b) The two dibromopentane stereoisomers are diastereomers. They are illustrated in Problem 9.12(a).

9.15   Two manipulations of Fischer projections are allowable.
1. A Fischer projection may be rotated on the page by 180°, but not by 90° or 270°.
2. Holding one group of a Fischer projection steady, we may rotate the other three groups either clockwise or counterclockwise.
If we hold the –COOH group of projection A steady and rotate the other three groups by 120°, we arrive at a projection identical with projection B. Thus A and B are identical.

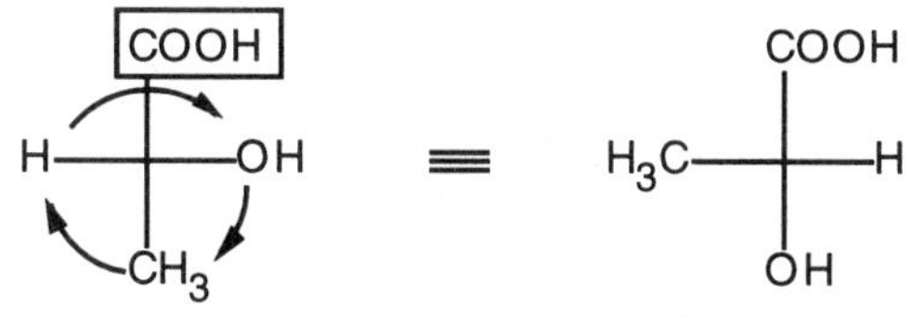

If projection C is rotated 180°, the resulting projection has two groups superimposable with projection B. However, no manipulation can make all four groups superimposable; C is thus an enantiomer of A and B.

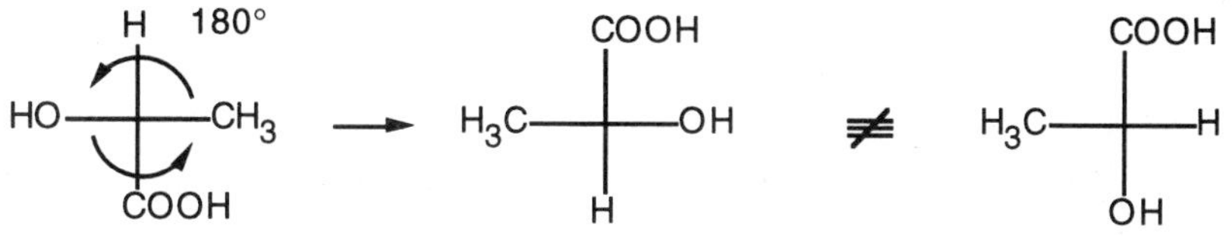

Likewise, projection D is not superimposable on projection B but is identical to projection C. Thus, A and B are identical and enantiomeric with C and D, which are also identical.

9.16 a) Manipulate two groups in the first structure to the positions they occupy in the second structure.

For example:

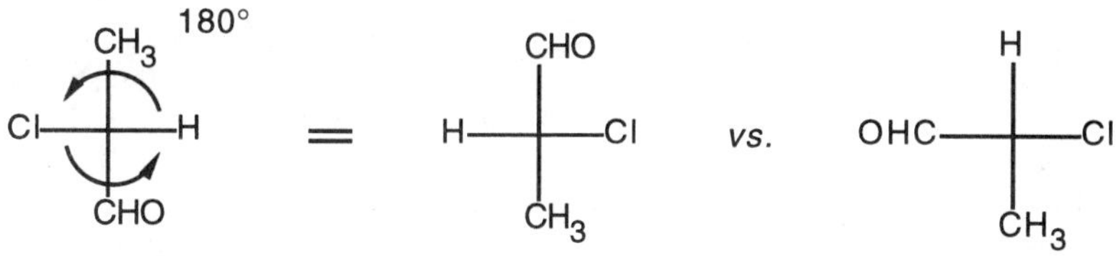

Although this projection resembles the second one in placement of –CH$_3$ and –Cl, –H and –CHO are interchanged. The two projections thus represent enantiomers.

b) As above, rotate the first projection by 180°.

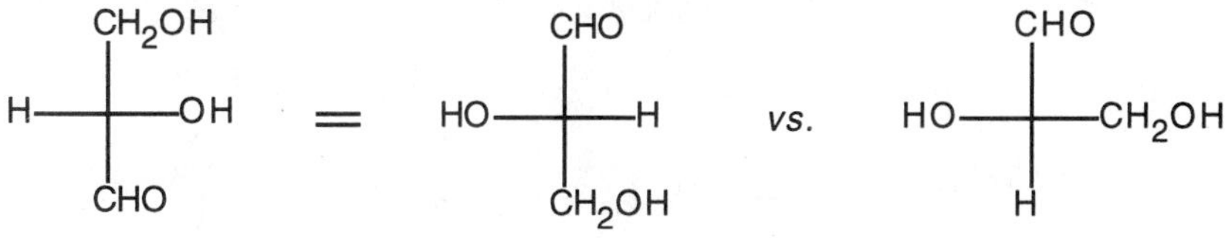

If we now compare this structure to the second structure of the pair, two of the groups occupy the same relative position. There is, however, no rotation that can make these two structures superimposable; they are enantiomers.

9.17 One of the hardest spatial problems in organic chemistry is to visualize two-dimensional drawings as three-dimensional chemical structures. This difficulty becomes particularly troublesome when it is necessary to assign R,S configurations to structures, especially if they are drawn as Fischer projections. Working out these assignments is easier when models are used, but it is still possible to determine a configuration from a two-dimensional drawing.

The following system may be used for assigning R,S configurations. Part (a) is used as an example.

Step 1. Rank substituents in priority order.

Substituents	Priority	
–Br	1	high
–COOH	2	
–CH$_3$	3	
–H	4	low

Step 2. Use either of the allowable manipulations of Fischer projections to bring the group of lowest priority to the top of the projection. In this case, hold –Br steady and rotate the other three groups clockwise.

Step 3. Indicate on the rotated projection the direction of the arrows that proceed from group 1 to group 2 to group 3. If the direction of the arrows is clockwise, the configuration is R; if the direction is counterclockwise, the configuration is S. Here the configuration is S.

c)

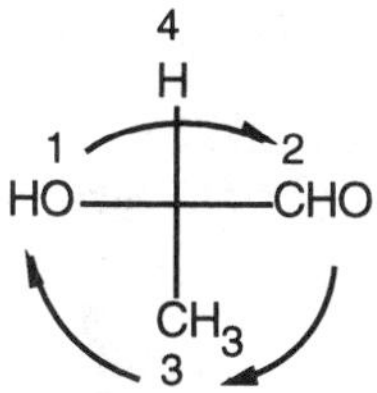

R

9.18 For simplicity, consider only top-side attack of bromine.

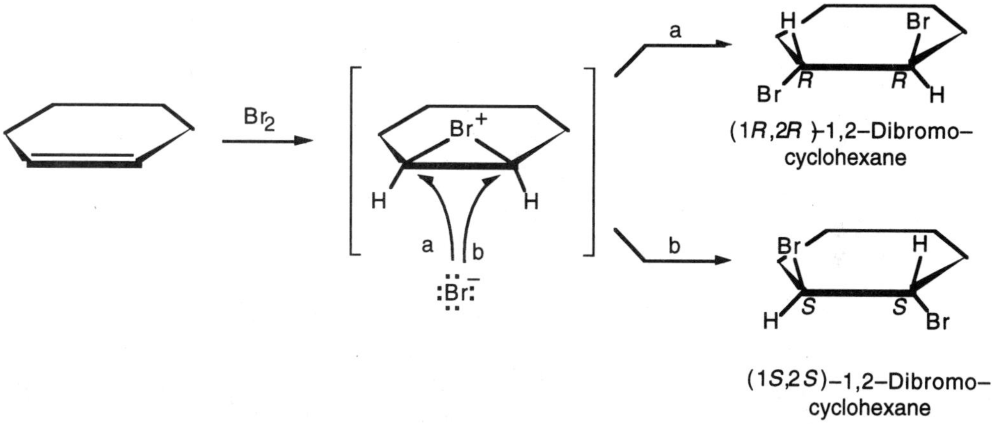

(1R,2R)–1,2–Dibromo-
cyclohexane

(1S,2S)–1,2–Dibromo-
cyclohexane

The product of addition of Br_2 to cyclohexene is a racemic mixture of the (1R,2R) and (1S,2S) enantiomers. The products of bottom-side attack are identical to those of top-side attack.

9.19 Possible bromonium ion intermediates are shown below. In this problem, assume that attack of Br^- is somewhat more likely at carbon 2.

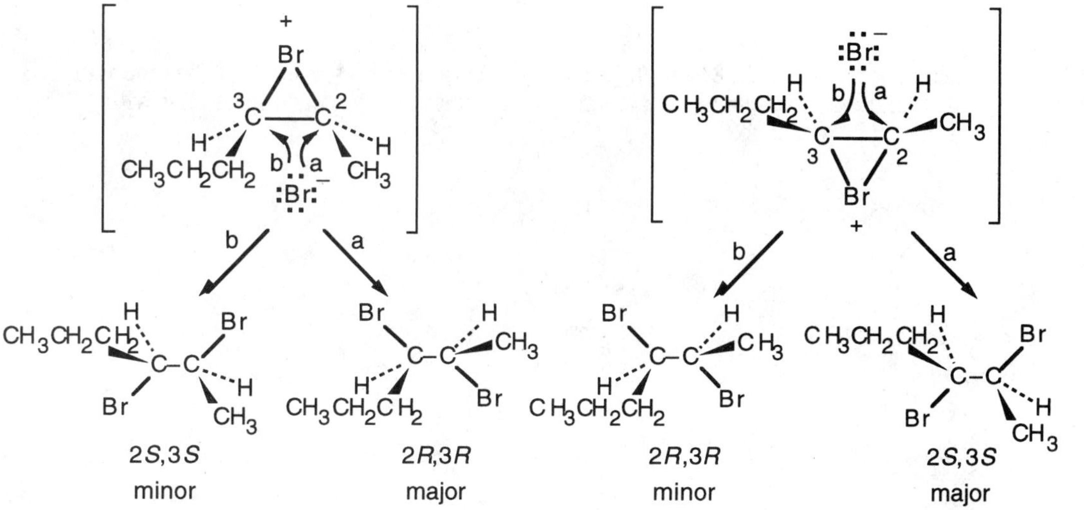

2S,3S
minor

2R,3R
major

2R,3R
minor

2S,3S
major

The products of attack of bromide ion on each bromonium ion are shown above. Notice that the major products are enantiomers of each other, as are the minor products. Because the bromonium ions are formed in a 50:50 mixture, and because the percent attack at carbon 2 is the same for each bromonium ion, the amount of $(2R,3R)$ and $(2S,3S)$–dibromohexanes is equal, and the product is a racemic mixture.

9.20   As in Problem 9.19, let's assume that attack at carbon 2 is be more likely.

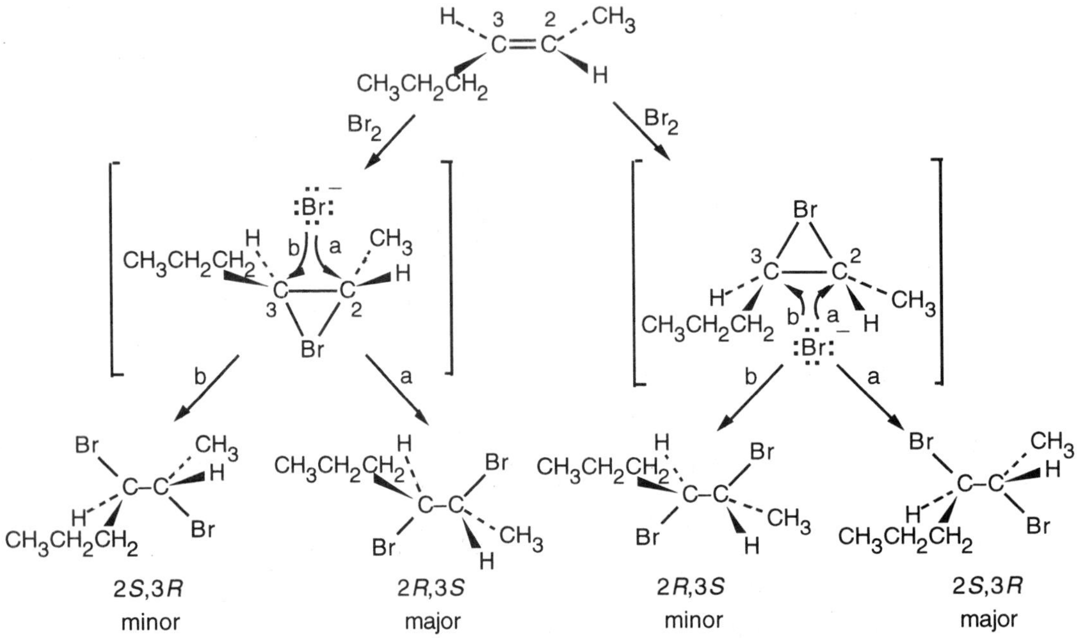

The product of bromination of *trans*–2–hexene is a racemic mixture of $(2S,3R)$–2,3–dibromohexane and $(2R,3S)$–2,3–dibromohexane. The reasoning is explained in Problem 9.19.

9.21   Look back to Figure 9.18, which shows the reaction of R–4–methyl–1–hexene with HBr. In a similar way, we can write a reaction mechanism for the reaction of HBr with S–4–methyl–1–hexene.

(2S,4S)–2–Bromo–4–methylhexane   (2R,4S)-2-Bromo–4–methylhexane

The 2S,4S stereoisomer is the enantiomer of the 2R,4R isomer, and the transition states leading to the formation of these two isomers are enantiomeric and of equal energy. Thus, the 2S,4S and 2R,4R enantiomers are formed in equal amounts. A similar argument can be used to show that the 2R,4S and 2S,4R isomers are formed in equal amounts. The product of the reaction of HBr with racemic starting material is thus a racemic mixture of the four possible stereoisomers and is optically inactive.

Note that the ratio of (2R,4R + 2S,4S):(2R,4S + 2S,4R) is not 50:50. Nevertheless, the product mixture is optically inactive because the enantiomers are formed in equal amounts.

9.22

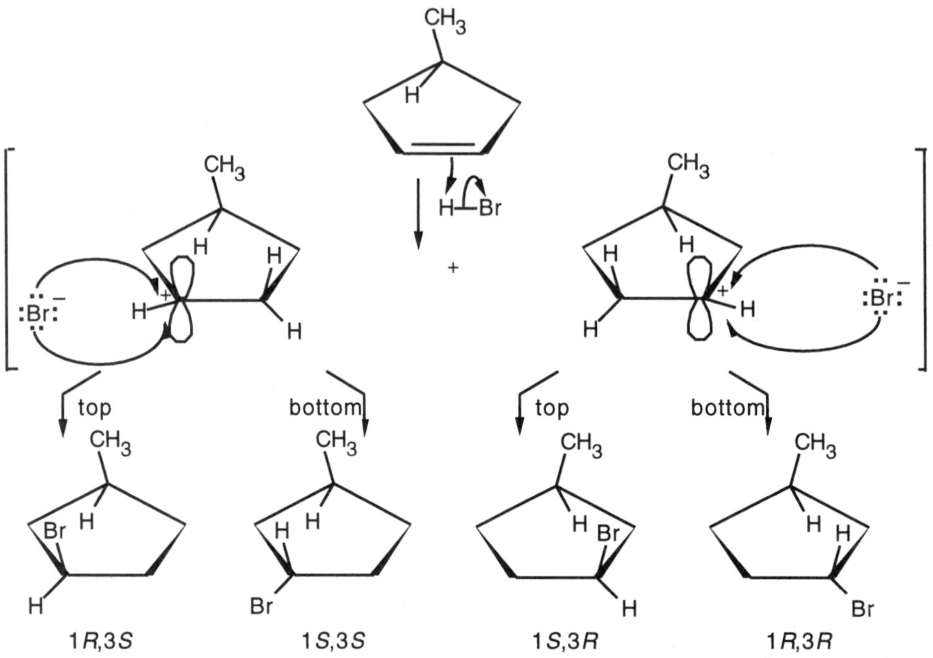

Two enantiomeric carbocations are formed. Each carbocation can be attacked by bromide ion from either the top or the bottom to yield four stereoisomers of 1–bromo–3–methylcyclopentane. The same argument used in Problem 9.21 can be used to show that the 1*S*,3*R* and 1*R*,3*S* enantiomers are formed in equal amounts, and the 1*S*,3*S* and 1*R*,3*R* isomers are formed in equal amounts. The product mixture is optically inactive (racemic).

9.23

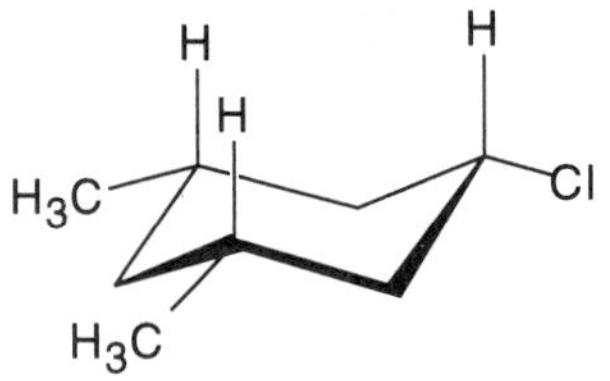

There are four stereoisomers of 1–chloro–3,5–dimethylcyclohexane. It is possible for (1) both methyl groups to be *cis* to chlorine; (2) both methyl groups to be *trans* to chlorine; (3) one methyl group to be *cis* and the other to be *trans* to chlorine (two isomers). The geometric isomer in which all three groups are *cis* is the most stable because the three groups are equatorial.

9.24

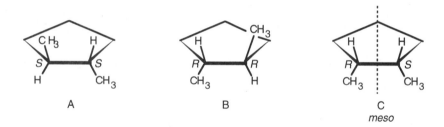

A and B are enantiomers and are chiral. Compound C is their diastereomer and is a *meso* compound.

9.25

for Cholic acid; $\quad [\alpha]_D = \dfrac{+2.22°}{0.100 \text{ dm} \times 0.600 \text{ g/mL}} = \dfrac{+2.22°}{0.0600} = +37.0°$

9.26

for Ecdysone; $\quad [\alpha]_D = \dfrac{+0.087°}{0.200 \text{ dm} \times 0.00700 \text{ g/mL}} = +62°$

9.27 Observed rotation is directly proportional to concentration. Thus, if the concentration of a sample is halved, the observed rotation is also halved. In this problem, halving the concentration of a solution with observed rotation of +90° would give a solution with a rotation of +45°. If, instead, the actual rotation were –270°, dilution would produce a solution with a rotation of –135°. The value of observed rotation after dilution allows us to know the original sign of rotation.

9.28
a) *Chirality* is the property of "handedness" — the property of a molecule or object that causes it to be nonsuperimposable on its mirror image.
b) A *stereogenic center* of a molecule is an atom that is bonded to four different atoms or groups of atoms.
c) *Optical activity* is the property of a substance that causes it to rotate the plane of polarization of plane-polarized light.
d) A *diastereomer* is a stereoisomer that is not the mirror image of another stereoisomer.
e) An *enantiomer* is one of a pair of stereoisomers that have a mirror image relationship.
f) A *racemate* is a 50:50 mixture of (+) and (–) enantiomers that behaves as if it were a pure compound and that is optically inactive.

9.29

a) $CH_3CH_2CH_2\overset{*}{C}H(CH_3)CH_2CH(CH_3)_2$. 2,4–Dimethylheptane has one stereogenic center.

b) $CH_3CH_2C(CH_3)_2CH_2CH(CH_2CH_3)_2$. 3–Ethyl–5,5–dimethylheptane is achiral.

c)

Cl——⟨ ⟩——Cl   *cis*–1,4–Dichlorocyclohexane is *achiral*. Notice the plane of symmetry that passes through the –Cl groups.

d) $CH_3C{\equiv}C\overset{*}{C}H(CH_3)\overset{*}{C}H(CH_3)C{\equiv}CCH_3$. 4,5–Dimethyl–2,6–octadiyne has two stereogenic centers.

The chirality of this compound depends on the configuration at each of the stereogenic centers. The R,R and S,S isomers are chiral enantiomers; the R,S isomer is an achiral *meso* compound.

9.30

CI
|
CH₃CH₂CH₂CHCH₃
*

$CH_3CH_2CH_2\overset{*}{C}HCH_3$

2–Chloropentane

b) $CH_3CH_2CH_2CH_2\overset{*}{C}HCH_3$

(OH)

2–Hexanol

c) $CH_3CH_2\overset{*}{C}HCH=CH_2$ (CH₃)

3–Methyl–1–pentene

d) $CH_3CH_2CH_2CH_2\overset{*}{C}HCH_2CH_3$ (CH₃)

3–Methylheptane

9.31

$CH_3CH_2CH_2CH_2CH_2OH$

achiral

$CH_3CH_2CH_2\overset{*}{C}HCH_3$ (OH)

chiral

$CH_3CH_2CHCH_2CH_3$ (OH)

achiral

$CH_3CH_2\overset{*}{C}HCH_2OH$ (CH₃)

chiral

$CH_3CH_2\overset{|}{C}CH_3$ (OH, CH₃)

achiral

$CH_3\overset{*}{C}H-CHCH_3$ (OH, CH₃)

chiral

$HOCH_2CH_2CHCH_3$ (CH₃)

achiral

$CH_3\overset{|}{C}CH_2OH$ (CH₃, CH₃)

achiral

9.32   Draw the five C_6H_{14} hexanes.

$CH_3CH_2CH_2CH_2CH_2CH_3$

3 kinds of –H

$CH_3CH_2CH_2CH(CH_3)_2$

5 kinds of –H

$CH_3CH_2CH(CH_3)CH_2CH_3$

4 kinds of –H

$(CH_3)_2CHCH(CH_3)_2$

2 kinds of –H

$CH_3CH_2C(CH_3)_3$

3 kinds of –H

17 monobromohexanes can be formed from the hexane isomers. You may need to draw all the bromohexanes to find the nine that are chiral:

$CH_3CH_2CH_2CH_2\overset{*}{C}HBrCH_3$ $CH_3CH_2CH_2\overset{*}{C}HBrCH_2CH_3$ $CH_3CH_2CH_2\overset{*}{C}H(CH_3)CH_2Br$

$CH_3CH_2\overset{*}{C}HBrCH(CH_3)_2$ $CH_3\overset{*}{C}HBrCH_2CH(CH_3)_2$ $CH_3CH_2\overset{*}{C}H(CH_3)CH_2CH_2Br$

$CH_3CH_2\overset{*}{C}H(CH_3)\overset{*}{C}HBrCH_3$ $(CH_3)_2CH\overset{*}{C}H(CH_3)CH_2Br$ $CH_3\overset{*}{C}HBrC(CH_3)_3$

9.33

a) $CH_3CH_2\overset{*}{C}H(OH)CH_3$

b) $CH_3CH_2\overset{*}{C}H(COOH)CH_3$

This carboxylic acid has no rings or carbon-carbon double or triple bonds. (The formula $C_5H_{10}O_2$ indicates that there is one double bond present, but it is in the carboxylic acid functional group.)

c) $CH_3\overset{*}{C}HBr\overset{*}{C}H(OH)CH_3$

d) $CH_3\overset{*}{C}HBrCHO$

9.34 Chiral: golf club, monkey wrench.
Achiral: basketball, fork, wine glass, snowflake.

9.35

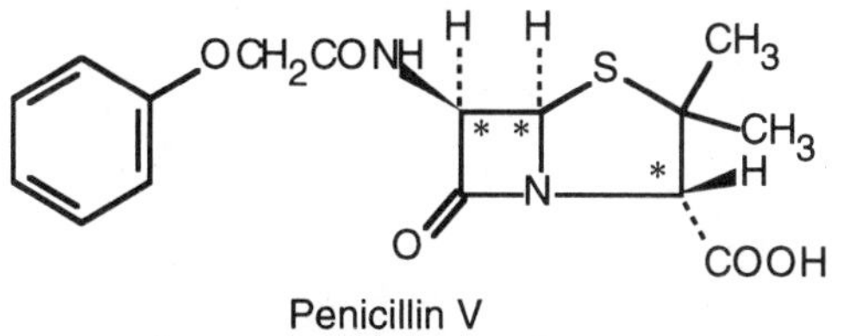

Penicillin V
three stereogenic carbons

9.36

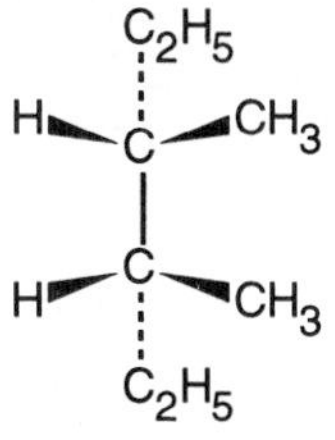

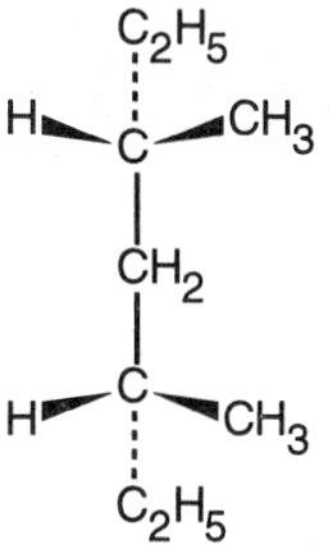

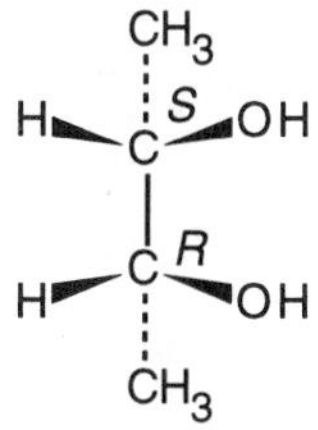

9.37 The specific rotation of (2R,3R)–dichloropentane is equal in magnitude and opposite in sign to the specific rotation of (2S,3S)–dichloropentane because the compounds are enantiomers. Because they are diastereomers, there is no predictable relationship between the specific rotations of the 2R,3S and 2R,3R isomers.

9.38–9.39

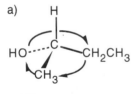

Br—C(S)—CH₃ / H₃C—C(R)—Br ... structures

   2S,4R   2R,4S   2S,4S   2R,4R

|———————————————| |———————————————|
      enantiomers   enantiomers

The 2R,4S stereoisomer is the enantiomer of the 2S,4R stereoisomer.
The 2S,4S and 2R,4R stereoisomers are diastereomers of the 2S,4R stereoisomer.

9.40

Highest priority ——————→ Lowest priority

a) $-C(CH_3)_3$, $-CH=CH_2$, $-CH(CH_3)_2$, $-CH_2CH_3$

b) ⬡ , $-C{\equiv}CH$, $-C(CH_3)_3$, $-CH=CH_2$

c) $-COOCH_3$, $-COCH_3$, $-CH_2OCH_3$, $-CH_2CH_3$

d) $-Br$, $-CH_2Br$, $-CN$, $-CH_2CH_2Br$

9.41

a) H OH
         ⟍S⟍

b) Cl H
         ⟍S⟍

c) H OCH₃
         ⟍S⟍
    HOCH₂   COOH

9.42

a) OH
       ⟍S⟍H
        S
          ⟍Cl
         H

b) H⟍S   S⟍CH₃
    CH₃CH₂⟍   ⟍H

9.43

a) H
    HO----C—CH₂CH₃
       ⟍CH₃

    (S)–2–Butanol

b) H
    Cl----C—CH₂CH₃
       ⟍CH=CH₂

    (R)–3–Chloro–1–pentene

9.44

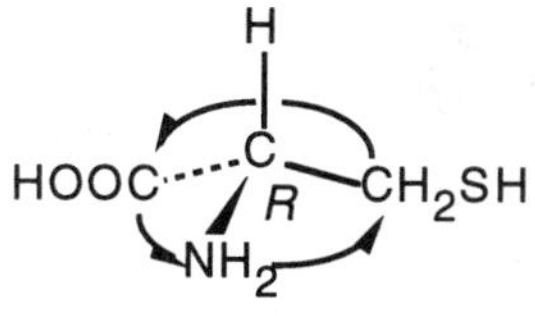

(R)–Cysteine 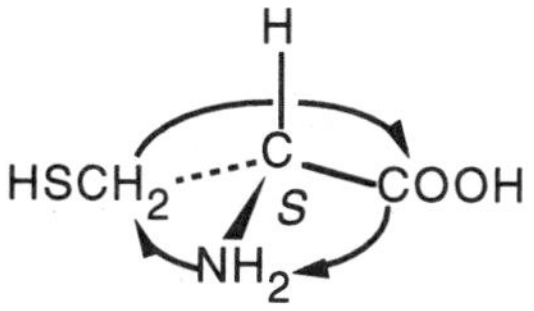   (S)–Cysteine

9.45   Identical molecules: b, c, d.
Pair of enantiomers: a.

9.46

a)

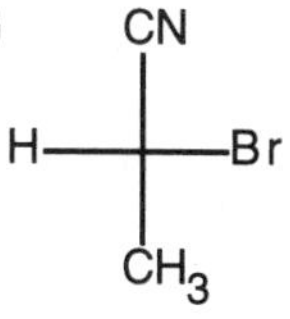

R

b)

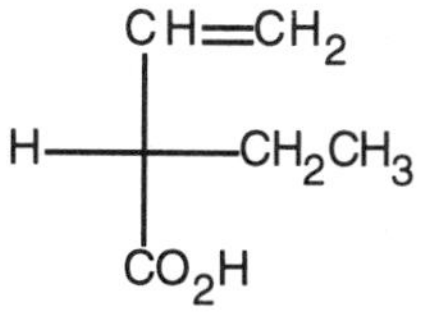

S

c)

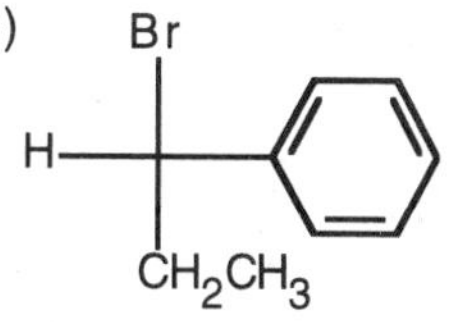

S

9.47

a)

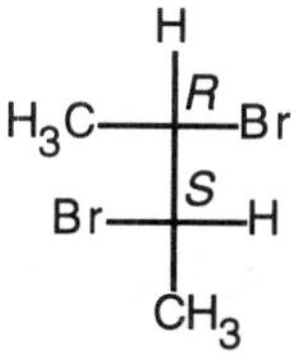

b)

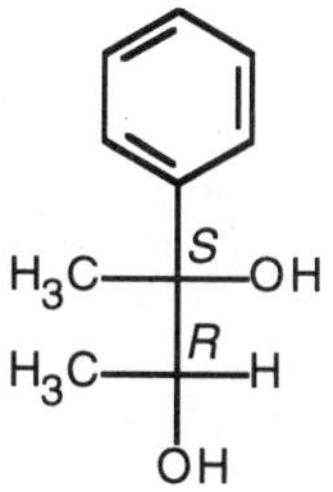

c)

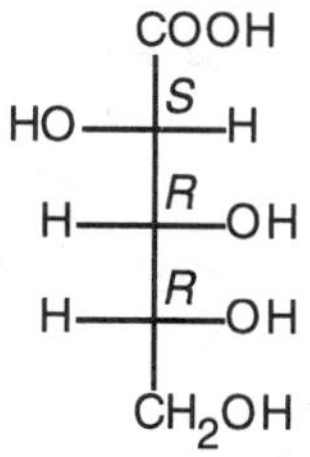

d)

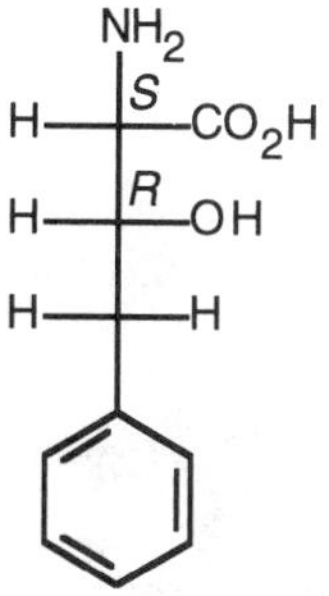

9.48

a)

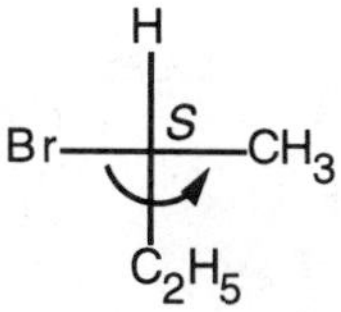

(S)–2–Bromobutane
$CH_3CH_2CHBrCH_3$

b)

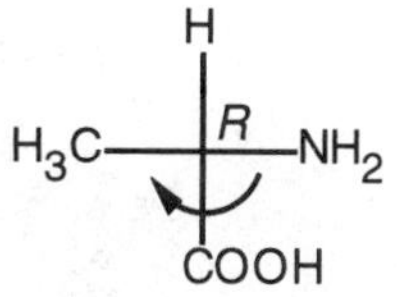

(R)–Alanine
$CH_3CH(NH_2)CO_2H$

c)

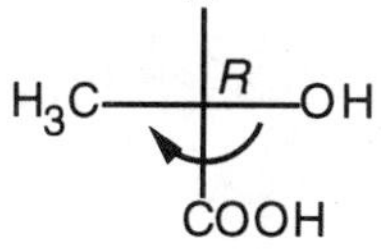

(R)–2–Hydroxypropanoic acid
$CH_3CH(OH)COOH$

d)

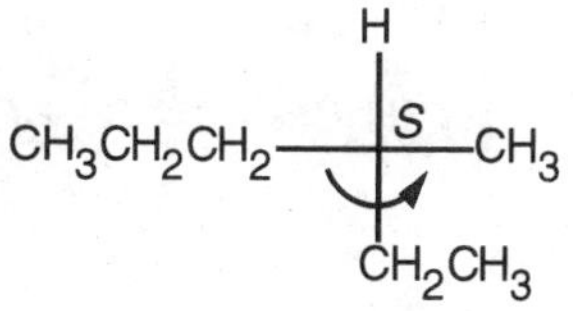

(S)–3–Methylhexane
$CH_3CH_2CH_2CH(CH_3)CH_2CH_3$

9.49

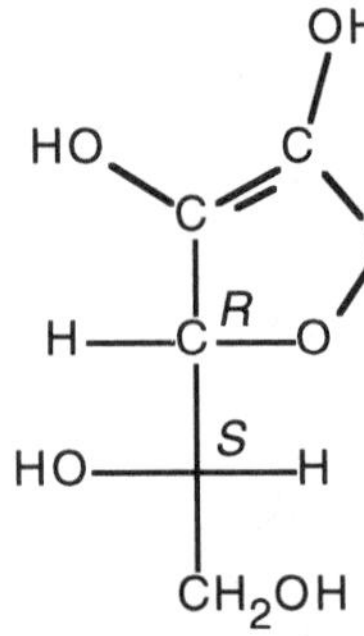

Ascorbic acid

9.50

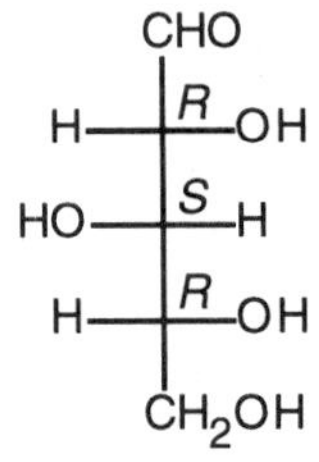

(+)–Xylose

9.51–9.52   The initial product of hydroxylation of a double bond is a cyclic *osmate*. Drawings of the osmates for *cis*–2–butene and *trans*–2–butene are shown below. No carbon-oxygen bonds are broken in the cleavage step; cleavage occurs at osmium-oxygen bonds. The final stereochemistry, therefore, is the same as that of the initial adduct.

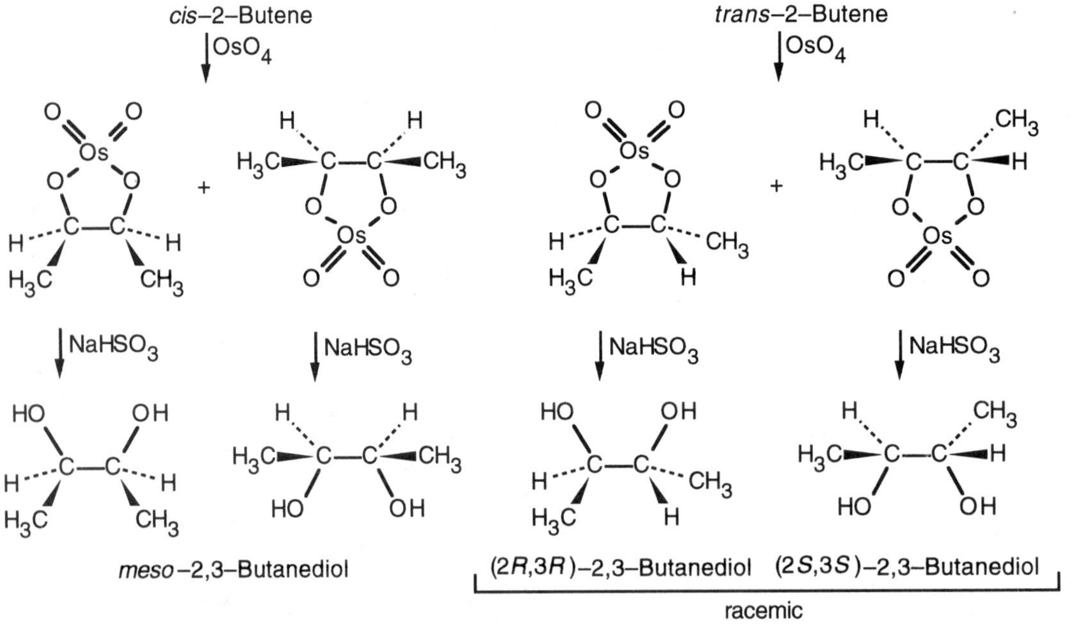

cis–2–Butene

OsO_4

trans–2–Butene

OsO_4

$NaHSO_3$ $NaHSO_3$ $NaHSO_3$ $NaHSO_3$

meso–2,3–Butanediol

(2*R*,3*R*)–2,3–Butanediol   (2*S*,3*S*)–2,3–Butanediol

racemic

9.53

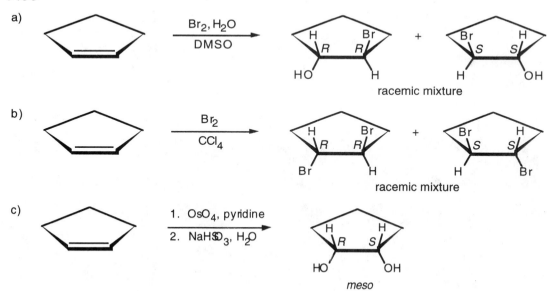

Peroxycarboxylic acids can attack either the "top" side or the "bottom" side of a double bond. The epoxide resulting from "top-side" attack on *cis*–4–octene has two stereogenic centers, but because it has a plane of symmetry, it is a *meso* compound. The two epoxides are identical.

9.54

The epoxide formed by "top-side" attack of a peroxyacid on *trans*–4–octene is pictured. This epoxide has two stereogenic centers of R configuration. The epoxide formed by "bottom-side" attack has S,S configuration. The two epoxide enantiomers are formed in equal amounts and constitute a racemic mixture.

9.55

a)

b)

c)

9.56

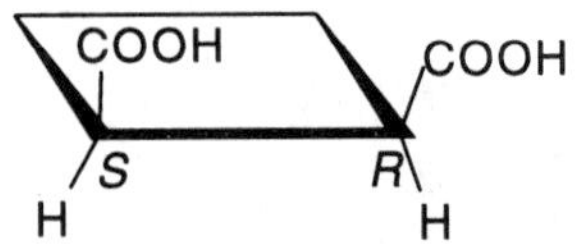
COOH
COOH
S
R
H
H

A

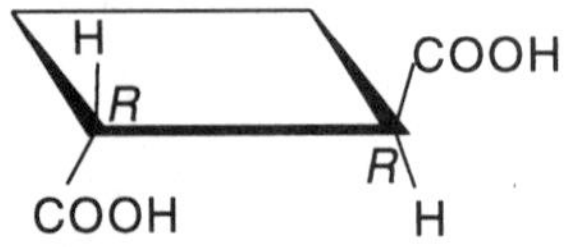
H
R
COOH
COOH
R
H

B

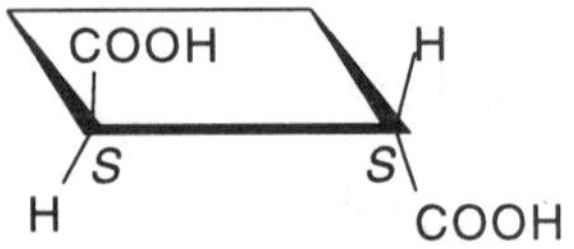
COOH
H
S
S
H
COOH

C

B and C are enantiomers and are optically active. Compound A is their diastereomer and is a *meso* compound.

The two isomeric cyclobutane–1,3–dicarboxylic acids are diastereomers and are both *meso* compounds.

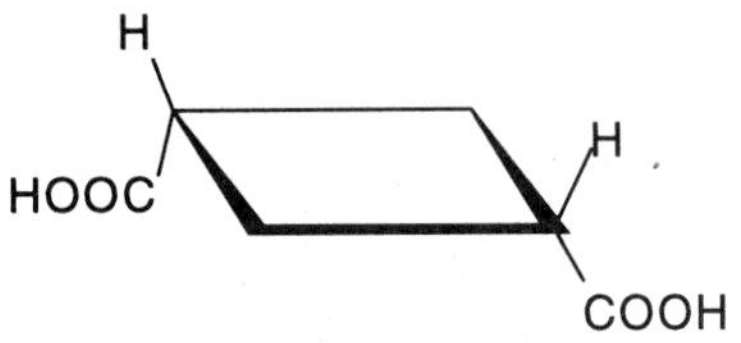
H
HOOC
H
COOH

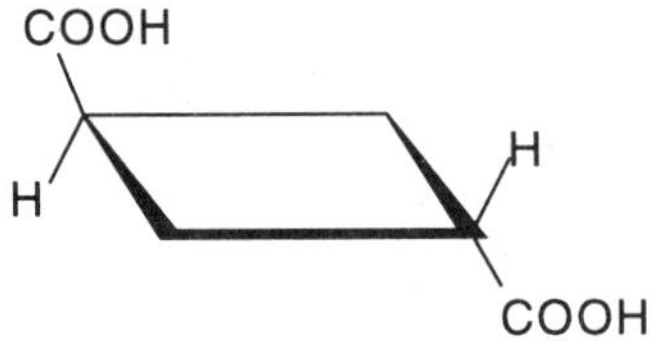
COOH
H
H
COOH

9.57

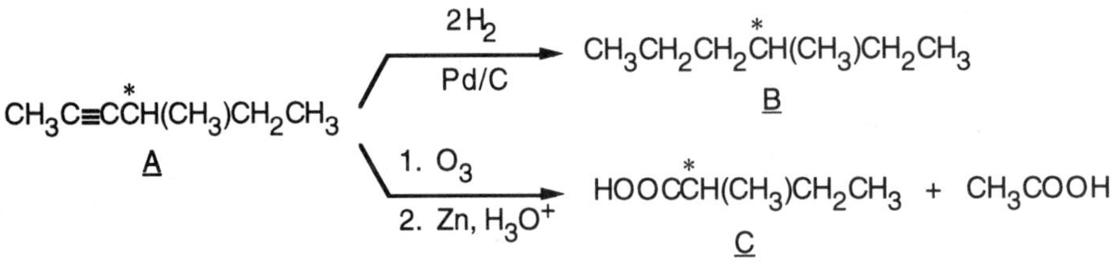
CH3C≡C*CH(CH3)CH2CH3
A
2H2
Pd/C
CH3CH2CH2*CH(CH3)CH2CH3
B
1. O3
2. Zn, H3O+
HOOC*CH(CH3)CH2CH3 + CH3COOH
C

9.58 A has four multiple bonds/rings.

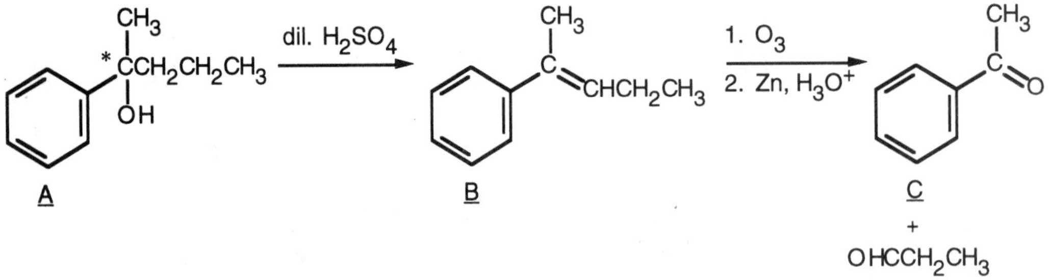
CH3
*CCH2CH2CH3
OH
A
dil. H2SO4
CH3
C
CHCH2CH3
B
1. O3
2. Zn, H3O+
CH3
C
O
C
+
OHCCH2CH3

2–Phenyl–3–pentanol is also a satisfactory answer.

9.59

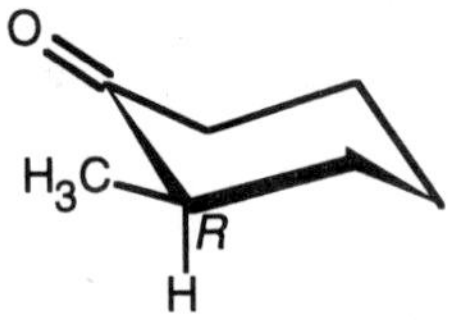
O
H3C
R
H

(*R*)–2–Methylcyclohexanone

9.60

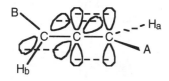

	CH₃				CH₃				CH₃				CH₃	

A   B   C   D

There are four stereoisomers of 2,4–dibromo–3–chloropentane. C and D are enantiomers and are optically active. A and B are optically inactive meso compounds and are diastereomers.

9.61 A tetrahedrane can be chiral. Notice that the orientations of the four substituents in space are the same as the orientations of the four substituents of a tetrasubstituted carbon atom. If you were able to make a model of a tetrasubstituted tetrahedrane (without having your models fall apart), you would also be able to make a model of its mirror image.

9.62 Mycomycin contains no stereogenic carbon atoms, yet is chiral. To see why, make a model of mycomycin. For simplicity, call –CH=CHCH=CHCH₂COOH "A" and –C≡CC≡CH "B". Remember from Chapter 6 that the carbon atoms of an allene are linear and that the pi bonds formed are perpendicular to each other. Attach substituents at the sp^2 carbons.

Notice that the substituents $\underline{A}$, H_a, and all carbon atoms lie in a plane that is perpendicular to the plane that contains B, H_b, and all carbon atoms.

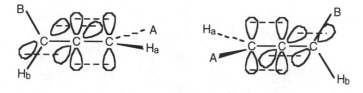

Now, make another model identical to the first, except for an exchange of $\underline{A}$ and H_a. This new allene is not superimposable on the original allene. The two allenes are enantiomers and are chiral because they possess no plane of symmetry.

9.63 4–Methylcyclohexylideneacetic acid is chiral for the same reason that mycomycin (Problem 9.62) is chiral: It possesses no plane of symmetry and is not superimposable on its mirror image. As in the case of allenes, the two functional groups at one end of the molecule lie in a plane perpendicular to the plane that contains the two functional groups at the other end.

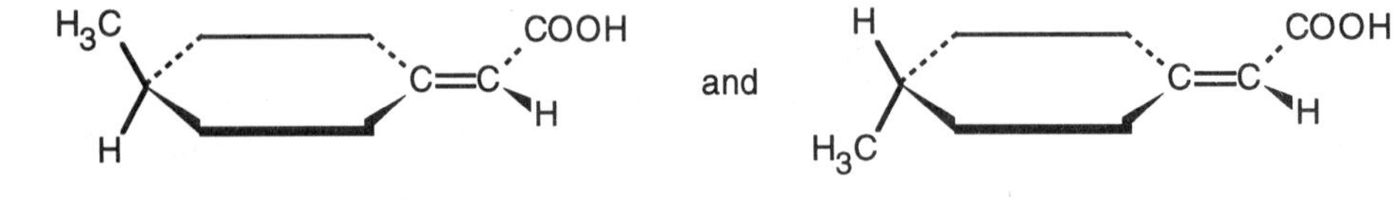

9.64

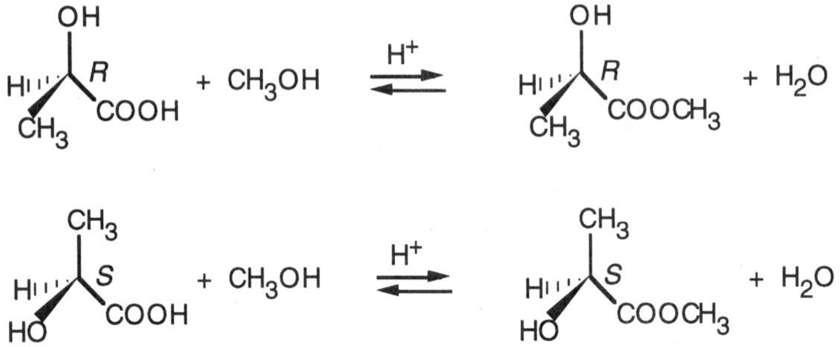

The product is a racemic mixture of *R* and *S* methyl lactates.

9.65

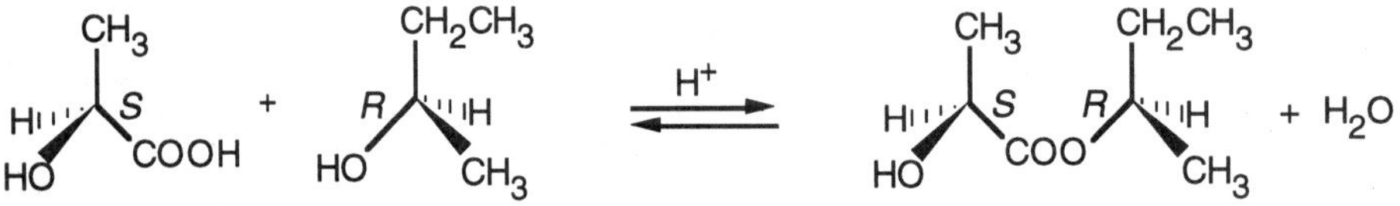

The product is *R*–*sec*–butyl *S*–lactate. As in the previous problem, the stereochemistry at the stereogenic centers of the product is the same as the stereochemistry of the reactants because no bonds were formed or broken at the stereogenic center.

9.66

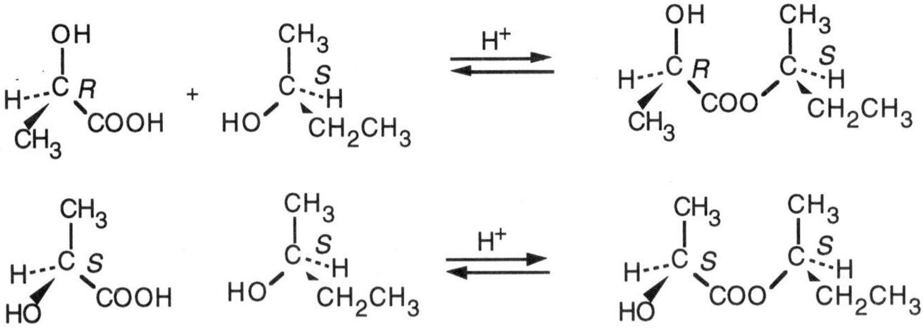

The product esters are diastereomers and differ in physical properties and chemical behavior. It should be possible to separate them by a technique such as distillation, fractional crystallization or chromatography. After separation, the esters can be converted back into (*R*) or (*S*)–lactic acid and (*S*)–2–butanol.

9.67

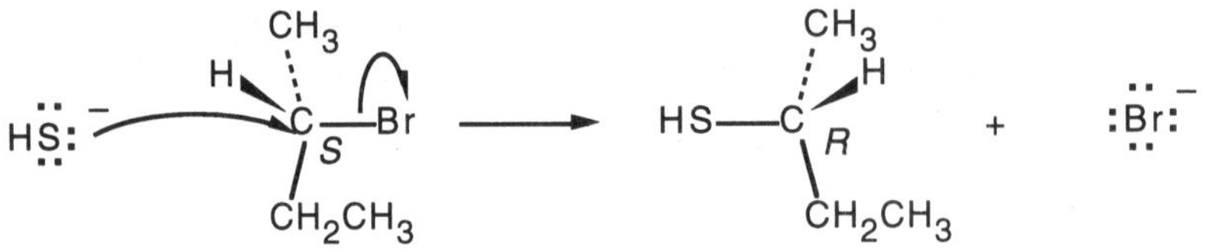

9.68

a)

| S–1–Chloro–
2–methylbutane | S–1,4–Dichloro–
2–methylbutane | R–1,2–Dichloro–
2–methylbutane | S–1,2–Dichloro–
2–methylbutane |

50:50 mixture

b) Chlorination at carbon 4 yields optically active product; chlorination at carbon 2 yields optically inactive product.

c) Radical chlorination reactions taking place at a stereogenic center occur with racemization; radical chlorination reactions at a site other than the stereogenic center do not affect the stereochemistry of the stereogenic center.

9.69

a) Reaction of a Grignard reagent with an achiral starting material, such as propanal, yields racemic product.

b) The product consists of a 50:50 mixture of R–2–butanol and its enantiomer, S–2–butanol.

9.70

(2S,3R)–3–Phenyl–
2–butanol

(2R,3R)–3–Phenyl–
2–butanol

a) Reaction of a Grignard reagent with a chiral starting material yields chiral products; the product mixture is optically active.

b) The two products are a mixture of the 2S,3R and 2R,3R diastereomers of 3–phenyl–2–butanol. The product ratio can't be predicted, but it is not 50:50.

Study Guide for Chapter 9

After studying this chapter, you should be able to:

(1)   Calculate the specific rotation of an optically active compound (9.5, 9.25, 9.26, 9.27).

(2)   Determine if an object or a molecule is chiral (9.1, 9.2, 9.34).

(3)   Locate stereogenic centers in molecules (9.3, 9.13, 9.35).

(4)   Assign priorities to substituents around a stereogenic carbon (9.6, 9.40).

(5)   Assign *R,S* designations to stereogenic centers (9.7, 9.9, 9.10, 9.24, 9.41, 9.42).

(6)   Given a stereoisomer, draw its enantiomer and/or diastereomers (9.38, 9.39, 9.56, 9.60).

(7)   Decide if a stereoisomer is a meso compound and locate its plane of symmetry (9.11, 9.12).

(8)   Manipulate Fischer projections to see if they are identical (9.15, 9.16, 9.45).

(9)   Assign *R,S* configurations to Fischer projections (9.17, 9.46, 9.47, 9.49, 9.50).

(10)   Draw Fischer projections of chiral compounds (9.48).

(11)   Draw chiral molecules corresponding to a given formula (9.8, 9.30, 9.31, 9.32, 9.33, 9.36, 9.43, 9.44, 9.57, 9.58, 9.59).

(12)   Predict the stereochemistry of reaction products (9.18, 9.19, 9.20, 9.21, 9.22, 9.51, 9.52, 9.53, 9.54, 9.55, 9.68, 9.69, 9.70).

(13)   Understand how a compound without a stereogenic carbon may be chiral (9.62, 9.63).

(14)   Define the important terms in this chapter (9.28).

10.1

(a) $CH_3CH_2CH_2CH_2I$

1-Iodobutane

(b) 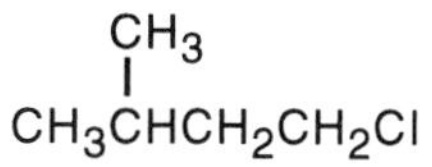
$$CH_3$$
$$|$$
$$CH_3CHCH_2CH_2Cl$$

1-Chloro-3-methylbutane

(c) 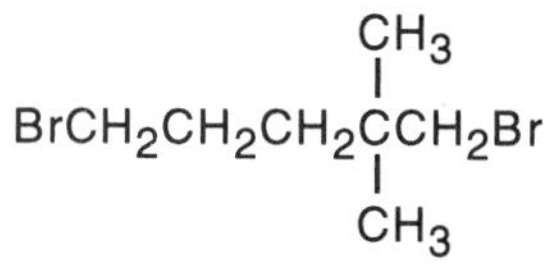
$$CH_3$$
$$|$$
$$BrCH_2CH_2CH_2CCH_2Br$$
$$|$$
$$CH_3$$

1,5-Dibromo-2,2-dimethylpentane

(d) 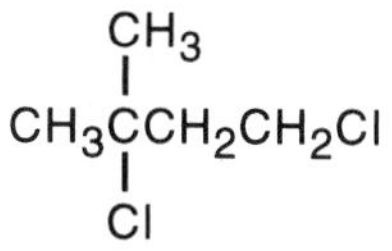
$$CH_3$$
$$|$$
$$CH_3CCH_2CH_2Cl$$
$$|$$
$$Cl$$

1,3-Dichloro-3-methylbutane

(e) 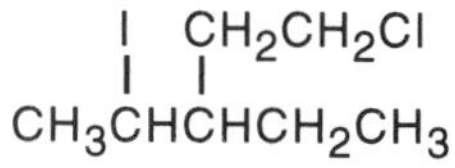
$$I \quad CH_2CH_2Cl$$
$$| \quad |$$
$$CH_3CHCHCH_2CH_3$$

1-Chloro-3-ethyl-4-iodopentane

(f) 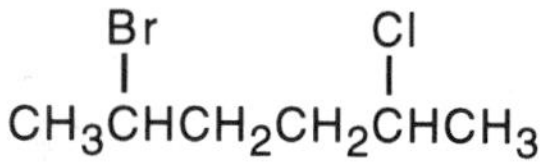
$$Br \qquad Cl$$
$$| \qquad |$$
$$CH_3CHCH_2CH_2CHCH_3$$

2-Bromo-5-chlorohexane

10.2

a) 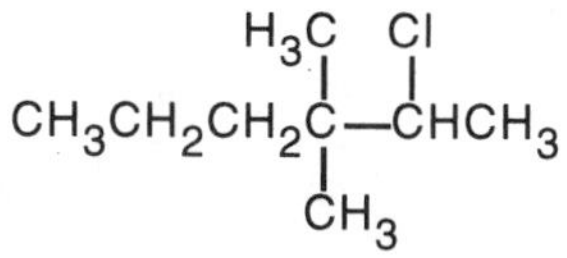
$$H_3C \quad Cl$$
$$| \quad |$$
$$CH_3CH_2CH_2C—CHCH_3$$
$$|$$
$$CH_3$$

2–Chloro–3,3–dimethylhexane

b) 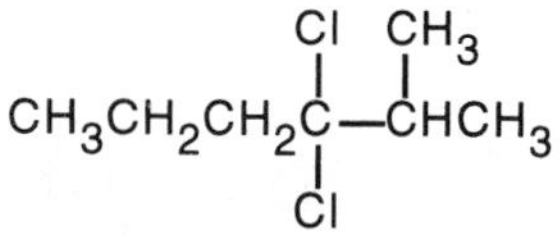
$$Cl \quad CH_3$$
$$| \quad |$$
$$CH_3CH_2CH_2C—CHCH_3$$
$$|$$
$$Cl$$

3,3–Dichloro–2–methylhexane

c) 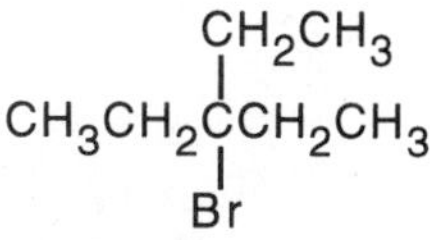
$$CH_2CH_3$$
$$|$$
$$CH_3CH_2CCH_2CH_3$$
$$|$$
$$Br$$

3–Bromo–3–ethylpentane

d)

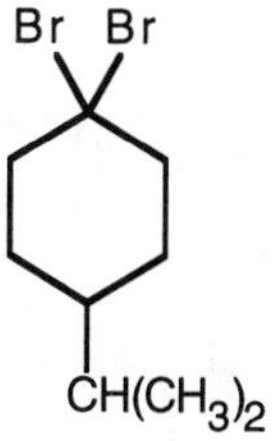

1,1–Dibromo–4–isopropylcyclohexane

e) 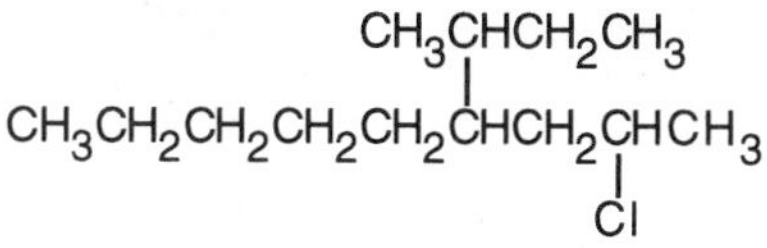
$$CH_3CHCH_2CH_3$$
$$|$$
$$CH_3CH_2CH_2CH_2CH_2CHCH_2CHCH_3$$
$$|$$
$$Cl$$

4–sec–Butyl–2–chlorononane

f)

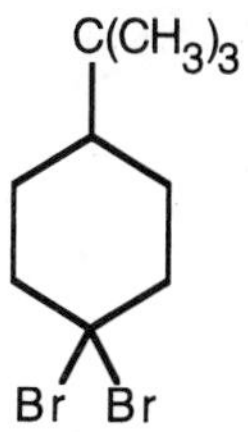

1,1–Dibromo–4–tert–butylcyclohexane

10.3

	Product	Site of Chlorination
	CH$_3$CH$_2$CH$_2$CH(CH$_3$)CH$_2$Cl 1–Chloro–2–methylpentane	a
	CH$_3$CH$_2$CH$_2$C(CH$_3$)$_2$Cl 2–Chloro–2–methylpentane	b
	CH$_3$CH$_2$CHClCH(CH$_3$)$_2$ 3–Chloro–2–methylpentane	c
	CH$_3$CHClCH$_2$CH(CH$_3$)$_2$ 2–Chloro–4–methylpentane	d
	CH$_2$ClCH$_2$CH$_2$CH(CH$_3$)$_2$ 1–Chloro–4–methylpentane	e

CH$_3$CH$_2$CH$_2$CH(CH$_3$)$_2$ $\xrightarrow[hv]{Cl_2}$
e   d   c   b   a
2–Methylpentane

Chlorination at sites b and e yields achiral products. The products of chlorination at sites a, c and d are chiral; each product is formed as a racemic mixture of enantiomers.

10.4

a
CH$_3$
|
CH$_3$—CH$_2$—C—CH$_3$
d   c | a
H
b
2–Methylbutane

Type of H	a	b	c	d
Number of H of each type	6	1	2	3
Relative reactivity	1.0	5.0	3.5	1.0
Number times reactivity	6.0	5.0	7.0	3.0
Percent chlorination	29%	24%	33%	14%

CH$_3$
|
CH$_3$CH$_2$CHCH$_3$ $\xrightarrow[hv]{Cl_2}$

CH$_3$
|
CH$_3$CH$_2$CHCH$_2$–Cl   +

CH$_3$
|
CH$_3$CH$_2$CCH$_3$
|
Cl

29%   24%

CH$_3$
|
+ CH$_3$CHCHCH$_3$ +
|
Cl   33%

CH$_3$
|
Cl—CH$_2$CH$_2$CHCH$_3$

14%

10.5 (CH$_3$)$_2$CH–H   +   Cl• →   (CH$_3$)$_2$CH• + H–Cl
D = 401 kJ/mol D = 432 kJ/mol

ΔH° = 401 kJ/mol – 432 kJ/mol = –31 kJ/mol.

$(CH_3)_2CH–H$ + Br• → $(CH_3)_2CH•$ + H–Br
D= 401 kJ/mol D = 366 kJ/mol

$\Delta H° = 401$ kJ/mol $- 366$ kJ/mol $= +35$ kJ/mol.

The reaction of Br• with a secondary hydrogen atom is more selective. In the endothermic reaction of $(CH_3)_2CH_2$ with Br•, the transition state resembles the isopropyl radical. Because the secondary isopropyl radical is much more stable than the primary radical, the reaction is much more likely to proceed by secondary hydrogen abstraction and thus to be more selective.

10.6

Abstraction of hydrogen by a bromine radical yields an allylic radical.

The allylic radical reacts with Br_2 to produce A and B.

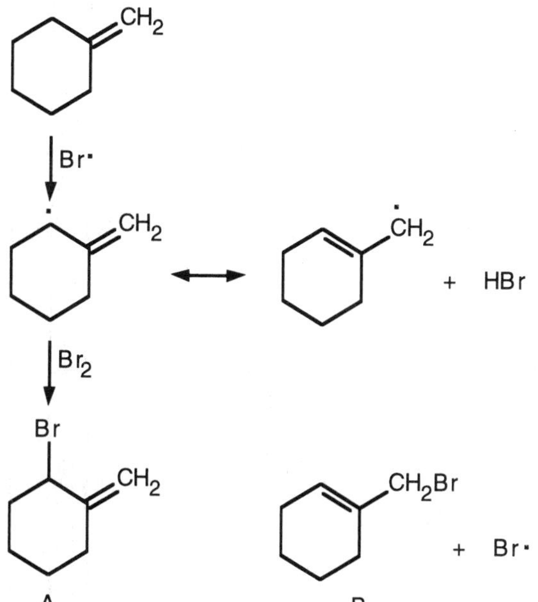

Product B, which has a trisubstituted double bond, forms in preference to product A, which has a disubstituted double bond.

10.7

a)

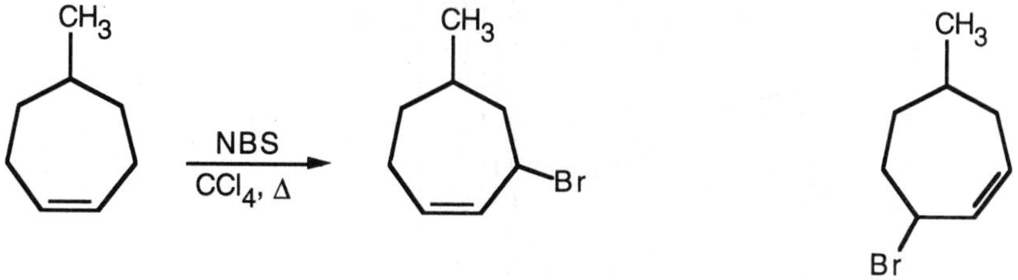

5–Methylcycloheptene   3–Bromo–5–methylcycloheptene   3–Bromo–6–methylcycloheptene

b)

$$CH_3\underset{.}{\overset{CH_3}{C}}-CH=CHCH_2CH_3 \updownarrow CH_3\overset{CH_3}{C}=CH-\underset{.}{C}HCH_2CH_3$$

$$CH_3\underset{.}{\overset{CH_3}{C}}H-CH=CHCHCH_3 \updownarrow CH_3\overset{CH_3}{C}H-\underset{.}{C}HCH=CHCH_3$$

$$CH_3\overset{CH_3}{\underset{.}{C}}HCH=CHCH_2CH_3 \xrightarrow{\text{Br·} \atop \text{radical}} \left[\quad \right] \xrightarrow[\text{CCl}_4,\ \Delta]{\text{NBS}}$$

$$CH_3\underset{Br}{\overset{CH_3}{C}}-CH=CHCH_2CH_3$$
+
$$CH_3\overset{CH_3}{C}=CH\underset{Br}{C}HCH_2CH_3$$
+
$$CH_3\overset{CH_3}{C}H-CH=CH\underset{Br}{C}HCH_3$$
+
$$CH_3\overset{CH_3}{C}H-\underset{Br}{C}HCH=CHCH_3$$

Two different allylic radicals can form, and four different bromohexenes can be produced.

10.8

a)

b)

$$\overset{:O:}{\underset{H-C-\overset{\cdot\cdot}{C}H_2}{\|}} \longleftrightarrow \overset{:\overset{\cdot\cdot}{O}:^-}{\underset{H-C=CH_2}{\|}}$$

c)

10.9

a) $$CH_3\underset{OH}{\overset{CH_3}{C}}CH_3 \xrightarrow{\text{HCl}} CH_3\underset{Cl}{\overset{CH_3}{C}}CH_3$$

HCl is a good reagent for converting a tertiary alcohol to a tertiary chloride.

b) $$CH_3\overset{CH_3}{C}HCH_2\overset{OH}{C}HCH_3 \xrightarrow[\text{ether}]{\text{PBr}_3} CH_3\overset{CH_3}{C}HCH_2\overset{Br}{C}HCH_3$$

Use PBr$_3$ for converting secondary alcohols to alkyl bromides.

c) $HOCH_2CH_2CH_2CH_2CH(CH_3)_2$ $\xrightarrow[\text{ether}]{PBr_3}$ $BrCH_2CH_2CH_2CH_2CH(CH_3)_2$

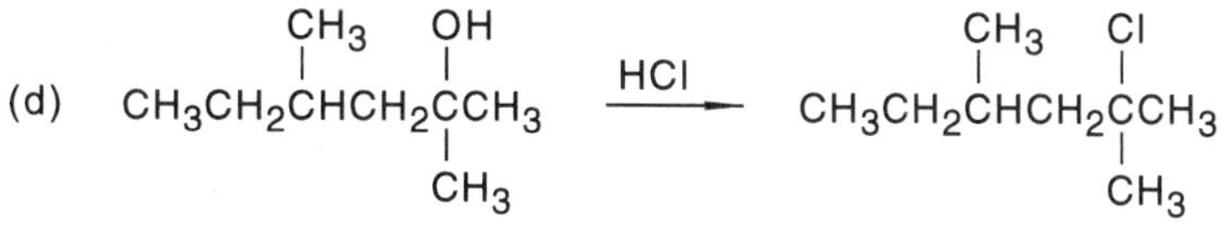

(d) $\underset{\underset{CH_3}{|}}{CH_3CH_2\overset{\overset{CH_3}{|}}{C}HCH_2\overset{\overset{OH}{|}}{C}CH_3}$ $\xrightarrow{HCl}$ $\underset{\underset{CH_3}{|}}{CH_3CH_2\overset{\overset{CH_3}{|}}{C}HCH_2\overset{\overset{Cl}{|}}{C}CH_3}$

10.10 Table 8.1 shows that the pK_a of CH_3–H is 60. Since CH_4 is a very weak acid, $^-$:CH_3 is a very strong base. Alkyl Grignard reagents are similar in base strength to $^-$:CH_3, but alkynyl Grignard reagents are somewhat weaker. Both reactions (a) and (b) occur as written.

a) CH_3MgBr + $H-C{\equiv}C-H$ $\longrightarrow$ CH_4 + $H-C{\equiv}C-MgBr$

| stronger base | stronger acid | | weaker acid | weaker base |

b) CH_3MgBr + NH_3 $\longrightarrow$ CH_4 + $H_2N-MgBr$

| stronger base | stronger acid | | weaker acid | weaker base |

10.11 Just as Grignard reagents react with *proton* donors to convert R–MgX into R–H, they also react with *deuterium* donors to convert R–MgX into R–D. In this case:

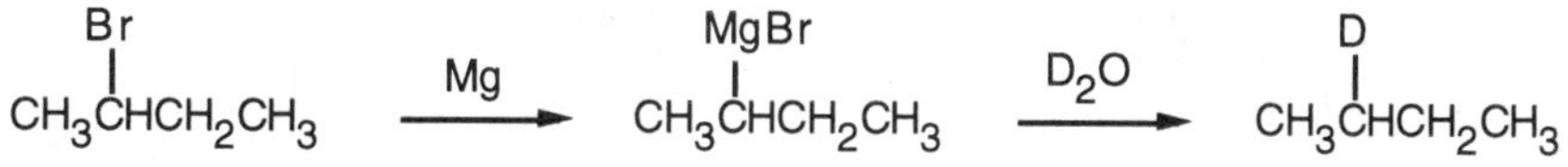

$\underset{CH_3\overset{\overset{Br}{|}}{C}HCH_2CH_3}{}$ $\xrightarrow{Mg}$ $\underset{CH_3\overset{\overset{MgBr}{|}}{C}HCH_2CH_3}{}$ $\xrightarrow{D_2O}$ $\underset{CH_3\overset{\overset{D}{|}}{C}HCH_2CH_3}{}$

10.12 A Grignard reagent can't be prepared from a compound containing a functional group that is a good proton donor because the Grignard reagent is immediately quenched by the proton source. For example, the –COOH, –OH, –NH$_2$, and RC≡CH functional groups are too acidic to be used for preparation of a Grignard reagent. Thus, $BrCH_2CH_2CH_2NH_2$ is another compound that does not form a Grignard reagent.

10.13

a)

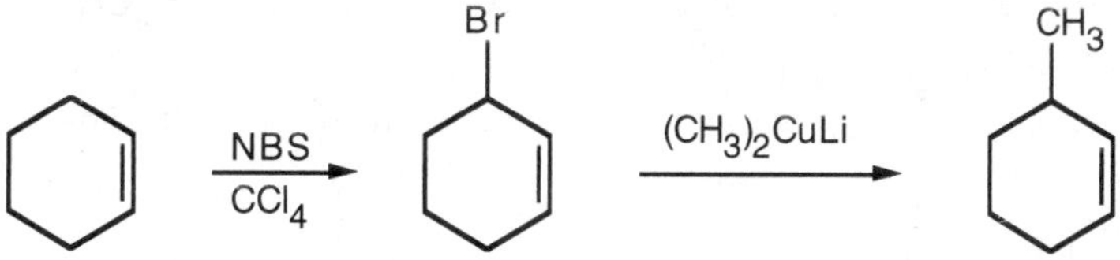

3–Methylcyclohexene

b) $2 \ CH_3CH_2CH_2CH_2Br \xrightarrow[\text{pentane}]{4 \ Li} 2 \ CH_3CH_2CH_2CH_2Li$

$+ \ 2 \ LiBr$ $\Big\downarrow$ CuI
ether

$$CH_3CH_2CH_2CH_2CH_2CH_2CH_2CH_3 \xleftarrow[\text{ether}]{CH_3CH_2CH_2CH_2Br} (CH_3CH_2CH_2CH_2)_2CuLi$$

Octane

c) $CH_3CH_2CH_2CH{=}CH_2 \xrightarrow[\text{peroxides}]{HBr} CH_3CH_2CH_2CH_2CH_2Br$

$2 \ CH_3CH_2CH_2CH_2CH_2Br \xrightarrow[\text{pentane}]{4 \ Li} 2 \ CH_3(CH_2)_3CH_2Li \ + \ 2 \ LiBr$

$\Big\downarrow$ CuI
ether

$$CH_3(CH_2)_8CH_3 \xleftarrow[\text{ether}]{CH_3(CH_2)_3CH_2Br} [CH_3(CH_2)_3CH_2]_2CuLi$$

Decane

10.14

(a)

(b) $CH_3CH_2NH_2 \ < \ H_2NCH_2CH_2NH_2 \ < \ CH_3CN$

10.15

a)   H₃C Br Br   CH₃

3,4–Dibromo–2,6–dimethylheptane

b)
$CH_3CH{=}CHCH_2CHCH_3$

5–Iodo–2–hexene

c)   Br   Cl CH₃

2–Bromo–4–chloro–2,5–dimethylhexane

d)   CH₂Br
$CH_3CH_2CHCH_2CH_2CH_3$

3–(Bromomethyl)hexane

e) $ClCH_2CH_2CH_2C{\equiv}CCH_2Br$

1–Bromo–6–chloro–2–hexyne

10.16

a)   CH₃ Cl
$CH_3CH_2CHCHCHCH_3$
       Cl

2,3–Dichloro–4–methylhexane

b)   CH₂CH₃
$CH_3CH_2CCH_2CHCH_3$
       Br   CH₃

4–Bromo–4–ethyl–2–methylhexane

c)

$$H_3C \quad CH_3$$
$$CH_3CCHCCH_3$$
$$H_3CI \quad CH_3$$

3–Iodo–2,2,4,4–tetramethylpentane

d)

cis–1-Bromo–2-ethylcyclopentane

10.17 Abstraction of hydrogen by Br• can produce either of two allylic radicals. The first radical, resulting from abstraction of a secondary hydrogen, is more likely to be formed.

$$CH_3\overset{\bullet}{C}HCH=CHCH_3 \longleftrightarrow CH_3CH=CH\overset{\bullet}{C}HCH_3 \qquad \text{(identical resonance forms)}$$

and

$$CH_3CH_2CH=\overset{\bullet}{C}HCH_2 \longleftrightarrow CH_3CH_2\overset{\bullet}{C}HCH=CH_2$$

Reaction of the radical intermediates with a bromine source leads to a mixture of products:

$$CH_3CH_2CHBrCH=CH_2 \qquad \qquad \text{3–Bromo–1–pentene}$$

and

$$CH_3CHBrCH=CHCH_3 \qquad \qquad cis- \text{ and } trans-4-Bromo-2-pentene$$

and

$$CH_3CH_2CH=CHCH_2Br \qquad \qquad cis- \text{ and } trans-1-Bromo-2-pentene$$

The major product is 4–bromo–2–pentene, instead of the desired product, 1–bromo–2–pentene.

10.18 Three different allylic radical intermediates can be formed. Bromination of these intermediates can yield as many as five bromoalkenes. This is definitely not a good reaction to use in a synthesis.

(allylic; secondary hydrogen abstracted)

NBS

3–Bromo–2–methylcyclohexene

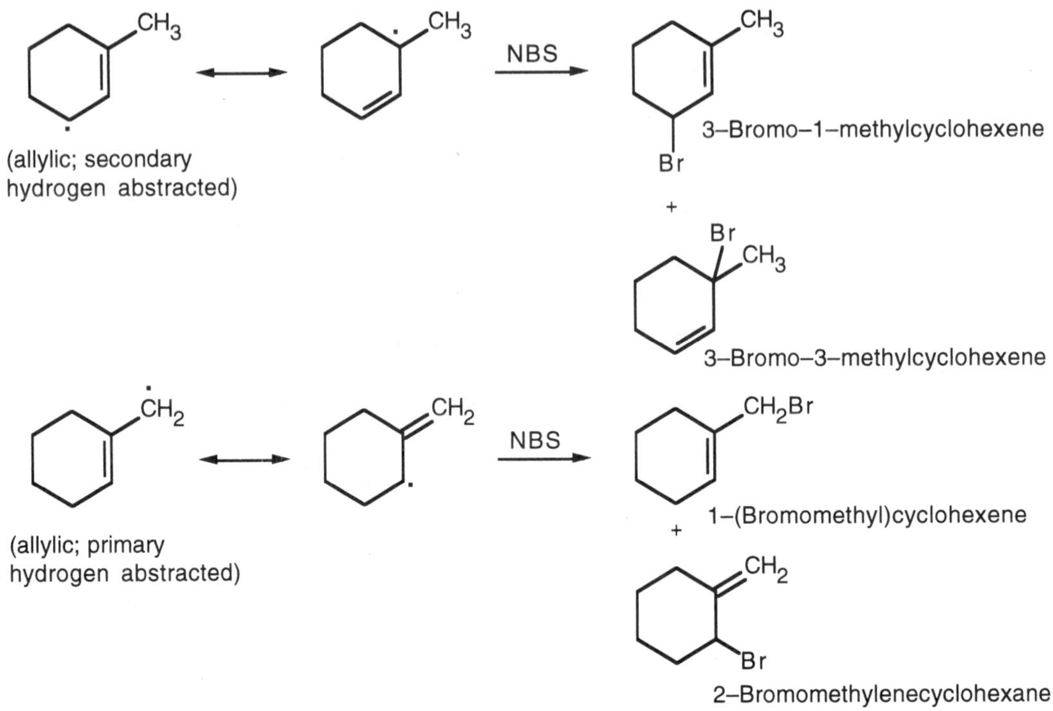

CH_3
(allylic; secondary hydrogen abstracted)
CH_3
NBS
CH_3
Br
3–Bromo–1–methylcyclohexene
+
Br
CH_3
3–Bromo–3–methylcyclohexene
CH_2
(allylic; primary hydrogen abstracted)
CH_2
NBS
CH_2Br
1–(Bromomethyl)cyclohexene
+
CH_2
Br
2–Bromomethylenecyclohexane

10.19

a) 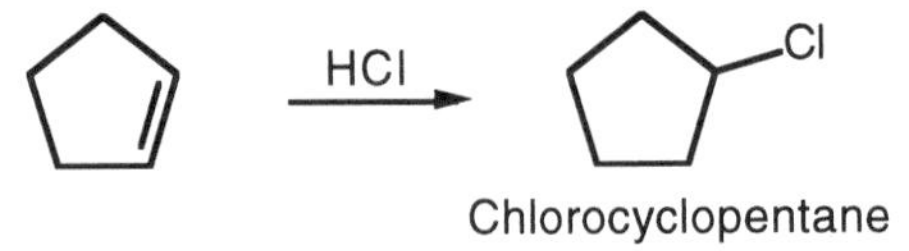

HCl
Cl
Chlorocyclopentane

b) 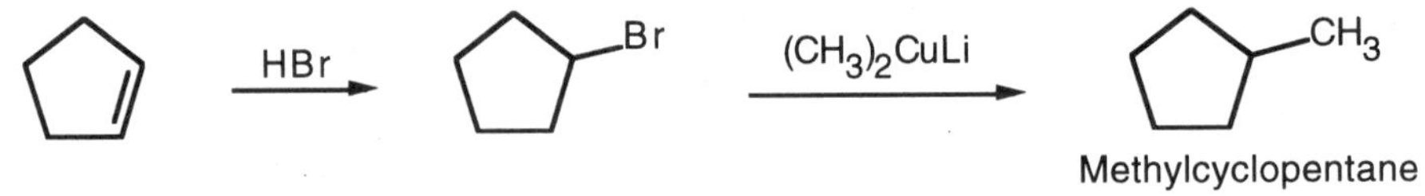

HBr
Br
$(CH_3)_2CuLi$
CH_3
Methylcyclopentane

c) 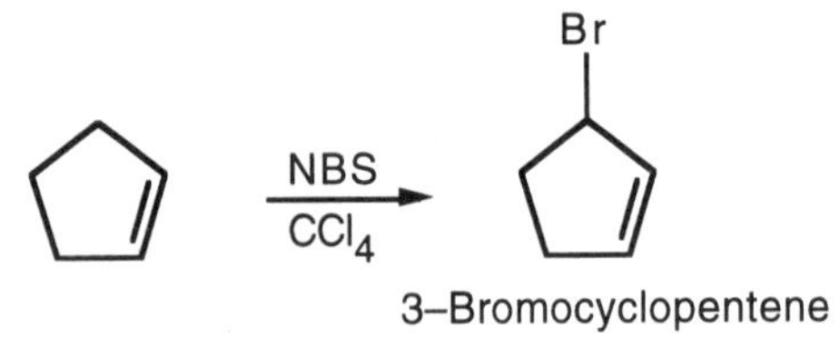

NBS
CCl_4
Br
3–Bromocyclopentene

d) 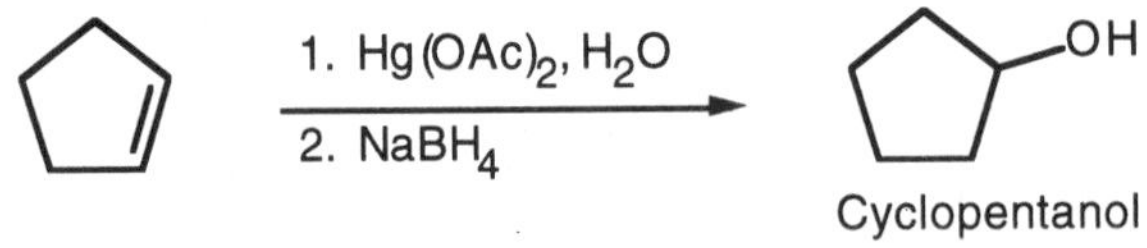

1. $Hg(OAc)_2$, H_2O
2. $NaBH_4$
OH
Cyclopentanol

e)

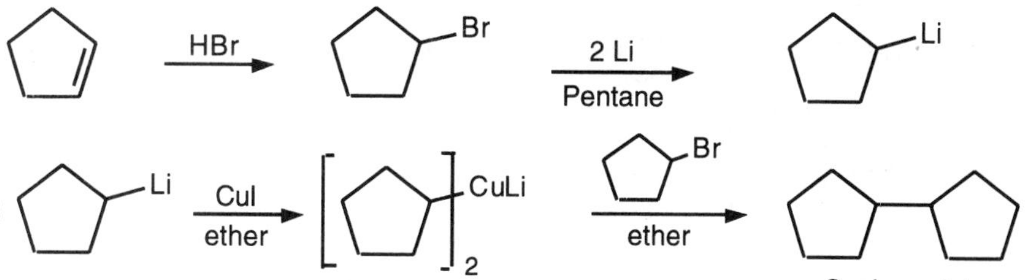

2

Cyclopentyl–
cyclopentane

f)

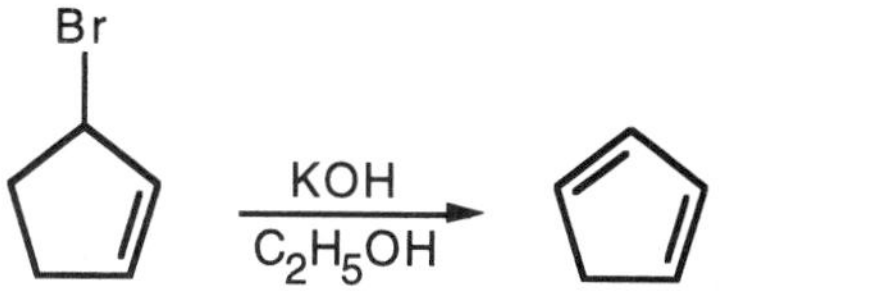

(from c)   1,3–Cyclopentadiene

10.20

a) 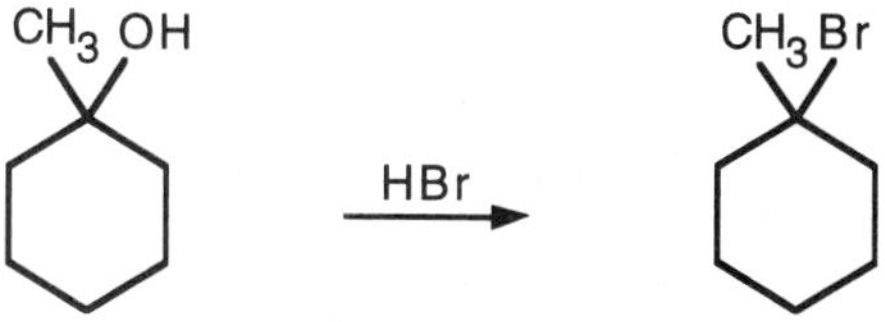 CH_3OH CH_3Br

This is a good method for converting a tertiary alcohol to a bromide.

b) $CH_3CH_2CH_2CH_2OH$ $\xrightarrow[\text{pyridine}]{SOCl_2}$ $CH_3CH_2CH_2CH_2Cl$

c)

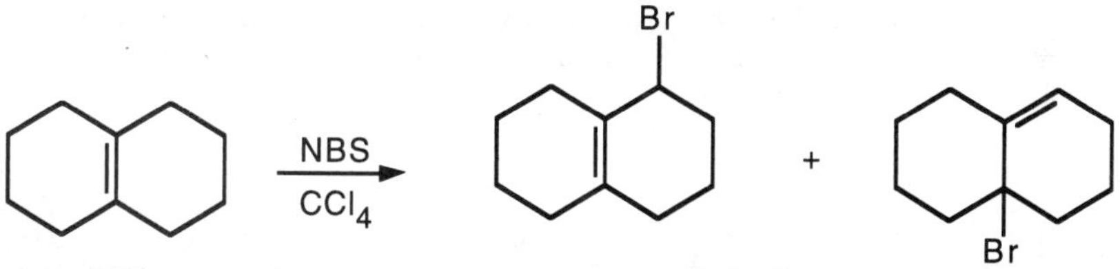

The major product contains a tetrasubstituted double bond; the minor product contains a trisubstituted double bond.

d)

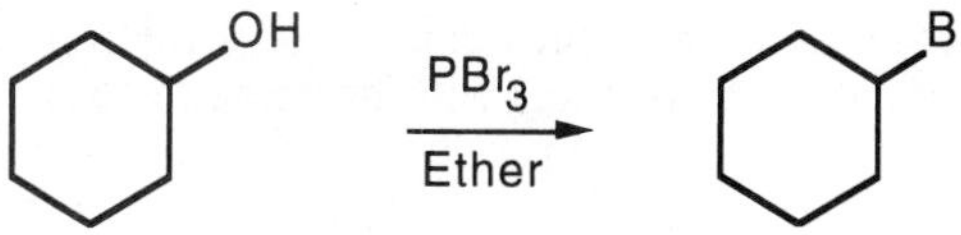

This is a good method for converting a primary or secondary alcohol to a bromide.

e) 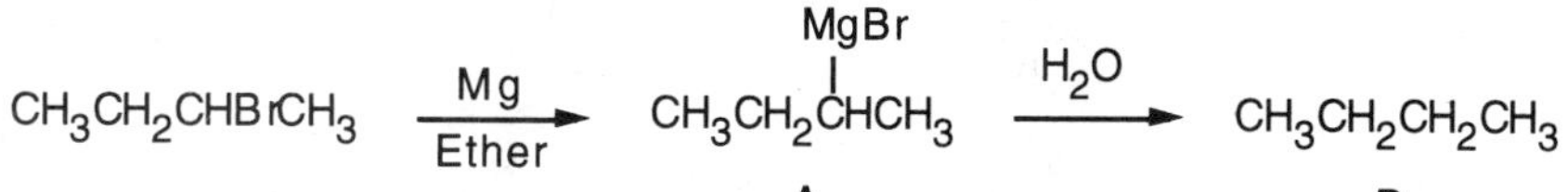 $CH_3CH_2CHBrCH_3$ $\xrightarrow[\text{Ether}]{Mg}$ $CH_3CH_2\overset{\overset{\displaystyle MgBr}{|}}{C}HCH_3$ $\xrightarrow{H_2O}$ $CH_3CH_2CH_2CH_3$

                                              A   B

f) $2 CH_3CH_2CH_2CH_2Br$ $\xrightarrow[\text{Pentane}]{4 Li}$ $2 CH_3CH_2CH_2CH_2Li$ $\xrightarrow{CuI}$ $(CH_3CH_2CH_2CH_2)_2CuLi$

$\underset{A}{}$ $\qquad\qquad\qquad\qquad\qquad\qquad$ $\underset{B}{}$

g) $CH_3CH_2CH_2CH_2Br + (CH_3)_2CuLi$ $\xrightarrow{\text{Ether}}$ $CH_3CH_2CH_2CH_2CH_3 + CH_3Cu + LiBr$

10.21

Three of the above products are chiral (stereogenic centers are starred). None of the products are optically active; each chiral product is a racemic mixture.

10.22

Abstraction of a hydrogen atom from the stereogenic center of S–3–methyl–hexane produces an achiral radical intermediate, which reacts with bromine to form a 1:1 mixture of R and S enantiomeric, chiral bromoalkanes. The product mixture is optically inactive.

10.23

Abstraction of a hydrogen atom from carbon 4 yields a chiral radical intermediate. Reaction of this intermediate with chlorine does not occur with equal probability from each side, and so the two diastereomeric products are not formed in 1:1 ratio. The first product is optically active, and the second product is a meso compound.

10.24

for Cl (1) Cl• + CH$_3$–H → CH$_3$• + H–Cl
          438 kJ/mol   432 kJ/mol

$\Delta H° = 438$ kJ/mol $- 432$ kJ/mol $= +6$ kJ/mol

(2) CH$_3$• + Cl–Cl → Cl• + CH$_3$–Cl
          243 kJ/mol   351 kJ/mol

$\Delta H° = 243$ kJ/mol $- 351$ kJ/mol $= -108$ kJ/mol

$\Delta H°_{total} = +6$ kJ/mol $- 108$ kJ/mol $= -102$ kJ/mol

for Br• (1) Br• + CH$_3$–H → CH$_3$• + H–Br
            438 kJ/mol   366 kJ/mol

$\Delta H° = 438$ kJ/mol $- 366$ kJ/mol $= +72$ kJ/mol

(2) CH$_3$• + Br–Br → Br• + CH$_3$–Br
         193 kJ/mol   293 kJ/mol

$\Delta H° = 193$ kJ/mol $- 293$ kJ/mol $= -100$ kJ/mol

$\Delta H°_{total} = +72$ kJ/mol $- 100$ kJ/mol $= -28$ kJ/mol

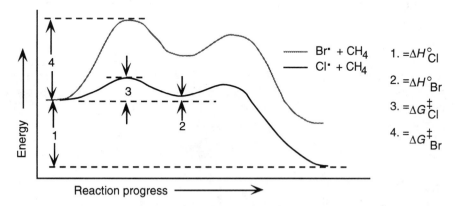

Reaction between Cl• and CH$_4$ is likely to be faster. $\Delta H°$ for formation of the CH$_3$• intermediate is lower for chlorination than for bromination, and thus $\Delta G^{\ddagger}$ is likely to be lower also, according to the Hammond Postulate.

10.25

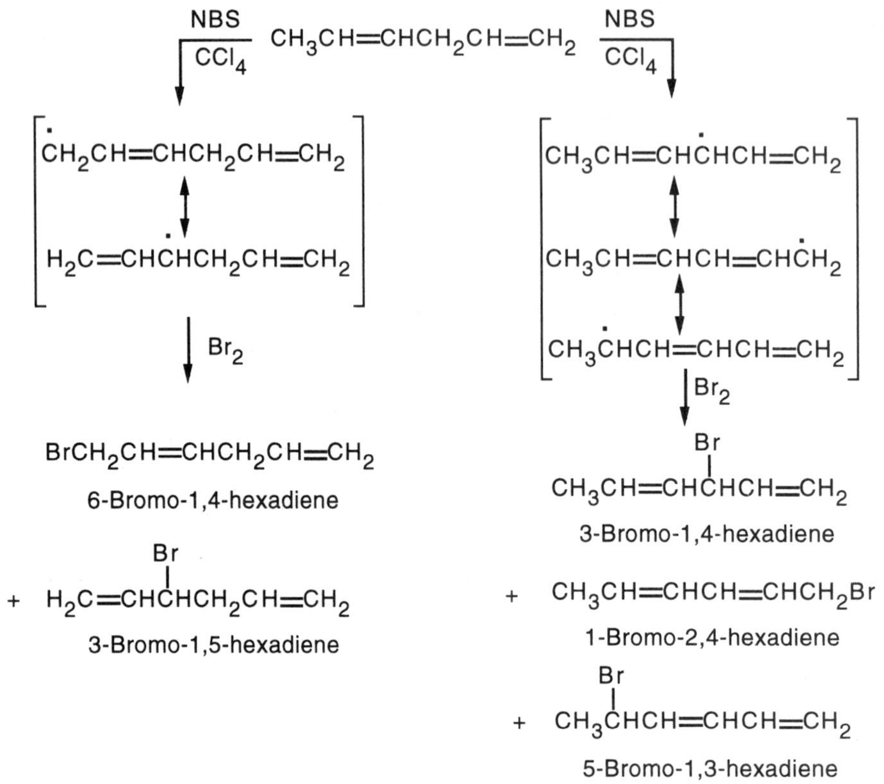

The intermediate on the right is the more stable, because the unpaired electron is delocalized over more atoms.

10.26 Table 5.4 shows that the bond dissociation energy for $C_6H_5CH_2$–H is 368 kJ/mol. This value is even smaller than the bond dissociation energy for allylic hydrogens, and thus it is relatively easy to form the $C_6H_5CH_2\cdot$ radical. The high bond dissociation energy for formation of $C_6H_5\cdot$, 464 kJ/mol, indicates the bromination on the benzene ring will not occur. The only product of reaction with NBS is $C_6H_5CH_2Br$.

10.27

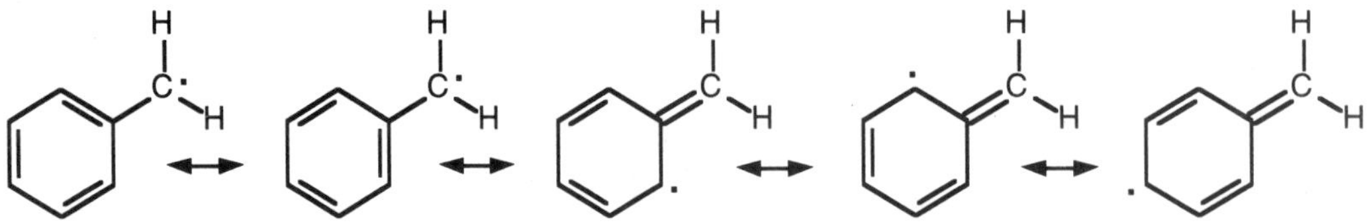

10.28 Two allylic radicals can form:

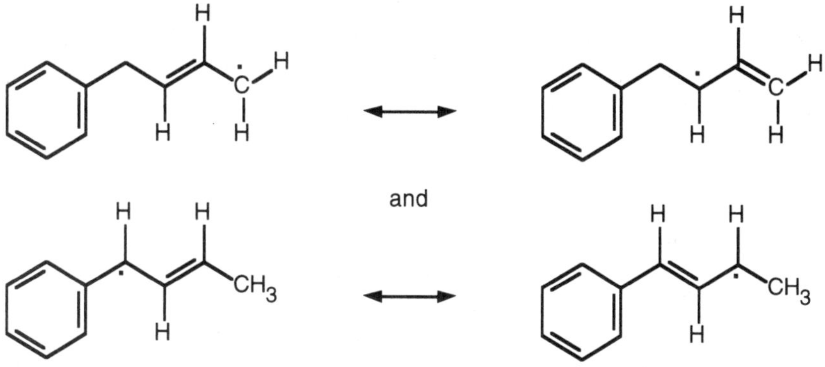

and

The second radical is much more likely to form because it is both allylic and benzylic, and it yields the following products:

+

10.29

a)

$$CH_3CH=CHCH=CHCH=CH\overset{+}{C}H_2$$

$$\updownarrow$$

$$CH_3CH=CHCH=\overset{+}{C}HCHCH=CH_2$$

$$\updownarrow$$

$$CH_3CH=CH\overset{+}{C}HCH=CHCH=CH_2$$

$$\updownarrow$$

$$CH_3\overset{+}{C}HCH=CHCH=CHCH=CH_2$$

b)

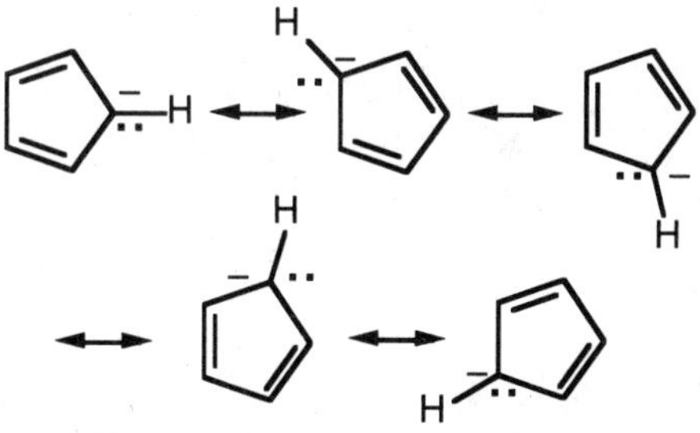

c)

$$CH_3C\equiv\overset{+}{N}-\overset{..}{\underset{..}{O}}:^- \longleftrightarrow CH_3\overset{..}{\underset{..}{C}}=\overset{+}{N}=\overset{..}{O}: \longleftrightarrow CH_3\overset{+}{C}=N-\overset{..}{\underset{..}{O}}:^-$$

10.30 In order of increasing oxidation level:

a) $CH_3CH=CHCH_3$, $CH_3CH_2CH=CH_2$ < $CH_3CH_2CH_2\overset{O}{\overset{\|}{C}}H$ < $CH_3CH_2CH_2\overset{O}{\overset{\|}{C}}OH$

b) $CH_3CH_2CH_2NH_2$, $CH_3CH_2CH_2Br$ < $BrCH_2CH_2CH_2Cl$ < $CH_3\overset{O}{\overset{\|}{C}}CH_2Cl$

10.31 Use Figure 10.4 if you need help.
   a) oxidation   b) neither oxidation nor reduction   c) reduction

10.32 All these reactions involve addition of a dialkylcopper reagent ($(CH_3CH_2CH_2CH_2)_2CuLi$) to an alkyl halide. The dialkylcopper is prepared by treating 1–bromobutane with lithium, followed by addition of CuI:

$$2\ CH_3CH_2CH_2CH_2Br \xrightarrow[\text{Pentane}]{2\ Li} 2\ CH_3CH_2CH_2CH_2Li \xrightarrow[\text{Ether}]{CuI} (CH_3CH_2CH_2CH_2)_2CuLi$$

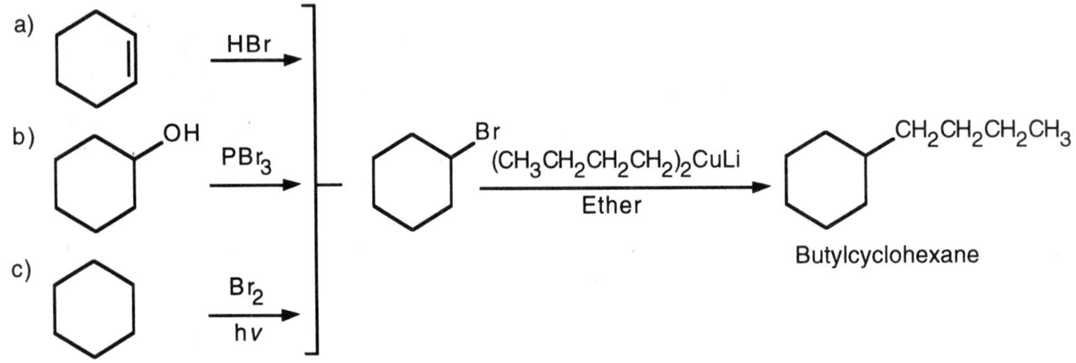

a)

b)

c)

Butylcyclohexane

10.33 a) Fluoroalkanes do not normally form Grignard reagents.

   b) Two allylic radicals can be produced.

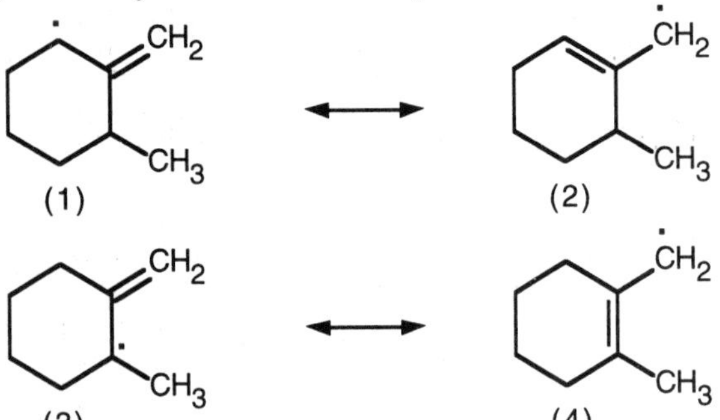

(1)   (2)

(3)   (4)

Instead of a single product, as many as four bromide products may result.

   c) Dialkylcopper reagents do not react with fluoroalkanes.

10.34

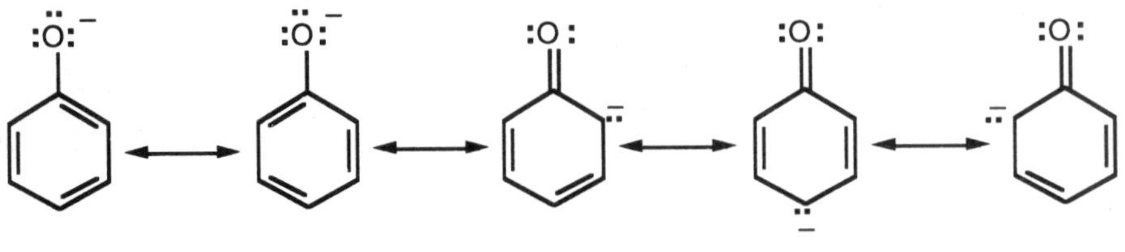

10.35

$$(C_4H_9)_3SnH \xrightarrow{h\nu} (C_4H_9)_3Sn\cdot + H\cdot$$

$$(C_4H_9)_3Sn\cdot + RX \longrightarrow (C_4H_9)_3SnX + R\cdot$$

$$(C_4H_9)_3SnH + R\cdot \longrightarrow (C_4H_9)_3Sn\cdot + RH$$

10.36 As we saw in Chapter 6, tertiary carbocations (R_3C+) are more stable than either secondary or primary carbocations, due to the ability of the three alkyl groups to stabilize positive charge. If the substrate is also allylic, as in the case of $H_2C=CHC(CH_3)_2Br$, positive charge can be further delocalized. Thus, $H_2C=CHC(CH_3)_2Br$ should form a carbocation faster than $(CH_3)_3CBr$ because the resulting carbocation is more stable.

Study Guide for Chapter 10

After studying this chapter, you should be able to:

(1)   Draw and name alkyl halides (10.1, 10.2, 10.15, 10.16, 10.21).

(2)   Understand the mechanism of radical halogenation and the stability order of radicals (10.3, 10.4, 10.5, 10.6, 10.7, 10.17, 10.18, 10.21, 10.22, 10.23, 10.24, 10.25, 10.26, 10.27, 10.28, 10.35).

(3)   Draw resonance structures (10.8, 10.27, 10.29, 10.34).

(4) ·   Prepare alkyl halides (10.9, 10.19, 10.20).

(5)   Prepare Grignard reagents and dialkylcopper reagents and use them in syntheses (10.11, 10.12, 10.13, 10.14, 10.32, 10.33).

(6)   Predict the oxidation level of a compound (10.14, 10.30, 10.31).

11.1

11.2

11.3   If back-side attack were necessary for S$_N$2 reaction, a molecule with a hindered back-side would not be able to react by an S$_N$2 mechanism. In this problem approach by the hydroxide ion from the back side of the bromoalkane is blocked by the rigid ring system, and displacement can't occur.

11.4

a) $CH_3CH_2CH_2CH_2Br$ $+ NaI$ $\longrightarrow$ $CH_3CH_2CH_2CH_2I$

b) $CH_3CH_2CH_2CH_2Br$ $+ KOH$ $\longrightarrow$ $CH_3CH_2CH_2CH_2OH$

c) $CH_3CH_2CH_2CH_2Br$ $+ HC\equiv C^-\,Li^+$ $\longrightarrow$ $CH_3CH_2CH_2CH_2C\equiv CH$

d) $CH_3CH_2CH_2CH_2Br$ $+ NH_3$ $\longrightarrow$ $CH_3CH_2CH_2CH_2\overset{+}{N}H_3\,Br^-$

11.5   a) $(CH_3)_2N^-$ is more nucleophilic. A negatively charged reagent is more nucleophilic than its conjugate acid.
b) $(CH_3)_3N$ is more nucleophilic than $(CH_3)_3B$. $(CH_3)_3B$ is non-nucleophilic because it has no lone electron pair.
c) Because nucleophilicity increases in going down a column of the periodic table, H_2S is more nucleophilic than H_2O.

11.6   In this problem, we are comparing two effects -- the effect of the substrate and the effect of the leaving group.

Most reactive $\longrightarrow$ Least reactive

$CH_3OTos > CH_3Br > (CH_3)_2CHCl \gg (CH_3)_3CCl$

11.7

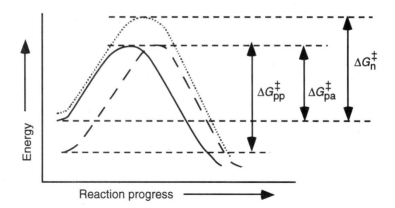

Polar protic solvents (dashed line) stabilize the charged transition state by solvation and also stabilize the nucleophile by hydrogen bonding.

Polar aprotic solvents (solid line) stabilize the charged transition state by solvation, but do not hydrogen-bond to the nucleophile. Since the energy level of the nucleophile is higher, $\Delta G^{\ddagger}$ is smaller and the reaction is faster in polar aprotic solvents than in polar protic solvents.

Nonpolar solvents (dotted line) stabilize neither the nucleophile nor the transition state. $\Delta G^{\ddagger}$ is therefore higher in nonpolar solvents than in polar solvents, and the reaction rate is slower.

11.8

Attack by acetate can occur on either side of the planar, achiral carbocation intermediate, resulting in a racemic mixture of acetate product

11.9 If reaction had proceeded with complete inversion, the product would have had a specific rotation of +53.6°. If complete racemization had occurred, $[\alpha]_D$ would have been zero. The observed rotation was +5.3°. Since $\dfrac{+5.3°}{+53.6°} = 0.099$, 9.9% of the original tosylate was inverted. The remaining 90.1% of the product must have been racemized.

11.10 S_N1 reactivity is related to carbocation stability. Thus, substrates that form the most stable carbocations are the most reactive in S_N1 reactions.

Most Reactive ⟶ Least Reactive

$H_2C=CHCHBrCH_3$ > $CH_3CHBrCH_3$ > CH_3CH_2Br > $H_2C=CHBr$

    allylic   secondary   primary   vinylic

11.11

$$CH_3CHBrCH=CH_2 \;\rightleftharpoons\; \left[CH_3\overset{+}{C}HCH=CH_2 \longleftrightarrow CH_3CH=CH\overset{+}{C}H_2 \right] \;\rightleftharpoons\; CH_3CH=CHCH_2Br$$

The two bromobutenes undergo S_N1 reaction at the same rate because they form the same allylic carbocation.

11.12

1–Chloro–1,2–diphenylethane

Nucleophilic substitution of 1–chloro–1,2–diphenylethane proceeds via an S_N1 mechanism because of stabilization of the carbocation intermediate by the phenyl group at C1. In an S_N1 reaction, the rate-limiting step is carbocation formation; all subsequent steps occur at a faster rate and do not affect the rate of reaction. After carbocation formation, the rate does not depend on the identity of the nucleophile that combines with the carbocation. In this example, $F:^-$ and $(CH_3CH_2)_3N:$ react at the same rate.

11.13 The first step in an S_N1 displacement is dissociation of the substrate to form a planar, sp^2–hybridized carbocation and a leaving group. The carbocation that would form from dissociation of this haloalkane can't become planar because of the rigid structure of the rest of the molecule. Because it's not possible to form the necessary carbocation, an S_N1 reaction can't occur.

11.14 The major elimination product in each case has the most substituted double bond.

a) $CH_3CH_2\overset{\overset{\displaystyle Br}{|}}{C}H\overset{\overset{\displaystyle CH_3}{|}}{C}HCH_3$ ⟶ $CH_3CH_2CH=\overset{\overset{\displaystyle CH_3}{|}}{C}CH_3$ + $CH_3CH=CH\overset{\overset{\displaystyle CH_3}{|}}{C}HCH_3$

                                          major   minor
                           (trisubstituted double bond) (disubstituted double bond)

b)

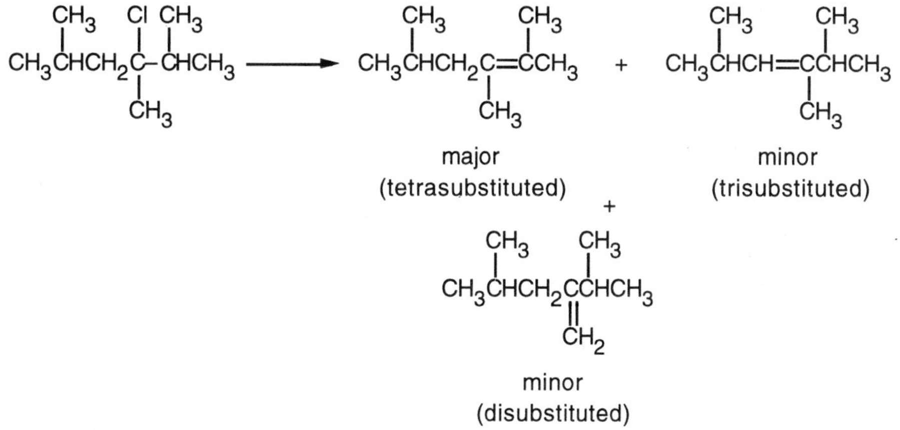

CH₃ Cl CH₃
 | | |
CH₃CHCH₂C-CHCH₃ ⟶
          |
          CH₃

CH₃   CH₃
 | |
CH₃CHCH₂C=CCH₃   +
          |
          CH₃

major
(tetrasubstituted)

CH₃   CH₃
 | |
CH₃CHCH=CCHCH₃
          |
          CH₃

minor
(trisubstituted)

+

CH₃   CH₃
 | |
CH₃CHCH₂CCHCH₃
        ‖
        CH₂

minor
(disubstituted)

c)

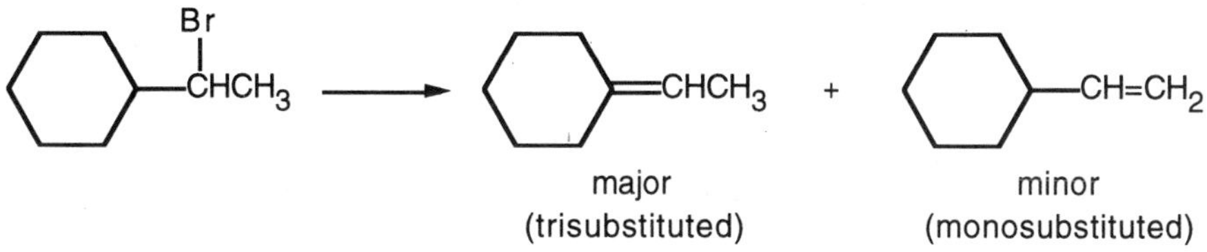

major
(trisubstituted)

minor
(monosubstituted)

11.15

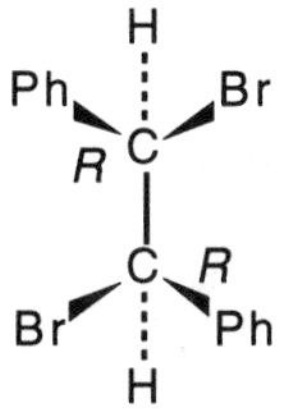

(1R,2R)–1,2–Dibromo–1,2–diphenylethane

Convert this drawing into a Newman projection, and draw the conformation having anti-periplanar geometry for –H and –Br.

The alkene resulting from dehydrohalogenation is (Z)–1–bromo–1,2–diphenylethylene.

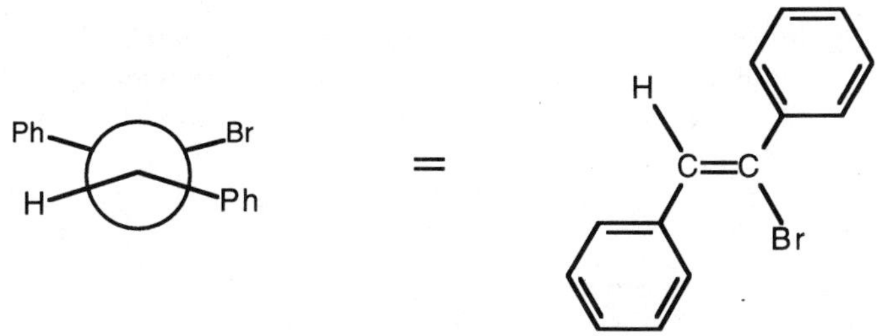

11.16

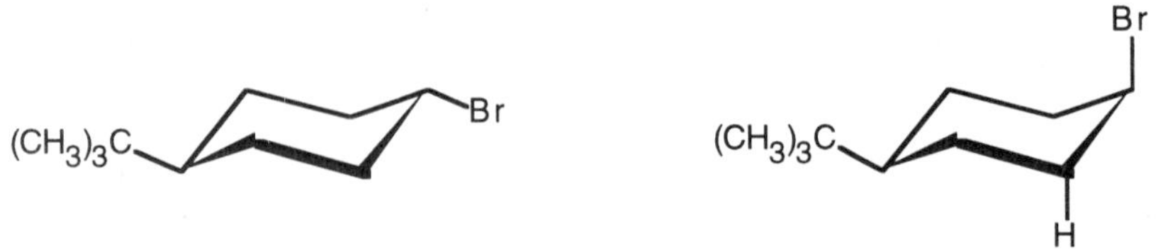

trans   cis

The more stable conformations of the two isomers are pictured above; the *tert*–butyl group is always equatorial in the more stable conformation. The cis isomer will react faster under E2 conditions because –Br and –H are in the anti-periplanar arrangement that favors E2 elimination.

11.17

a) $CH_3CH_2CH_2CH_2Br$ + NaN_3 $\longrightarrow$ $CH_3CH_2CH_2CH_2N_3$

       primary   substitution product

Since the halide is primary and a substitution product is formed, this is an S_N2 reaction.

b) $CH_3CH_2\overset{\underset{\textstyle |}{Cl}}{C}HCH_2CH_3$ +   KOH $\longrightarrow$ $CH_3CH_2CH=CHCH_3$

       secondary   strong base   elimination product

This is an E2 reaction since a secondary halide reacts with a strong base to yield an elimination product.

c)   + CH_3COOH $\longrightarrow$

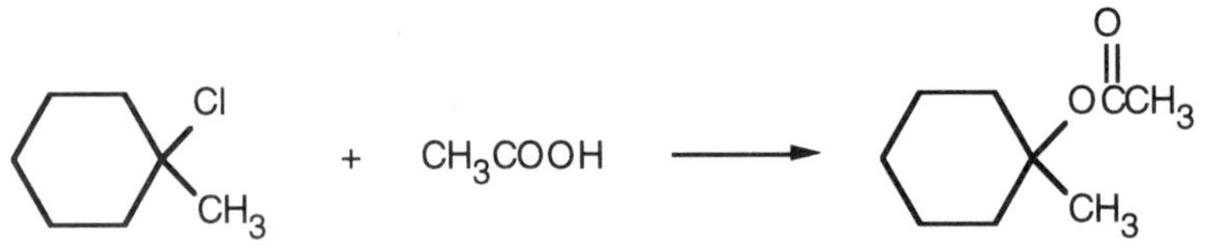

       tertiary   substitution product

This is an S_N1 reaction. Tertiary halides form substitution products only by the S_N1 route.

11.18 a) CH_3I reacts faster than CH_3Br because I^- is a better leaving group than Br^-.
   b) CH_3CH_2I reacts faster with OH^- in dimethylsulfoxide (DMSO) than in ethanol. Ethanol, a protic solvent, hydrogen-bonds with hydroxide ion and decreases its reactivity.
   c) Under the S_N2 conditions of this reaction, CH_3Cl reacts faster than $(CH_3)_3CCl$. Approach of the nucleophile to the bulky $(CH_3)_3CCl$ molecule is hindered.
   d) $H_2C=CHCH_2Br$ reacts faster because vinylic halides such as $H_2C=CHBr$ are unreactive to displacement reactions.

11.19

$$CH_3CH_2\overset{\overset{\displaystyle CH_3}{|}}{C}HCH_2I + {}^-{:}CN \longrightarrow CH_3CH_2\overset{\overset{\displaystyle CH_3}{|}}{C}HCH_2CN + {:}\overset{..}{\underset{..}{I}}{:}^-$$

(primary halide)

This is a S_N2 reaction, so the reaction rate depends on the concentration of both alkyl halide and nucleophile.

a) Halving the concentration of cyanide ion and doubling the concentration of alkyl halide won't change the reaction rate.
b) Tripling the concentrations of both cyanide ion and alkyl halide will cause a ninefold increase in reaction rate.

11.20

$$CH_3CH_2\overset{\overset{\displaystyle CH_3}{|}}{\underset{\underset{\displaystyle I}{|}}{C}}CH_3 + CH_3CH_2OH \longrightarrow CH_3CH_2\overset{\overset{\displaystyle CH_3}{|}}{\underset{\underset{\displaystyle OCH_2CH_3}{|}}{C}}CH_3 + HI$$

(tertiary halide)

This is an S_N1 reaction, so the reaction rate depends only on the concentration of 2–iodo–2–methylbutane. Tripling the concentration of alkyl halide will triple the rate of reaction.

11.21

a) $CH_3Br + Na^+ {}^-C{\equiv}CCH(CH_3)_2 \longrightarrow CH_3C{\equiv}CCH(CH_3)_2$

NOT $CH_3C{\equiv}C^-Na^+ + BrCH(CH_3)_2$. The strong base $CH_3C{\equiv}C{:}$ brings about E2 elimination producing $CH_3C{\equiv}CH$ and $H_2C{=}CHCH_3$.

b) $CH_3CH_2CH_2CH_2Br + NaCN \longrightarrow CH_3CH_2CH_2CH_2CN$

c) $H_3C{-}Br + {:}\overset{..}{O}C(CH_3)_3 \longrightarrow H_3C{-}OC(CH_3)_3$. The reaction of $CH_3\overset{..}{\underset{..}{O}}{:}^-$ with $Br{-}C(CH_3)_3$ results in elimination, not substitution.

d) $CH_3CH_2CH_2Br + NH_3 \longrightarrow CH_3CH_2CH_2NH_2 + HBr$

e)

f)

11.22 a) The difference in this pair of reactions is in the leaving group. Since ⁻OTos is a better leaving group than ⁻Cl (see Section 11.5), S_N2 displacement by iodide on CH_3–OTos proceeds faster.

b) The substrates in these two reactions are different. Bromoethane is a primary bromoalkane, and bromocyclohexane is a secondary bromoalkane. Since S_N2 reactions proceed faster at primary than secondary carbon atoms, S_N2 displacement on bromoethane is a faster reaction.

c) Ethoxide ion and cyanide ion are different nucleophiles. Since CN^- is more reactive than $CH_3CH_2O^-$ in S_N2 reactions, S_N2 displacement on 2-bromopropane by CN^- proceeds at a faster rate.

d) The solvent in each reaction is different. S_N2 reactions in hexamethylphosphoramide (HMPA) proceed faster than those in other solvents. Thus, S_N2 displacement by acetylide ion on bromomethane proceeds faster in HMPA than in benzene.

11.23 Because 1–bromopropane is a primary haloalkane, the mode of reaction is either S_N2 or E2, depending on the basicity and the amount of steric hindrance in the nucleophile.

a) $CH_3CH_2CH_2Br$ + $NaNH_2$ ⟶ $CH_3CH_2CH_2NH_2$

b) $CH_3CH_2CH_2Br$ + $\overset{+}{K}:\overset{-}{O}C(CH_3)_3$ ⟶ $CH_3CH=CH_2$

   K^+ ⁻O-t-Butyl is a hindered, strong base that causes elimination, not substitution.

c) $CH_3CH_2CH_2Br$ + NaI ⟶ $CH_3CH_2CH_2I$

d) $CH_3CH_2CH_2Br$ + $NaCN$ ⟶ $CH_3CH_2CH_2CN$

e) $CH_3CH_2CH_2Br$ + $Na^+\ ^-C{\equiv}CH$ ⟶ $CH_3CH_2CH_2C{\equiv}CH$

f) $CH_3CH_2CH_2Br$ + Mg ⟶ $CH_3CH_2CH_2MgBr$ $\xrightarrow{H_2O}$ $CH_3CH_2CH_3$

11.24 Remember two rules used to predict nucleophilicity:
  1) In comparing nucleophiles that have the same attacking atom, nucleophilicity parallels basicity. In other words, a more basic nucleophile is a more effective nucleophile.
  2) Nucleophilicity increases in going down a column of the periodic table.

	More Nucleophilic	Less Nucleophilic	Reason
a)	⁻$\ddot{N}H_2$	$:NH_3$	Rule 1
b)	$CH_3CO\ddot{O}:^-$	$H_2\ddot{O}:$	Rule 1
c)	$:\ddot{F}:^-$	BF_3	BF_3 is not a nucleophile
d)	$(CH_3)_3P:$	$(CH_3)_3N:$	Rule 2
e)	$:\ddot{I}:^-$	$:\ddot{C}l:^-$	Rule 2
f)	⁻$:C{\equiv}N$	⁻$:\ddot{O}CH_3$	Table 11.1

11.25 An alcohol is converted to an ether by two different routes in this series of reactions. The two resulting ethers have identical structural formulas but differ in sign of specific rotation. Therefore, at some step or steps in these reaction sequences, inversion of configuration at the stereogenic carbon must have occurred. Let's study each step of the Phillips and Kenyon series to find where inversion is occurring.

In step 1, the alcohol reacts with potassium metal to produce a potassium alkoxide. Since the bond between carbon and oxygen has not been broken, no inversion occurs in this step.

The potassium alkoxide acts as a nucleophile in the S_N2 displacement on CH_3CH_2Br in step 2. It is the C–Br bond of bromoethane, however, not the C–O bond of the alkoxide, that is broken. No inversion at the stereogenic carbon occurs in step 2.

The starting alcohol reacts with tosyl chloride in step 3. Again, because the O–H rather than the C–O bond of the alcohol is broken, no inversion occurs at this step.

Inversion does occur at step 4 when the ⁻OTos group is displaced by CH_3CH_2OH. The C–O bond of the tosylate (OTos) is broken, and a new C–O bond is formed.

Notice the specific rotations of the two enantiomeric products. The product of steps 1 and 2 should be enantiomerically pure because neither reaction has affected the C–O bond. Reaction 4 proceeds with some racemization at the stereogenic center to give a smaller value of $[\alpha]_D$.

11.26 a) Substitution does not take place with secondary alkyl halides when a strong, bulky base is used. Elimination occurs instead, and produces $H_2C=CHCH_2CH_3$ and $CH_3CH=CHCH_3$.

b) Reaction of this secondary fluoroalkane with hydroxide yields both elimination and substitution products.

c) $SOCl_2$ in pyridine converts primary and secondary alcohols to chlorides by an S_N2 mechanism. 1–Methyl–1–cyclohexanol is a tertiary alcohol, and does not undergo S_N2 substitution. Instead, E2 elimination occurs to give 1–methylcyclohexene.

11.27 S_N1 reactivity:

Most reactive ———————————————————→ Least reactive

a)

$C_6H_5{-}C(CH_3)_2Cl$ > $(CH_3)_3C{-}Cl$ >> $CH_3CH_2\overset{\displaystyle NH_2}{\underset{|}{C}H}CH_3$

(most stable carbocation)

b) $(CH_3)_3C{-}Br$ > $(CH_3)_3C{-}F$ > $(CH_3)_3C{-}OH$

(best leaving group)

c)

$[C_6H_5]_3{-}CBr$ > $C_6H_5{-}CH(CH_3)Br$ > $C_6H_5{-}CH_2Br$

(most stable carbocation)

11.28 S_N2 reactivity:

Most reactive ———————————————————→ Least reactive

a) $CH_3CH_2CH_2Cl$ > $CH_3CH_2CHClCH_3$ > $(CH_3)_3CCl$

(primary carbon atom)

b) $(CH_3)_2CHCH_2Br$ > $(CH_3)_2CH\overset{\displaystyle Br}{\underset{|}{C}H}CH_3$ > $(CH_3)_3CCH_2Br$

(least sterically hindered carbon atom)

c) $CH_3CH_2CH_2OTos$ > $CH_3CH_2CH_2Br$ > $CH_3CH_2CH_2OCH_3$

(best leaving group)

11.29

R–2–Bromooctane

(R)–2–Bromooctane is a secondary bromoalkane, which undergoes S_N2 substitution. Since S_N2 reactions proceed with inversion of configuration, the configuration at the stereogenic carbon atom is inverted. (This does not necessarily mean that all R isomers become S isomers after an S_N2 reaction. The R–S designation refers to the priorities of groups, which may change when the nucleophile is varied.)

$$Nu: \quad + \quad \underset{n-C_6H_{13}}{\overset{H}{\underset{H_3C}{C}}}C-Br \quad \longrightarrow \quad Nu-\underset{n-C_6H_{13}}{\overset{H}{C}}CH_3 \quad + \quad :Br:^-$$

Nucleophile		Product	
a)	$^-:CN$	$NC-\underset{n-C_6H_{13}}{\overset{H}{C}}CH_3$	S
b)	$CH_3\overset{O}{\overset{\|}{C}}\overset{..}{O}:^-$	$CH_3\overset{O}{\overset{\|}{C}}O-\underset{n-C_6H_{13}}{\overset{H}{C}}CH_3$	S
c)	$CH_3\overset{..}{S}:^-$	$CH_3S-\underset{n-C_6H_{13}}{\overset{H}{C}}CH_3$	S
d)	$:\overset{..}{Br}:^-$	$Br-\underset{n-C_6H_{13}}{\overset{H}{C}}CH_3 \quad S \qquad + \qquad \underset{n-C_6H_{13}}{\overset{H}{\underset{H_3C}{C}}}C-Br \quad R$	

2-Bromooctane is 100% racemized after 50% of the original (*R*)-2-bromooctane has reacted with Br⁻.

11.30 a) The rates of both S$_N$1 and S$_N$2 reactions are affected by the use of polar solvents. S$_N$1 reactions are accelerated because polar solvents stabilize developing charges in the transition state. Most S$_N$2 reactions, however, are slowed down by polar *protic* solvents because these solvents hydrogen-bond to the nucleophile and decrease its reactivity. Polar aprotic solvents solvate nucleophiles without hydrogen bonding and increase nucleophile reactivity in S$_N$2 reactions.

b) Good leaving groups (weak bases whose negative charge can be delocalized) increase the rates of S$_N$1 and S$_N$2 reactions.

c) A good attacking nucleophile accelerates the rate of an S$_N$2 reaction. Since the nucleophile is involved in the rate-limiting step of an S$_N$2 reaction, a good attacking nucleophile lowers the energy of the transition state and increases the rate of reaction. Choice of nucleophile has no effect on the rate of a S$_N$1 reaction because attack of the nucleophile occurs after the rate-limiting step.

d) Because the rate-limiting step in an S$_N$2 reaction involves attack of the nucleophile on the substrate, any factor that makes approach of the nucleophile more difficult slows down the rate of reaction. Especially important is the degree of crowding at the reacting carbon atom. Tertiary carbon atoms are too crowded to allow S$_N$2 substitution to occur. Even steric hindrance one carbon atom away from the reacting site causes a drastic slowdown in rate of reaction.

The rate-limiting step in an S$_N$1 reaction involves formation of a carbocation. Any structural factor in the substrate that stabilizes a carbocation will increase the rate of reaction. Substrates that are tertiary, allylic, or benzylic react the fastest.

11.31

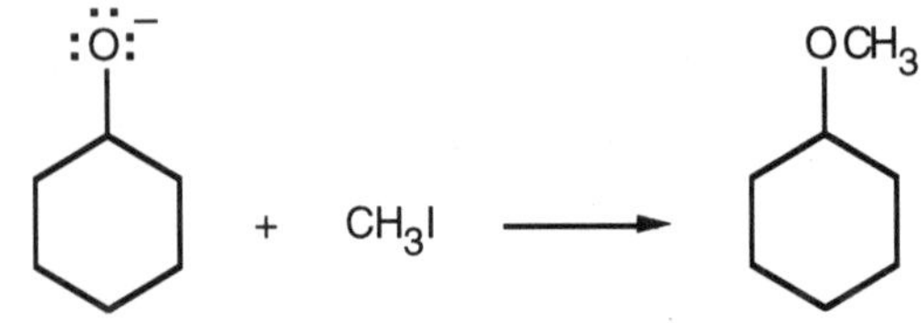

This is an excellent method of ether preparation because iodomethane is very reactive in S_N2 displacements.

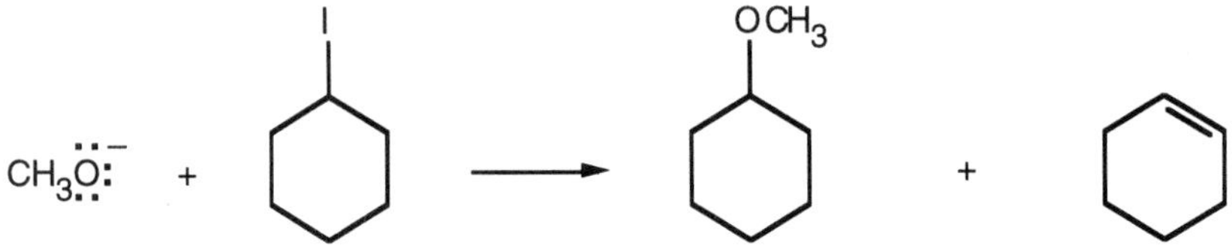

Reaction of a secondary haloalkane with a basic nucleophile yields both substitution and elimination products. This is a less satisfactory method of ether preparation.

11.32

Methoxide removes a proton from the hydroxyl group of 4–bromo–1–butanol.

S_N2 displacement of :Br⁻ by the alkoxide oxygen yields the cyclic ether tetrahydrofuran.

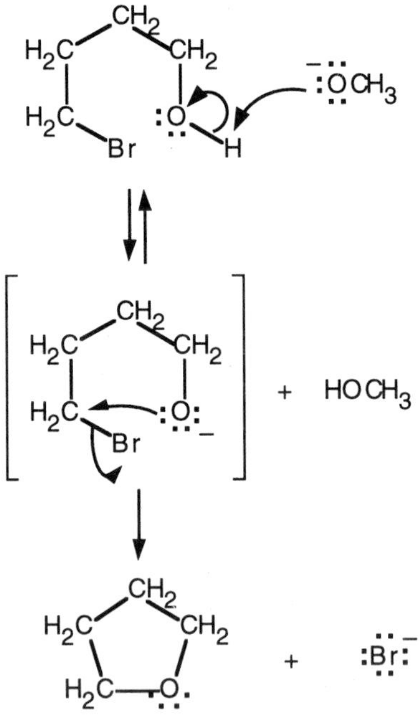

Tetrahydrofuran

$CH_3OCH_2CH_2CH_2CH_2OH$ is also produced.

11.33

$$BrCH_2CH_2Br \ + \ 2\,NaOH \longrightarrow HOCH_2CH_2OH$$

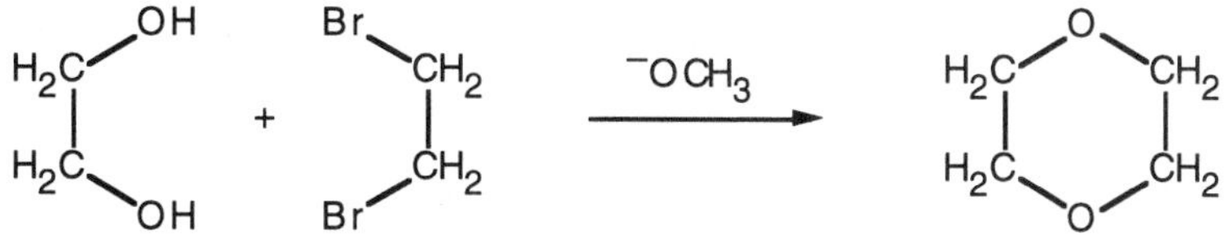

11.34

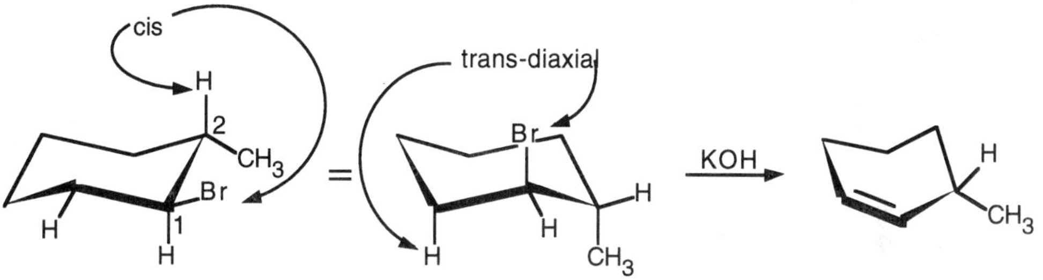

E2 reactions require that the two atoms to be eliminated have a trans-diaxial relationship. Since it's impossible for bromine and the hydrogen at C2 to be trans-diaxial, elimination occurs in the opposite direction to yield 3–methylcyclohexene, the non-Zaitsev product.

11.35

a)

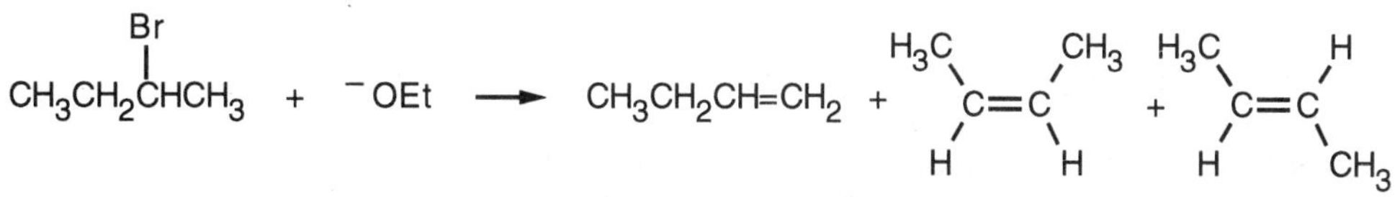

b)

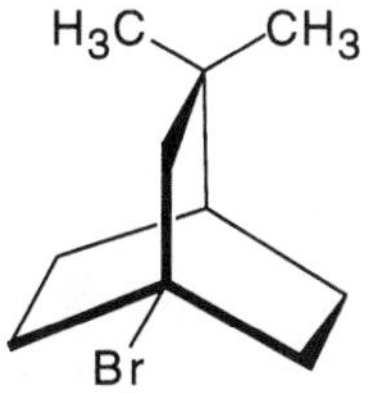

does not undergo nucleophilic substitution
(see Problems 11.3 and 11.13)

c)

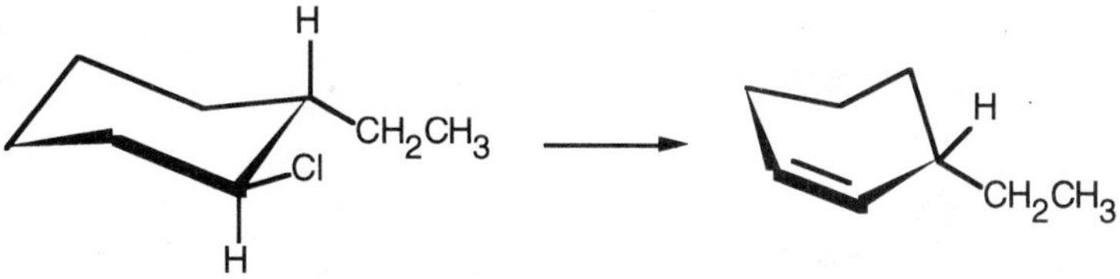

This alkyl halide gives the less substituted cycloalkene (non-Zaitsev product). Elimination to form Zaitsev product is not likely to occur because the –Cl and –H involved cannot assume the anti-periplanar geometry preferred for E2 elimination.

d) $(CH_3)_3C–OH \ + \ HCl \ \xrightarrow{\ 0°\ } \ (CH_3)_3C–Cl$

11.36

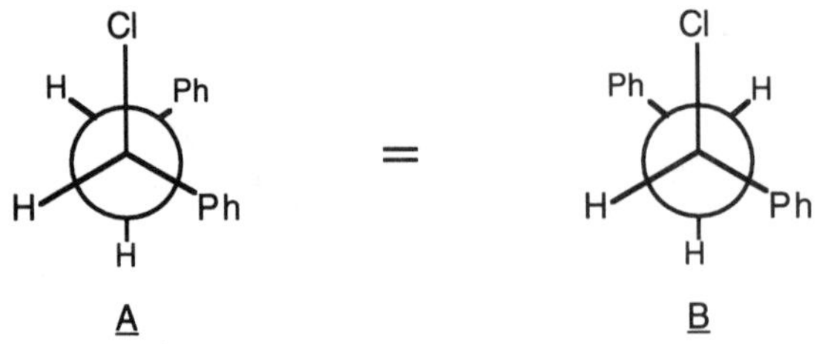

A   =   B

Both Newman projections place –H and –Cl in the correct anti-periplanar geometry for E2 elimination.

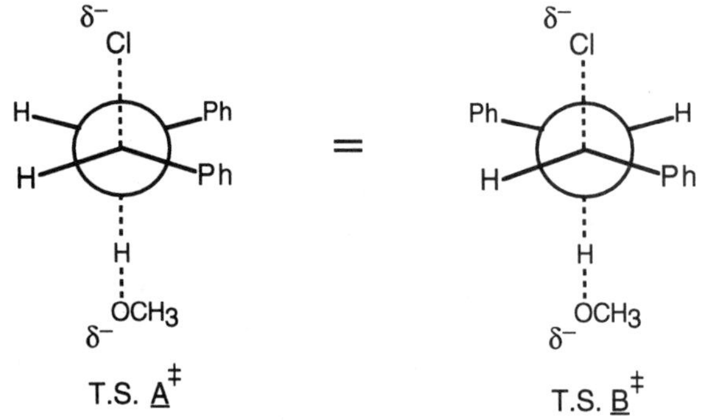

T.S. A‡   =   T.S. B‡

Either transition state $\underline{A}^{\ddagger}$ or $\underline{B}^{\ddagger}$ can form when 1–chloro–1,2–diphenylethane undergoes E2 elimination. Steric interactions of the two phenyl groups in T.S. $\underline{A}^{\ddagger}$ make this transition state (and the product resulting from it) of higher energy than transition state $\underline{B}^{\ddagger}$. Formation of the product from $\underline{B}^{\ddagger}$ is therefore favored, and *trans*–1,2–diphenylethylene is the major product.

11.37

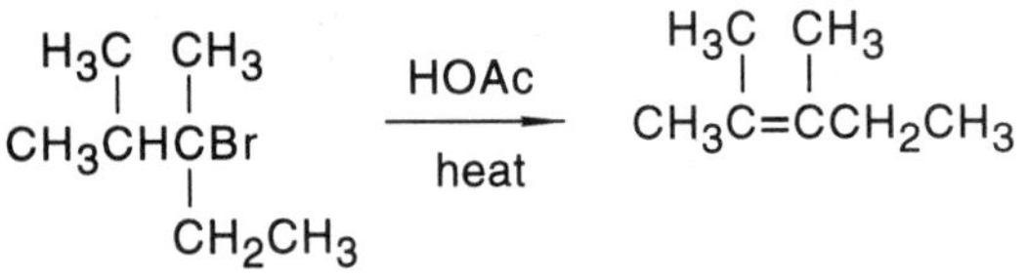

This alkene has the most highly substituted double bond.

11.38

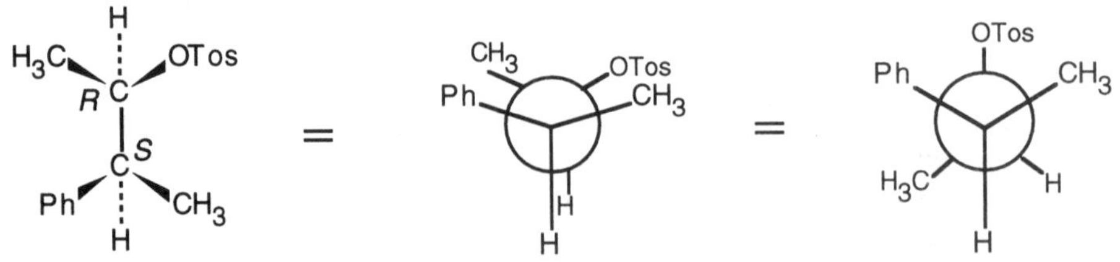

Draw a Newman projection of the tosylate of (2R,3S)–3–phenyl–2–butanol, and then rotate the projection until the –OTos and the –H on the adjoining carbon atom are anti–periplanar. Even though this conformation has several gauche interactions, it is the only conformation in which –OTos and –H are 180° apart.

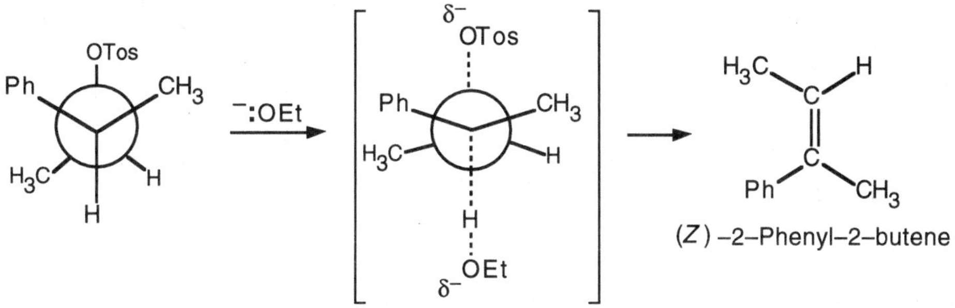

(Z)–2–Phenyl–2–butene

Elimination yields the Z isomer of 2–phenyl–2–butene. Refer to Chapter 6 for the method of assigning E, Z designation.

11.39 By the same argument used in Problem 11.38, you can show that elimination from (2R,3R)–3– phenyl–2–butyl tosylate give the E–alkene.

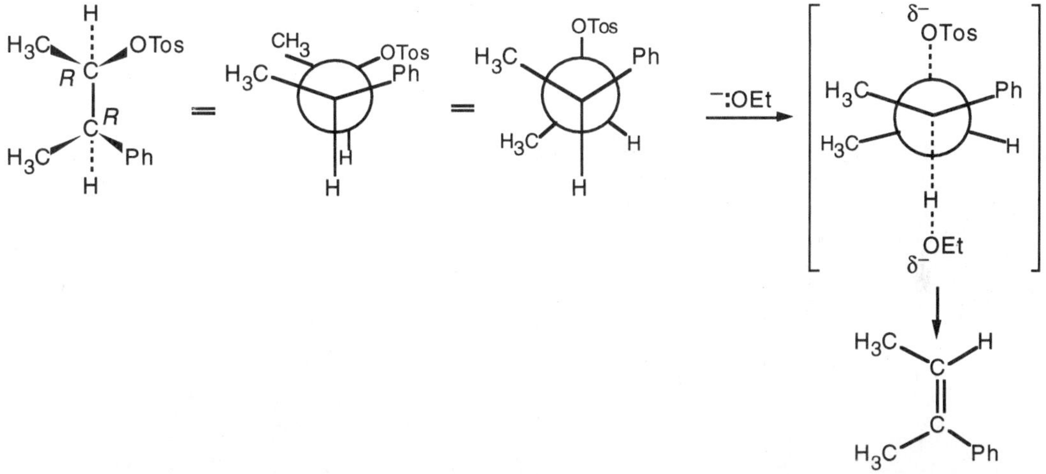

(E)–2–Phenyl–2–butene

The 2S,3S isomer also forms the E–alkene; the 2S,3R isomer yields Z–alkene.

11.40

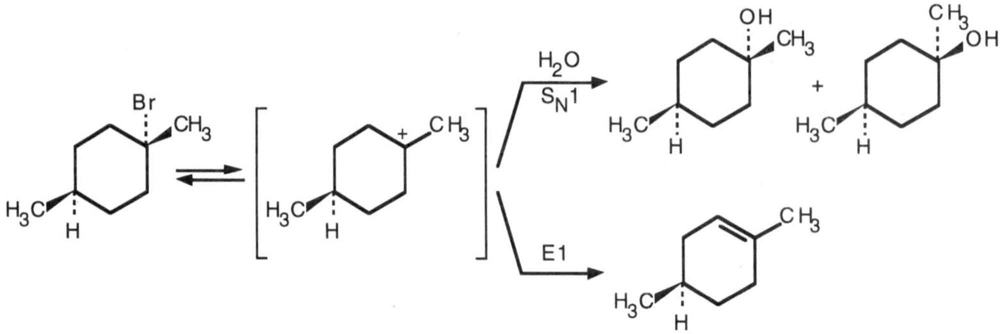

This tertiary bromoalkane reacts by S_N1 and E1 routes to yield alcohol and alkene products.

11.41 S_N2 reactivity:

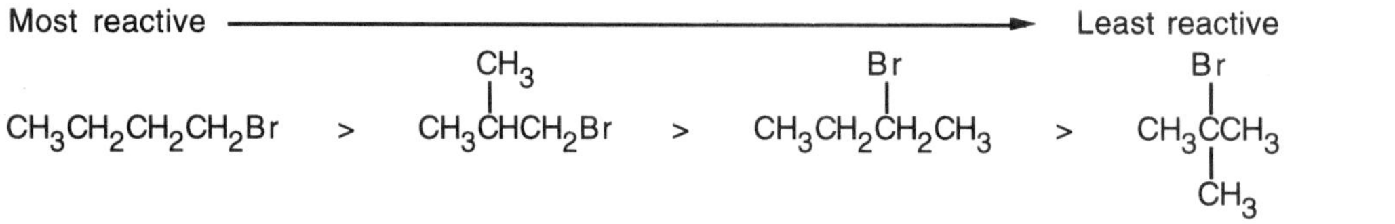

Most reactive ⟶ Least reactive

$CH_3CH_2CH_2CH_2Br$ > CH_3CHCH_2Br (with CH_3) > $CH_3CH_2CH_2CH_3$ (with Br) > CH_3CCH_3 (with Br and CH_3)

1–Bromobutane   1–Bromo–2–methyl–propane   2–Bromobutane   2–Bromo–2–methyl–propane

11.42

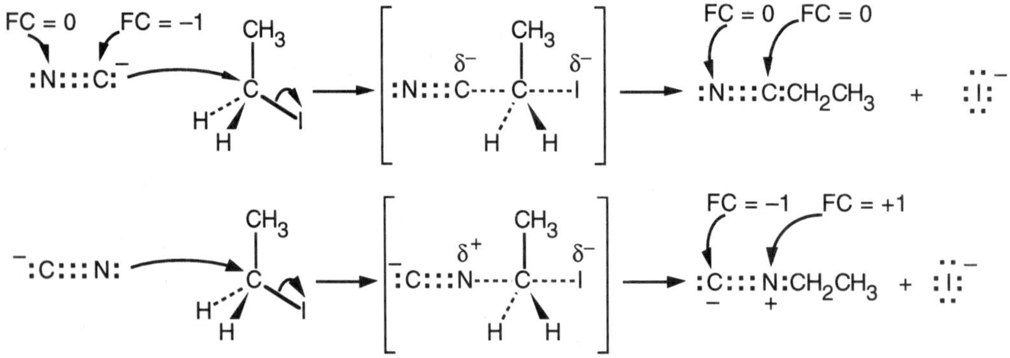

S_N2 attack by the lone pair electrons associated with carbon gives the nitrile product. Attack by the lone pair electrons associated with nitrogen yields isonitrile product.

11.43

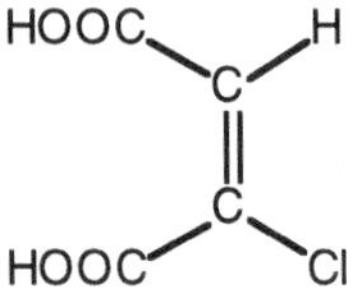

(E)–2–Chloro–2–butene–1,4–dioic acid 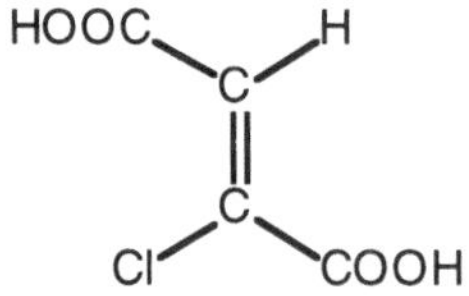   (Z)–2–Chloro–2–butene–1,4–dioic acid

Hydrogen and chlorine are anti to each other in the Z isomer and are syn in the E isomer. Since the Z isomer reacts fifty times faster than the E isomer, elimination must proceed more favorably when the substituents to be eliminated are anti to one another. This is the same stereochemical result as occurs in E2 eliminations of alkyl halides.

11.44 Since 2–butanol is a secondary alcohol, substitution can occur by either an S_N1 or S_N2 route, depending on reaction conditions. Two factors favor an S_N1 mechanism in this case. (1) The reaction is run under solvolysis (solvent as nucleophile) conditions in a polar, protic solvent. (2) Dilute acid converts a poor leaving group (^-OH) into a good leaving group (OH_2), which dissociates easily.

Protonation of oxygen . . .

. . . is followed by loss of water to form a planar carbocation.

Attack of water from either side of the planar cation yields racemic product.

11.45 The chiral tertiary alcohol (R)–3–methyl–3–hexanol reacts with HBr by an S_N1 pathway. HBr protonates the hydroxyl group, which dissociates to yield a planar, achiral carbocation. Attack by the nucleophilic bromide anion can occur from either side of the carbocation to produce (±)3–bromo–3–methylhexane.

11.46 Since carbon-deuterium bonds are slightly stronger than carbon-hydrogen bonds, more energy is required to break a C–D bond than to break a C–H bond. In a reaction where either a carbon-deuterium or a carbon-hydrogen bond can be broken in the rate-limiting step, a higher percentage of C–H bond-breaking will occur because the energy of activation for C–H breakage is lower.

Transition state $\underline{A}^{\ddagger}$ is of higher energy than transition state $\underline{B}^{\ddagger}$ because more energy is required to break the C–D bond. The product that results from transition state $\underline{B}^{\ddagger}$ is thus formed in greater abundance.

11.47 One of the steric requirements of E2 elimination is the need for periplanar geometry, which optimizes orbital overlap in the transition state leading to alkene product. Two types of periplanar arrangements of substituents are possible — syn and anti.

A model of the deuterated bromo compound shows that the deuterium, bromine, and the two carbon atoms that will constitute the double bond all lie in a plane. This arrangement of atoms leads to syn elimination. Even though anti elimination is usually preferred, it does not occur for this compound because the bromine, hydrogen, and two carbons can't occur can't achieve the necessary geometry.

11.48

We concluded in Problem 11.48 that E2 elimination in compounds of this bicyclic structure occurs with syn-periplanar geometry. In compound $\underline{A}$, –H and –Cl can be eliminated via the syn-periplanar route. Since neither syn nor anti-periplanar elimination is possible for $\underline{B}$, elimination occurs by a slower, E1 route.

11.49

Diastereomer 8 reacts much more slowly than other isomers in an E2 reaction. No pair of hydrogen and chlorine atoms can assume the anti-periplanar orientation preferred for E2 elimination.

11.50 Build molecular models of triethylamine and quinuclidine. A model of the most stable conformation of triethylamine shows that the ethyl groups interfere with approach of the nitrogen lone pair electrons to iodomethane. In quinuclidine, however, the hydrocarbon framework is rigidly held back from the nitrogen lone pair. It is sterically easier for quinuclidine to approach methyl iodide, and reaction therefore occurs at a faster rate.

11.51 The two pieces of evidence indicate that the reaction proceeds by an S_N2 mechanism: S_N2 reactions proceed much faster in polar aprotic solvents such as DMF, and methyl esters react faster than ethyl esters. This reaction is an S_N2 displacement on a methyl ester by iodide ion.

Other experiments can provide additional evidence for an S_N2 mechanism. We can determine if the reaction is second-order by varying the concentration of LiI. We can also vary the type of nucleophile to distinguish an S_N2 mechanism from an S_N1 mechanism, which does not depend on the identity of the nucleophile.

11.52 Because Cl⁻ is a relatively poor leaving group and acetate is a relatively poor nucleophile, a displacement involving these two groups proceeds at a very slow rate. I⁻, however, is both a good nucleophile and a good leaving group. 1–Chlorooctane therefore reacts preferentially with iodide to form 1–iodooctane. Only a small amount of 1–iodooctane is formed (because of the low concentration of iodide ion), but 1–iodooctane is more reactive than 1–chlorooctane toward substitution by acetate. Reaction with acetate produces 1–octyl acetate and regenerates iodide ion. The whole process can now be repeated with another molecule of 1–chlorooctane. The net result is production of 1–octyl acetate; no iodide is consumed.

11.53 Two optically inactive structures are possible for compound X. Any other structure consistent with the series of reactions is optically active.

11.54

(2R,3S)–2–Bromo–3–methyl–
2–phenylpentane

(E)–3–Methyl–
2–phenyl–2–pentene

The 2S,3R isomer also yields E product.

11.55

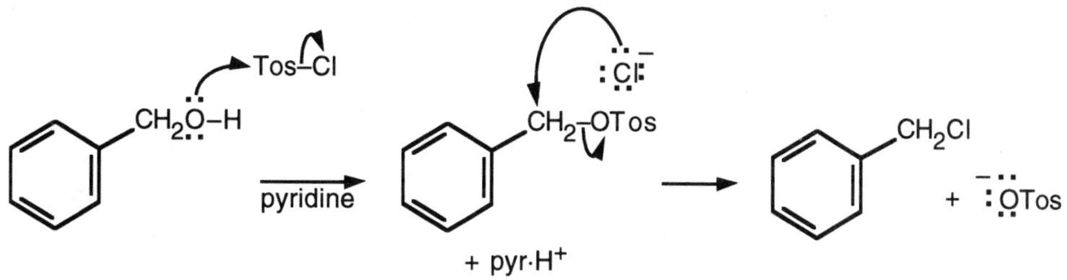

11.56

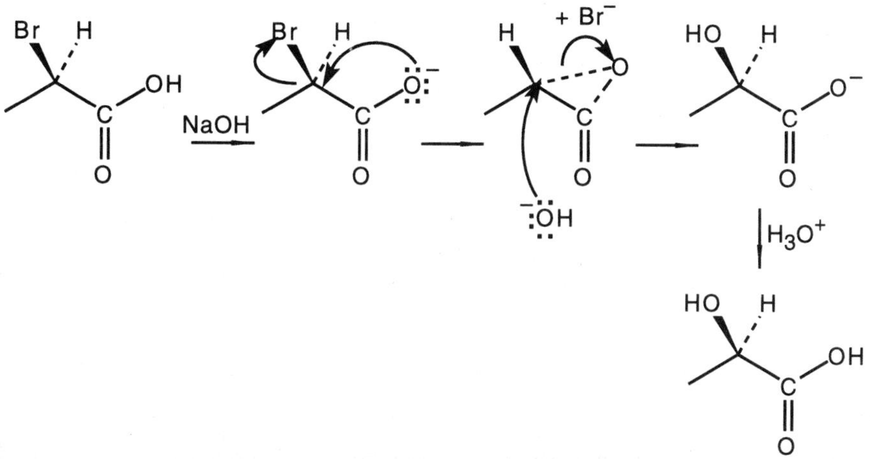

Two inversions of configuration equal a net retention of configuration.

11.57

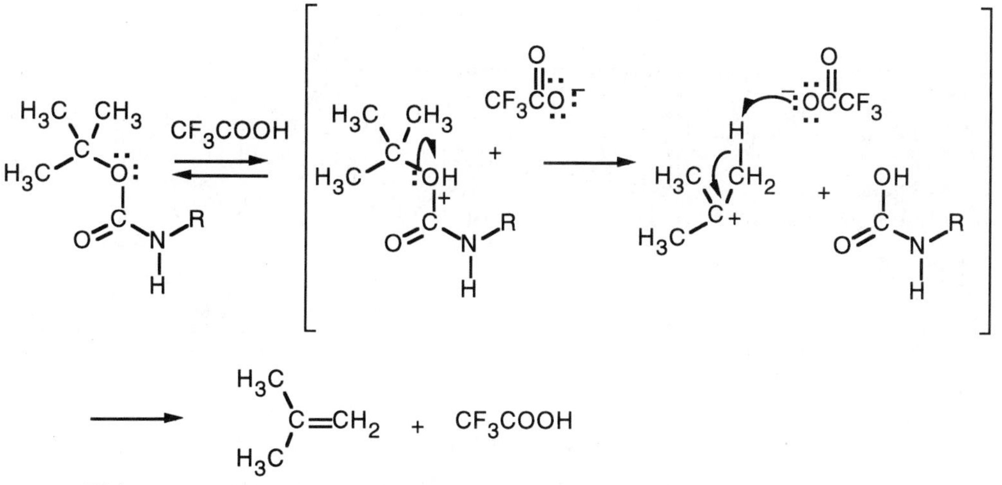

This process is an E1 reaction.

11.58

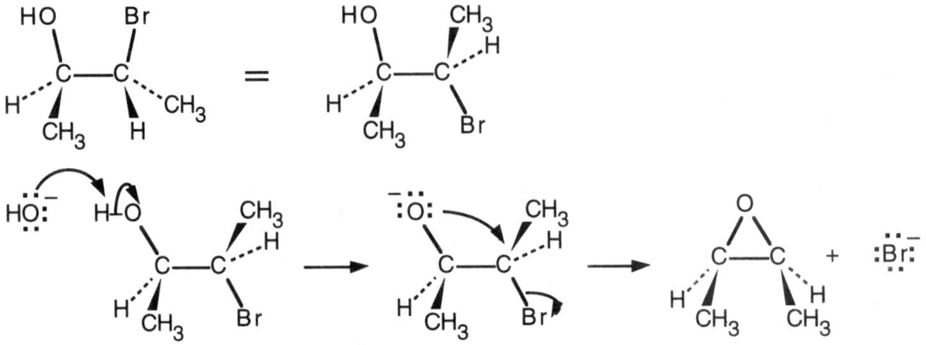

This reaction is an intramolecular S_N2 displacement.

11.59

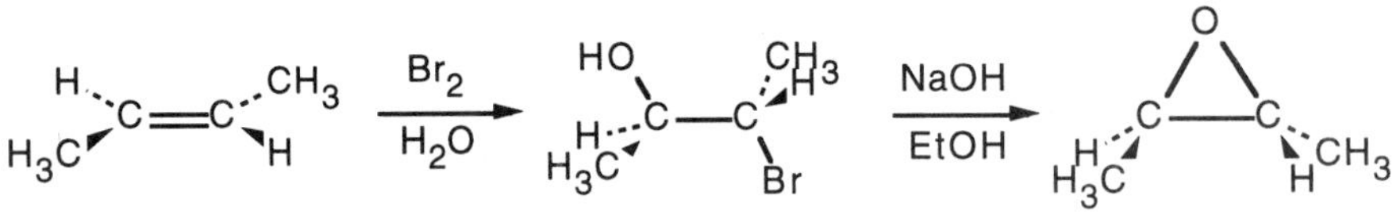

11.60

$$CH_3CH_2CH_2CH_2CH_2NH_2 \xrightarrow[CH_3I]{\text{excess}} CH_3CH_2CH_2CH_2CH_2\overset{+}{N}(CH_3)_3 \quad I^-$$

$$\xrightarrow[H_2O]{Ag_2O} CH_3CH_2CH_2CH{=}CH_2 \quad + \quad N(CH_3)_3$$

The intermediate is a charged quaternary ammonium compound that results from S_N2 substitutions on three CH_3I molecules by the amine nitrogen. E2 elimination occurs because the neutral $N(CH_3)_3$ molecule is a good leaving group.

Study Guide for Chapter 11

After studying this chapter, you should be able to:

(1) Formulate the mechanism of:
(a) S_N2 reactions (11.1, 11.2, 11.3, 11.25, 11.42, 11.51, 11.55, 11.56, 11.58, 11.59)
(b) S_N1 reactions (11.8, 11.9, 11.11, 11.44, 11.45)
(c) Elimination reactions (11.15, 11.16, 11.36, 11.38, 11.39, 11.43, 11.46, 11.57)

(2) Predict the effect of substrate, nucleophile or base:
(a) S_N2 reactions (11.6, 11.18, 11.19, 11.22, 11.24, 11.28, 11.30, 11.41)
(b) S_N1 reactions (11.10, 11.12, 11.13, 11.20, 11.27, 11.30)
(c) Elimination reactions (11.47, 11.48, 11.49)

(3) Predict the products of:
(a) Substitution reactions (11.4, 11.21, 11.23, 11.29, 11.31)
(b) Elimination reactions (11.14, 11.34, 11.37, 11.40, 11.53, 11.54)

(4) Classify reactions as S_N1, S_N2, E1 or E2 (11.17, 11.26).

(5) Use substitution and elimination reactions in synthetic sequences (11.32, 11.33).

Chapter 12 – Structure Determination.
Mass Spectrometry and Infrared Spectroscopy

12.1 The following systematic approach may be helpful.

a) $M^+ = 86$
  1. Find the compound of molecular weight 86 that contains only C and H. Remember that a hydrocarbon with n carbon atoms can have no more than $2n + 2$ hydrogen atoms. Here, C_6H_{14} is the correct formula.
  2. Find the formula corresponding to $M^+ = 86$ that contains carbon, hydrogen, and one oxygen atom. If one oxygen atom (atomic weight = 16) is added to the base formula from step 1, one carbon atom (atomic weight 12) and four hydrogen atoms (atomic weight 4) must be removed. This formula is $C_5H_{10}O$.
  3. Proceed to find the remaining molecular formulas. Each time one oxygen is added, one carbon and four hydrogens must be removed. The remaining formulas for $M^+ = 86$ are $C_4H_6O_2$ and $C_3H_2O_3$.

b) $M^+ = 128$. The procedure is the same as in part a). Two hydrocarbons having $M^+ = 128$ are C_9H_{20} and $C_{10}H_8$. The formulas containing one oxygen are $C_8H_{16}O$ and C_9H_4O. The remaining formulas are $C_7H_{12}O_2$, $C_6H_8O_3$, $C_5H_4O_4$.

c) $M^+ = 156$. Possible formulas are $C_{11}H_{24}$, $C_{12}H_{12}$, $C_{11}H_8O$, $C_{10}H_{20}O$, $C_{10}H_4O_2$, $C_9H_{16}O_2$, $C_8H_{12}O_3$, $C_7H_8O_4$, $C_6H_4O_5$.

12.2 Use the method described in Problem 12.1(a). The hydrocarbons having $M^+ = 218$ are $C_{16}H_{26}$ and $C_{17}H_{14}$. Since nootkatone also contains oxygen, we must consider only those formulas that include oxygen. Using the previous procedure, we can determine that $C_{15}H_{22}O$, $C_{14}H_{18}O_2$, $C_{13}H_{14}O_3$, $C_{12}H_{10}O_4$, $C_{11}H_6O_5$, $C_{16}H_{10}O$ and $C_{15}H_6O_2$ are possible formulas for nootkatone. The actual formula of nootkatone is $C_{15}H_{22}O$.

12.3 Each carbon atom has a 1.10% probability of being ^{13}C and a 98.90% probability of being ^{12}C. The ratio of the height of the ^{13}C peak to the height of the ^{12}C peak for a one-carbon compound is $(1.10/98.9) \times 100\% = 1.11\%$. For a six-carbon compound, the contribution to $(M+1)^+$ from ^{13}C is $6 \times (1.10/98.9) \times 100\% = 6.66\%$. For benzene, the relative height of $(M+1)^+$ is 6.66% of the height of M^+.

A similar line of reasoning can be used to calculate the contribution to $(M+1)^+$ from ^2H. The natural abundance of ^2H is 0.015%, so the ratio of a ^2H peak to a ^1H peak for a one-hydrogen compound is 0.015%. For a six-hydrogen compound, the contribution to $(M+1)^+$ from ^2H is $6 \times 0.015\% = 0.09\%$.

For benzene, $(M+1)^+$ is 6.75% of M^+. Notice that ^2H contributes very little to the size of $(M+1)^+$.

12.4 The structural formula of 2,2–dimethylpropane and of its molecular ion are given below.

When the molecular ion fragments, neutral and positively charged species are produced. The fragment of $m/z = 57$ corresponds to $C_4H_9^+$. The base peak usually represents the cation best able to stabilize positive charge. Since tertiary carbocations are relatively stable, $C_4H_9^+$ is most likely to be the *tert*–butyl cation.

12.5

CH$_3$CH$_2$CH=C(CH$_3$)CH$_3$

2–Methyl–2–pentene

$CH_3CH_2CH_2CH=CHCH_3$

2–Hexene

Fragmentation occurs to a greater extent at the weakest carbon-carbon bonds, and the positive charge remains with the fragment that is more able to stabilize it. A table of bond-dissociation energies (Table 5.4) shows that allylic bonds have lower bond-dissociation energies than the other bonds in these two compounds. Thus, the principal fragmentations of these compounds yield allylic cations.

$^+CH_2CH=C(CH_3)CH_3$

$m/z = 69$

$^+CH_2CH=CHCH_3$

$m/z = 55$

Spectrum (b), which has $m/z = 55$ as its base peak, corresponds to 2–hexene. Spectrum (a), which has an abundant peak at $m/z = 69$, corresponds to 2–methyl–2–pentene.

12.6

$E = h\nu = hc/\lambda$; $h = 6.62 \times 10^{-34}$ J·s; $c = 3.00 \times 10^{10}$ cm/s

for $\lambda = 10^{-4}$ cm (infrared radiation):

$$E = \frac{(6.62 \times 10^{-34} \text{ J·s})(3.00 \times 10^{10} \text{ cm/s})}{1.0 \times 10^{-4} \text{ cm}} = 2.0 \times 10^{-19} \text{ J}$$

for $\lambda = 3.0 \times 10^{-7}$ cm (X radiation):

$$E = \frac{(6.62 \times 10^{-34} \text{ J·s})(3.00 \times 10^{10} \text{ cm/s})}{3.0 \times 10^{-7} \text{ cm}} = 7.0 \times 10^{-17} \text{ J}$$

Thus, an X ray is of higher energy than infrared radiation.

12.7 First, convert radiation in cm to radiation in Hz by the equation:

$$\nu = \frac{c}{\lambda} = \frac{3.00 \times 10^{10} \text{ cm/s}}{9.0 \times 10^{-4} \text{ cm}} = 3.0 \times 10^{13} \text{ Hz}$$

The equation $E = h\nu$ says that the greater the value of ν, the greater the energy. Thus, radiation with $\nu = 3.0 \times 10^{13}$ Hz ($\lambda = 9.0 \times 10^{-4}$ cm) is higher in energy than radiation with $\nu = 4.0 \times 10^9$ Hz.

12.8

a) $E = \dfrac{1.20 \times 10^{-2} \text{ kJ/mol}}{\lambda \text{ (in cm)}} = \dfrac{1.20 \times 10^{-2} \text{ kJ/mol}}{5.0 \times 10^{-9}}$

     $= 2.4 \times 10^{6} \text{kJ/mol for gamma rays}$

b) $E = 4.0 \times 10^{4} \text{kJ/mol for X rays}$

c) $\upsilon = \dfrac{c}{\lambda}$; $\lambda = \dfrac{c}{\upsilon} = \dfrac{3.0 \times 10^{10} \text{cm/s}}{6.0 \times 10^{15} \text{ Hz}} = 5.0 \times 10^{-6} \text{cm}$

     $E = \dfrac{1.20 \times 10^{-2} \text{ kJ/mol}}{5.0 \times 10^{-6}} = 2.4 \times 10^{3} \text{kJ/mol for ultraviolet light}$

d) $E = 2.8 \times 10^{2} \text{kJ/mol for visible light}$

e) $E = 6.0 \text{ kJ/mol for infrared radiation}$

f) $E = 4.0 \times 10^{-2} \text{kJ/mol for microwave radiation}$

12.9

Wavenumber $= \dfrac{1}{\text{wavelength}}$; wavenumber has units of cm^{-1}. $1 \, \mu\text{m} = 10^{-4} \text{ cm}$.

a) $3.10 \, \mu\text{m} = 3.10 \times 10^{-4} \text{ cm}$; $\dfrac{1}{3.1 \times 10^{-4} \text{ cm}} = 3225 \text{ cm}^{-1}$

b) $5.85 \, \mu\text{m}$; 1710 cm^{-1}

c) $\dfrac{1}{2250 \text{ cm}^{-1}} = 4.44 \times 10^{-4} \text{ cm} = 4.44 \, \mu\text{m}$

d) 970 cm^{-1}; $10.3 \, \mu\text{m}$

12.10 a) A compound with a strong absorption at 1710 cm^{-1} contains a carbonyl group and is either a ketone or aldehyde.

     b) A nitro compound has a strong absorption at 1540 cm^{-1}.

     c) A compound showing both carbonyl (1720 cm^{-1}) and $-$OH ($2500\text{-}3000 \text{ cm}^{-1}$ broad) absorptions is a carboxylic acid.

12.11 To use IR spectroscopy to distinguish between isomers, find a strong IR absorption that is present in one isomer but absent in the other.

     a) CH_3CH_2OH CH_3OCH_3

            Strong hydroxyl band   No band in the
            at $3400\text{-}3640 \text{ cm}^{-1}$.   region $3400\text{-}3640 \text{ cm}^{-1}$.

b) $CH_3CH_2CH_2CH_2CH=CH_2$

Alkene bands at
$3020–3100$ cm^{-1} and
at $1650–1670$ cm^{-1}.

No bands in alkene region

c) CH_3CH_2COOH

Strong, broad band
at $2500–3100$ cm^{-1}.

$HOCH_2CH_2CHO$

Strong band
at $3400–3640$ cm^{-1}.

12.12 Based on what we know at present, we can identify four absorptions in this spectrum.

a) Absorptions in the region 1450 cm^{-1} - 1600 cm^{-1} are due to aromatic ring $–C=C–$ motions.

b) The absorption at 2100 cm^{-1} is due to a $–C\equiv C–$ stretch.

c) Absorptions in the range 3000 cm^{-1} - 3100 cm^{-1} are due to aromatic ring $=C–H$ stretches.

d) The absorption at 3300 cm^{-1} is due to a $\equiv C–H$ stretch.

12.13 a) An ester next to a double bond absorbs at 1715 cm^{-1}. The alkene double bond absorbs at 1650-1670 cm^{-1}.

b)

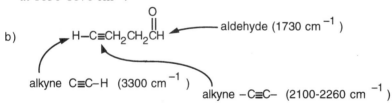

12.14 In this problem, all formulas must represent hydrocarbons.

a) For $M^+ = 64$, the only possible molecular formula is C_5H_4.
b) For $M^+ = 186$, possible formulas are $C_{14}H_{18}$ and $C_{15}H_6$.
c) For $M^+ = 158$, the only reasonable formula is $C_{12}H_{14}$.
d) Three formulas for $M^+ = 220$ are possible: $C_{16}H_{28}$, $C_{17}H_{16}$, and $C_{18}H_4$.

12.15

M$^+$		Molecular Formula	Degree of Unsaturation
a)	86	C_6H_{14}	0
b)	110	C_8H_{14}	2
c)	146	$C_{11}H_{14}$	5
d)	190	$C_{14}H_{22}$	4
		$C_{15}H_{10}$	11

12.16

M^+	Molecular Formula	Degree of Unsaturation	Possible Structure
a) 132	$C_{10}H_{12}$	5	
b) 166	$C_{13}H_{10}$	9	
	$C_{12}H_{22}$	2	
c) 84	C_6H_{12}	1	

12.17 Remember that compounds in this problem may contain carbon, hydrogen, oxygen, and nitrogen. In addition, the molecular ions of many compounds may have the same value of M^+. Some of the less likely molecular formulas -- those with few carbon or hydrogen atoms -- have been omitted.

a) $M^+ = 74$. Any nitrogen-containing compound that has a molecular ion at $M^+ = 74$ must have an even number of nitrogen atoms.
Compounds containing:

C, H;	C_6H_2
C, H, O;	$C_4H_{10}O$, $C_3H_6O_2$, $C_2H_2O_3$
C, H, N;	$C_3H_{10}N_2$, CH_6N_4
C, H, N, O;	$C_2H_6N_2O$, $CH_2N_2O_2$

b) $M^+ = 131$ has an odd number of nitrogen atoms; no hydrocarbons correspond to this molecular ion.

C, H, N;	C_9H_9N, $C_7H_5N_3$, $C_6H_{17}N_3$, $C_4H_{13}N_5$
C, H, N, O;	$C_7H_{17}NO$, C_8H_5NO, $C_6H_{13}NO_2$, $C_5H_9NO_3$, $C_4H_5NO_4$, $C_5H_{13}N_3O$, $C_4H_9N_3O_2$, $C_3H_5N_3O_3$, $C_3H_9N_5O$

12.18 Reasonable molecular formulas for camphor are $C_{10}H_{16}O$, $C_9H_{12}O_2$, $C_8H_8O_3$. The actual formula, $C_{10}H_{16}O$, corresponds to three degrees of unsaturation. The ketone functional group accounts for one of these. Since camphor is a saturated compound, the other two degrees of unsaturation are due to two rings.

Camphor

12.19 Carbon is tetravalent, and nitrogen is trivalent. If a C–H unit (formula weight 13) is replaced by an N atom (formula weight 14), the molecular weight of the resulting compound increases by one. Since all neutral hydrocarbons have even molecular weights (C_nH_{2n+2}, C_nH_{2n}, and so forth) the resulting nitrogen-containing compounds have odd molecular weights. If two C–H units are replaced by two N atoms, the molecular weight of the resulting compound increases by two and remains an even number.

12.20 Because M^+ is an odd number, pyridine contains an odd number of nitrogen atoms. If pyridine contained one nitrogen atom (atomic weight 14) the remaining atoms would have a formula weight of 65, corresponding to $-C_5H_5$. C_5H_5N is, in fact, the molecular formula of pyridine.

12.21 The molecular formula of nicotine is $C_{10}H_{14}N_2$. To find the equivalent hydrocarbon formula, subtract the number of nitrogens from the number of hydrogens. The equivalent hydrocarbon formula of nicotine, $C_{10}H_{12}$, indicates five degrees of unsaturation — two of them due to the two rings and the other three due to three double bonds.

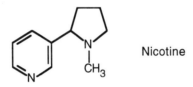

Nicotine

12.22 In order to simplify this problem, neglect the ^{13}C and ^2H isotopes in determining the molecular ions of these compounds.

a) The formula weight of $-CH_3$ is 15, and the atomic masses of the two bromine isotopes are 79 and 81. The two molecular ions of bromoethane occur at $M^+ = 96$ (49.3%) and $M^+ = 94$ (50.7%).

b) The formula weight of $-C_6H_{13}$ is 85, and the atomic masses of the two chlorine isotopes are 35 and 37. The two molecular ions of 1-chlorohexane occur at $M^+ = 122$ (24.2%) and $M^+ = 120$ (75.8%).

12.23 Again, neglect ^{13}C and ^2H in these calculations.

a) Finding the molecular ions of chloroform is a statistical exercise.
  1) The probability that all three chlorine atoms are ^{37}Cl is $(0.242)^3 = 0.014$.
  2) The probability that two chlorine atoms are ^{37}C and one is ^{35}Cl is $3(0.242)(0.242)(0.758) = 3(0.0444) = 0.133$. The factor 3 enters the calculations because three permutations of two ^{37}Cl's and one ^{35}Cl are possible.
  3) The probability that one chlorine atom is ^{37}Cl and two are ^{35}Cl is $3(0.242)(0.758)(0.758) = 3(0.139) = 0.417$.
  4) The probability that all chlorine atoms are ^{35}Cl is $(0.758)^3 = 0.436$.
  5) The mass of: $CH^{37}Cl^{37}Cl^{37}Cl = 124$
$CH^{37}Cl^{37}Cl^{35}Cl = 122$
$CH^{37}Cl^{35}Cl^{35}Cl = 120$
$CH^{35}Cl^{35}Cl^{35}Cl = 118$.
  6) Thus, molecular ions for chloroform occur at:

M^+	124	122	120	118
Abundance	1.4%	13.3%	41.7%	43.6%

b) The molecular ions for Freon 12:

M^+	124	122	120
Abundance	5.9%	36.7%	57.4%

12.24

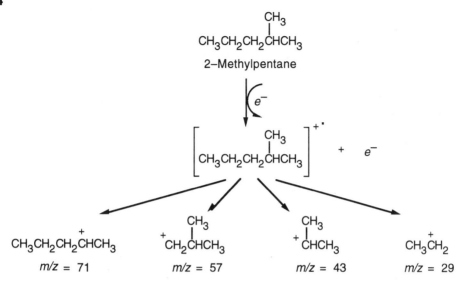

The molecular ion, at $m/z = 86$, is present in very low abundance. The base peak, at $m/z = 43$, represents a stable secondary carbocation.

12.25 Before doing the hydrogenation, familiarize yourself with the mass spectra of cyclohexene and cyclohexane. Note that M^+ is different for each compound. After the reaction is underway, inject a sample from the reaction mixture into the mass spectrometer. If the reaction is finished, the mass spectrum of the reaction mixture should be superimposable with the mass spectrum of cyclohexane.

12.26 See Problem 12.9 for the method of solution.
a) 3360 cm^{-1} b) 1720 cm^{-1} c) 2030 cm^{-1}

12.27 a) 5.70 μm   b) 3.08 μm   c) 5.80 μm   d) 5.62 μm

12.28 $CH_3CH_2C{\equiv}CH$ shows absorptions at 2100-2260 cm^{-1} and at 3300 cm^{-1} that are due to the terminal alkyne bond.

$CH_2{=}CHCH{=}CH_2$ has absorptions in the regions 1650-1670 cm^{-1} and 3020-3100 that are due to the double bonds. No absorptions occur in the alkyne region.

$CH_3C{\equiv}CCH_3$. For reasons we won't discuss, symmetrically substituted alkynes such as 2–butyne do not show a C≡C bond absorption in the IR. This alkyne is distinguished from the other isomers in that it shows no absorptions in either the alkyne or alkene regions.

12.29 Two enantiomers have identical physical properties (other than the sign of specific rotation). Thus, their IR spectra are also identical.

12.30 Diastereomers have different physical properties and chemical behavior. Hence the IR spectra of two diastereomers are also different.

12.31 a) Absorptions at 3300 cm^{-1} and 2150 cm^{-1} are due to a terminal triple bond. Possible structures:

$$CH_3CH_2CH_2C{\equiv}CH \qquad\qquad (CH_3)_2CHC{\equiv}CH$$

b) An IR absorption at 3400 cm^{-1} is due to a hydroxyl group. Since no double bond absorption is present, the compound must be a cyclic alcohol.

c) An absorption at 1715 cm^{-1} is due to a ketone. The only possible structure is $CH_3CH_2COCH_3$.

d) Absorptions at 1600 cm^{-1} and 1500 cm^{-1} are due to an aromatic ring. Possible structures:

12.32 a) $HC{\equiv}CCH_2NH_2$
Alkyne absorptions at
3300 cm^{-1}, 2100-2260 cm^{-1}
Amine absorption at
3310-3500 cm^{-1}

$CH_3CH_2C{\equiv}N$
Nitrile absorption at
2210-2260 cm^{-1}

b) CH_3COCH_3
Strong ketone absorption
at 1715 cm^{-1}

CH_3CH_2CHO
Strong aldehyde absorption
at 1725 cm^{-1}

12.33 Spectrum (b) differs from spectrum (a) in several respects. Note in particular the absorbances at 715 cm^{-1} (strong), 1140 cm^{-1} (strong), 1650 cm^{-1} (medium), and 3000 cm^{-1} (medium) in spectrum (b). The absorbances at 1650 cm^{-1} (C=C stretch) and 3000 cm^{-1} (=C–H stretch) can be found in Table 12.1. They allow us to assign spectrum (b) to cyclohexene and spectrum (a) to cyclohexane.

12.34 a) $CH_3C{\equiv}CCH_3$ does not exhibit a terminal ≡C–H stretching vibration at 3300 cm^{-1}, as $CH_3CH_2C{\equiv}CH$ does.

b) $CH_3COCH{=}CHCH_3$, a conjugated unsaturated ketone, shows a strong ketone absorbance at 1690 cm^{-1}; $CH_3COCH_2CH{=}CH_2$, a nonconjugated ketone, shows a ketone absorption at 1710 cm^{-1}.

c) CH_3CH_2CHO exhibits an aldehyde band at 1725 cm^{-1}; $H_2C{=}CHOCH_3$ shows characteristic alkene absorbances, as well as a C–O stretch near 1200 cm^{-1}.

12.35 If the isotopic masses of the atoms C, H, and O had integral values of 12 amu, 1 amu and 16 amu, many molecular formulas would correspond to a molecular weight of 360 amu. Because isotopic masses are not integral, however, only one molecular formula is associated with a molecular ion at 360.1937 amu.

To reduce the number of possible formulas, assume that the difference in molecular weight between 360 and 360.1937 is due mainly to hydrogen. Divide 0.1937 by 0.00783, the amount by which the atomic weight of one ^{1}H atom differs from 1. The answer, 24.8, gives a "ballpark" estimate of the number of hydrogens in cortisone. Then make a list of molecular formulas containing C, H and O whose mass is 360 and which contain 20-30 hydrogens. Tabulate these, and calculate their exact masses using the values given in the text.

Isotopic mass x

Molecular formula	Mass of carbons	Mass of hydrogens	Mass of oxygens	Mass of molecular ion
$C_{27}H_{20}O$	324.0000 amu	20.1566 amu	15.9949 amu	360.1515 amu
$C_{25}H_{28}O_2$	300.0000	28.2192	31.9898	360.2090
$C_{24}H_{24}O_3$	288.0000	24.1879	47.9847	360.1726
$C_{21}H_{28}O_5$	252.0000	28.2192	79.9745	360.1937

The molecular weight of $C_{21}H_{28}O_5$ corresponds to the observed molecular weight of cortisone.

12.36

1–Methylcyclohexanol   1–Methylcyclohexene

The infrared spectrum of the starting alcohol shows a broad absorption at 3400-3640 cm^{-1}, due to an O–H stretch, and another strong absorption at 1050-1100 cm^{-1}, due to a C–O stretch. The alkene product exhibits medium intensity absorbances at 1645-1670 cm^{-1} and at 3000-3100 cm^{-1}. Monitoring the disappearance of the alcohol absorptions allows one to decide when the alcohol is gone. It is also possible to monitor the *appearance* of the alkene absorbances.

12.37

3–Bromo–3–methylpentane   3–Methyl–2–pentene   2–Ethyl–1–butene

The IR spectra of both products show the characteristic absorptions of alkenes in the regions 3020-3100 cm^{-1} and 1650 cm^{-1}. However, in the region 700-1000 cm^{-1}, 2–ethyl–1–butene shows a strong absorption at 890 cm^{-1} that is typical of

2,2–disubstituted $R_2C=CH_2$ alkenes. The presence or absence of this peak should help to identify the product. (3–Methyl–2–pentene is the major product of the dehydrobromination reaction.)

12.38

Compound	Distinguishing Absorption	Due to
a) $CH_3CH_2\overset{\overset{\displaystyle O}{\|}}{C}CH_3$	1715 cm^{-1}	$-\overset{\overset{\displaystyle O}{\|}}{C}-$ (ketone)
b) $(CH_3)_2CHCH_2C\equiv CH$	2140 cm^{-1} 3300 cm^{-1}	$-C\equiv C-$ $-C\equiv C-H$
c) $(CH_3)_2CHCH_2CH=CH_2$	910 cm^{-1} , 990 cm^{-1} 1650-1670 cm^{-1} 3020-3100 cm^{-1}	$R-CH=CH_2$ $-\overset{\|}{C}=\overset{\|}{C}-$ $=\overset{\|}{C}-H$
d) $CH_3CH_2CH_2\overset{\overset{\displaystyle O}{\|}}{C}OCH_3$	1735 cm^{-1}	$R-\overset{\overset{\displaystyle O}{\|}}{C}-OR$ (ester)
e) 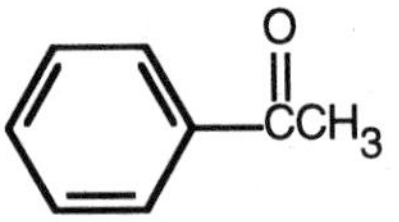 (phenyl)$-\overset{\overset{\displaystyle O}{\|}}{C}CH_3$	1690 cm^{-1}	$-\overset{\overset{\displaystyle O}{\|}}{C}-$ (ketone next to aromatic ring)

12.39 The following expressions are needed:

$E = h\nu = hc/\lambda = hc\,\tilde{\nu}$ where $\tilde{\nu}$ is the wavenumber. The last expression shows that, as $\tilde{\nu}$ increases, the energy needed to cause IR absorption increases, indicating greater bond strength. Thus, an ester C=O bond ($\tilde{\nu}$ = 1735 cm^{-1}) is stronger than a ketone C=O bond ($\tilde{\nu}$ = 1715 cm^{-1}).

12.40 Possible molecular formulas containing carbon, hydrogen, and oxygen and having M$^+$ = 150 are $C_{10}H_{14}O$, $C_9H_{10}O_2$, and $C_8H_6O_3$. The first formula has four degrees of unsaturation, the second has five degrees of unsaturation, and the third has six degrees of unsaturation. Since carvone has three double bonds (including the ketone) and one ring, $C_{10}H_{14}O$ is the correct molecular formula for carvone.

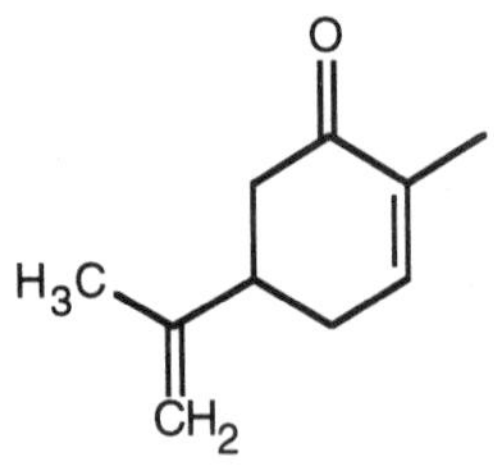

Carvone

12.41 The intense absorption at 1690 cm^{-1} is due to a carbonyl group next to a double bond.

12.42 The peak of maximum intensity (base peak) in the mass spectrum occurs at $m/z = 67$. This peak does *not* represent the molecular ion, however, because M^+ of a hydrocarbon must be an even number. Careful inspection reveals the molecular ion peak at $m/z = 68$. $M^+ = 68$ corresponds to a hydrocarbon of molecular formula C_5H_8 with a degree of unsaturation of two.

Fairly intense peaks in the mass spectrum occur at $m/z = 67, 53, 40, 39$, and 27. The peak at $m/z = 67$ corresponds to loss of one hydrogen atom, and the peak at $m/z = 53$ represents loss of a methyl group. The unknown hydrocarbon thus contains a methyl group.

Significant IR absorptions occur at 2130 cm^{-1} (–C≡C– stretch) and at 3320 cm^{-1} (≡C–H stretch). These bands indicate that the unknown hydrocarbon is a terminal alkyne. Possible structures for C_5H_8 are $CH_3CH_2CH_2C≡CH$ and $(CH_3)_2CHC≡CH$. [In fact, 1–pentyne is correct.]

12.43 The molecular ion, $M^+ = 70$, corresponds to the molecular formula C_5H_{10}. This compound has one double bond or ring.

The base peak in the mass spectrum occurs at $m/z = 55$. This peak represents loss of a methyl group from the molecular ion and indicates the presence of a methyl group in the unknown hydrocarbon. All other peaks occur with low intensity.

In the IR spectrum, it is possible to distinguish absorbances at 1660 cm^{-1} and at 3000 cm^{-1} due to a double bond. (The 2960 cm^{-1} absorption is rather hard to detect because it occurs as a shoulder on the alkane C–H stretch at 2850-2960 cm^{-1}.)

Since no absorptions occur in the region 890 cm^{-1} – 990 cm^{-1}, we can exclude terminal alkenes as possible structures. The remaining possibilities for C_5H_{10} are $CH_3CH_2CH=CHCH_3$ and $(CH_3)_2C=CHCH_3$. [2–Methyl–2–butene is correct.]

12.44

a) $CH_3CH_2\underset{\underset{CH_3}{|}}{C}H\overset{\overset{O}{||}}{C}H$

b) $H_3C\underset{\underset{CH_3}{|}}{\overset{\overset{CH_3}{|}}{C}}-C≡N$ $CH_3\underset{\underset{CH_3}{|}}{C}HCH_2C≡N$

$CH_3CH_2CH_2CH_2C≡N$

12.45

1. CH₃MgBr
2. H₃O⁺

The absorption at 3400 cm^{-1} is due to a hydroxyl group.

12.46

$CH_3CH_2\overset{\overset{O}{||}}{C}CH_3$ → (1. NaBH₄ 2. H₃O⁺) $CH_3CH_2\underset{\underset{}{\overset{\overset{OH}{|}}{}}}{C}HCH_3$

$M^+ = 74$

12.47

$$CH_3CH_2C{\equiv}N \xrightarrow[\text{heat}]{H_3O^+} CH_3CH_2\overset{\displaystyle O}{\overset{\|}{C}}OH$$

Study Guide for Chapter 12

After studying this chapter, you should be able to:

(1)   Write molecular formulas corresponding to a given molecular ion (12.1, 12.2, 12.14, 12.15, 12.16, 12.17, 12.18, 12.19, 12.20, 12.21, 12.40).

(2)   Use the natural abundance of isotopes to calculate molecular ions (12.3, 12.22, 12.23).

(3)   Use mass spectra to determine molecular weights and base peaks, and to distinguish between hydrocarbons (12.4, 12.5, 12.24, 12.35, 12.42, 12.43).

(4)   Calculate the energy of electromagnetic radiation (12.6, 12.7, 12.8, 12.39).

(5)   Convert from wavelength to wavenumber, and *vice versa* (12.9, 12.26, 12.27).

(6)   Identify functional groups by their infrared absorptions (12.10, 12.11, 12.12, 12.13, 12.28, 12.29, 12.30, 12.31, 12.32, 12.33, 12.34, 12.38, 12.41, 12.42, 12.43, 12.44).

(7)   Use IR and MS information to monitor reaction progress (12.25, 12.36, 12.37, 12.45, 12.46).

13.1

$$E = \frac{1.20 \times 10^{-2} \text{ kJ/mol}}{\lambda \text{ (in cm)}}$$

$$\lambda = \frac{c}{\upsilon} = \frac{3.0 \times 10^{10} \text{ cm/s}}{\upsilon}$$

here $\upsilon = 56$ MHz, or 5.6×10^7 Hz

so $\lambda = \dfrac{3.0 \times 10^{10} \text{cm/s}}{5.6 \times 10^7 \text{Hz}} = 0.54 \times 10^3 \text{ cm}$

$$E = \frac{1.20 \times 10^{-2} \text{kJ/mol}}{0.54 \times 10^3} = 2.2 \times 10^{-5} \text{kJ/mol}$$

Compare this value with $E = 2.4 \times 10^{-5}$ kJ/mol for ^{1}H. It takes less energy to spin-flip a ^{19}F nucleus than to spin-flip a ^{1}H nucleus.

13.2

$$\lambda = \frac{c}{\upsilon} = \frac{3.0 \times 10^{10} \text{ cm/s}}{\upsilon}$$

here $\upsilon = 100$ MHz $= 100 \times 10^6$ Hz, or 1.00×10^8 Hz

so $\lambda = \dfrac{3.0 \times 10^{10} \text{cm/s}}{1.00 \times 10^8 \text{ Hz}} = 3.0 \times 10^2 \text{cm}$

$$E = \frac{1.20 \times 10^{-2} \text{ kJ/mol}}{3.0 \times 10^2} = 4.0 \times 10^{-5} \text{kJ/mol}$$

Increasing the spectrometer frequency increases the amount of energy needed for resonance.

13.3

a)

This alkene has two different types of carbon and shows two signals in its ^{13}C NMR. Since all protons are equivalent, only one ^{1}H NMR signal appears.

b)   ¹H: ¹³C:

H_3C CH_3 ← a

H_2C CH_2 ← b

H_2C CH_2 ← c

CH_2 ← d

H_3C CH_3 ← a

← b

← c

← d

← e

At room temperature, 1,1-dimethylcyclohexane shows four ¹H NMR signals and five ¹³C NMR signals.

c)

$$CH_3\overset{\overset{\displaystyle O}{\|}}{C}CH_3$$

Acetone shows two ¹³C NMR signals and one ¹H NMR signal.

d) $(CH_3)_3C\overset{\overset{\displaystyle O}{\|}}{C}OCH_3$

↑   ↑↑ ↑

a   b c d

Four signals appear in the ¹³C NMR spectrum of this ester because four different kinds of carbon atoms are present. The ¹H NMR shows two signals.

e)

H_3C—⟨benzene ring⟩—CH_3

¹³C: Three signals
¹H: Two signals

f) H_3C CH_3

⟨cyclopropane⟩

¹³C: Three signals
¹H: Two signals

13.4

b H CH_3 a

C=C

c H   Cl

2–Chloropropene has three kinds of protons. Protons b and c differ because one is cis to the chlorine and the other is trans.

13.5

a)

$$\delta = \frac{\text{Observed chemical shift (\# Hz away from TMS)}}{\text{Spectrometer frequency (MHz)}}$$

δ = Parts per million. Here, δ = 2.1 ppm

$$2.1 \text{ ppm} = \frac{\text{Observed chemical shift}}{60 \text{ (MHz)}}$$

126 Hz = Observed chemical shift

b) If the ^{1}H NMR spectrum of acetone were recorded at 100 MHz, the position of absorption would still be 2.1 δ because measurements given in ppm or δ units are independent of the operating frequency of the NMR spectrometer.

c) $2.1 \delta = \dfrac{\text{Observed chemical shift}}{100 \text{ MHz}}$; observed chemical shift = 210 Hz

13.6

$$\delta = \frac{\text{Observed chemical shift (in Hz)}}{60 \text{ MHz}}$$

a) $\delta = \dfrac{436 \text{ Hz}}{60 \text{ MHz}} = 7.27 \, \delta$ for $CHCl_3$

b) $\delta = \dfrac{183 \text{ Hz}}{60 \text{ MHz}} = 3.05 \, \delta$ for CH_3Cl

c) $\delta = \dfrac{208 \text{ Hz}}{60 \text{ MHz}} = 3.47 \, \delta$ for CH_3OH

d) $\delta = \dfrac{318 \text{ Hz}}{60 \text{ MHz}} = 5.30 \, \delta$ for CH_2Cl_2

13.7

	Compound	Kinds of non-equivalent protons
a)	$\overset{1}{C}H_3\overset{2}{C}H_2Br$	Two
b)	$\overset{1}{C}H_3O\overset{2}{C}H_2\overset{3}{C}H{\overset{4}{C}H_3 / \overset{4}{C}H_3}$	Four
c)	$\overset{1}{C}H_3\overset{2}{C}H_2\overset{3}{C}H_2NO_2$	Three
d)		Four
e)		Five

The two protons attached to the double bond are non-equivalent.

f)

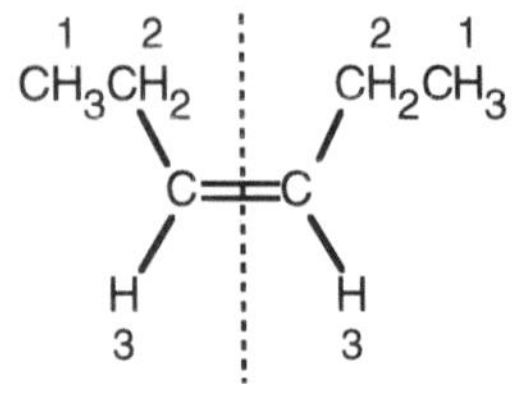

CH$_3$CH$_2$ (1 2)   CH$_2$CH$_3$ (2 1)

C=C

H (3)   H (3)

plane of symmetry

Three

13.8

Compound	δ	Kind of proton
a) Cyclohexane	1.43	secondary alkyl
b) CH$_3$COCH$_3$	2.17	methyl ketone
c) C$_6$H$_6$	7.37	aromatic
d) Glyoxal	9.70	aldehyde
e) CH$_2$Cl$_2$	5.30	protons adjacent to two halogens
f) (CH$_3$)$_3$N	2.12	methyl protons adjacent to nitrogen

13.9

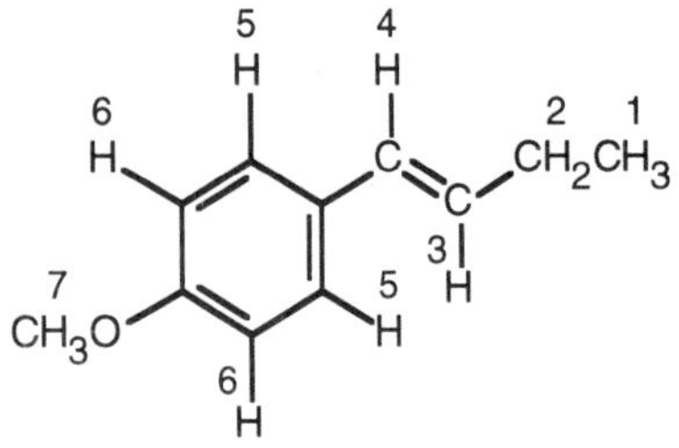

This compound has seven different kinds of protons.

Proton	δ	Kind of proton
1	1.0	primary alkyl
2	1.8	allylic
3	6.1	vinylic
4	6.2	vinylic
	(different from proton 3)	
5	7.2	aromatic
6	6.7	aromatic
7	3.8	ether

Note: The two "5" protons are equivalent to each other, as are the two "6" protons, because of free rotation around the bond joining the aromatic ring and the alkenyl side chain.

13.10

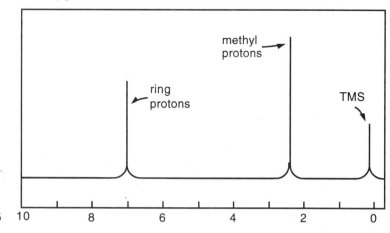

H₃C—⟨benzene ring⟩—CH₃   *p*–Xylene

There are two absorbances in the ^{1}H NMR spectrum of *p*–xylene. The four ring protons absorb at 7.0 δ, and the six methyl-group protons absorb at 2.3 δ. The peak ratio of methyl protons:ring protons is 3:2.

13.11

Compound	Proton	Number of Adjacent Protons	Splitting
a) $\overset{1}{C}HB\overset{2}{r_2}CH_3$	1	3	quartet
	2	1	doublet
b) $\overset{1}{C}H_3O\overset{2}{C}H_2\overset{3}{C}H_2Br$	1	0	singlet
	2	2	triplet
	3	2	triplet
c) $Cl\overset{1}{C}H_2\overset{2}{C}H_2\overset{1}{C}H_2Cl$	1	2	triplet
	2	4	quintet
d) $\overset{1}{C}H_3\overset{2}{C}H_2O\overset{\overset{\displaystyle O}{\|}}{\overset{3}{C}}\overset{4}{C}H(CH_3)_2$	1	2	triplet
	2	3	quartet
	3	6	septet
	4	1	doublet

13.12

a) CH_3OCH_3

b) $CH_3CHClCH_3$

c) $ClCH_2CH_2OCH_2CH_2Cl$

d) $CH_3CH_2\overset{\overset{\displaystyle O}{\|}}{C}OCH_3$

13.13 The ^{1}H NMR spectrum shows two signals, corresponding to two types of hydrogens in the ratio 33:50, or 2:3. Since the unknown contains 10 hydrogens, four protons are of one type and six are of the other type.

The upfield signal at 1.2 δ is due to saturated primary protons. The downfield signal at 3.5 δ is due to protons on carbon adjacent to an electronegative atom — in this case, oxygen.

The signal at 1.1 δ is a triplet, indicating two neighboring protons. The signal at 3.5 δ is a quartet, indicating three neighboring protons. The compound is diethyl ether, $CH_3CH_2OCH_2CH_3$.

13.14

3–Bromo–1–phenyl–1–propene

Coupling of the C2 proton to the Cl vinylic proton occurs with $J = 16$ Hz and causes the signal of the C2 proton to be split into a doublet. The C2 proton is also coupled to the two C3 protons with $J = 8$ Hz. This splitting causes each leg of the C2 proton doublet to be split into a triplet, producing six lines in all. Because of the size of the coupling constants, two of the lines coincide, and a quintet is observed.

$J_{1-2} = 16$ Hz

$J_{2-3} = 8$ Hz

13.15

HCl

or

Focus on the ^{1}H NMR methyl group absorption. In the first product, the methyl group signal is unsplit; in the other product, it appears as a doublet. In addition, the second product shows a downfield absorption in the 2.5 δ – 4.0 δ region due to the proton adjacent to an electronegative atom. If you were to take the ^{1}H NMR spectrum of the reaction product, you would find an unsplit methyl group, and you could conclude that the product was 1-chloro-1-methylcyclohexane.

13.16

$$\overset{4}{CH_3}\overset{3}{CH_2}\overset{2}{CO_2}\overset{1}{CH_3}$$

$\delta\,(ppm)$ *Assignment*

9.3	4
27.6	3
51.4	1
174.6	2

13.17

a)

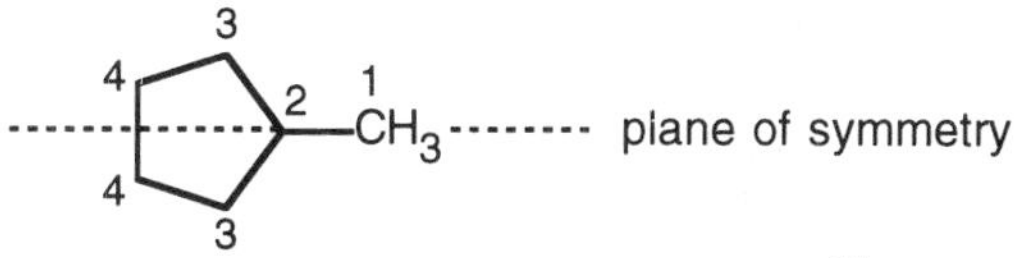

Four resonance lines are observed in the ^{13}C NMR spectrum of methylcyclopentane.

b)

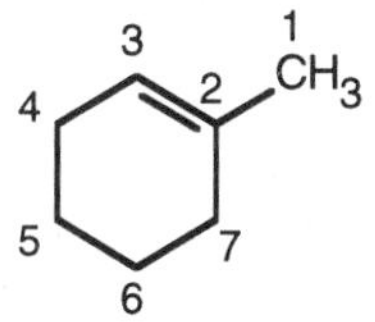

Seven resonance lines are seen. No two carbon atoms in 1–methylcyclohexene are equivalent because no plane of symmetry is present.

c)

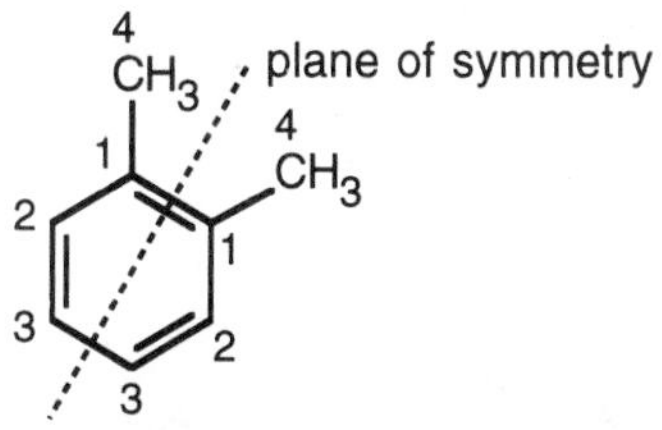

Four resonance lines are observed. A plane of symmetry causes one half of the carbon atoms to be equivalent to the other half.

d)

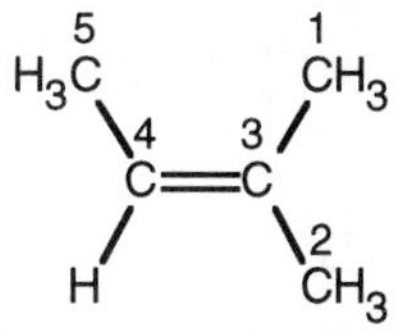

Five resonance lines are observed. Carbons 1 and 2 are nonequivalent because of the double bond stereochemistry.

13.18 Each part of this problem has several correct answers.

a) 1–Methylcyclohexene (see Problem 13.17b) and 1,3–dimethylcyclopentene show seven resonance lines.

b) H_3C \
    $CHCH_2CH_2CH_3$ Two of the six carbons are equivalent \
    H_3C

c) H_3C \
    $CHCH_2Cl$ The two methyl groups are equivalent. \
    H_3C

13.19

Carbon	Chemical Shift (δ)
1	18
2	68
3	24 (negative)
4	39 (negative)
5	124
6	132
7, 8	23, 26

13.20 Either of two products may result from radical addition of HBr to 2–methylpropene: $(CH_3)_2CHCH_2Br$ (non-Markovnikov product) and $(CH_3)_3CBr$ (Markovnikov product). ^{13}C NMR can easily distinguish between them.
$(CH_3)_2CHCH_2Br$ shows three peaks in its ^{13}C NMR spectrum, whereas $(CH_3)_3CBr$ shows only two. If you ran a ^{13}C NMR spectrum of the product, you would see that the spectrum has three absorptions, and you could identify the product as 1-bromo-2-methylpropane, the non-Markovnikov product.

13.21 See Problem 13.6 for the method of solution.

a) 2.18 δ b) 4.78 δ c) 7.52 δ d) 9.05 δ

13.22 See Problem 13.5 for the method of solution.

a) 168 Hz   b) 276 Hz   c) 504 Hz   d) 616 Hz

13.23 a) Since the symbol "δ" indicates ppm downfield from TMS, chloroform absorbs at 7.3 ppm.

b)
$$\delta = \frac{\text{Observed chemical shift (\# Hz from TMS)}}{\text{Spectrometer frequency (in MHz)}}$$

7.3 ppm $= \dfrac{\text{Chemical shift}}{360 \text{ MHz}}$; 7.3 ppm x 360 MHz = chemical shift

Chemical shift = 2600 Hz

c) δ is still 7.3 because the chemical shift measured in δ is independent of the operating frequency of the spectrometer.

13.24–13.25

Compound	Number of ^{13}C Absorptions	Carbons Showing Peaks in DEPT-135 ^{13}C NMR Spectrum		
		Positive Peaks	Negative Peaks	No Peaks
a) 1,1-dimethylcyclohexane (H₃C—C(CH₃), carbons 1, 2, 3, 4, 5)	5	carbon 1	carbons 3, 4, 5	carbon 2
b) $CH_3CH_2OCH_3$ (1, 2, 3)	3	carbons 1, 3	carbon 2	
c) 1-tert-butylcyclohexane (H_3C—C—CH_3 with CH_3, carbons 1,2,3,4,5,6)	6	carbons 1, 3	carbons 4, 5, 6	carbon 2
d) $CH_3CH_2CHC\equiv CH$ with CH_3 branch (carbons 6 5 2 1, CH₃=4, 3)	6	carbons 1, 3, 4, 6	carbon 5	carbon 2
e) 1,2-dimethylcyclohexane (H_3C, CH_3, carbons 1,2,3,4)	4	carbons 1, 2	carbons 3, 4	
f) cyclohexanone (=O, carbons 1,2,3,4)	4		carbons 2, 3, 4	carbon 1

13.26 ^{13}C NMR absorptions occur over a range of 250 ppm, while ^1H NMR absorptions generally occur over a range of only 10 ppm. The spread of peaks in ^{13}C NMR is therefore much greater, so accidental overlap is less likely. In addition, normal ^{13}C NMR spectra are uncomplicated by spin-splitting, and the total number of lines is smaller.

13.27 a) The *chemical shift* is the position at which a nucleus absorbs rf energy in an NMR spectrum.

b) *Spin-spin splitting* is the splitting of a single NMR resonance into multiple lines. Spin-spin splitting occurs when the effective magnetic field felt by a nucleus is influenced by the small magnetic moments of adjacent nuclei. In ^{1}H NMR, the signal of a proton with n equivalent neighboring protons is split into $n + 1$ peaks. The magnitude of spin-spin splitting is given by the coupling constant J.

c) The *applied magnetic field* is the magnetic field that is externally applied to a sample by an NMR spectrometer.

d) The *spectrometer operating frequency* is the frequency of applied rf energy used by the spectrometer to bring a magnetic nucleus into resonance. The rf energy required depends on the magnetic field strength and on the nature of the nucleus being observed.

e) If the NMR signal of nucleus $\underline{A}$ is split by the spin of adjacent nucleus $\underline{B}$, there is reciprocal splitting of the signal of nucleus $\underline{B}$ by the spin of nucleus $\underline{A}$. The spins of the two nuclei are said to be coupled. The distance between two individual peaks within the multiplet of $\underline{A}$ is the same as the distance between two individual peaks within the multiplet of $\underline{B}$. This distance, measured in Hz, is called the *coupling constant*.

f) The right side of an NMR spectrum is the *upfield* side; the left side is the *downfield* side. Nuclei that absorb upfield are more shielded and absorb at a higher applied magnetic field. Nuclei that absorb downfield are less shielded and absorb at a lower applied magnetic field.

13.28

Compound	Non-equivalent protons

a)   4

b) CH$_3$CH$_2$CH$_2$OCH$_3$ 4

c)   2

d)   6

e)

13.29

Lowest Chemical Shift ⟶ Highest Chemical Shift

CH_4 < Cyclohexane < CH_3COCH_3 < CH_2Cl_2, $H_2C=CH_2$ < benzene

0.23   1.43   2.17   5.30   5.33   7.37

13.30

Compound	Number of peaks	Peak Assignment	Splitting Pattern	
a) $(CH_3)_3CH$	2	1	doublet	(9H)
		2	multiplet (dectet)	(1H)
b) $CH_3CH_2COOCH_3$	3	1	triplet	(3H)
		2	quartet	(2H)
		3	singlet	(3H)
c)	2	1	doublet	(6H)
		2	quartet	(2H)

13.31

^{1}H: $CH_3CH_2COOCH(CH_3)_2$

Peak Assignment	Splitting Pattern	
1	triplet	(3H)
2	quartet	(2H)
3	septet	(1H)
4	doublet	(6H)

13.32 Use of ^{13}C NMR to distinguish between the two isomers has been described in the text. ^{1}H NMR can also be useful.

A   B

Isomer <u>A</u> has only four kinds of protons because of symmetry. Its vinylic proton absorption (4.5 – 6.5 δ) represents two hydrogens. Isomer <u>B</u> contains six different kinds of protons. Its ^{1}H NMR shows an unsplit methyl group signal and one vinylic proton signal of relative area 1. These differences make it possible to distinguish between <u>A</u> and <u>B</u>.

13.34 First, examine each isomer for structural differences that are obviously recognizable in the NMR spectrum. If it is not possible to pick out distinguishing features immediately, it may be necessary to sketch an approximate spectrum of each isomer for comparison.

a) $CH_3CH=CHCH_2CH_3$ has two vinylic protons with chemical shifts at $5.4 - 5.5$ δ. Because ethylcyclopropane shows no signal in this region, it should be easy to distinguish one isomer from the other.

b) $CH_3CH_2OCH_2CH_3$ has two kinds of protons, so its ^1H NMR spectrum consists of two resonances — a triplet and a quartet. $CH_3OCH_2CH_2CH_3$ has four different types of protons, and its spectrum is more complex. In particular, the methyl group bonded to oxygen shows an unsplit singlet absorption.

c) Each compound shows three peaks in its ^1H NMR spectrum. The ester, however, shows a downfield absorption due to the $-CH_2-$ hydrogens next to oxygen. No comparable peak shows in the spectrum of the ketone.

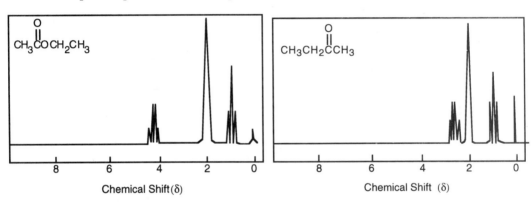

d) Each isomer contains four different kinds of protons — two kinds of methyl protons and two kinds of vinylic protons. For the first isomer, the methyl resonances are both singlets, whereas for the second isomer, one resonance is a singlet and one is a doublet.

13.34

a) $(CH_3)_4C$

b) ⬠

c) (structure with two O atoms, six-membered ring)

13.35 a,b) C_3H_6O contains one double bond or ring. Possible structures for C_3H_6O include:

H_2C-CH_2
H_2C-O
Cyclic ether

$H_2C-CHCH_3$ (with O bridging)
Cyclic ether

$H_2C=CHOCH_3$
Ether, double bond

$H_2C=CHCH_2OH$
Alcohol, double bond

H_2C / H_2C — C with H and OH
Cyclic alcohol

CH_3CCH_3 (with =O)
Ketone (acetone)

CH_3CH_2CH (with =O)
Aldehyde

c) Carbonyl functional groups (usually ketones) absorb at 1715 cm^{-1} in the infrared. Only the last two compounds above show an infrared absorption in this region.

d) Because the aldehyde from part b) has three different kinds of protons, its ^{1}H NMR spectrum has three resonances. The ketone, however, shows only one resonance. Since the unknown compound of this problem shows only one ^{1}H NMR absorption (in the methyl ketone region), it must be acetone.

13.36 Either ^{1}H NMR or ^{13}C NMR can be used to distinguish among these isomers. In either case, it is first necessary to find the number of different kinds of protons or carbon atoms.

Compound	H₂C—CH₂ / H₂C—CH₂	H₂C=CHCH₂CH₃	CH₃CH=CHCH₃	(CH₃)₂C=CH₂
Kinds of protons	1	5	2	2
Kinds of carbon atoms	1	4	2	3
Number of ^{1}H NMR peaks	1	5	2	2
Number of ^{13}C NMR peaks	1	4	2	3

^{13}C NMR is the simplest method for identifying these compounds because each isomer differs in the number of absorptions in its ^{13}C NMR spectrum. ^{1}H NMR can also be used to distinguish among the isomers because the two isomers that show two ^{1}H NMR peaks differ in their splitting patterns.

13.37

	Number of peaks	Distinguishing Absorptions
^{1}H	5	Unsplit vinylic peak, relative area 1
^{13}C	7	Two vinylic peaks
^{1}H	4	Split vinylic peak, relative area 2
^{13}C	5	One vinylic peak

The two isomers have different numbers of peaks in both ^{1}H NMR and ^{13}C NMR. In addition, the distinguishing absorptions in the vinylic region of both the ^{1}H and ^{13}C spectra make it possible to identify each isomer by its NMR spectrum.

13.38 The ketone IR absorption of 3–methyl–2–cyclohexenone occurs near 1690 cm^{-1} because the double bond is next to the ketone group. The ketone IR absorption of 4–cyclopentenyl methyl ketone occurs near 1715 cm^{-1}, the usual position for ketone absorption.

13.39 BrCH₂CH₂CH₂Br

13.40

a) $(CH_3)_2CHCCH_3$
   (with O double-bonded to the carbonyl carbon)

b)
$$\begin{array}{ccc} CH_3 & & H \\ & C=C & \\ Br & & H \end{array}$$

13.41 Possible structures for $C_4H_7ClO_2$ are $CH_3CH_2CO_2CH_2Cl$ and $ClCH_2CO_2CH_2CH_3$. Chemical-shift data can distinguish between them.

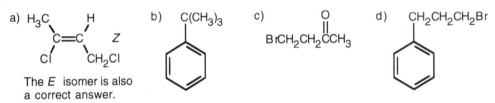

CH₃CH₂COCH₂Cl   ClCH₂COCH₂CH₃
        I   II

In I, the protons attached to the carbon bonded to both oxygen and chlorine (–OCH₂Cl) absorb far downfield (5.0 – 6.0 δ). Because no signal is present in this region of the ¹H NMR spectrum given, the unknown must be II.

13.42

a)
$$\begin{array}{ccc} H_3C & & H \\ & C=C & \\ Cl & & CH_2Cl \end{array}\ Z$$

The *E* isomer is also a correct answer.

b)   C(CH₃)₃

c)   BrCH₂CH₂CCH₃
   (with O double-bonded to the carbonyl carbon)

d)   CH₂CH₂CH₂Br

13.43 Let's try to differentiate between the isomers using both ¹H NMR and ¹³C NMR.

A   B

¹H NMR. Isomer <u>A</u> has two kinds of protons — those in four-membered rings and those in the eight-membered ring. Each group of eight protons absorbs in the allylic region of the spectrum (C=C–C–H; 1.5 - 2.5 δ). Two kinds of protons are also present in isomer <u>B</u>, but you will have to build a model to see the difference between them. One group of eight protons points toward a double bond; the other group of eight protons points away from a double bond. Both groups of protons are also allylic and absorb in the region of 1.5 - 2.5 δ. ¹H NMR thus cannot be used to distinguish between <u>A</u> and <u>B</u>.

^{13}C NMR. Isomer $\underline{A}$ has three different kinds of carbon atoms:

Carbon atom		Quantity	Chemical shift
=C̣–		four	100–150 δ
–CH$_2$–	(in 4-membered rings)	four	15–55 δ
–CH$_2$–	(in 8-membered rings)	four	15–55 δ

Isomer $\underline{B}$ contains only two different kinds of carbon atoms:

	Quantity	Chemical shift
=C̣–	four	100–150 δ
–CH$_2$–	eight	15–55 δ

Three absorptions should appear in the ^{13}C NMR spectrum of $\underline{A}$, but only two resonances should appear in the ^{13}C NMR spectrum of $\underline{B}$. Although ^1H NMR cannot distinguish between $\underline{A}$ and $\underline{B}$, ^{13}C NMR will solve the problem.

13.44

a) $(CH_3)_2CHCH_2Br$

b) $CH_3CHCH_2CH_2Cl$
            |
            Cl

13.45

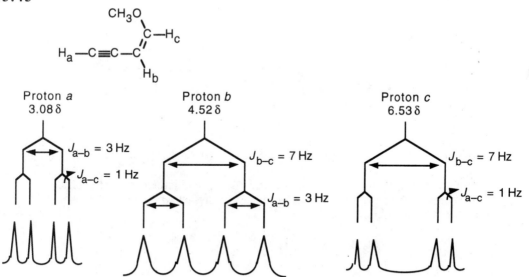

The absorptions in the ^1H NMR spectrum can be identified by comparison with the tree diagrams. H_a absorbs at 3.08 δ, H_b absorbs at 4.52 δ, and H_c absorbs at 6.35 δ.

13.46

Carbon	δ (ppm)
1	14
2	61
3	166
4	
5	
6	127–133 (four absorptions)
7	

Ethyl benzoate

13.47 Compound $\underline{A}$ (4 multiple bonds and/or rings) must be symmetrical because it exhibits only six peaks in its ^{13}C NMR spectrum. Saturated carbons account for two of these peaks (δ = 15, 28 ppm), and unsaturated carbons account for the other four (δ = 119, 129, 131, 143 ppm).

^{1}H NMR shows a triplet (3 H at 1.1 δ), and a quartet (2 H at 2.5 δ), indicating the presence of an ethyl group. The other signals (4 H at 6.9 – 7.3 δ are due to aromatic protons.

Compound $\underline{A}$

13.48

a)

$CH_3CH_2CH_2\overset{\overset{\displaystyle O}{\|}}{C}CH_3$

b)

c)

13.49 The peak in the mass spectrum at m/z = 84 is probably the molecular ion of the unknown compound and corresponds to a molecular weight of 84 (C_6H_{12} — one double bond or ring).

^{13}C NMR shows three different kinds of carbons and indicates a symmetrical hydrocarbon. The absorption at 132 δ is due to a vinylic carbon atom. A reasonable structure for the unknown is:

$$CH_3CH_2CH=CHCH_2CH_3$$

3–Hexene

13.50 Compound A, a hydrocarbon having M^+ = 96, has the formula C_7H_{12}, indicating two degrees of unsaturation. From reaction with BH_3, we can see that Compound A contains a double bond. From the broadband decoupled ^{13}C NMR spectrum, we can see that C_7H_{12} is symmetrical, since it shows only five peaks.

The DEPT-135 spectrum of Compound A indicates three different $-CH_2-$carbons, one $=CH_2$ carbon and one $-C=$ carbon; the last two carbons are shown to be sp^2 hybridized by their chemical shifts. In the DEPT-135 spectrum of Compound B, the absorptions due to double bond carbons have been replaced by a C-H carbon and a $-CH_2-$ carbon bonded to an electronegative group.

$$\underset{\text{Compound A}}{\text{(structure)}} \xrightarrow[\text{2. } H_2O_2,\ OH^-]{\text{1. } BH_3,\ THF} \underset{\text{Compound B}}{\text{(structure)}}$$

Compound A		Compound B	
1.	106.9 δ	1.	68.2 δ
2.	149.7 δ	2.	40.5 δ
3.	35.7 δ	3.	29.9 δ
4,5.	26.8 δ, 28.7 δ	4,5.	26.1 δ, 26.9 δ

13.51

$$HOCH_2CH_2CH_2CH=CH_2$$

13.52

$$\overset{\displaystyle OH}{\underset{\displaystyle |}{CH_3CH_2CHCH=CH_2}}$$

13.53

$$H_2C=CH\overset{O}{\overset{\|}{C}}-OCH_2\overset{CH_3}{\overset{|}{C}HCH_3}$$

13.54

Compound F   Compound G

13.55 Make a model of one enantiomer of 3-methyl-2-butanol and orient it as a staggered Newman projection along the C2-C3 bond. The *S* enantiomer is pictured.

Because of the stereogenic center at C2, the two methyl groups aren't equivalent. Prove it to yourself: no rotation can place the two methyl groups on C3 in an identical environment. Since the methyl groups aren't equivalent, their carbons show slightly different signals in the ^{13}C NMR.

13.56 Commercial 2,4-pentanediol is a mixture of three stereoisomers - (*R,R*), (*S,S*), and (*R,S*). The *meso* isomer shows three signals in its ^{13}C NMR spectrum. Its diastereomers, the *R,R* and *S,S* enantiomeric pair, also show three signals, but two of these signals occur at different δ values from the *meso* isomer. This is expected, because diastereomers differ in physical and chemical properties.

13.57

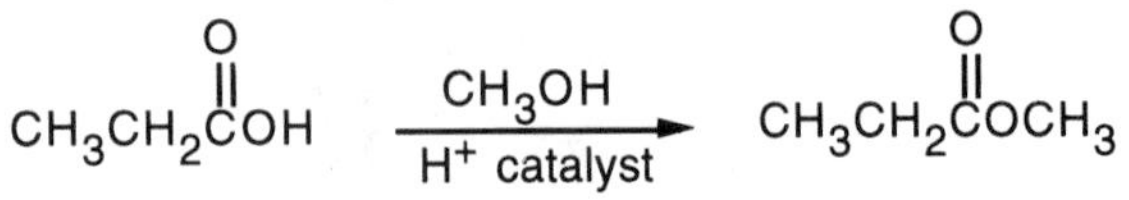

$$CH_3CH_2\overset{\overset{\displaystyle O}{\|}}{C}OH \quad \xrightarrow[\text{H}^+ \text{ catalyst}]{CH_3OH} \quad CH_3CH_2\overset{\overset{\displaystyle O}{\|}}{C}OCH_3$$

13.58

$$\underset{\overset{\displaystyle |}{CH_3}}{CH_3CHC\equiv N} \quad \xrightarrow[\text{2. H}_3O^+]{\text{1. CH}_3\text{MgBr}} \quad \underset{\overset{\displaystyle |}{CH_3}}{CH_3CH\overset{\overset{\displaystyle O}{\|}}{C}CH_3}$$

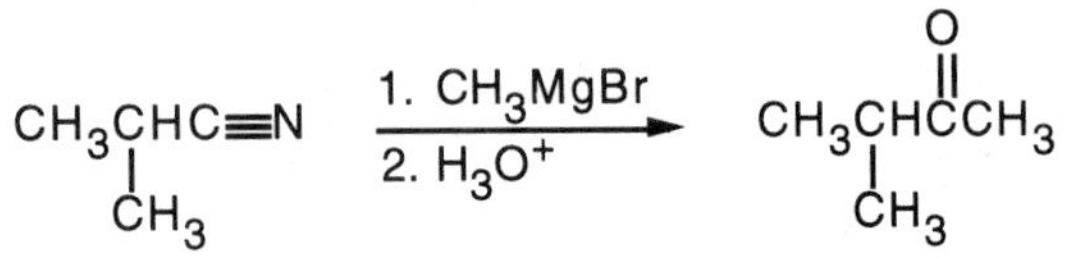

Study Guide for Chapter 13

After studying this chapter, you should be able to:

(1)   Understand the principle of NMR and be able to define important terms (13.1, 13.2, 13.26, 13.27).

(2)   Calculate the relationship between delta value, chemical shift and spectrometer operating frequency (13.5, 13.6, 13.21, 13.22, 13.23).

(3)   Identify non-equivalent carbons and hydrogens, and predict the number of signals appearing in the ^{1}H NMR and ^{13}C NMR spectra of compounds (13.3, 13.4, 13.7, 13.17, 13.24, 13.25, 13.28).

(4)   Assign resonances to specific carbons or hydrogens of a given structure (13.16, 13.29, 13.46, 13.47).

(5)   Propose structures for compounds, given their NMR spectrum (13.12, 13.13, 13.18, 13.34, 13.39, 13.40, 13.41, 13.42, 13.44, 13.47, 13.48, 13.49).

(6)   Describe or sketch spectra corresponding to a given compound (13.8, 13.9, 13.10).

(7)   Predict ^{1}H splitting patterns, using tree diagrams if necessary (13.11, 13.14, 13.30, 13.31, 13.45).

(8)   Use NMR to distinguish between isomers (13.33, 13.35, 13.36, 13.37, 13.43).

(9)   Use NMR to identify reaction products (13.15, 13.20, 13.32, 13.54, 13.57, 13.58).

(10)   Use DEPT-NMR to identify compounds (13.19, 13.50, 13.51, 13.52, 13.53).

14.1

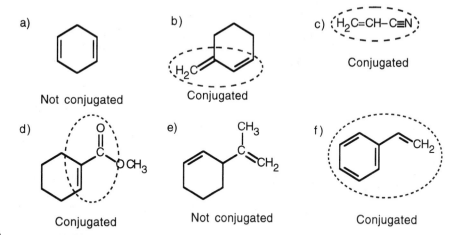

a) Not conjugated

b) H_2C Conjugated

c) $H_2C=CH-C\equiv N$ Conjugated

d) Conjugated

e) Not conjugated

f) Conjugated

14.2

We would expect $\Delta H_{hydrog} = -126 + (-126) = -252$ kJ/mol for allene if the heat of hydrogenation for each double bond were the same as that for an isolated double bond. The measured ΔH_{hydrog}, -298 kJ/mol, is 46 kJ/mol more negative than the expected value. Thus, allene is higher in energy (less stable) than a nonconjugated diene, which in turn is less stable than a conjugated diene.

14.3

$CH_3CH=CHCH=CH_2$ 1,3–Pentadiene

Product	Name	Results from:
(1) $CH_3CH=CHCHClCH_3$	4–Chloro–2–pentene	1,2 Addition 1,4 Addition
(2) $CH_3CH_2CHClCH=CH_2$	3–Chloro–1–pentene	1,2 Addition
(3) $CH_3CH_2CH=CHCH_2Cl$	1–Chloro–2–pentene	1,4 addition

14.4

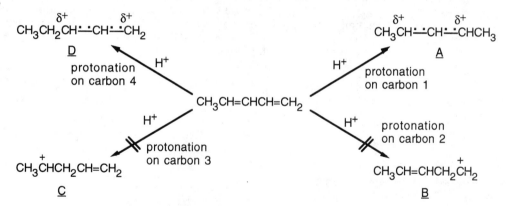

$$CH_3CH_2CH\overset{\delta+}{\cdots}CH\overset{\delta+}{\cdots}CH_2$$
D
protonation on carbon 4

$$CH_3CH\overset{\delta+}{\cdots}CH\overset{\delta+}{\cdots}CHCH_3$$
A
protonation on carbon 1

$CH_3CH=CHCH=CH_2$

H^+ protonation on carbon 3

H^+ protonation on carbon 2

$CH_3\overset{+}{C}HCH_2CH=CH_2$
C

$CH_3CH=CHCH_2\overset{+}{C}H_2$
B

The positive charge of allylic carbocation $\underline{A}$ is delocalized over two secondary carbons, while the positive charge of carbocation $\underline{D}$ is delocalized over one secondary and one primary carbon. We therefore predict that carbocation $\underline{A}$ is the major intermediate formed, and that 4-chloro-2-pentene predominates. Note that this product results from both 1,2– and 1,4–addition.

14.5　Figure 14.3 shows the three pi molecular orbitals of an allylic pi system. An allyl radical has three pi electrons. Two of them occupy the bonding molecular orbital, and the third electron occupies the nonbonding orbital.

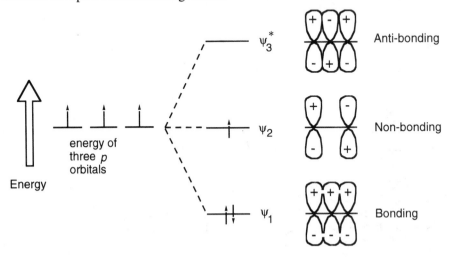

14.6

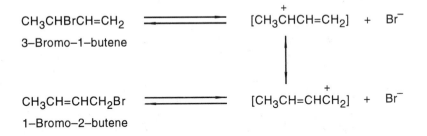

Allylic halides can undergo slow dissociation to form stabilized carbocations. Both 3–bromo–1–butene and 1–bromo–2–butene form the same allylic carbocation, pictured above, on dissociation. Addition of bromide ion to the allylic carbocation then occurs to form a mixture of bromobutenes. Since the reaction is run under equilibrium conditions, the thermodynamically more stable 1–bromo–2–butene predominates.

14.7

monosubstituted double bond

disubstituted double bond

1,4–adducts are more stable than 1,2–adducts because disubstituted double bonds are more stable than monosubstituted double bonds (see Chapter 6).

14.8

14.9

14.10

Good Dienophiles: (a) $H_2C=CHCCl$, with O double bonded   (d)

Poor Dienophiles: (b) $H_2C=CHCH_2CH_2COOCH_3$,   (c)   ,   (e)

Compound (a) and (d) are good dienophiles because they have electron-withdrawing groups conjugated with a double bond. Alkene (c) is a poor dienophile because it has no electron-withdrawing functional group. Compounds (b) and (e) are poor dienophiles because their electron-withdrawing groups are not conjugated with the double bond.

14.11 a) This diene has an s-cis conformation and should undergo Diels-Alder cycloaddition.

     b) This diene has an s-trans conformation. Because the double bonds are in a fused ring system, it is not possible for them to rotate to an s-cis conformation.

c) Rotation can also occur about the single bond of this *s*-trans diene. The resulting *s*-cis conformation, however, has an unfavorable steric interaction of the interior methyl group with a hydrogen at carbon 1. Rotation to the *s*-cis conformation is therefore not favored energetically.

s–trans
(more stable)

s–cis
(less stable)

14.12

$$200 \text{ nm} = 200 \times 10^{-7} \text{ cm} = 2 \times 10^{-5} \text{ cm}$$
$$400 \text{ nm} = 400 \times 10^{-7} \text{ cm} = 4 \times 10^{-5} \text{ cm}$$

$$E = \frac{1.20 \times 10^{-2} \text{ kJ/mol}}{\lambda \text{ (in cm)}}$$

for $\lambda = 2 \times 10^{-5}$ cm

$$E = \frac{1.20 \times 10^{-2} \text{ kJ/mol}}{2 \times 10^{-5}} = 6 \times 10^2 \text{ kJ/mol}$$

for $\lambda = 4 \times 10^{-5}$ cm

$$E = \frac{1.20 \times 10^{-2} \text{ kJ/mol}}{4 \times 10^{-5}} = 3 \times 10^2 \text{ kJ/mol}$$

The energy of electromagnetic radiation occurs over the range of 300–600 kJ/mol.

14.13

Energy (in kJ/mol)	UV	IR	^{1}H NMR (at 60 MHz)
	300 – 600	4.7 – 47	2.4×10^{-5}

The energy required for UV transitions is greater than the energy required for IR or ^{1}H NMR transitions.

14.14

$$\varepsilon = \frac{A}{C \cdot l}$$

Where ε = molar absorptivity
A = absorbance
l = sample path length (in cm)
C = concentration (in M)

In this problem

$$\varepsilon = 50{,}100 = 5.01 \times 10^4$$
$$l = 1.00 \text{ cm}$$
$$A = 0.735$$

$$C = \frac{A}{\varepsilon \times l} = \frac{0.735}{(5.01 \times 10^4)(1.00)} = 1.47 \times 10^{-5} \text{ M}$$

14.15 All compounds having alternating single and multiple bonds should show ultraviolet absorption in the range 200–400 nm. Only compound (a) is not UV-active. All of the compounds pictured below are UV active.

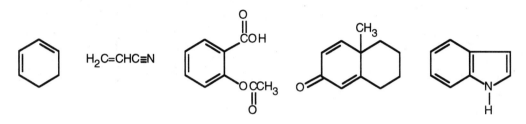

14.16 a) 3–Methyl–2,4–hexadiene   b) 1,3,5–Heptatriene

c) 2,3,5–Heptatriene   d) 3–Propyl–1,3–pentadiene

14.17

a) 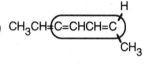   b) 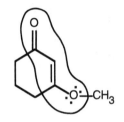   c)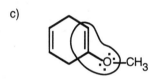

d) 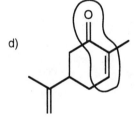   e) 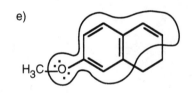   f)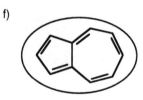

14.18

a)

$\xrightarrow[\text{CCl}_4]{\text{1 mole Br}_2}$

b)

$\xrightarrow[\text{2. Zn, H}_3\text{O}^+]{\text{1. O}_3}$

c)

$\xrightarrow{\text{1 mole HCl}}$

d)

1 mole DCl

e)

f)

1. OsO$_4$
2. NaHSO$_3$

enantiomers

14.19

Conjugated Dienes:	CH$_3$CH=CHCH=CH$_2$	H$_2$C=CHĊ=CH$_2$
	1,3–Pentadiene	2–Methyl–1,3–butadiene

Conjugated Dienes: CH$_3$CH=CHCH=CH$_2$ H$_2$C=CHC=CH$_2$ (with CH$_3$)

1,3–Pentadiene   2–Methyl–1,3–butadiene

Cumulated Dienes: CH$_3$CH$_2$CH=C=CH$_2$ CH$_3$CH=C=CHCH$_3$

1,2–Pentadiene   2,3–Pentadiene

H$_2$C=C=C(CH$_3$)$_2$

3–Methyl–1,2–butadiene

Non-conjugated Diene: H$_2$C=CHCH$_2$CH=CH$_2$

1,4–Pentadiene

14.20

CH$_3$CH$_2$C≡CCH$_2$CH$_3$ CH$_3$CH=CHCH=CHCH$_3$ CH$_3$CH$_2$CH=C=CHCH$_3$

3–Hexyne   2,4–Hexadiene   2,3–Hexadiene

^{1}H:	2 peaks triplet, quartet below 2.0 δ	3 peaks two in region 4.5–6.5 δ	5 peaks two in region 4.5–6.5 δ
^{13}C:	3 peaks 8–55 δ : 2 65–85 δ : 1	3 peaks 8–30 δ : 1 100–150 δ : 2	6 peaks 8–55 δ : 3 100–150 δ : 2 ~ 200 δ : 1 (sp carbon)
UV Absorp-tion?	No	Yes	No

2,4–Hexadiene can easily be distinguished from the other two isomers because it is the only isomer that absorbs in the UV region. The other two isomers show significant differences in their ^{1}H and ^{13}C NMR spectra and can be identified by either technique.

14.21

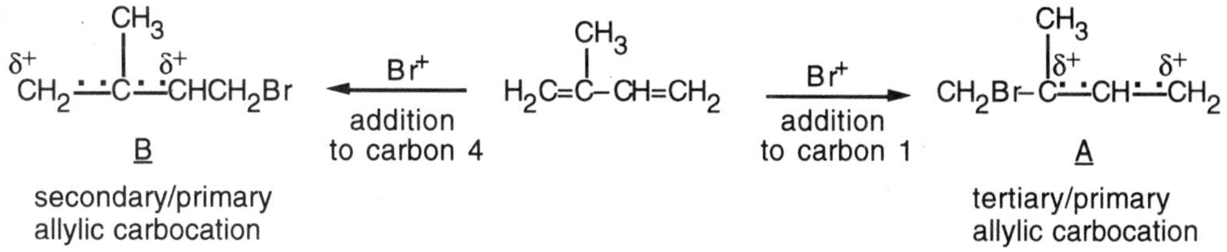

Tertiary/primary allylic carbocation A is more stable than secondary/primary allylic carbocation B. Since the products formed from the more stable intermediate predominate, 3,4–dibromo–3–methyl– 1–butene is the major product of 1,2 addition of bromine to isoprene.

14.22 A vinyl branch in a diene polymer is the result of an occasional 1,2–double bond addition to the polymer chain.

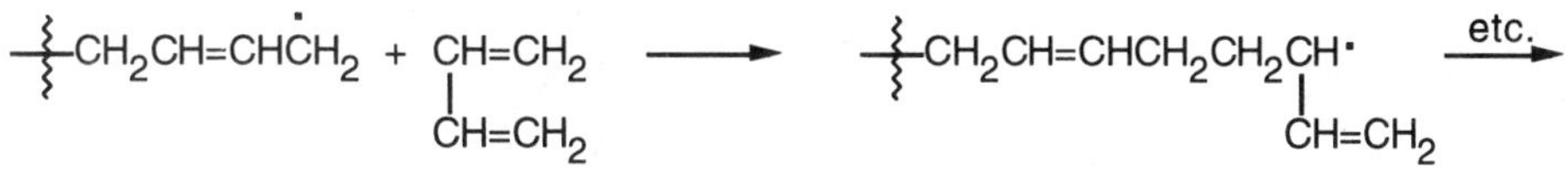

14.23

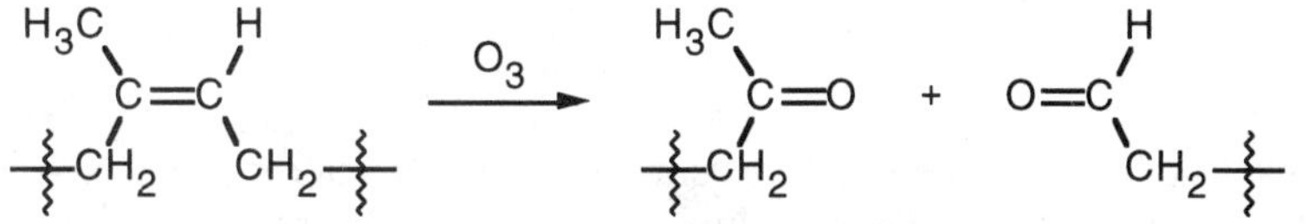

Ozone causes oxidative cleavage of the double bonds in rubber and breaks the polymer chain.

14.24 To absorb in the 200–400 nm range, an alkene must be conjugated. Since the double bonds of allene are not conjugated, allene does not absorb light in the UV region.

14.25

a)

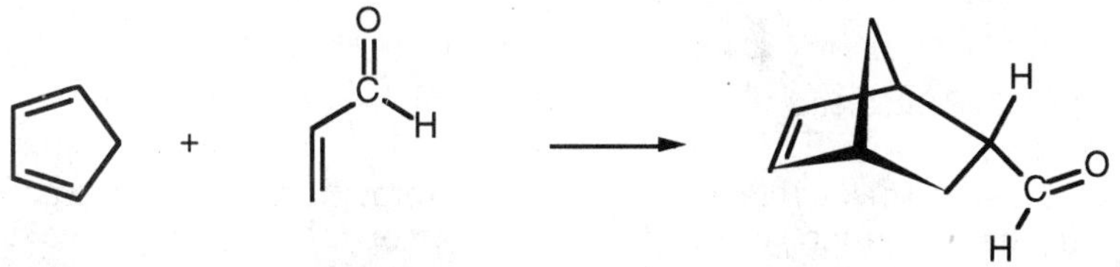

b)

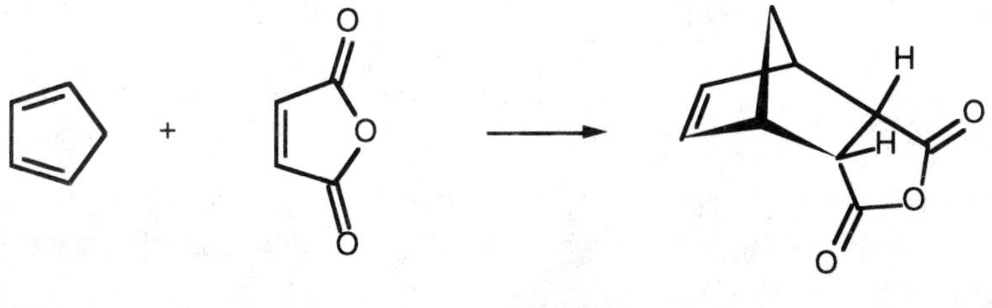

c)

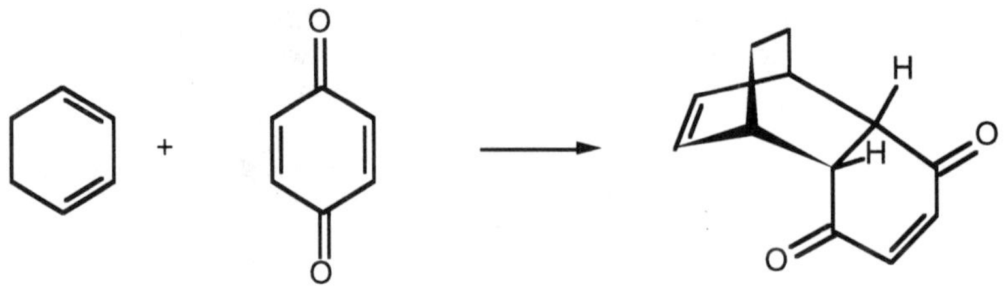

If two moles of cyclohexadiene are present for each mole of dienophile, you can also obtain a second product:

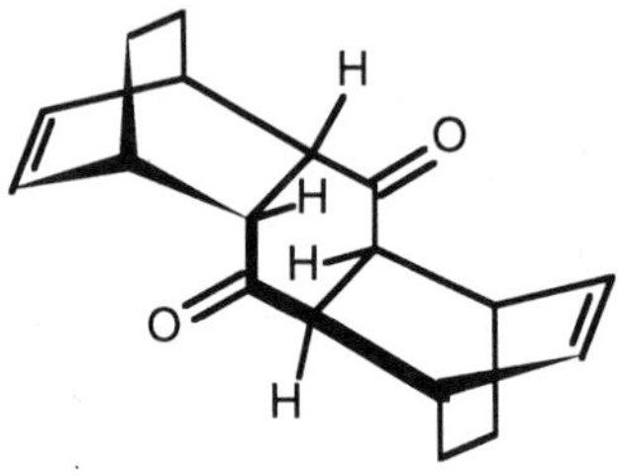

14.26

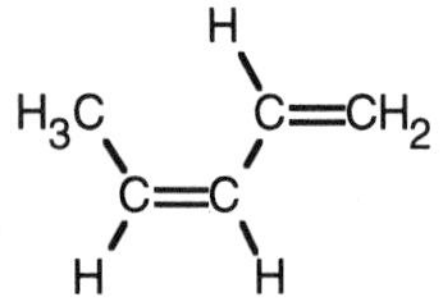

cis–1,3–Pentadiene 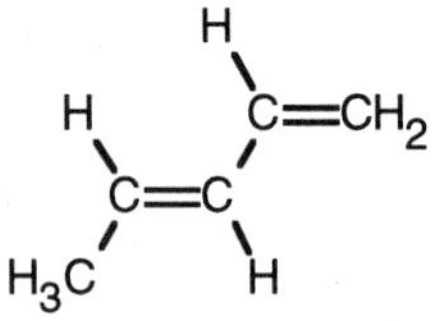   trans–1,3–Pentadiene

Both pentadienes are more stable in *s*–trans conformations. To undergo Diels-Alder reactions, however, they must rotate about the single bond between the double bonds to assume *s*–cis conformations.

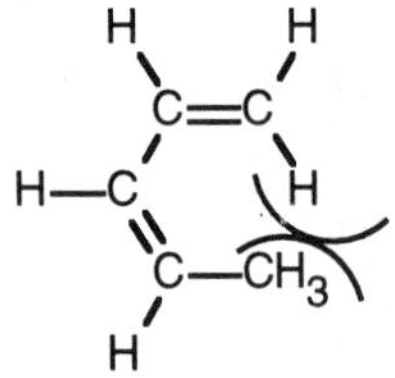

cis–1,3–Pentadiene 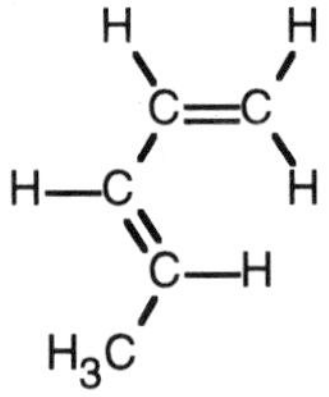   trans–1,3–Pentadiene

When *cis*–1,3–pentadiene rotates to the *s*–cis conformation, a steric interaction occurs between the methyl-group protons and a hydrogen on C1. Since it's more difficult for *cis*–1,3–pentadiene to assume the *s*–cis conformation, it is less reactive in the Diels-Alder reaction.

14.27 Only compounds having alternating multiple bonds show $\pi \rightarrow \pi^*$ ultraviolet absorptions in the 200–400 nm range. Of the compounds shown, only pyridine (b) absorbs in this range.

14.28 HC≡CC≡CH can't be used as a Diels-Alder diene because it is linear. The end carbons are too far apart to be able to react with a dienophile in a cyclic transition state.

14.29 Among the possible structures:

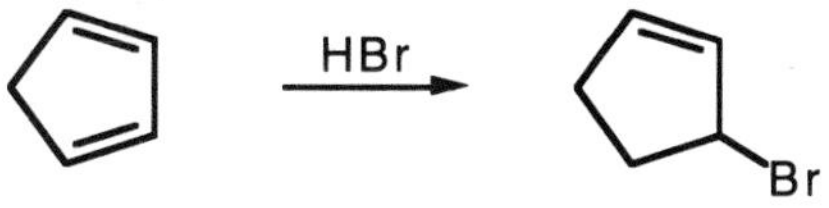

(cyclopentadiene) $\xrightarrow{\text{HBr}}$ (3-bromocyclopentene)

14.30 Protonation on carbon 1:

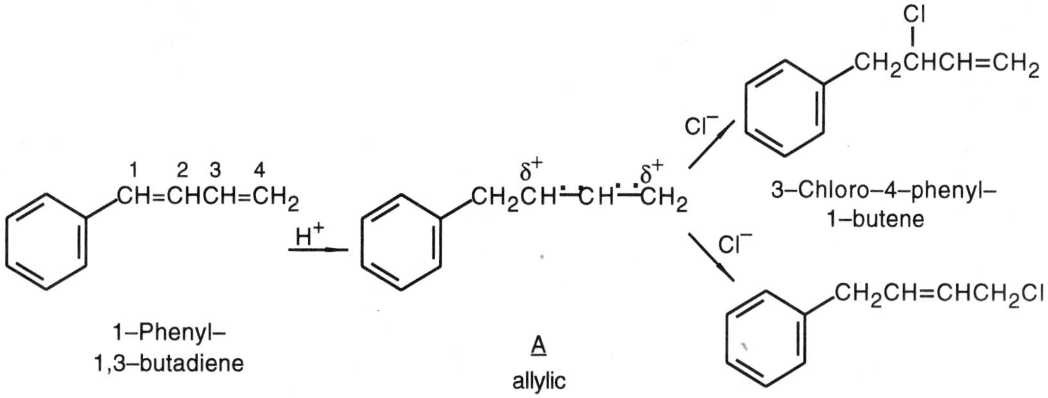

1   2   3   4
$CH=CHCH=CH_2$ (with phenyl) $\xrightarrow{H^+}$ $CH_2\overset{\delta^+}{CH}\text{---}CH\text{---}\overset{\delta^+}{CH_2}$ (with phenyl)

1–Phenyl–
1,3–butadiene

A
allylic

$\overset{Cl}{|}$
$CH_2CHCH=CH_2$ (with phenyl)

Cl^-

3–Chloro–4–phenyl–
1–butene

$CH_2CH=CHCH_2Cl$ (with phenyl)

Cl^-

1–Chloro–4–phenyl–
2–butene

Protonation on carbon 2:

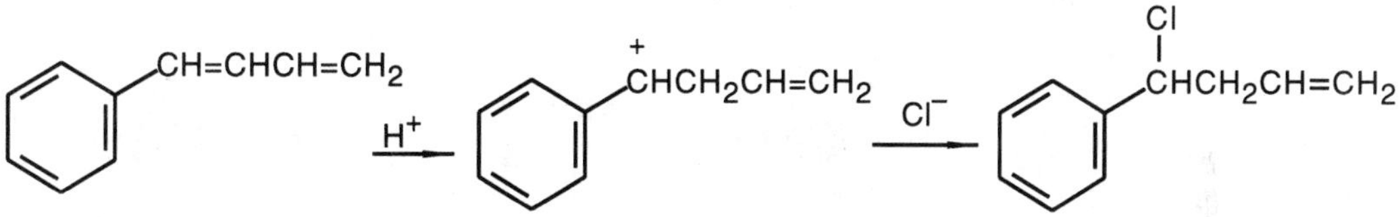

$CH=CHCH=CH_2$ (with phenyl) $\xrightarrow{H^+}$ $\overset{+}{CH}CH_2CH=CH_2$ (with phenyl)

B

Cl^-

$\overset{Cl}{|}$
$CHCH_2CH=CH_2$ (with phenyl)

4–Chloro–4–phenyl–
1–butene

Protonation on carbon 3:

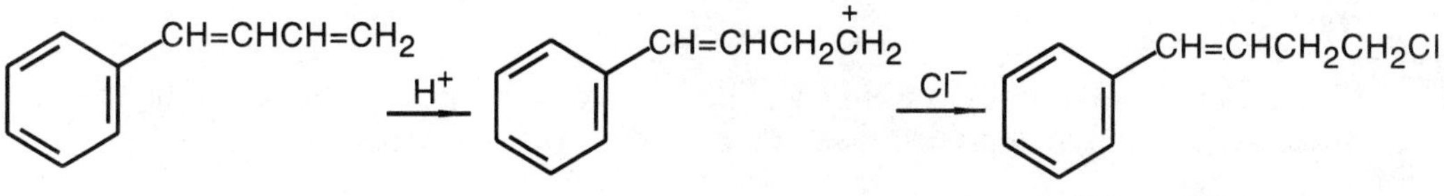

$CH=CHCH=CH_2$ (with phenyl) $\xrightarrow{H^+}$ $CH=CHCH_2\overset{+}{CH_2}$ (with phenyl)

C

Cl^-

$CH=CHCH_2CH_2Cl$ (with phenyl)

4–Chloro–1–phenyl–
1–butene

Protonation on carbon 4:

3–Chloro–1–phenyl–1–butene

1–Chloro–1–phenyl–2–butene

D
allylic

Carbocation D is most stable because it can use the pi systems of both the benzene ring and the side chain to further delocalize positive charge. 3–Chloro–1–phenyl–1–butene is the major product because it results from cation D and because its double bond can conjugate with the benzene ring to provide extra stability.

14.31

Two different orientations of the dienophile ester group are possible in the cyclic transition state, and two different products can form.

14.32 The most reactive dienophiles contain electron-withdrawing groups.

Most reactive ——————————————————————→ Least reactive

$(NC)_2C=C(CN)_2$ > $H_2C=CHCHO$ > $H_2C=CHCH_3$ > $(CH_3)_2C=C(CH_3)_2$

Four electron-withdrawing groups   One electron-withdrawing group   Four electron-*donating* groups

14.33 The difference in reactivity of the three cyclic dienes is due to steric factors. As the "non-diene" part of the molecule becomes larger, the carbon atoms at the end of the diene portion of the ring are forced farther apart. Overlap with the pi system of the dienophile in the pericyclic transition state is poorer, and reaction is slower.

14.34 First, find the cyclohexene ring formed by the Diels-Alder reaction. After you locate the new bonds, you should then be able to identify the diene and the dienophile.

a)

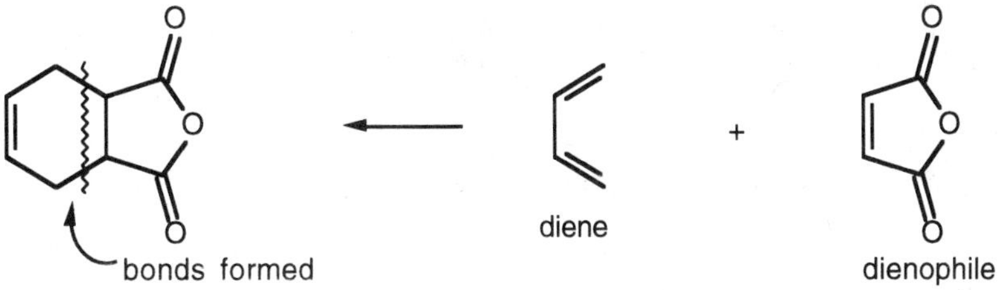

bonds formed

diene

dienophile

b)

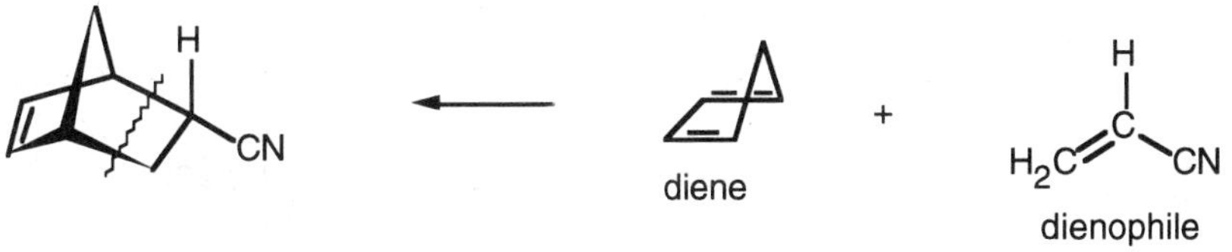

diene

dienophile

c)

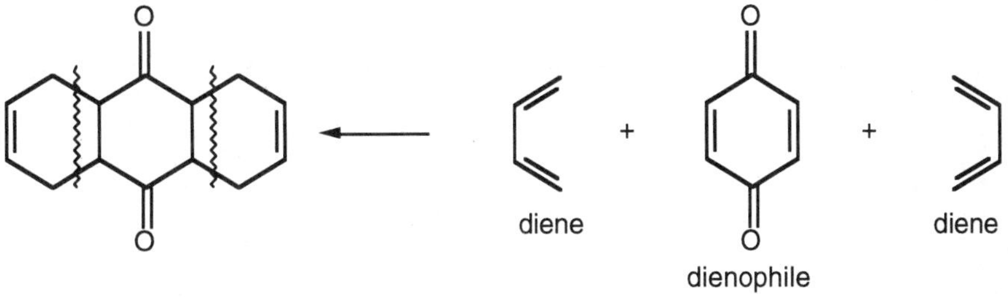

diene

dienophile

diene

d)

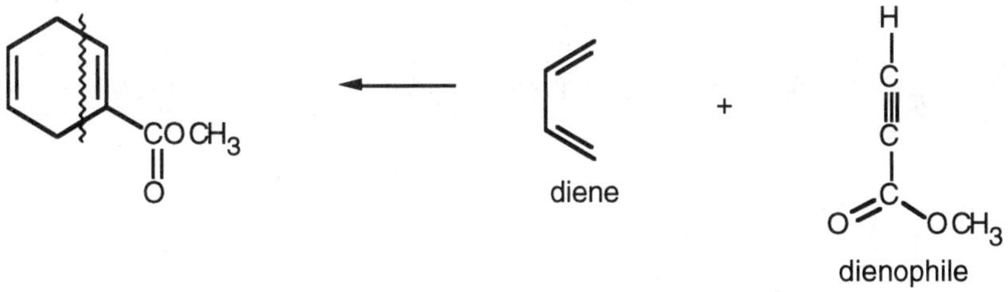

diene

dienophile

14.35

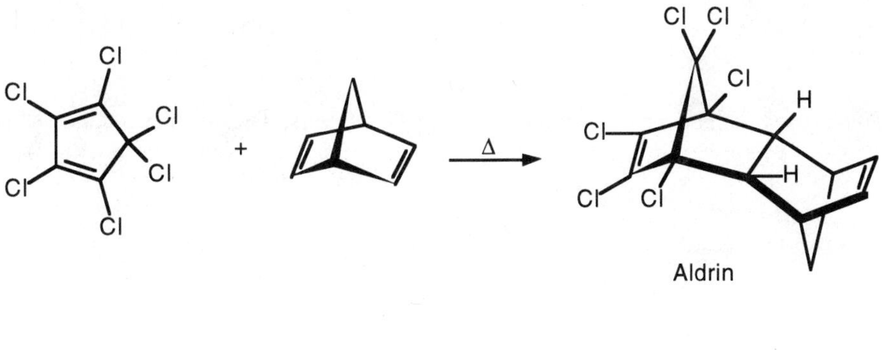

Aldrin

14.36

Diels-Alder reaction   E2 elimination

14.37 Diels-Alder reaction are reversible when the products are much more stable (of lower energy) than the reactants. In this case, the reactant is a nonconjugated diene, and the products are benzene (a stable, conjugated molecule) and ethylene.

14.38 A Diels-Alder reaction between α–pyrone (diene) and the alkyne dienophile yields the following product.

The double bonds in this product are not conjugated, and a more stable product can be formed by loss of CO_2.

This process can occur in a manner similar to the reverse Diels-Alder reaction of the previous problem.

14.39 The value of λ_{max} in the ultraviolet spectrum of dienes becomes larger with increasing alkyl substitution. Since energy is inversely related to λ_{max}, the energy needed to produce ultraviolet absorption decreases with increasing substitution.

Diene	# of $-CH_3$ groups	λ_{max}	$\lambda_{max} - \lambda_{max}$ (butadiene)
H C=CH₂ H₂C=C H	0	217 nm	0
H₃C C=CH₂ H₂C=C H	1	220	3
H H C=CH₂ C=C H₃C H	1	223	6
H₃C C=CH₂ H₂C=C CH₃	2	226	9
H CH₃ H C=C C=C H H₃C H	2	227	10
H₃C H₃C C=CH₂ C=C H₃C H	3	232	15
H CH₃ H₃C C=C C=C CH₃ H₃C H	4	240	23

The average increase in λ_{max} is 5 nm per methyl group.

14.40

H_2C=CH—CH=CH—CH=CH_2

1,3,5–Hexatriene

λ_{max} = 258 nm

H_2C=CH—CH=C—CH₃—C=CH_2 with CH₃ groups

2,3–Dimethyl–1,3,5–hexatriene

λ_{max} ~ 268 nm

In Problem 14.39, we concluded that one alkyl group increases λ_{max} of a conjugated diene by 5 nm. Since 2,3–dimethyl–1,3,5–hexatriene has two methyl substituents, its UV λ_{max} should be about 10 nm longer than the λ_{max} of 1,3,5–hexatriene.

14.41 a) ß–Ocimene, $C_{10}H_{16}$, has three degrees of unsaturation. Catalytic hydrogenation yields a hydrocarbon of formula $C_{10}H_{22}$. ß–Ocimene thus contains three double bonds and no rings.

   b) The ultraviolet absorption at 232 nm indicates that ß–ocimene is conjugated.

   c) The carbon skeleton, as determined from hydrogenation, is:

$$CH_3CH_2\overset{\overset{\displaystyle CH_3}{|}}{C}HCH_2CH_2CH_2\overset{\overset{\displaystyle CH_3}{|}}{C}HCH_3$$

2,6–Dimethyloctane

Ozonolysis data are used to determine the location of the double bonds. The acetone fragment, which comes from carbon atoms 1 and 2 of 2,6–dimethyloctane, fixes the position of one double bond. Formaldehyde results from ozonolysis of a double bond at the other end of ß–ocimene. Placement of the other fragments to conform to the carbon skeleton yields the following structural formula for ß–ocimene.

$$H_2C=CH\overset{\overset{\displaystyle CH_3}{|}}{C}=CHCH_2CH=\overset{\overset{\displaystyle CH_3}{|}}{C}CH_3$$

β-Ocimene

d)

$$H_2C=O \ + \ O=CH\overset{\overset{\displaystyle CH_3}{|}}{C}=O \ + \ O=CHCH_2CH=O \ + \ O=\overset{\overset{\displaystyle CH_3}{|}}{C}CH_3$$

Formaldehyde   Pyruvaldehyde   Malonaldehyde   Acetone

$$H_2C=CH\overset{\overset{\displaystyle CH_3}{|}}{C}=CHCH_2CH=\overset{\overset{\displaystyle CH_3}{|}}{C}CH_3$$

β-Ocimene

     1. O_3
2. Zn, H_3O^+

H_2/Pd

$$CH_3CH_2\overset{\overset{\displaystyle CH_3}{|}}{C}HCH_2CH_2CH_2\overset{\overset{\displaystyle CH_3}{|}}{C}HCH_3$$

14.42 Much of what was proven for ß–ocimene is also true for myrcene, since both hydrocarbons have the same carbon skeleton and contain conjugated double bonds. The difference between the two isomers is in the placement of double bonds.

    The ozonolysis fragments from myrcene are 2–oxopentanedial (five carbon atoms), acetone (three carbon atoms), and two equivalents of formaldehyde (one carbon atom each). Putting these fragments together in a manner consistent with the data gives the following structural formula for myrcene:

1. O_3
2. Zn, H_3O^+

$H_2C=CHCCH_2CH_2CH=CCH_3$

CH$_2$ CH$_3$

Myrcene

→ $2 H_2C=O$ + $O=CHCCH_2CH_2CH=O$ + $O=CCH_3$

O   CH$_3$

H_2/Pd $CH_3CH_2CHCH_2CH_2CH_2CHCH_3$

CH$_3$ CH$_3$

14.43

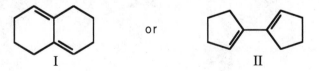

Conjugation with
the oxygen non-
bonding electrons
makes the double
bond more
nucleophilic. (Fig. 14.10)

Reaction with HCl
yields a cation
intermediate that
can be stabilized by
the oxygen electrons.

Addition of
Cl⁻ leads to
the observed
product.

There are two reasons why the other regioisomer is not formed: (1) Carbon 1 is less nucleophilic than carbon 2; (2) The cation intermediate that would result from protonation at carbon 1 can't be stabilized by the oxygen electrons.

14.44 a) Hydrocarbon **A** must have two double bonds and two rings, since no carbons are lost on ozonolysis and a diketone–dialdehyde is formed.

I   or   II

b) Rotation about the central single bond of II allows the double bond to assume the *s*–cis conformation necessary for a Diels-Alder reaction. Rotation is not possible for I.

c)

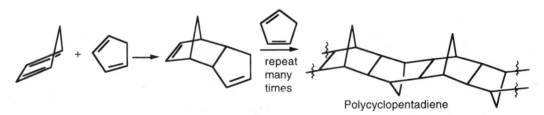

14.45

$$H_2C=CHCH=CH_2 \xrightarrow[\text{1,4--Addition}]{\text{Br}_2,\ \text{CCl}_4} BrCH_2CH=CHCH_2Br \xrightarrow{\text{2 NaCN}} NCCH_2CH=CHCH_2CN$$

$$\downarrow \begin{array}{l} H_2 \\ Pd/C \end{array}$$

$$NCCH_2CH_2CH_2CH_2CN$$

Adiponitrile

14.46

$$C = \frac{A}{\varepsilon \times l} = \frac{0.065}{11,900 \times 1.00\ \text{cm}} = \frac{6.5 \times 10^{-2}}{1.19 \times 10^4} = 5.5 \times 10^{-6}\ \text{M}$$

14.47

Polycyclopentadiene is the product of successive Diels-Alder additions of cyclopentadiene to a growing polymer chain. Strong heat causes depolymerization of the chain and reversion to cyclopentadiene monomer units.

14.48

14.49

The stereochemistry of the product resulting from Diels-Alder reaction of the (2*E*,4*Z*) diene differs at the starred carbon from that of the (2*E*,4*E*) diene. Not only is the stereochemistry of the dienophile maintained during the Diels-Alder reaction, the stereochemistry of the diene is also maintained.

14.50 Double bonds can be conjugated not only with other multiple bonds but also with the lone-pair electrons of atoms such as oxygen and nitrogen. *p*–Toluidine has the same number of double bonds as benzene, yet its λ_{max} is 31 nm greater. The electron pair of the nitrogen atom can conjugate with the pi electrons of the three double bonds of the ring, extending the pi system and increasing λ_{max}.

14.51 Hydrogen ion protonates the nitrogen atom of *p*-toluidine and prevents its lone pair of electrons from conjugating with the ring double bonds. λ_{max} is therefore lowered to a value very close to that of benzene.

14.52 Dilute NaOH removes the proton from the –OH group, leaving the phenoxide anion.

The increased electron density at oxygen increases conjugation with the pi electrons of the ring double bonds. The extended conjugation increases λ_{max} in a manner similar to *p*–toluidine (Problem 14.50).

Study Guide for Chapter 14

After studying this chapter, you should be able to:

(1)   Locate conjugated portions of molecules (14.1, 14.17).

(2)   Draw and name conjugated dienes (14.16, 14.19).

(3)   Understand the reasons for the stability of conjugated molecules (14.2, 14.5).

(4)   Predict the products of electrophilic additions to conjugated molecules (14.3, 14.4, 14.18, 14.21, 14.29, 14.30, 14.43, 14.45).

(5)   Understand the concept of kinetic *vs* thermodynamic control of reactions (14.6, 14.7).

(6)   Recognize diene polymers, and draw a representative segment of a diene polymer (14.8, 14.9, 14.22, 14.47).

(7)   Predict the products of the Diels-Alder reaction. You should be able to identify compounds that are good dienes and dienophiles (14.10, 14.11, 14.25, 14.26, 14.28, 14.31, 14.32, 14.33, 14.34, 14.35, 14.36, 14.37, 14.38, 14.48, 14.49).

(8)   Calculate the energy required for ultraviolet absorption, and use molar absorptivity to calculate concentration (14.12, 14.13, 14.14, 14.46).

(9)   Predict if and where a compound absorbs radiation in the ultraviolet region (14.15, 14.24, 14.27, 14.39, 14.40, 14.50, 14.51, 14.52).

(10)   Deduce the structure of an unknown diene from structural and spectral data (14.20, 14.41, 14.42, 14.44).

Chapter 15 – Benzene and Aromaticity

15.1 An ortho disubstituted benzene has two substituents in a 1,2 relationship. A meta disubstituted benzene has its two substituents in a 1,3 relationship. A para disubstituted benzene has its two substituents in a 1,4 relationship.

a) 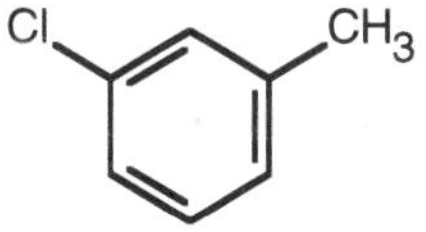
Cl
CH_3

meta disubstituted

b) 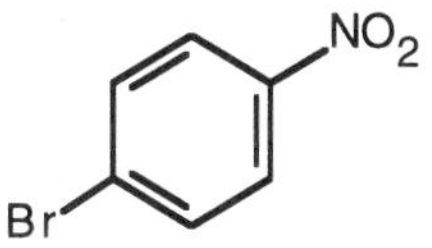
NO_2
Br

para disubstituted

c) 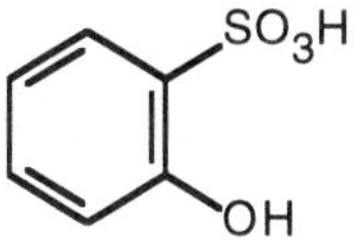
SO_3H
OH

ortho disubstituted

15.2

a) 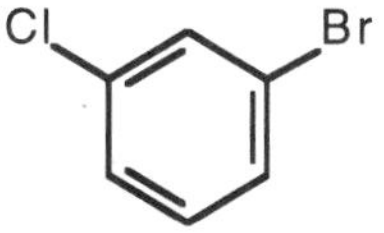
Cl
Br

m–Bromochlorobenzene

b) 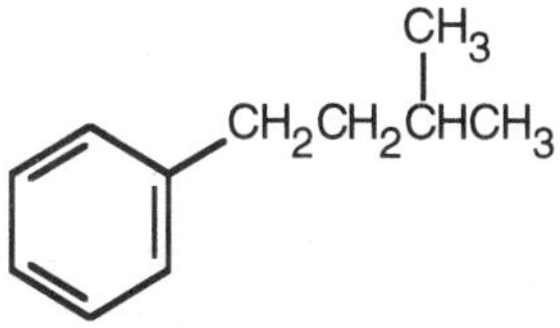
CH_3
$CH_2CH_2CHCH_3$

(3–Methylbutyl)benzene

c) 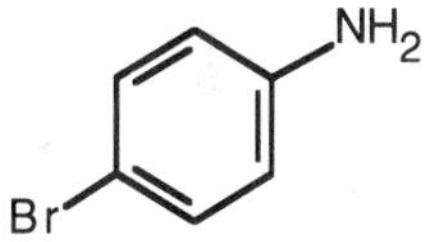
NH_2
Br

p–Bromoaniline

d) 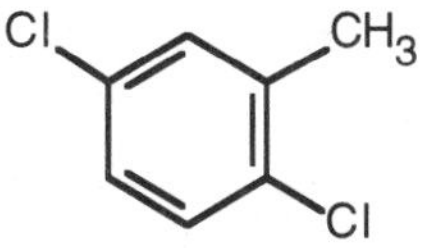
Cl
CH_3
Cl

2,5–Dichlorotoluene

e) 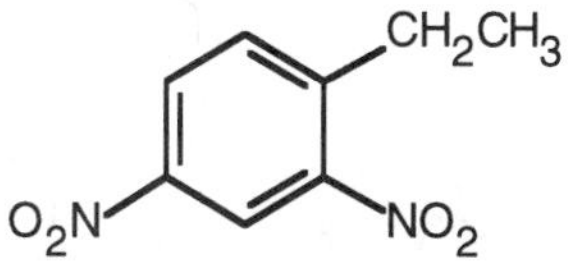
CH_2CH_3
O_2N
NO_2

1–Ethyl–2,4–dinitro–benzene

f) 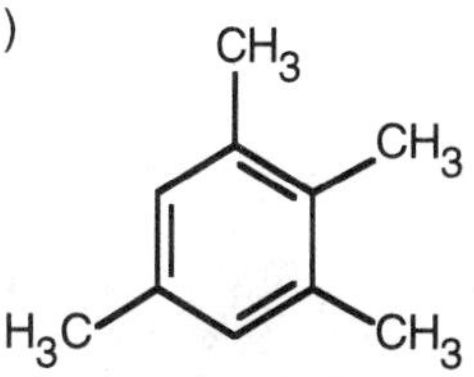
CH_3
CH_3
H_3C
CH_3

1,2,3,5–Tetra-methylbenzene

15.3

a) 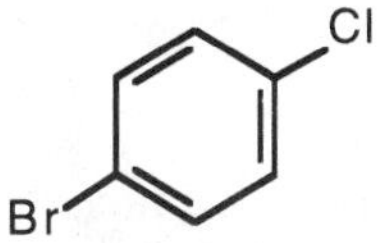
Cl
Br

p–Bromochlorobenzene

b) 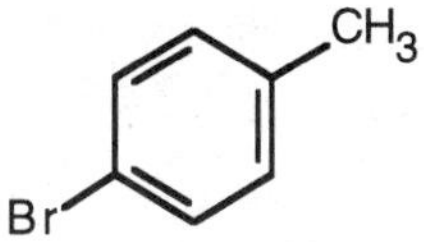
CH_3
Br

p–Bromotoluene

c) 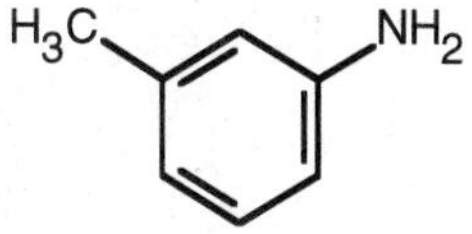
H_3C
NH_2

m–Chloroaniline

d) 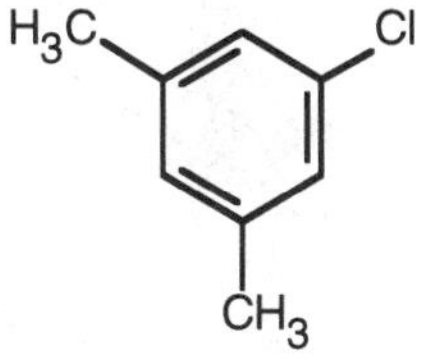
H_3C
Cl
CH_3

1–Chloro–3,5–dimethylbenzene

15.4

a)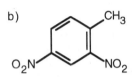

The correct name is 1–Bromo–2–chlorobenzene. You must use the lowest possible combination of numbers. You may also use the *ortho* designation.

b)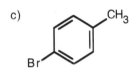

The correct name is 2,4–dinitrotoluene. This is the same mistake as in a).

c)

The correct name is *p*-bromotoluene or 1-bromo-4-methylbenzene.

d)

The correct name is 2-chloro-1,4-dimethylbenzene. *p*-Xylene is not a parent name.

15.5

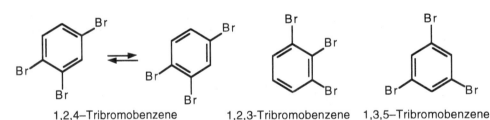

1,2,4–Tribromobenzene   1,2,3-Tribromobenzene   1,3,5–Tribromobenzene

According to Kekulé theory, four tribromobenzenes are possible. Kekulé would say that the two 1,2,4–tribromobenzenes rapidly interconvert, and that only three isomers can be isolated.

15.6   Since all carbon atoms are equivalent in Ladenburg benzene, only one monobromo derivative is possible. Four dibromo derivatives, including a pair of enantiomers, are possible.

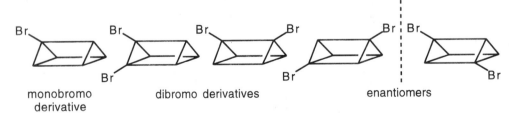

monobromo   dibromo derivatives   enantiomers
derivative

Dewar benzene, which is a bent molecule, has two different kinds of carbons. Three monobromo derivatives, including a pair of enantiomers, are possible.

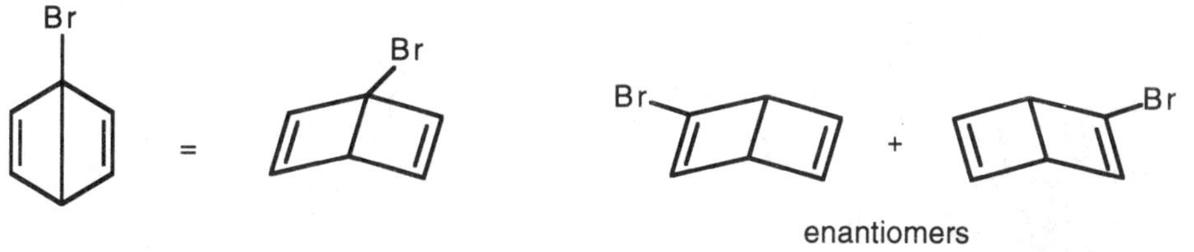

enantiomers

The dibromo derivatives include three pairs of enantiomers and three other dibromo Dewar benzenes.

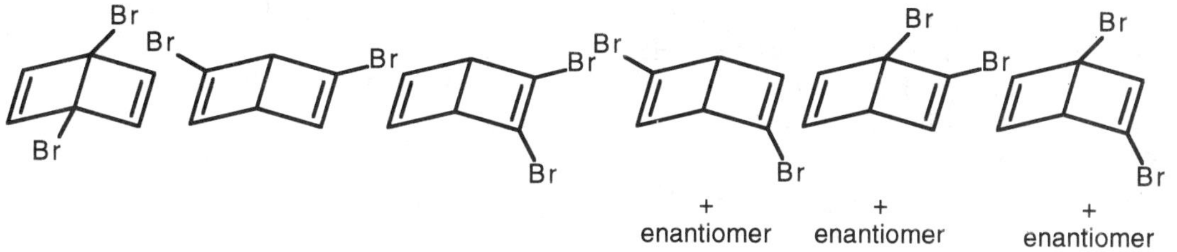

+   +   +
enantiomer   enantiomer   enantiomer

15.7

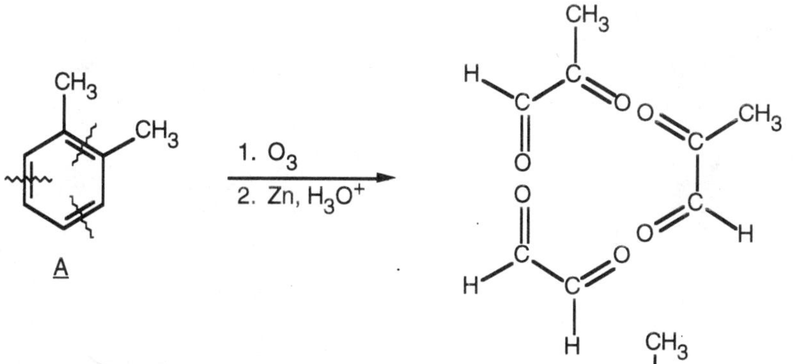

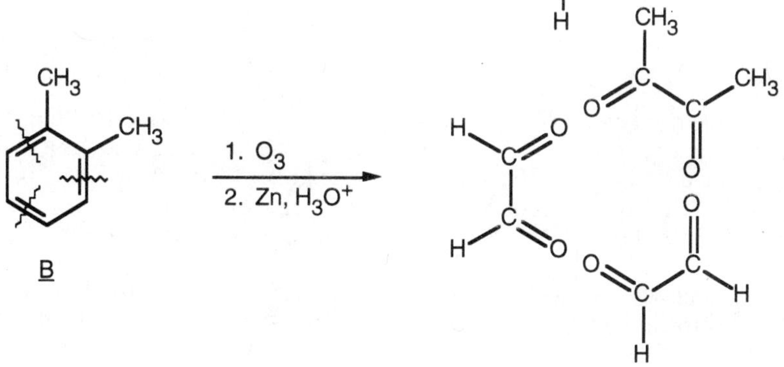

If *o*–xylene exists only as a structure <u>A</u>, ozonolysis would cause cleavage at the bonds indicated and would yield two moles of pyruvaldehyde and one mole of glyoxal for each mole of <u>A</u> consumed. If *o*–xylene exists only as structure <u>B</u>, ozonolysis would yield one mole of 2,3–butanedione and two moles of glyoxal. If *o*–xylene exists as a resonance hybrid of <u>A</u> and <u>B</u>, the ratio of ozonolysis products would be glyoxal : pyruvaldehyde :

2,3–butanedione = 3:2:1. Since this ratio is identical to the experimentally determined ratio, we know that $\underline{A}$ and $\underline{B}$ contribute equally to the structure of o–xylene.

15.8

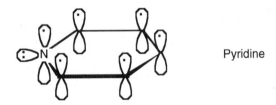

Pyridine

The electronic descriptions of pyridine and benzene are very similar. The pyridine ring is formed by the sigma overlap of carbon and nitrogen sp^2 orbitals. In addition, six p orbitals, perpendicular to the plane of the ring, hold six electrons. These six p orbitals form six π molecular orbitals that allow electrons to be delocalized over the π system of the pyridine ring. The lone pair of nitrogen electrons occupies an sp^2 orbital that lies in the plane of the ring.

15.9

Cyclodecapentaene has $4n + 2$ π electrons (n = 2), but it is not flat. If cyclodecapentaene were flat, the hydrogen atoms circled would crowd each other across the ring. To avoid this interaction, cyclodecapentaene twists so that it is neither planar nor aromatic.

15.10

A compound that can be described by several resonance forms has a structure that can be represented by no one form. The structure of the cyclopentadienyl anion is a combination of all of the above structures and contains only one kind of carbon atom and one kind of hydrogen atom. All carbon-carbon bond lengths are equivalent, as are all carbon-hydrogen bonds lengths. Both the ^{1}H NMR and ^{13}C NMR spectra show only one absorption.

15.11 When cyclooctatetraene accepts two electrons, it becomes a $(4n + 2)$ π electron aromatic ion. Cyclooctatetraenyl dianion is planar with a carbon-carbon bond angle of 135° (a regular octagon).

15.12

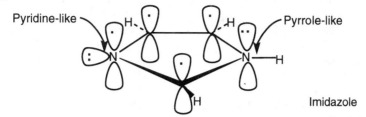

The aromatic heterocycle imidazole contains six π electrons. Each carbon contributes one electron, the nitrogen bonded to hydrogen (pyrrole-like) contributes two electrons, and the remaining nitrogen (pyridine-like) contributes one electron. Both nitrogens are sp^2 hybridized.

15.13

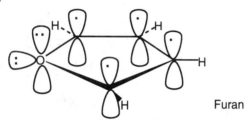

15.14

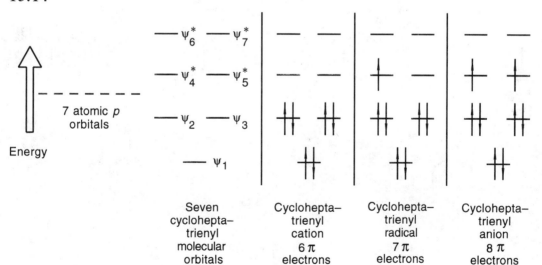

The cycloheptatrienyl cation has six π electrons (a Hückel number) and is aromatic.

15.15

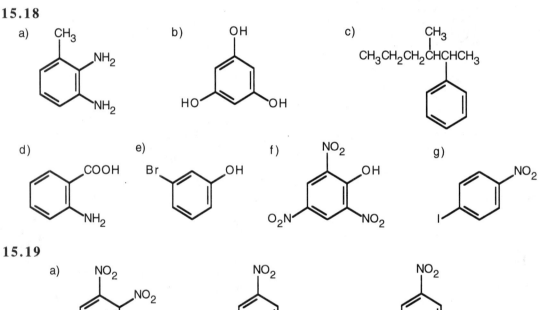

Azulene

Azulene is aromatic because it has a conjugated cyclic π electron system containing ten π electrons (a Hückel number).

15.16 Naphthalene is a ten π electron compound; the circle in each ring represents five electrons.

15.17 a) 2–Methyl–5–phenylhexane   b) *m*–Bromobenzoic acid
c) 1–Bromo–3,5–dimethylbenzene   d) *o*–Bromopropylbenzene
e) 1–Fluoro–2,4–dinitrobenzene   f) *p*–Chloroaniline

15.18

a)   b)   c)

d)   e)   f) NO₂   g)

15.19

a) NO₂   NO₂   NO₂

o–Dinitrobenzene   *m*–Dinitrobenzene   *p*–Dinitrobenzene

b)

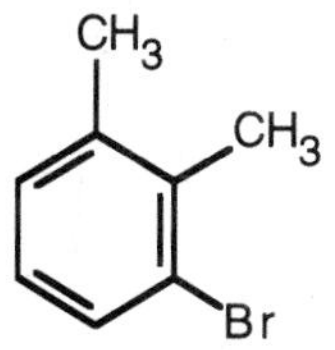

1–Bromo–2,3–dimethyl-
benzene

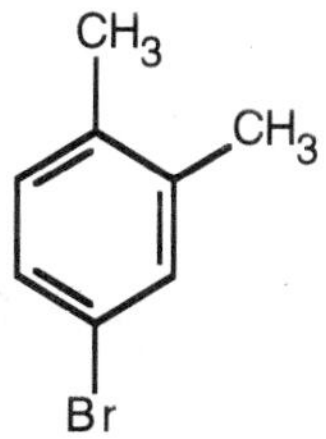

4–Bromo–1,2–dimethyl-
benzene

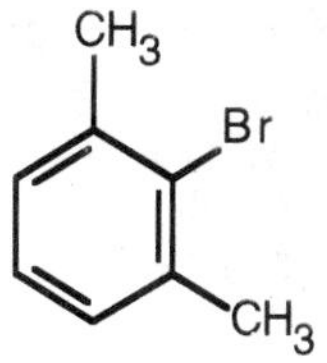

2–bromo–1,3–dimethyl-
benzene

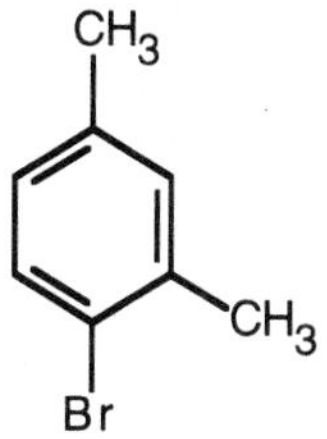

1–Bromo–2,4–dimethyl-
benzene

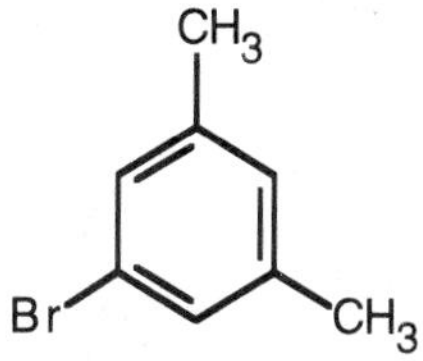

1–Bromo–3,5–dimethyl-
benzene

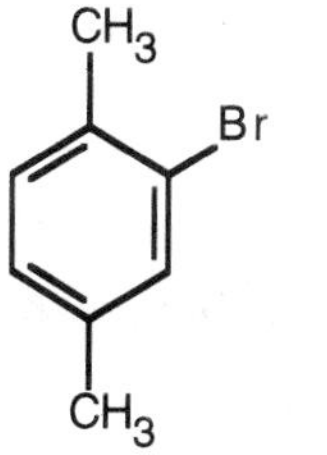

2–Bromo–1,4–dimethyl-
benzene

c)

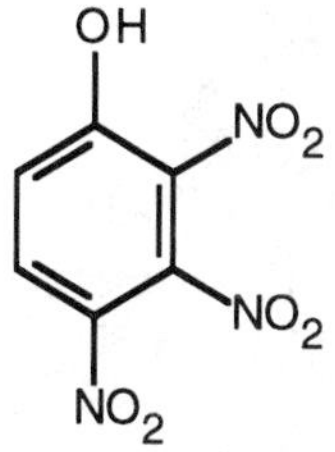

2,3,4–Trinitrophenol

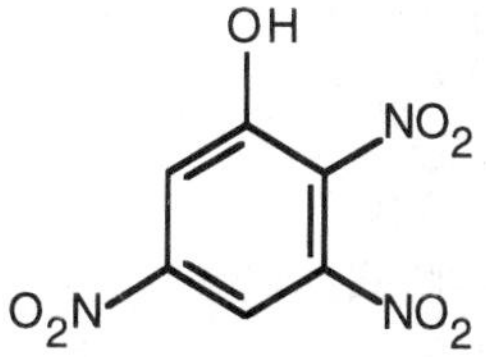

2,3,5–Trinitrophenol

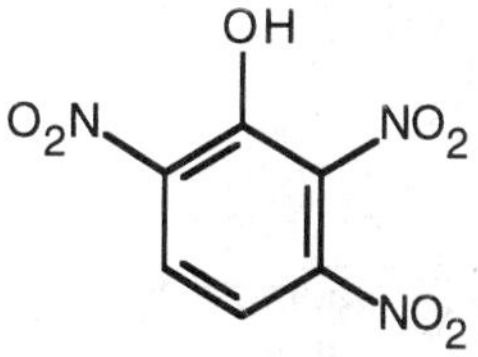

2,3,6–Trinitrophenol

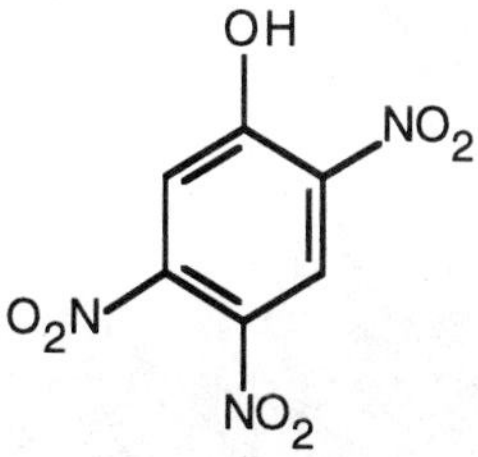

2,4,5–Trinitrophenol

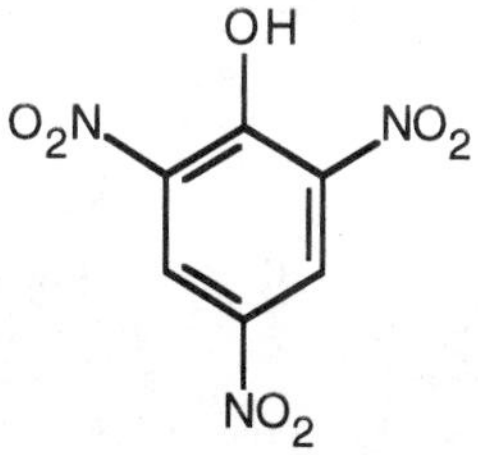

2,4,6–Trinitrophenol

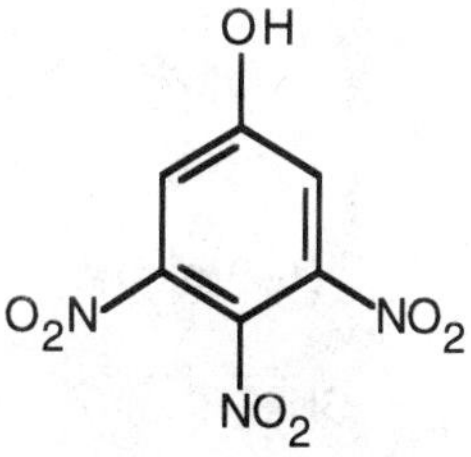

3,4,5–Trinitrophenol

15.20 All aromatic compounds of formula C_7H_7Cl have one ring and three double bonds.

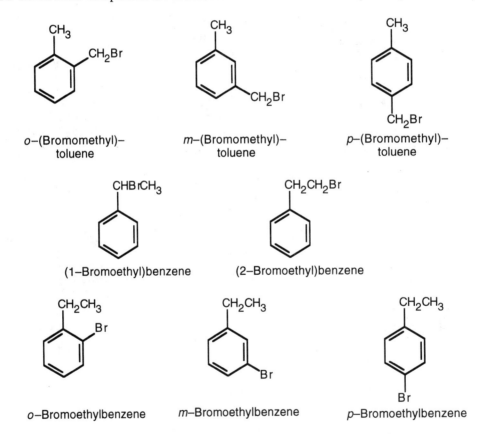

o–Chlorotoluene   m–Chlorotoluene   p–Chlorotoluene   Benzyl chloride
or
(Chloromethyl)–
benzene

15.21 Six of these compounds are illustrated and named in Problem (15.19b).

o–(Bromomethyl)–
toluene

m–(Bromomethyl)–
toluene

p–(Bromomethyl)–
toluene

(1–Bromoethyl)benzene

(2–Bromoethyl)benzene

o–Bromoethylbenzene   m–Bromoethylbenzene   p–Bromoethylbenzene

15.22 All compounds in this problem have four double bonds and/or rings and must be substituted benzenes, if they are to be aromatic. They may be substituted by methyl, ethyl, propyl, or butyl groups.

a)

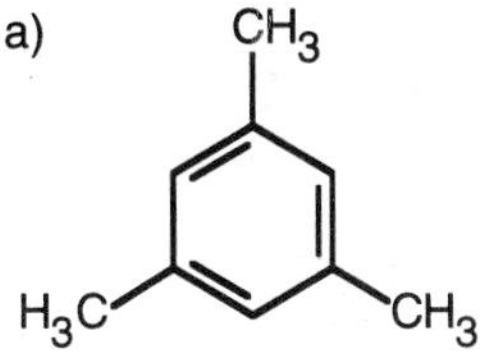

b)

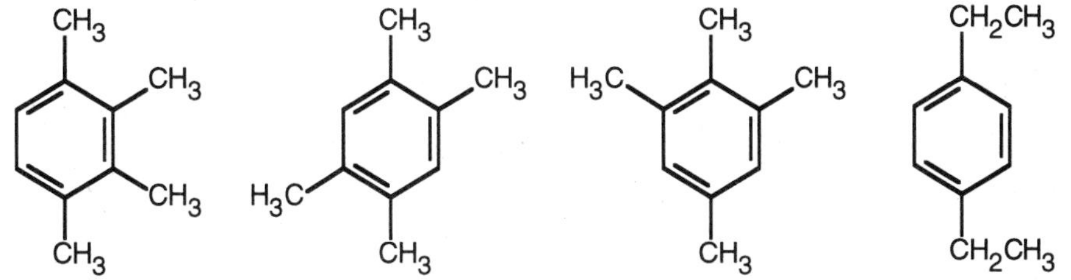

c)

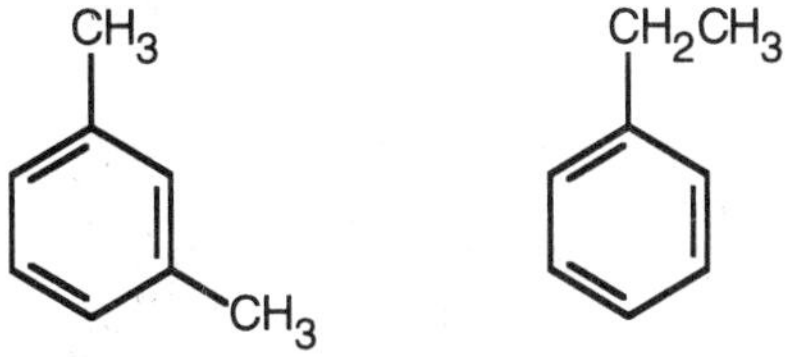

d)

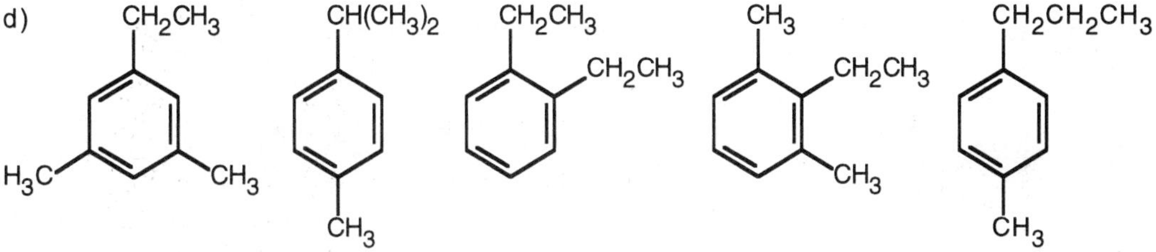

15.23

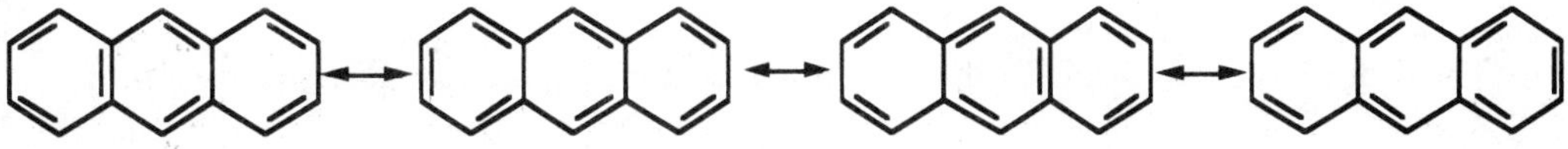

15.24

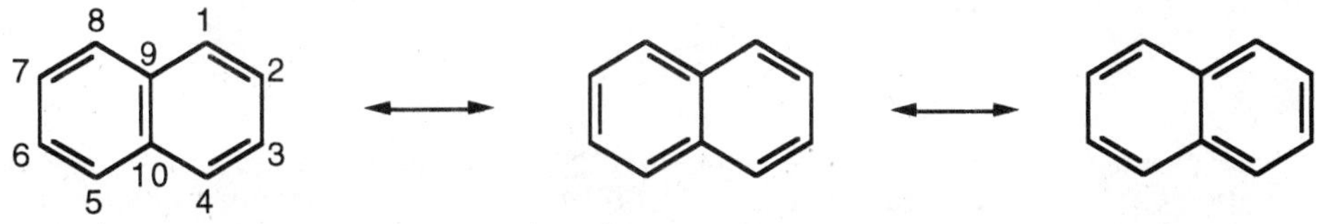

The bond between carbons 1 and 2 is represented as a double bond in two of the three resonance structures, but the bond between carbons 2 and 3 is represented as a double bond in only one resonance structure. The C1-C2 bond thus has more double bond character in the resonance hybrid, and it is shorter than the C2-C3 bond. The C3-C4, C5-C6, and C7-C8 bonds also have more double bond character than the remaining bonds.

15.25, 15.26

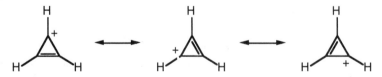

The circled bond is represented as a double bond in four of the five resonance forms of phenanthrene. This bond has more double-bond character and thus is shorter than the other carbon-carbon bonds of phenanthrene.

15.27 a) *Aromaticity* is a property of cyclic conjugated compounds having $(4n + 2)$ π electrons. Aromatic compounds are usually stable and unreactive, and undergo substitution rather than addition reactions.

b) A system of alternating single and multiple bonds having overlapping p orbitals is said to be *conjugated*. Compounds having conjugated bonds are stabilized by their ability to delocalize π electrons.

c) The *Hückel 4n + 2 rule* predicts that compounds having $(4n + 2)$ π electrons (where $n = 0, 1, 2...$) in a planar, cyclic conjugated system will be aromatic.

d) For some compounds it is possible to draw two or more Kekulé structures that differ only in the placement of electrons and not of atoms. These structures are called resonance forms. The actual structure of the compound can't be represented by any one of the resonance forms but is a *resonance hybrid* of all of them.

15.28 The heat of hydrogenation is the amount of heat liberated when a compound reacts with hydrogen.

(1) Benzene + 3 H_2 → Cyclohexane ΔH_{hydrog} = –206 kJ/mol

(2) 1,3–Cyclohexadiene + 2 H_2 → Cyclohexane ΔH_{hydrog} = –230 kJ/mol

---

Benzene + H_2 → 1,3–Cyclohexadiene ΔH_{hydrog} = –206 kJ/mol – (–230 kJ/mol)
                                                 = 24 kJ/mol

The heat of hydrogenation for this reaction is positive, so the reaction is endothermic.

15.29

The product of the reaction of 3–chlorocyclopropene with $AgBF_4$ is the cyclopropenyl cation $C_3H_3^+$. The resonance structures of the cation indicate that all hydrogen atoms are equivalent, and the ^1H NMR spectrum, which shows only one type of hydrogen atom, confirms this equivalence. The cyclopropenyl cation contains two π electrons and is aromatic according to Hückel's rule. (Here, $n = 0$.)

15.30

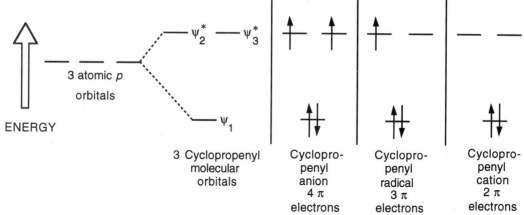

3 Cyclopropenyl molecular orbitals

Cyclopropenyl anion 4 π electrons

Cyclopropenyl radical 3 π electrons

Cyclopropenyl cation 2 π electrons

The cyclopropenyl cation is aromatic according to Hückel's rule.

15.31 The circle in the cyclopropenyl cation represents two π electrons.

15.32

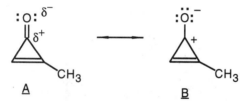

In resonance structure <u>A</u>, methylcyclopropenone is a cyclic conjugated compound with three π electrons in its ring. Because the electronegative oxygen attracts the π electrons of the carbon-oxygen π bond, however, a second resonance structure <u>B</u> is possible in which both carbonyl π electrons are located on oxygen, leaving only two π electrons in the ring. Since 2 is a Hückel number, the methylcyclopropenone ring fulfills the criteria of aromaticity.

15.33

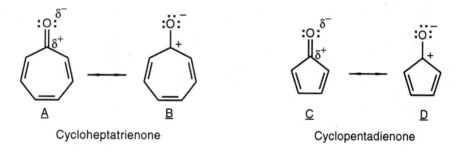

A   B
Cycloheptatrienone

C   D
Cyclopentadienone

As in the previous problem, we can draw resonance forms in which both carbonyl π electrons are located on oxygen. The cycloheptatrienone ring in <u>B</u> contains six π electrons and is aromatic according to Hückel's rule. The cyclopentadienone ring in D contains four π electrons and is antiaromatic.

15.34 Check the number of electrons in the π system of each compound. The species with a Hückel (4n + 2) number of π electrons is the most stable.

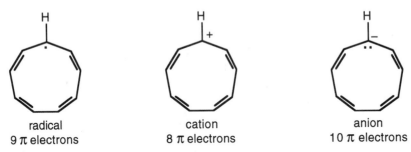

radical	cation	anion
9 π electrons	8 π electrons	10 π electrons

The ten π electron anion is the most stable.

15.35

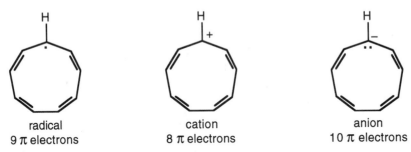

15.36 Compound A has four multiple bonds and/or rings. Possible structures that yield three monobromo substitution products are:

Only structure I shows a six-proton singlet at 2.30 δ, because it contains two identical methyl groups unsplit by other protons. The presence of four protons in the aromatic region of the ^{1}H NMR spectrum confirms that I is the correct structure.

15.37

Molecules with dipole moments are polar because electron density is drawn from one part of the molecule to another. In azulene, electron density is drawn from the seven-membered ring to the five-membered ring, satisfying Hückel's rule for both rings and producing a dipole moment. The five-membered ring resembles the cyclopentadienyl anion in having six pi electrons, while the seven-membered ring resembles the cycloheptatrienyl cation.

15.38 The molecular weight of the hydrocarbon (120) corresponds to the structural formula C_9H_{12}, which indicates four double bonds and/or rings. The ^1H NMR singlet at 7.25 δ indicates the presence of five aromatic ring protons. The septet at 2.90 δ is due to a benzylic proton that has six neighboring protons.

Isopropylbenzene

15.39

a) o–Diethylbenzene

b) p–Isopropyltoluene

The IR values are essential for deciding if the rings are *o, m* or *p*–substituted.

15.40

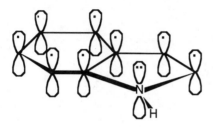

Indole, like naphthalene, has ten π electrons in two rings and is aromatic. Two π electrons come from the nitrogen atom.

15.41

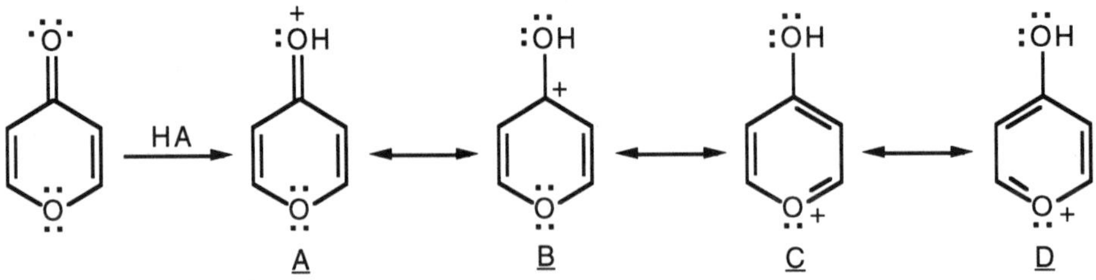

Protonation of 4–pyrone gives structure A, which has the resonance forms B, C and D. In C and D, a lone pair of electrons of the ring oxygen is delocalized into the ring to produce a six π electron system, which should be aromatic according to Hückel's rule.

15.42

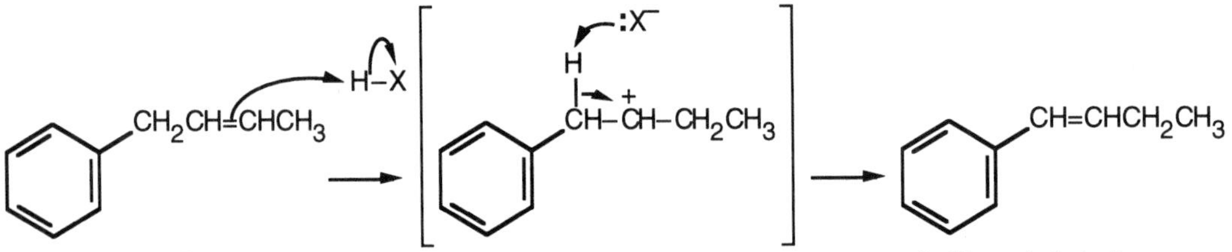

1–Phenyl–2–butene

1–Phenyl–1–butene

The alkene double bond is protonated to yield an intermediate carbocation, which loses a proton to give a product in which the double bond is conjugated with the aromatic ring, as shown by the increased value of λ_{max}.

15.43

a)

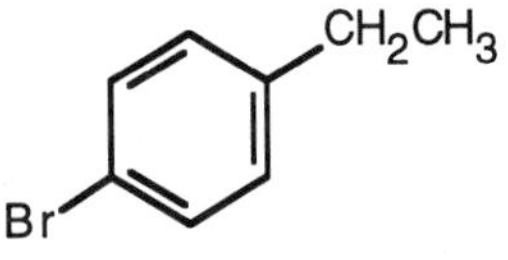

p–Bromoethylbenzene

b)

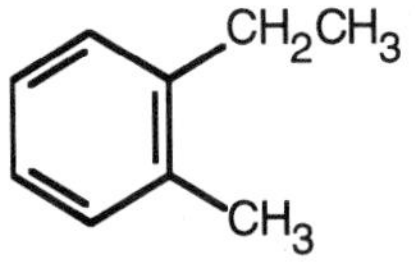

o–Ethyltoluene

c)

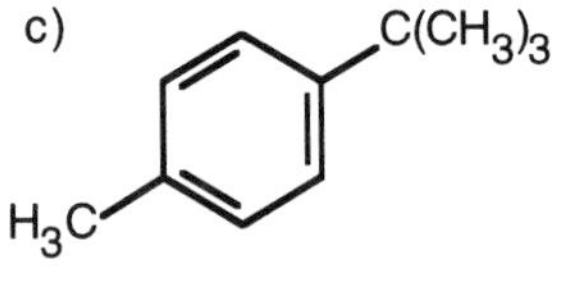

p–tert–Butyltoluene

15.44

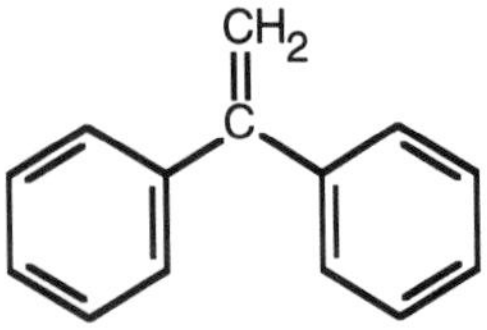

15.45 Pentalene has eight π electrons and is antiaromatic. Pentalene dianion, however, has ten π electrons and is a stable, aromatic ion.

15.46

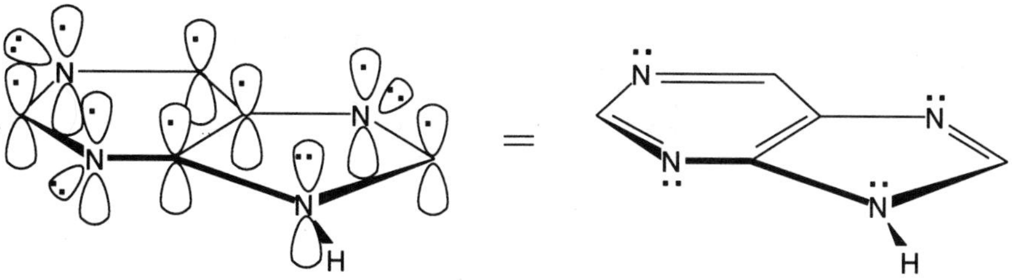

Purine is a ten-π-electron aromatic molecule. The N–H nitrogen atom in the five-membered ring donates *both* electrons of its lone pair to the π electron system, and each of the other three nitrogens donates one electron to the π electron system.

15.47

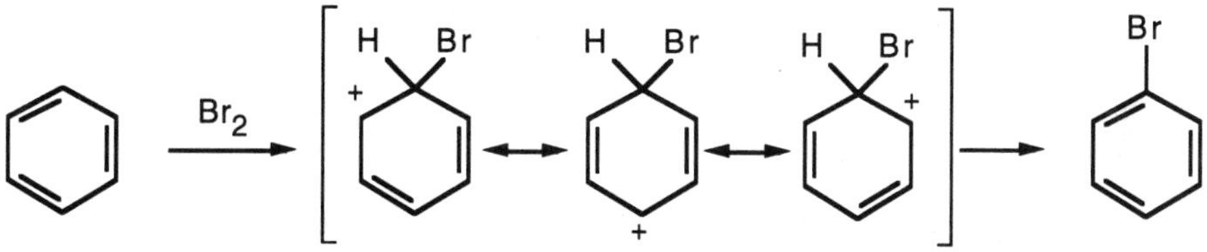

15.48

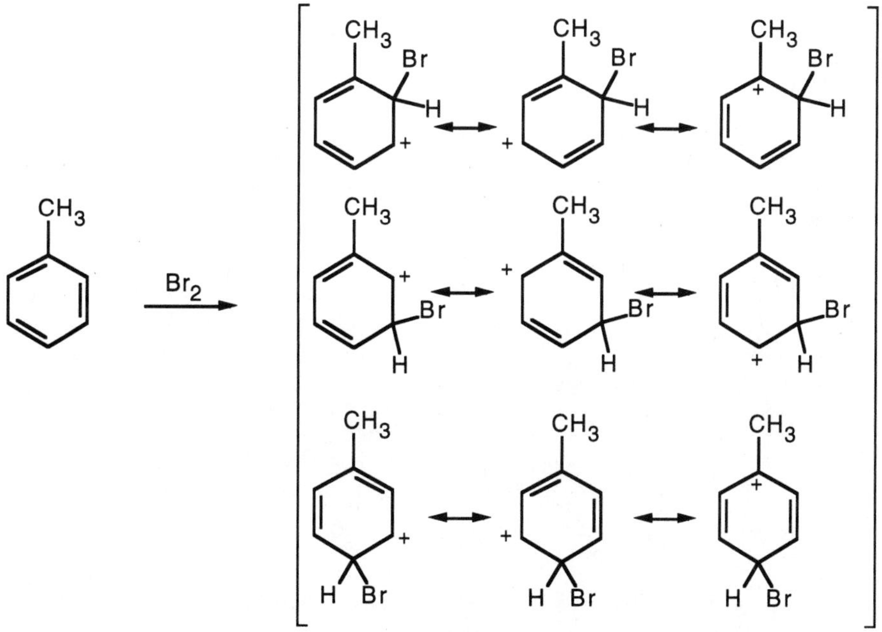

The ortho and para products predominate because the intermediate carbocation is more stabilized. The third resonance form drawn for ortho-para attack places the positive charge at the methyl-substituted carbon, which can form a more stable tertiary carbocation.

Study Guide for Chapter 15

After studying this chapter, you should be able to:

(1)   Recognize aromatic compounds; name and draw substituted benzene compounds (15.2, 15.3, 15.4, 15.5, 15.17, 15.18, 15.19, 15.20, 15.21, 15.22).

(2)   Draw resonance structures and molecular orbital diagrams for benzene and other cyclic conjugated molecules (15.5, 15.7, 15.8, 15.14, 15.15, 15.23, 15.24, 15.25, 15.26, 15.30).

(3)   Use Hückel's rule to predict aromaticity (15.9, 15.10, 15.11, 15.29, 15.31, 15.32, 15.33, 15.34, 15.35, 15.37, 15.41, 15.45, 15.46).

(4)   Draw orbital pictures of cyclic conjugated molecules (15.8, 15.12, 15.13, 15.40).

(5)   Use NMR and IR data to predict structures of aromatic compounds (15.36, 15.38, 15.39, 15.43, 15.44).

Chapter 16 – Chemistry of Benzene.
Electrophilic Aromatic Substitution

16.1

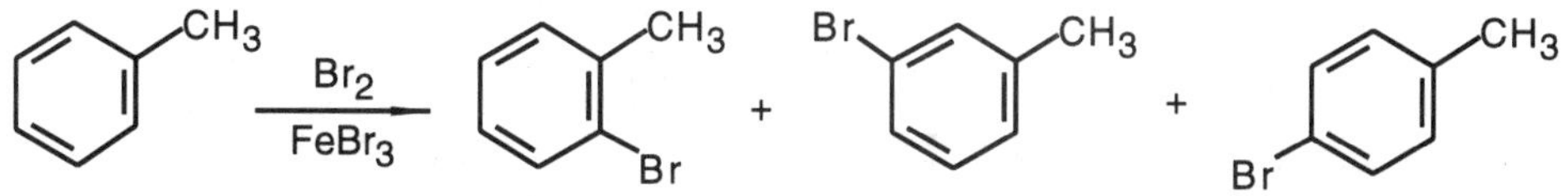

Toluene   *o*–Bromotoluene   *m*–Bromotoluene   *p*–Bromotoluene

16.2

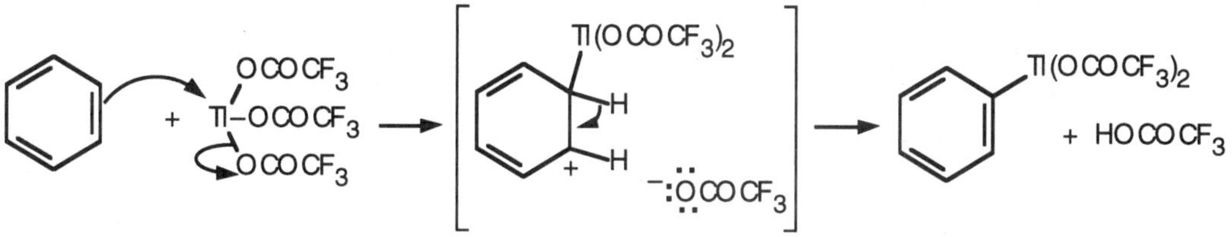

Thallium acts as an electrophile and substitutes on the aromatic ring.

16.3

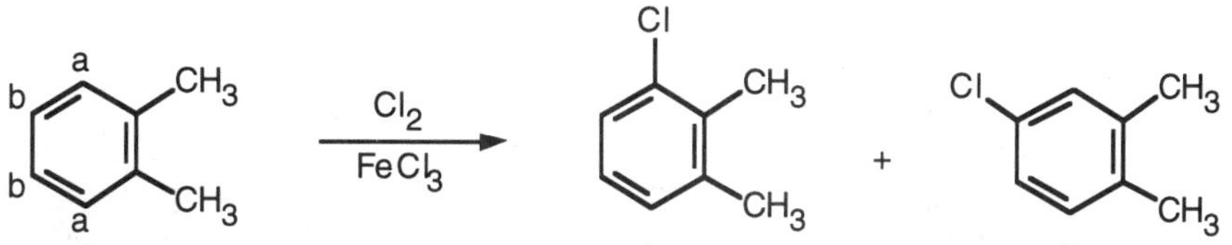

o–Xylene   A   B

Chlorination at position "a" of *o*–xylene yields product A; chlorination at position "b" yields product B.

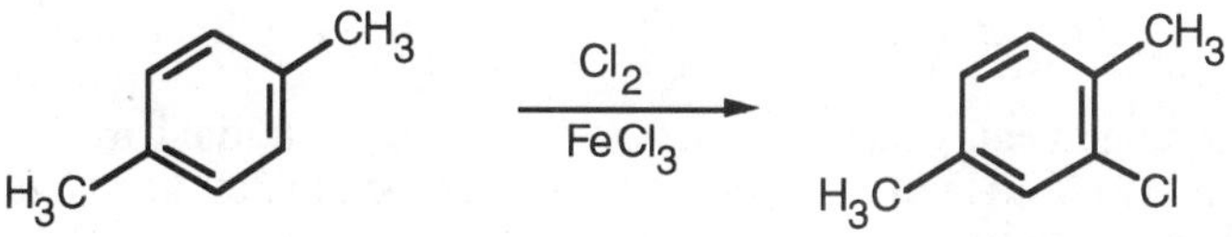

p–Xylene

Only one product results from chlorination of *p*–xylene because all sites for chlorination are equivalent.

16.4

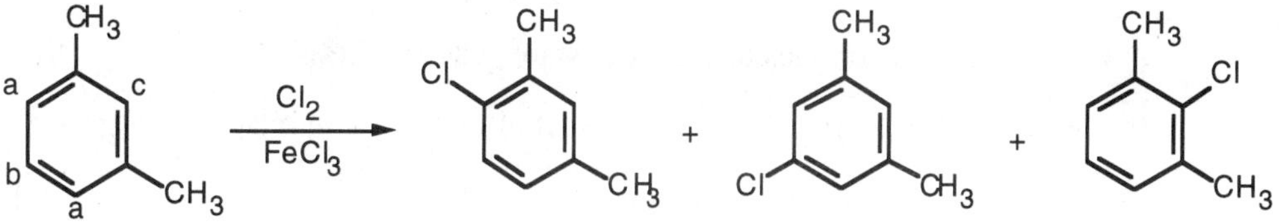

m–Xylene   A   B   C

Three products might be expected to form on chlorination of *m*–xylene. Product C is unlikely to form because substitution rarely occurs between two meta substituents. Product B is also unlikely to form for reasons to be explained later in the chapter.

16.5

carbocation intermediate

Benzene can be protonated by strong acids. The resulting intermediate can lose either deuterium or hydrogen. If H$^+$ is lost, deuterated benzene is produced. Attack by D$^+$ can occur at all positions of the ring and leads to eventual replacement of all hydrogens by deuterium.

16.6

Isobutyl carbocation (primary)

tert–Butyl carbocation (tertiary)

The isobutyl carbocation is initially formed when 1–chloro–2–methylpropane and AlCl$_3$ react. This carbocation rearranges via a hydride shift to give the more stable *tert*–butyl carbocation, which can then alkylate benzene to form *tert*–butylbenzene.

16.7   Carbocation rearrangements of alkyl halides occur: (1) if the initial carbocation is primary or secondary, and (2) if it is possible for the initial carbocation to rearrange to a more stable secondary or tertiary cation.

a)   Although CH$_3$CH$_2^+$ is a primary carbocation, it can't rearrange to a more stable cation.

b)   CH$_3$CH$_2$CHClCH$_3$ forms a secondary carbocation that doesn't rearrange.

c)   CH$_3$CH$_2$CH$_2^+$ rearranges to the more stable CH$_3$CHCH$_3^+$.

d)   (CH$_3$)$_3$CCH$_2^+$ (primary) undergoes an alkyl shift to yield (CH$_3$)$_2$CCH$_2$CH$_3^+$ (tertiary).

e)   The cyclohexyl carbocation doesn't rearrange.

In summary:
No rearrangement: CH$_3$CH$_2$Cl, CH$_3$CH$_2$CHClCH$_3$, chlorocyclohexane
Rearrangement: CH$_3$CH$_2$CH$_2$Cl, (CH$_3$)$_3$CCH$_2$Cl

16.8 Refer to Figure 16.10 in the text for the directing effects of substituents. You should memorize the effects of the most important groups.

a)

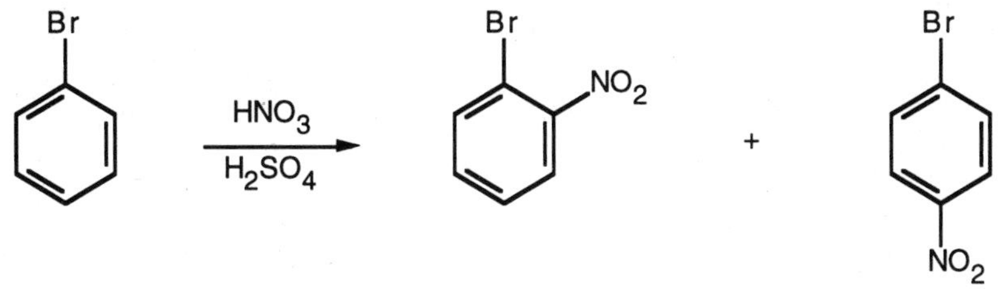

Even though bromine is a deactivator, it is an ortho–para director.

b)

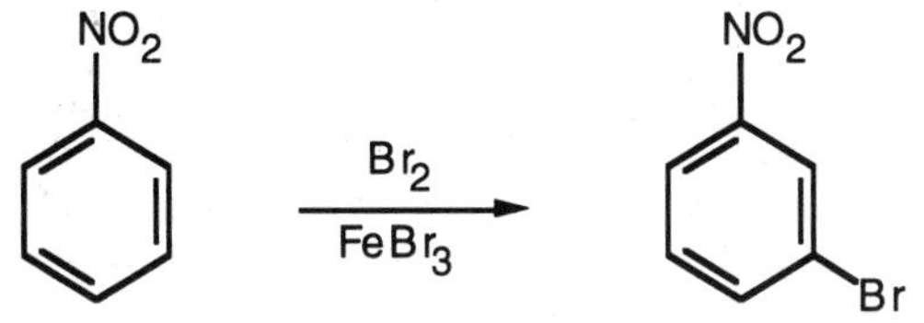

The $-NO_2$ group is a meta–director.

c)

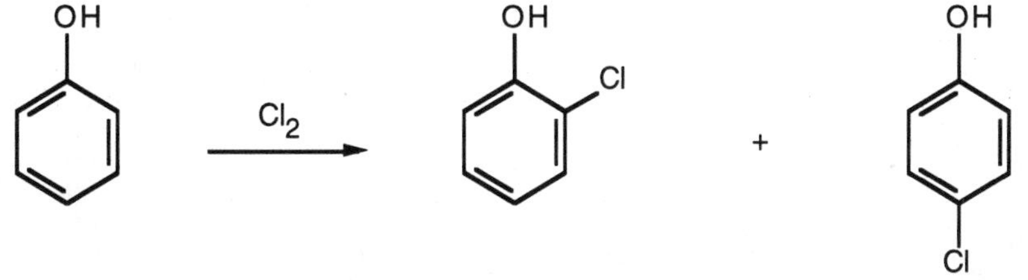

d)

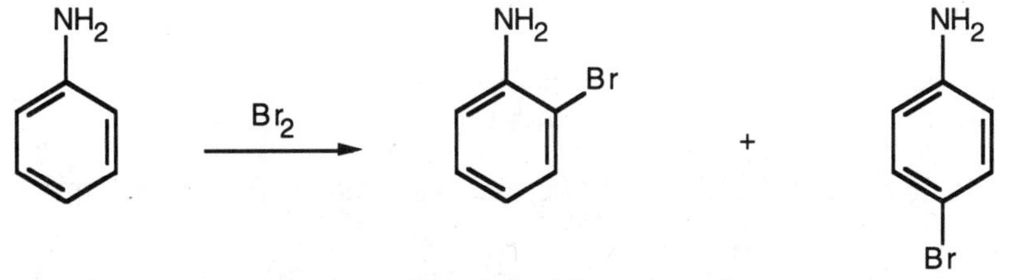

No catalyst is necessary because aniline is highly activated.

16.9

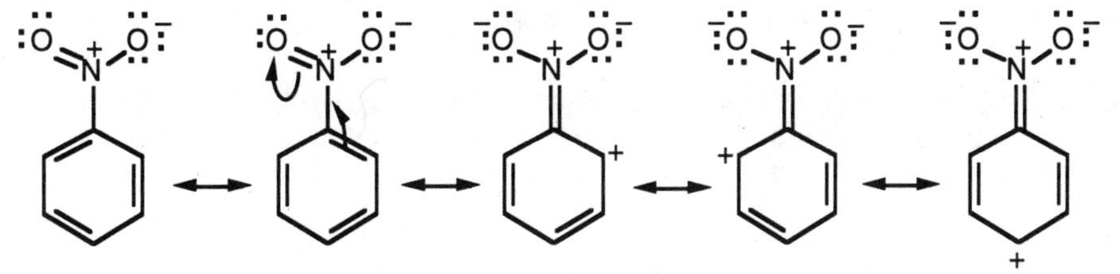

16.10

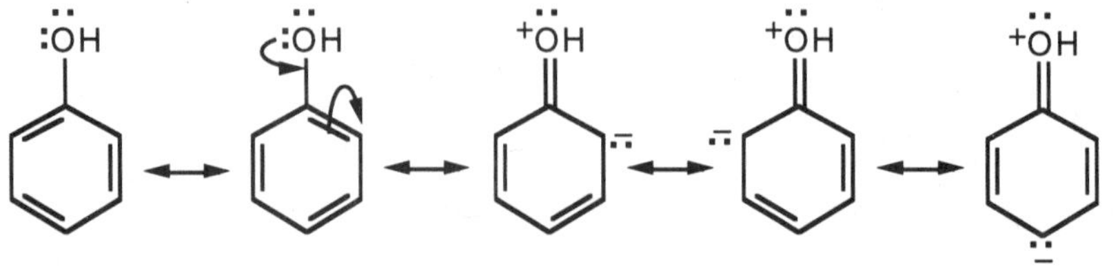

16.11

a) *Para* attack:

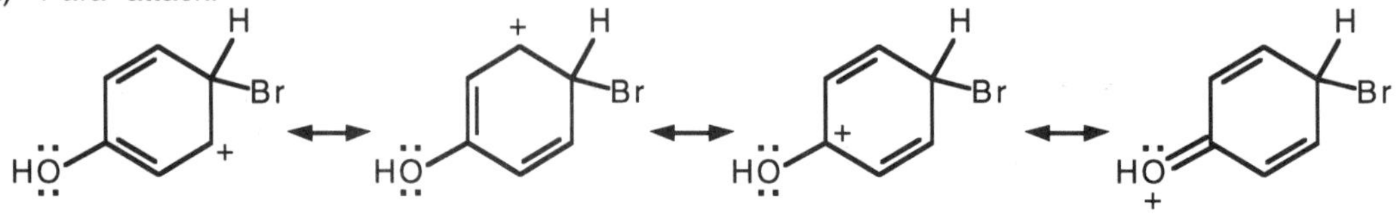

b) *Meta* attack:

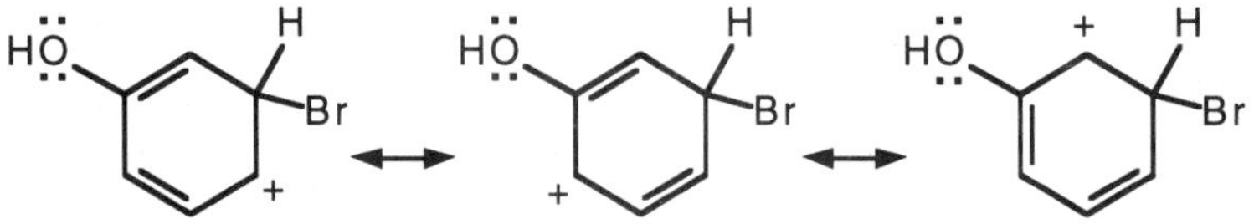

c) *Ortho* attack:

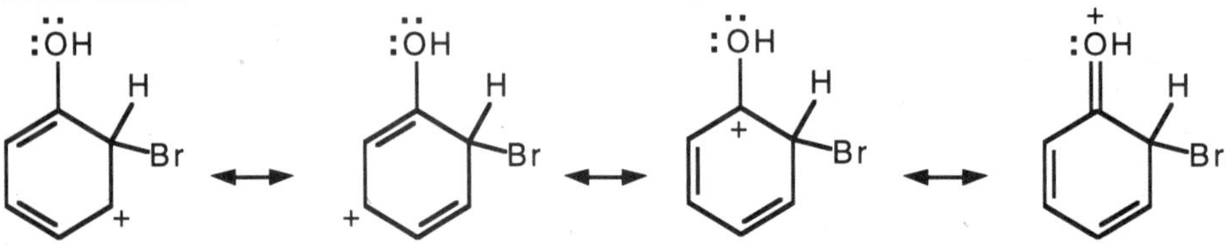

The more resonance forms that can be drawn, the greater is the extent of charge delocalization. Since the intermediates from ortho and para attack can be written as four resonance forms each, these intermediates are of lower energy and are more stable.

16.12 Use Figure 16.10 to find the activating and deactivating effects of groups.

Most Reactive ——————————→ Least Reactive

a)   Phenol > toluene > benzene > nitrobenzene.
b)   Phenol > benzene > chlorobenzene > benzoic acid.
c)   Aniline > benzene > bromobenzene > benzaldehyde.

16.13 An acyl substituent is deactivating. Once an aromatic ring has been acylated, it is much less reactive to further substitution. An alkyl substituent is activating, however, so an alkyl-substituted ring is more reactive than an unsubstituted ring, and polysubstitution occurs readily.

16.14 Toluene is more reactive toward electrophilic substitution than (trifluoromethyl)benzene. The electronegativity of the three fluorine atoms causes the trifluoromethyl group to be electron-withdrawing and deactivating toward electrophilic substitution.

16.15

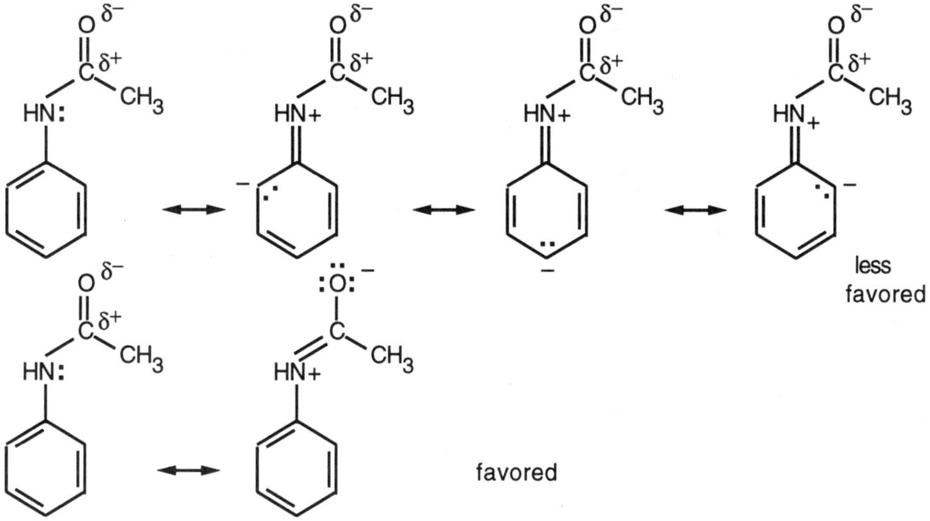

For acetanilide, resonance donation of the nitrogen lone pair electrons to the aromatic ring is less favored because the positive charge on nitrogen is next to the positively polarized carbonyl group. Resonance donation to the carbonyl oxygen *is* favored because of the electronegativity of oxygen. Since the nitrogen lone pair electrons are less available to the ring, the reactivity of the ring toward electrophilic substitution is decreased, and acetanilide is less reactive than aniline toward electrophilic substitution.

16.16

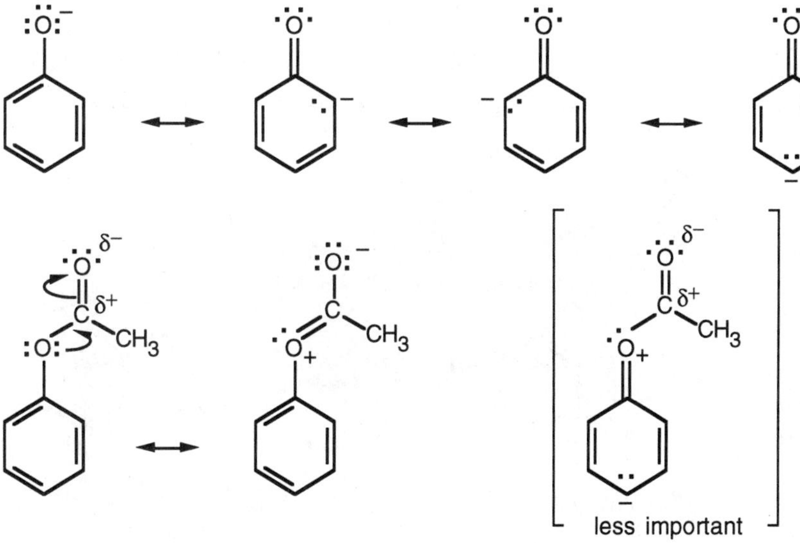

Phenoxide ion is the most reactive and phenyl acetate is the least reactive towards electrophilic substitution. The full negative charge of the phenoxide anion can be delocalized into the ring, which becomes electron rich and strongly activated toward electrophilic aromatic substitution. For phenyl acetate, delocalization of an electron pair onto the electronegative carbonyl oxygen makes the ring less reactive.

16.17 The aromatic ring is deactivated toward electrophilic aromatic substitution by the combined electron-withdrawing inductive effect of electronegative nitrogen and oxygen. The lone pair of electrons of nitrogen can, however, stabilize by resonance the ortho and para substituted intermediates but not the meta intermediate

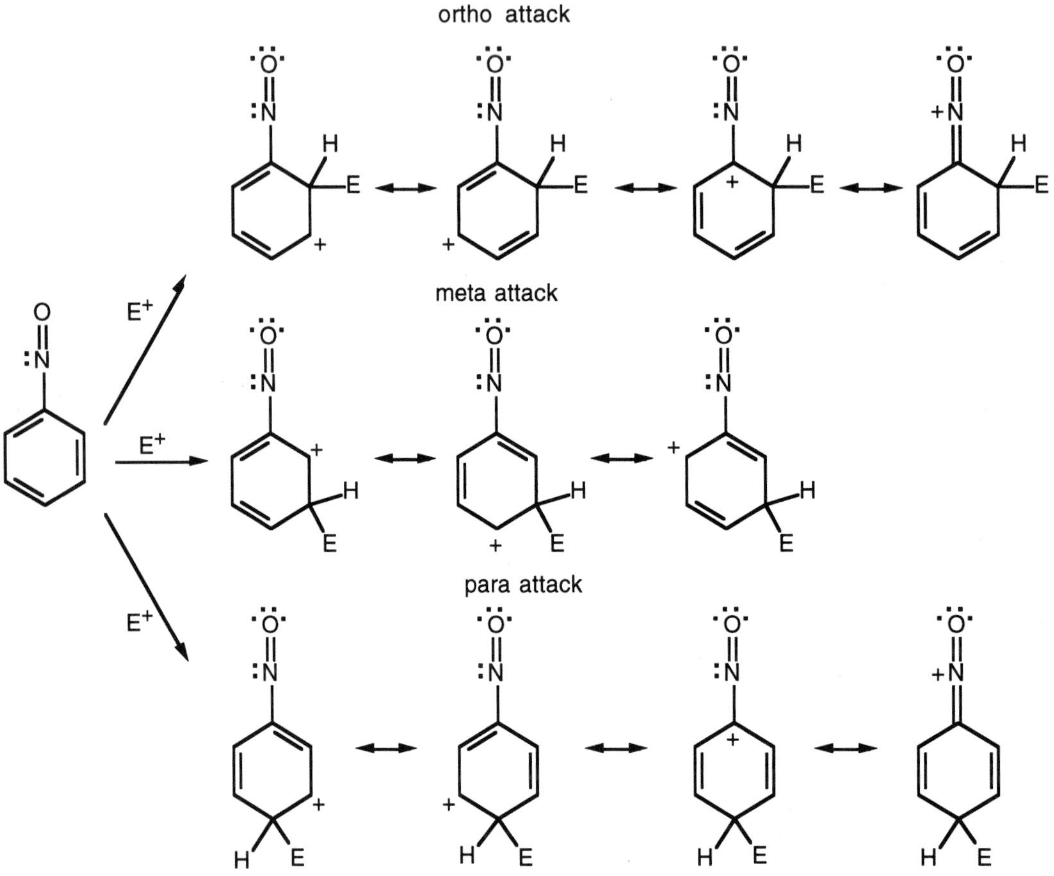

ortho attack

meta attack

para attack

16.18 Ortho attack:

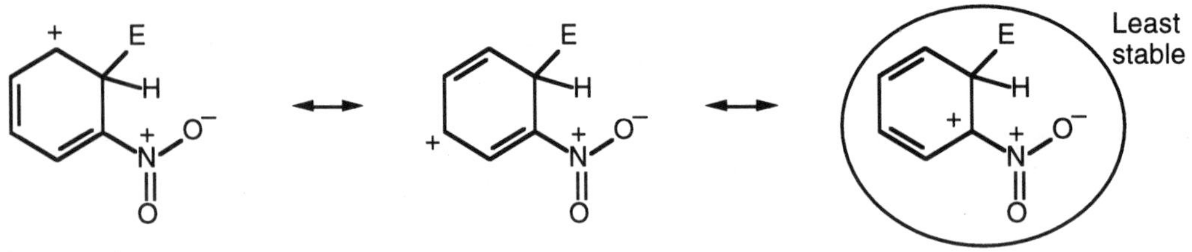

Least
stable

Meta attack:

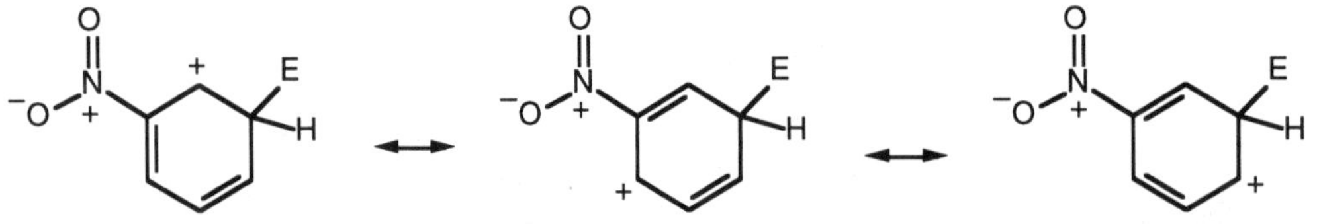

Para attack:

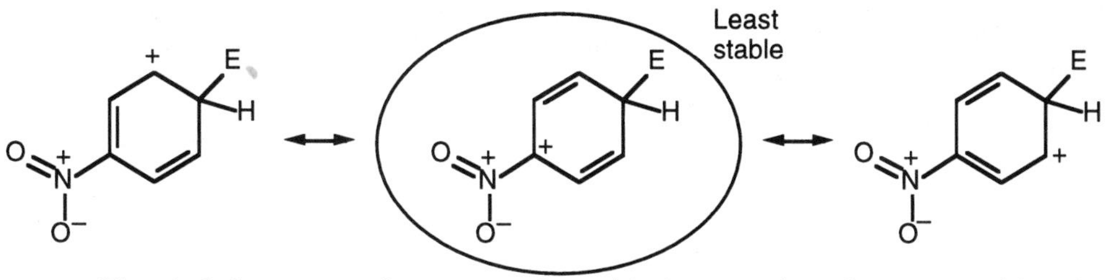

The circled resonance forms are unfavorable, because they place two positive charges adjacent to each other. The intermediate from *meta* attack is thus favored.

16.19

a)

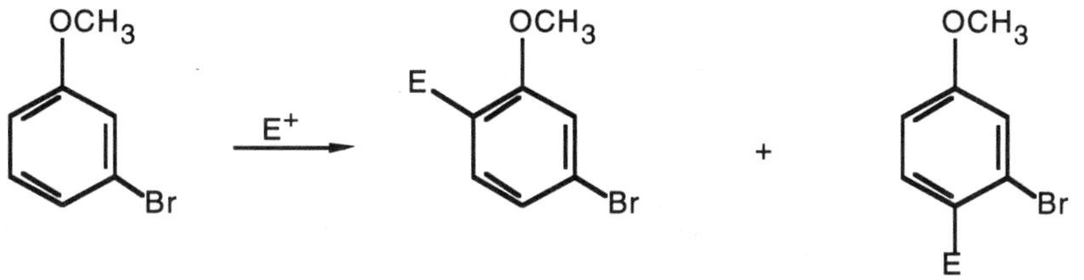

Both groups are ortho-para directors and direct substitution to the same positions. Attack doesn't occur between the two groups for steric reasons.

b)

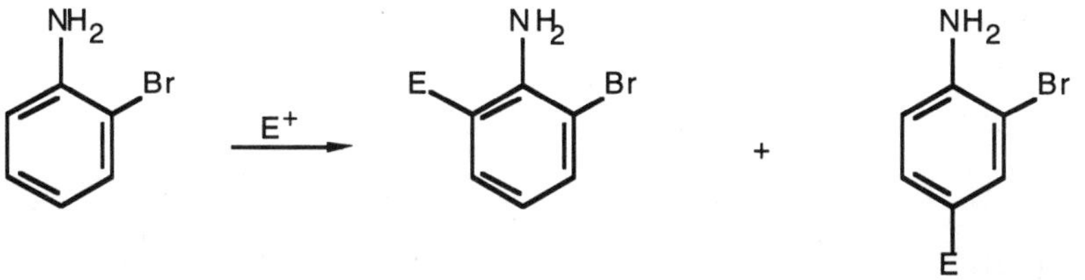

Both groups are ortho-para directors but direct substitution to different positions. Because –NH$_2$ group is a more powerful activator, substitution occurs ortho and para to it.

c)

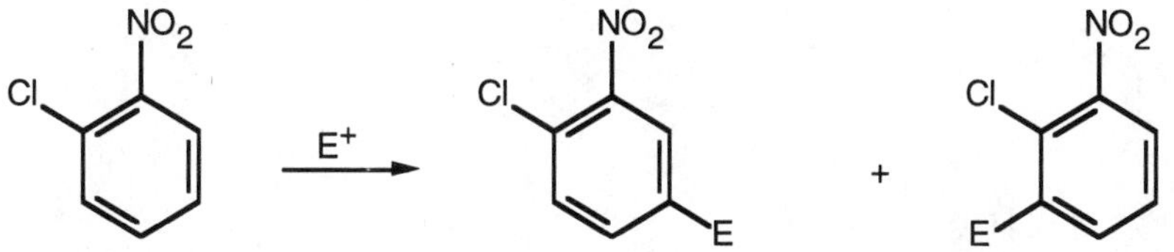

Both groups are deactivating, but they orient substitution toward the same positions.

16.20

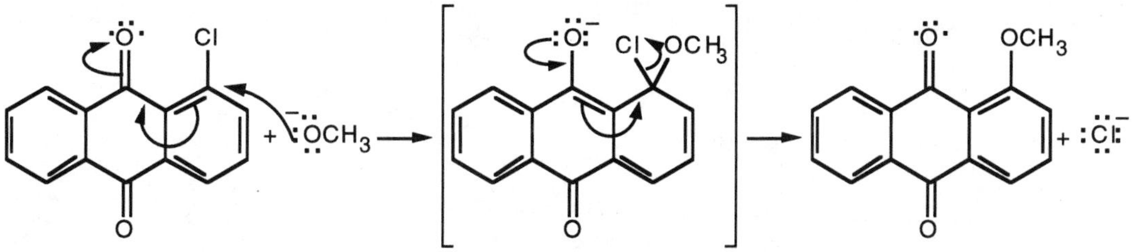

The carbonyl oxygens make the chlorine–containing ring electron-poor and vulnerable to attack by the nucleophile ⁻OCH₃. They also stabilize the negatively charged Meisenheimer complex. This nucleophilic aromatic substitution occurs by an addition-elimination pathway.

16.21

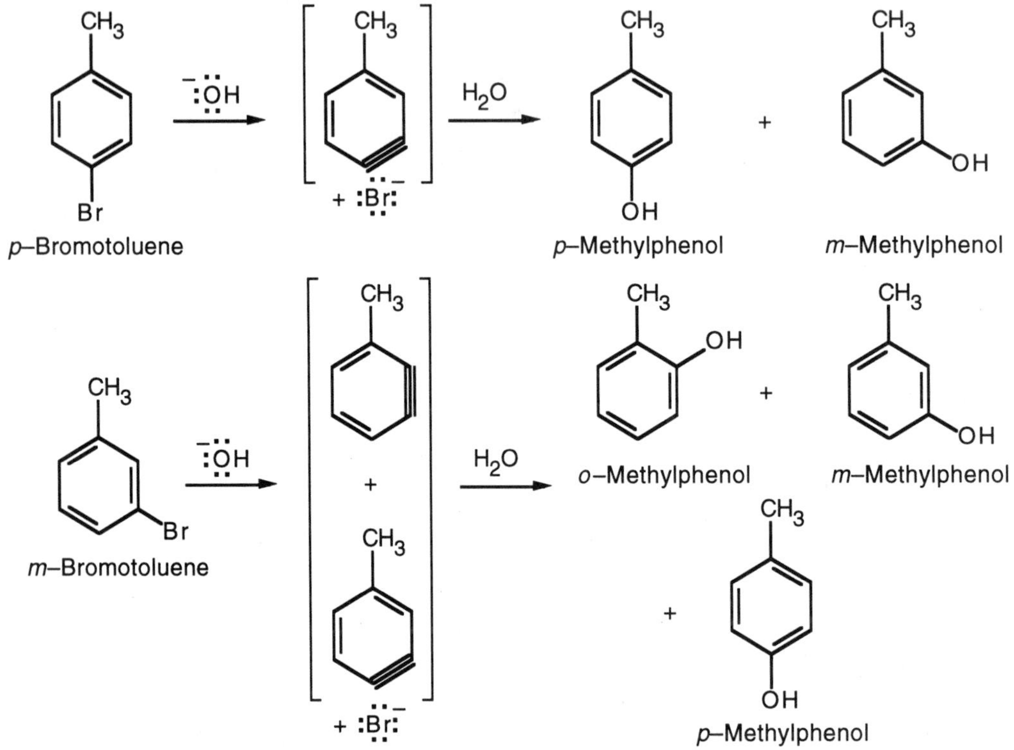

p–Bromotoluene   *p*–Methylphenol   *m*–Methylphenol

m–Bromotoluene   *o*–Methylphenol   *m*–Methylphenol

p–Methylphenol

Treatment of *m*–bromotoluene with NaOH leads to two benzyne intermediates, which react with water to yield three methylphenol products.

16.22

a) $\xrightarrow[\text{H}_2\text{O}]{\text{KMnO}_4}$

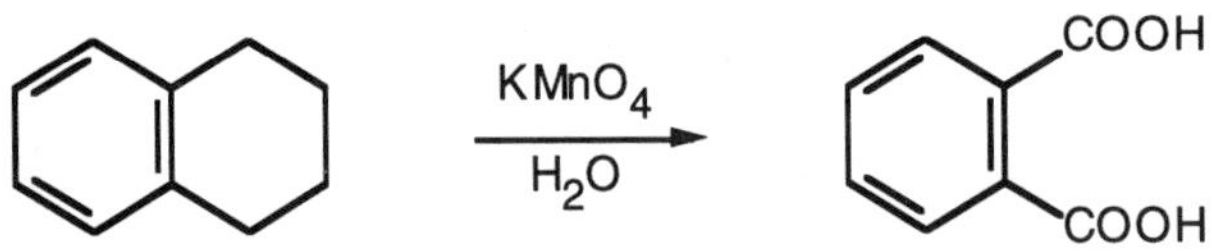

b) $\xrightarrow[\text{H}_2\text{O}]{\text{KMnO}_4}$

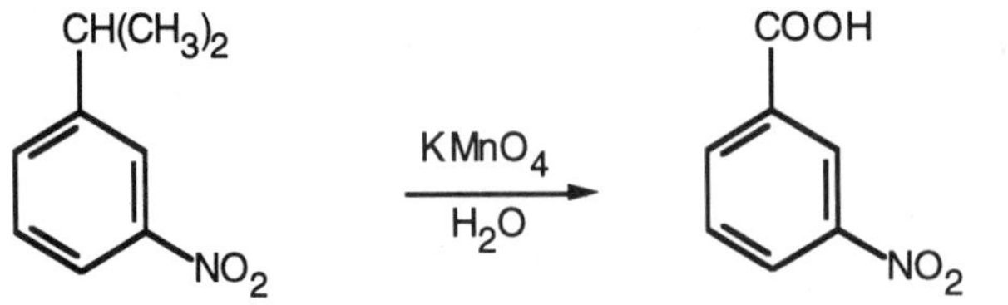

16.23

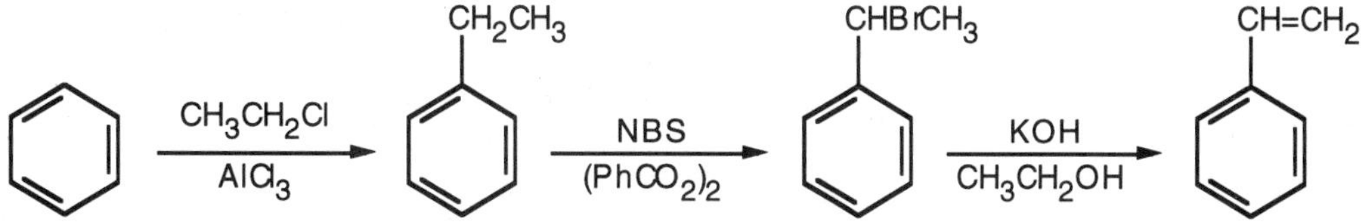

16.24

Bond	$CH_3CH_2–H$	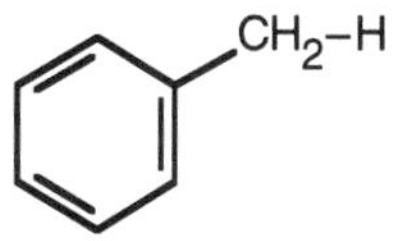	$H_2C{=}CHCH_2–H$
Bond dissociation energy	420 kJ/mol	368 kJ/mol	361 kJ/mol

Bond dissociation energies measure the amount of energy that must be supplied to cleave a bond into two radical fragments. A radical is thus higher in energy and less stable than the compound it came from. Since the C–H bond dissociation energy is 420 kJ/mol for ethane and 368 kJ/mol for a methyl group C–H bond of toluene, less energy is required to form a benzyl radical than to form an ethyl radical. A benzyl radical is thus more stable than a primary alkyl radical by 52 kJ/mol. The bond dissociation energy of an allyl C–H bond is 361 kJ/mol, indicating that a benzyl radical is nearly as stable as an allyl radical.

16.25

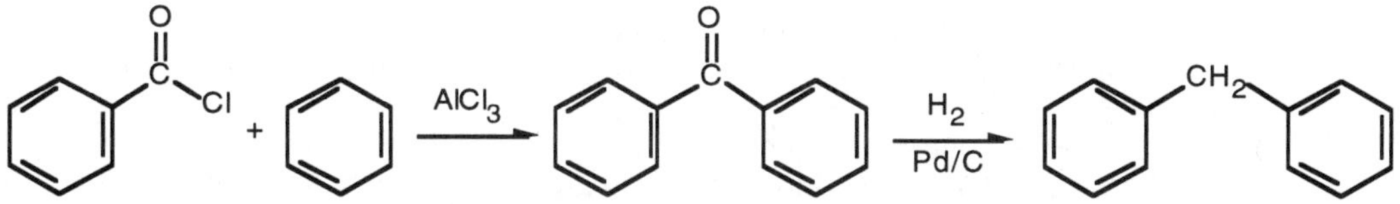

16.26

a)

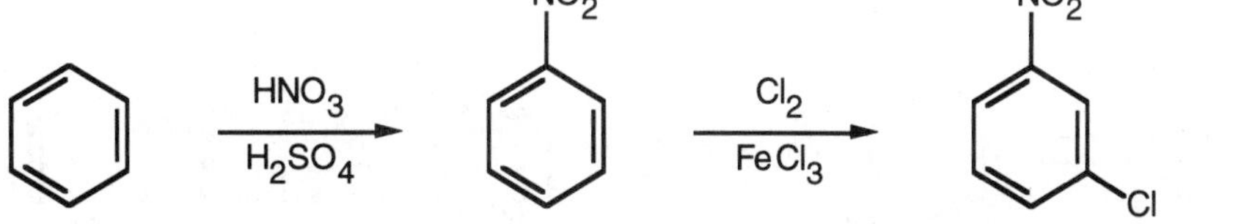

m–Chloronitrobenzene

b)

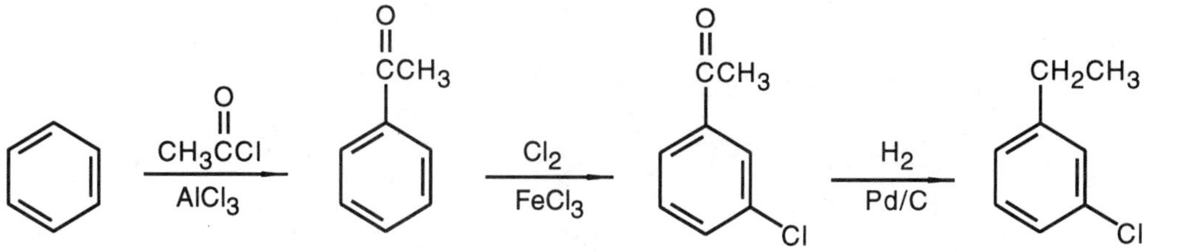

m–Chloroethyl–
benzene

c) Two routes are possible:

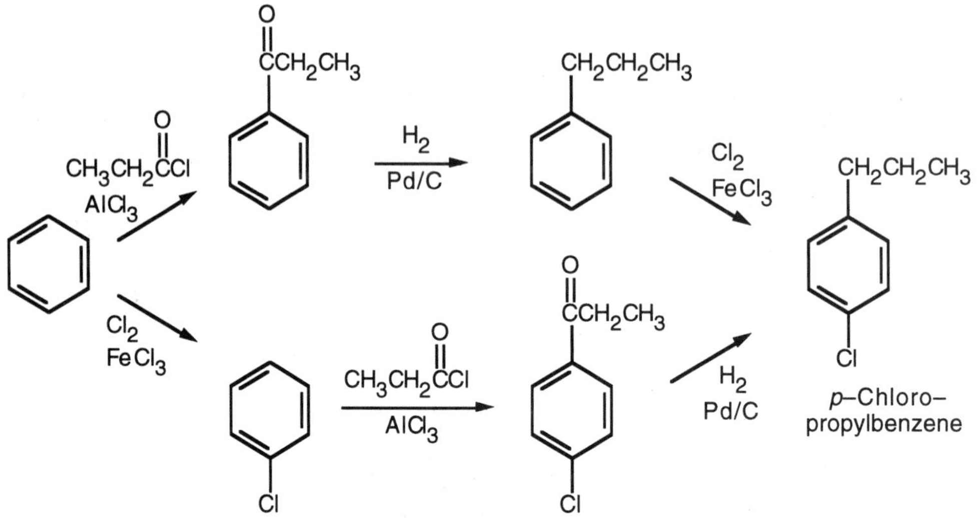

p–Chloro–propylbenzene

16.27 a) Friedel-Crafts acylation, like Friedel-Crafts alkylation, does not occur at an aromatic ring carrying a meta-directing group. We will learn in a later chapter how to synthesize this compound by another route.

b) There are two problems with this synthesis as it is written:
1. Rearrangement often occurs during Friedel-Crafts alkylations using primary halides.
2. Even if *p*–chloropropylbenzene could be synthesized, introduction of the second –Cl group would occur ortho to the alkyl group.

A possible route to this compound:

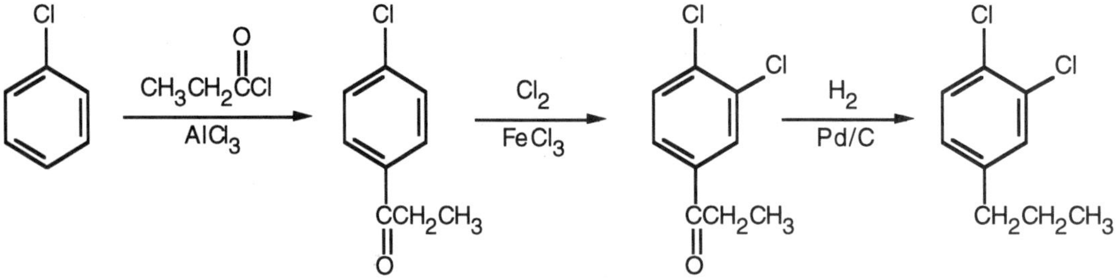

16.28

a)

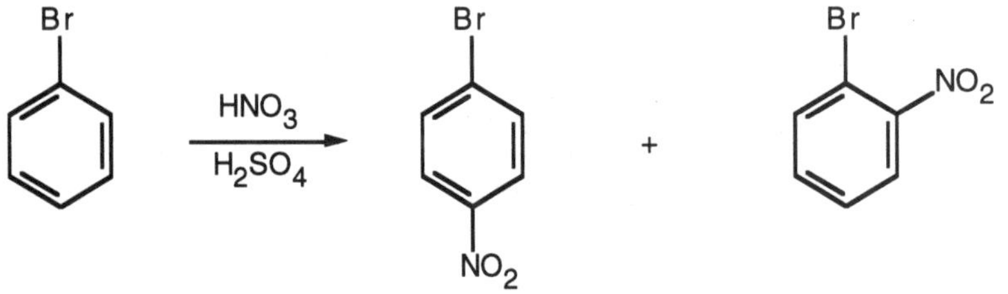

p–Bromonitrobenzene *o*–Bromonitrobenzene

b)

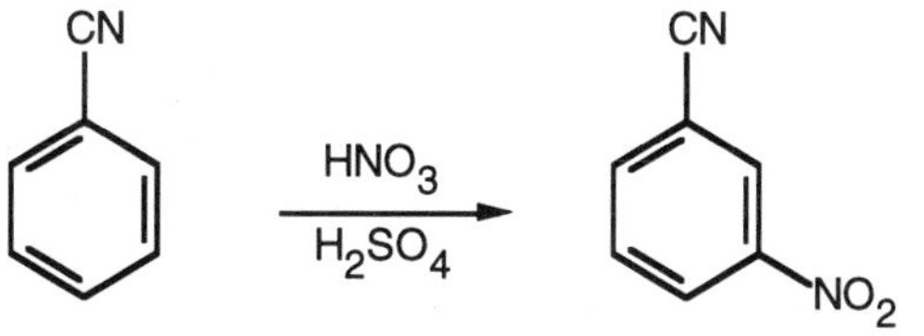

m–Nitrobenzonitrile

c)

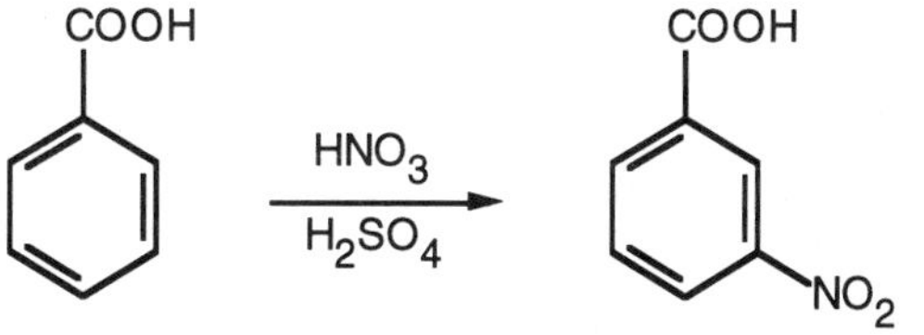

m–Nitrobenzoic acid

d)

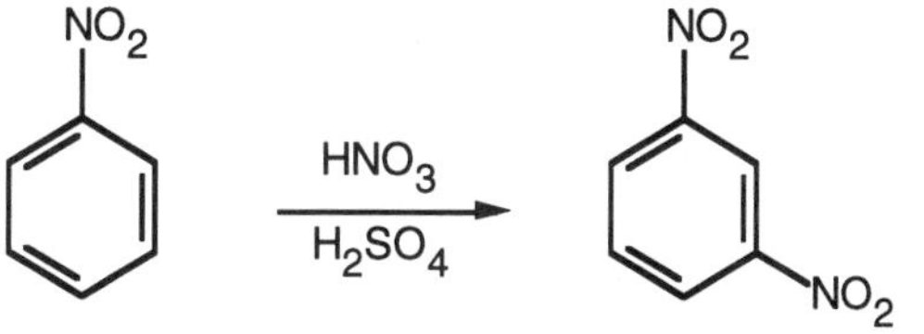

m–Dinitrobenzene

e)

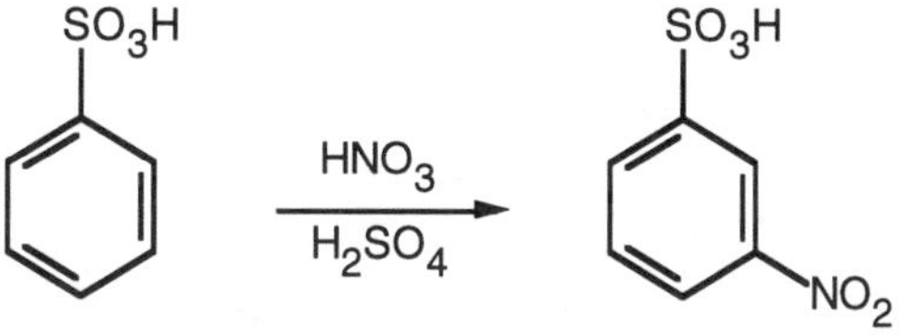

m–Nitrobenzenesulfonic acid

f)

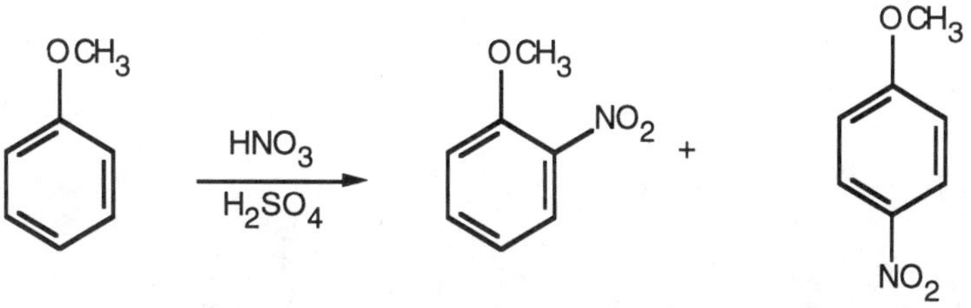

o–Methoxynitrobenzene   *p*–Methoxynitrobenzene

Only methoxybenzene reacts faster than benzene.

16.29   Most reactive ————————————————————> Least reactive

(a) Benzene > Chlorobenzene > *o*-Dichlorobenzene
(b) Phenol > Nitrobenzene > *p*-Bromonitrobenzene
(c) *o*-Xylene > Fluorobenzene > Benzaldehyde
(d) *p*-Methoxybenzonitrile > *p*-Methylbenzonitrile > Benzonitrile

16.30

a)

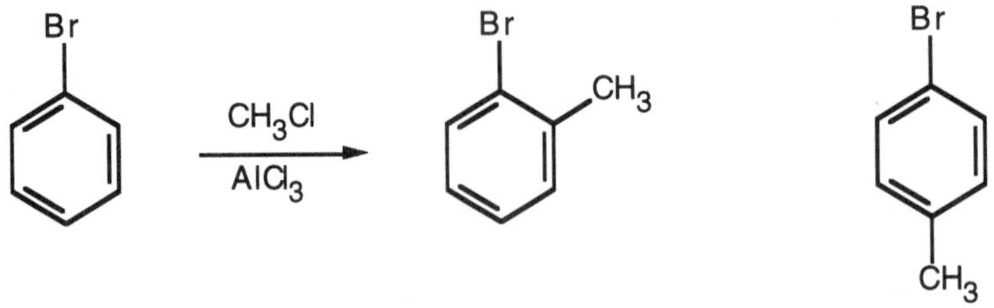

o–Bromotoluene   p–Bromotoluene

b)

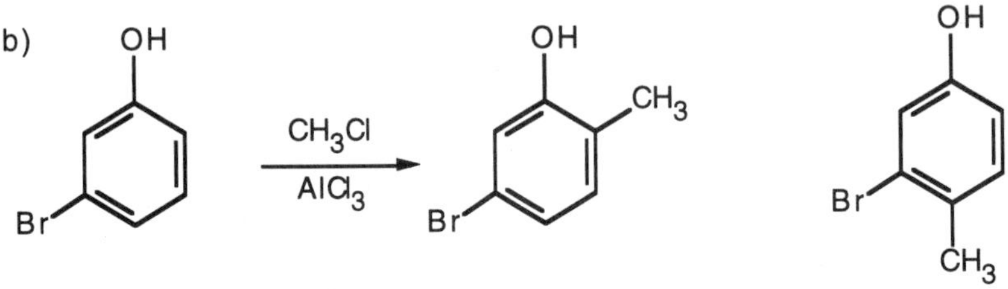

5–Bromo–2–methyl–   3–Bromo–4–methyl
phenol   phenol

c)

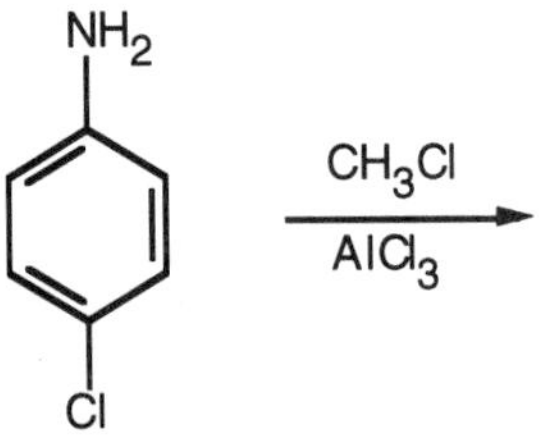

No reaction. AlCl$_3$ combines with $-\ddot{N}H_2$ to form a complex that deactivates the ring toward Friedel-Crafts alkylation.

d)

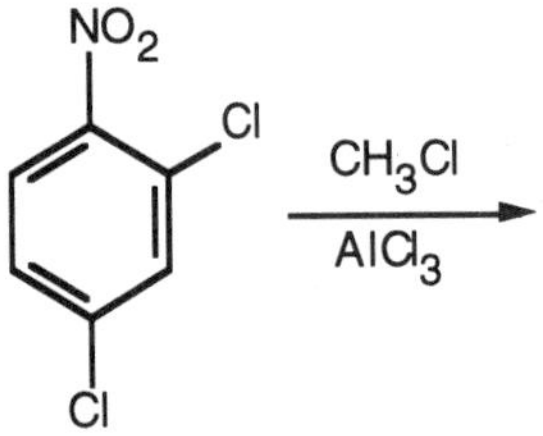

No reaction. The ring is deactivated

e)

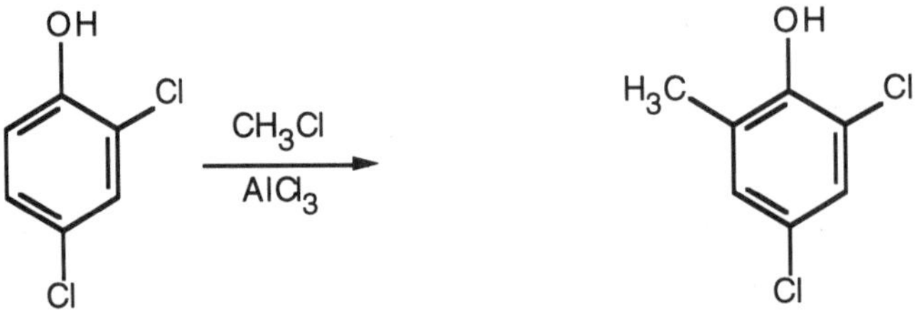

2,4–Dichloro–6–methylphenol

f)

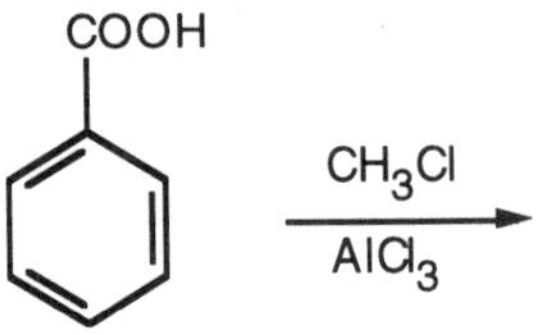

No reaction.

g)

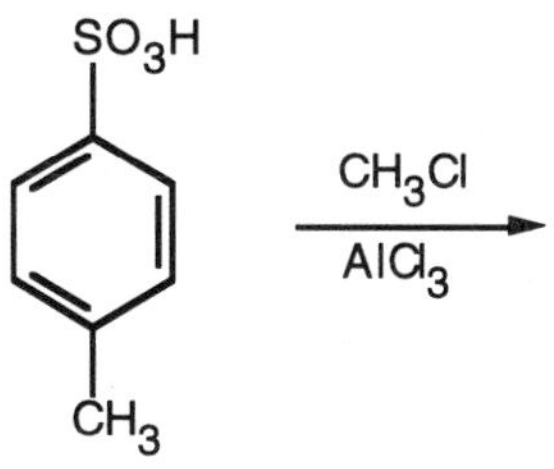

CH₃Cl / AlCl₃

No reaction. The ring is deactivated

h)

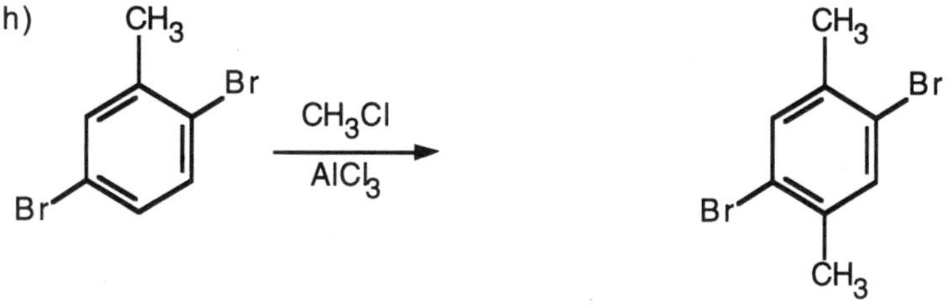

CH₃Cl / AlCl₃

1,4–Dibromo–2,5–dimethylbenzene

16.31

a)

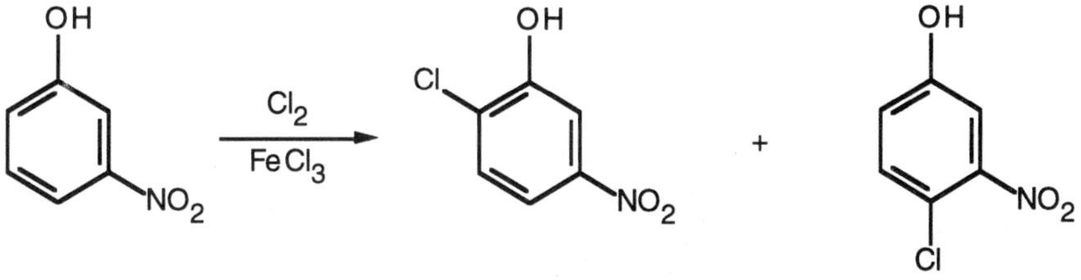

Cl₂ / FeCl₃

2–Chloro–5–nitrophenol   +   4–Chloro–3–nitrophenol

b)

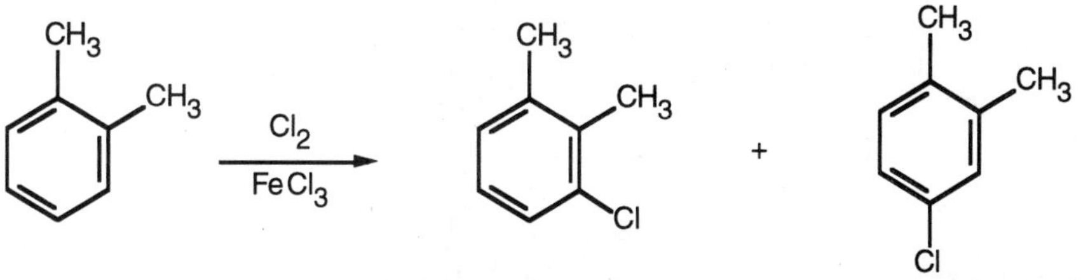

Cl₂ / FeCl₃

1–Chloro–2,3–dimethylbenzene   +   4–Chloro–1,2–dimethylbenzene

c)

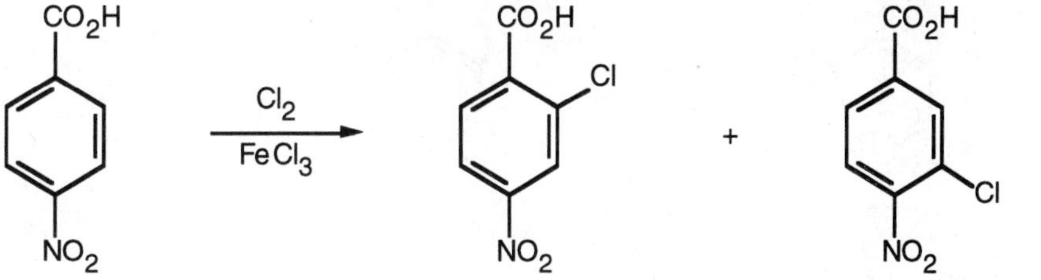

Cl₂ / FeCl₃

2–Chloro–4–nitro-benzoic acid   +   3–Chloro–4–nitro-benzoic acid

d)

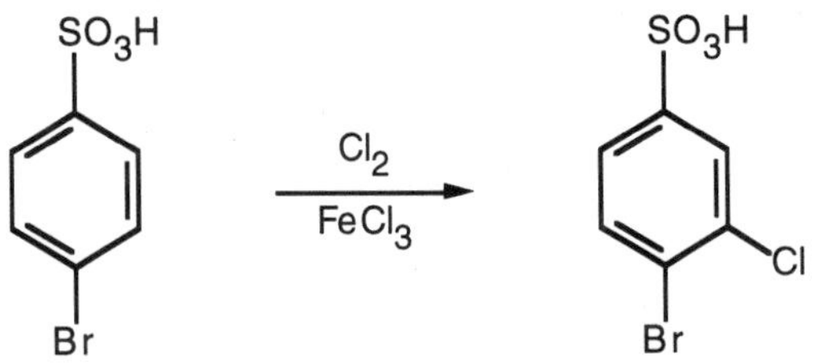

4–Bromo–3–chloro-
benzenesulfonic acid

16.32

a)

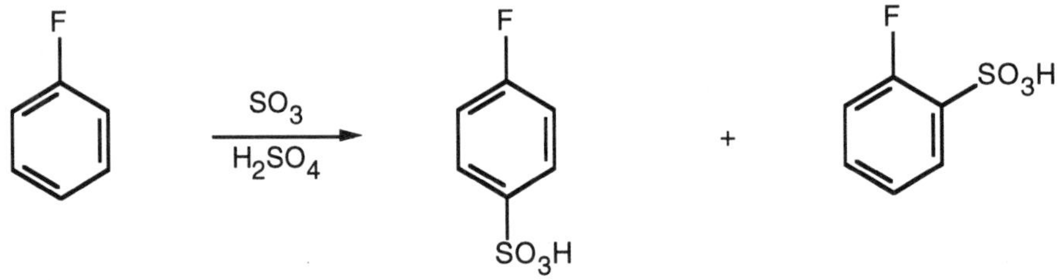

b)

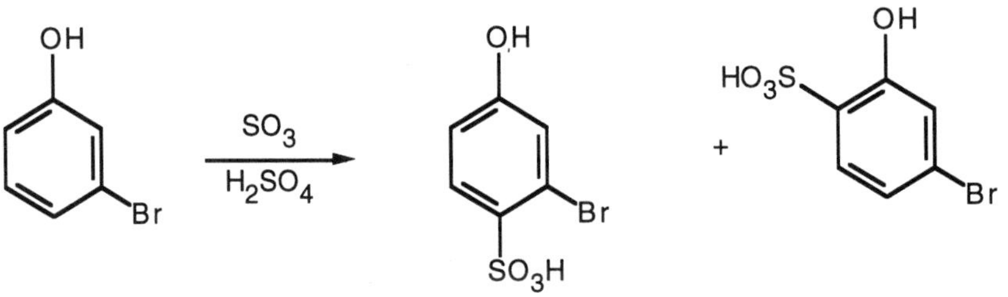

c)

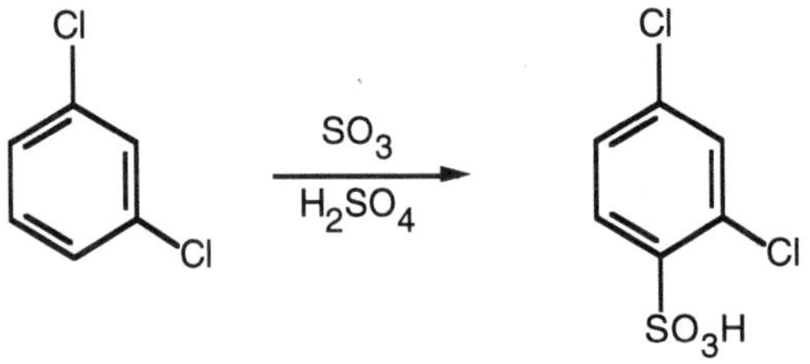

d)

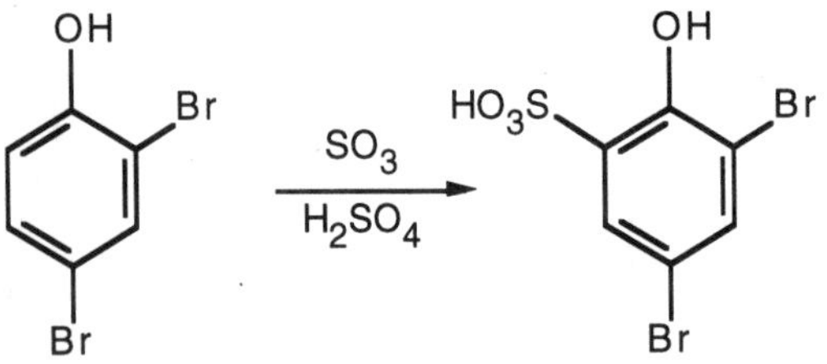

16.33 Most reactive ──────────────────────→ Least reactive

Phenol > Toluene > *p*-Bromotoluene > Bromobenzene

Aniline and nitrobenzene do not undergo Friedel-Crafts alkylations.

16.34

Resonance structures show that bromination occurs in the ortho and para positions of the rings. The positively charged intermediate formed from ortho or para attack can be stabilized by resonance contributions from the second ring of biphenyl, but this stabilization is not possible for meta attack.

16.35

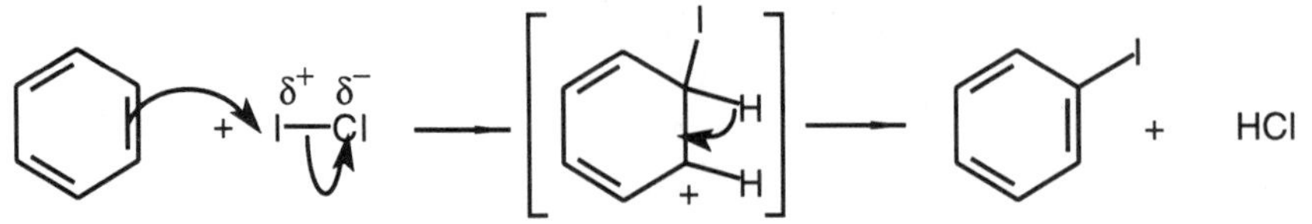

$$\delta^+ \quad \delta^-$$
ICl can be represented as I—Cl because chlorine is a more electronegative element than iodine. The iodine atom can act as an electrophile in electrophilic aromatic substitution reactions.

16.36

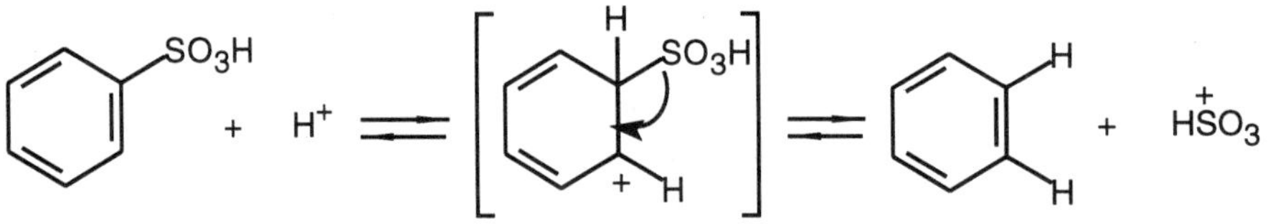

This mechanism is the reverse of the sulfonation mechanism illustrated in the text. H^+ is the electrophile in this reaction.

16.37

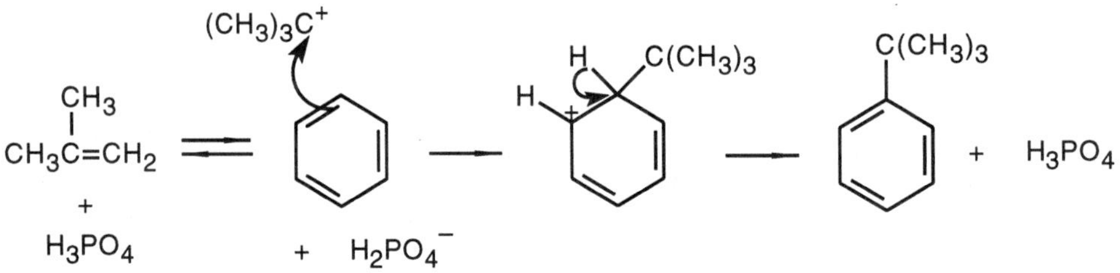

16.38
Ortho attack:

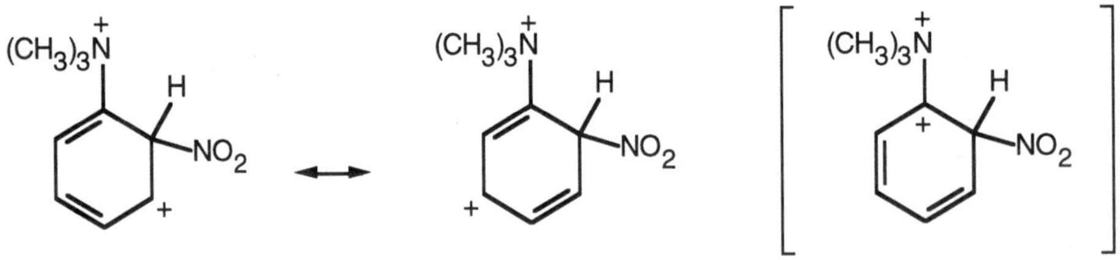

This form is not an important resonance form because two positive charges are next to each other.

Meta attack -- *MOST FAVORED*

Para attack:

This form is not important.

The *N,N,N*-trimethylammonium group has no electron-withdrawing resonance effect because it has no vacant *p* orbitals to overlap with the π orbital system of the aromatic ring. The $(CH_3)_3N^+-$ group is inductively deactivating, however, because it is positively charged. It is meta-directing because the cationic intermediate resulting from meta attack is somewhat more stable than those resulting from ortho or para attack.

16.39 Attack occurs on the unsubstituted ring because bromine is a deactivating group. Attack occurs at the ortho and para positions of the ring because the positively charged intermediate can be stabilized by resonance contributions from bromine and from the second ring (Problem 16.34).

16.40 When directly attached to a ring, the –CN group is a meta-directing deactivator for both inductive and resonance reasons. In 3–phenylpropanenitrile, however, the saturated side chain does not allow resonance interactions of –CN with the aromatic ring, and the –CN group is too far from the ring for its inductive effect to be strongly felt. The side chain acts as an alkyl substituent, and ortho–para substitution is observed.

In 3–phenylpropenenitrile, the –CN group interacts with the ring through the π electrons of the side chain. Resonance forms show that –CN deactivates the ring toward electrophilic substitution, and substitution occurs at the meta position.

16.41

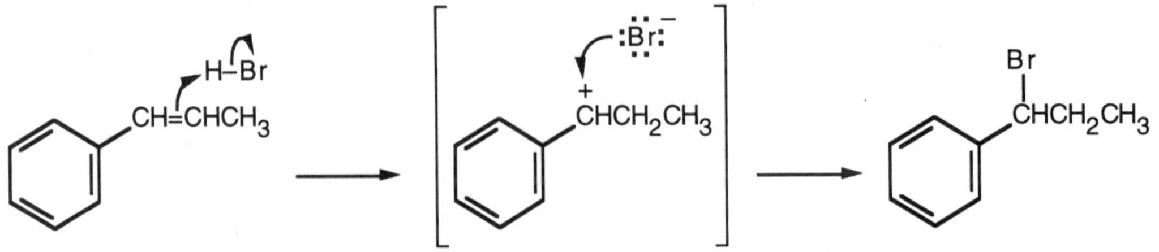

Protonation of the double bond at carbon 2 of 1–phenylpropene leads to an intermediate that can be stabilized by resonance involving the benzene ring.

16.42

$$CHCl_3 + AlCl_3 \rightleftharpoons {}^+CHCl_2 \ {}^-AlCl_4$$

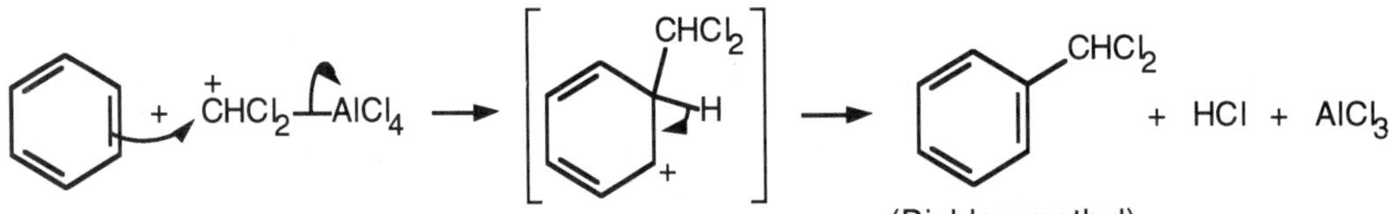

(Dichloromethyl)-
benzene

(Dichloromethyl)benzene can react with two additional equivalents of benzene to produce triphenylmethane.

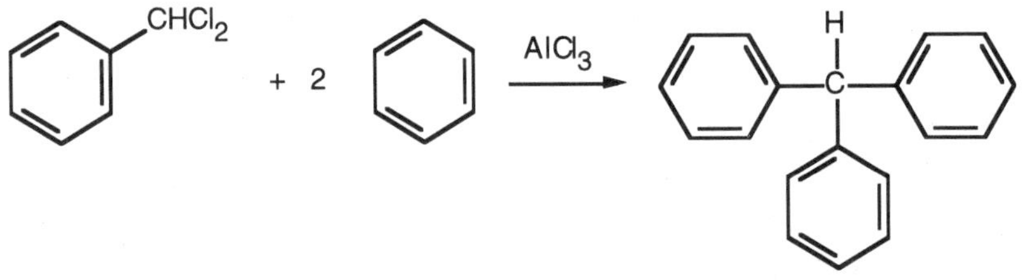

Triphenylmethane

16.43

a)

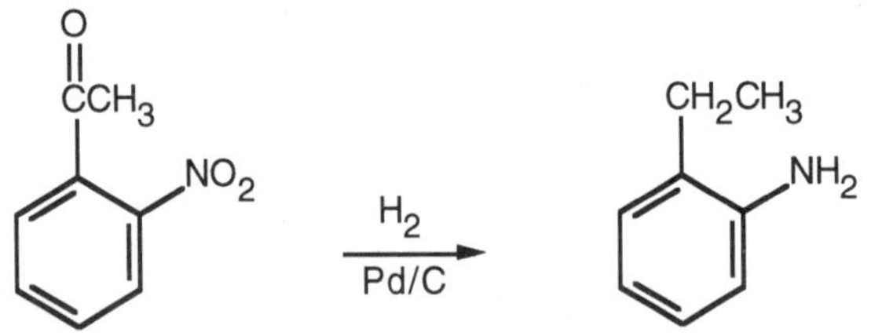

b) No reaction. Since this compound has no benzylic protons, it can't be oxidized by $KMnO_4$.

c)

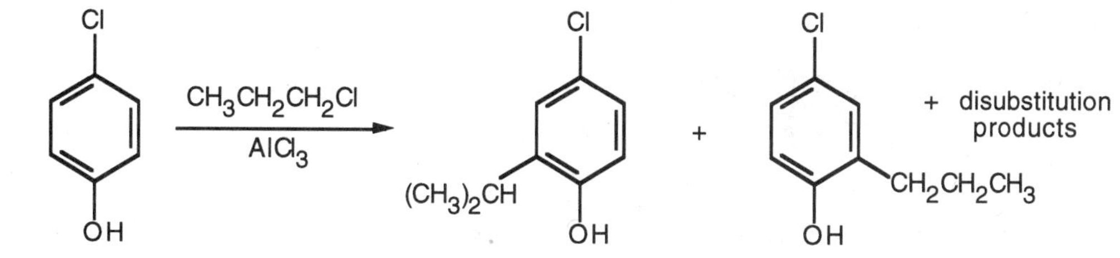

major product

16.44

a)

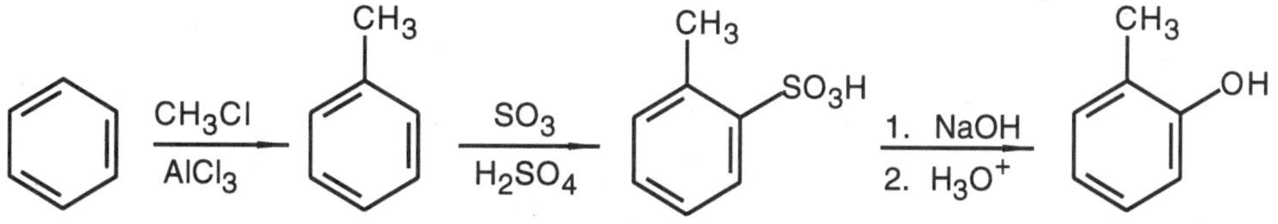

o–Methylphenol

b)

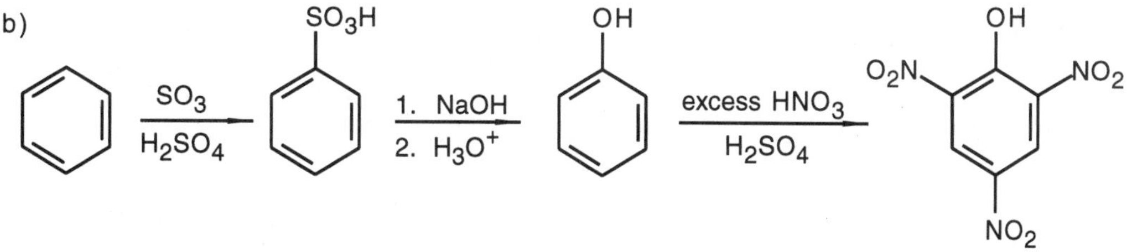

2,4,6–Trinitrophenol

c)

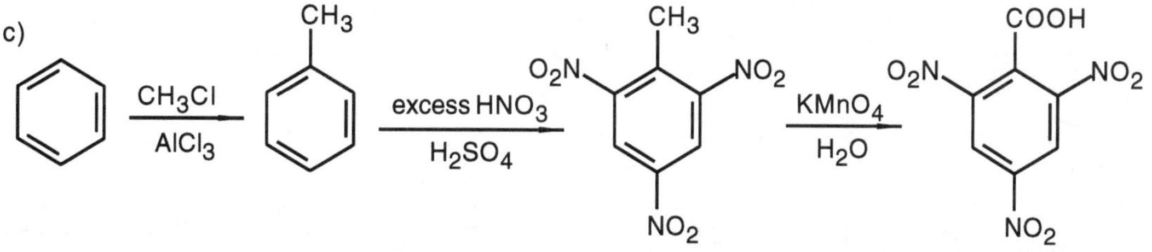

2,4,6–Trinitrobenzoic acid

d)

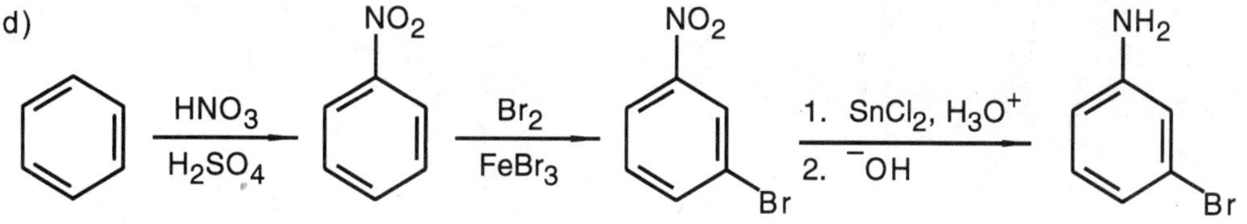

m–Bromoaniline

16.45

a)

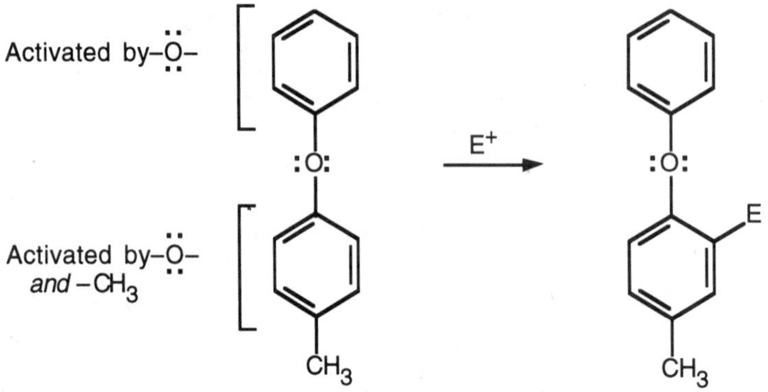

Activated by –Ö–

Activated by –Ö–
and –CH₃

E^+

Substitution occurs in the more activated ring. The position of substitution is determined by the more powerful activating group – in this case, the ether oxygen.

b)

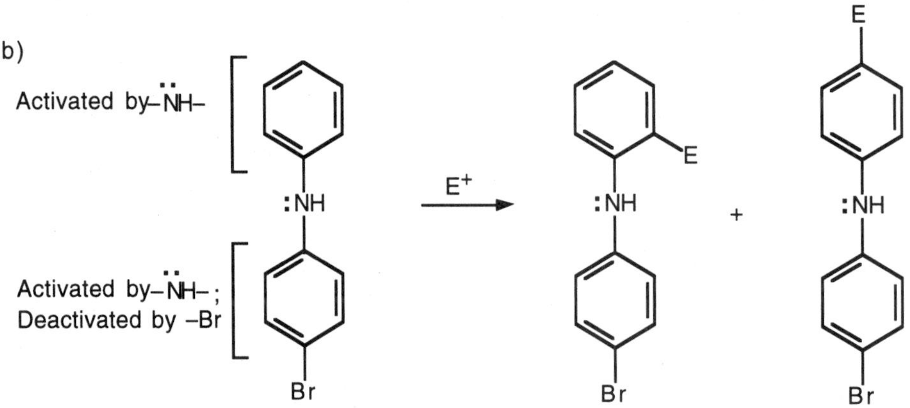

Activated by –N̈H–

Activated by –N̈H– ;
Deactivated by –Br

E^+

+

The upper ring is more activated than the lower ring. –NH– is an *ortho-para* director.

c)

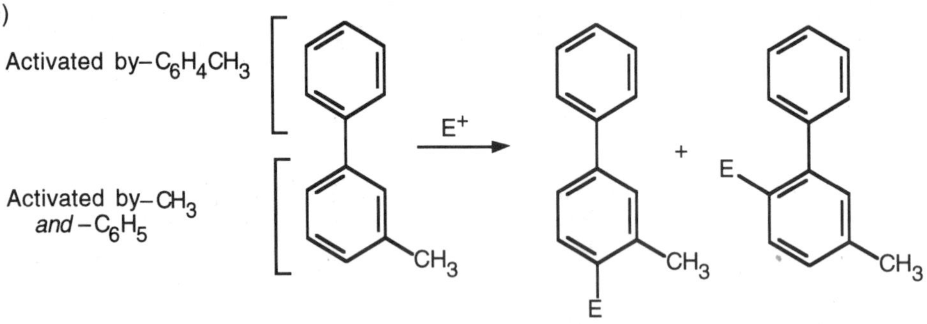

Activated by –C₆H₄CH₃

Activated by –CH₃
and –C₆H₅

E^+

+

Substitution occurs in the ortho and para positions of the more activated ring. Substitution doesn't occur between –C₆H₅ and –CH₃ for steric reasons.

16.46

Attack occurs in the activated ring and yields ortho and para bromination products. The intermediate is resonance-stabilized by overlap of the nitrogen lone pair electrons with the π electrons of the substituted ring.

Similar drawings can be made of the resonance forms of the intermediate resulting from ortho attack.

16.47 Reaction of (R)-2-chlorobutane with $AlCl_3$ produces an ion pair [$CH_3{}^+CHCH_2CH_3$ $^-AlCl_4$]. The planar, sp^2-hybridized carbocation is achiral, and its reaction with benzene yields racemic product.

16.48 When synthesizing substituted aromatic rings, it is necessary to introduce substituents in the proper order. A group that is introduced out of order will not have the proper directing effect. Remember that in many of these reactions a mixture of ortho and para isomers may be formed.

a)

C_6H_6 $\xrightarrow[\text{H}_2\text{SO}_4]{\text{SO}_3}$ [benzene with SO₃H] $\xrightarrow[\text{2. H}_3\text{O}^+]{\text{1. NaOH}}$ [phenol OH] $\xrightarrow[\text{FeCl}_3]{\text{Cl}_2}$ [p-chlorophenol OH with Cl]

p–Chlorophenol

b)

C_6H_6 $\xrightarrow[\text{H}_2\text{SO}_4]{\text{HNO}_3}$ [nitrobenzene NO₂] $\xrightarrow[\text{FeBr}_3]{\text{Br}_2}$ [m-bromonitrobenzene]

m–Bromonitrobenzene

c)

C_6H_6 $\xrightarrow[\text{FeBr}_3]{\text{Br}_2}$ [bromobenzene Br] $\xrightarrow[\text{H}_2\text{SO}_4]{\text{SO}_3}$ [o-Bromobenzene-sulfonic acid] + [p-bromobenzenesulfonic acid Br / SO₃H]

o–Bromobenzene-
sulfonic acid

d)

C_6H_6 $\xrightarrow[\text{H}_2\text{SO}_4]{\text{SO}_3}$ [benzenesulfonic acid SO₃H] $\xrightarrow[\text{FeCl}_3]{\text{Cl}_2}$ [m-Chlorobenzene-sulfonic acid SO₃H / Cl]

m–Chlorobenzene-
sulfonic acid

16.49

a)

[toluene CH₃] $\xrightarrow[\text{H}_2\text{SO}_4]{\text{HNO}_3}$ [p-nitrotoluene CH₃ / NO₂] $\xrightarrow[\text{FeBr}_3]{\text{Br}_2}$ [2-Bromo-4-nitrotoluene CH₃ / Br / NO₂]

2–Bromo–4–nitrotoluene

b)

[benzene] $\xrightarrow[\text{H}_2\text{SO}_4]{\text{3 HNO}_3}$ [1,3,5-Trinitrobenzene with NO₂, O₂N, NO₂]

1,3,5–Trinitrobenzene

c)

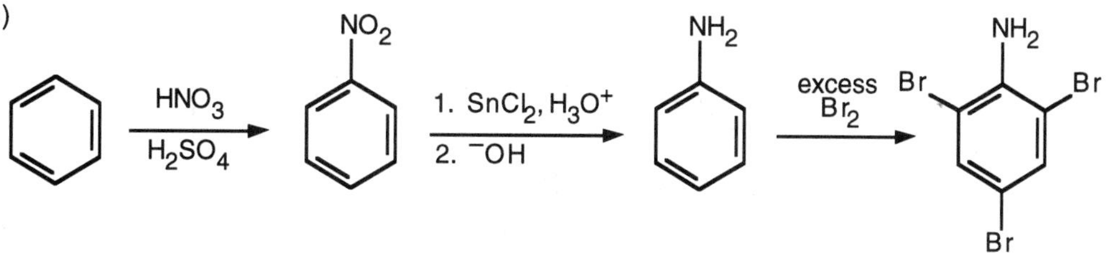

2,4,6–Tribromoaniline

No catalyst is needed for bromination because aniline is very activated toward substitution.

d)

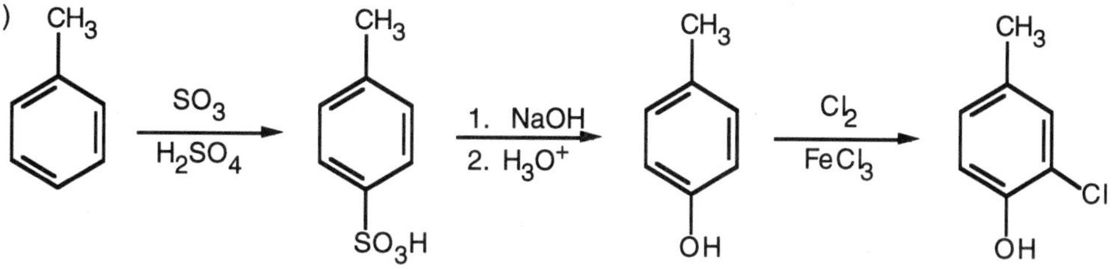

plus ortho
isomer

2–Chloro–4–
methylphenol

The –OH group is a stronger director than –CH$_3$.

16.50 a) Chlorination of toluene occurs at the ortho and para positions. To synthesize the given product, first oxidize toluene to benzoic acid and then chlorinate.

b) *p*–Nitrochlorobenzene is inert to Friedel-Crafts alkylation because the nitrochlorobenzene ring is deactivated.

c) The first two steps in the sequence are correct, but H$_2$/Pd reduces the nitro group as well as the ketone.

16.51

a)

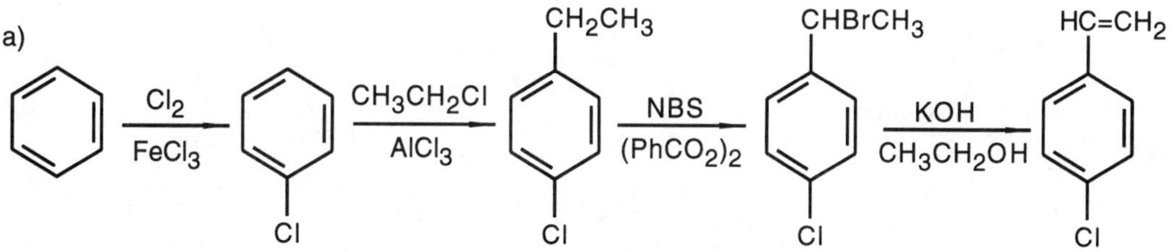

b)

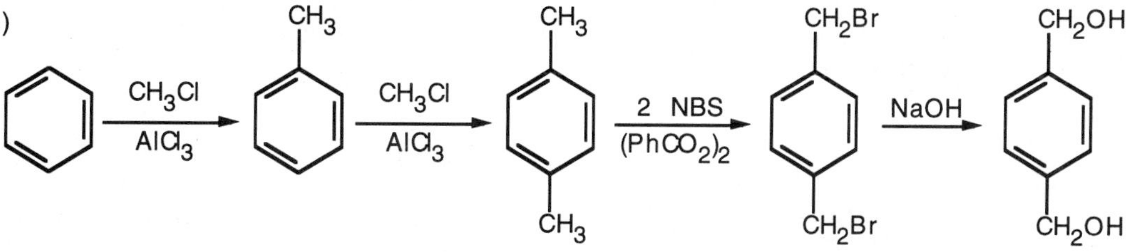

c)

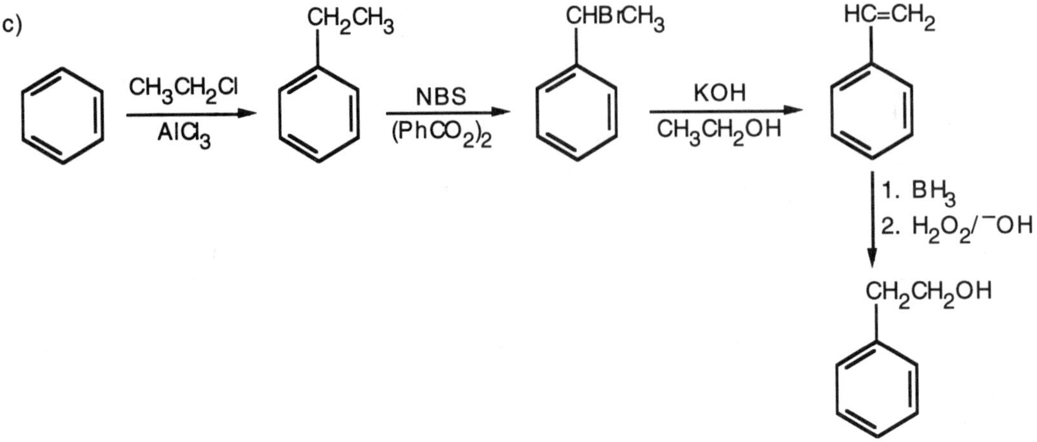

16.52

a)

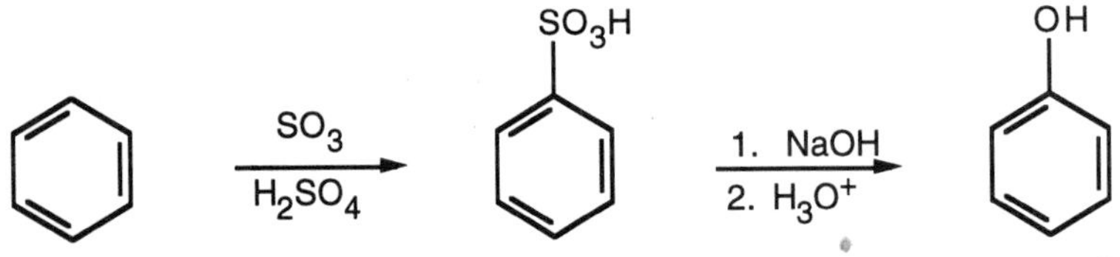

Phenol

b)

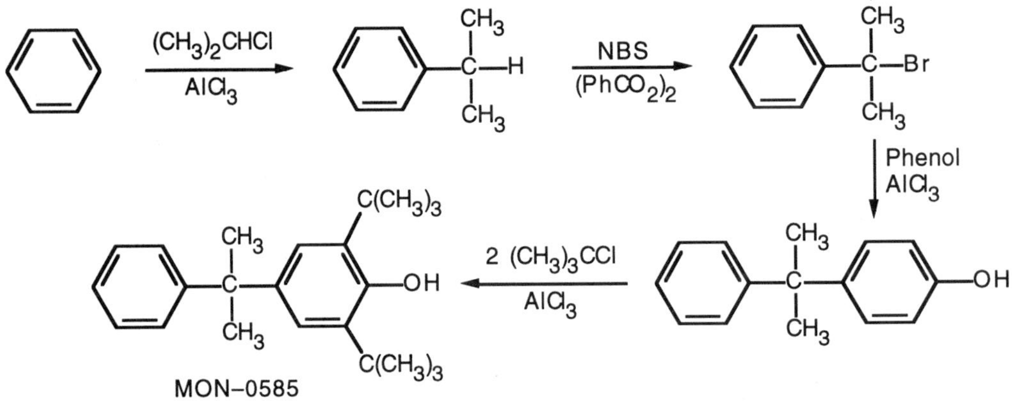

MON–0585

16.53

1)

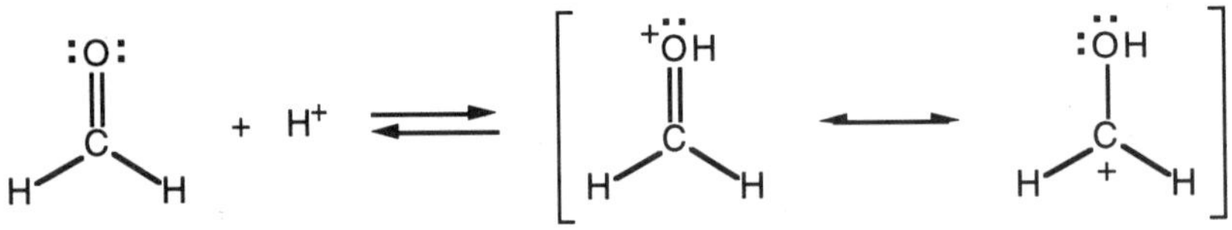

Formaldehyde is protonated to form a carbocation.

2)

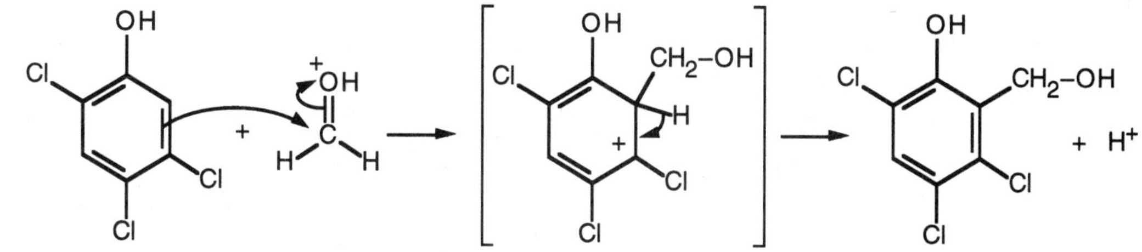

The formaldehyde cation acts as the electrophile in a substitution reaction at the 6 position of 2,4,5–trichlorophenol.

3)

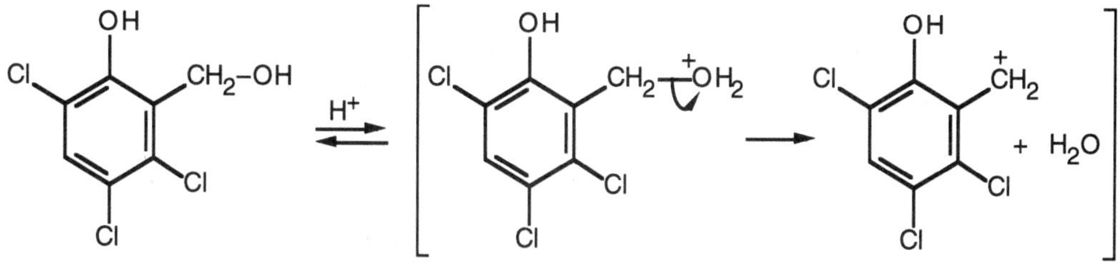

The product from step 2 is protonated by strong acid to produce a carbocation.

4)

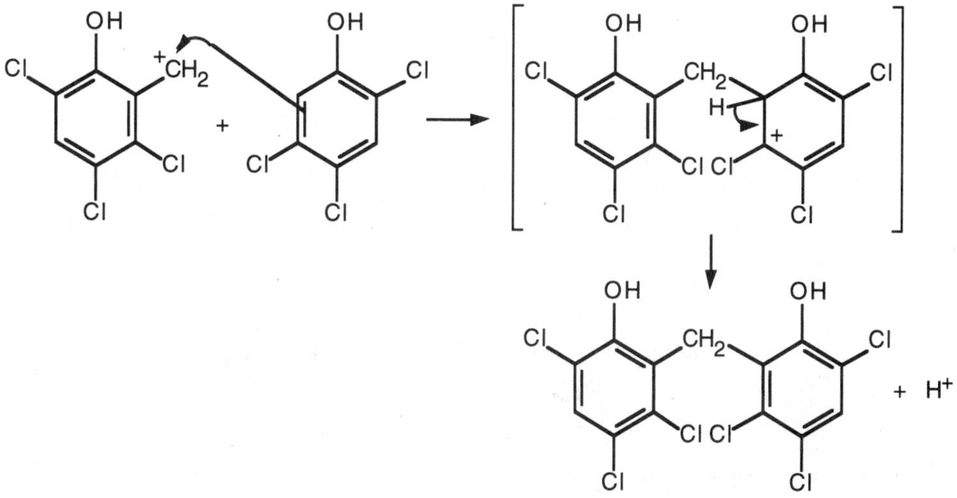

Hexachlorophene

This carbocation is attacked by a second molecule of 2,4,5–trichlorophenol to produce hexachlorophene.

16.54

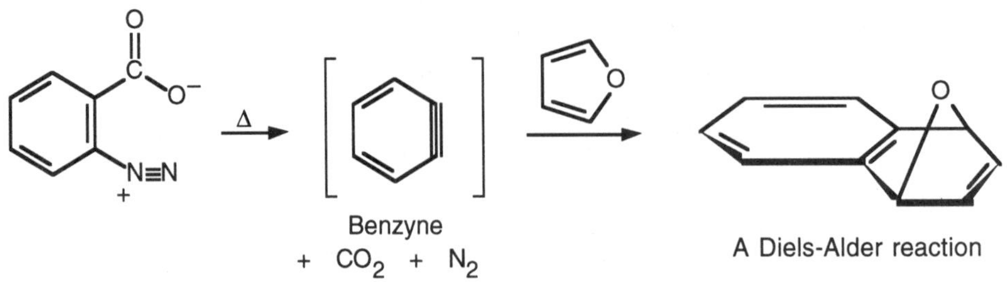

Benzyne

+ CO$_2$ + N$_2$

A Diels-Alder reaction

16.55

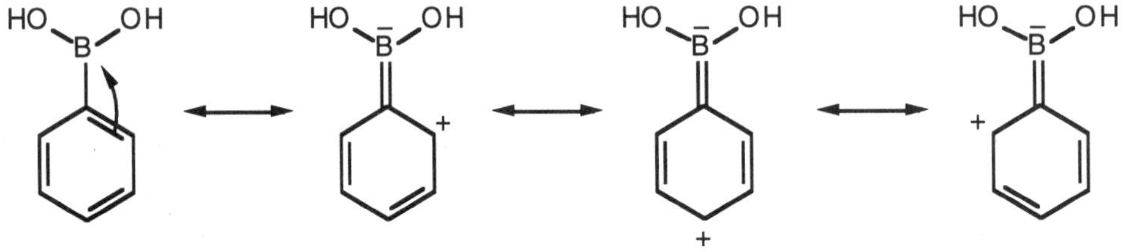

The trivalent boron in phenylboronic acid is a Lewis acid (electron pair acceptor). It is possible to write resonance forms for phenylboronic acid in which an electron pair from the phenyl ring is delocalized onto boron. In these resonance forms, the ortho and para positions of phenylboronic acid are the most electron-deficient, and substitutions occur primarily at the meta position.

16.56 Resonance forms from the intermediate from attack at C1:

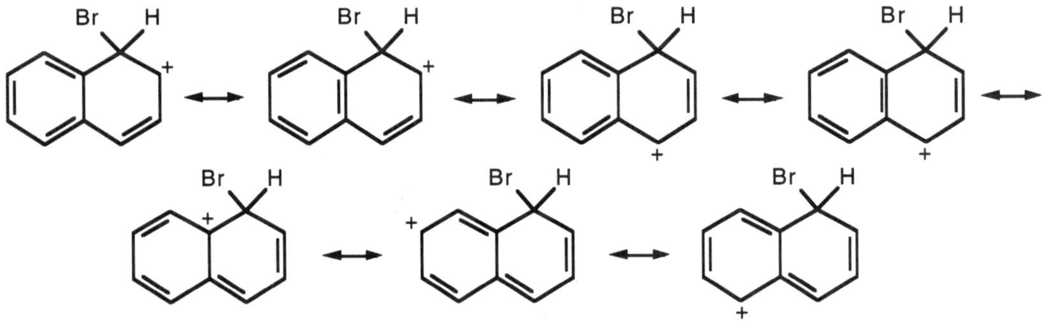

Resonance forms for the intermediate from attack at C2:

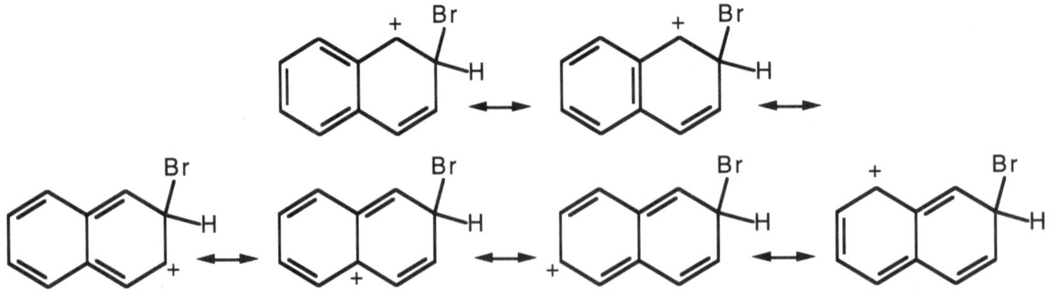

There are seven resonance forms for attack at C1 and six for attack at C2. Look carefully at the forms, however. In the first four resonance structures for C1 attack, the second ring is still fully aromatic. In the other three forms, however, the positive charge has been delocalized into the second ring, destroying the ring's aromaticity. For C2 attack, only the first two resonance structures have a fully aromatic second ring. Since stabilization is lost when aromaticity is disrupted, the intermediate from C2 attack is less stable than the intermediate from C1 attack, and C1 attack is favored.

16.57

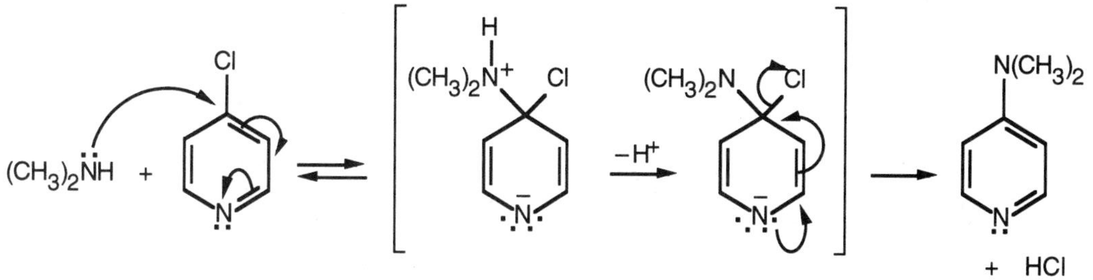

This reaction is an example of nucleophilic aromatic substitution. Dimethylamine is a nucleophile, and the pyridine ring acts as an electron-withdrawing group that can stabilize the negatively-charged intermediate.

16.58

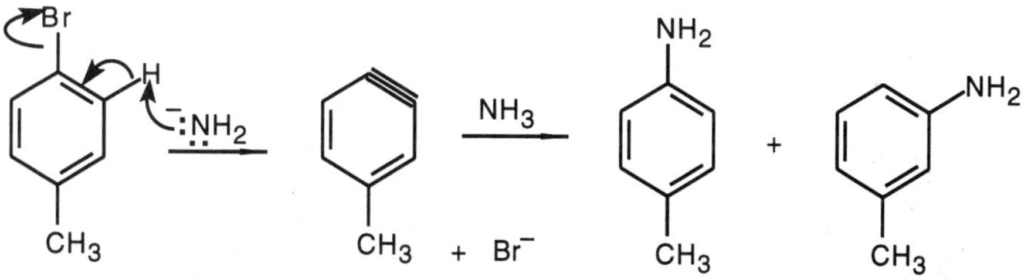

The reaction of an aryl halide with potassium amide proceeds through a benzyne intermediate. Ammonia can then add to either end of the triple bond to produce the two methylanilines observed.

16.59

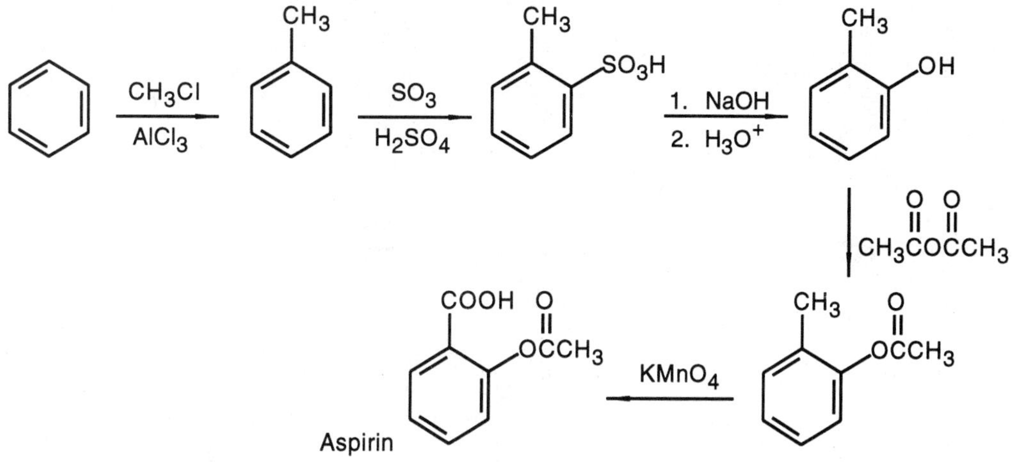

16.60

a)

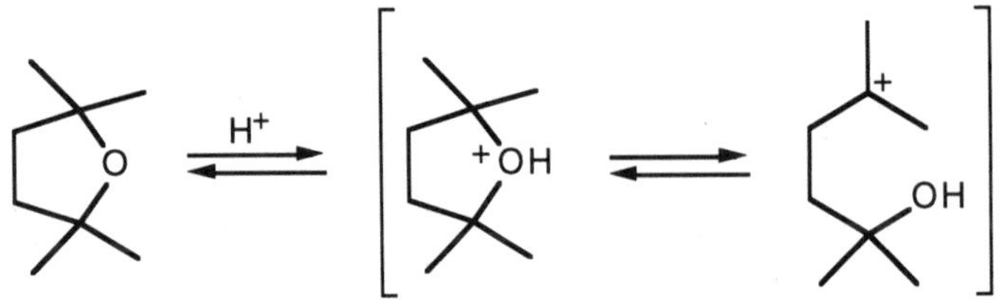

b)

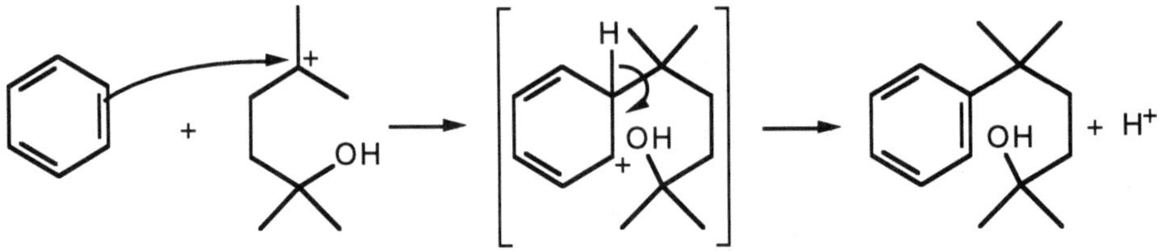

c)

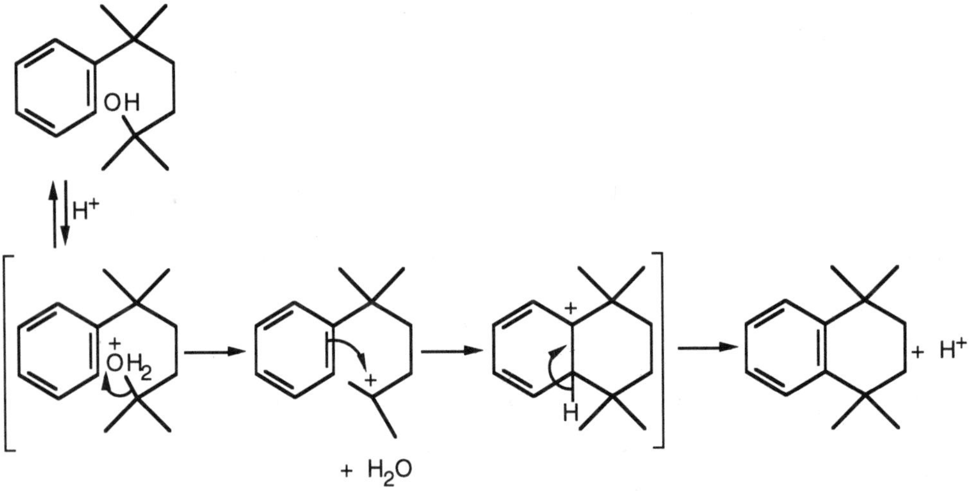

+ H₂O

16.61

1)

2)

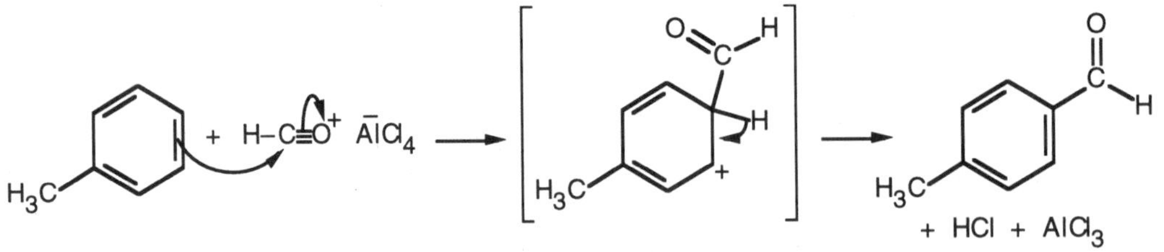

16.62

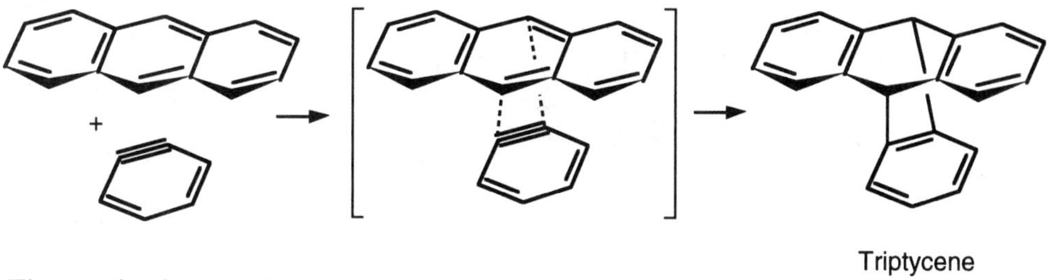

Triptycene

The reaction between benzyne and anthracene is a Diels-Alder reaction.

16.63

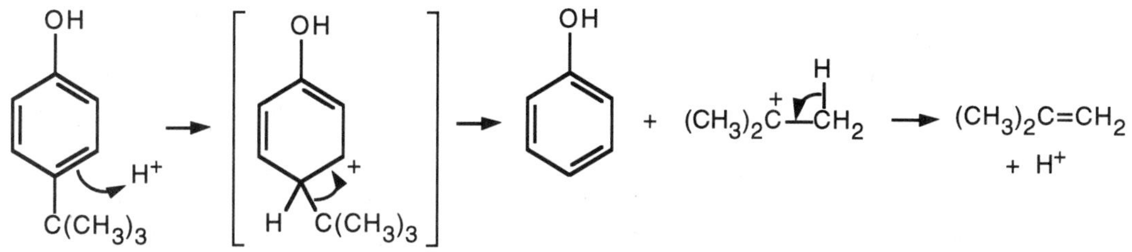

16.64

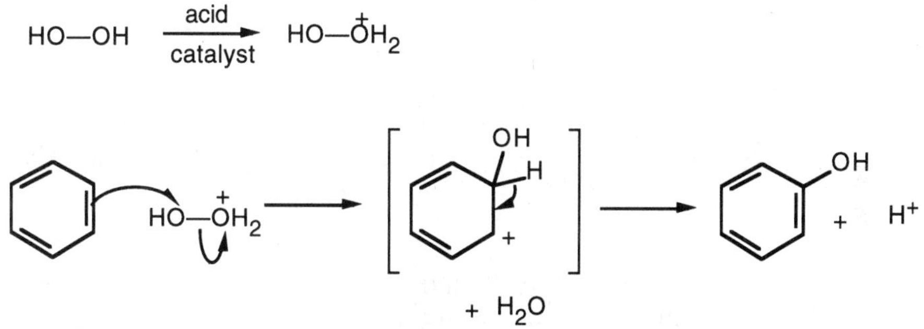

$$HO-OH \xrightarrow[\text{catalyst}]{\text{acid}} HO-\overset{+}{O}H_2$$

$$+ H_2O$$

The reactive electrophile (protonated H_2O_2) is equivalent to ^+OH.

16.65

a)

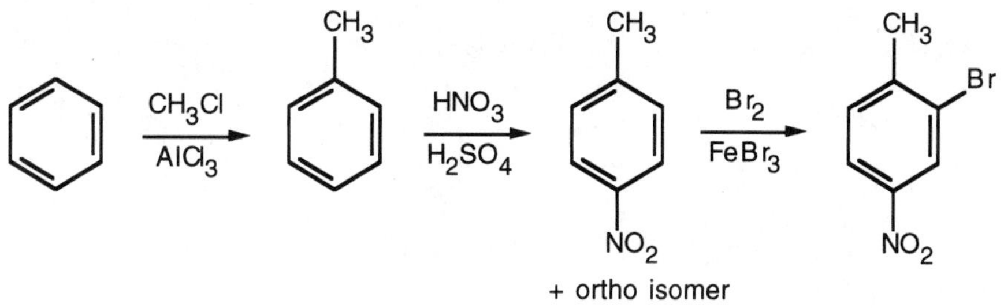

+ ortho isomer

b)

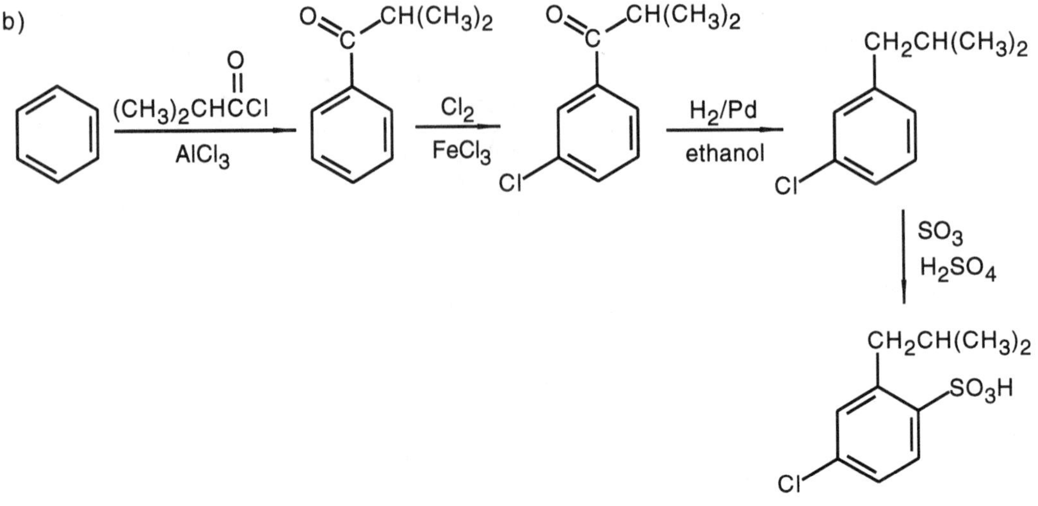

16.66 Problem 16.41 shows the mechanism of the addition of HBr to 1–phenylpropene and shows how the aromatic ring stabilizes the carbocation intermediate. For the methoxy-substituted styrene, an additional form can be drawn in which the cation is stabilized by the electron-donating resonance effect of the oxygen atom. For the nitro-substituted styrene, the cation is destabilized by the electron-withdrawing effect of the nitro group.

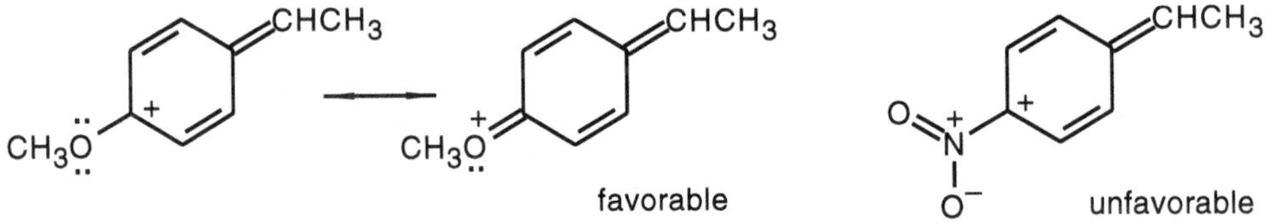

Thus, the intermediate resulting from addition of HBr to the methoxy-substituted styrene is more stable, and reaction of p–methoxystyrene is faster.

16.67

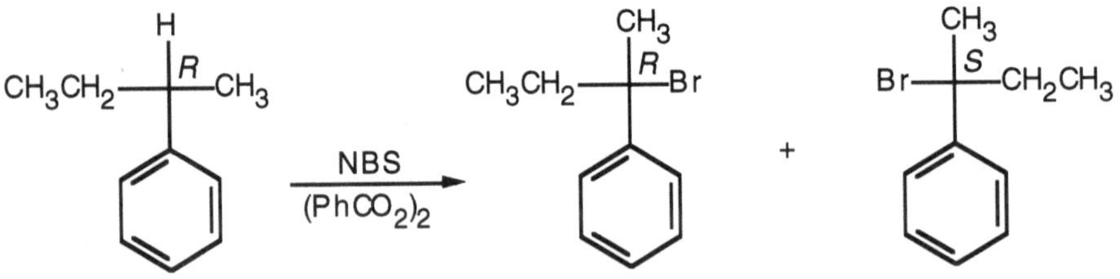

The intermediate benzyl radical is planar and achiral, and it reacts with bromine to produce a racemic mixture.

16.68

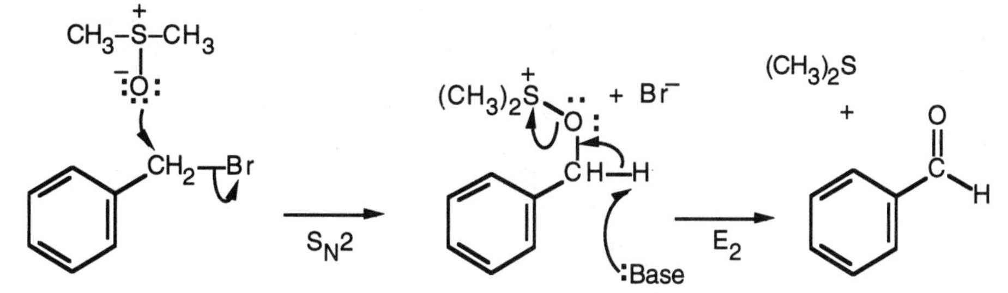

16.69

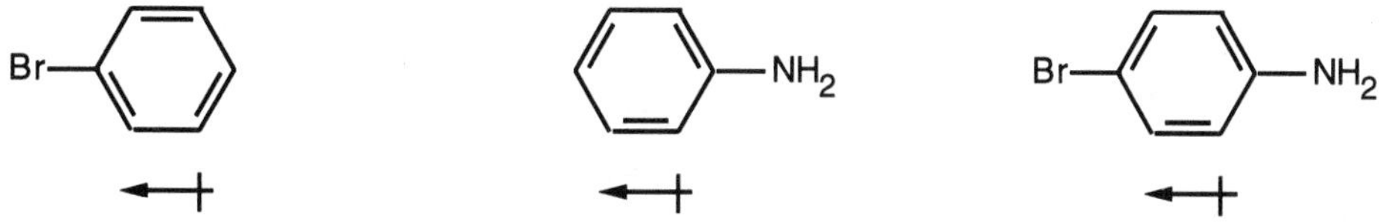

μ = 1.53 D

−Br has a strong electron-withdrawing inductive effect.

μ = 1.52 D

−NH$_2$ has a strong electron-donating resonance effect.

μ = 2.91 D

The polarities of the two groups are additive and produce a net dipole moment almost equal to the sum of the individual moments.

16.70

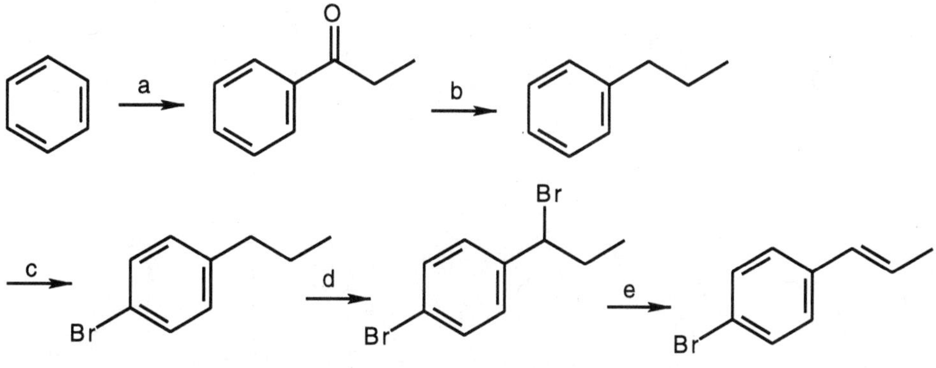

(a) CH_3CH_2COCl, $AlCl_3$; (b) H_2, Pd; (c) Br_2, $FeBr_3$; (d) NBS, $h\nu$; (e) KOH, ethanol

16.71 An electron-withdrawing substituent *destabilizes* a positively charged intermediate (as in electrophilic aromatic substitution) but *stabilizes* a negatively charged intermediate. In the case of the dissociation of a phenol, an −NO$_2$ group stabilizes the phenoxide anion by resonance, thus lowering ΔG° and pK_a.

16.72 For the same reason as in the previous problem, a methyl group destabilizes the negatively charged intermediate, thus raising ΔG° and pK_a.

Study Guide for Chapter 16

After studying this chapter, you should be able to:

(1)   Predict the products of electrophilic aromatic substitution reactions (16.1, 16.3, 16.4, 16.6, 16.7, 16.28, 16.30, 16.31, 16.32).

(2)   Formulate the mechanisms of electrophilic aromatic substitution reactions (16.2, 16.5, 16.35, 16.36, 16.37, 16.42, 16.47, 16.53, 16.60, 16.61, 16.63).

(3)   Understand the activating and directing effects of substituents on aromatic rings (16.9, 16.10, 16.12, 16.13, 16.15, 16.16, 16.17, 16.18, 16.55, 16.56, 16.69, 16.70, 16.71).

(4)   Use inductive and resonance arguments to predict orientation and reactivity in electrophilic substitution reactions (16.8, 16.11, 16.14, 16.29, 16.33, 16.34, 16.38, 16.40, 16.66).

(5)   Predict the position of electrophilic aromatic substitution of polysubstituted aromatic compounds (16.19, 16.39, 16.45, 16.46).

(6)   Understand and predict the products of other reactions of benzenes.
(a) Nucleophilic aromatic substitution (16.20, 16.57).
(b) Benzyne (16.21, 16.54, 16.58, 16.62).
(c) Oxidation of alkylbenzene side-chains (16.22, 16.43).
(d) Bromination of alkylbenzene side-chains (16.23, 16.24, 16.67).
(e) Reduction of aromatic compounds (16.25, 16.43).
(f) Birch reduction (16.27).

(7)   Synthesize substituted benzenes (16.29, 16.30, 16.43, 16.47, 16.48, 16.49, 16.50, 16.51, 16.58, 16.63).

Chapter 17 – Alcohols and Thiols

17.1

a)

OH OH
| |
CH₃CHCH₂CHCH(CH₃)₂

5–Methyl–2,4–hexanediol

b)

2–Methyl–4–phenyl–2–butanol

c)

4,4–Dimethylcyclohexanol

d)

trans–2–Bromocyclopentanol

17.2

a)

CH₂CH₃
/
CH₃CH=C
\
CH₂OH

2–Ethyl–2–buten–1–ol

b)

3–Cyclohexen–1–ol

c)

trans–3–Chlorocycloheptanol

d)

OH
|
CH₃CHCH₂CH₂CH₂OH

1,4–Pentanediol

17.3 In general, the boiling points of a series of isomers decrease with branching. The more nearly spherical a compound becomes, the less surface area it has relative to a straight chain compound of the same molecular weight and functional group type. A smaller surface area allows fewer van der Waals interactions, the weak forces that cause covalent molecules to be attracted to each other.

In addition, branching in alcohols makes it more difficult for hydroxyl groups to approach each other to form hydrogen bonds. A given volume of 2–methyl–2–propanol therefore contains fewer hydrogen bonds than the same volume of 1–butanol, so less energy is needed to break them in boiling.

17.4

Least acidic ⟶ Most acidic

$HC\equiv CH$ < $(CH_3)_2CHOH$ < CH_3OH < $(CF_3)_2CHOH$

alkyne   hindered   alcohol   alcohol with
                 alcohol   electron-withdrawing
                                                      groups

17.5   We saw in Chapter 16 that a nitro group is electron-withdrawing. Since electron-withdrawing groups stabilize alkoxide anions, *p*–nitrobenzyl alcohol is more acidic than benzyl alcohol.

17.6

a)

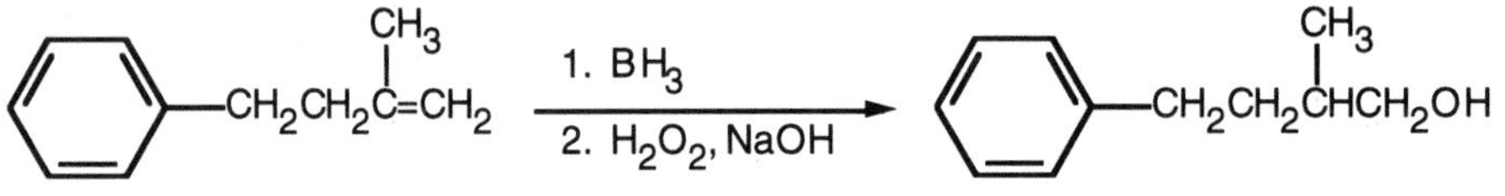

2–Methyl–4–phenyl–1–butanol

b)

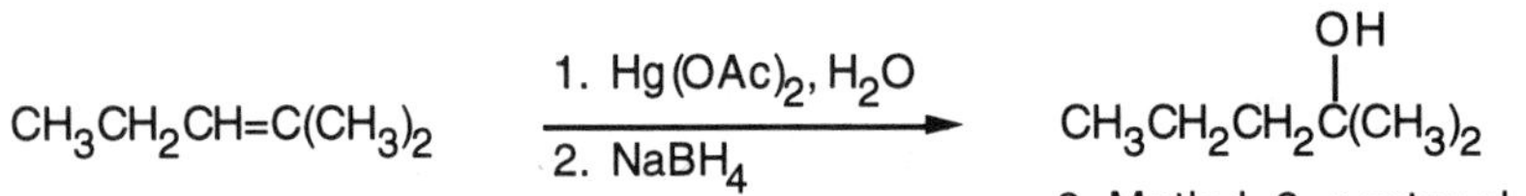

2–Methyl–2–pentanol

c)

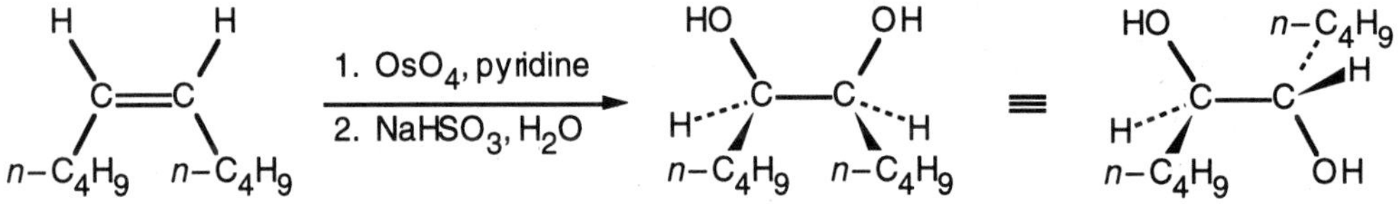

meso–5,6–Decanediol

17.7

a)

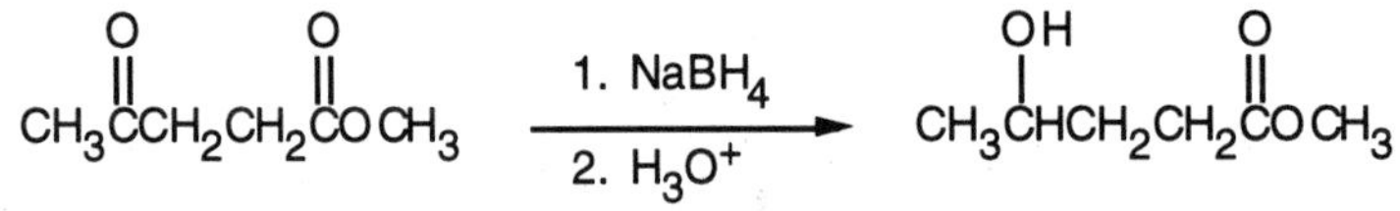

NaBH$_4$ reduces aldehydes and ketones without disturbing other functional groups.

b)

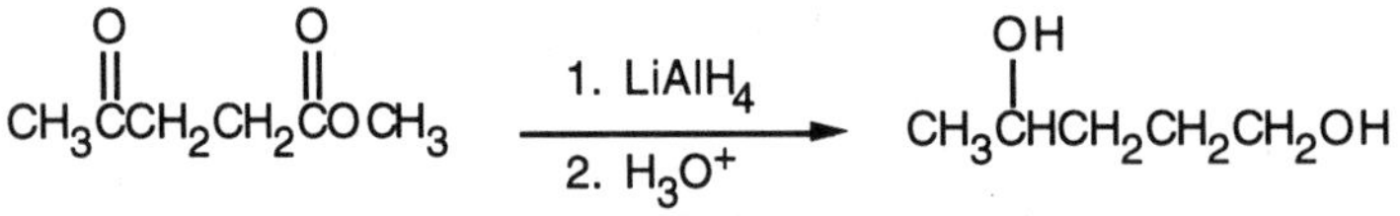

LiAlH$_4$ reduces both ketones and esters.

c)

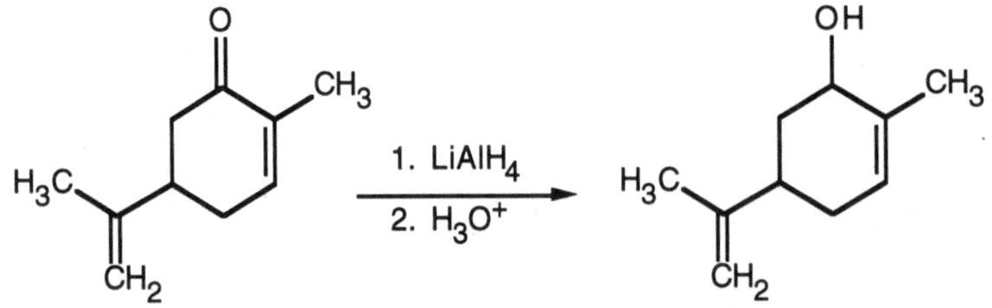

LiAlH$_4$ reduces carbonyl functional groups without reducing double bonds.

d)

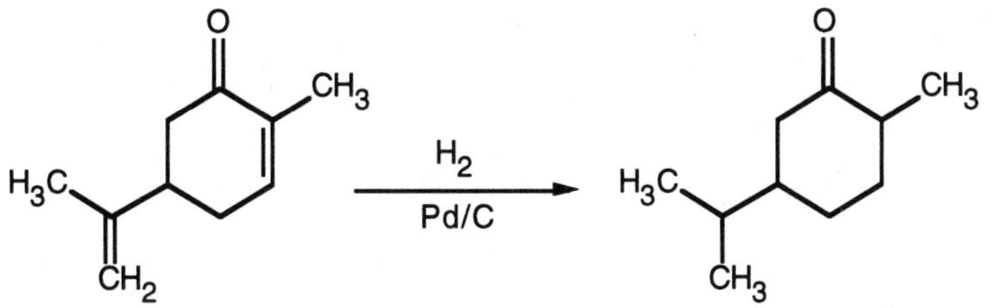

H₂, Pd/C reduces double bonds without reducing carbonyl functional groups.

17.8

a)

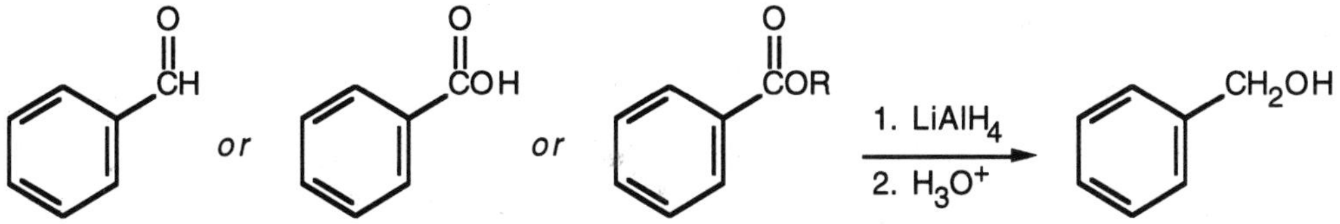

b)

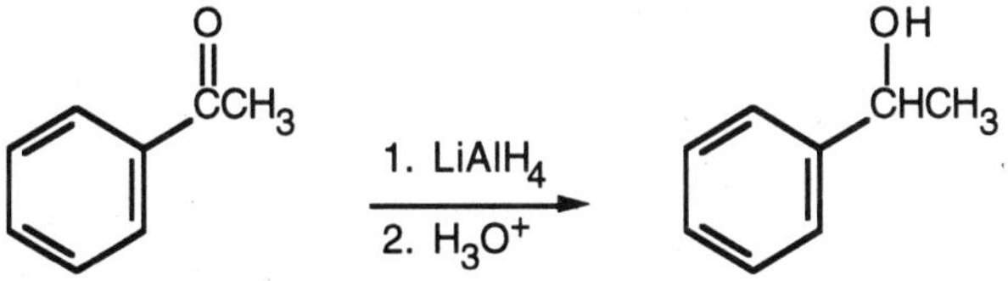

c)

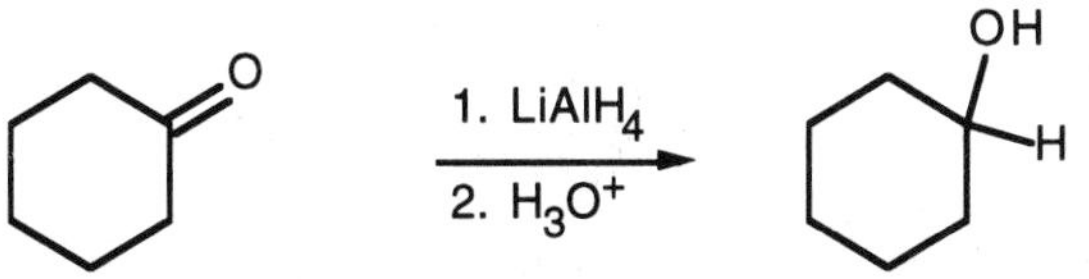

d)   (CH₃)₂CHCHO   or   (CH₃)₂CHCOOH   or   (CH₃)₂CHCOOR

17.9

a)

b)

c)   CH₃CH₂CH₂CCH₂CH₃

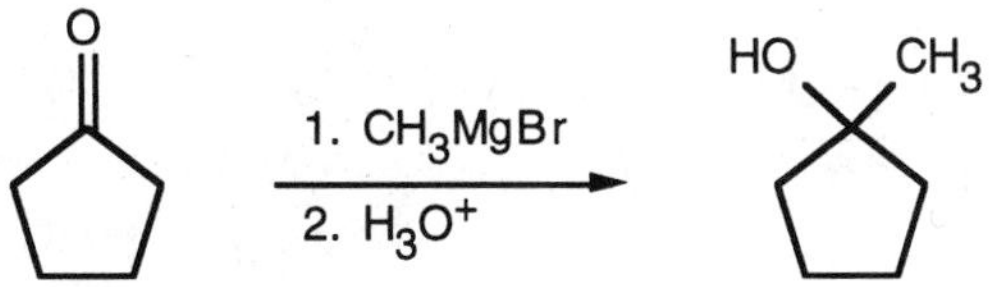

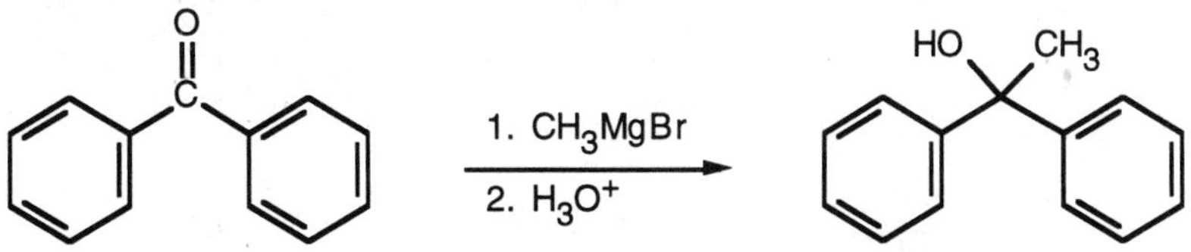

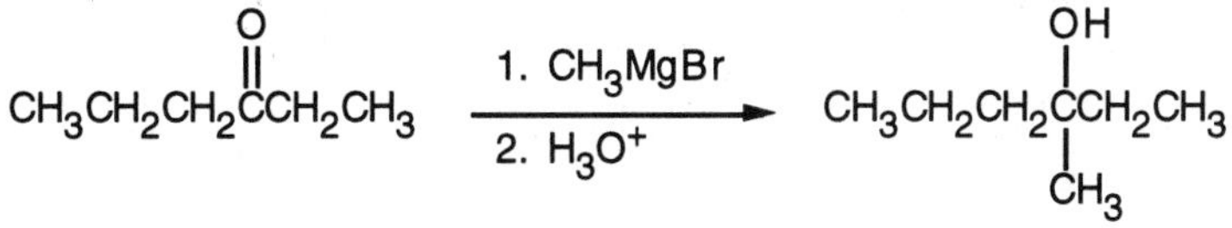

17.10 a) 2–Methyl–2–propanol is a tertiary alcohol. To synthesize a tertiary alcohol, start with a ketone.

$$CH_3\overset{\overset{\displaystyle O}{\|}}{C}CH_3 \quad \xrightarrow[\text{2. }H_3O^+]{\text{1. }CH_3MgBr} \quad CH_3\underset{\underset{\displaystyle CH_3}{|}}{\overset{\overset{\displaystyle OH}{|}}{C}}CH_3$$

If two or more alkyl groups bonded to the carbon bearing the –OH group are the same, an alcohol can be synthesized from an ester and a Grignard reagent.

$$CH_3\overset{\overset{\displaystyle O}{\|}}{C}OR \quad \xrightarrow[\text{2. }H_3O^+]{\text{1. 2 }CH_3MgBr} \quad CH_3\underset{\underset{\displaystyle CH_3}{|}}{\overset{\overset{\displaystyle OH}{|}}{C}}CH_3$$

2–Methyl–2–propanol

b) Since 1–methylcyclohexanol is a tertiary alcohol, start with a ketone.

1–Methylcyclohexanol

c) 3–Methyl–3–pentanol is a tertiary alcohol. Either a ketone or an ester can be used as a starting material.

$$CH_3CH_2\overset{\overset{\displaystyle O}{\|}}{C}CH_2CH_3 \quad \xrightarrow[\text{2. }H_3O^+]{\text{1. }CH_3MgBr}$$

or

$$CH_3CH_2\overset{\overset{\displaystyle O}{\|}}{C}CH_3 \quad \xrightarrow[\text{2. }H_3O^+]{\text{1. }CH_3CH_2MgBr} \qquad CH_3CH_2\underset{\underset{\displaystyle CH_3}{|}}{\overset{\overset{\displaystyle OH}{|}}{C}}CH_2CH_3$$

3–Methyl–3–pentanol

or

$$CH_3\overset{\overset{\displaystyle O}{\|}}{C}OR \quad \xrightarrow[\text{2. }H_3O^+]{\text{1. 2 }CH_3CH_2MgBr}$$

d) Three possible combinations of ketone plus Grignard reagent can be used to synthesize this tertiary alcohol.

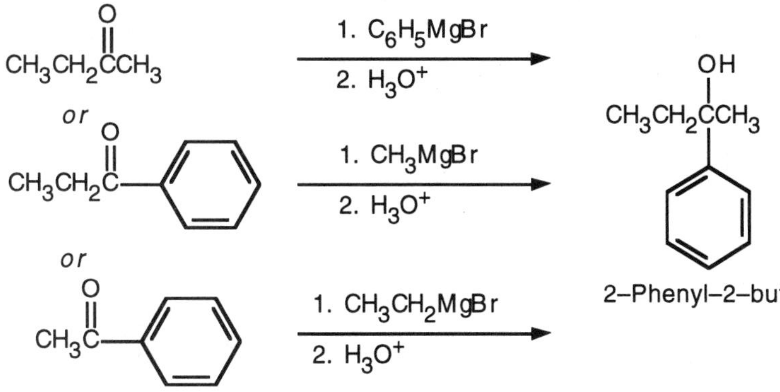

2–Phenyl–2–butanol

e) Formaldehyde must be used to synthesize this primary alcohol.

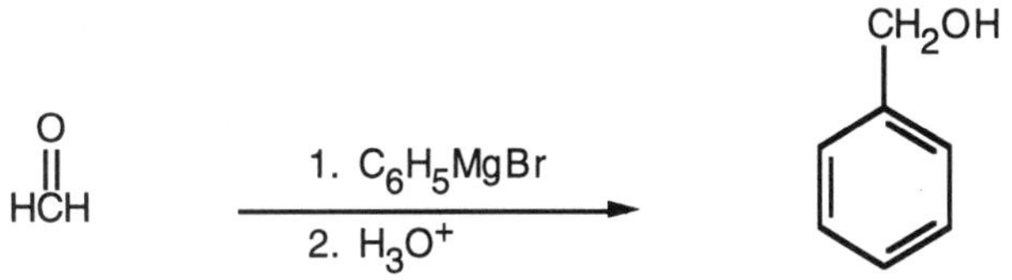

Benzyl alcohol

17.11

a)

b)

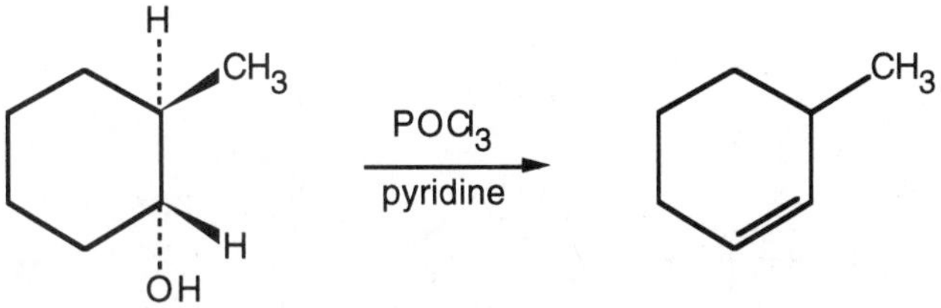

In E2 elimination, dehydration proceeds most readily when the two groups to be eliminated have a trans–diaxial relationship. In this compound, the only hydrogen with the proper stereochemical relationship to the –OH group is at C6. Thus the non-Zaitsev product, 3–methylcyclohexene, is formed.

c)

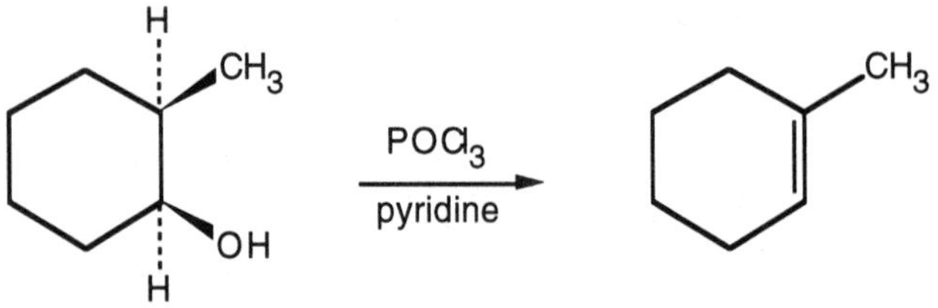

POCl₃
pyridine

Here, the hydrogen at C2 is trans to the hydroxyl, and dehydration yields the Zaitsev product, 1–methylcyclohexene.

17.12

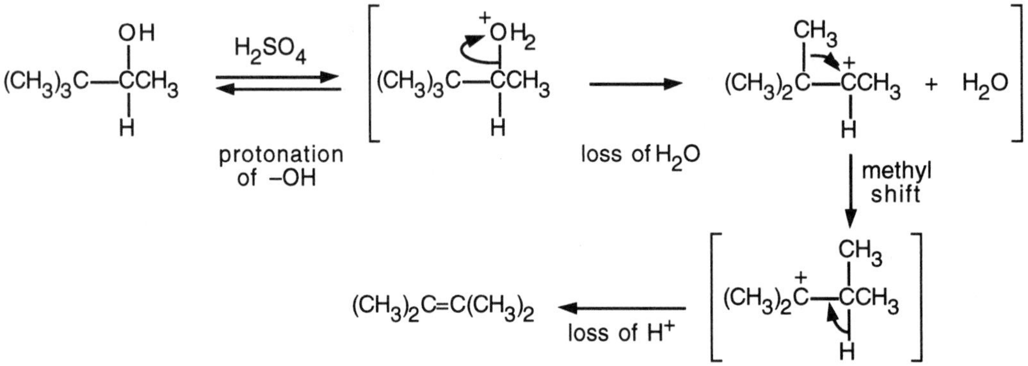

17.13

a)

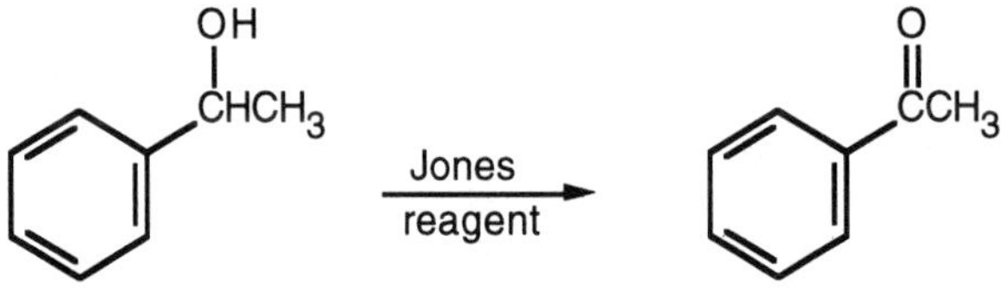

Jones
reagent

b)   CH₃CHCH₂OH   PCC / CH₂Cl₂   CH₃CHCHO

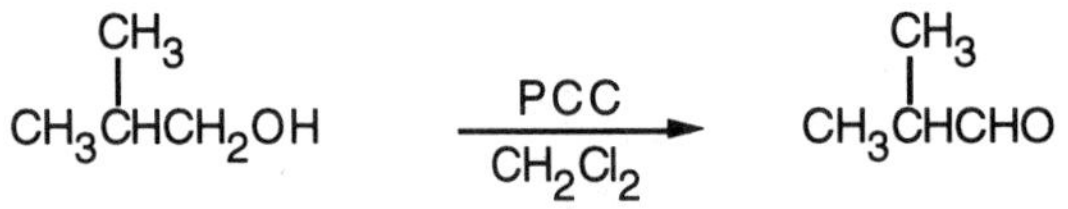

c)

Jones
reagent

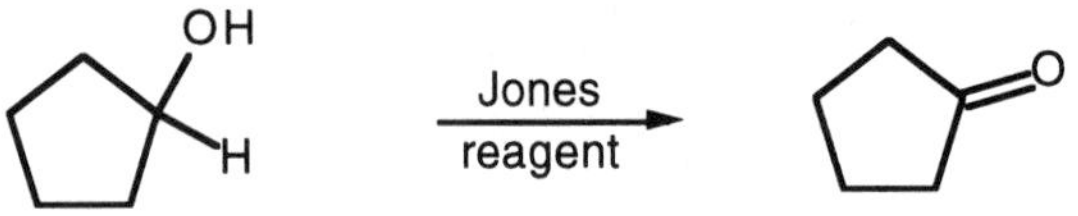

17.14

Starting Material	Jones	PCC
a) $CH_3CH_2CH_2CH_2CH_2CH_2OH$	$CH_3CH_2CH_2CH_2CH_2COOH$	$CH_3CH_2CH_2CH_2CH_2CHO$
b) $CH_3CH_2CH_2CH_2CHCH_3$ (OH)	$CH_3CH_2CH_2CH_2CCH_3$ (O)	$CH_3CH_2CH_2CH_2CCH_3$ (O)
c) $CH_3CH_2CH_2CH_2CH_2CHO$	$CH_3CH_2CH_2CH_2CH_2COOH$	no reaction

17.15

This is an S_N2-like reaction in which the nucleophile F^- attacks silicon and displaces an alkoxide ion as leaving group.

17.16 The infrared spectra of cholesterol and 5–cholesten–3–one each exhibit a unique absorption that makes it easy to distinguish between them. Cholesterol shows an –OH stretch at 3300-3600 cm^{-1}; 5–cholesten–3–one shows a C=O stretch at 1710 cm^{-1}. In the Jones oxidation of cholesterol to 5–cholesten–3–one, the –OH band will disappear and will be replaced by a C=O band. When oxidation is complete no –OH absorption should be visible.

17.17 In general, the –OH signal of a tertiary alcohol (R_3COH) is unsplit, the signal of a secondary alcohol (R_2CHOH) is split into a doublet, and the signal of a primary alcohol (RCH_2OH) is split into a triplet.

a) 2–Methyl–2–propanol is a tertiary alcohol; its –OH signal is unsplit.
b) Cyclohexanol is a secondary alcohol; its –OH absorption is a doublet.
c) Ethanol is a primary alcohol; its –OH signal appears as a triplet.
d) 2-Propanol is a secondary alcohol; its –OH absorption is split into a doublet.
e) Cholesterol is a secondary alcohol; its –OH absorption is split into a doublet.
f) 1–Methylcyclohexanol is a tertiary alcohol; its –OH signal is unsplit.

17.18

a)
$$\underset{\text{2–Butanethiol}}{CH_3CH_2\overset{\overset{\displaystyle SH}{|}}{C}HCH_3}$$

b)
$$\underset{\text{2,2,6–Trimethyl–4–heptanethiol}}{(CH_3)_3CCH_2\overset{\overset{\displaystyle SH}{|}}{C}HCH_2CH(CH_3)_2}$$

c)
2–Cyclopentene–1–thiol

17.19

a)
$$\underset{\text{Methyl 2–butenoate}}{CH_3CH=CHCO_2CH_3} \xrightarrow[\text{2. } H_3O^+]{\text{1. LiAlH}_4} CH_3CH=CHCH_2OH \xrightarrow{PBr_3} CH_3CH=CHCH_2Br$$

$$\downarrow \begin{array}{l} \text{1. } (H_2N)_2C=S \\ \text{2. } ^-OH, H_2O \end{array}$$

$$\underset{\text{2–Butene–1–thiol}}{CH_3CH=CHCH_2SH}$$

b)
$$H_2C=CHCH=CH_2 \xrightarrow[\Delta]{HBr} CH_3CH=CHCH_2Br \xrightarrow[\text{2. } ^-OH, H_2O]{\text{1. } (H_2N)_2C=S} CH_3CH=CHCH_2SH$$

17.20 a) 2–Methyl–1,4–butanediol   b) 3–Ethyl–2–hexanol
c) *cis*–1,3–Cyclobutanediol   d) *cis*–2–Methyl–4–cyclohepten–1–ol
e) *cis*–3–Phenylcyclopentanol   f) 2,3–Dimethyl–3–hexanethiol

17.21 None of these alcohols has multiple bonds or rings.

$CH_3CH_2CH_2CH_2CH_2OH$

1–Pentanol

$CH_3CH_2CH_2\overset{\displaystyle OH}{\underset{|}{C}}HCH_3$

2–Pentanol

$CH_3CH_2\overset{\displaystyle OH}{\underset{|}{C}}HCH_2CH_3$

3–Pentanol

$CH_3CH_2\overset{\displaystyle CH_3}{\underset{|}{C}}HCH_2OH$

2–Methyl–1–butanol

$CH_3CH_2\overset{\displaystyle CH_3}{\underset{|}{\underset{|}{\underset{OH}{C}}}}CH_3$

2–Methyl–2–butanol

$CH_3\overset{\displaystyle CH_3}{\underset{|}{C}}H\overset{}{\underset{|}{C}}HCH_3$ with OH below

3–Methyl–2–butanol

$HOCH_2CH_2\overset{\displaystyle CH_3}{\underset{|}{C}}HCH_3$

3–Methyl–1–butanol

$CH_3\overset{\displaystyle CH_3}{\underset{|}{\underset{|}{\underset{CH_3}{C}}}}CH_2OH$

2,2–Dimethyl–1–propanol

17.22 Primary alcohols react with Jones' reagent to form carboxylic acids, secondary alcohols yield ketones, and tertiary alcohols are unreactive to oxidation. Of the eight alcohols in the previous problem, only 2–methyl–2–butanol is unreactive to Jones oxidation.

$CH_3CH_2CH_2CH_2CH_2OH \xrightarrow{\text{Jones}} CH_3CH_2CH_2CH_2COOH$

$CH_3CH_2CH_2\overset{\displaystyle OH}{\underset{|}{C}}HCH_3 \xrightarrow{\text{Jones}} CH_3CH_2CH_2\overset{\displaystyle O}{\overset{\|}{C}}CH_3$

$CH_3CH_2\overset{\displaystyle OH}{\underset{|}{C}}HCH_2CH_3 \xrightarrow{\text{Jones}} CH_3CH_2\overset{\displaystyle O}{\overset{\|}{C}}CH_2CH_3$

$CH_3CH_2\overset{\displaystyle CH_3}{\underset{|}{C}}HCH_2OH \xrightarrow{\text{Jones}} CH_3CH_2\overset{\displaystyle CH_3}{\underset{|}{C}}HCOOH$

$CH_3\overset{\displaystyle CH_3}{\underset{|}{C}}H\overset{}{\underset{|}{C}}HCH_3 \xrightarrow{\text{Jones}} CH_3\overset{\displaystyle CH_3}{\underset{|}{\underset{|}{\underset{O}{\overset{\|}{C}}}}}HCH_3$ (OH below)

$HOCH_2CH_2\overset{\displaystyle CH_3}{\underset{|}{C}}HCH_3 \xrightarrow{\text{Jones}} HOOCCH_2\overset{\displaystyle CH_3}{\underset{|}{C}}HCH_3$

$CH_3\overset{\displaystyle CH_3}{\underset{|}{\underset{|}{\underset{CH_3}{C}}}}CH_2OH \xrightarrow{\text{Jones}} CH_3\overset{\displaystyle CH_3}{\underset{|}{\underset{|}{\underset{CH_3}{C}}}}COOH$

17.23

a)

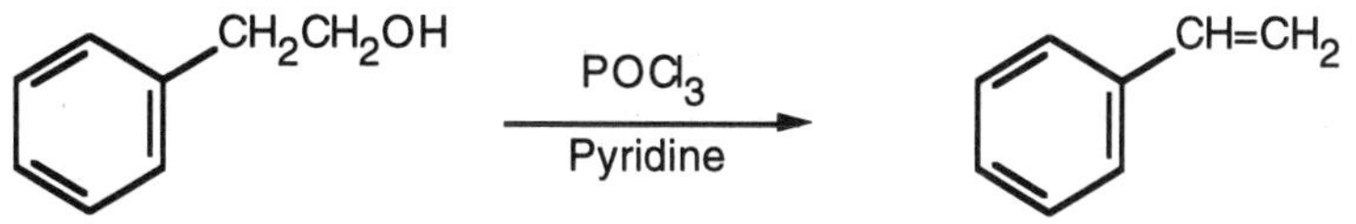

2-Phenylethanol → (POCl₃ / Pyridine) → Styrene

$$\text{2-Phenylethanol} \xrightarrow[\text{Pyridine}]{\text{POCl}_3} \text{Styrene}$$

b)

Ar-CH₂CH₂OH →(PCC* / CH₂Cl₂)→ Ar-CH₂CHO

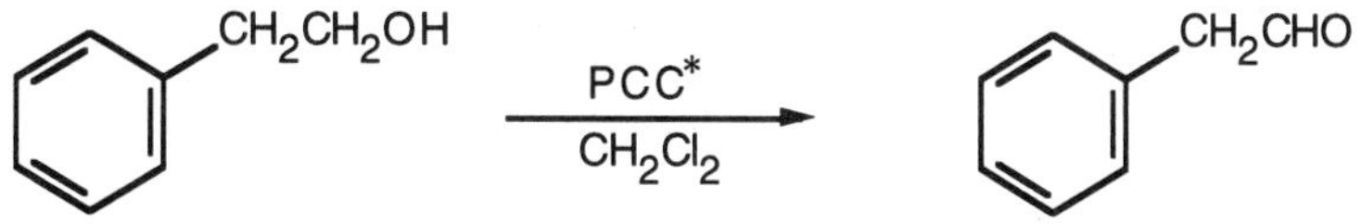

Phenylacetaldehyde

* $C_5H_6NCrO_3Cl$ (pyridinium chlorochromate)

c)

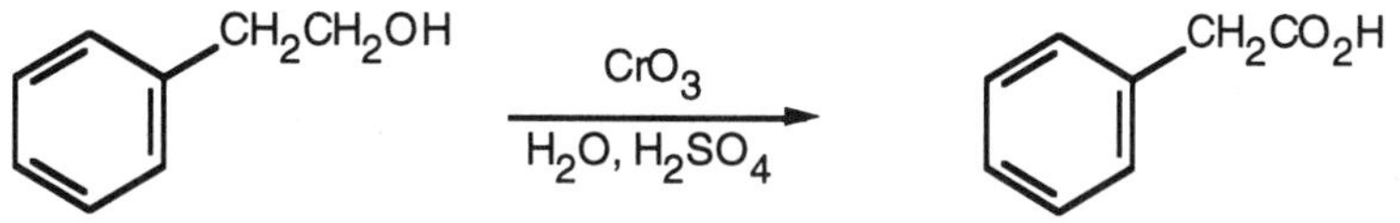

Ar-CH₂CH₂OH →(CrO₃ / H₂O, H₂SO₄)→ Ar-CH₂CO₂H

Phenylacetic acid

d)

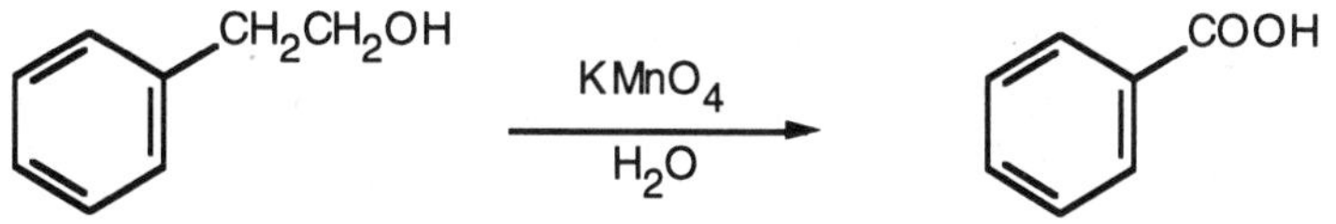

Ar-CH₂CH₂OH →(KMnO₄ / H₂O)→ Ar-COOH

Benzoic acid

e)

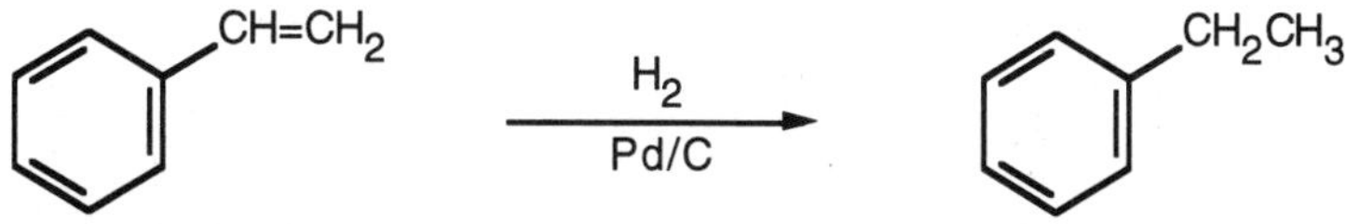

Ar-CH=CH₂ →(H₂ / Pd/C)→ Ar-CH₂CH₃

(from a)   Ethylbenzene

f)

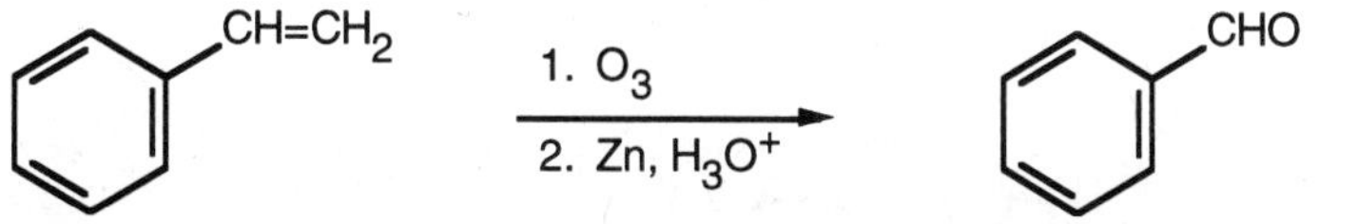

Ar-CH=CH₂ →(1. O₃ 2. Zn, H₃O⁺)→ Ar-CHO

(from a)   Benzaldehyde

g)

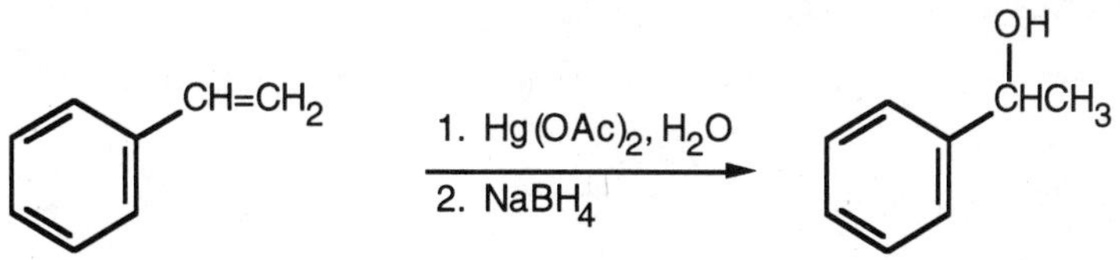

Ar-CH=CH₂ →(1. Hg(OAc)₂, H₂O 2. NaBH₄)→ Ar-CH(OH)CH₃

(from a)   1-Phenylethanol

h)

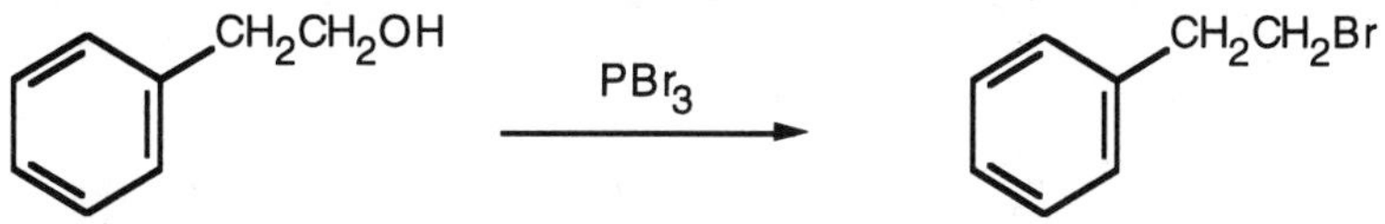

Ar-CH₂CH₂OH →(PBr₃)→ Ar-CH₂CH₂Br

1-Bromo-2-phenylethane

17.24

a)

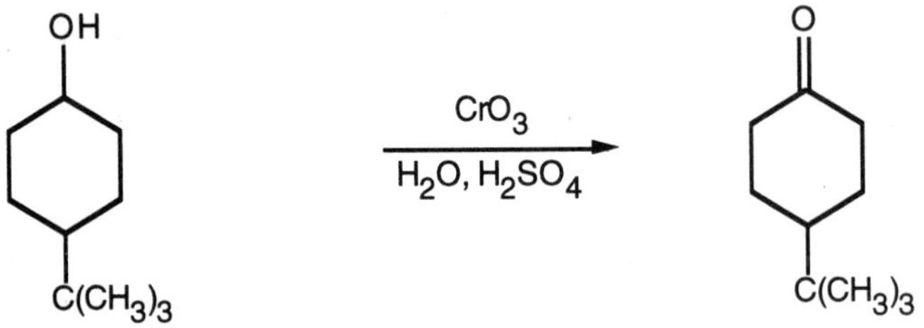

$$\xrightarrow[\text{H}_2\text{O, H}_2\text{SO}_4]{\text{CrO}_3}$$

b)

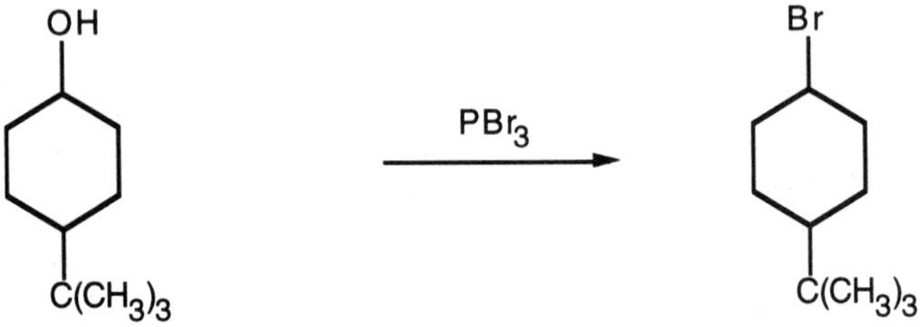

$$\xrightarrow{\text{PBr}_3}$$

c)

$$\xrightarrow[\text{Pyridine}]{\text{POCl}_3}$$

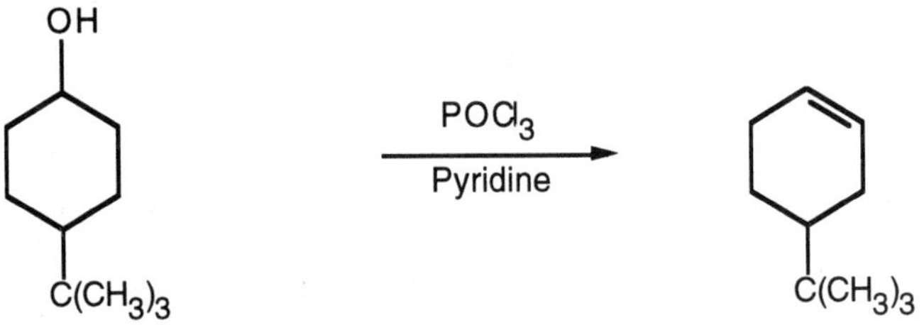

d) $CH_3CH_2CH_2OH$ $\xrightarrow[\text{CH}_2\text{Cl}_2]{\text{PCC}}$ CH_3CH_2CHO

e) $CH_3CH_2CH_2OH$ $\xrightarrow[\text{H}_2\text{O, H}_2\text{SO}_4]{\text{CrO}_3}$ CH_3CH_2COOH

f)

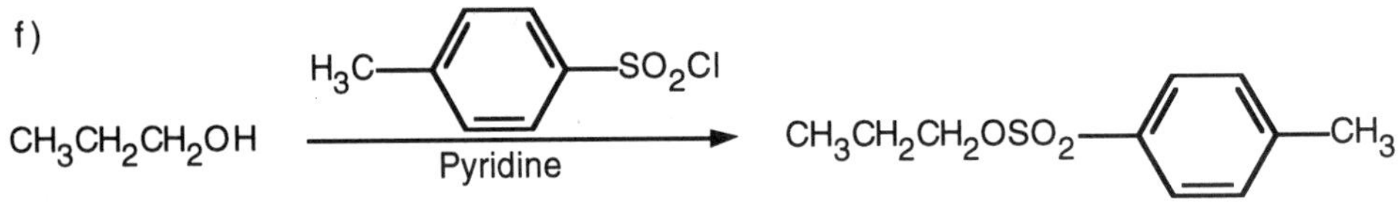

$CH_3CH_2CH_2OH$ $\xrightarrow{\text{Pyridine}}$ $CH_3CH_2CH_2OSO_2$—

g) $CH_3CH_2CH_2OH$ $\xrightarrow{\text{SOCl}_2}$ $CH_3CH_2CH_2Cl$

h) $CH_3CH_2CH_2OH$ $\xrightarrow{\text{NaH}}$ $CH_3CH_2CH_2O^-\ Na^+\ +\ H_2$

17.25 In some of these problems, different combinations of Grignard reagent and carbonyl compound are possible. Remember that aqueous acid is added to the initial Grignard adduct to yield the alcohol.

a) CH₃MgBr + CH₃CH₂CHO⎤
   or   ⎥ ⟶
   CH₃CHO + CH₃CH₂MgBr⎦

$$\underset{\text{2–Butanol}}{\overset{\overset{\displaystyle OH}{|}}{CH_3CHCH_2CH_3}}$$

b)

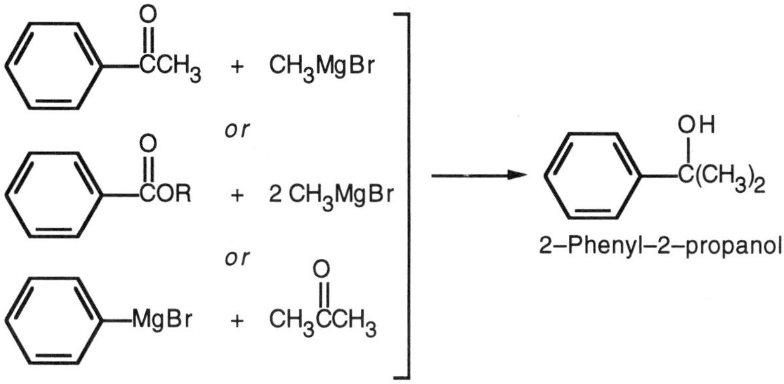

2–Phenyl–2–propanol

c)
$$\underset{}{\overset{\overset{\displaystyle CH_3}{|}}{H_2C=C-MgBr}} + CH_2O \longrightarrow \underset{\text{2–Methyl–2–propen–1–ol}}{\overset{\overset{\displaystyle CH_3}{|}}{H_2C=CCH_2OH}}$$

d)

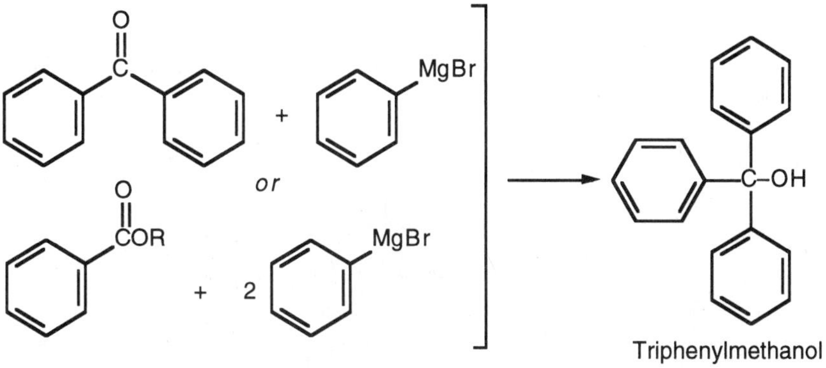

Triphenylmethanol

e) CH₃MgBr + HCCH₂CH₂CH₂Br ⟶
$$\underset{\text{5–Bromo–2–pentanol}}{\overset{\overset{\displaystyle OH}{|}}{CH_3CHCH_2CH_2CH_2Br}}$$

17.26

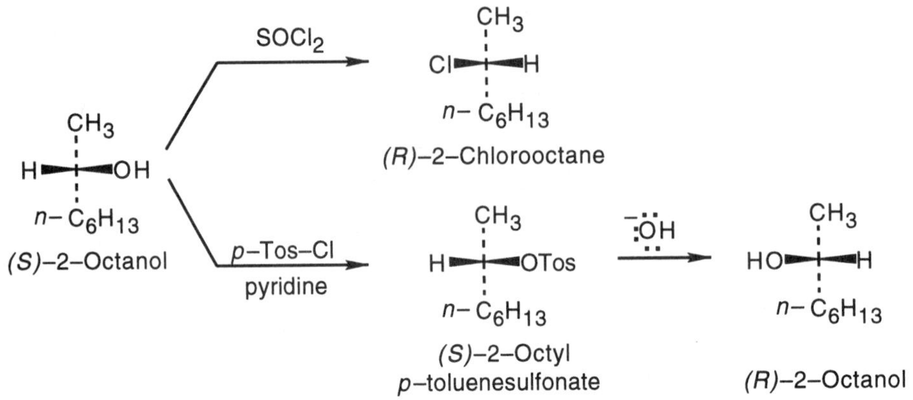

CH3

Cl—H

n-C6H13

(R)-2-Chlorooctane

SOCl2

CH3

H—OH

n-C6H13

(S)-2-Octanol

p-Tos-Cl
pyridine

CH3

H—OTos

n-C6H13

(S)-2-Octyl
p-toluenesulfonate

:OH

CH3

HO—H

n-C6H13

(R)-2-Octanol

To form optically pure (R)–2–octanol, the poor hydroxide leaving group must be converted to the very good toluenesulfonate leaving group. Reaction of (S)–2–octyl–p–toluenesulfonate with hydroxide ion proceeds with inversion (S_N2 mechanism) to give (R)–2–octanol.

17.27

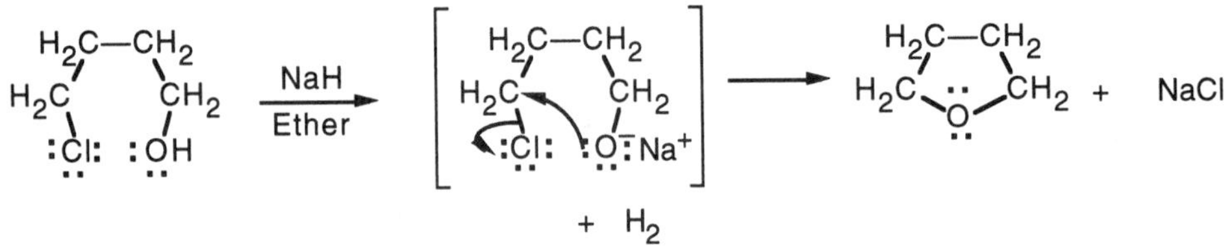

H2C—CH2

H2C   CH2

:Cl: :OH

NaH
Ether

H2C—CH2

H2C   CH2

:Cl: :O: Na+

+ H2

H2C—CH2

H2C   CH2
   O

+   NaCl

17.28

Compound	Carbonyl precursors

a) 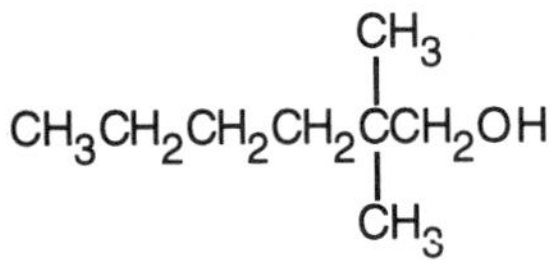 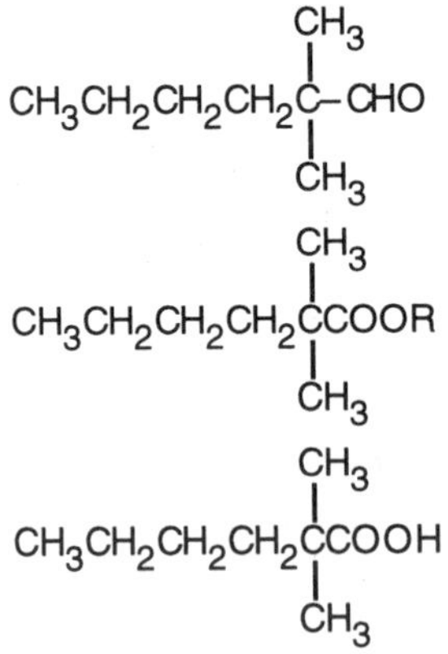   CH3CH2CH2CH2C(CH3)(CH3)CH2OH

    CH3CH2CH2CH2C(CH3)(CH3)-CHO

    CH3CH2CH2CH2C(CH3)(CH3)COOR

    CH3CH2CH2CH2C(CH3)(CH3)COOH

b) 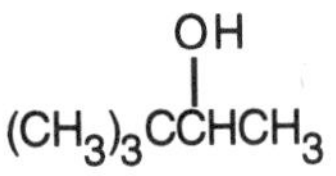 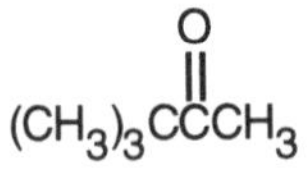   (CH3)3CCH(OH)CH3

    (CH3)3CCCH3 (O)

c) 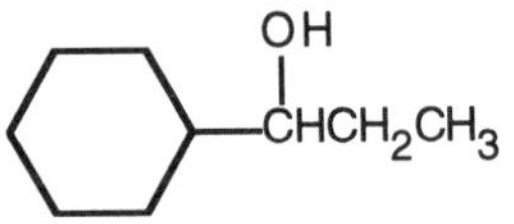 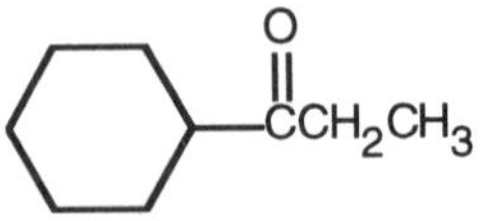   cyclohexyl-CH(OH)CH2CH3

    cyclohexyl-CCH2CH3 (O)

d)

-CH$_2$OH   -CHO   -COOH   -COOR

,   ,

17.29 In these compounds you want to reduce some, but not all, of the functional groups present. To do this, you must select the correct reducing agent.

a)

CO$_2$H

$\xrightarrow{\text{H}_2 \atop \text{Rh/C}}$

CO$_2$H

b)

CO$_2$H

$\xrightarrow{\text{1. LiAlH}_4 \atop \text{2. H}_3\text{O}^+}$

CH$_2$OH

17.30

Grignard Reagent +   Carbonyl Compound $\longrightarrow$ Product

a)   CH$_3$MgBr   +   CH$_3\overset{\text{O}}{\overset{||}{\text{C}}}CH_3$

or

2 CH$_3$MgBr   +   CH$_3\overset{\text{O}}{\overset{||}{\text{C}}}$OR $\longrightarrow$ (CH$_3$)$_3$COH

b) CH$_3$CH$_2$MgBr   +

$\longrightarrow$

OH

-CH$_2$CH$_3$

c)   CH$_3$CH$_2$MgBr   +   CH$_3$CH$_2\overset{\text{O}}{\overset{||}{\text{C}}}$-

or

2 CH$_3$CH$_2$MgBr   +   RO-$\overset{\text{O}}{\overset{||}{\text{C}}}$-

or

-MgBr   +   CH$_3$CH$_2\overset{\text{O}}{\overset{||}{\text{C}}}CH_2CH_3$

$\longrightarrow$

OH

CH$_3$CH$_2\overset{|}{\text{C}}$CH$_2$CH$_3$

d)

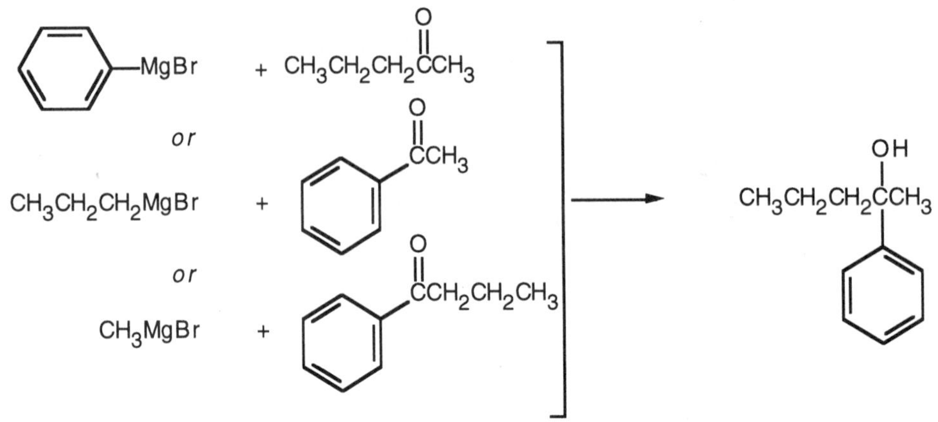

e)

f)

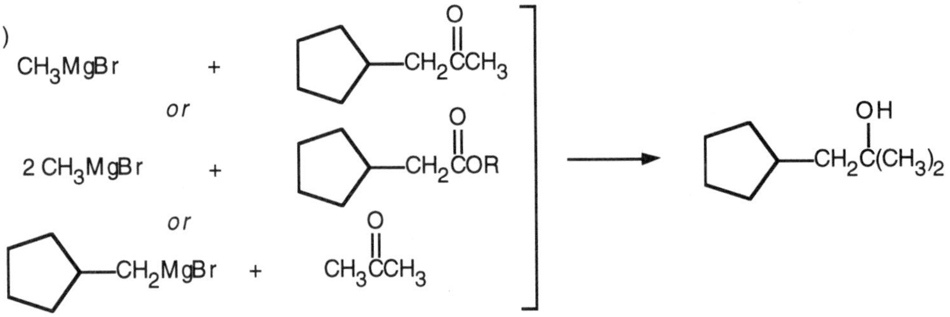

17.31

a) $CH_3CH_2CH_2CH_2CH_2OH$ $\xrightarrow{\ PBr_3\ }$ $CH_3CH_2CH_2CH_2CH_2Br$

b) $CH_3CH_2CH_2CH_2CH_2OH$ $\xrightarrow{\ SOCl_2\ }$ $CH_3CH_2CH_2CH_2CH_2Cl$

c) $CH_3CH_2CH_2CH_2CH_2OH$ $\xrightarrow{\ CrO_3,\ H_2O,\ H_2SO_4\ }$ $CH_3CH_2CH_2CH_2COOH$

d) $CH_3CH_2CH_2CH_2CH_2OH$ $\xrightarrow{\ PCC\ }$ $CH_3CH_2CH_2CH_2CHO$

17.32

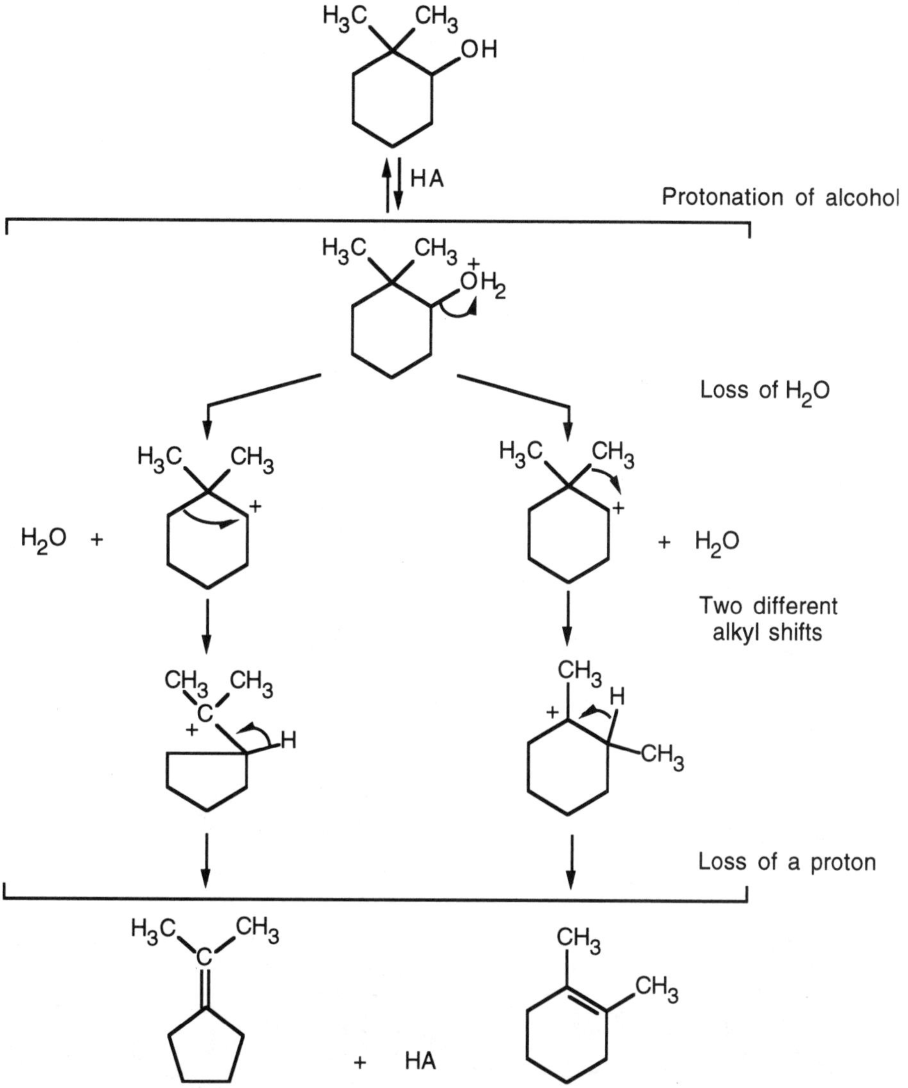
H3C
CH3
OH
HA
Protonation of alcohol
+OH2
Loss of H2O
H2O +
+
+ H2O
Two different
alkyl shifts
C
H
CH3
Loss of a proton
+ HA

17.33

a)

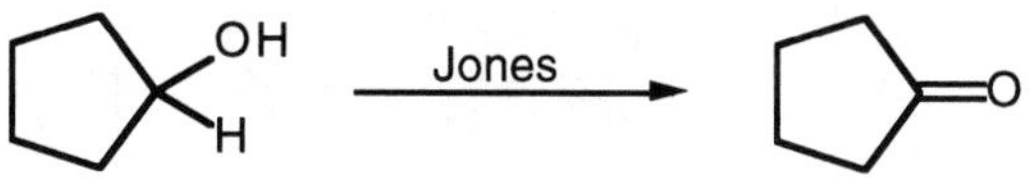
OH
H
Jones
O

b)

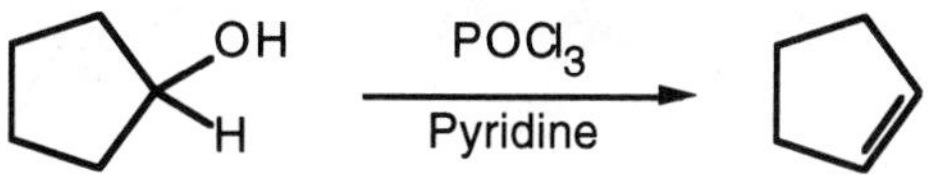
OH
H
POCl3
Pyridine

c)

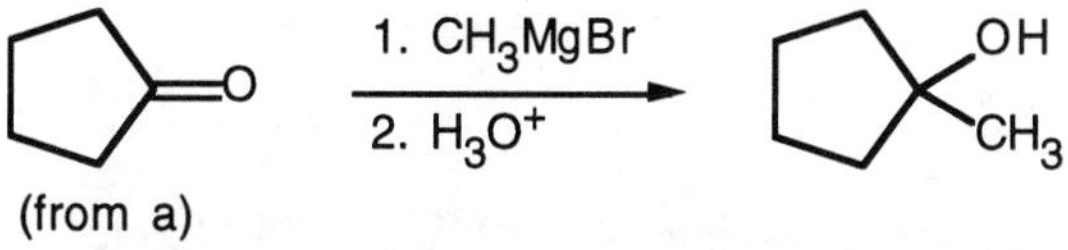
O
(from a)
1. CH3MgBr
2. H3O+
OH
CH3

d)

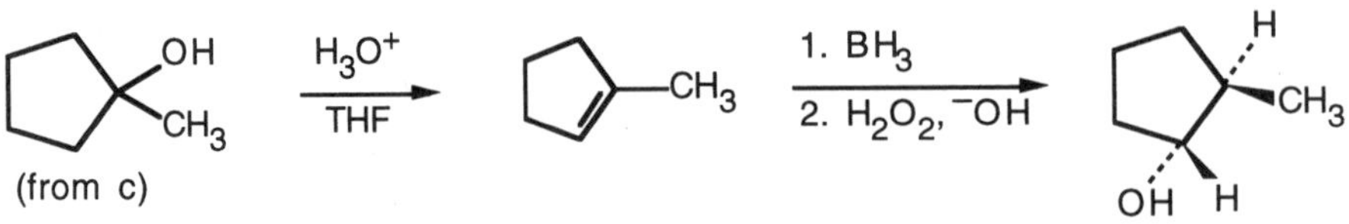

(from c)

17.34

a)

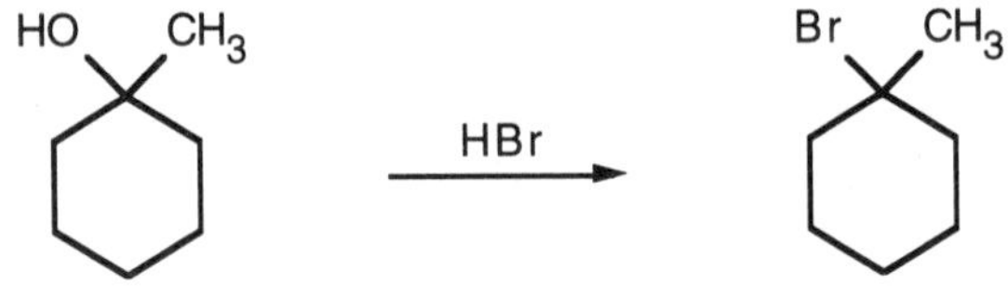

HBr

b)

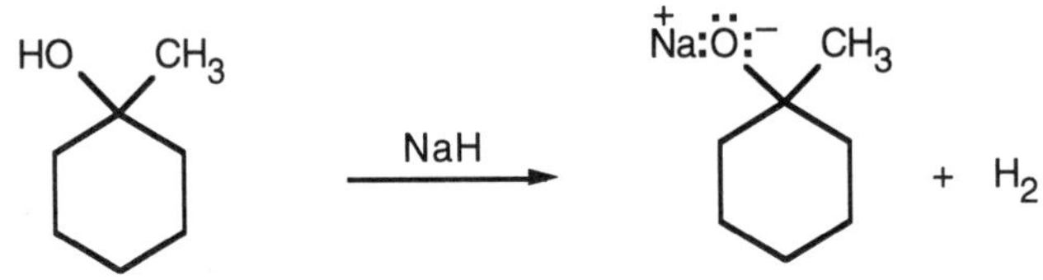

NaH

+ H$_2$

c)

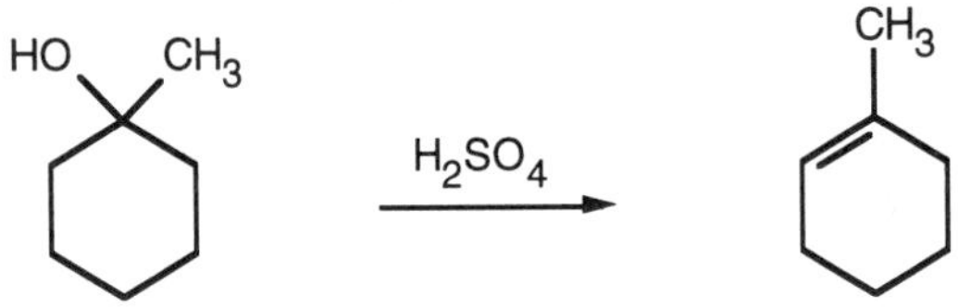

H$_2$SO$_4$

d)

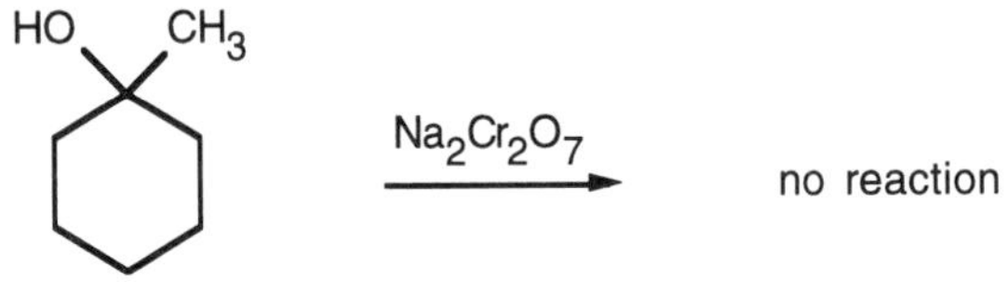

Na$_2$Cr$_2$O$_7$

no reaction

17.35

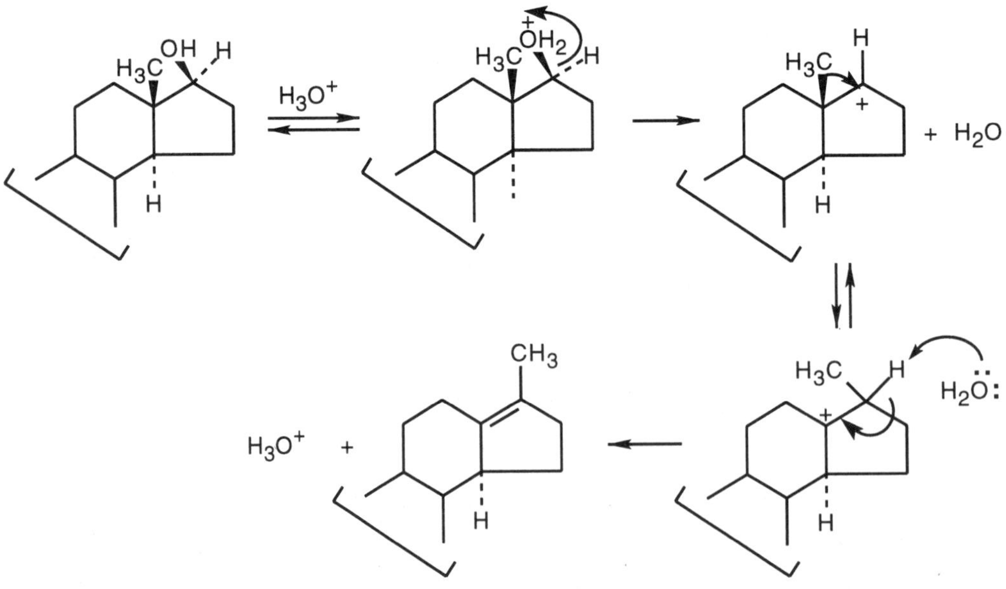

+ H$_2$O

This is a carbocation rearrangement involving the shift of an alkyl group.

17.36

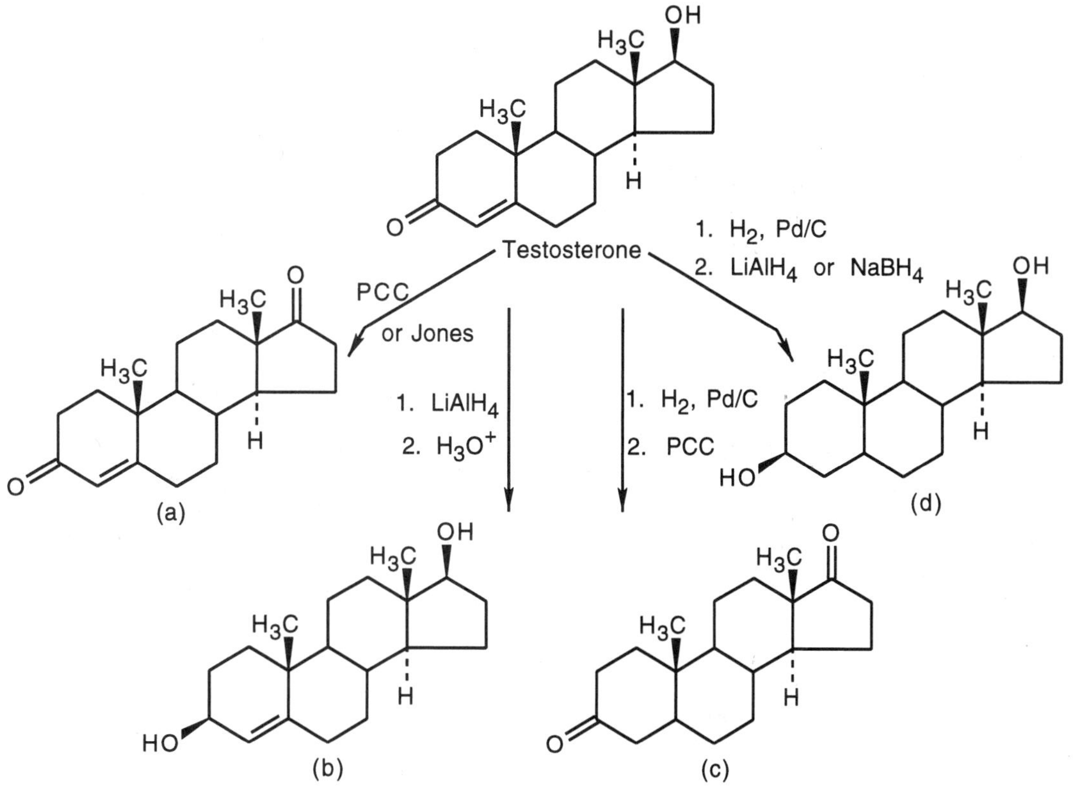

(a)

(b)

(c)

(d)

17.37

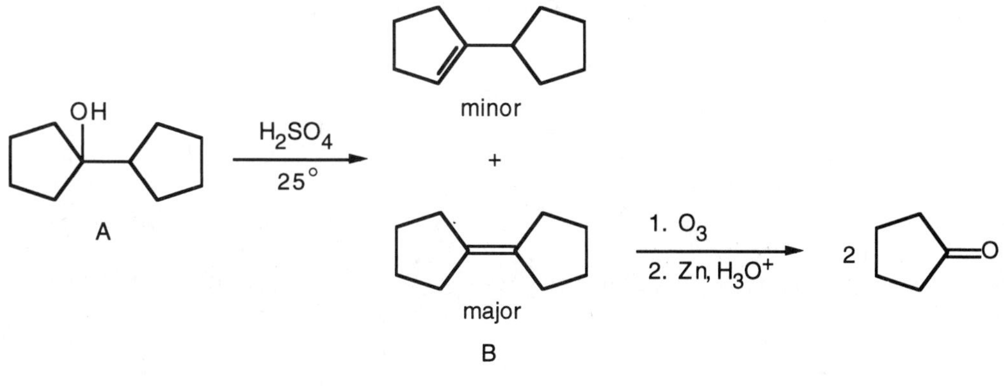

17.38

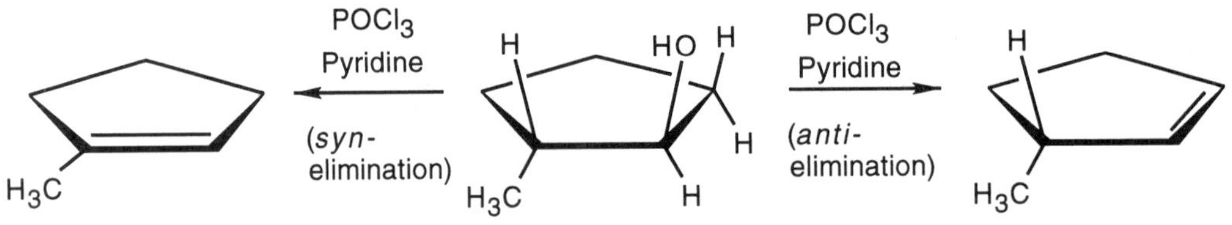

1–Methylcyclopentene | *trans*–2–Methylcyclopentanol | 3–Methylcyclopentene

The more stable product of dehydration of *trans*–2–methylcyclopentanol is 1–methylcyclopentene, which can be formed only via syn periplanar elimination. The product of anti periplanar elimination is 3–methylcyclopentene. Since this product predominates, the requirement of anti periplanar geometry must be more important than formation of the more stable product.

17.39

a)

$$CH_3CH_2OH \xrightarrow{PBr_3} CH_3CH_2Br \xrightarrow[\text{ether}]{Mg} CH_3CH_2MgBr$$

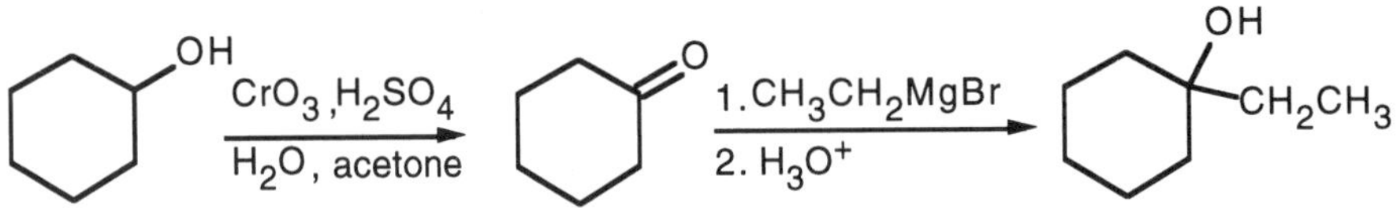

b)

$$CH_3CH_2CH_2\overset{\displaystyle CH_3}{\overset{|}{C}}HCH_2OH \xrightarrow{PBr_3} CH_3CH_2CH_2\overset{\displaystyle CH_3}{\overset{|}{C}}HCH_2Br$$

$$\xrightarrow[\text{ether}]{Mg} CH_3CH_2CH_2\overset{\displaystyle CH_3}{\overset{|}{C}}HCH_2MgBr$$

$$CH_3OH \xrightarrow{PCC} H_2C{=}O$$

$$CH_3CH_2CH_2\overset{\displaystyle CH_3}{\overset{|}{C}}HCH_2MgBr + H_2C{=}O \xrightarrow[2. H_3O^+]{1.\ \text{ether}} CH_3CH_2CH_2\overset{\displaystyle CH_3}{\overset{|}{C}}HCH_2CH_2OH$$

c)

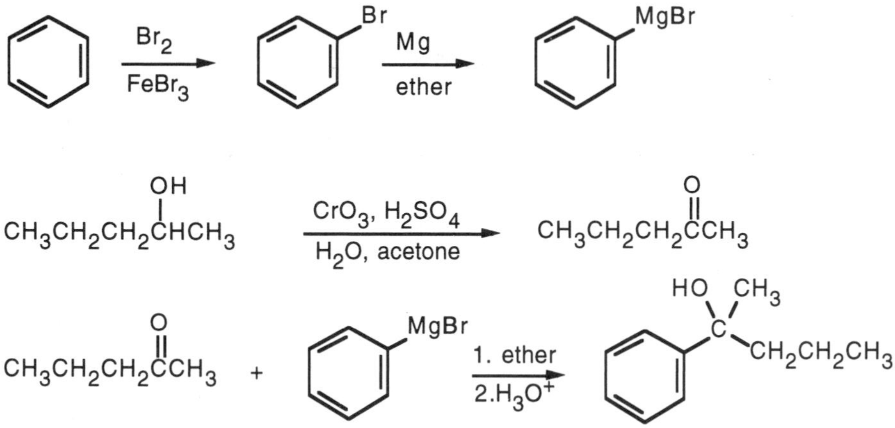

d)

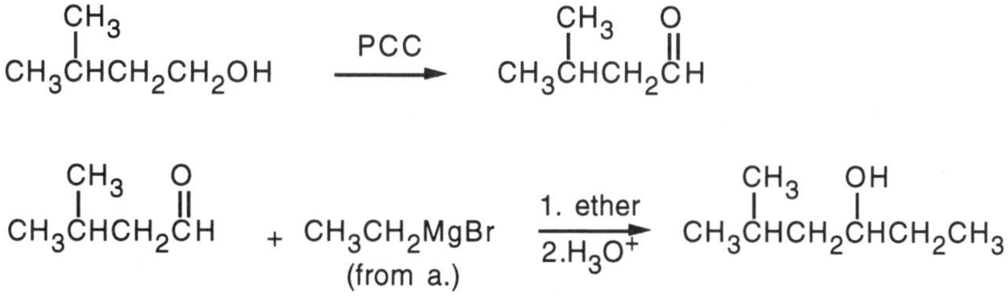

17.40

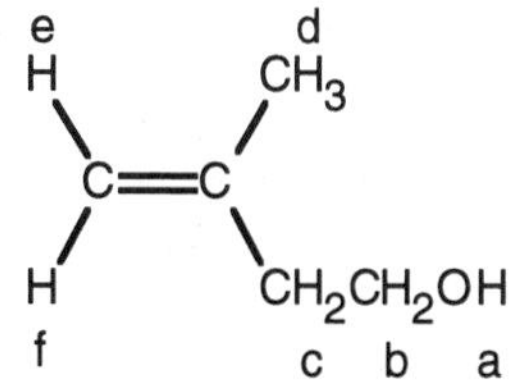

3–Methyl–3–buten–1–ol

The peak absorbing at 1.75 δ (3 H) is due to the d protons. This peak, which occurs in the allylic region of the spectrum, is unsplit.

The peak absorbing at 2.13 δ (1 H) is due to the –OH proton a.

The peak absorbing at 2.30 δ (3 H) is due to protons c. The peak is a triplet because of splitting by the adjacent b protons.

The peak absorbing at 3.70 δ (2 H) is due to the b protons. The adjacent oxygen causes the peak to be downfield, and the adjacent –CH₂– group splits the peak into a triplet.

The peaks at 4.78 δ and 4.85 δ (2 H) are due to protons e and f.

17.41 1. $C_8H_{18}O_2$ has *no* double bonds or rings.
2. The IR band at 3350 cm^{-1} shows the presence of a hydroxyl group.
3. The compound is probably symmetrical (simple NMR).
4. There is no splitting.

A structure that meets all these criteria:

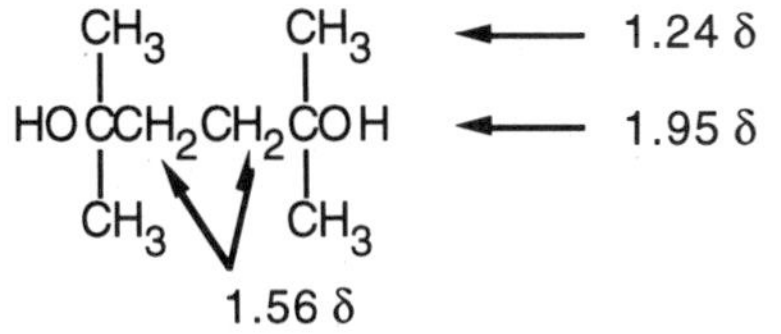

$$\text{HO\overset{CH_3}{\underset{CH_3}{C}}CH_2CH_2\overset{CH_3}{\underset{CH_3}{C}}OH}$$

1.24 δ

1.95 δ

1.56 δ

2,5–Dimethyl–2,5–hexanediol

17.42

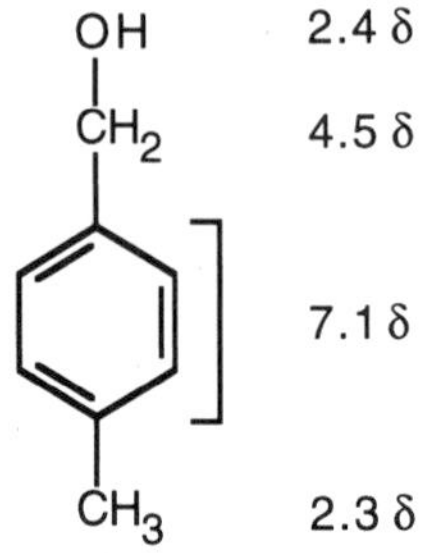

OH   2.4 δ

CH₂   4.5 δ

7.1 δ

CH₃   2.3 δ

p–Methylbenzyl alcohol

17.43

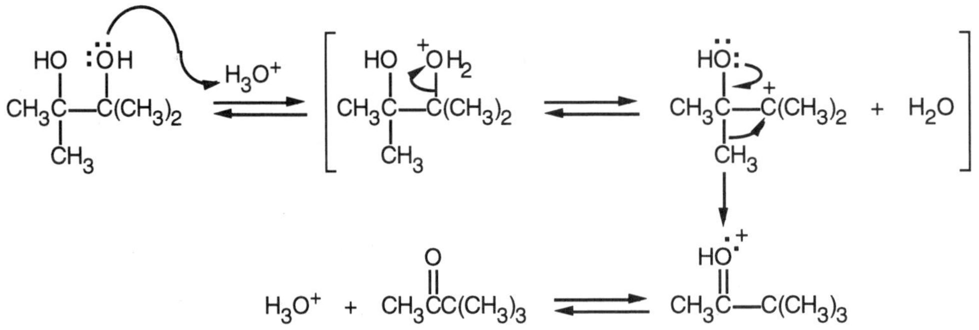

17.44 a) Compound A has one double bond or ring.
b) The infrared absorption at 3400 cm^{-1} indicates the presence of an alcohol. (The weak absorption at 1640 cm^{-1} is due to a C=C stretch.)
c) (1) The absorptions at 1.63 δ and 1.70 δ are due to unsplit methyl protons. Because the absorptions are shifted slightly downfield, the protons are adjacent to an unsaturated center.
  (2) The broad singlet at 3.83 δ is due to an alcohol proton.
  (3) The doublet at 4.15 δ is due to two protons attached to a carbon bearing an electronegative atom (oxygen, in this case).
  (4) The proton absorbing at 5.70 δ is a vinylic proton.

d)

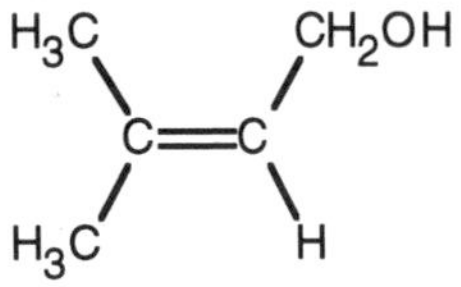

3–Methyl–2–buten–1–ol

Compound A

17.45

a)

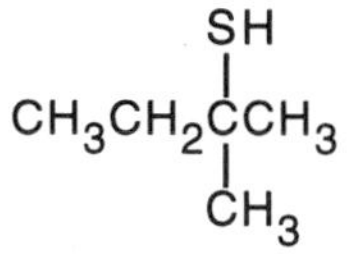

b)

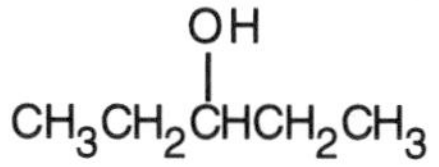

c)

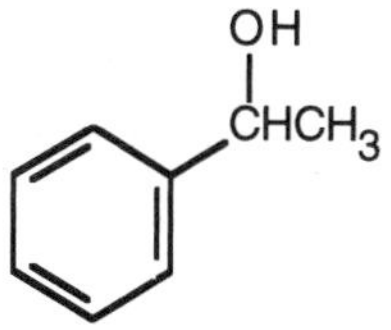

17.46

a)

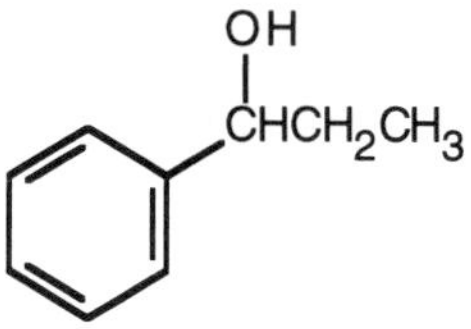

b)

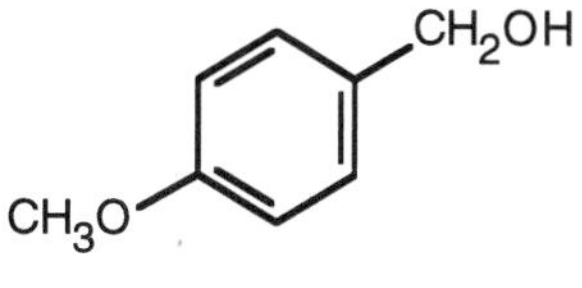

17.47

trans–4–tert–Butylcyclohexanol

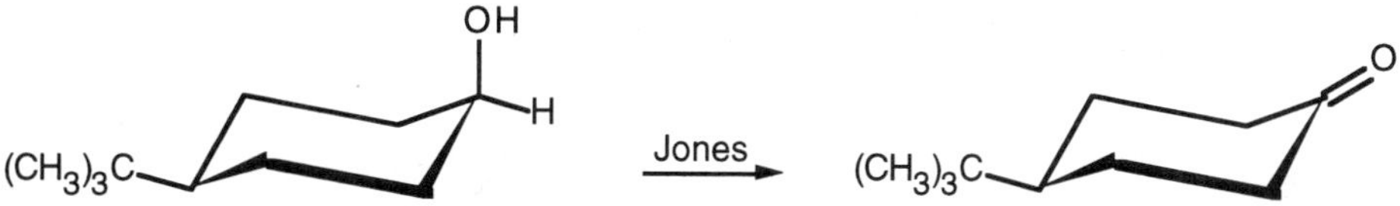

cis–4–tert–Butylcyclohexanol   →(Jones)→

Recall that the bulky *tert*-butyl group occupies the equatorial position. The cis isomer should oxidize faster than the trans isomer.

17.48

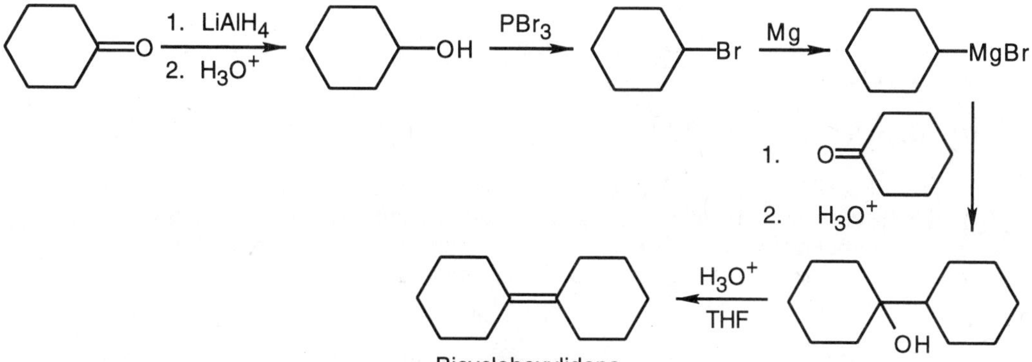

Bicyclohexylidene

17.49 CH₃SSCH₃

17.50 An alcohol adds to an aldehyde by a mechanism that we will study in a later chapter. The hydroxyl group of the intermediate undergoes oxidation (as shown in Section 17.9), and an ester is formed.

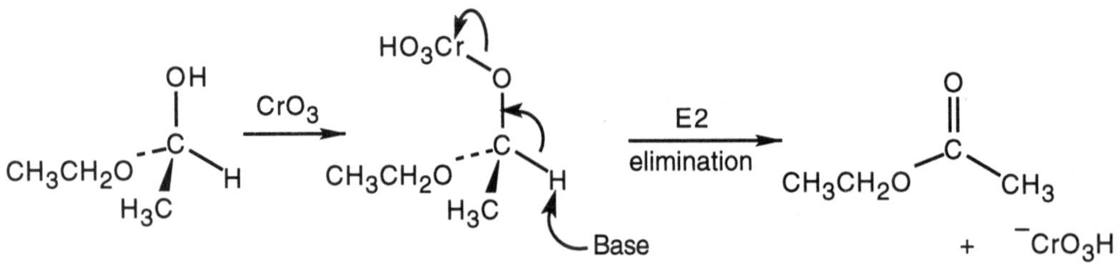

17.51 The first equivalent of CH$_3$MgBr reacts with the hydroxyl hydrogen of 4–hydroxycyclohexanone.

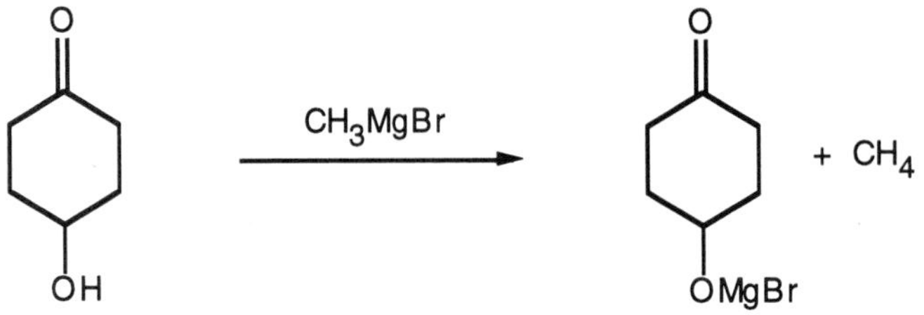

A second equivalent of CH$_3$MgBr adds to the ketone in the expected manner to yield 1–methyl–1,4–cyclohexanediol.

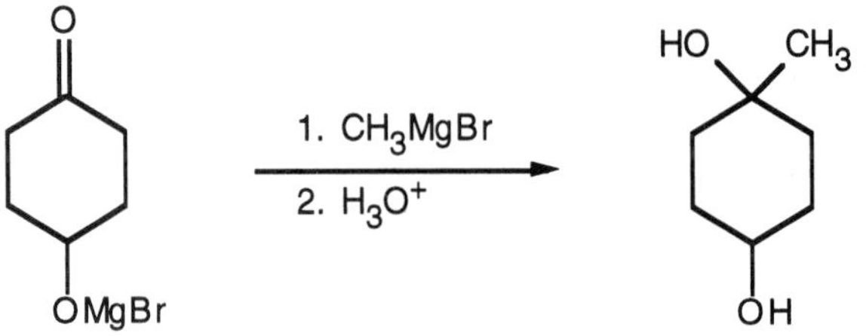

17.52 (a) NaBH$_4$, ethanol; (b) PBr$_3$; (c) Mg, ether, then CH$_2$O; (d) PCC, CH$_2$Cl$_2$; (e) C$_6$H$_5$CH$_2$MgBr; (f) POCl$_3$, pyridine.

17.53 a) C$_5$H$_{12}$O, C$_4$H$_8$O$_2$, C$_3$H$_4$O$_3$

b) The ^{1}H NMR data show that the compound has twelve protons.
c) The IR absorption at 3600 cm^{-1} shows that the compound is an alcohol.
d) The compound contains five carbons, but two of them are identical.
e) C$_5$H$_{12}$O is the structural formula of the compound.
f)

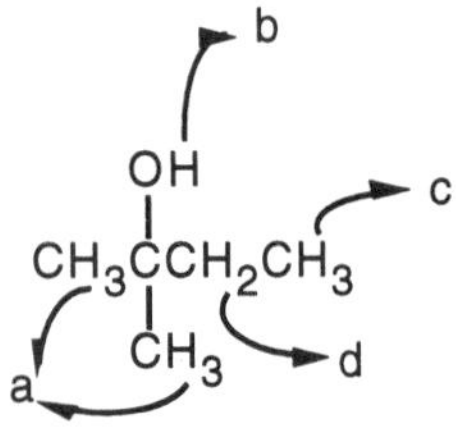

a. 1.2 δ (6 H, singlet)
b. 1.0 δ (1 H, singlet)
c. 0.9 δ (3 H, triplet)
d. 1.4 δ (2 H, quartet)

17.54

17.55 Phenol is a stronger acid than cyclohexanol because the phenoxide anion is stabilized by resonance involving the aromatic ring. No such stabilization is possible for the cyclohexoxide anion.

Study Guide for Chapter 17

After studying this chapter, you should be able to:

(1) Name and draw structures of alcohols and thiols (17.1, 17.2, 17.18, 17.20, 17.21).

(2) Understand the properties and acidity of alcohols (17.3, 17.4, 17.5, 17.54).

(3) Prepare alcohols:
a) From alkenes (17.6).
b) From reduction of carbonyl compounds (17.7, 17.8, 17.28, 17.29, 17.36).
c) From reaction of carbonyl compounds with Grignard reagents (17.9, 17.10, 17.25, 17.30, 17.39, 17.51).
d) Prepare thiols (17.19).

(4) Predict the products of reactions involving alcohols (17.11, 17.12, 17.13, 17.14, 17.22, 17.23, 17.24, 17.26, 17.31, 17.33, 17.34, 17.37, 17.47, 17.48, 17.53).

(5) Formulate mechanisms of reactions of alcohols (17.15, 17.27, 17.32, 17.35, 17.38, 17.43, 17.50).

(6) Use chemical and spectroscopic data to identify alcohols (17.16, 17.17, 17.40, 17.41, 17.42, 17.44, 17.45, 17.46, 17.49, 17.52).

18.1

a) $(CH_3)_2CHOCH(CH_3)_2$

2–Isopropoxypropane *or*
Diisopropyl ether

b)

Propoxycyclopentane *or*
Cyclopentyl propyl ether

c)

p–Bromoanisole *or*
4–Bromo–1–methoxybenzene

d)

1–Methoxycyclohexene

e) $(CH_3)_2CHCH_2OCH_2CH_3$

1–Ethoxy–2–methylpropane *or*
Ethyl isobutyl ether

f) $H_2C{=}CHCH_2OCH{=}CH_2$

Allyl vinyl ether

18.2 The first step of the dehydration procedure is protonation of an alcohol; water is then displaced by another molecule of alcohol to form an ether. If two different alcohols are present, either one can be protonated and either one can displace water, yielding a mixture of products.

   If this procedure were used with ethanol and 1–propanol, the products would be diethyl ether, ethyl propyl ether, and dipropyl ether. If there were equimolar amounts of the alcohols, and if they were of equal reactivity, the product ratio would be diethyl ether : ethyl propyl ether : dipropyl ether = 1:2:1.

18.3

a)

Formation of alkoxide ion.

b)

S_N2 displacement of iodide by alkoxide to yield an ether.

18.4

a)
$$CH_3CH_2CH_2O^- + CH_3Br$$
or
$$CH_3CH_2CH_2Br + CH_3O^- \, Na^+$$
$$\longrightarrow \quad CH_3CH_2CH_2OCH_3 \quad + \quad NaBr$$
Methyl propyl ether

b)

+ CH₃Br ⟶ Anisole + NaBr

c)

$$\underset{CH_3CHO^- \, Na^+}{\overset{CH_3}{|}} \quad + \quad$$

⟶ Benzyl isopropyl ether + NaBr

d)
$$\underset{\underset{CH_3}{|}}{\overset{\overset{CH_3}{|}}{CH_3CCH_2O^-}} Na^+ \; + \; CH_3CH_2Br \quad \longrightarrow \quad \underset{\underset{CH_3}{|}}{\overset{\overset{CH_3}{|}}{CH_3CCH_2OCH_2CH_3}} \; + \; NaBr$$

Ethyl 2,2-dimethyl
propyl ether

18.5 The compounds most reactive in the Williamson ether synthesis are also most reactive in any S_N2 process (review Chapter 11 if necessary).

Most reactive ⟶ Least reactive

a) CH_3CH_2Br > $CH_3CHBrCH_3$ >> C_6H_5Br

    primary   secondary   aromatic halide
                                  (not reactive)

b) CH_3CH_2Br > CH_3CH_2Cl >> $ICH=CHCH_3$

    better   poorer   vinylic
    leaving group   leaving group   (not reactive)

18.6

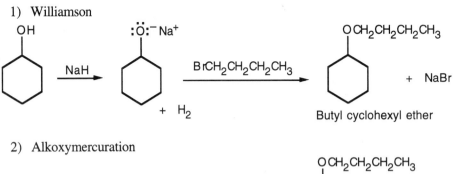

The reaction mechanism of alkoxymercuration/demercuration of an alkene is similar to other electrophilic additions we have studied. First, the cyclopentene π electrons attack Hg^{2+} with formation of a mercurinium ion. Next, the nucleophilic alcohol displaces mercury. Markovnikov addition occurs because the carbon bearing the methyl group is better able to stabilize the partial positive charge arising from cleavage of the carbon-mercury bond. The ethoxyl and the mercuric groups are trans to each other. Finally, removal of mercury by $NaBH_4$ by a mechanism that is not fully understood results in the formation of 1–ethoxy–1–methylcyclopentane.

18.7 a) Either method of synthesis is appropriate.

1) Williamson

2) Alkoxymercuration

b) Either method is possible, but the Williamson synthesis is simpler.

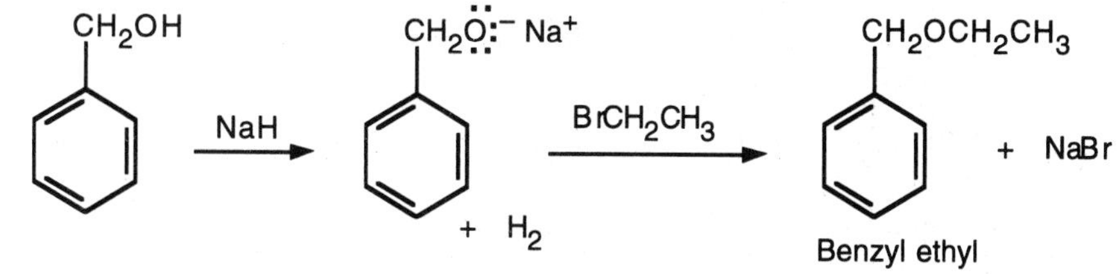

Benzyl ethyl ether

c) Because both parts of the ether are somewhat hindered, use alkoxymercuration.

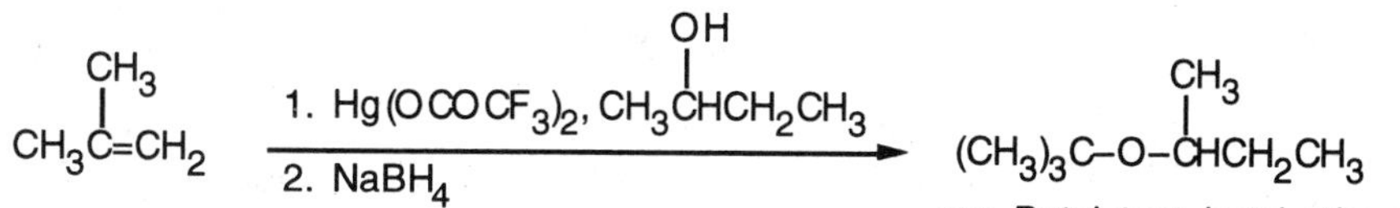

sec–Butyl *tert*–butyl ether

d) The Williamson synthesis is better.

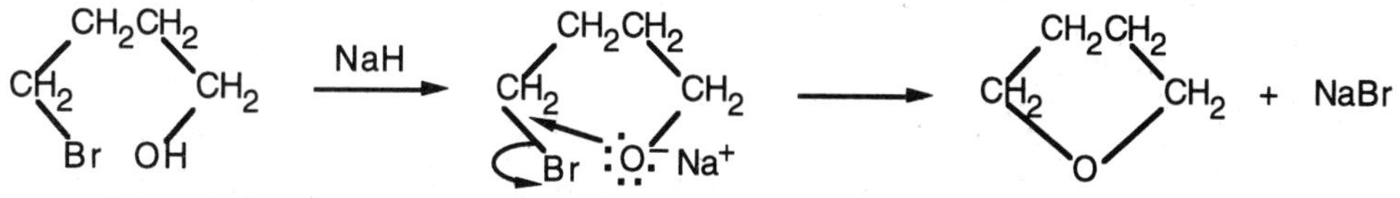

Tetrahydrofuran

18.8 Let R be any other alkyl function.

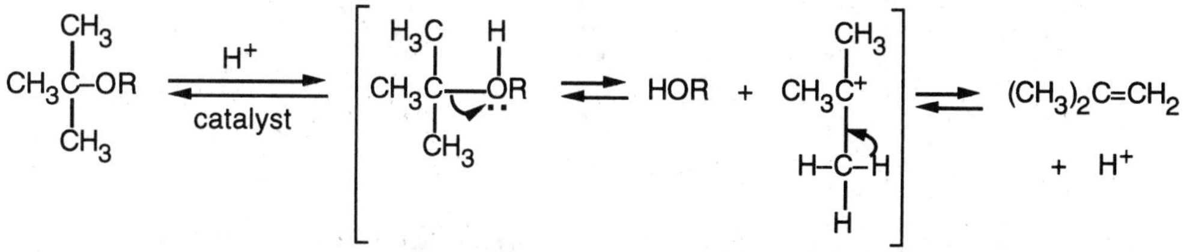

The first step of trifluoroacetic ether cleavage is protonation of the ether oxygen to give an intermediate, which undergoes spontaneous dissociation to form an alcohol and a tertiary carbocation. The carbocation then loses a proton to form an alkene, 2-methylpropene. This is an example of E_1 elimination.

18.9

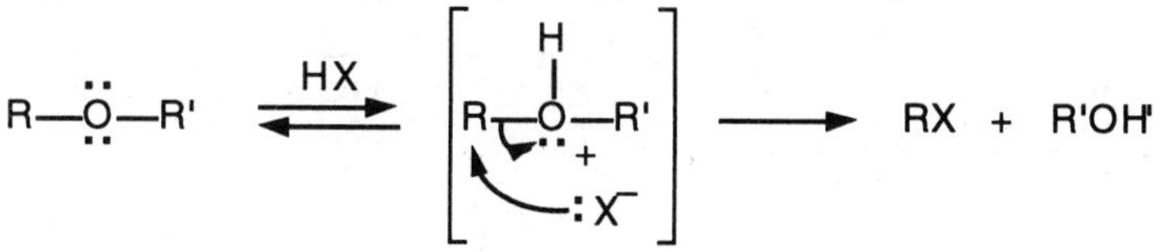

HX first protonates the oxygen atom, and halide then effects nucleophilic displacement to form an alcohol and an organic halide. The better the nucleophile, the more effective the displacement. Since I^- and Br^- are more nucleophilic than Cl^-, ether cleavage proceeds more smoothly with HI or HBr than with HCl.

18.10 Epoxidation by use of *m*-chloroperoxybenzoic acid (RCO₃H) is a syn addition of oxygen to a double bond; the original bond stereochemistry is retained.

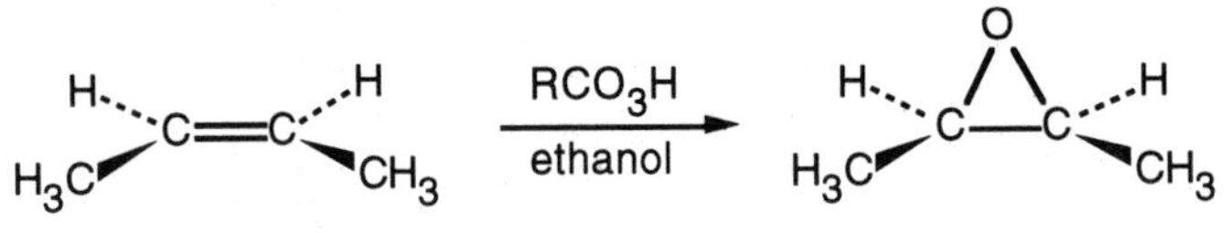

cis–2–Butene   cis–2,3–Epoxybutane

In the epoxide product, as in the alkene starting material, the methyl groups are cis.

18.11

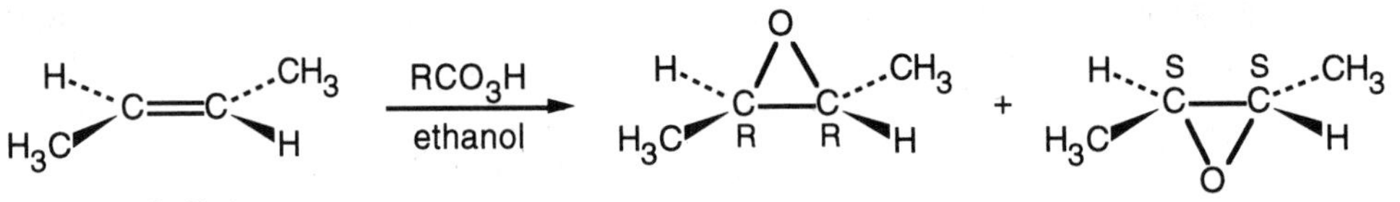

trans–2–Butene   trans–2,3–Epoxybutane

The argument in the previous problem can be used to show that reaction of *trans*–2–butene with *m*-chloroperoxybenzoic acid yields *trans*–2,3–epoxybutane. A mixture of enantiomers is formed because the peroxyacid can attack either the top or bottom of the double bond.

18.12

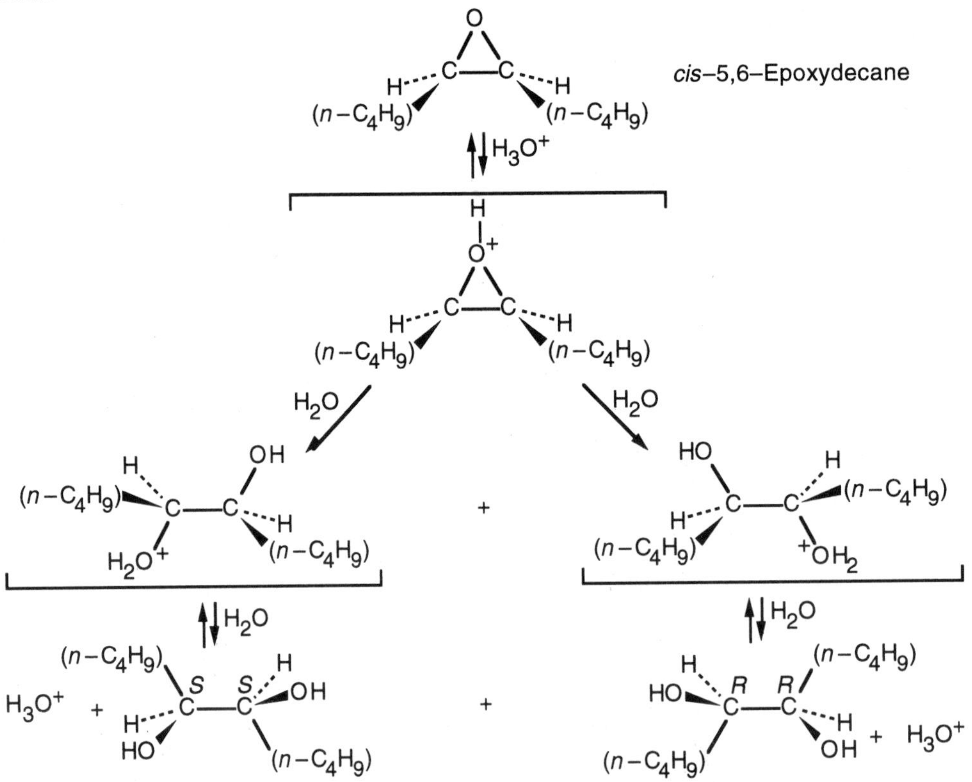

cis–5,6–Epoxydecane

The product of acid hydrolysis of *cis*-5,6-epoxydecane is a racemic mixture.

18.13

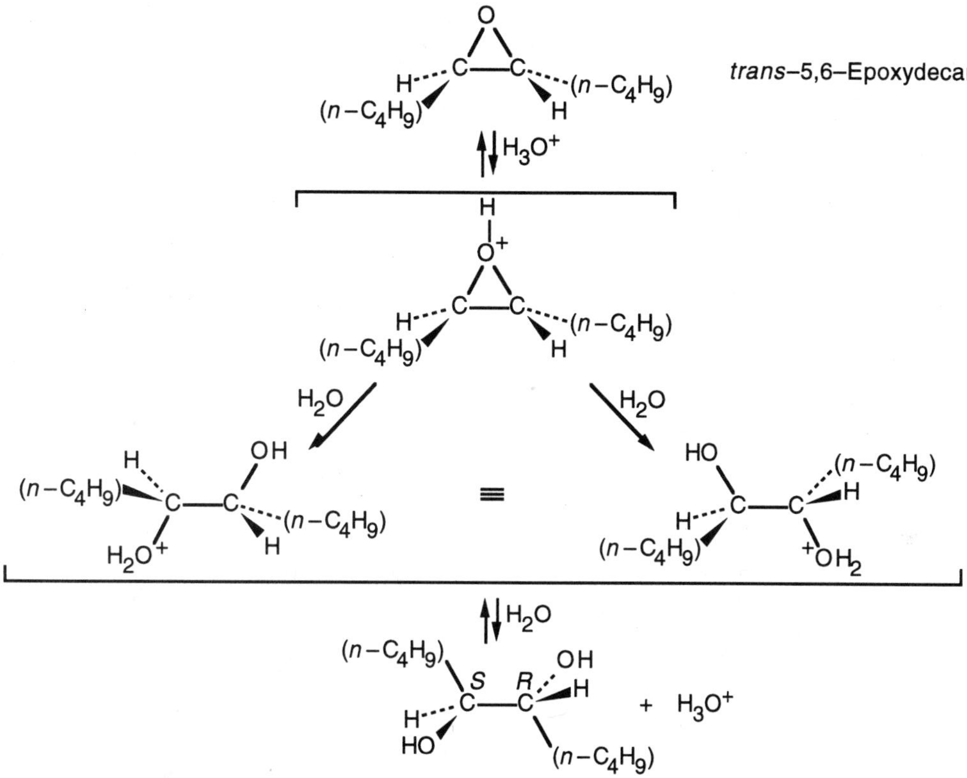

trans–5,6–Epoxydecane

The product of acid hydrolysis of *trans*–5,6–epoxydecane is a meso compound that is a diastereomer of the one formed in the previous problem.

18.14

a)

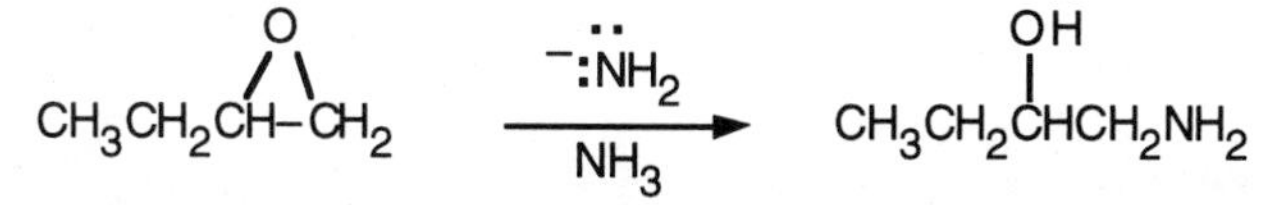

Attack of the basic nucleophile occurs at the less substituted carbon atom.

b)

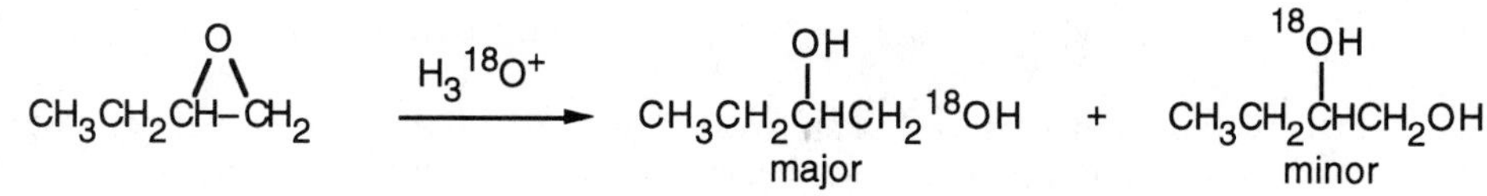

Attack of the nucleophile under acidic conditions occurs at the less substituted carbon atom when the carbons are primary or secondary.

18.15

15–Crown–5   12–Crown–4

The ion-to-oxygen distance in 15-crown-5 is about 40% longer than the ion-to-oxygen distance in 12-crown-4.

18.16

$$1.5 \, \delta$$

CH$_3$CH$_2$CH–C

1.0 δ

~2.9 δ

$\begin{bmatrix} 2.5 \, \delta \\ \text{and} \\ 2.7 \, \delta \end{bmatrix}$

1,2–Epoxybutane

18.17

a)   CH$_3$CH$_2$SCH$_3$

Ethyl methyl sulfide

b)   (CH$_3$)$_3$CSCH$_2$CH$_3$

tert–Butyl ethyl sulfide

c)   SCH$_3$

SCH$_3$

o–(Dimethylthio)benzene

d)   S

CH$_3$

Phenyl *p*–tolyl sulfide *or*
p–(Phenylthio)toluene

18.18

O$^-$

CH$_3$SCH$_3$
   $+$

Dimethyl sulfoxide

CH$_3$SCH$_3$

Dimethyl sulfide

The boiling point of dimethyl sulfoxide is high because it is a dipolar compound. Dimethyl sulfoxide is miscible with water because it can hydrogen-bond with water.

18.19

a)   CH$_2$CH$_3$

CH$_3$CH$_2$OCH

CH$_2$CH$_3$

b)   Cl–⟨⟩–O–⟨⟩–Cl

c)

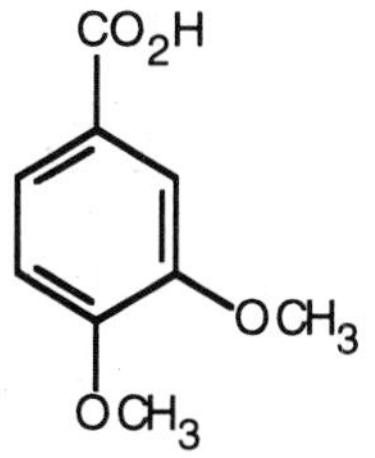

d)

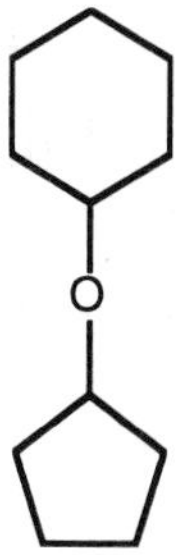

e)

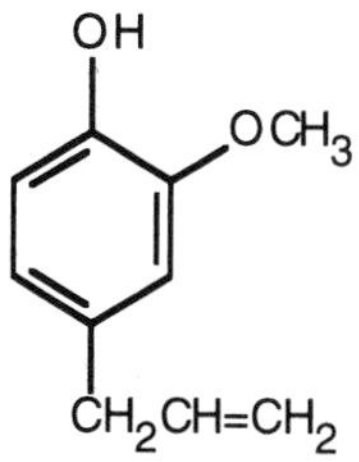

18.20 a) *o*–Dimethoxybenzene

b) Cyclopropyl isopropyl ether *or* Isopropoxycyclopropane

c) 2–Methyltetrahydrofuran

d) Cyclohexyl cyclopropyl sulfide

e) 2,2–Dimethoxypropane

f) 1,2–Epoxycyclopentane

g) *p*–Nitroethoxybenzene *or* Ethyl *p*–nitrophenyl ether

h) 1,1–(Dimethylthio)cyclohexane

18.21

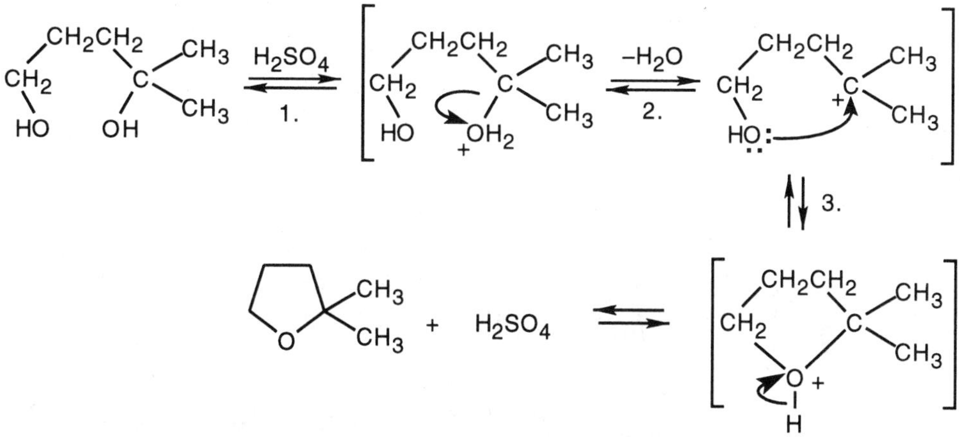

The reaction involves: (1) protonation of the tertiary hydroxyl group; (2) loss of water to form a tertiary carbocation; (3) nucleophilic attack on the cation by the other hydroxyl group. The tertiary hydroxyl group is more likely to be eliminated because the resulting carbocation is more stable.

18.22

a)

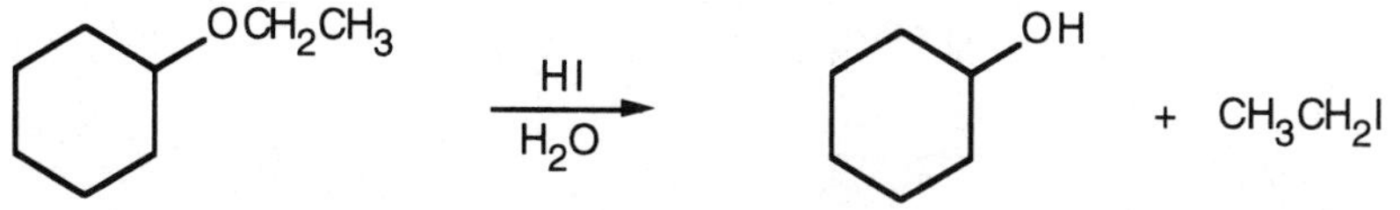

b)

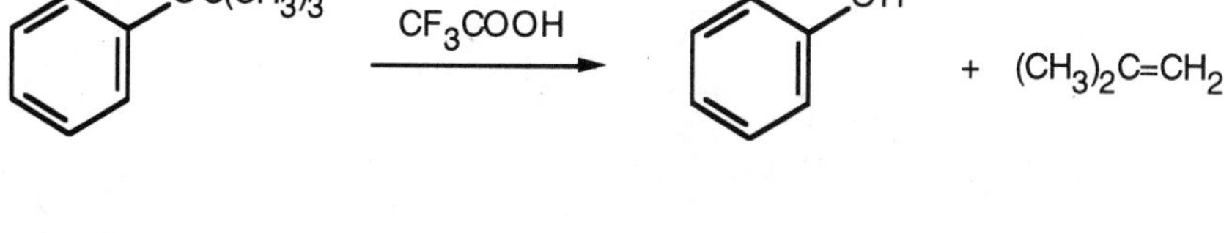

c)

OCH_2CH_3

$H_2C=CH$ $\xrightarrow[H_2O]{HI}$ CH_3CH_2I + $\left[\begin{array}{c} OH \\ H_2C=CH \end{array} \right]$ ⟶ CH_3CHO

d) $(CH_3)_3CCH_2OCH_2CH_3$ $\xrightarrow[H_2O]{HI}$ $(CH_3)_3CCH_2OH$ + CH_3CH_2I

18.23

$$\frac{1.06 \text{ g vanillin}}{152 \text{ g/mol}} = 6.97 \times 10^{-3} \text{ mol vanillin}$$

$$\frac{1.60 \text{ g AgI}}{234.8 \text{ g/mol}} = 6.81 \times 10^{-3} \text{ mol AgI}$$

6.81×10^{-3} mol ⟶ 6.81×10^{-3} mol ⟶ 6.81×10^{-3} mol ⟶ 6.81×10^{-3} mol
    AgI   :I⁻ CH_3I $-OCH_3$

Thus, 6.97×10^{-3} moles of vanillin contains 6.81×10^{-3} moles of methoxyl groups. Since the ratio of moles vanillin to moles methoxyl is approximately 1:1, each vanillin contains one methoxyl group.

Vanillin

18.24

a)

$\xrightarrow{\text{NaH}}$ ... $\xrightarrow{CH_3CH_2Br}$ ...

b) $CH_3CH=CH_2$ $\xrightarrow[\text{2. NaBH}_4]{\text{1. } C_6H_5OH, Hg(OCOCF_3)_2}$

c)

H_3C ... H / $C=C$ / H ... CH_3 $\xrightarrow[\text{ethanol}]{RCO_3H}$ H_3C···C—C···H / H ... CH_3

d)

—OH $\xrightarrow[\text{2. NaBH}_4]{\text{1. Hg(OCOCF}_3)_2, (CH_3)_2C=CH_2}$ ... $-OCCH_3$ with CH3 groups

e)

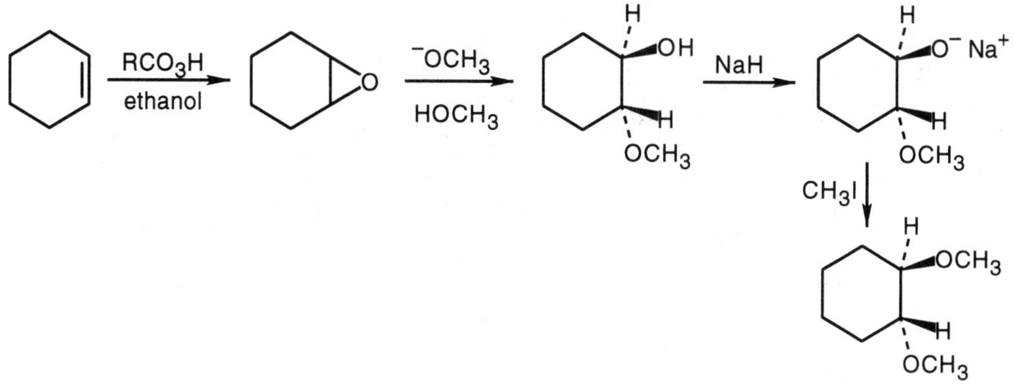

f)

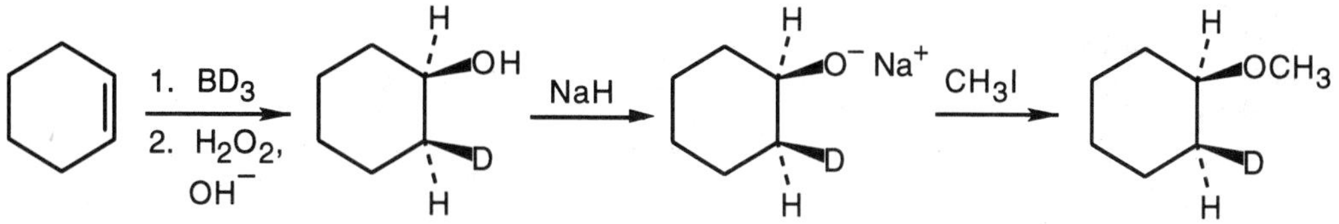

18.25

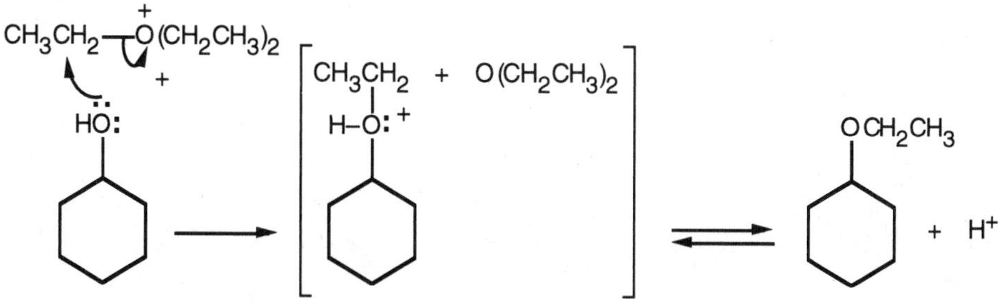

Trialkyloxonium salts are more reactive alkylating agents than alkyl iodides because a neutral ether is a better leaving group than an iodide ion.

18.26

a)

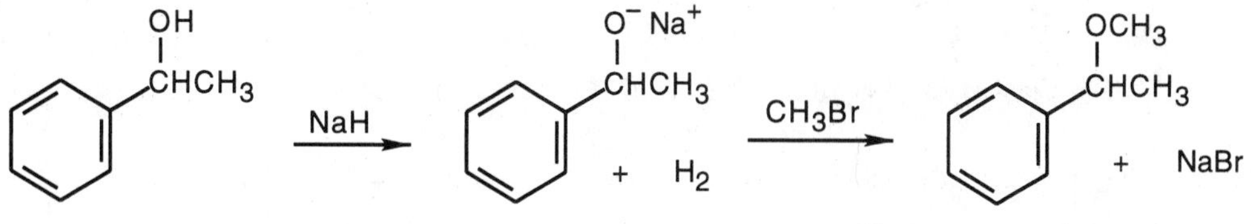

Methyl 1–phenylethyl ether

b)

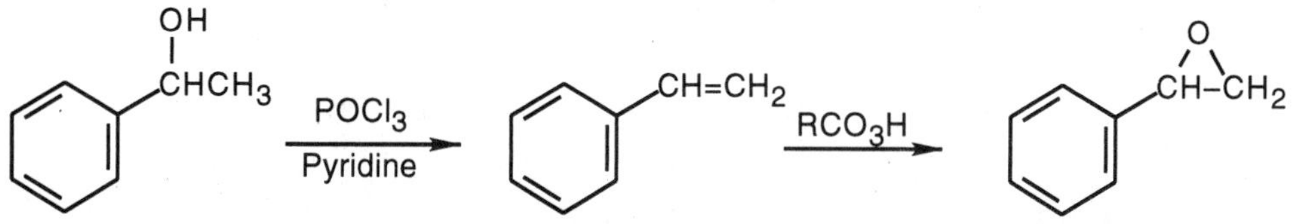

Styrene   Phenylepoxyethane

c)

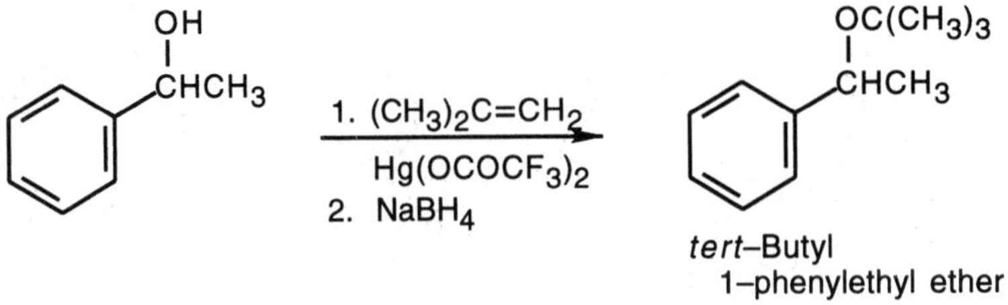
OH
CHCH3
1. (CH3)2C=CH2
Hg(OCOCF3)2
2. NaBH4
OC(CH3)3
CHCH3
tert–Butyl
1–phenylethyl ether

18.27

a)

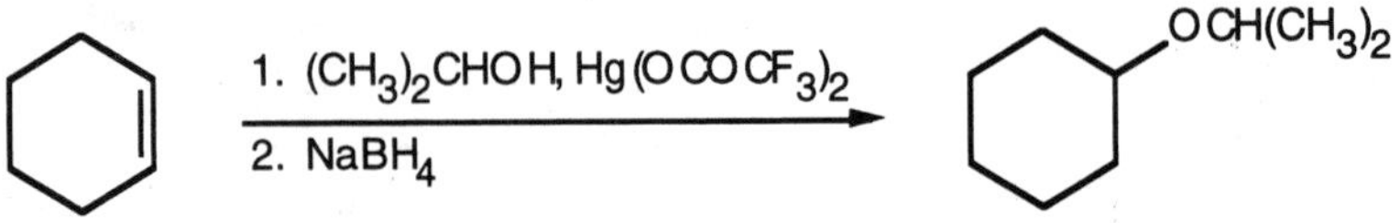
1. (CH3)2CHOH, Hg(OCOCF3)2
2. NaBH4
OCH(CH3)2

b)

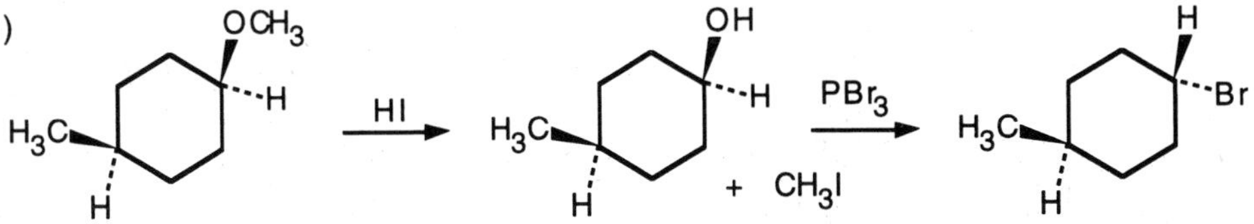
OCH3
H
H3C
H
HI
OH
H
H3C
H
+ CH3I
PBr3
H
Br
H3C
H

c)

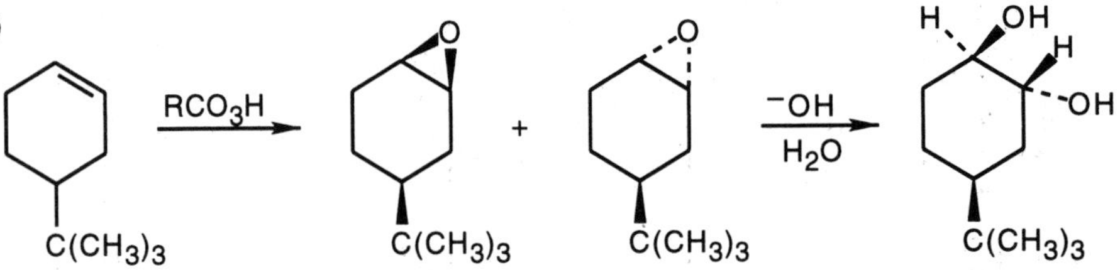
C(CH3)3
RCO3H
O
C(CH3)3
+
O
C(CH3)3
−OH
H2O
H
OH
H
OH
C(CH3)3

d)

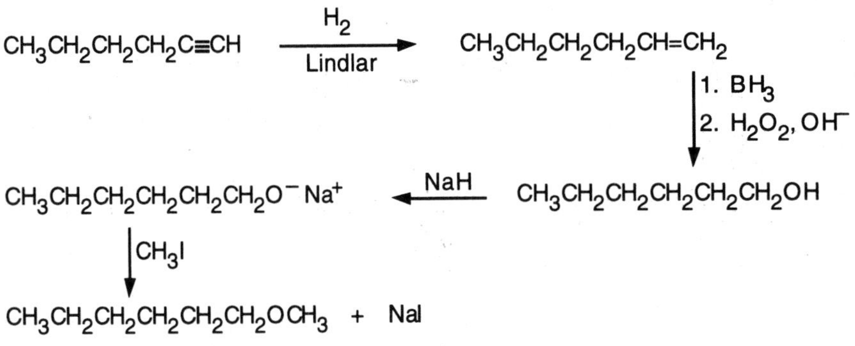
CH3CH2CH2CH2C≡CH
H2
Lindlar
CH3CH2CH2CH2CH=CH2
1. BH3
2. H2O2, OH−
CH3CH2CH2CH2CH2CH2OH
NaH
CH3CH2CH2CH2CH2CH2O− Na+
CH3I
CH3CH2CH2CH2CH2CH2OCH3 + NaI

e)

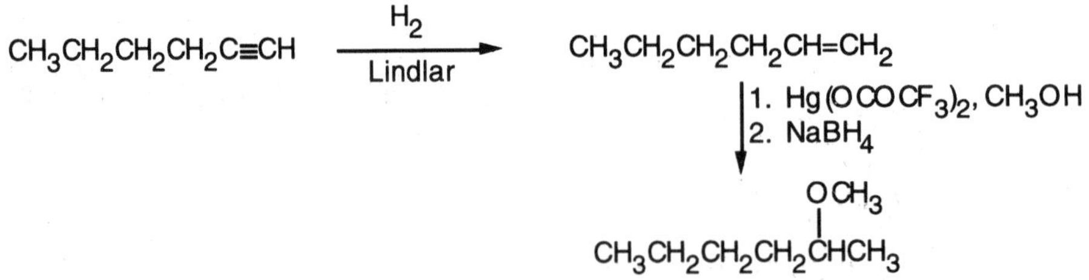
CH3CH2CH2CH2C≡CH
H2
Lindlar
CH3CH2CH2CH2CH=CH2
1. Hg(OCOCF3)2, CH3OH
2. NaBH4
OCH3
CH3CH2CH2CH2CHCH3

18.28

a)

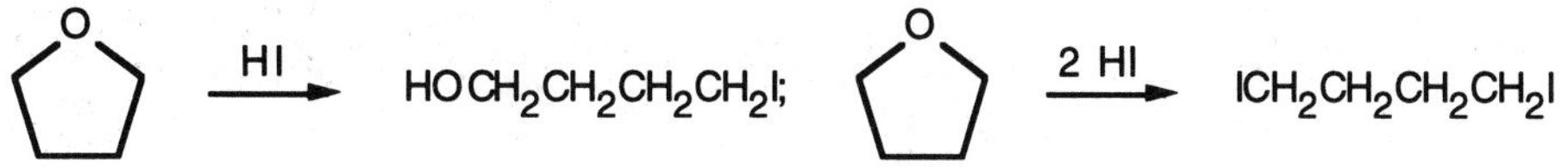

HO CH$_2$CH$_2$CH$_2$CH$_2$I;

ICH$_2$CH$_2$CH$_2$CH$_2$I

18.29

a)

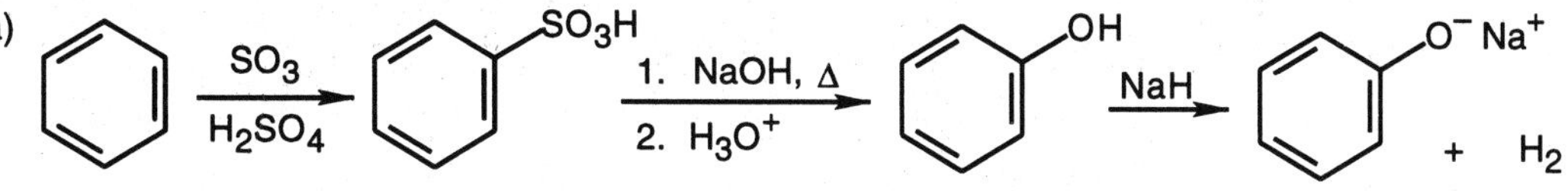

b)

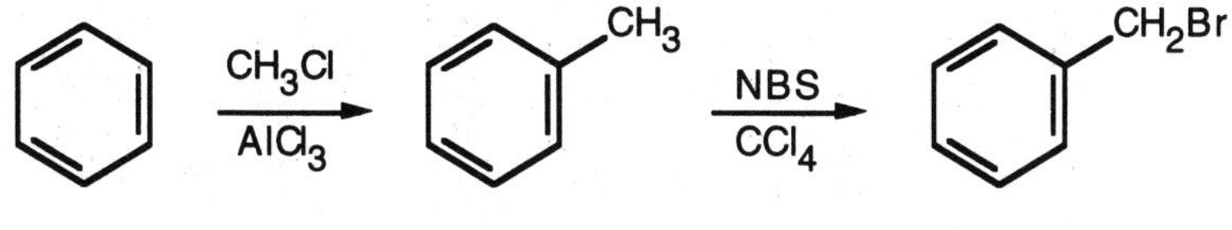

c)

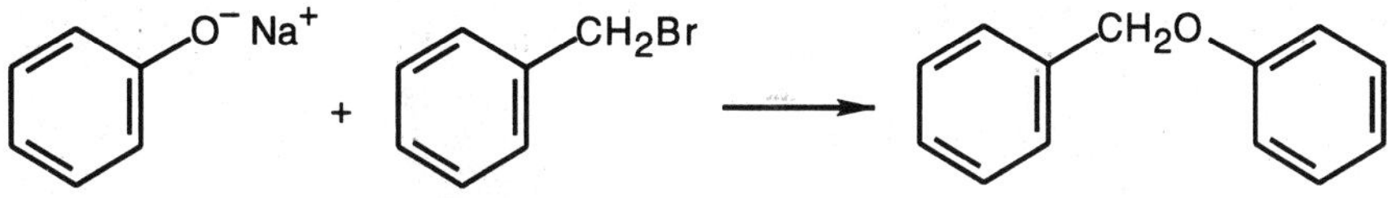

from a   from b   Benzyl phenyl ether

18.30

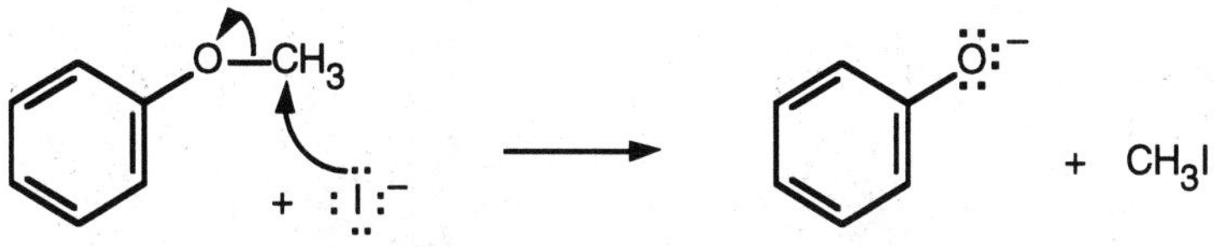

This reaction is an S$_N$2 displacement and can't occur at an aryl carbon.

18.31

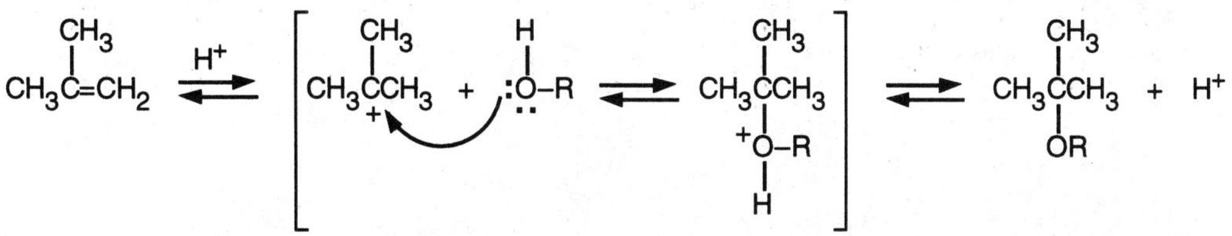

Notice that this reaction is the reverse of acid-catalyzed cleavage of a tertiary ether.

18.32

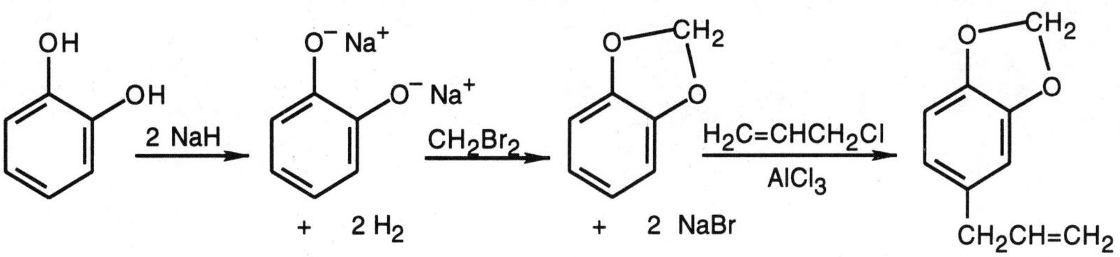

+   2 H$_2$ +   2 NaBr

CH$_2$CH=CH$_2$

Safrole

18.33

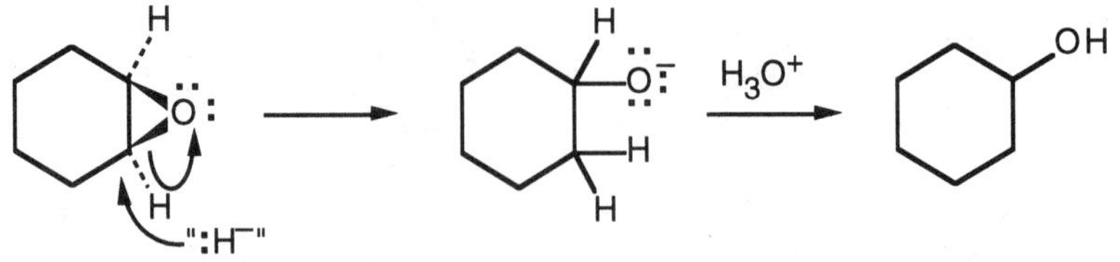

The reaction is an S_N2 epoxide cleavage by attack of hydride ion nucleophile.

18.34

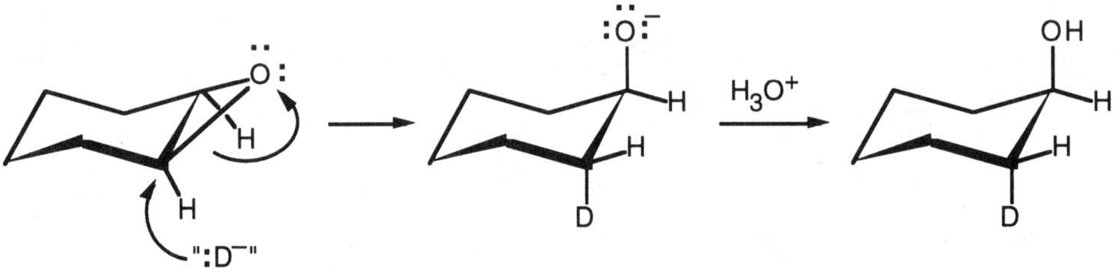

18.35

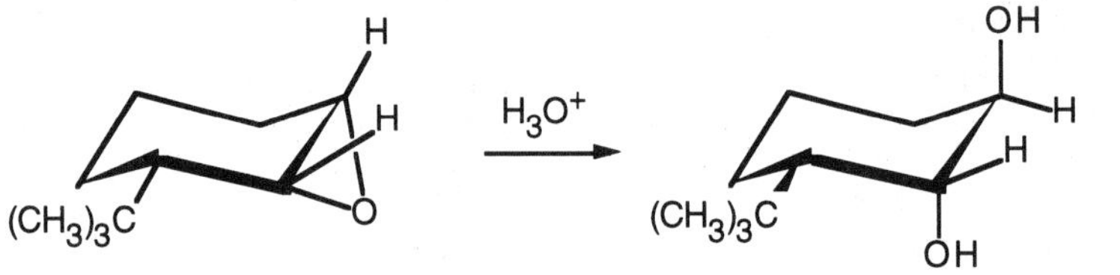

cis–3–*tert*–Butyl–1,2–epoxycyclohexane

The hydroxyl groups in the product have a trans diaxial relationship.

18.36 The mechanism of Grignard addition to oxetane is the same as the mechanism of Grignard addition to epoxides, described in Section 18.8. The reaction proceeds at a reduced rate because oxetane is less reactive than ethylene oxide. The four-membered ring oxetane is less strained, and therefore more stable, than the three-membered ethylene oxide ring.

18.37

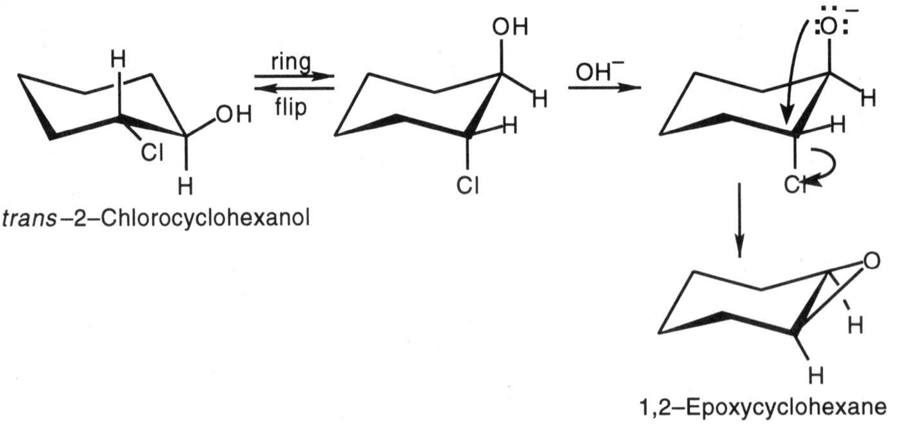

trans–2–Chlorocyclohexanol

1,2–Epoxycyclohexane

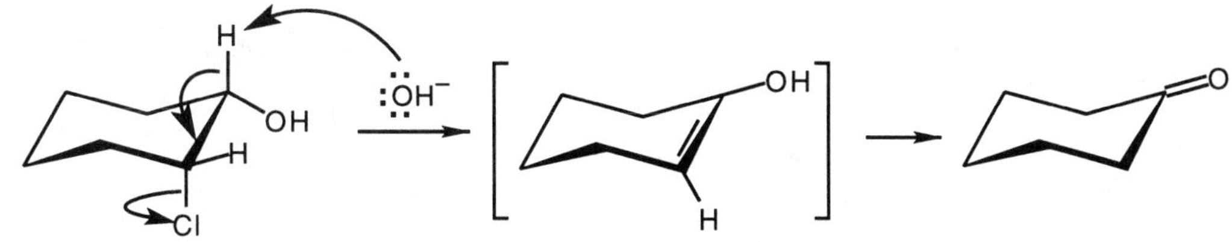

cis–2–Chlorocyclohexanol   enol   Cyclohexanone

In the trans isomer, the –OH and –Cl are in the trans orientation that allows epoxide formation to occur as described in Section 18.7. Epoxidation can't occur for the cis isomer, however. Instead, the base ⁻OH brings about E2 elimination, producing an enol, which rearranges to cyclohexanone.

18.38

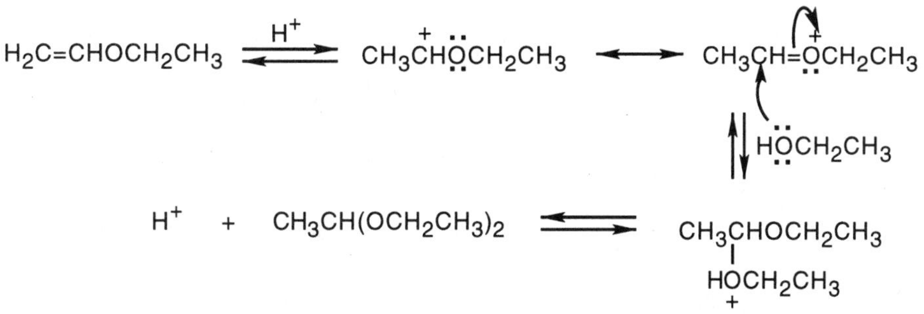

Addition of ethanol occurs with the observed regiochemistry because the initial carbocation is stabilized by resonance involving the ether oxygen.

18.39

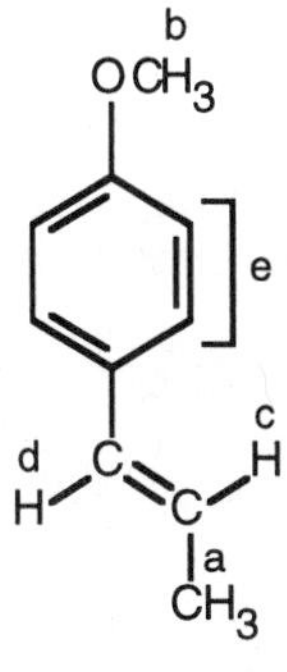

Anethole

Protons	δ	Multiplicity	Split by
a.	1.83	doublet	c
b.	3.75	singlet	
c.	6.08	two quartets	a,d
d.	6.28	doublet	c
e.	6.80, 7.23	multiplet	

18.40

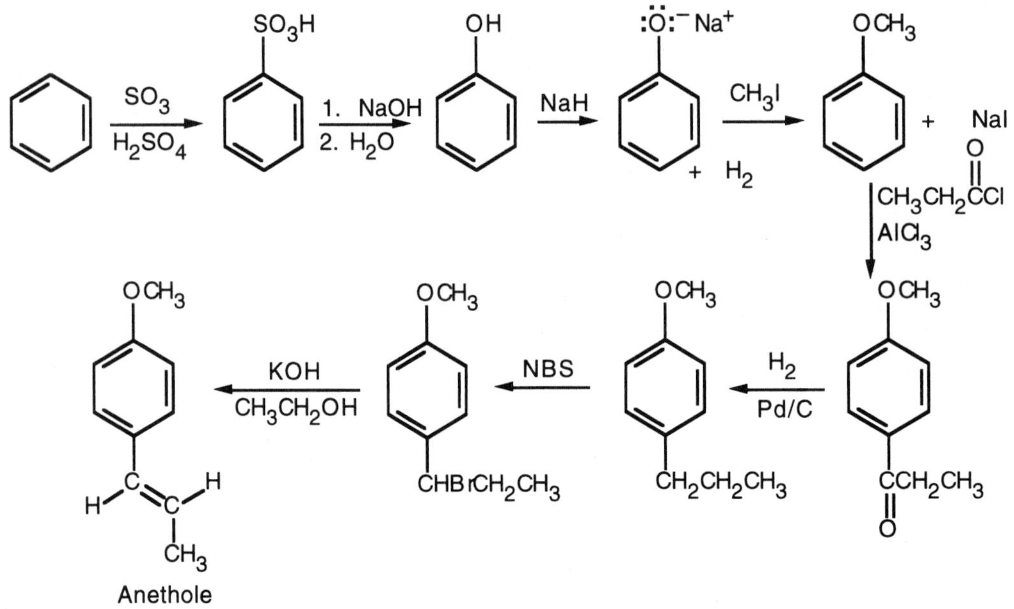

Anethole

18.41 $M^+ = 116$ corresponds to a sulfide of molecular formula $C_6H_{12}S$, indicating one degree of unsaturation.

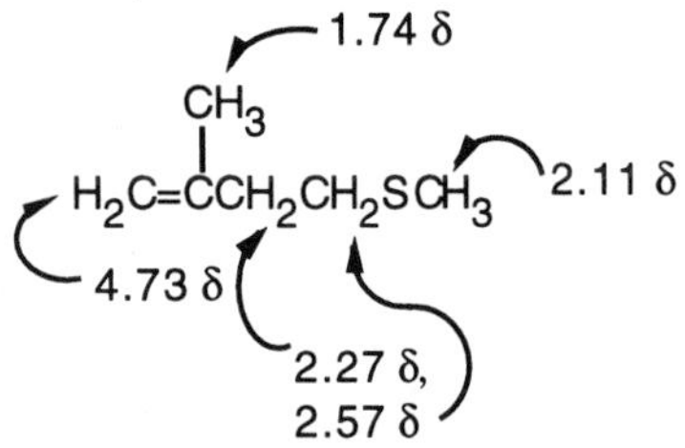

18.42 A molecular model of bornene shows that approach to the upper face of the double bond is hindered by a methyl group. Reaction with RCO_3H occurs at the lower face of the double bond to produce epoxide $\underline{A}$.

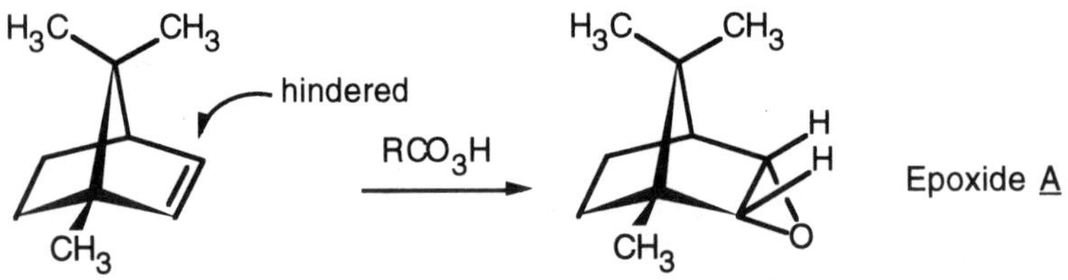

Epoxide $\underline{A}$

In the reaction of Br_2 and H_2O with bornene, the intermediate bromonium ion also forms at the lower face. Reaction with water yields a bromohydrin which, when treated with base, forms epoxide **B**.

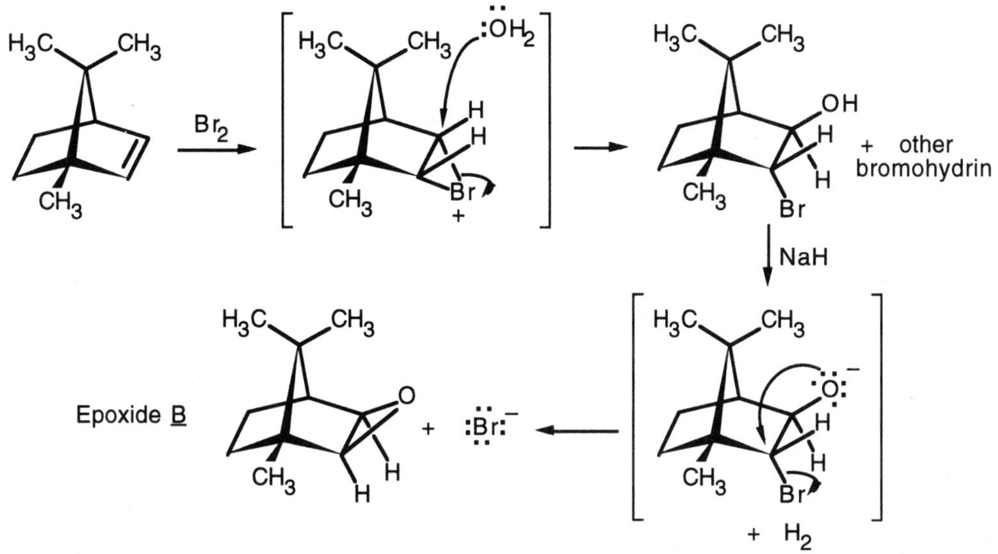

18.43 Disparlure, $C_{19}H_{38}O$, contains one degree of unsaturation, which the ${}^1H$ NMR absorption at 2.8 δ identifies as an epoxide ring.

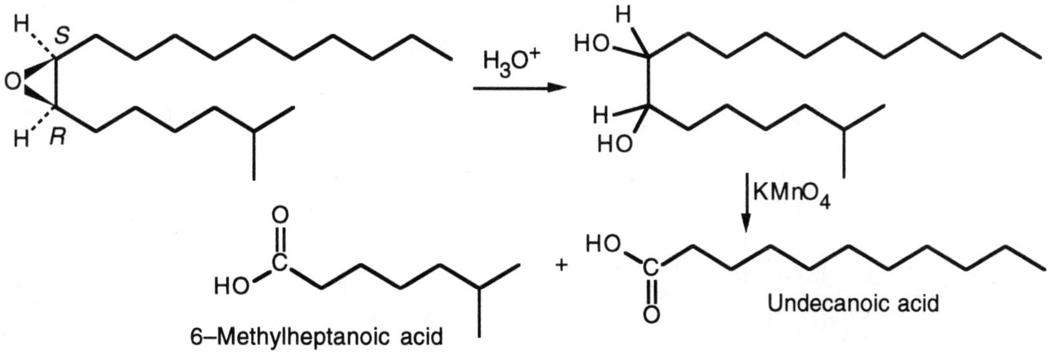

6–Methylheptanoic acid

Undecanoic acid

18.44

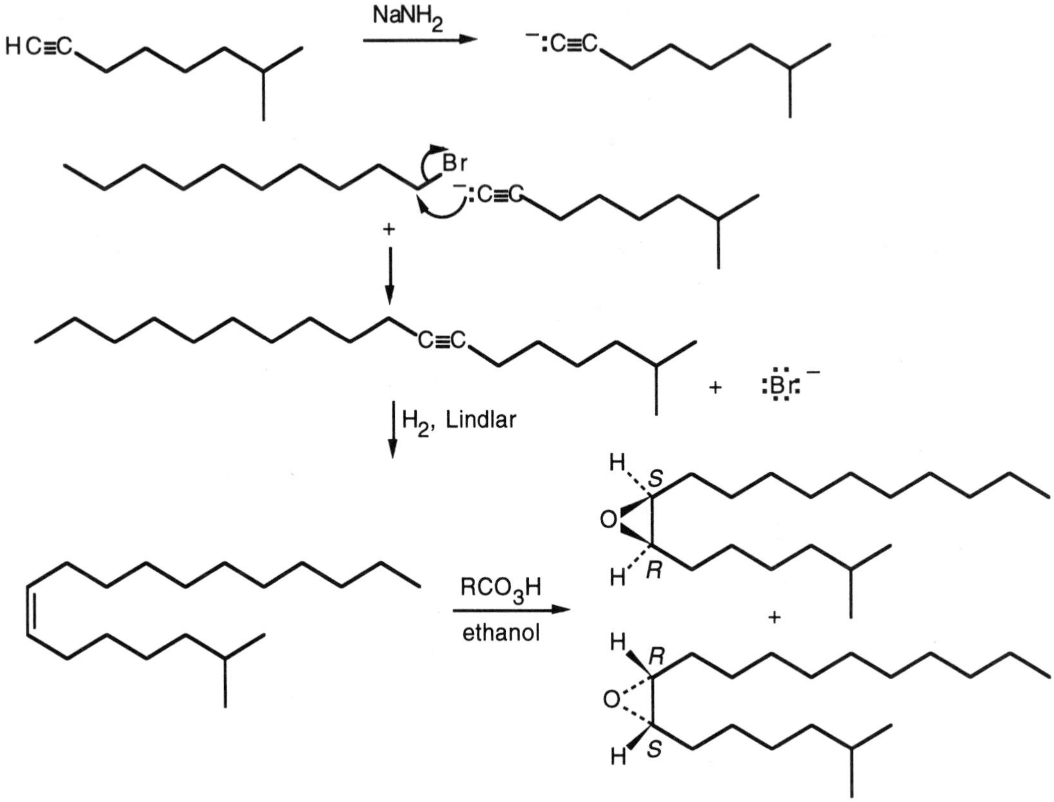

$HC\equiv C$ ~~~~ $\xrightarrow{\text{NaNH}_2}$ $^-:C\equiv C$ ~~~~

$+$:B̈r: $^-$

$\Big\downarrow$ H_2, Lindlar

$\xrightarrow[\text{ethanol}]{\text{RCO}_3\text{H}}$

$+$

18.45

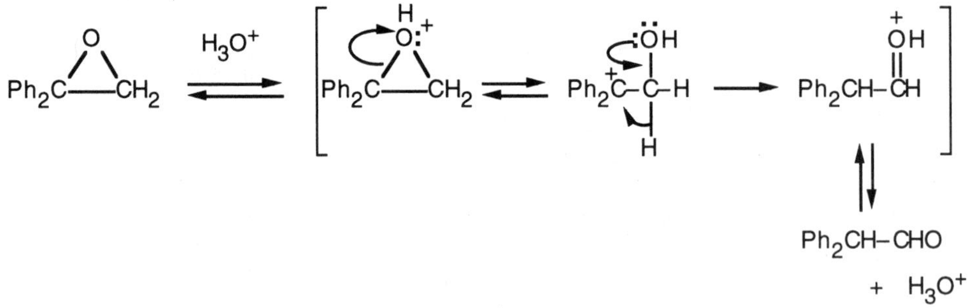

$Ph_2C\text{---}CH_2$ with O (epoxide) $\xrightleftharpoons{H_3O^+}$ $\Big[$ $Ph_2C\text{---}CH_2$ $\xrightleftharpoons{}$ $Ph_2\overset{+}{C}\text{-}\overset{..}{C}\text{-H}$ $\longrightarrow$ $Ph_2CH\text{-}\overset{+}{\overset{OH}{C}}H$ $\Big]$

$\Big\updownarrow$

$Ph_2CH\text{-}CHO$

$+$ H_3O^+

18.46

$\xrightarrow{H_3O^+}$

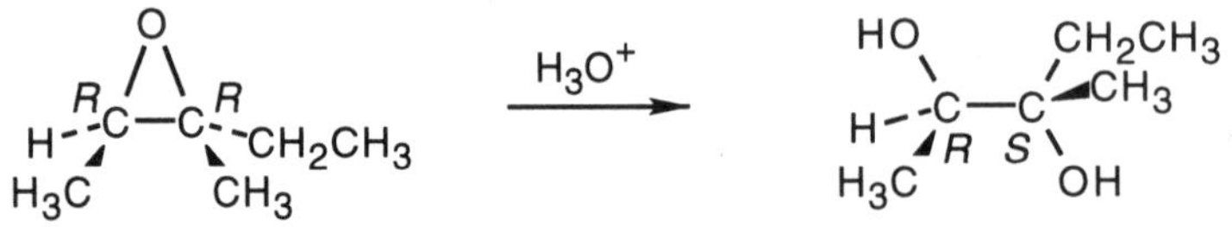

(2R,3R)-2,3-Epoxy-3-methylpentane   (2R, 3S)-3-Methyl-2,3-pentanediol

Reaction with aqueous acid causes ring opening to occur at C3 because the positive
charge of the cationic intermediate is more stabilized at the tertiary carbon.

c) If ring opening occurs exclusively at C3, the product is the 2R,3S isomer and is chiral. (If ring opening occurred equally at either carbon, the product would be a mixture of chiral enantiomers).

d) The product is optically active because one chiral enantiomer is produced.

18.47

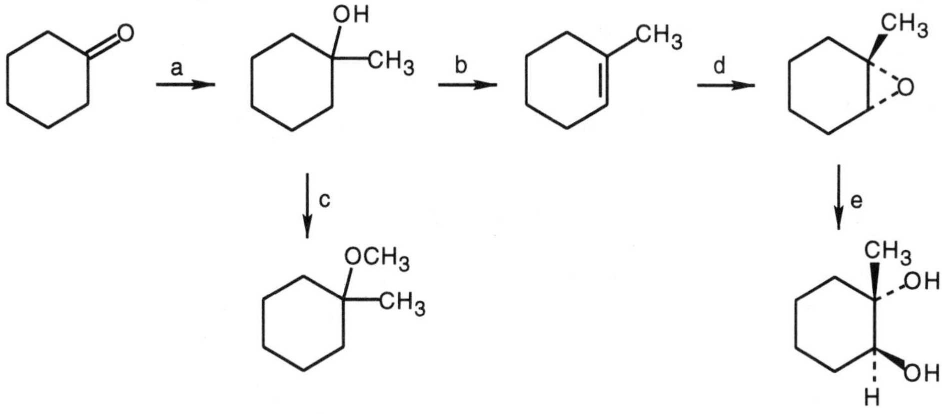

(a) CH₃MgBr, ether; (b) H₂SO₄, H₂O; (c) NaH, CH₃I; (d) m-ClC₆H₄CO₃H; (e) H₃O⁺

18.48

a)

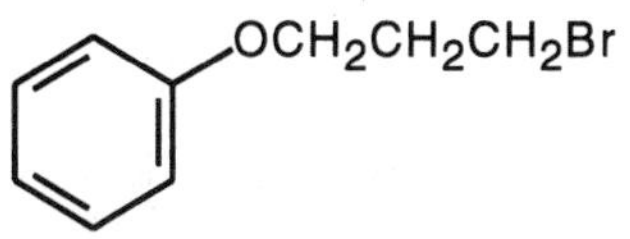

b) CH₃CH(OCH₃)₂

c)

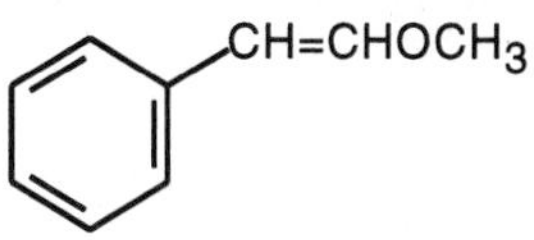

18.49

a)

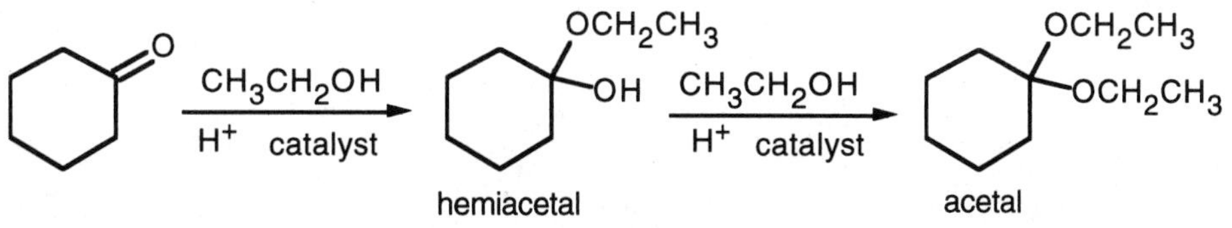

b)

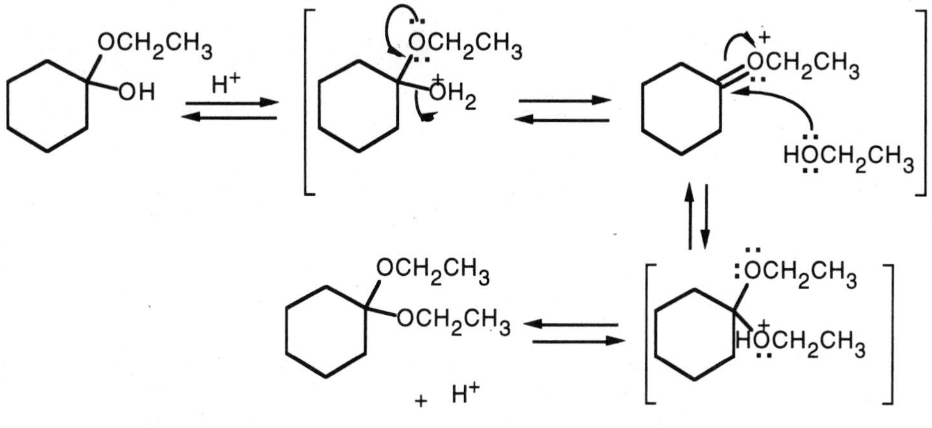

18.50

The intermediate resulting from addition of H:⁻ is like the intermediate in a Williamson ether synthesis. Intramolecular reaction occurs to form the epoxide.

Study Guide for Chapter 18

After studying this chapter, you should be able to:

(1)   Name and draw ethers, epoxides and sulfides (18.1, 18.15, 18.17, 18.19, 18.20).

(2)   Prepare ethers and epoxides (18.2, 18.3, 18.4, 18.5, 18.7, 18.10, 18.11, 18.24, 18.26, 18.29, 18.32, 18.40, 18.44, 18.49).

(3)   Predict the products of reactions involving ethers (18.22, 18.27, 18.28).

(4)   Formulate mechanisms for preparation and cleavage of ethers (18.6, 18.8, 18.9, 18.21, 18.25, 18.30, 18.31, 18.36, 18.37, 18.38, 18.48).

(5)   Formulate mechanisms and predict products of ring-opening reactions of epoxides (18.12, 18.13, 18.14, 18.33, 18.34, 18.35, 18.42, 18.45, 18.46).

(6)   Identify ethers, epoxides and sulfides by chemical and spectroscopic techniques (18.16, 18.23, 18.39, 18.41, 18.43, 18.47).

Chapter 19 – Aldehydes and Ketones. Nucleophilic Addition Reactions

19.1

a)

$$CH_3CH_2\overset{\overset{\displaystyle O}{\|}}{C}CH(CH_3)_2$$

2–Methyl–3–pentanone

b)

3–Phenylpropanal

c)

$$CH_3\overset{\overset{\displaystyle O}{\|}}{C}CH_2CH_2CH_2\overset{\overset{\displaystyle O}{\|}}{C}CH_2CH_3$$

2,6–Octanedione

d)

trans–2–Methylcyclohexane-
carbaldehyde

e)

$$H\overset{\overset{\displaystyle O}{\|}}{C}CH_2CH_2CH_2\overset{\overset{\displaystyle O}{\|}}{C}H$$

Pentanedial

f)

cis–2,5–Dimethylcyclohexanone

g)

$$CH_3CH_2\overset{\overset{\displaystyle CH_3}{|}}{C}H\overset{\overset{\displaystyle O}{\|}}{C}H\overset{\displaystyle}{C}CH_3$$
$$\underset{CH_2CH_2CH_3}{|}$$

4–Methyl–3–propyl–2–hexanone

h)

$$CH_3CH=CHCH_2CH_2\overset{\overset{\displaystyle O}{\|}}{C}H$$

4–Hexenal

19.2

a)

$$(CH_3)_2CHCH_2\overset{\overset{\displaystyle O}{\|}}{C}H$$

3–Methylbutanal

b)

$$CH_3\overset{\overset{\displaystyle Cl}{|}}{C}HCH_2\overset{\overset{\displaystyle O}{\|}}{C}CH_3$$

4–Chloro–2–pentanone

c)

Phenylacetaldehyde

d)

cis–3–tert–Butylcyclohexane-
carbaldehyde

e)

$$H_2C=\overset{\overset{\displaystyle CH_3}{|}}{C}CH_2\overset{\overset{\displaystyle O}{\|}}{C}H$$

3–Methyl–3–butenal

f)

$$CH_3CH_2\overset{\overset{\displaystyle CH_3}{|}}{C}HCH_2CH_2\overset{\displaystyle}{C}H\overset{\overset{\displaystyle O}{\|}}{C}H$$
$$\underset{CHClCH_3}{|}$$

2–(1–Chloroethyl)–5–methylheptanal

19.3

a) $CH_3CH_2CH_2CH_2CH_2OH$ $\xrightarrow[\text{CH}_3\text{Cl}_2]{\text{PCC}}$ $CH_3CH_2CH_2CH_2CHO$

         1–Pentanol   Pentanal

b) $CH_3CH_2CH_2CH_2CH=CH_2$ $\xrightarrow[\text{2. Zn, H}_3O^+]{\text{1. O}_3}$ $CH_3CH_2CH_2CH_2CHO$

1–Hexene

c) $CH_3CH_2CH_2CH_2COOCH_3$ $\xrightarrow[\text{2. H}_3O^+]{\text{1. DIBAH}}$ $CH_3CH_2CH_2CH_2CHO$

19.4

a) 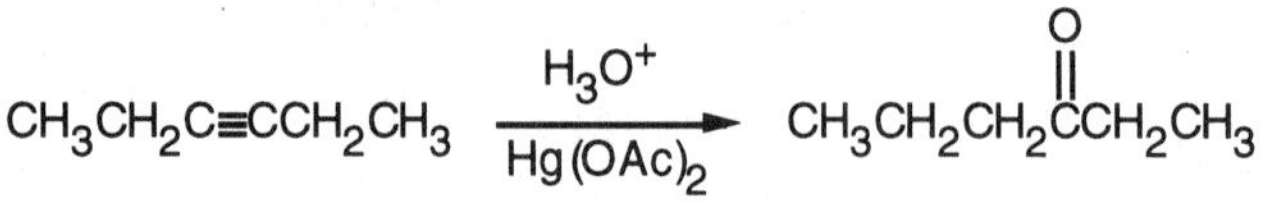 $CH_3CH_2C≡CCH_2CH_3$ $\xrightarrow[\text{Hg(OAc)}_2]{\text{H}_3O^+}$ $CH_3CH_2CH_2\overset{\overset{\displaystyle O}{\|}}{C}CH_2CH_3$

b)

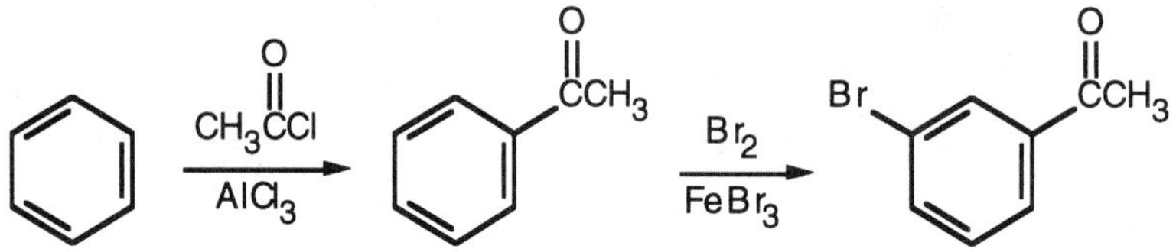

c)

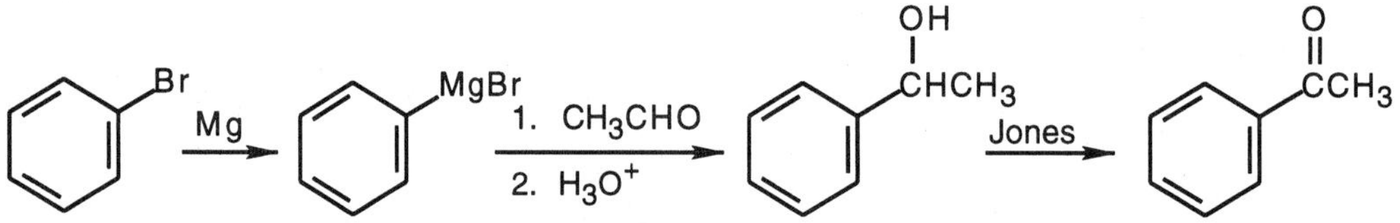

d)

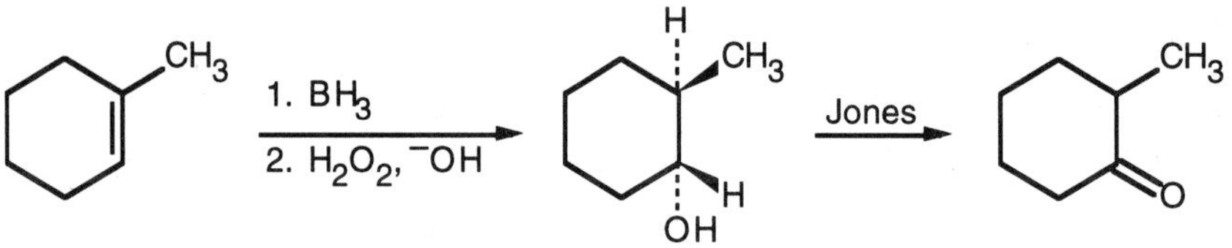

19.5

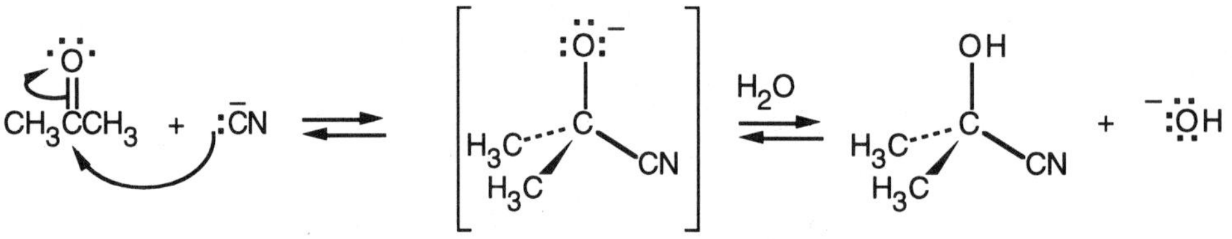

19.6   An aromatic aldehyde is less reactive than an aliphatic aldehyde toward nucleophilic addition because the partial positive charge of the carbonyl carbon can be delocalized into the aromatic ring. The electron-donating aromatic ring thus makes benzaldehyde less reactive toward nucleophiles.

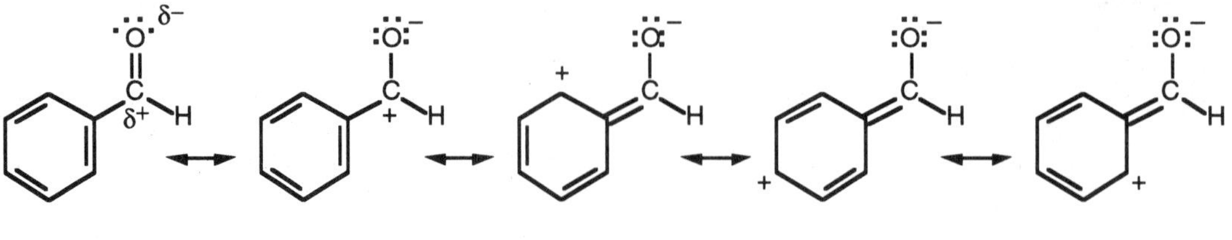

19.7

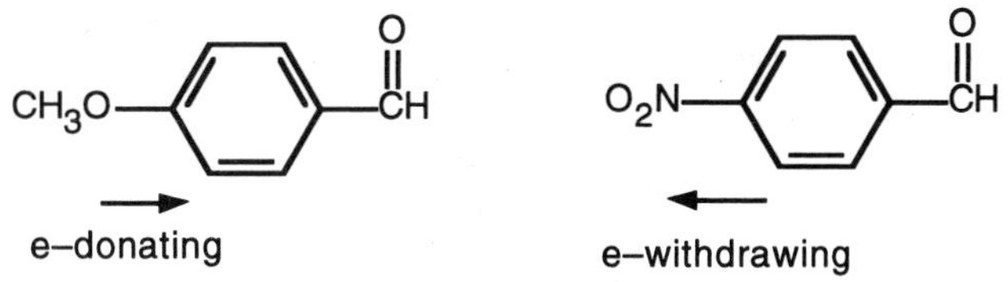

e–donating   e–withdrawing

The electron-withdrawing nitro group makes the aldehyde carbon of *p*-nitrobenzaldehyde more electron-poor (more electrophilic) and more reactive toward nucleophiles than the aldehyde carbon of *p*–methoxybenzaldehyde.

19.8

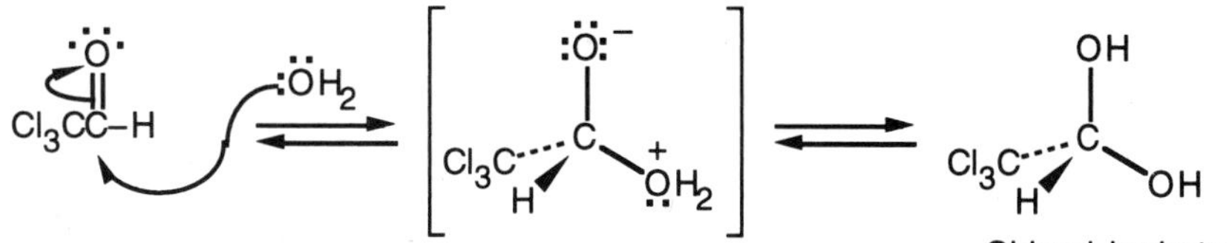

Chloral hydrate

19.9

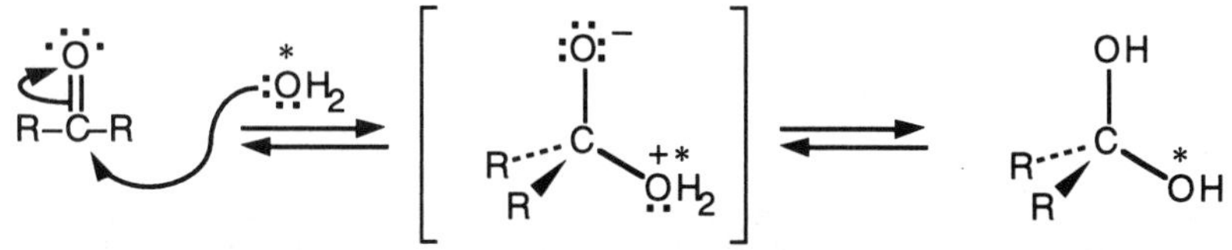

The above mechanism is similar to other nucleophilic addition mechanisms we have studied. Since all steps are reversible, we can write the above mechanism in reverse to show how labeled oxygen is incorporated into an aldehyde or ketone.

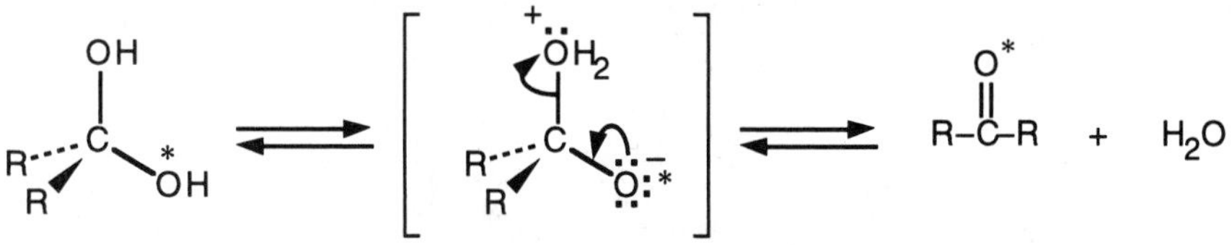

This exchange is very slow in water but proceeds more rapidly when either acid or base is present.

19.10

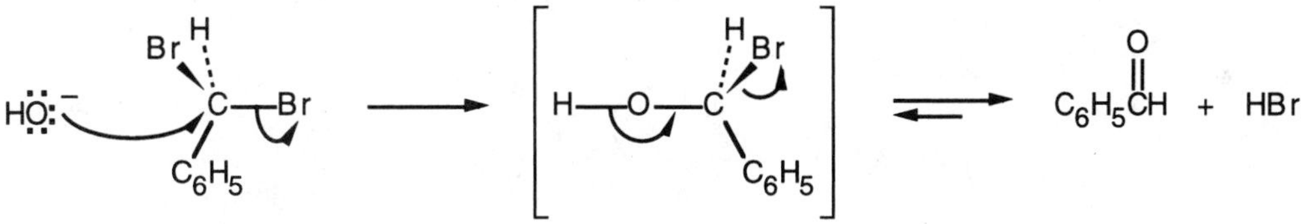

S_N2 addition of hydroxide to $C_6H_5CHBr_2$ yields an unstable bromoalcohol intermediate, which loses HBr to yield benzaldehyde.

19.11

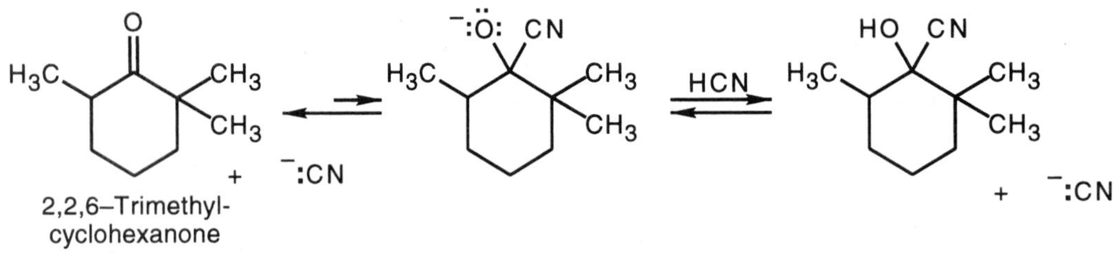

2,2,6–Trimethyl-
cyclohexanone

Cyanohydrin formation is an equilibrium process. Because the product of addition of HCN to 2,2,6–trimethylcyclohexanone is sterically hindered by the three methyl groups, the equilibrium lies toward the side of the unreacted ketone.

19.12

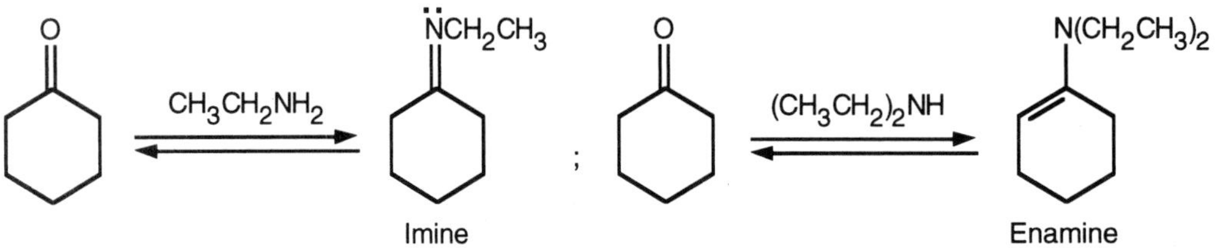

Imine   ;   Enamine

19.13

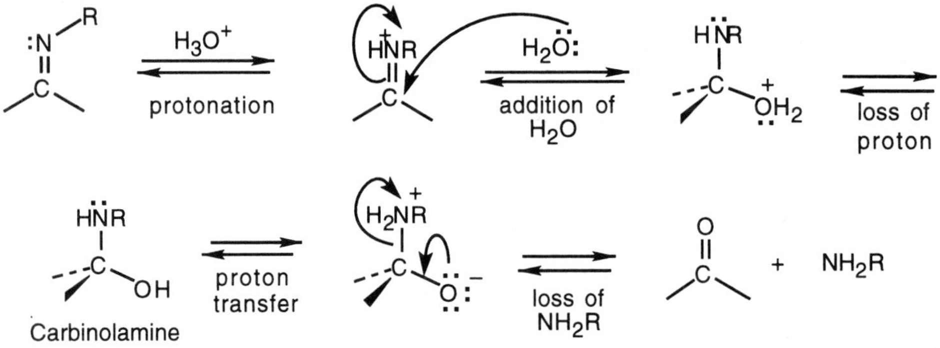

Carbinolamine

19.14

a)

Hemiacetal

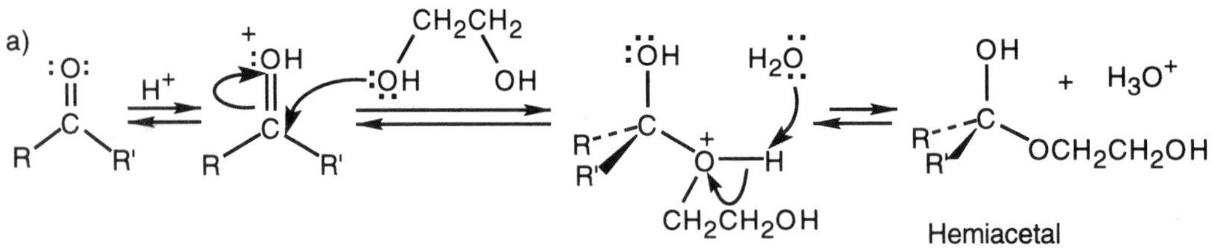

Formation of the hemiacetal is the first step.

b)

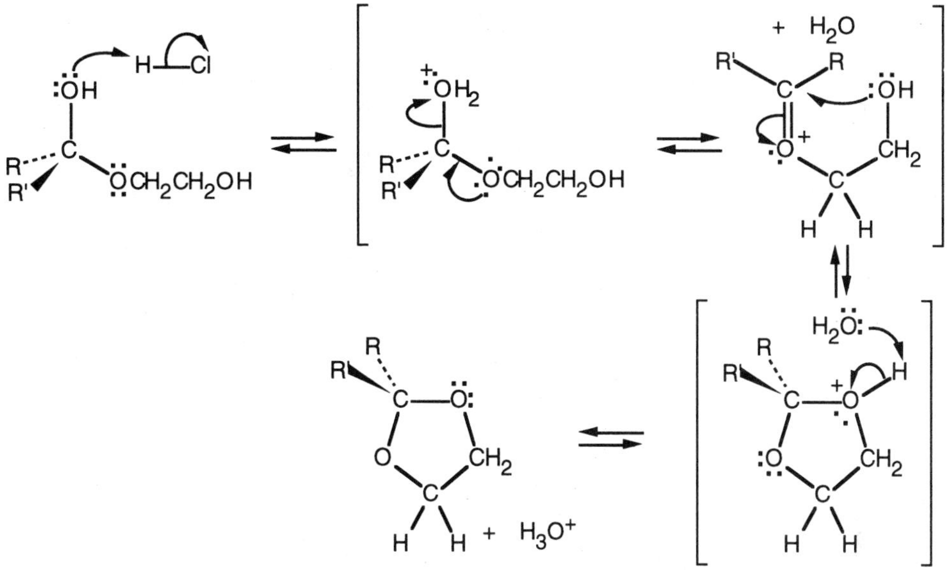

Protonation of the hemiacetal hydroxyl group is followed by loss of water. Attack by the second hydroxyl group of ethylene glycol forms the cyclic acetal ring.

19.15

a)

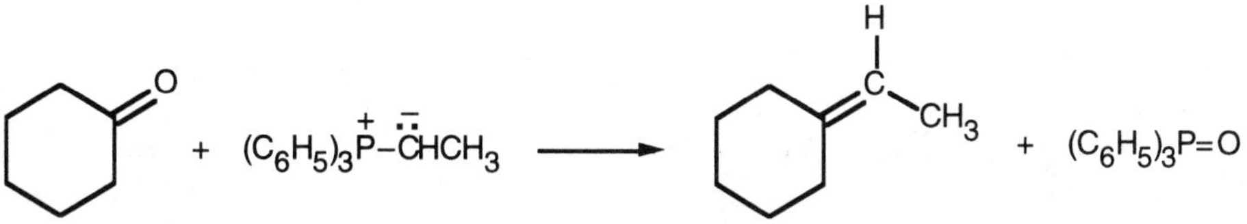

b)

c)

$$CH_3\overset{O}{\underset{\|}{C}}CH_3 \ + \ (C_6H_5)_3\overset{+}{P}-\overset{..}{\underset{..}{C}}HCH_2CH_2CH_3 \longrightarrow (CH_3)_2C=CHCH_2CH_2CH_3 \ + \ (C_6H_5)_3P=O$$

d)

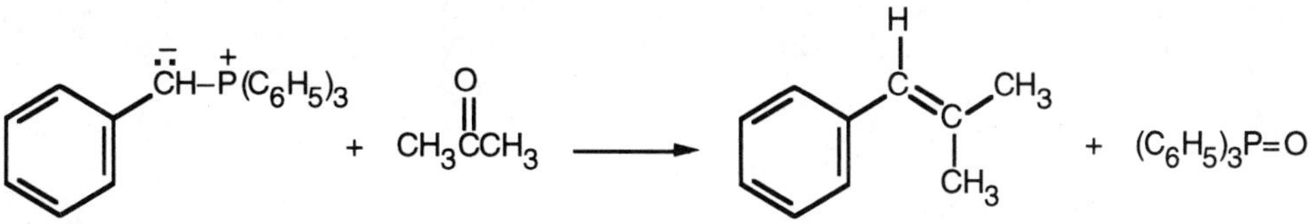

e)

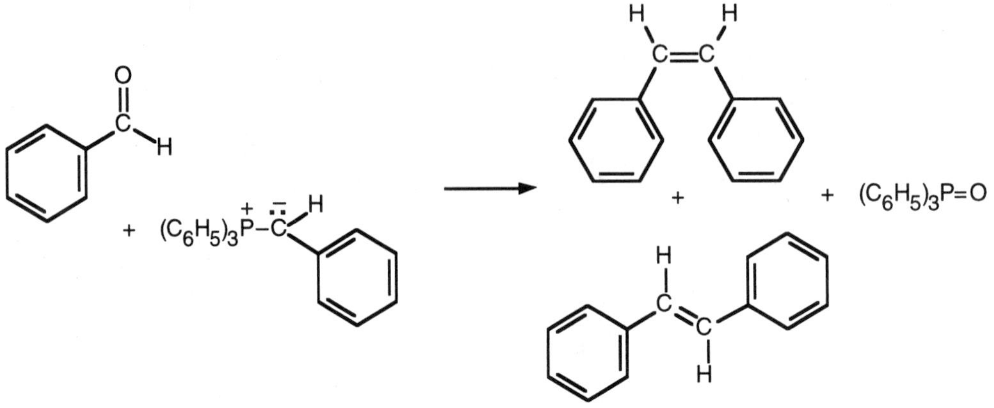

19.16

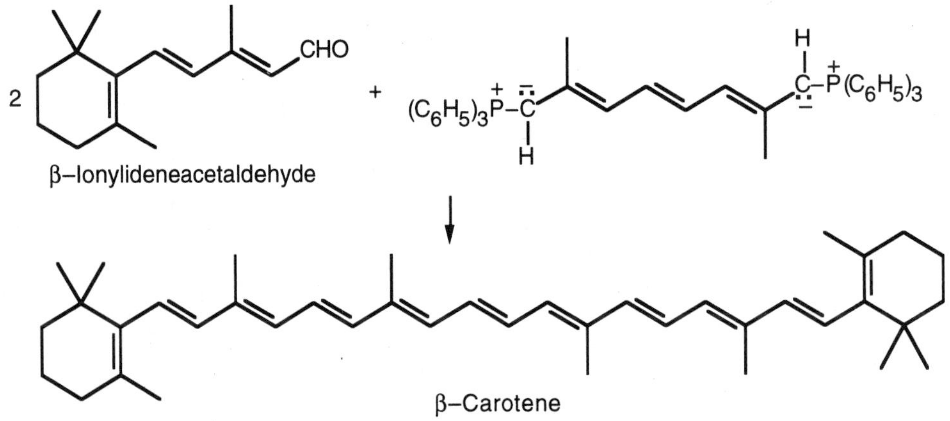

β−Ionylideneacetaldehyde

β−Carotene

19.17

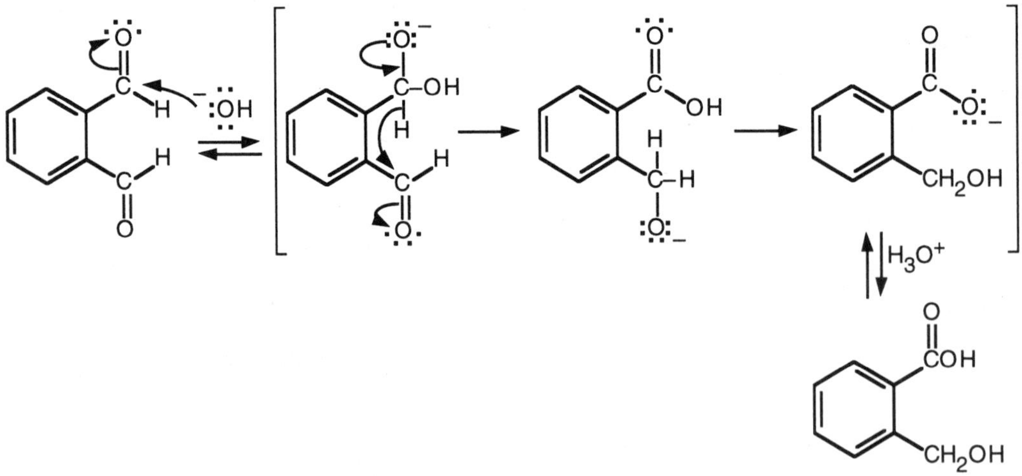

This is an internal Cannizzaro reaction.

19.18

a)

$H_2C=CHCCH_3$ $\xrightarrow{\begin{array}{c}1.\ Li(CH_3CH_2CH_2)_2Cu\\2.\ H_3O^+\end{array}}$ $CH_3CH_2CH_2CH_2CH_2CCH_3$

2–Heptanone

b)

$\xrightarrow{\begin{array}{c}1.\ Li(CH_3)_2Cu\\2.\ H_3O^+\end{array}}$

3,3–Dimethylcyclohexanone

c)

$\xrightarrow{\begin{array}{c}1.\ Li(CH_3CH_2)_2Cu\\2.\ H_3O^+\end{array}}$

4–*tert*–Butyl–3–ethylcyclohexanone

d)

$\xrightarrow{\begin{array}{c}1.\ Li(CH_2=CH)_2Cu\\2.\ H_3O^+\end{array}}$

19.19

2–Cyclohexenone $\xrightarrow{\begin{array}{c}1.\ Li(CH_3)_2Cu\\2.\ H_3O^+\end{array}}$

A
1–Methyl–2–
cyclohexen–1–ol

or

B
3–Methylcyclo-
hexanone

2–Cyclohexenone is a cyclic α,ß–unsaturated ketone whose carbonyl IR absorption occurs at 1685 cm^{-1}. If direct addition product $\underline{A}$ is formed, the carbonyl absorption will vanish and a hydroxyl absorption will appear at 3300 cm^{-1}. If conjugate addition produces $\underline{B}$, the carbonyl absorption will shift to 1715 cm^{-1}, where 6-membered-ring saturated ketones absorb.

19.20 a) $H_2C=CHCH_2COCH_3$ absorbs at 1715 cm^{-1}. (4–Penten–2–one is not an α,ß–unsaturated ketone.)

b) $CH_3CH=CHCOCH_3$ is an α,ß–unsaturated ketone and absorbs at 1685 cm^{-1}.

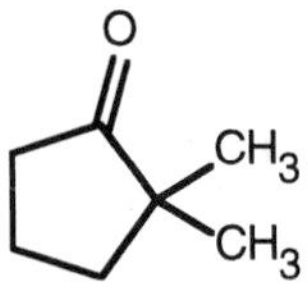

c

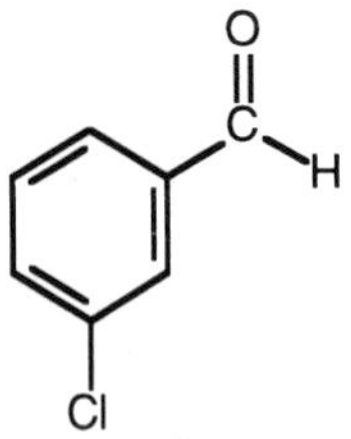

d

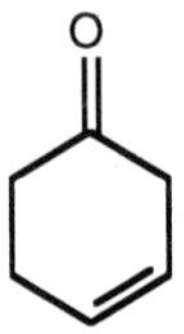

e

c) 2,2–Dimethylcyclopentanone, a five-membered-ring ketone, absorbs at 1750 cm^{-1}.

d) *m*–Chlorobenzaldehyde shows a singlet absorption at 1705 cm^{-1} and a doublet at 2720 cm^{-1} and 2820 cm^{-1}.

e) 3–Cyclohexenone absorbs at 1715 cm^{-1}.

f) $CH_3CH_2CH_2CH{=}CHCHO$ is an α,ß–unsaturated aldehyde and absorbs at 1705 cm^{-1}.

19.21

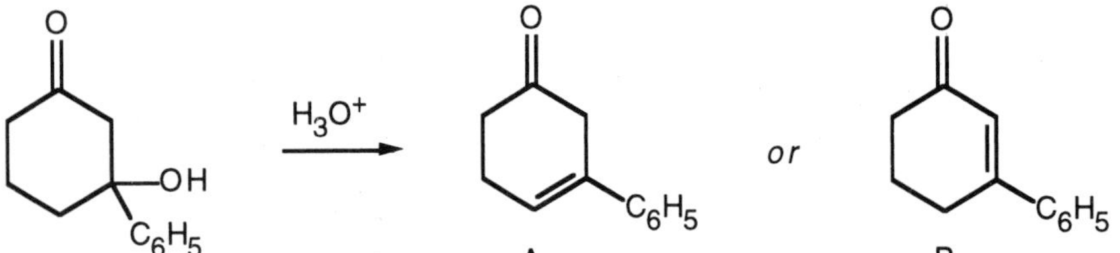

A B

Compound A is a cyclic, nonconjugated keto alkene whose carbonyl infrared absorption should occur at 1715 cm^{-1}. Compound B is an α,ß–unsaturated, cyclic ketone; additional conjugation with the phenyl ring should lower its IR absorption below 1685 cm^{-1}. Because the actual IR absorption occurs at 1670 cm^{-1}, B is the correct structure.

19.22

a)

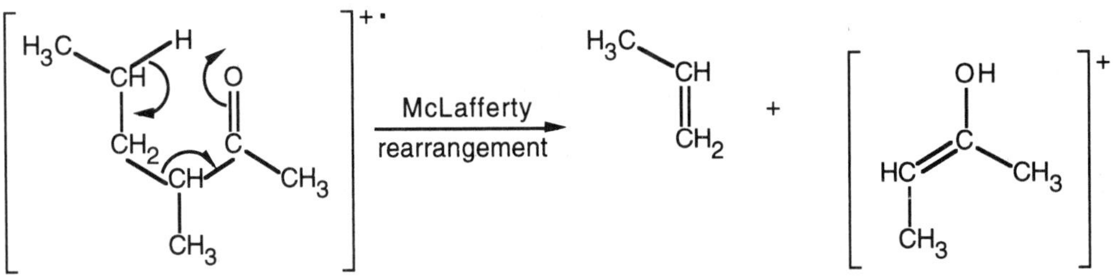

m/z = 72

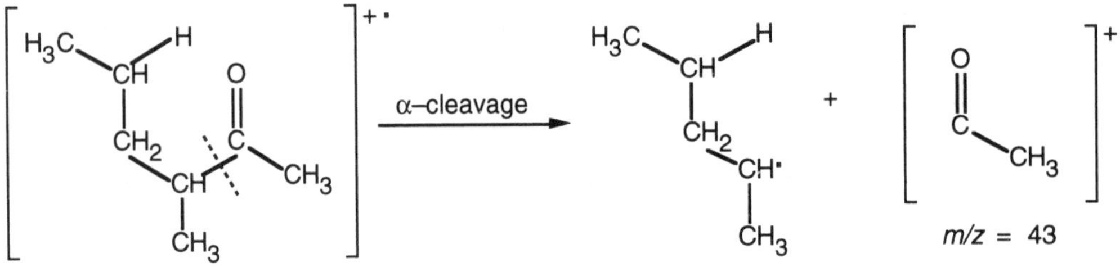

m/z = 114 m/z = 43

3–Methyl–2–hexanone

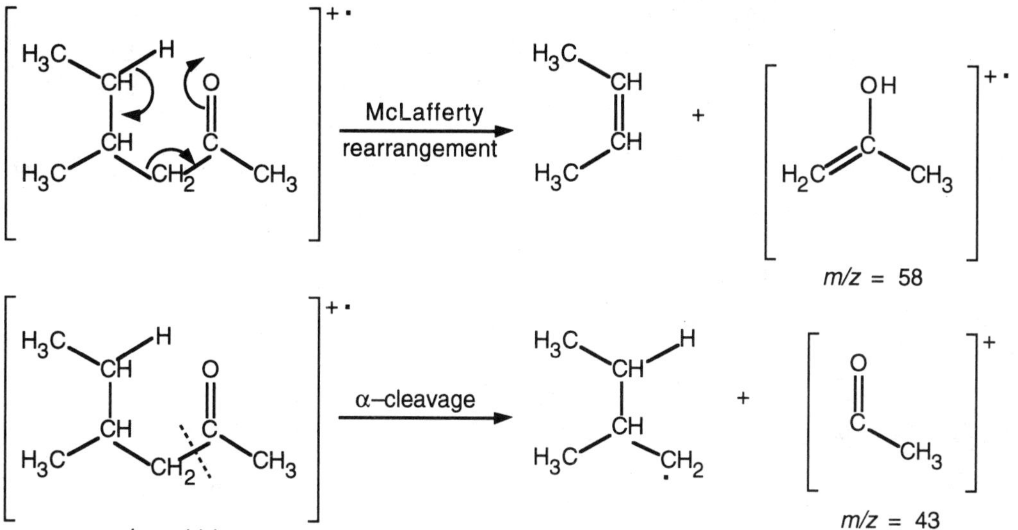

$m/z = 58$

$m/z = 114$
4–Methyl–2–hexanone

$m/z = 43$

Both isomers exhibit peaks at $m/z = 43$ due to α–cleavage. The products of McLafferty rearrangement, however, occur at different values of m/z.

b)

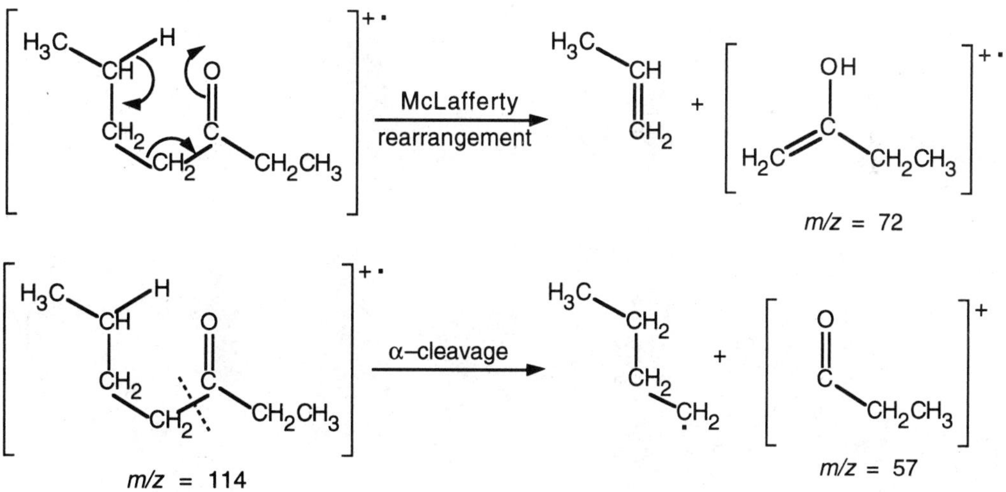

$m/z = 72$

$m/z = 114$
3–Heptanone

$m/z = 57$

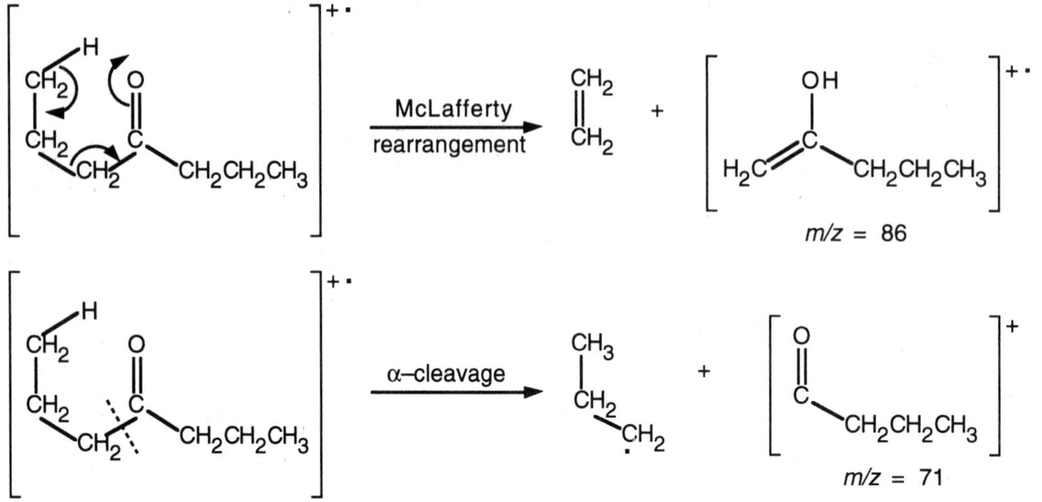

McLafferty rearrangement

CH_2
CH_2

$+$

OH

H_2C $CH_2CH_2CH_3$

$m/z = 86$

α–cleavage

CH_3
CH_2
CH_2

$+$

O

C $CH_2CH_2CH_3$

$m/z = 71$

$m/z = 114$
4–Heptanone

The isomers can be distinguished on the basis of both α–cleavage products ($m/z = 57$ vs $m/z = 71$) and McLafferty rearrangement products ($m/z = 72$ vs $m/z = 86$).

c)

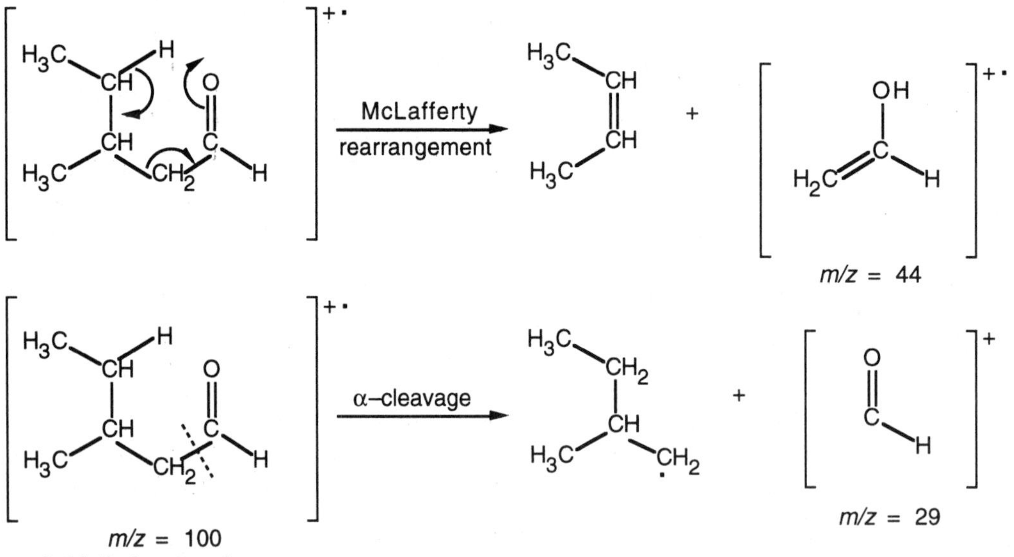

McLafferty rearrangement

H_3C CH
CH
H_3C

$+$

OH

H_2C H

$m/z = 44$

α–cleavage

H_3C CH_2
CH
H_3C CH_2

$+$

O

C   H

$m/z = 29$

$m/z = 100$
3–Methylpentanal

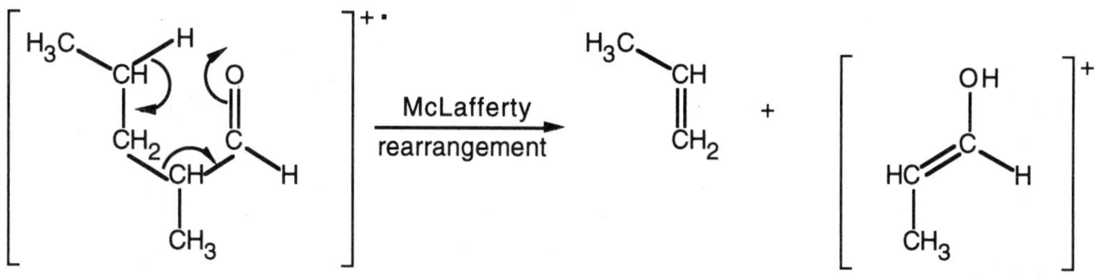

The fragments from McLafferty rearrangement, which occur at different values of m/z, serve to distinguish the two isomers.

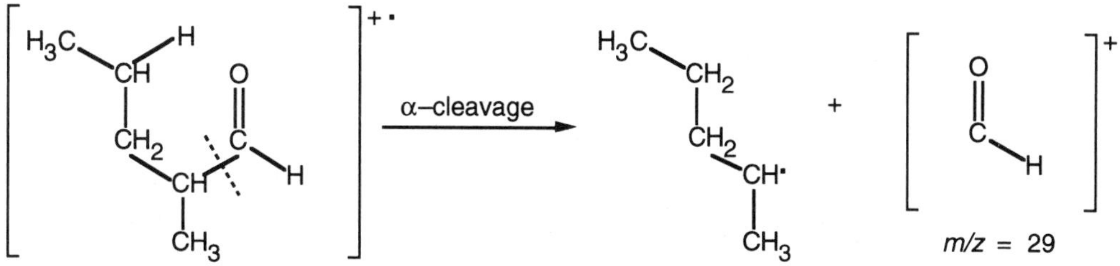

19.23

a) 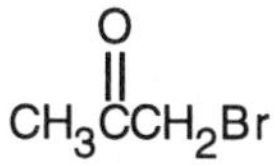 $CH_3\overset{O}{\overset{\|}{C}}CH_2Br$

b)

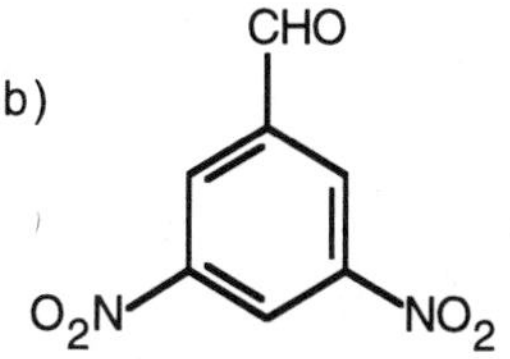

c) 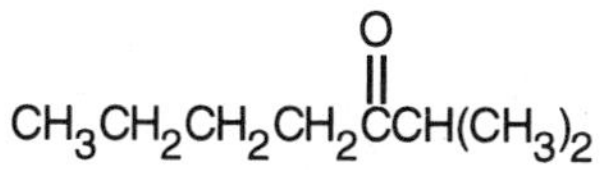 $CH_3CH_2CH_2CH_2\overset{O}{\overset{\|}{C}}CH(CH_3)_2$

d)

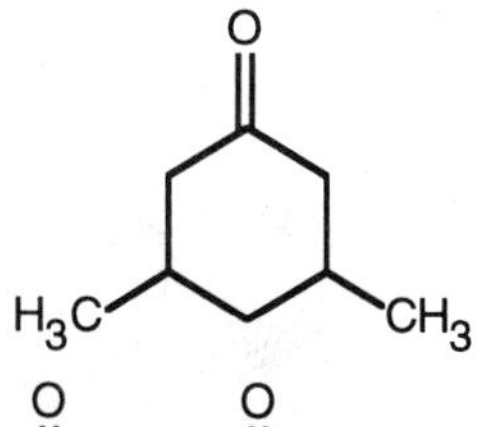

e) 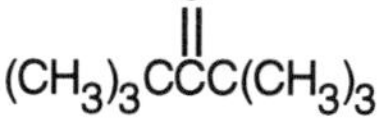 $(CH_3)_3C\overset{O}{\overset{\|}{C}}C(CH_3)_3$

f) 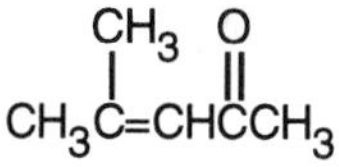 $CH_3\overset{CH_3}{\overset{|}{C}}=CH\overset{O}{\overset{\|}{C}}CH_3$

g) 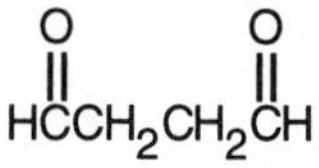 $H\overset{O}{\overset{\|}{C}}CH_2CH_2\overset{O}{\overset{\|}{C}}H$

h)

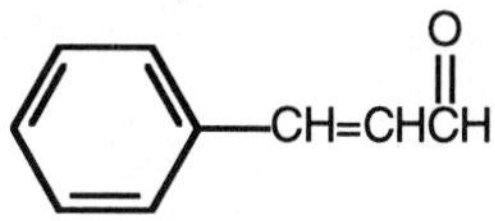

i)

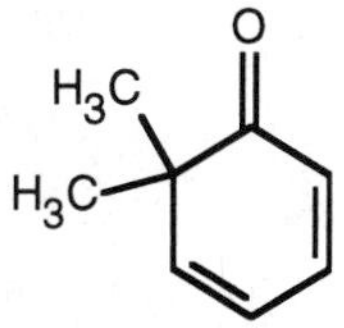

j)

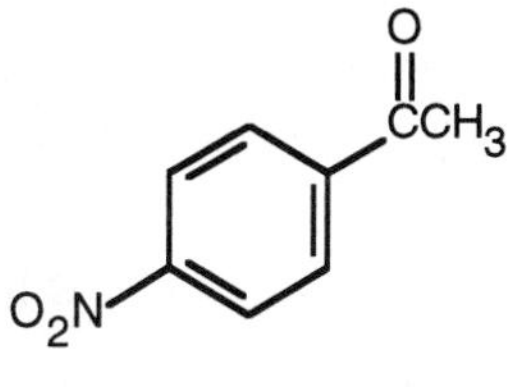

k)

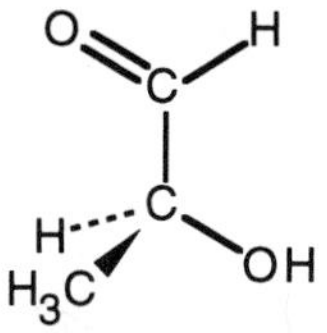

l)

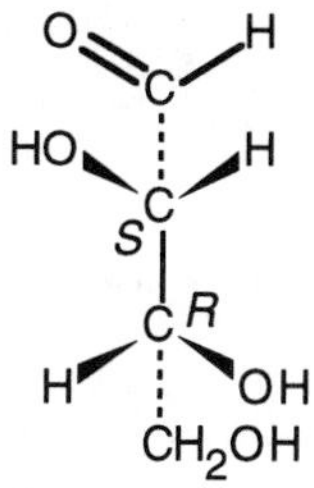

19.24

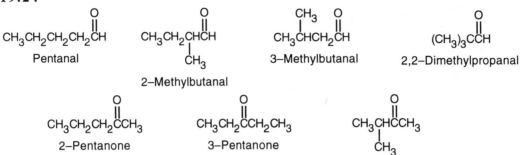

CH₃CH₂CH₂CH₂CH (Pentanal)

CH₃CH₂CHCH with CH₃ below (2–Methylbutanal)

CH₃CHCH₂CH with CH₃ (3–Methylbutanal)

(CH₃)₃CCH (2,2–Dimethylpropanal)

CH₃CH₂CH₂CCH₃ (2–Pentanone)

CH₃CH₂CCH₂CH₃ (3–Pentanone)

CH₃CHCCH₃ with CH₃ (3–Methyl–2–butanone)

19.25 a) 3–Methyl–3–cyclohexenone

b) (2*R*)–2,3–Dihydroxypropanal (D–Glyceraldehyde)

c) 5–Isopropyl–2–methyl–2–cyclohexenone

d) 2–Methyl–3–pentanone

e) 3–Hydroxybutanal

f) *p*–Benzenedicarbaldehyde

19.26 a) The α,ß–unsaturated ketone C₆H₈O contains one ring. Possible structures:

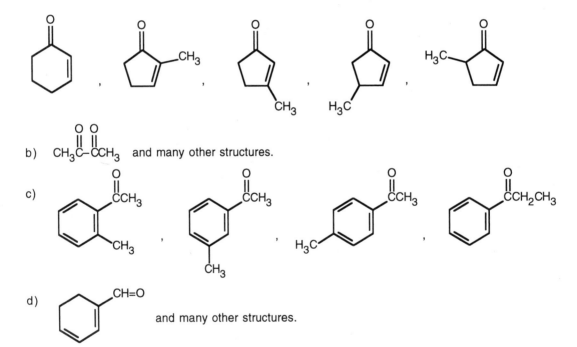

b)   CH₃C–CCH₃ (with two C=O)   and many other structures.

c)   [structures with aromatic rings bearing CCH₃ and CH₃ groups, and CCH₂CH₃]

d)   [cyclohexadiene with CH=O]   and many other structures.

19.27

a) [benzene ring with CH₂CH₂OH]

b) [benzene ring with CH₂COH, C=O]

c) [benzene ring with CH₂CH, N–OH]

d) [benzene ring with CH₂CHCH₃, OH]

e) 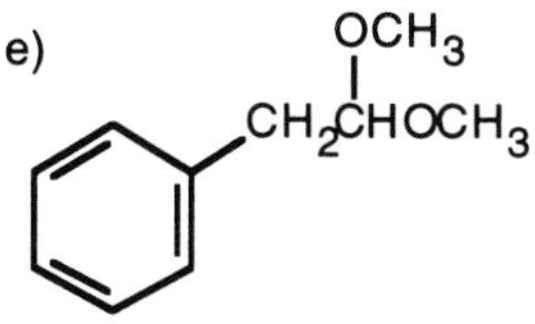 OCH_3 / CH_2CHOCH_3 f) 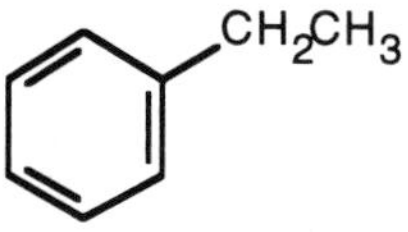 CH_2CH_3 g) 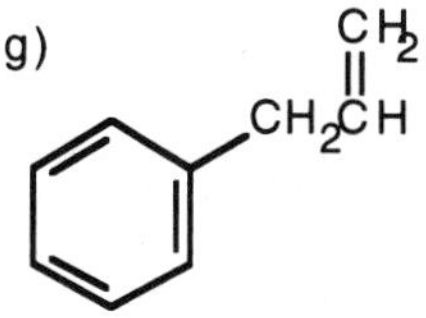 CH_2 / CH_2CH h) 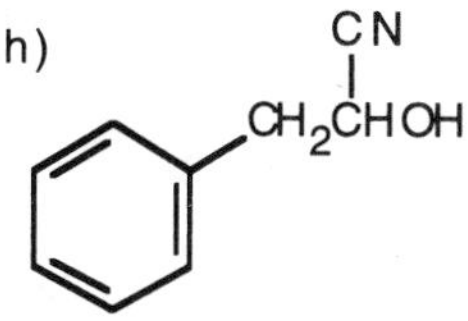 CN / CH_2CHOH

19.28

a) 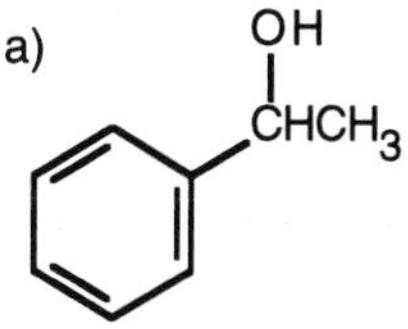 OH / $CHCH_3$ b) no reaction   c) 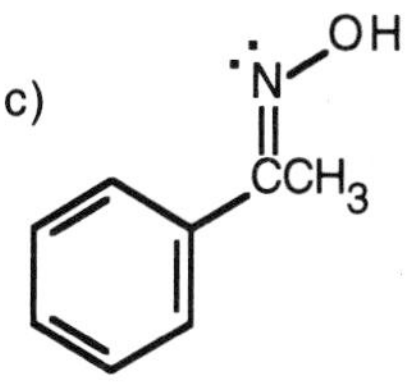 N—OH / CCH_3 d) 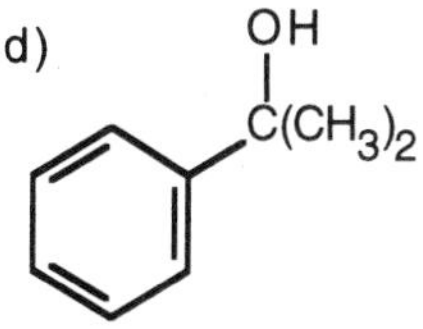 OH / $C(CH_3)_2$

e) 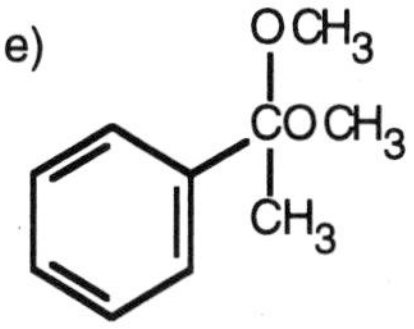 OCH_3 / $COCH_3$ / CH_3 f) 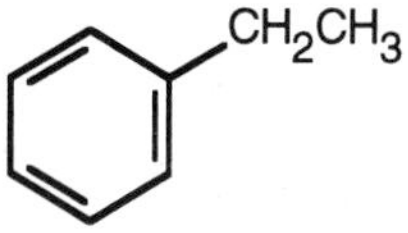 CH_2CH_3 g) 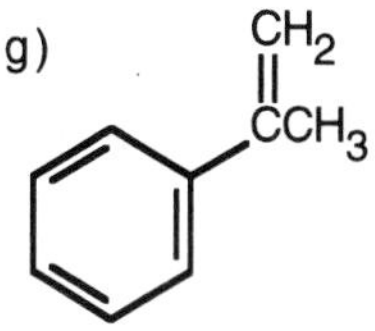 CH_2 / CCH_3 h) 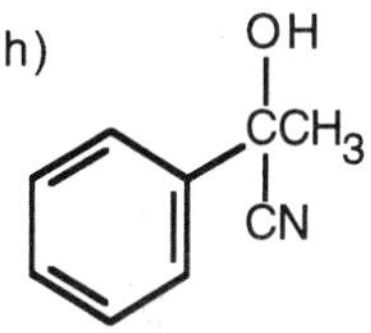 OH / CCH_3 / CN

19.29

a)

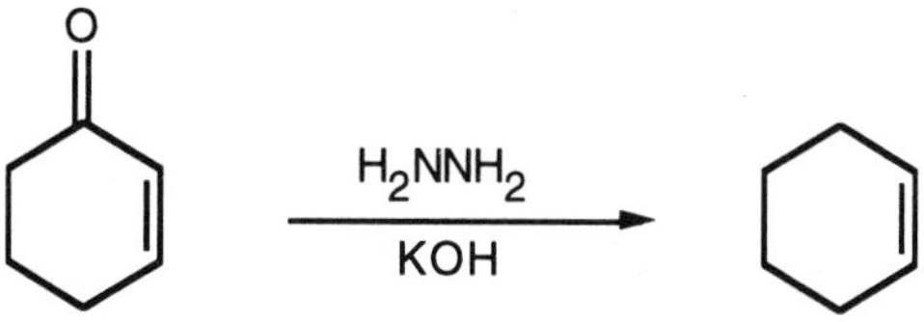

$$\underset{\text{KOH}}{\xrightarrow{H_2NNH_2}}$$

b)

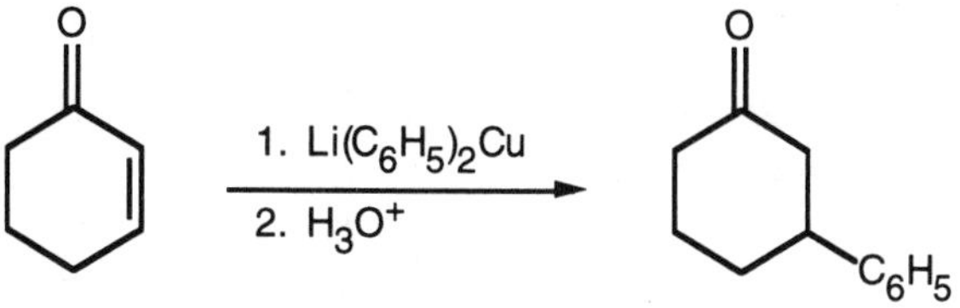

$$\xrightarrow[\text{2. } H_3O^+]{\text{1. } Li(C_6H_5)_2Cu} \quad C_6H_5$$

c)

$$\xrightarrow[\text{2. } H_3O^+]{\text{1. } Li(CH=CH_2)_2Cu} \quad CH=CH_2 \xrightarrow[\text{2. } H_3O^+]{\text{1. } KMnO_4} \quad COOH$$

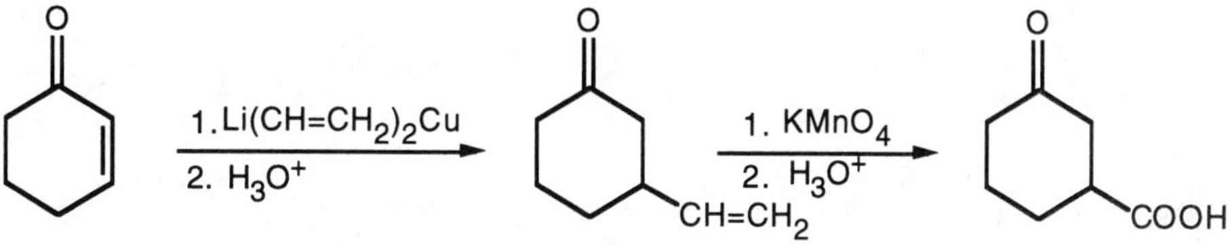

d)

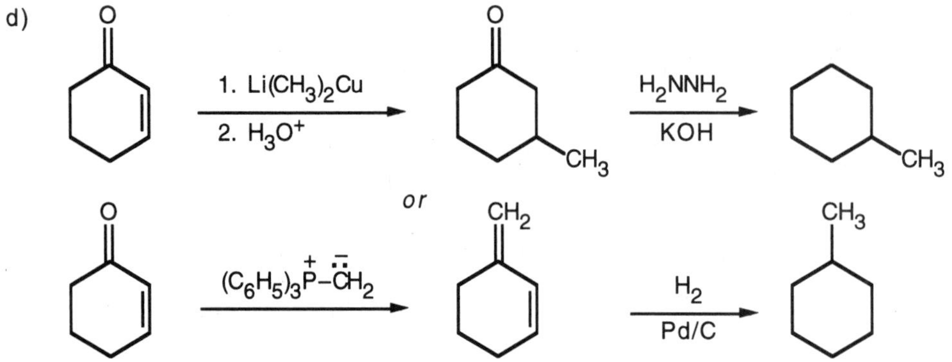

or

19.30 Glucose exists mainly as the cyclic hemiacetal, formed by the addition of the hydroxyl group of carbon atom 5 to the aldehyde group at carbon 1. A small amount of the open-chain aldehyde is in equilibrium with the cyclic hemiacetal. The open-chain aldehyde gives a positive Tollens' test, like any other aldehyde.

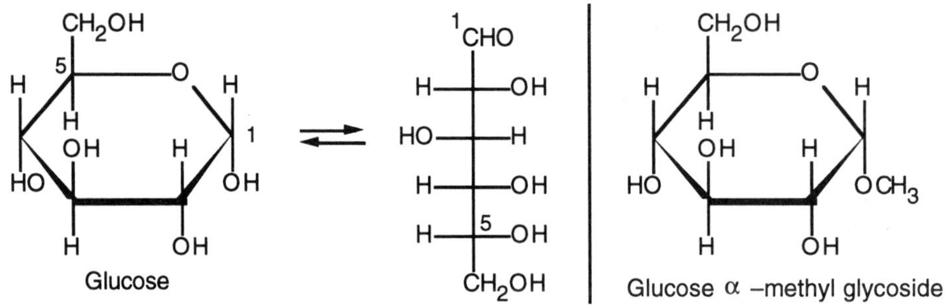

Glucose α–methyl glycoside, an acetal, is not in equilibrium with an open-chain aldehyde and doesn't react with Tollens reagent.

19.31

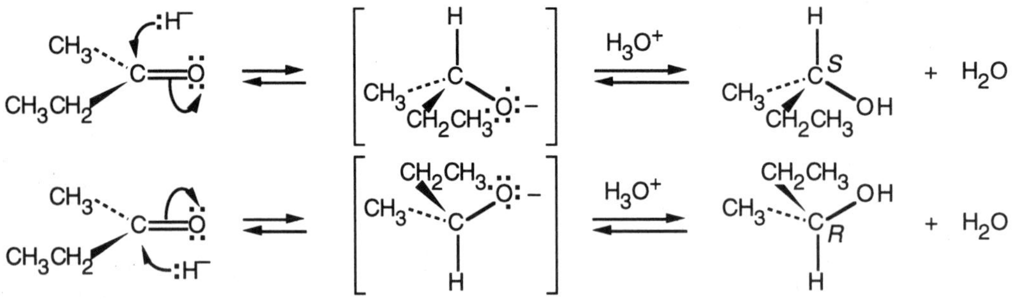

Attack can occur with equal probability from either side of the planar carbonyl group to yield a racemic product mixture.

19.32 Remember:

$$RCH_2-X \quad + \quad (C_6H_5)_3P: \quad \longrightarrow \quad (C_6H_5)_3\overset{+}{P}CH_2R \ X^-$$

Alkyl halide   Triphenyl phosphine   Phosphonium salt

$$(C_6H_5)_3\overset{+}{P}CH_2R \ X^- \quad + \quad CH_3CH_2CH_2\overset{..}{C}H_2Li^+ \quad \longrightarrow \quad (C_6H_5)_3\overset{+}{P}-\overset{..}{C}HR$$

Phosphonium salt   Butyllithium   Ylide

$$(C_6H_5)_3\overset{+}{P}-\overset{..}{C}HR \quad + \quad O=C \diagup \quad \longrightarrow$$

Ylide   Carbonyl   Alkene

	Alkyl halide	Carbonyl	Product
a	CH₂Br (benzyl)	O=HCCH=CH-(phenyl)	(phenyl)-CH=CHCH=CH-(phenyl)
b	CH₂Br (benzyl)	cyclohexanone	benzylidene cyclohexane
c	CH₃Br	cyclohexenone	methylene cyclohexene
d	CH₃Br	cyclohexene carbaldehyde	cyclohexenyl-CH=CH₂

19.33 Suppose that tri*methyl*phosphine were to react with alkyl halide.

$$(CH_3)_3P: \ + \ RCH_2CH_2X \quad \longrightarrow \quad (CH_3)_3\overset{+}{P}CH_2CH_2R \ X^-$$

phosphonium salt

Treatment of the phosphonium salt with strong base would yield two different ylides.

$$(CH_3)_3\overset{+}{P}CH_2CH_2R \ X^- \quad \xrightarrow[THF]{BuLi} \quad (CH_3)_3\overset{+}{P}-\overset{..}{C}HCH_2R \quad and \quad (CH_3)_2\overset{+}{P}CH_2CH_2R$$
$$\overset{-}{:}CH_2$$

Reaction of the ylides with a carbonyl compound would produce two different alkenes, a problem that can't happen when triphenylphosphine is used.

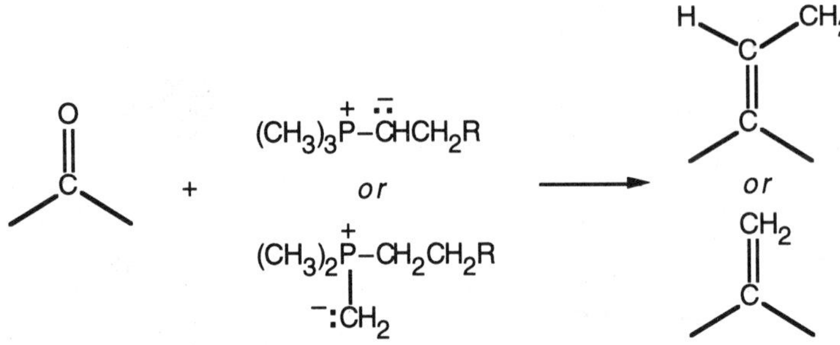

19.34

	Carbonyl Compound	Grignard Reagent	Product
a	CH_3CH (with O double bond)	$CH_3CH_2CH_2MgBr$	$CH_3CH_2CH_2CHCH_3$ with OH
	$CH_3CH_2CH_2CH$ (with O double bond)	CH_3MgBr	
b	CH_2O	$CH_3CH_2CH_2MgBr$	$CH_3CH_2CH_2CH_2OH$
c	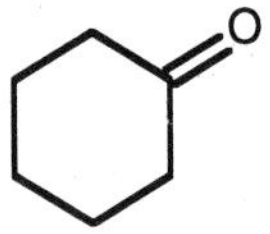 (cyclohexanone)	C_6H_5MgBr	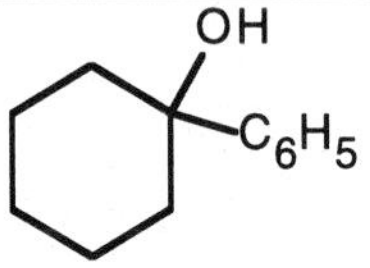 (1-phenylcyclohexanol) OH, $-C_6H_5$
d	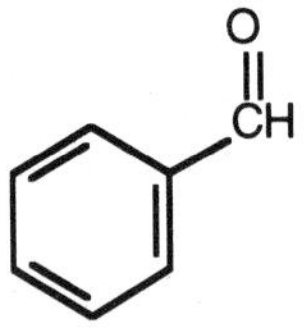 (benzaldehyde) CH with O	C_6H_5MgBr	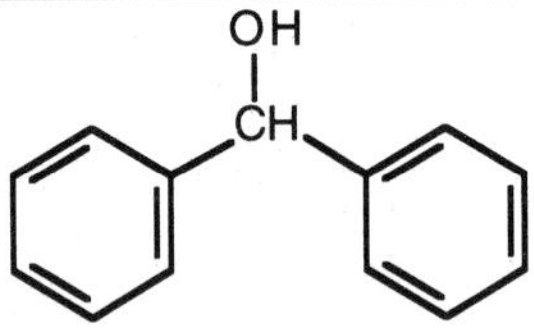 (diphenylmethanol) OH, CH

19.35

a) $(C_6H_5)_3P: + ClCH_2OCH_3 \longrightarrow (C_6H_5)_3\overset{+}{P}CH_2OCH_3 \ Cl^-$

$$\downarrow BuLi$$

$(C_6H_5)_3\overset{+}{P}-\overset{..}{\overset{-}{C}}HOCH_3 + LiCl$

b)

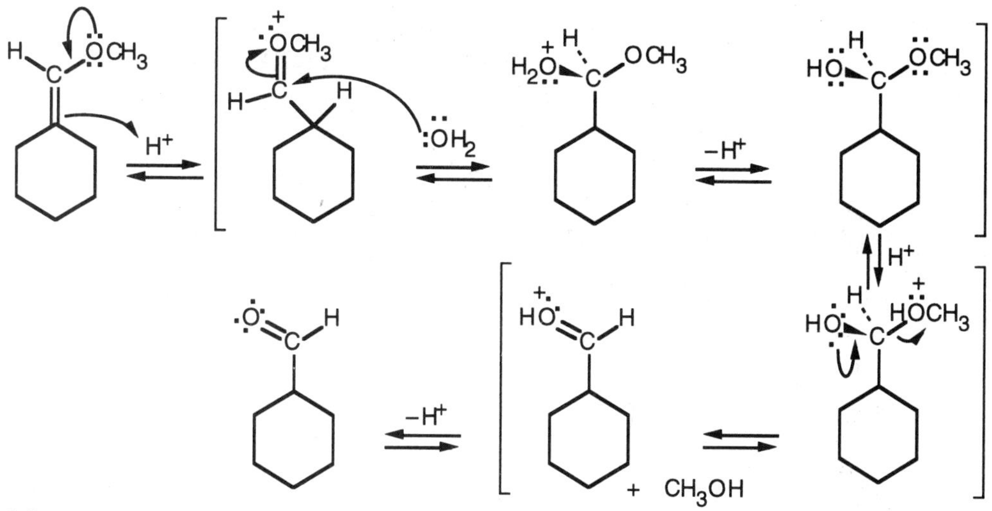

19.36

a)

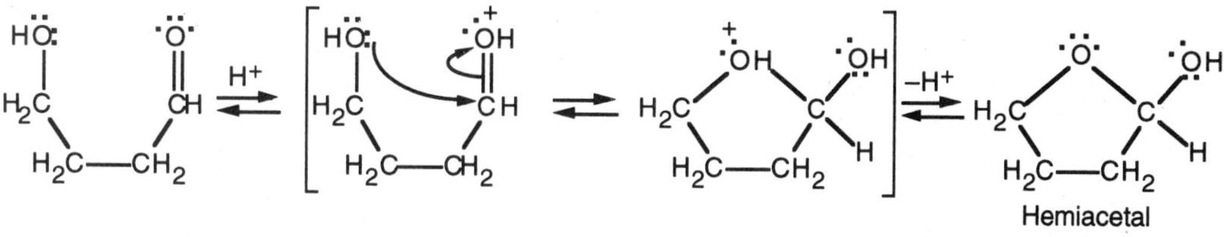

Hemiacetal

b)

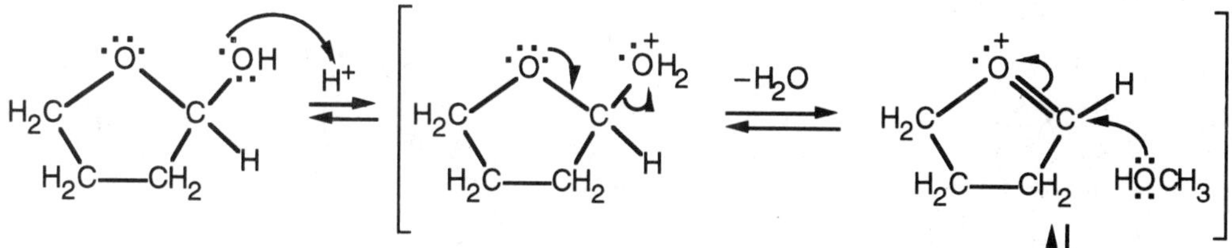

c)

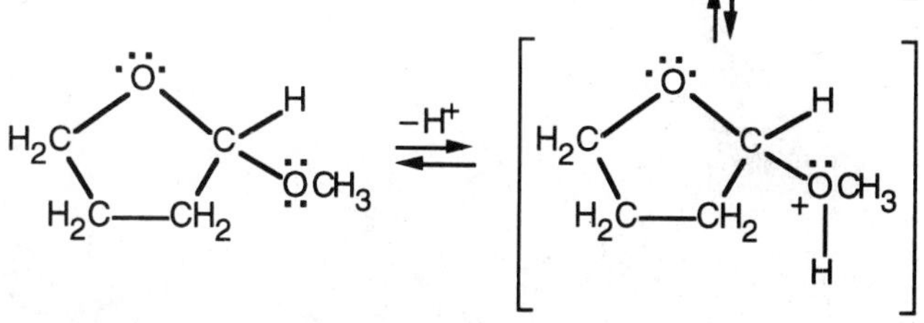

2–Methoxytetrahydrofuran is a cyclic acetal. The hydroxyl oxygen of 4–hydroxybutanal reacts with the aldehyde to form the cyclic ether linkage.

19.37 In general, ketones are less reactive than aldehydes for both steric (excess crowding) and electronic reasons. If the keto aldehyde in this problem were reduced with one equivalent of $NaBH_4$, the aldehyde functional group would be reduced in preference to the ketone.

For the same reason, reaction of the keto aldehyde with one equivalent of ethylene glycol selectively forms the acetal of the aldehyde functional group. The ketone can then be reduced with $NaBH_4$ and the acetal protecting group cleaved.

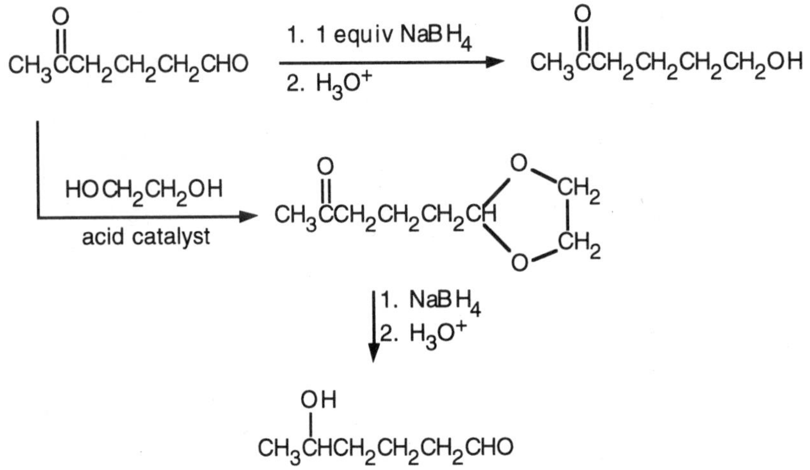

CH₃CCH₂CH₂CH₂CHO $\xrightarrow{\begin{array}{c}\text{1. 1 equiv NaBH}_4\\\text{2. H}_3\text{O}^+\end{array}}$ CH₃CCH₂CH₂CH₂CH₂OH

$\downarrow$ HOCH₂CH₂OH, acid catalyst

CH₃CCH₂CH₂CH₂CH(OCH₂CH₂O)

$\downarrow$ 1. NaBH₄ 2. H₃O⁺

OH
CH₃CHCH₂CH₂CH₂CHO

19.38

a)

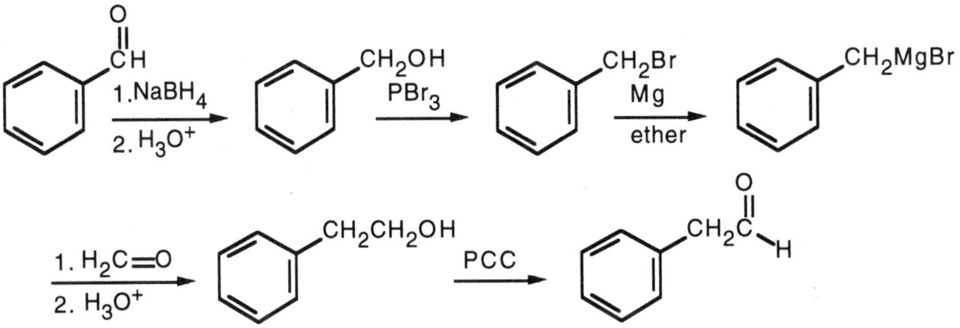

$\xrightarrow{\begin{array}{c}\text{1.NaBH}_4\\\text{2. H}_3\text{O}^+\end{array}}$ benzyl alcohol $\xrightarrow{\text{PBr}_3}$ benzyl bromide $\xrightarrow{\begin{array}{c}\text{Mg}\\\text{ether}\end{array}}$ benzyl magnesium bromide

$\xrightarrow{\begin{array}{c}\text{1. H}_2\text{C=O}\\\text{2. H}_3\text{O}^+\end{array}}$ phenethyl alcohol $\xrightarrow{\text{PCC}}$ phenylacetaldehyde

b)

benzaldehyde $\xrightarrow{\begin{array}{c}\text{1.CH}_3\text{MgBr}\\\text{2.H}_3\text{O}^+\end{array}}$ 1-phenylethanol $\xrightarrow{\begin{array}{c}\text{CrO}_3\\\text{H}_2\text{SO}_4\text{,H}_2\text{O}\end{array}}$ acetophenone

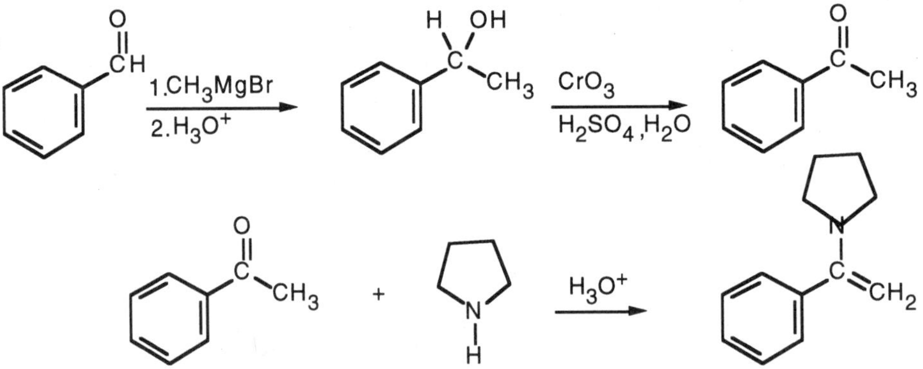

acetophenone + pyrrolidine $\xrightarrow{\text{H}_3\text{O}^+}$ enamine product

c)

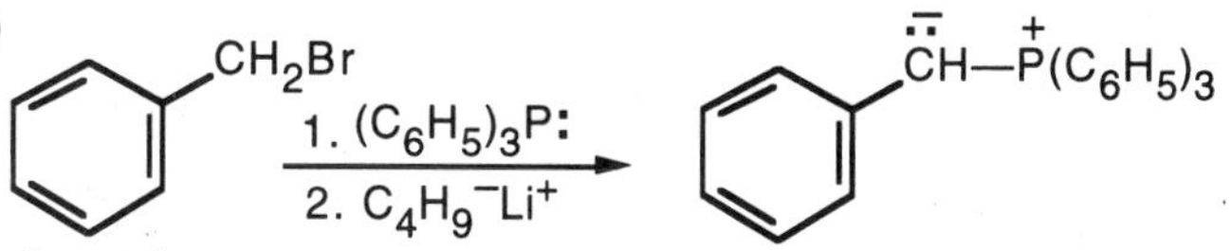

from a)

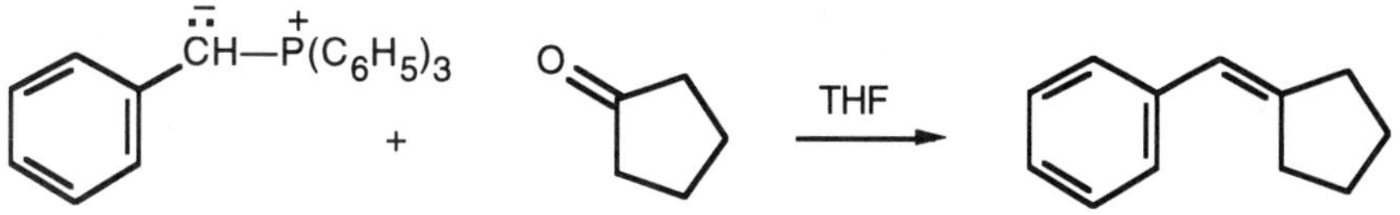

19.39

a)

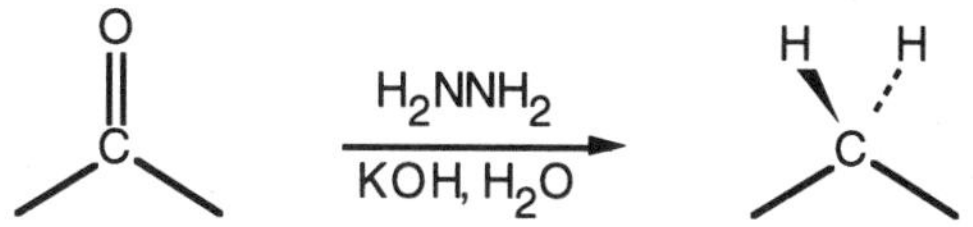

Advantage: reduction is one-step
Disadvantage: can't be used when base-sensitive functional groups are present

b)

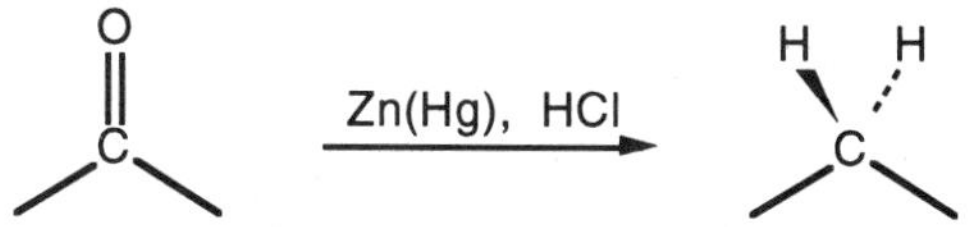

Advantage: reduction is one step
Disadvantage: can't be used when acid-sensitive functional groups are present

c)

d)

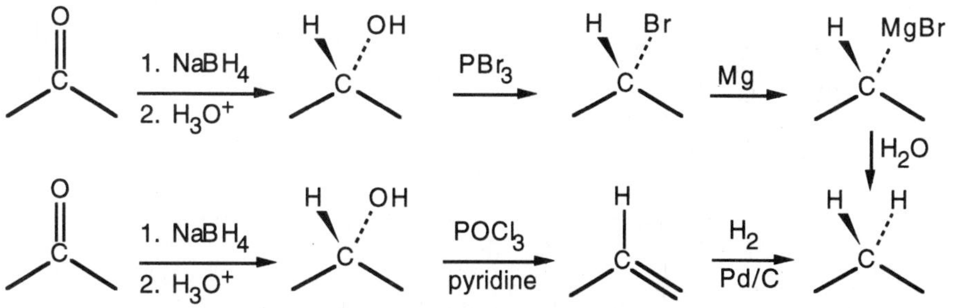

Disadvantage: these two methods require several steps.

19.40

a) 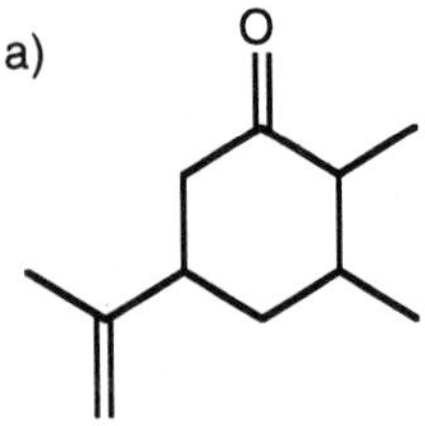
O

b) 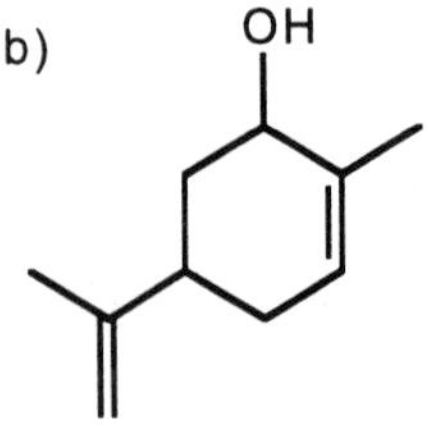
OH

c) 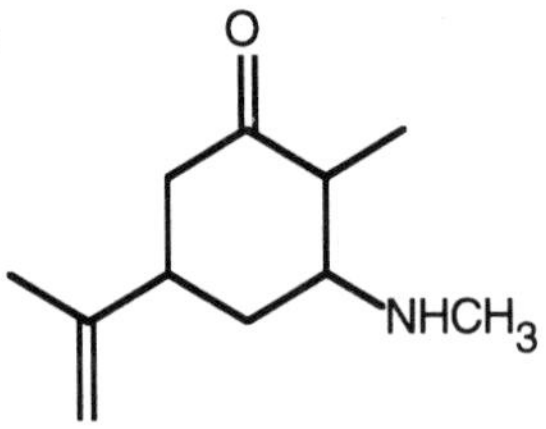
O
NHCH3

d) 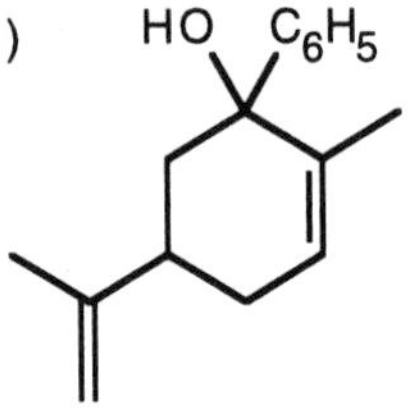
HO
C6H5

e) 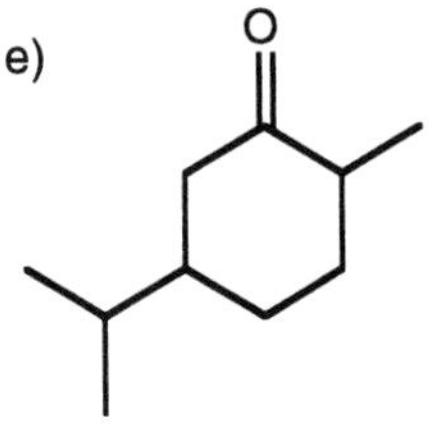
O

f) no reaction

g) 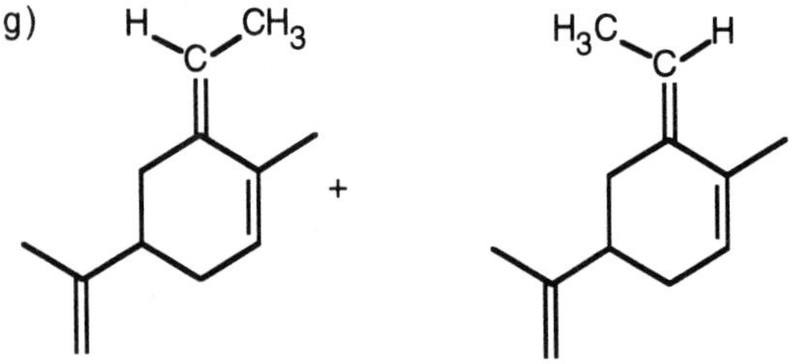
H
CH3
C
+
H3C
C
H

h) 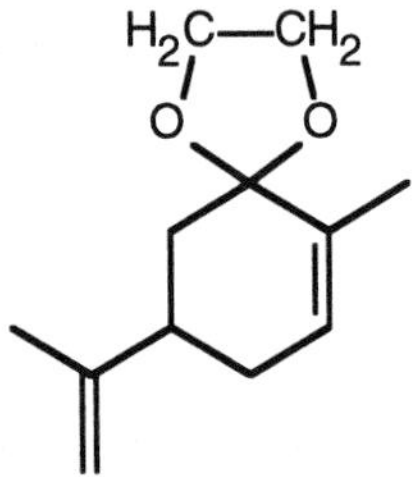
H2C—CH2
O
O

19.41

$$(CH_3)_3C\overset{O}{\overset{\|}{C}}H$$

2,2–Dimethylpropanal
Compound A

19.42

$$CH_3\underset{CH_3}{\underset{|}{C}}H\overset{O}{\overset{\|}{C}}CH_3$$

3–Methyl–2–butanone
Compound B

19.43

a) 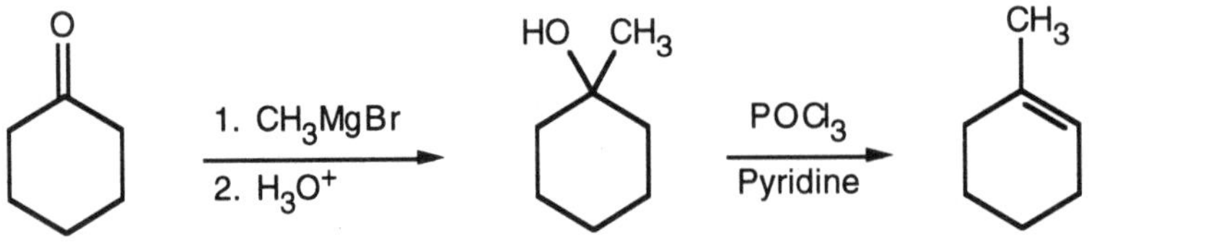
O
1. CH3MgBr
2. H3O+
HO
CH3
POCl3
Pyridine
CH3

1–Methylcyclohexene

b)

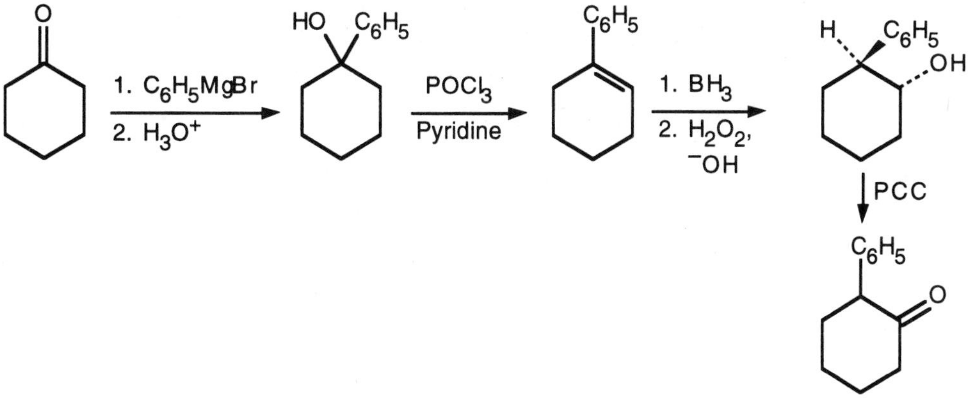

2–Phenylcyclohexanone

c)

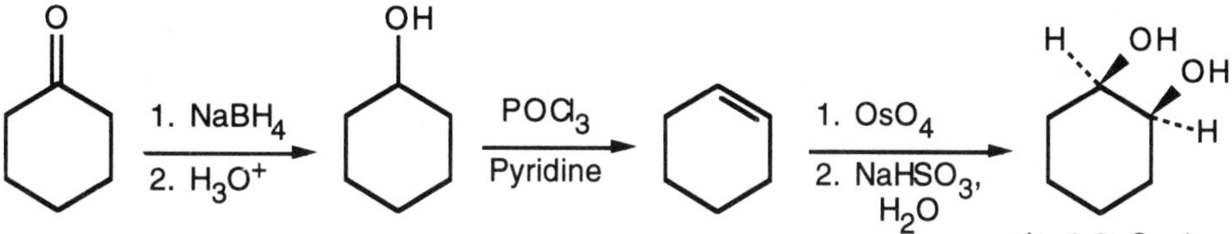

cis–1,2–Cyclo-
hexanediol

d)

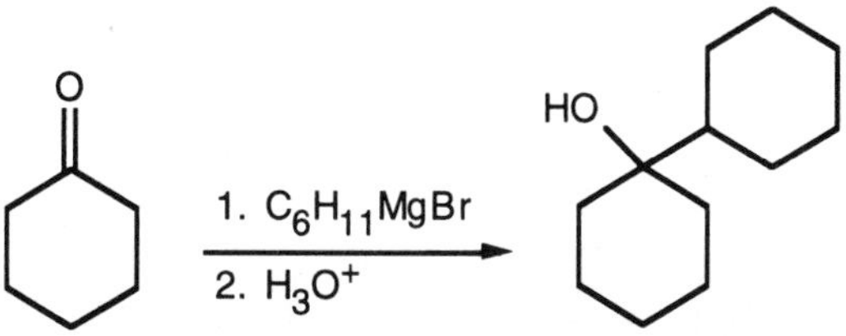

1–Cyclohexylcyclohexanol

19.44

Absorption:	Due to:
a) 1750 cm^{-1} 1685 cm^{-1}	5–membered ring ketone α,ß–unsaturated ketone
b) 1710 cm^{-1}	5–membered ring *and* α,ß–unsaturated ketone
c) 1750 cm^{-1}	5–membered ring ketone
d) 1705 cm^{-1}, 2720 cm^{-1}, 2820 cm^{-1} 1715 cm^{-1}	aromatic aldehyde aliphatic ketone

Compounds in parts b-d also show aromatic ring IR absorptions in the range 1450 cm^{-1} – 1600 cm^{-1} and in the range 690 – 900 cm^{-1}.

19.45 a) Basic silver ion does not oxidize secondary alcohols to ketones. Grignard addition to a conjugated ketone yields the 1,2 product, not the 1,4 product. The correct scheme:

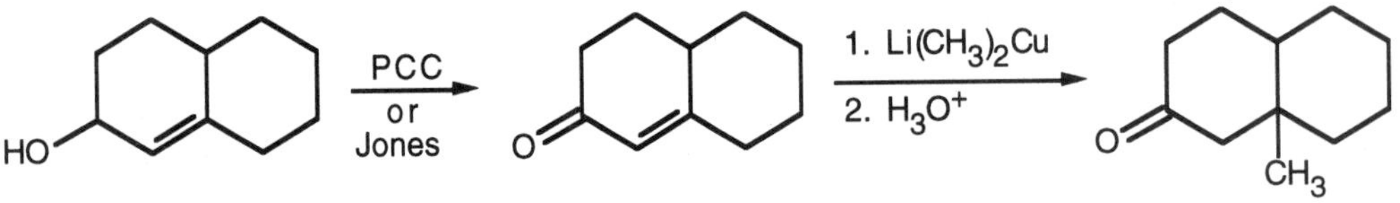

b) The Jones oxidation converts primary alcohols to carboxylic acids, not to aldehydes. The correct scheme:

$$C_6H_5CH=CHCH_2OH \xrightarrow{PCC} C_6H_5CH=CHCHO$$

$$\downarrow H^+, CH_3OH$$

$$C_6H_5CH=CHCH(OCH_3)_2$$

c) Treatment of a cyanohydrin with H_3O^+ produces a carboxylic acid, not an amine. The correct scheme:

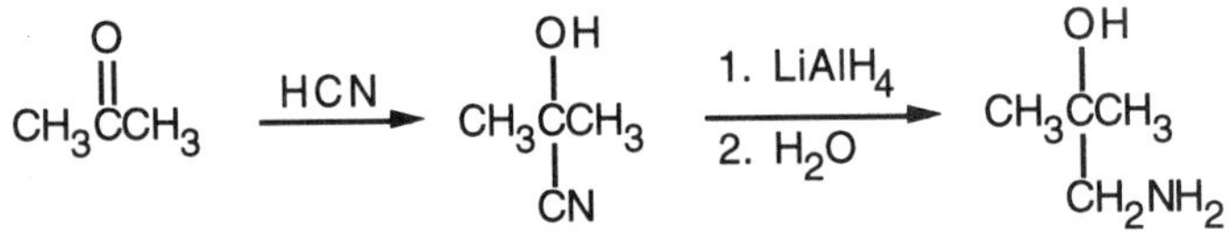

19.46

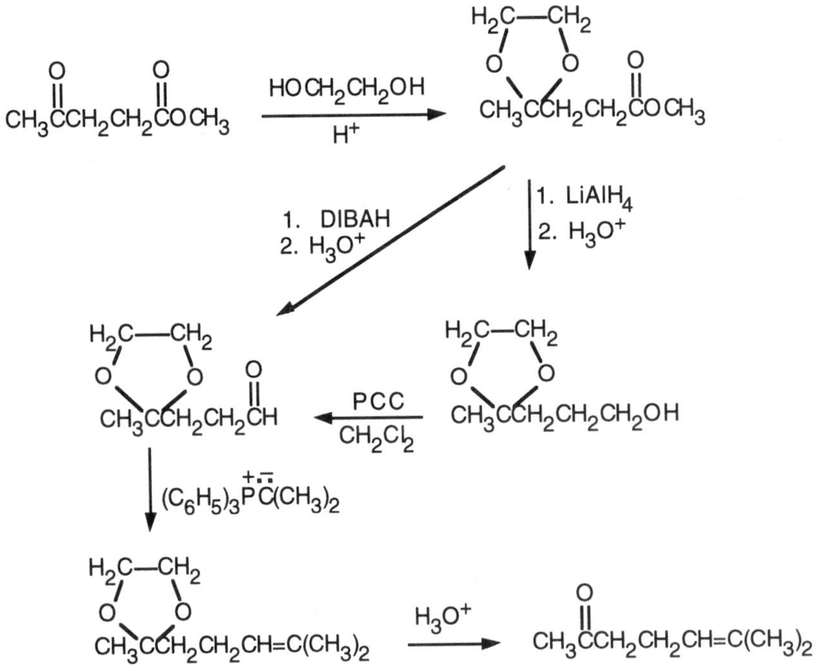

6-Methyl-5-hepten-2-one

19.47 Even though the product looks unusual, this reaction is made up of steps with which you are familiar.

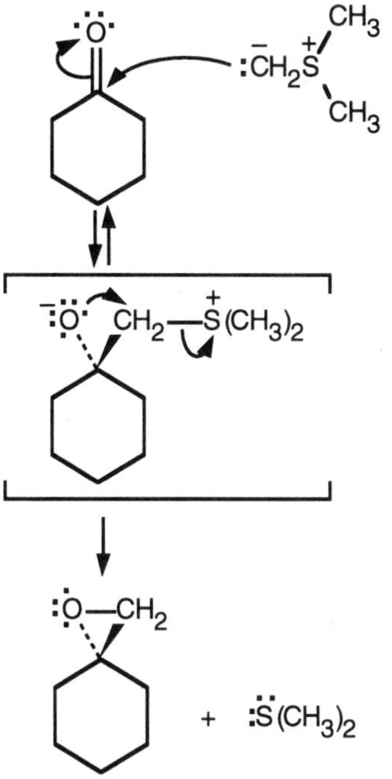

Nucleophilic addition of the ylide to a carbonyl group . . .

. . . is followed by S_N2 displacement of dimethyl sulfide by oxygen.

19.48

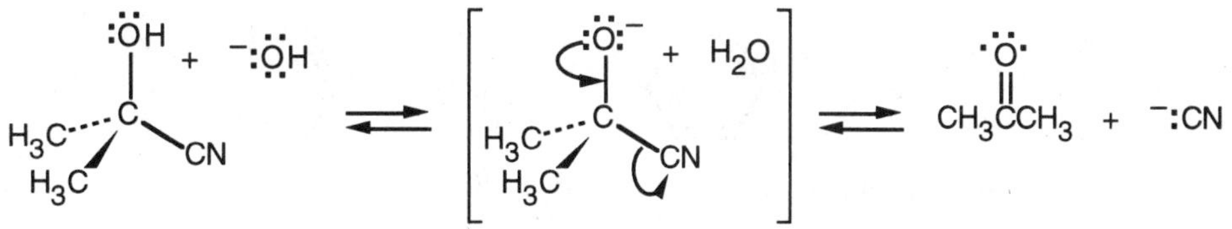

The above steps are the reverse of Problem 19.5.

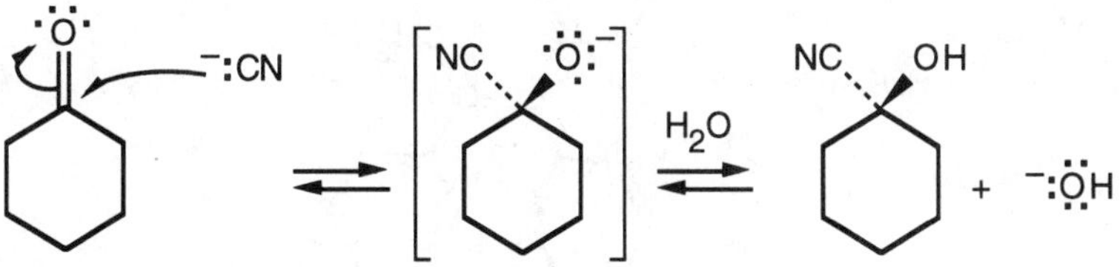

This step is a nucleophilic addition of cyanide.

19.49

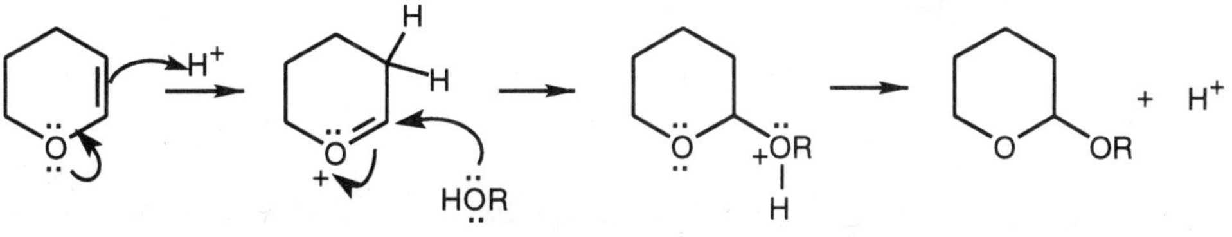

19.50

19.51

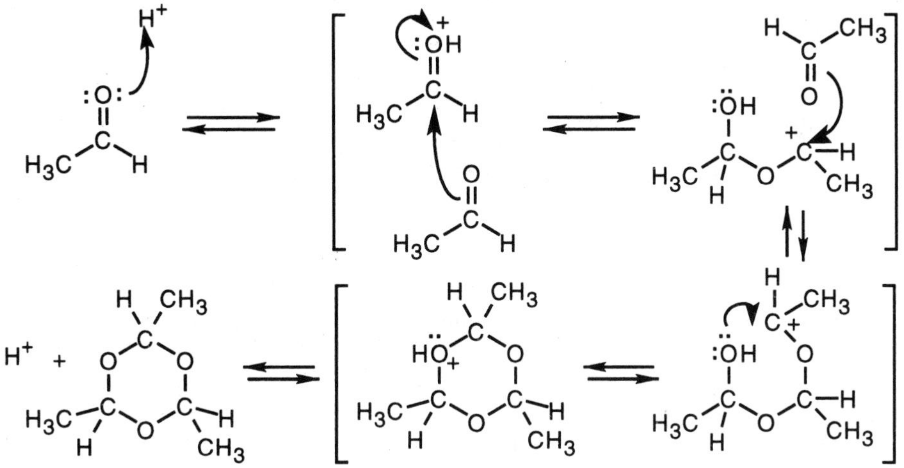

19.52

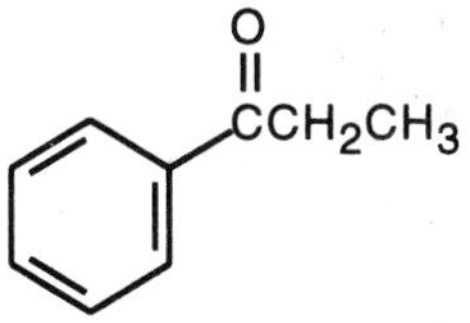

19.53

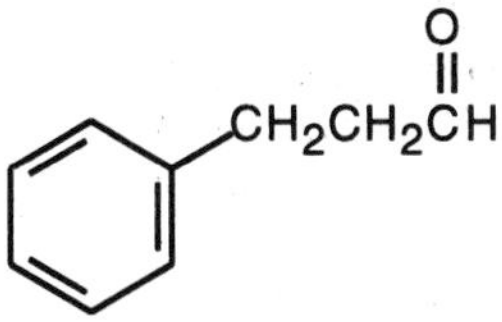

19.54

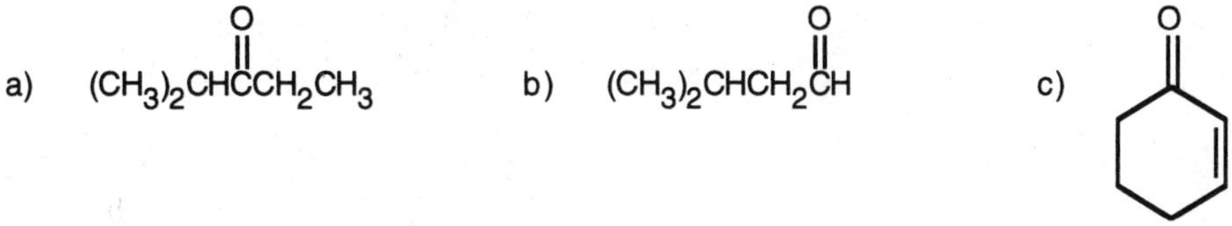

a) $(CH_3)_2CHCCH_2CH_3$ b) $(CH_3)_2CHCH_2CH$ c)

19.55

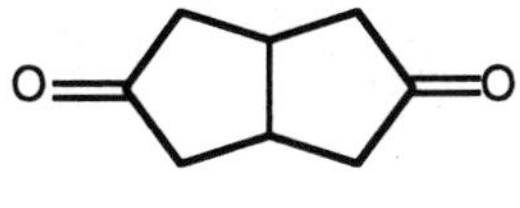

19.56 1) Aluminum, a Lewis Acid, complexes with the carbonyl oxygen.

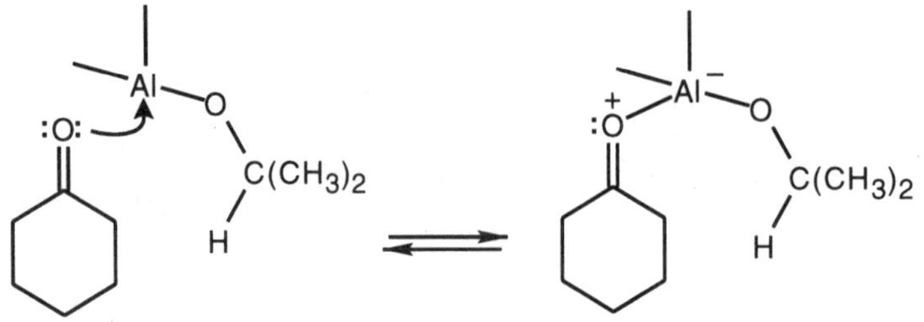

2) Complexation with aluminum makes the carbonyl functional group electrophilic and facilitates hydride transfer from isopropoxide.

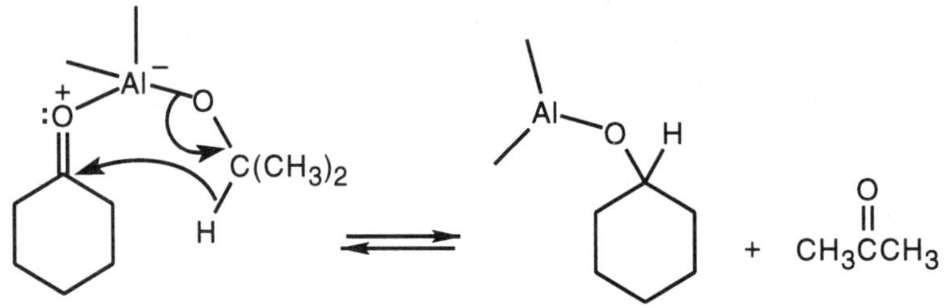

$$+ \quad CH_3CCH_3$$

3) Treatment of the reaction mixture with aqueous acid cleaves the aluminum-oxygen bond and produces cyclohexanol.

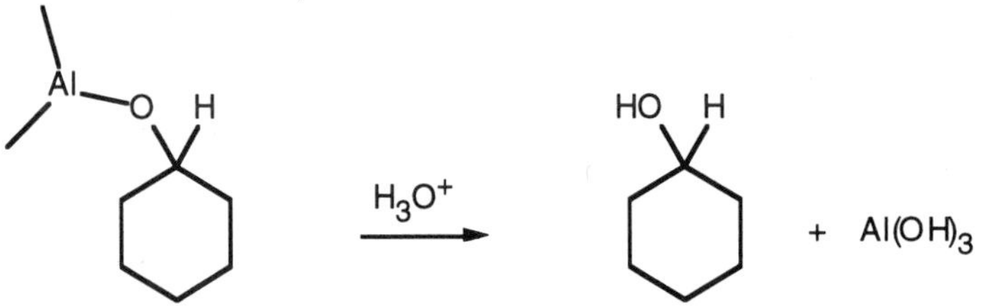

$$\xrightarrow{H_3O^+}$$

$$+ \quad Al(OH)_3$$

Both the MPV reaction and the Cannizzaro reaction are hydride transfers in which a carbonyl group is reduced by an alkoxide ion, which is oxidized.

19.57

a) 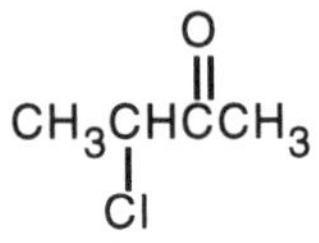 CH_3CHCCH_3 with Cl

b) 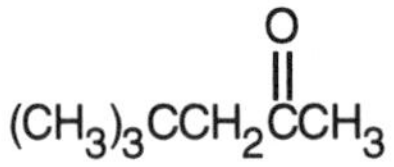 $(CH_3)_3CCH_2CCH_3$

c)

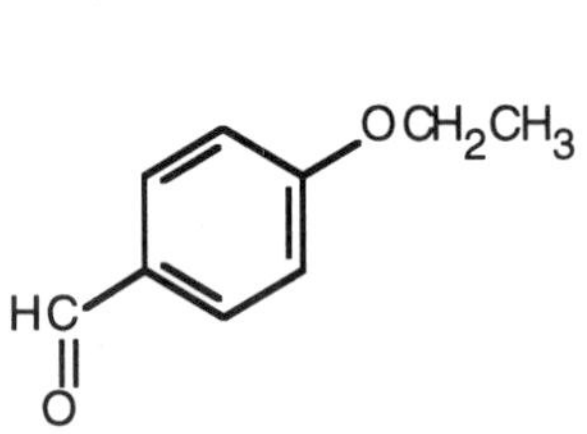

19.58

a)

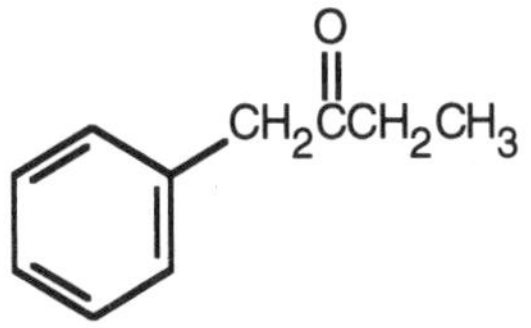

b) 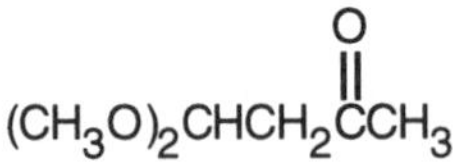 $(CH_3O)_2CHCH_2CCH_3$

c)

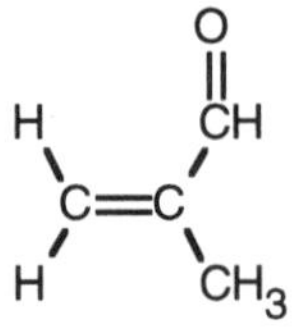

19.59 a) Nucleophilic addition of hydrazine to one of the carbonyl groups, followed by elimination of water, produces a hydrazone.

hydrazone

b) In a similar manner, the *other* nitrogen of hydrazine can add to the *other* carbonyl carbonyl group of 2,4–pentanedione to form the pyrazole.

19.60 The same sequence of steps used in the previous problem leads to the formation of 3,5–dimethylisoxazole when hydroxylamine is the reagent.

a)

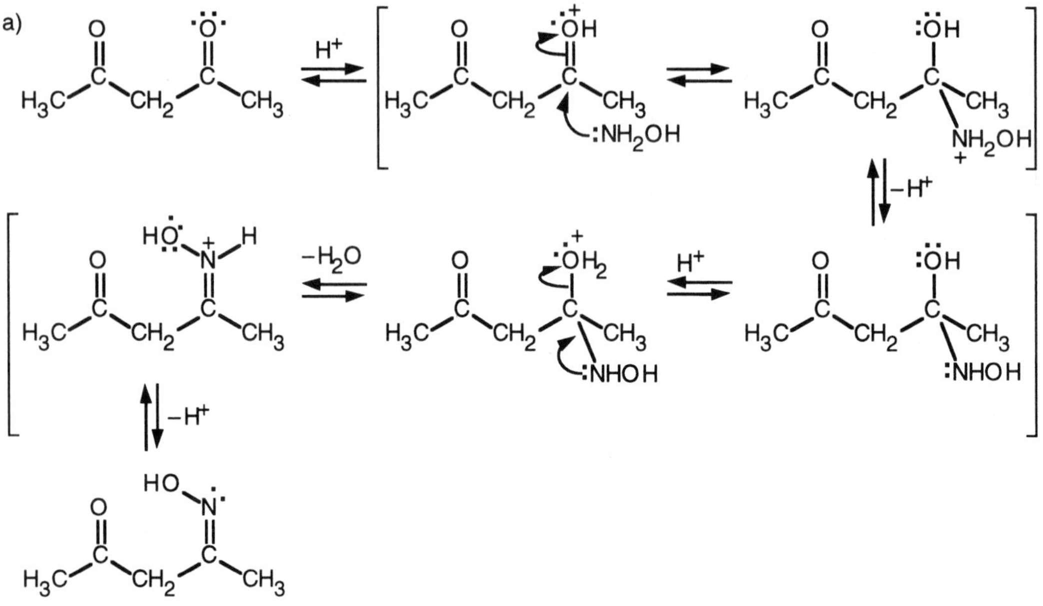

b)

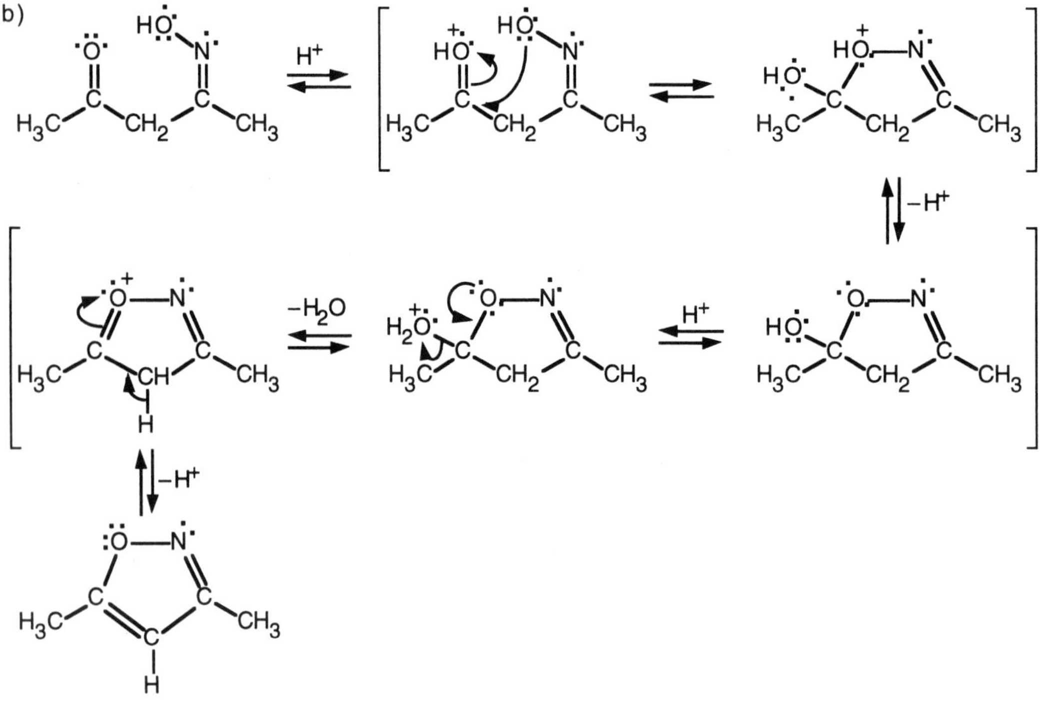

19.61

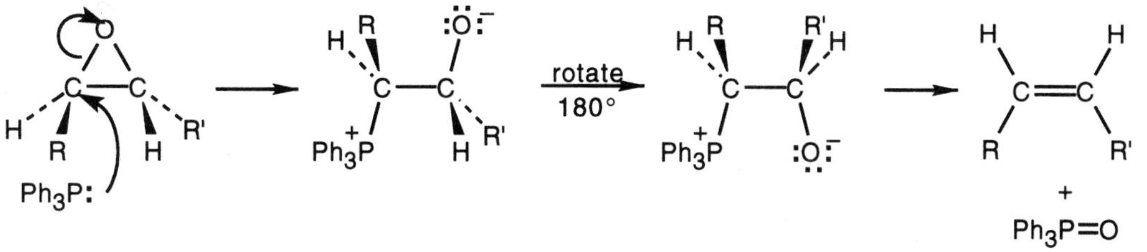

The final step is the same as the last step in a Wittig reaction.

19.62

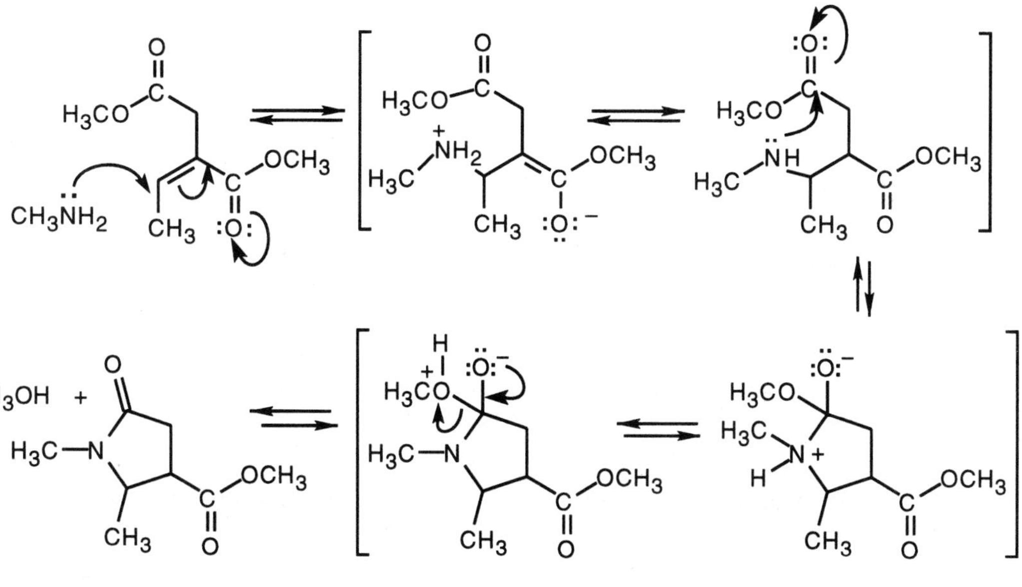

19.63

In this series of equilibrium steps, the hemiacetal ring of α–glucose opens to yield the free aldehyde. Bond rotation is followed by formation of the cyclic hemiacetal of ß–glucose. The reaction is catalyzed by both acid and base.

19.64 The free aldehyde form of glucose (Problem 19.63) is reduced in the same manner described in the text for other aldehydes to produce the polyalcohol sorbitol.

α-Glucose   Sorbitol

Study Guide for Chapter 19

After studying this chapter, you should be able to:

(1)   Name and draw aldehydes and ketones (19.1, 19.2, 19.23, 19.24, 19.25, 19.26).

(2)   Prepare aldehydes and ketones (19.3, 19.4).

(3)   Explain the difference in reactivity between aldehydes and ketones, and explain substituent effects on their reactivity (19.6, 19.7, 19.37).

(4)   Formulate mechanisms for nucleophilic addition reactions of aldehydes and ketones (19.5, 19.8, 19.9, 19.10, 19.11, 19.31, 19.35, 19.36, 19.47, 19.48, 19.49, 19.51, 19.56, 19.59, 19.60, 19.61, 19.62, 19.63).

(5)   Predict the products of reactions of aldehydes and ketones with:
(a) Amines (19.12, 19.13).
(b) Alcohols (19.14, 19.30).
(c) Phosphorus ylides (19.15, 19.16, 19.32, 19.33).
(d) Strong base (19.17).
(e) Organocopper reagents (19.18).
(f) Grignard reagents (19.34).
(g) A combination of reagents (19.27, 19.28, 19.29, 19.38, 19.39, 19.40).

(6)   Synthesize aldehydes and ketones (19.45, 19.47, 19.48).

(7)   Use spectroscopic information to determine the structure of aldehydes and ketones (19.19, 19.20, 19.21, 19.22, 19.41, 19.42, 19.44, 19.52, 19.53, 19.54, 19.55, 19.57, 19.58).

Chapter 20 – Carboxylic Acids

20.1

a) $(CH_3)_2CHCH_2COOH$

3–Methylbutanoic acid

b) $CH_3CHBrCH_2CH_2COOH$

4–Bromopentanoic acid

c) $CH_3CH=CHCH=CHCOOH$

2,4–Hexadienoic acid

d)
$$CH_3CH_2\overset{\overset{\displaystyle COOH}{|}}{C}HCH_2CH_2CH_3$$

2–Ethylpentanoic acid

e)

cis–1,3–Cyclopentane-
dicarboxylic acid

f)

2–Phenylpropanoic acid

20.2

a)
$$CH_3CH_2CH_2\overset{\overset{\displaystyle CH_3}{|}}{C}H\overset{\overset{\displaystyle}{}}{C}HCOOH$$
$$|$$
$$CH_3$$

2,3–Dimethylhexanoic acid

b) $(CH_3)_2CHCH_2CH_2COOH$

4–Methylpentanoic acid

c)

trans–1,2–Cyclobutane-
dicarboxylic acid

d)

o–Hydroxybenzoic acid

e)

(9Z, 12Z)–Octadecadienoic acid

20.3 $\Delta G° = -RT \ln K_a = -2.303\ RT \log K_a$. Here R = 8.314 J/mol • K; T = 300 K

$\Delta G° = (-2.303)(8.314\ \text{J/mol} • \text{K})(300\ \text{K})(\log K_a)$
  $= (-5.744 \times 10^3\ \text{J/mol})(\log K_a)$
  $= (-5.744\ \text{kJ/mol})(\log K_a)$

For ethanol: $pK_a = 16$; $\log K_a = -16$
   $\Delta G° = (-5.744\ \text{kJ/mol})(-16) = +92\ \text{kJ/mol}$

For acetic acid: $pK_a = 4.75$; $\log K_a = -4.75$
   $\Delta G° = (-5.744\ \text{kJ/mol})(-4.75) = +27.3\ \text{kJ/mol}$

Dissociation of acetic acid is more favored. Since $\Delta G°$ for acetic acid is a smaller number,
less energy is required for dissociation of acetic acid than for dissociation of ethanol.

20.4 Naphthalene is insoluble in water and benzoic acid is only slightly soluble. The *salt* of benzoic acid is very soluble in water, however, and we can take advantage of this solubility in separating naphthalene from benzoic acid.

Dissolve the mixture in an organic solvent, and extract with a dilute aqueous solution of sodium hydroxide or sodium bicarbonate, which will neutralize benzoic acid. Naphthalene will remain in the organic layer, and all benzoic acid, now converted to the benzoate salt, will be in the aqueous layer. To recover benzoic acid, remove the aqueous layer, acidify it with dilute mineral acid, and extract with an organic solvent.

20.5

$$Cl_2CHCOH + H_2O \ \overset{K_a}{\rightleftharpoons} \ Cl_2CHCO^- + H_3O^+$$

$$K_a = \frac{[Cl_2CHCO_2^-][H_3O^+]}{[Cl_2CHCOOH]} = 3.32 \times 10^{-2}$$

	Initial molarity	Molarity after dissociation
Cl_2CHCOH	0.10	0.10 − y
Cl_2CHCO^-	0	y
H_3O^+	0	y

$$K_a = \frac{y \cdot y}{(0.10 - y)} = 3.32 \times 10^{-2}$$

Using the quadratic formula to solve for y, we find that y = 0.0434.

$$\text{Percent dissociation} = \frac{0.0434}{0.1000} \times 100\% = 43.4\%.$$

20.6

Weaker acid ⎯⎯⎯⎯⎯⎯⟶ Stronger acid

a) $CH_3CH_2COOH < BrCH_2COOH < FCH_2COOH$

Fluoride is the most electronegative group and can stabilize the carboxylate anion the best.

b)

The electron-withdrawing nitro group stabilizes the carboxylate anion. The methoxyl group, which is a resonance electron donor, destabilizes the carboxylate anion.

c) $CH_3CH_2NH_2 < CH_3CH_2OH < CH_3CH_2COOH$

20.7

The pK_1 of oxalic acid is lower than that of a monocarboxylic acid because the carboxylate anion is stabilized both by resonance and by the inductive effect of the nearby second carboxylic acid group.

The pK_2 of oxalic acid is higher than the pK_1 for two reasons: (1) The first carboxylate group inductively destabilizes the negative charge resulting from dissociation of the second proton. (2) Electrostatic repulsion between the two adjacent negative charges destabilizes the dianion.

20.8 A pK_a of 4.45 indicates that *p*–cyclopropylbenzoic acid is a weaker acid than benzoic acid. This, in turn, indicates that a cyclopropyl group must be electron-donating. Since electron-donating groups increase reactivity in electrophilic substitution reactions, *p*–cyclopropylbenzene should be more reactive than benzene toward electrophilic bromination.

20.9

Least acidic ⟶ Most acidic

a)

b)

20.10 Both methods are simple two-step transformations, and both can be used with compounds containing base-sensitive functional groups. Advantages of the cyanide hydrolysis method: (1) it can be used with acid-sensitive compounds; (2) it produces an optically active carboxylic acid from an optically active halide; (3) it can be used with compounds containing functional groups that interfere with Grignard reagent formation. The advantage of the Grignard method is that it can be used with tertiary halides and also with aryl and vinylic halides, which do not undergo S$_N$2 reactions.

20.11

a)

$$\text{(bromobenzene)} \quad \xrightarrow[\text{3. H}_3\text{O}^+]{\begin{array}{l}\text{1. Mg, ether}\\ \text{2. CO}_2\text{, ether}\end{array}} \quad \text{(benzoic acid)}$$

b) $(CH_3)_3CCl$ $\xrightarrow[\text{3. H}_3\text{O}^+]{\begin{array}{l}\text{1. Mg, ether}\\ \text{2. CO}_2\text{, ether}\end{array}}$ $(CH_3)_3CCOOH$

c) $CH_3CH_2CH_2Br$ $\xrightarrow[\text{3. H}_3\text{O}^+]{\begin{array}{l}\text{1. Mg, ether}\\ \text{2. CO}_2\text{, ether}\end{array}}$ or $\xrightarrow[\text{2. H}_3\text{O}^+]{\text{1. NaCN}}$ $CH_3CH_2CH_2COOH$

20.12 a) 2,5–Dimethylhexanedioic acid   b) 2,2–Dimethylpropanoic acid
       c) 3–Propylhexanoic acid   d) *p*–Nitrobenzoic acid
       e) 1–Cyclodecenecarboxylic acid   f) 4,5–Dibromopentanoic acid

20.13

a)

b) $HOOCCH_2CH_2CH_2CH_2CH_2COOH$

c) $CH_3C\equiv CCH=CHCOOH$

d)

$$\overset{\displaystyle CH_2CH_3}{\underset{\displaystyle CH_2CH_2CH_3}{CH_3CH_2CH_2CH_2\overset{|}{C}HCH_2\overset{|}{C}HCOOH}}$$

e)

f) $(C_6H_5)_3CCOOH$

20.14 Acetic acid molecules are strongly associated because of hydrogen bonding. Molecules of the ethyl ester are much more weakly associated, and less heat is required to overcome the attractive forces between molecules of the ethyl ester. Even though the ethyl ester has a greater molecular weight, it boils at a lower temperature than the acid.

20.15

$CH_3CH_2CH_2CH_2CH_2COOH$

Hexanoic acid

$$\overset{\displaystyle CH_3}{CH_3CH_2CH_2\overset{|}{C}HCOOH}$$

2–Methylpentanoic acid

$$\overset{\displaystyle CH_3}{CH_3CH_2\overset{|}{C}HCH_2COOH}$$

3–Methylpentanoic acid

$$\overset{\displaystyle CH_3}{CH_3\overset{|}{C}HCH_2CH_2COOH}$$

4–Methylpentanoic acid

$$\overset{\displaystyle COOH}{CH_3CH_2\overset{|}{C}HCH_2CH_3}$$

2–Ethylbutanoic acid

$$\underset{\displaystyle CH_3}{\overset{\displaystyle CH_3}{CH_3CH_2\overset{|}{\underset{|}{C}}COOH}}$$

2,2–Dimethylbutanoic acid

$$CH_3CHCHCOOH$$

with CH₃ groups —

2,3–Dimethylbutanoic acid

$$CH_3CCH_2COOH$$

3,3–Dimethylbutanoic acid

20.16

Least acidic ⟶ Most acidic

a) CH_3COOH < $HCOOH$ < $HOOCCOOH$

  Acetic acid   Formic acid   Oxalic acid

b)   *p*–Bromobenzoic acid < *p*–Nitrobenzoic acid < 2,4-Dinitrobenzoic acid

c) $C_6H_5CH_2CH_2COOH$ < $C_6H_5CH_2COOH$, $(C_6H_5)_2CHCOOH$

d) FCH_2CH_2COOH < ICH_2COOH < FCH_2COOH

20.17 Remember that the conjugate base of a weak acid is a strong base. In other words, the stronger the acid, the weaker the base derived from that acid.

Least basic ⟶ Most basic

a) $\left[CH_3C{-}\ddot{O}{:}^- \right]_2 Mg^{2+}$ < $Mg(OH)_2$ < $H_3C{:}^- Mg^+Br$

b)

$<$ $<$ $HC{\equiv}C{:}^- Na^+$

c) $HC{-}\ddot{O}{:}^- Li^+$ < $H\ddot{O}{:}^- Li^+$ < $CH_3CH_2\ddot{O}{:}^- Li^+$

20.18 Two factors are responsible for the difference in pK_2 values between these benzenedicarboxylic acids. (1) The mono-anion of phthalic acid is stabilized by hydrogen bonding of the remaining –COOH proton with the adjacent carboxylate group. This type of hydrogen bonding is not possible for terephthalic acid.

(2) The dianion of phthalic acid has its two negative charges close to one another, resulting in electrostatic repulsion that destabilizes the phthalate dianion. Since the energy difference between mono- and dianion is greater for phthalic acid than for terephthalic acid, the pK_2 of phthalic acid is greater than the pK_2 of terephthalic acid.

20.19

a) $CH_3CH_2CH_2\overset{\displaystyle O}{\overset{\displaystyle \|}{C}}OH$ $\xrightarrow[\text{2. H}_3\text{O}^+]{\text{1. BH}_3 \text{ or LiAlH}_4}$ $CH_3CH_2CH_2CH_2OH$
1–Butanol

b) $CH_3CH_2CH_2CH_2OH$ $\xrightarrow{\text{PBr}_3}$ $CH_3CH_2CH_2CH_2Br$
from a   1–Bromobutane

c) $CH_3CH_2CH_2CH_2Br$ $\xrightarrow{\text{NaCN}}$ $CH_3CH_2CH_2CH_2C{\equiv}N$
from b

$\downarrow H_3O^+$

$CH_3CH_2CH_2CH_2COOH$
Pentanoic acid

d) $CH_3CH_2CH_2CH_2Br$ $\xrightarrow{(CH_3)_3CO^-K^+}$ $CH_3CH_2CH{=}CH_2$
from b   1–Butene

e) $2\,CH_3CH_2CH_2CH_2Br$ $\xrightarrow{\text{2Li}}$ $2\,CH_3CH_2CH_2\overset{\displaystyle ..}{C}H_2{}^-\,Li^+$
from b

$\downarrow \text{CuI}$

$(CH_3CH_2CH_2CH_2)_2Cu^-\,Li^+$

$(CH_3CH_2CH_2CH_2)_2CuLi \; + \; CH_3CH_2CH_2CH_2Br$

$\downarrow$

$CH_3CH_2CH_2CH_2CH_2CH_2CH_2CH_3$
Octane

20.20

a) $CH_3CH_2CH_2CH_2OH$ $\xrightarrow[\text{reagent}]{\text{Jones}}$ $CH_3CH_2CH_2COOH$

b) $CH_3CH_2CH_2CH_2Br$ $\xrightarrow{\text{NaOH}}$ $CH_3CH_2CH_2CH_2OH$ $\xrightarrow[\text{reagent}]{\text{Jones}}$ $CH_3CH_2CH_2COOH$

c) $CH_3CH_2CH{=}CH_2$ $\xrightarrow[\text{2. H}_2\text{O}_2,\,^-\text{OH}]{\text{1. BH}_3}$ $CH_3CH_2CH_2CH_2OH$ $\xrightarrow[\text{reagent}]{\text{Jones}}$ $CH_3CH_2CH_2COOH$

d) $CH_3CH_2CH_2Br$

1. NaCN
2. ‾OH, H_2O
3. H_3O^+

or

1. Mg
2. CO_2
3. H_3O^+

$CH_3CH_2CH_2COOH$

e) $CH_3CH_2CH_2CH=CHCH_2CH_2CH_3$ $\xrightarrow[\text{H}_3\text{O}^+]{\text{KMnO}_4}$ 2 $CH_3CH_2CH_2COOH$

20.21

a)

b)

c)

Alternatively, benzyl bromide can be treated with cyanide, and the resulting nitrile can be hydrolyzed.

20.22 a) $K_a = 8.4 \times 10^{-4}$ for lactic acid
$pK_a = -\log (8.4 \times 10^{-4}) = 3.08$

b) $K_a = 5.6 \times 10^{-6}$ for acrylic acid
$pK_a = -\log (5.6 \times 10^{-6}) = 5.25$

20.23 a) $pK_a = 3.14$ for citric acid
$K_a = 10^{-3.14} = 7.24 \times 10^{-4}$

b) $pK_a = 2.98$ for tartaric acid
$K_a = 10^{-2.98} = 1.05 \times 10^{-3}$

20.24 Inductive effects of functional groups are transmitted through bonds. As the length of the carbon chain increases, the effect of one functional group on another decreases. In this example, the influence of the second carboxyl group on the ionization of the first is barely felt by succinic and adipic acids.

20.25

a. 1. BH_3
   2. H_3O^+
→ CH_3—⬡—CH_2OH

b. NBS
   CCl_4
→ $BrCH_2$—⬡—COOH

CH_3—⬡—COOH

c. 1. CH_3MgBr
   2. H_3O^+
→ CH_3—⬡—COOH + CH_4

d. $KMnO_4$
   H_3O^+
→ HOOC—⬡—COOH

e. 1. $LiAlH_4$
   2. H_3O^+
→ CH_3—⬡—CH_2OH

20.26

a) CH_3CH_2Br $\xrightarrow{Mg}$ CH_3CH_2MgBr $\xrightarrow[\text{2. } H_3O^+]{\text{1. } ^{13}CO_2}$ $CH_3CH_2{}^{13}COOH$

b) CH_3Br $\xrightarrow{Mg}$ CH_3MgBr $\xrightarrow[\text{2. } H_3O^+]{\text{1. } ^{13}CO_2}$ $CH_3{}^{13}\overset{O}{\overset{||}{C}}OH$ $\xrightarrow[\text{2. } H_3O^+]{\text{1. } BH_3}$ $CH_3{}^{13}CH_2OH$

$\downarrow PBr_3$

$CH_3{}^{13}CH_2\overset{O}{\overset{||}{C}}OH$ $\xleftarrow[\text{2. } H_3O^+]{\text{1. } CO_2}$ $CH_3{}^{13}CH_2MgBr$ $\xleftarrow{Mg}$ $CH_3{}^{13}CH_2Br$

20.27

$(CH_3)_3CCH_2COOH$

3,3–Dimethylbutanoic acid

20.28

a)

$\xrightarrow[\text{peroxides}]{HBr}$ ... $\xrightarrow[\text{2. } H_3O^+]{\text{1. } {}^-:CN}$ or $\xrightarrow[\text{3. } H_3O^+]{\substack{\text{1. Mg} \\ \text{2. } CO_2}}$ ...

b)

CH_2 $\xrightarrow{HBr}$ H_3C Br $\xrightarrow{Mg}$ H_3C MgBr $\xrightarrow[2.\ H_3O^+]{1.\ CO_2}$ H_3C COOH

20.29 Either ^{13}C NMR or ^1H NMR can be used to distinguish among these three isomeric carboxylic acids.

Compound	Number of ^{13}C NMR signals	Number of ^1H NMR signals	splitting of ^1H NMR signals
$CH_3(CH_2)_3COOH$	5	5	1 triplet, peak area 3, 1.0 δ 1 triplet, peak area 2, 2.4 δ 2 multiplets, peak area 4, 1.5 δ 1 singlet, peak area 1, 12.0 δ
$(CH_3)_2CHCH_2COOH$	4	4	1 doublet, peak area 6, 1.0 δ 1 doublet, peak area 2, 2.4 δ 1 multiplet, peak area 1, 1.6 δ 1 singlet, peak area 1, 12.0 δ
$(CH_3)_3CCOOH$	3	2	1 singlet, peak area 9, 1.3 δ 1 singlet, peak area 1, 12.1 δ

20.30 a) Grignard carboxylation can't be used to prepare the carboxylic acid because of the acidic hydroxyl group. Use nitrile hydrolysis.

b) Either method produces the carboxylic acid. Grignard carboxylation is a better reaction for preparing a carboxylic acid from a secondary bromide. Nitrile hydrolysis produces an optically active carboxylic acid from an optically active bromide.

c) Neither method of acid synthesis yields the desired product. Any Grignard reagent formed will react with the carbonyl functional group present in the starting material. Reaction with cyanide occurs at the carbonyl functional group, producing a cyanohydrin, as well as at the halogen. However, if the ketone is first protected by forming an acetal, either method can be used.

d) Since the hydroxyl proton interferes with formation of the Grignard reagent, nitrile hydrolysis must be used to form the carboxylic acid.

20.31 2–Chloro–2–methylpentane is a tertiary alkyl halide and ‾CN is a base. Instead of the desired S_N2 reaction of cyanide with a halide, E2 elimination occurs and yields 2–methyl–2–pentene.

20.32 a) BH_3 is a reducing agent, not an oxidizing agent. To obtain benzoic acid from toluene, use $KMnO_4$.

b) Use CO_2 instead of NaCN to form the carboxylic acid, or eliminate Mg from this reaction scheme and form the acid by nitrile hydrolysis.

c) Reduction of a carboxylic acid with $LiAlH_4$ yields an alcohol, not an alkyl group.

d) Acid hydrolysis of the nitrile will also dehydrate the tertiary alcohol. Use basic hydrolysis to form the carboxylic acid.

20.33

Notice that the order of the reactions is very important. If toluene is oxidized first, the nitro group will be introduced in the meta position. If the nitro group is reduced first, oxidation to the carboxylic acid will reoxidize the $-NH_2$ group.

20.34 Before starting this type of problem, identify the functional groups present in the starting material. Lithocholic acid contains only alcohol and carboxylic acid functional groups. The given reagents can react with one, both, or neither functional group. Remember to keep track of stereochemistry.

20.35

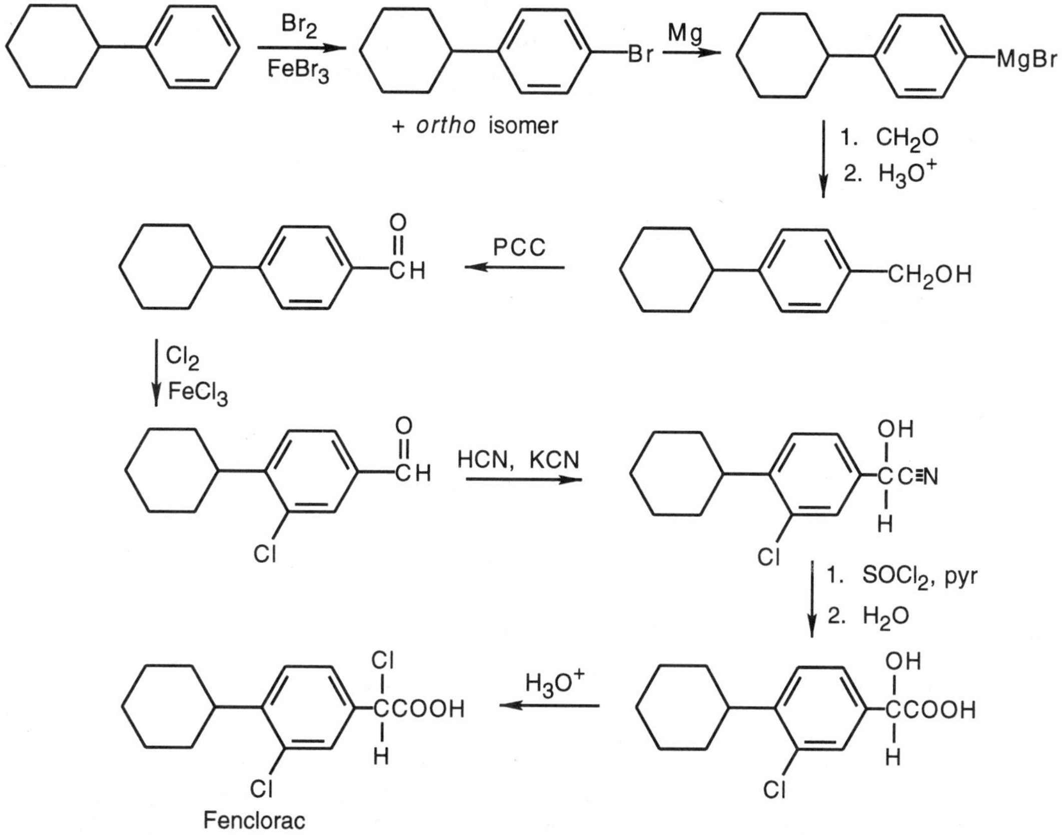

+ *ortho* isomer

Other routes to this compound are possible. Notice that the aldehyde functional group and the cyclohexyl group both serve to direct the aromatic chlorination to the correct position. Also, reaction of the hydroxy acid with $SOCl_2$ converts –OH to –Cl and –COOH to –COCl. Treatment with H_2O regenerates the carboxylic acid.

20.36 a), b) Use either ^{1}H NMR or ^{13}C NMR to distinguish between the isomers.

Compound	# of ^{13}C NMR Absorptions	# of ^{1}H NMR Absorptions
a) COOH (benzene with COOH meta)	5	4
COOH (benzene with COOH para)	3	2
b) HOOCCH$_2$CH$_2$COOH	2	2
CH$_3$CH(COOH)$_2$	3	3

c) Use ^{1}H NMR to distinguish between these two compounds. The carboxylic acid proton of the first compound absorbs near 12 δ; the aldehyde proton of the second compound absorbs near 10 δ and is split into a triplet.

d) The cyclic acid shows four absorptions in both its ^{1}H NMR and ^{13}C NMR spectra. The unsaturated acid shows six absorptions in its ^{13}C NMR and five in its ^{1}H NMR spectrum; one of the ^{1}H NMR signals occurs in the vinylic region (4.5 – 6.5 δ) of the spectrum.

20.37

Substituent	pK_a	Acidity	*E.A.S. Reactivity
–PCl$_2$	3.59	Most acidic	Least reactive (most deactivating)
–OSO$_2$CH$_3$	3.84		
–CH=CHCN	4.03		
–HgCH$_3$	4.10		
–H	4.19		
–Si(CH$_3$)$_3$	4.27	Least acidic	Most reactive (least deactivating)

*Electrophilic aromatic substitution

Recall from Section 20.5 that substituents that increase acidity also decrease reactivity in electrophilic aromatic substitution reactions. Of the above substituents, only –Si(CH$_3$)$_3$ is an activator.

20.38

a)

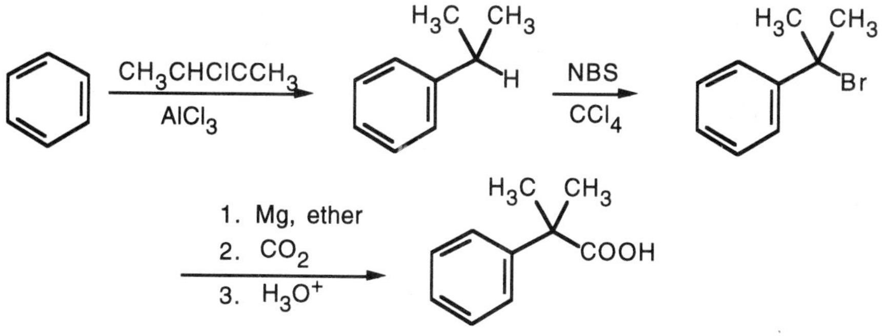

b)

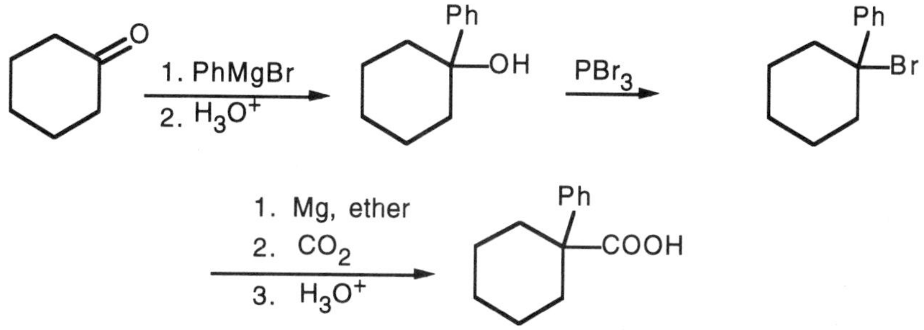

20.39 For hydroxyacetic acid, the negative charge of the carboxylate anion is stabilized by the electron-withdrawing *inductive* effect of –OH. For *p*-hydroxybenzoic acid, the negative charge of the anion is destabilized by the electron-donating *resonance* effect of –OH that acts over the π electron system of the ring.

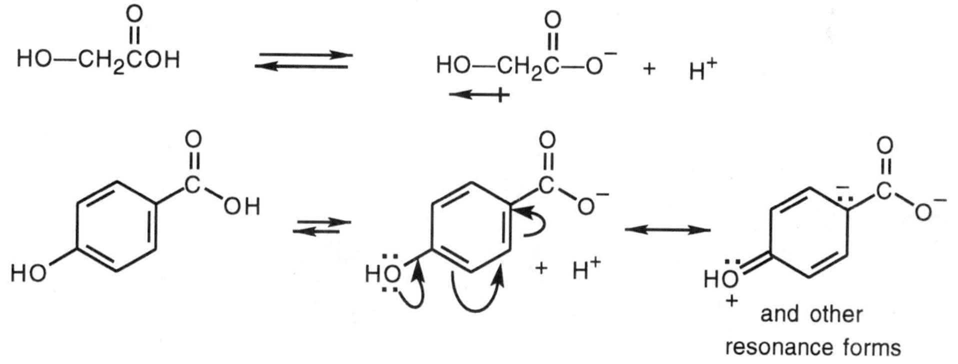

20.40

$$CH_3CH_2OCH_2\overset{\overset{\displaystyle O}{\|}}{C}OH$$

20.41 Both compounds contain four different kinds of protons (remember that the $H_2C=$ protons are nonequivalent). The carboxylic acid proton absorptions are easy to identify; the other three absorptions in each spectrum are more complex.

It is possible to assign the spectra by studying the methyl group absorptions. The methyl group peak of crotonic acid is split into a doublet by the geminal $(CH_3CH=)$ proton, while the methyl group absorption of methacrylic acid is a singlet. The first spectrum is that of crotonic acid, and the second spectrum is that of methacrylic acid.

20.42

a)

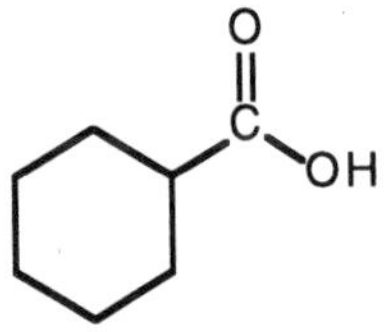

b)

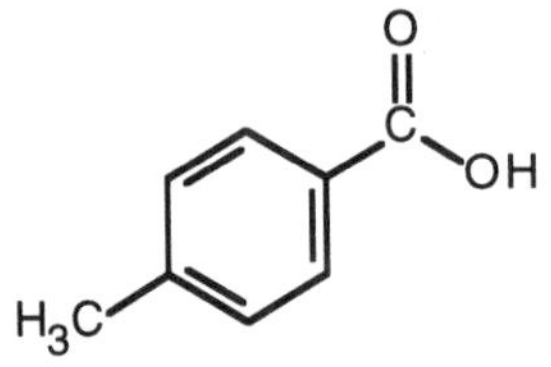

20.43

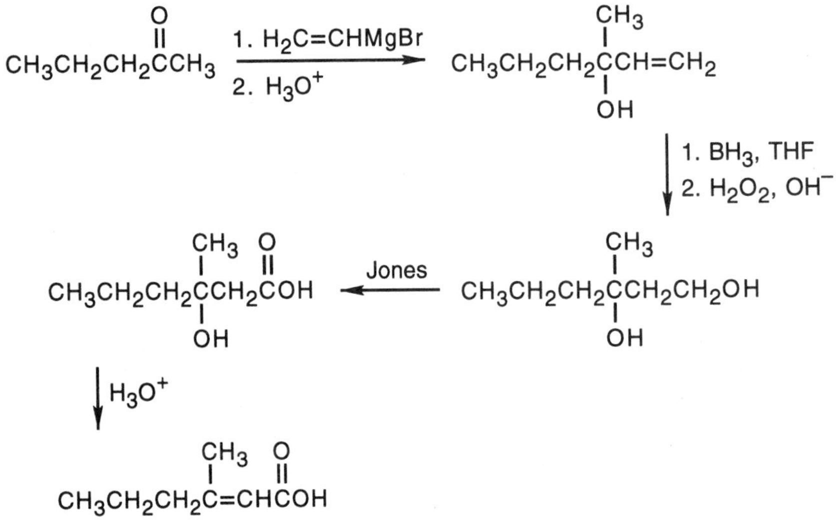

3-Methyl-2-hexenoic acid

Dehydration will occur in the indicated direction to produce a double bond conjugated with the carboxylic acid carbonyl group.

20.44

$$H_2O_2 + OH^- \rightleftharpoons OOH^- + H_2O$$

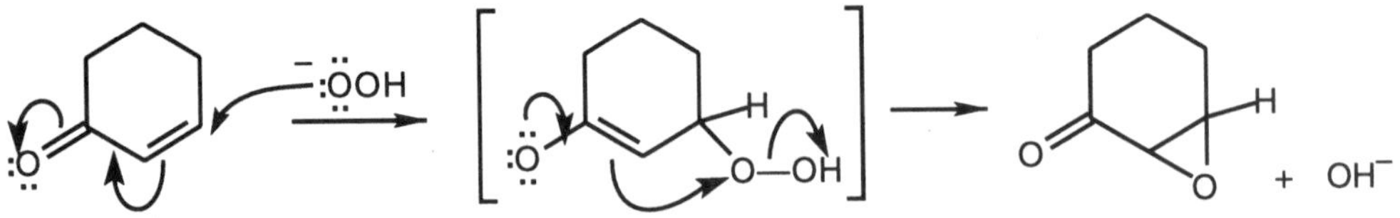

20.45

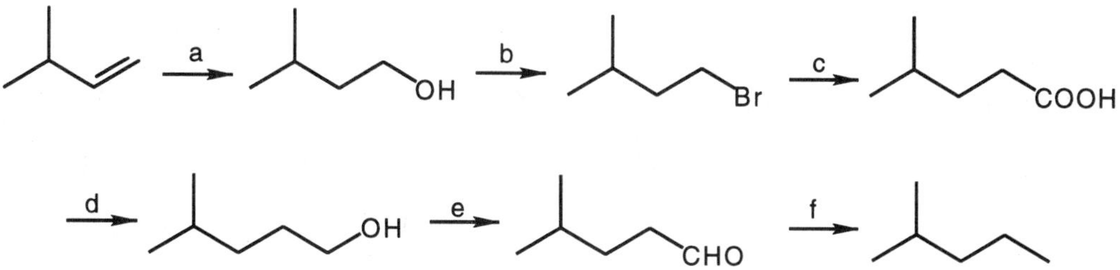

(a) BH$_3$, then H$_2$O$_2$, OH$^-$; (b) PBr$_3$; (c) Mg, then CO$_2$, or CN$^-$, then H$_3$O$^+$; (d) LiAlH$_4$; (e) PCC; (f) N$_2$H$_4$, KOH

20.46

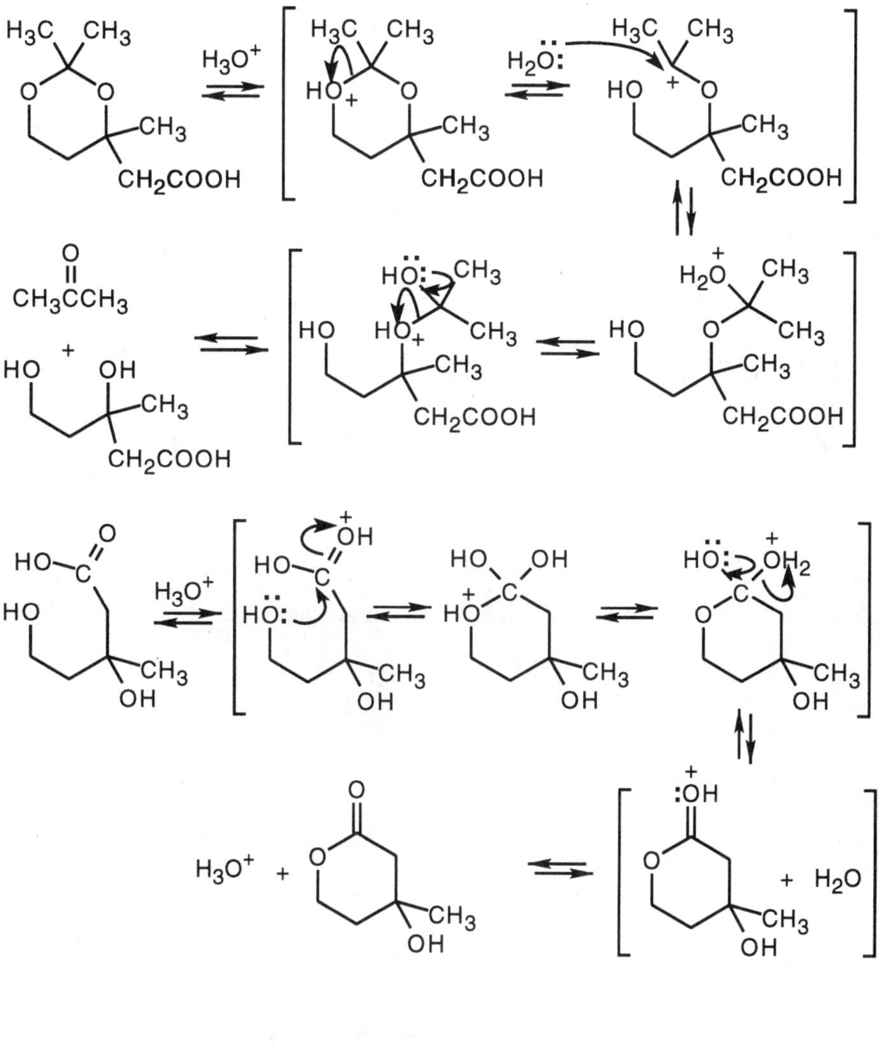

20.47

Study Guide for Chapter 20

After studying this chapter, you should be able to:

(1)   Name and draw carboxylic acids (20.1, 20.2, 20.12, 20.13, 20.15).

(2)   Calculate dissociation constants, pH, $\Delta G°$ and percent dissociation for carboxylic acids (20.3, 20.5, 20.22, 20.23).

(3)   Explain the physical properties of carboxylic acids (20.4, 20.14).

(4)   Predict the effects of substituents on carboxylic acid acidity (20.6, 20.7, 20.8, 20.9 20.16, 20.17, 20.18, 20.24, 20.37, 20.39).

(5)   Prepare carboxylic acids (20.10, 20.11, 20.20, 20.21, 20.26, 20.28, 20.30, 20.31, 20.32, 20.33, 20.35, 20.38, 20.43).

(6)   Recognize the types of reactions carboxylic acids undergo, and predict the products of these reactions (20.21, 20.25, 20.34).

(7)   Use spectroscopic techniques to identify carboxylic acids (20.27, 20.29, 20.36, 20.40, 20.41 20.42).

21.1

a)

$$(CH_3)_2CHCH_2CH_2\overset{\overset{\displaystyle O}{\|}}{C}Cl$$

4–Methylpentanoyl chloride

b)

Cyclohexylacetamide

c) $CH_3CH_2CH(CH_3)CN$

2–Methylbutanenitrile

d)

Benzoic anhydride

e)

Isopropyl
cyclopentanecarboxylate

f)

Cyclopentyl
2–methylpropanoate

g)

$$H_2C=CHCH_2CH_2\overset{\overset{\displaystyle O}{\|}}{C}NH_2$$

4–Pentenamide

h)

$$\overset{\displaystyle CN}{\underset{\displaystyle |}{}}$$
$$CH_3CH_2CHCH_2CH_3$$

2–Ethylbutanenitrile

i)

2,3–Dimethyl–2–butenoyl
chloride

j)

$$CF_3\overset{\overset{\displaystyle O}{\|}}{C}O\overset{\overset{\displaystyle O}{\|}}{C}CF_3$$

Bis(trifluoroacetic)anhydride

21.2

a) $CH_3CH_2CH=CHCN$ Correct name: 2–Pentenenitrile

The nitrile carbon is at position 1.

b) $CH_3CH_2CH_2CONHCH_3$ Correct name: *N*–Methylbutanamide

You must specify that the methyl group is bonded to nitrogen.

c) CH_3
                        |
     $(CH_3)_2CHCH_2CHCOCl$ Correct name: 2,4–Dimethylpentanoyl chloride

The prefix "di" must be put before "methyl".

d)

Correct name: Methyl 1–methylcyclohexanecarboxylate

The methyl group on the cyclohexane ring is at position 1. "Cyclohexanecarboxylate" is one word.

21.3

Most reactive ⎯⎯⎯⎯⎯⎯⎯⎯⎯⎯⎯⎯⎯⎯⎯⎯⎯→ Least reactive

a)

$$CH_3\overset{O}{\overset{\|}{C}}Cl \quad > \quad CH_3\overset{O}{\overset{\|}{C}}OCH_3 \quad > \quad CH_3\overset{O}{\overset{\|}{C}}NH_2$$

b)

$$CH_3\overset{O}{\overset{\|}{C}}OCH(CF_3)_2 \quad > \quad CH_3\overset{O}{\overset{\|}{C}}OCH_2CCl_3 \quad > \quad CH_3\overset{O}{\overset{\|}{C}}OCH_3$$

The most reactive acyl derivatives contain strongly electron-withdrawing groups in the alkyl portion of the structure.

21.4

The strongly electron-withdrawing trifluoromethyl group makes the carbonyl carbon more electron-poor and more reactive toward nucleophiles than the methyl acetate carbonyl group. Methyl trifluoroacetate is thus more reactive than methyl acetate in nucleophilic acyl substitution reactions.

21.5

a)

$$CH_3\overset{O}{\overset{\|}{C}}-OH \; + \; H-OCH_2CH_2CH_2CH_3 \; \underset{}{\overset{HCl}{\rightleftharpoons}} \; CH_3\overset{O}{\overset{\|}{C}}-OCH_2CH_2CH_2CH_3 \; + \; H_2O$$

Acetic acid   Butanol   Butyl acetate

b)

$$CH_3CH_2CH_2\overset{O}{\overset{\|}{C}}-OH \; + \; H-OCH_3 \; \underset{}{\overset{HCl}{\rightleftharpoons}} \; CH_3CH_2CH_2\overset{O}{\overset{\|}{C}}-OCH_3 \; + \; H_2O$$

Butanoic acid   Methanol   Methyl butanoate

21.6

5–Hydroxypentanoic acid   a lactone

21.7

a) 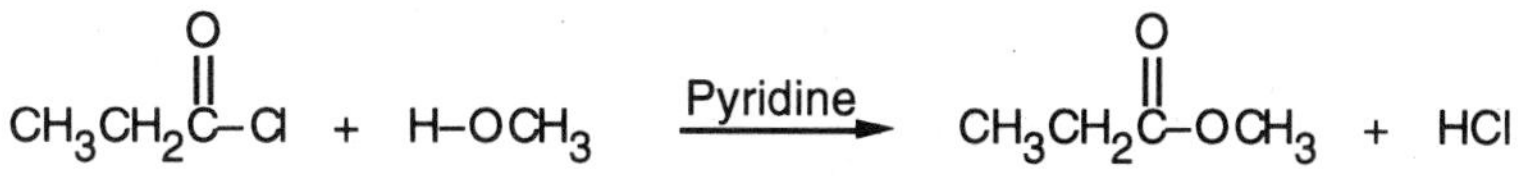
CH3CH2C(=O)-Cl + H-OCH3
Pyridine
CH3CH2C(=O)-OCH3 + HCl

Methyl propanoate

b) 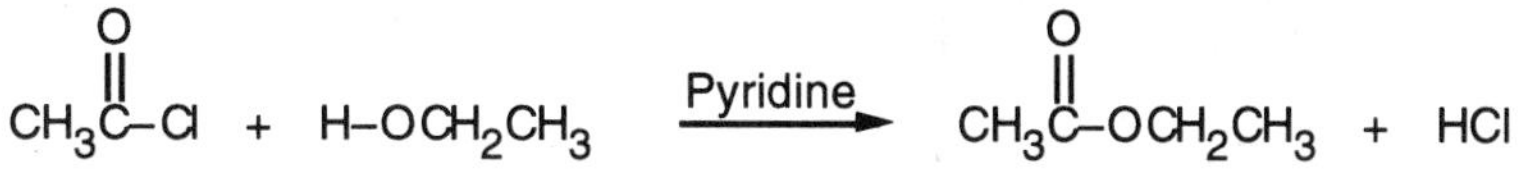
CH3C(=O)-Cl + H-OCH2CH3
Pyridine
CH3C(=O)-OCH2CH3 + HCl

Ethyl acetate

c) 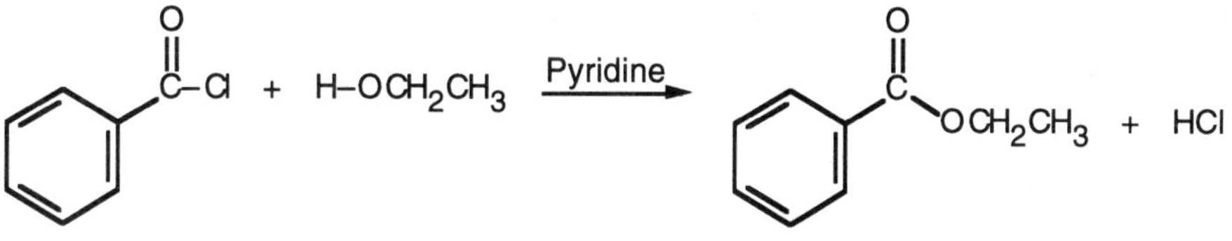
C-Cl + H-OCH2CH3
Pyridine
OCH2CH3 + HCl

Ethyl benzoate

21.8 Cyclohexanol is a secondary alcohol, which for steric reasons is less reactive in the Fischer esterification reaction. Thus, reaction of cyclohexanol with benzoyl chloride is the preferred method for preparing cyclohexyl benzoate.

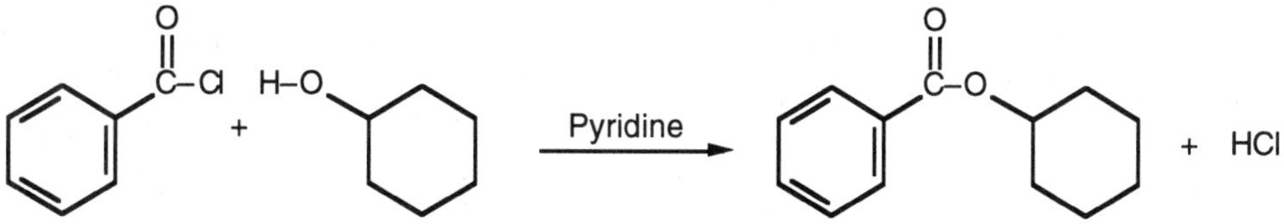
C-Cl H-O
+
Pyridine
C-O
+ HCl

21.9

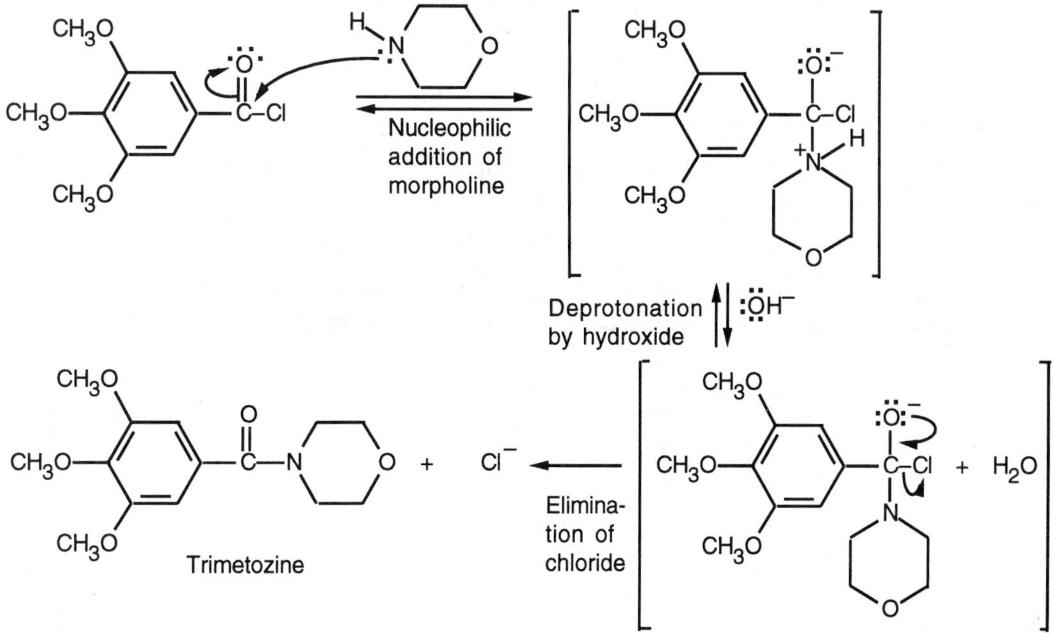
CH3O
CH3O
CH3O
C-Cl
Nucleophilic
addition of
morpholine
C-Cl
H
Deprotonation
by hydroxide
:OH⁻
Elimination of chloride
Cl⁻
+ H2O
Trimetozine

21.10 For primary and secondary amines:

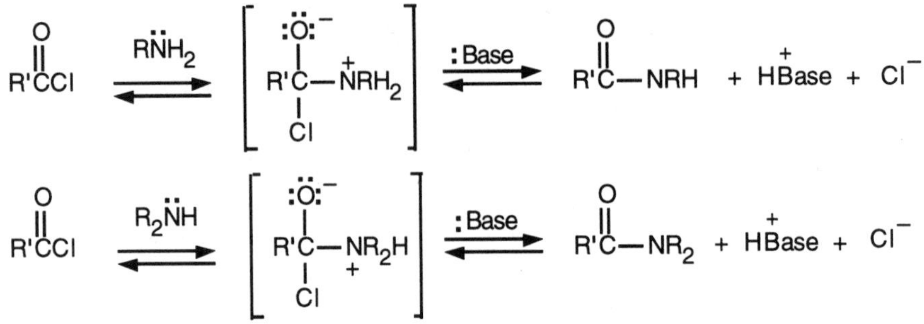

For triethylamine:

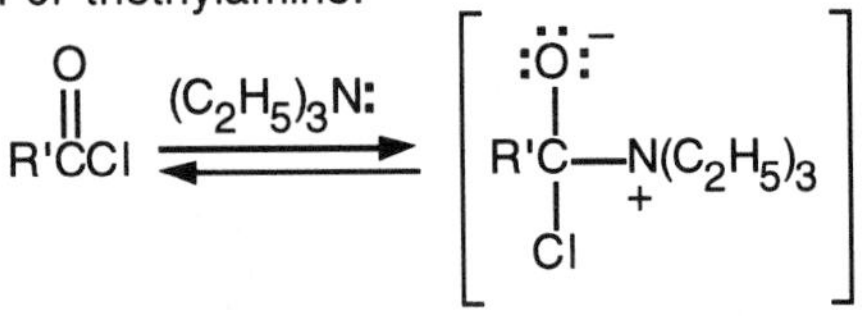

Triethylamine, like other amines, is a base whose lone pair of electrons can scavenge HCl. Unlike primary and secondary amines, triethylamine does not form an amide. There is no proton on the nitrogen atom, and so the tetrahedral adduct formed by nucleophilic addition of triethylamine to an acid chloride reverts to starting material instead of forming an amide.

21.11

a)

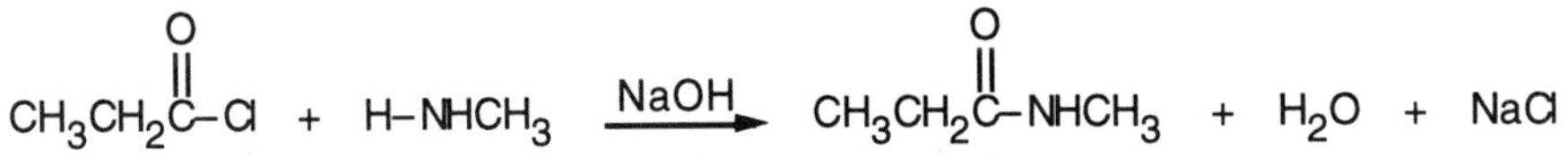

N–Methylpropanamide

b)

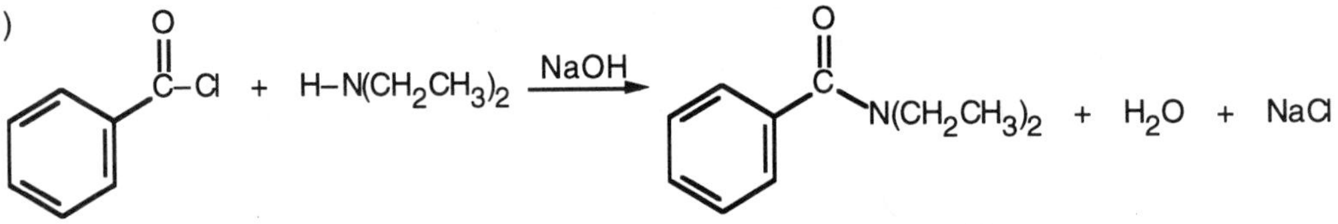

N,N–Dimethylbenzamide

c)

Propanamide

21.12

a)

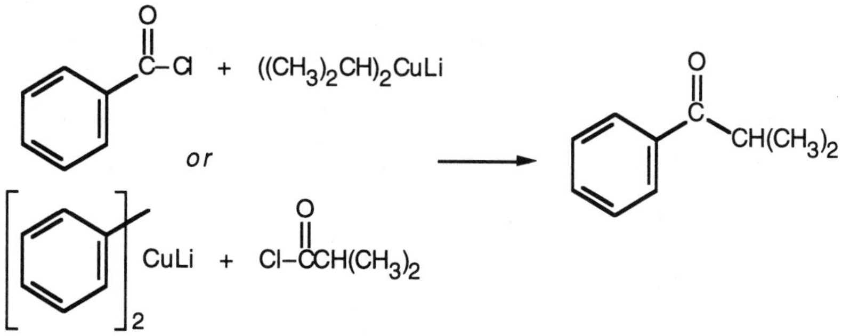

or

b)

or

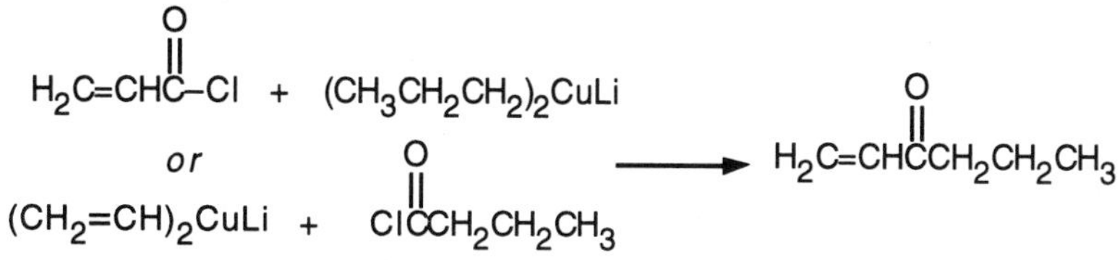

21.13 In the slow addition of Grignard reagent to a solution of an acid chloride, each drop of Grignard reagent is surrounded by a large amount of acid chloride. The Grignard reagent is consumed immediately in converting the reactive acid chloride to less reactive ketone, and no Grignard reagent remains to react with the ketone to form a tertiary alcohol. If the acid chloride were slowly added to a solution of Grignard reagent, however, excess reagent would convert the initially formed ketone into a tertiary alcohol.

21.14

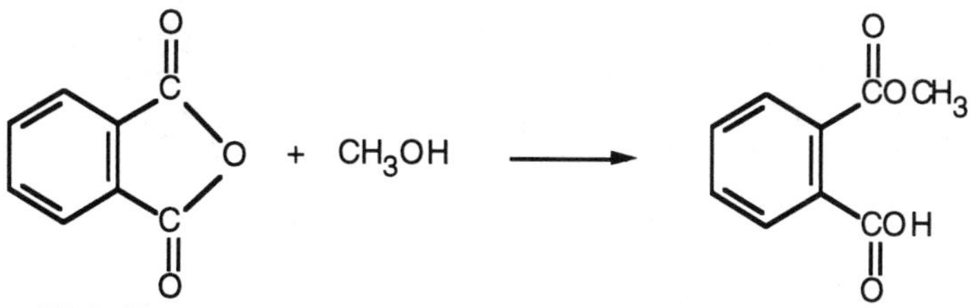

Phthalic
anhydride

The second half of the anhydride becomes a carboxylic acid.

21.15

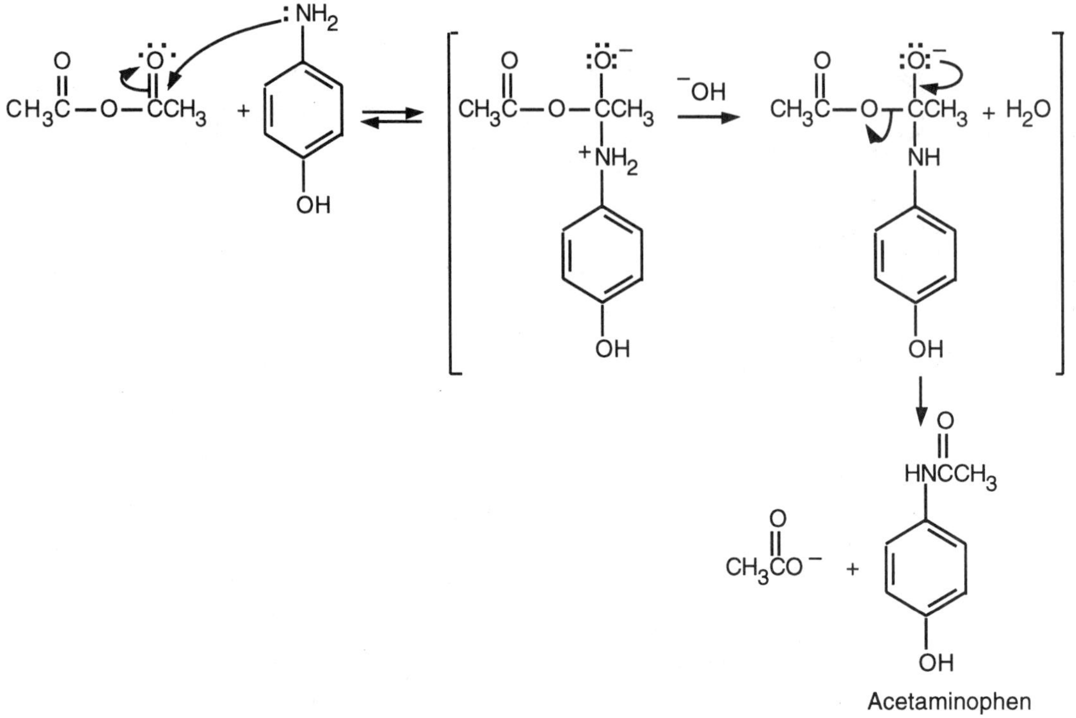
Acetaminophen

21.16 One equivalent of base must be added to the reaction mixture when an amine reacts with an anhydride. This base removes a proton from nitrogen after the formation of the initial tetrahedral intermediate. If base were not added, the amine starting material would serve as the base, and the reaction would stop when half of the starting amine was converted to amide. The rest of the amine would be protonated and would no longer be nucleophilic.

21.17

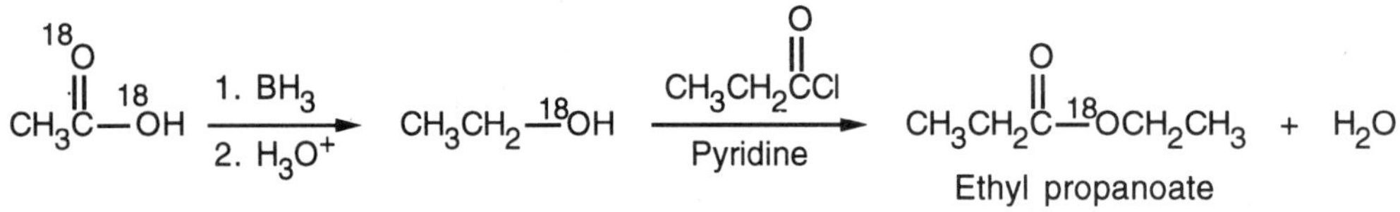
1. BH_3
2. H_3O^+
Pyridine
Ethyl propanoate

21.18 Acidic hydrolysis of an ester is a reversible reaction because the products are an alcohol and a carboxylic acid. Basic hydrolysis of an ester is irreversible because the products are an alcohol and a carboxylate anion, which has a negative charge and is not attacked by nucleophiles.

21.19

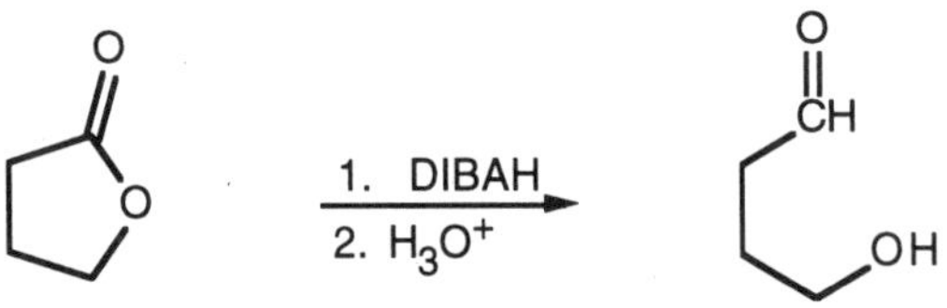
1. DIBAH
2. H_3O^+

21.20

a)

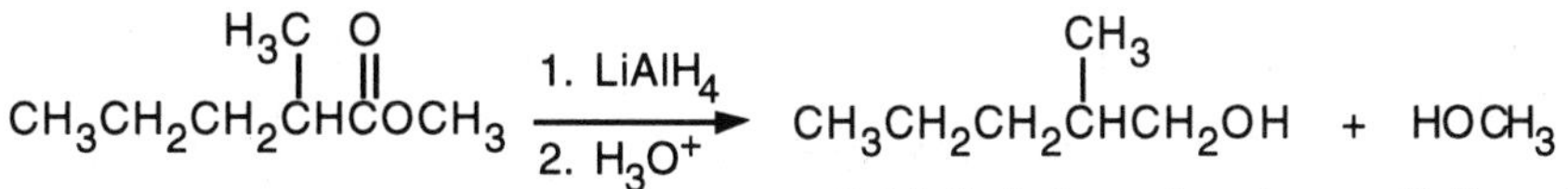

                           2-Methyl-1-pentanol   Methanol

b)

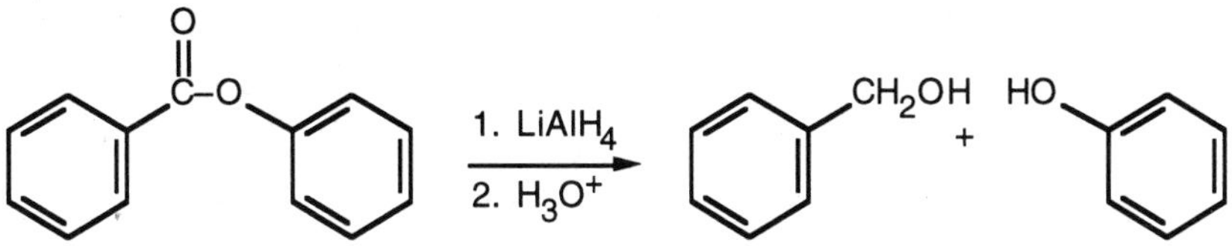

                                Benzyl   Phenol
                                alcohol

21.21

Ester	Grignard Reagent $\longrightarrow$	Tertiary alcohol

a)

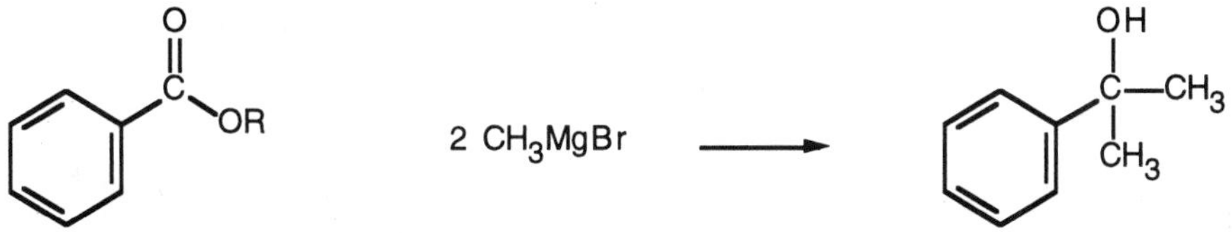

                                              2-Phenyl-2-propanol

b)

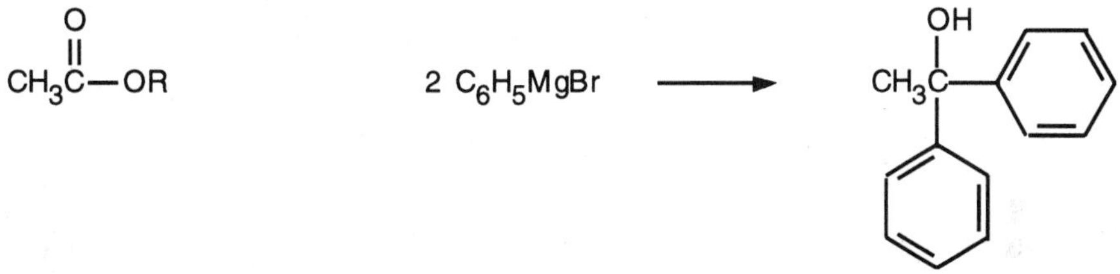

                                              1,1-Diphenylethanol

c)

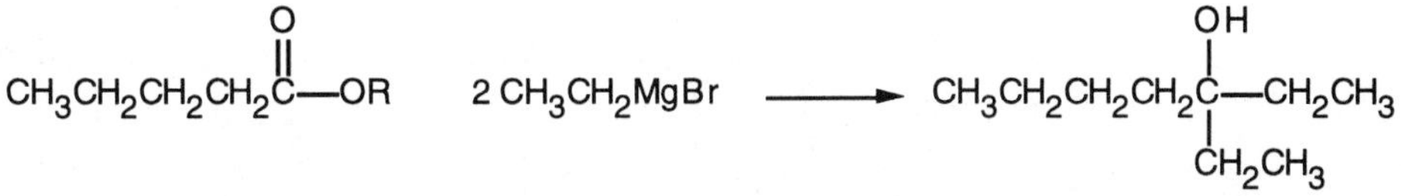

                                              3-Ethyl-3-heptanol

21.22

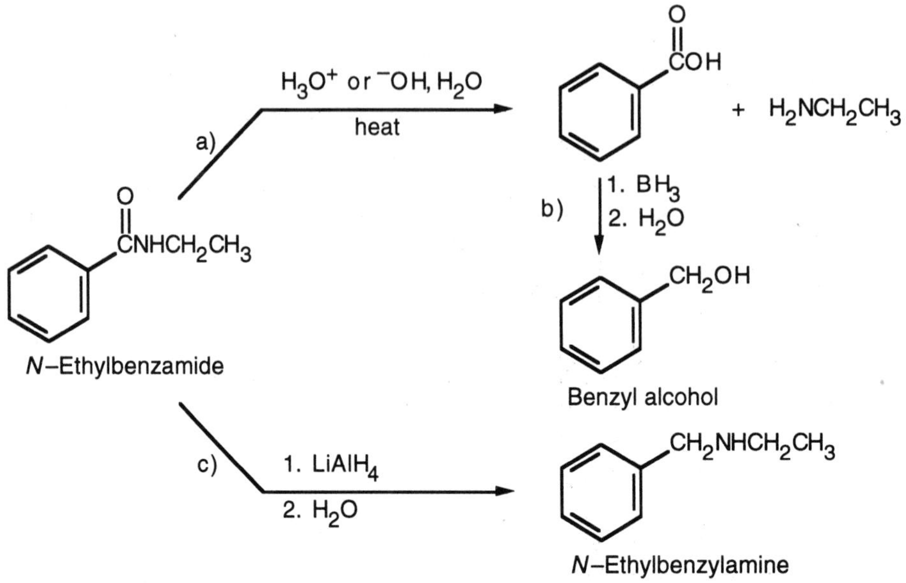

21.23

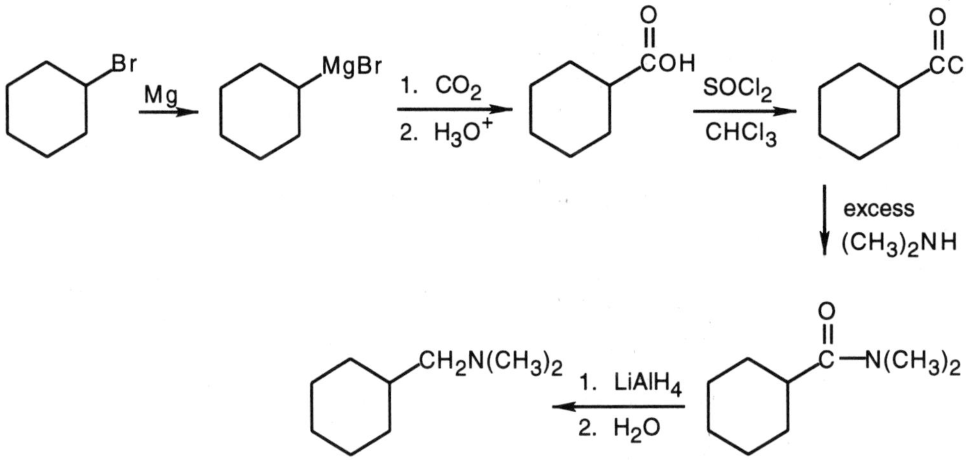

21.24

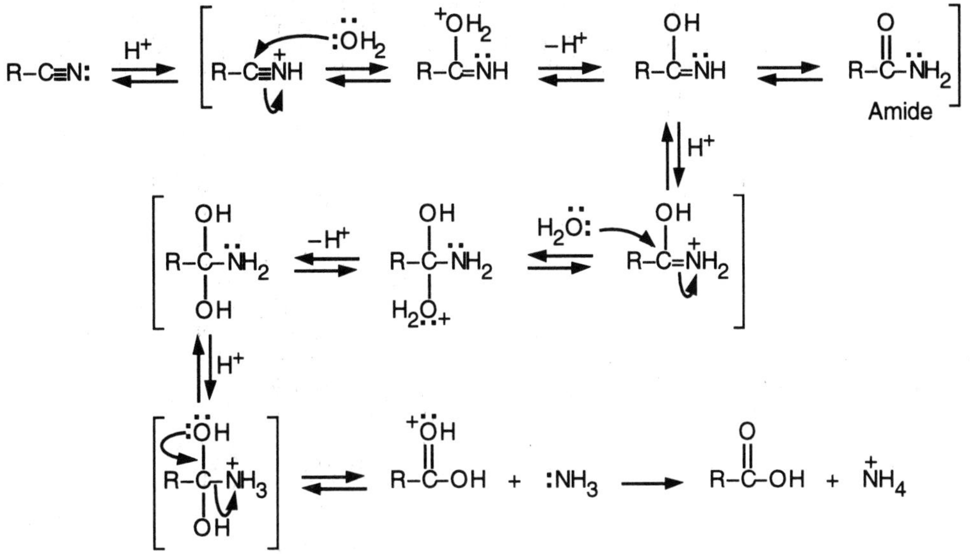

An amide is an intermediate in the acidic hydrolysis of nitrile to a carboxylic acid.

21.25

a) 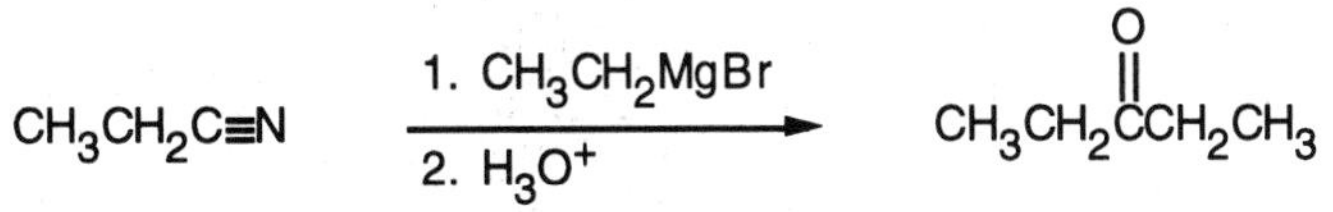 $CH_3CH_2C\equiv N$ $\xrightarrow[\text{2. } H_3O^+]{\text{1. } CH_3CH_2MgBr}$ $CH_3CH_2\overset{\overset{\displaystyle O}{\|}}{C}CH_2CH_3$

b) $CH_3CH_2C\equiv N$ $\xrightarrow[\text{2. } H_3O^+]{\text{1. } (CH_3)_2CHMgBr}$

or

$(CH_3)_2CHC\equiv N$ $\xrightarrow[\text{2. } H_3O^+]{\text{1. } CH_3CH_2MgBr}$

$(CH_3)_2CH\overset{\overset{\displaystyle O}{\|}}{C}CH_2CH_3$

c) $(CH_3)_2CHC\equiv N$ $\xrightarrow[\text{2. } H_2O]{\text{1. DIBAH}}$ $(CH_3)_2CH\overset{\overset{\displaystyle O}{\|}}{C}H$

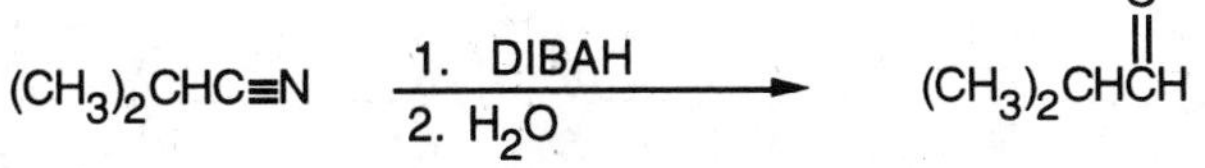

d) 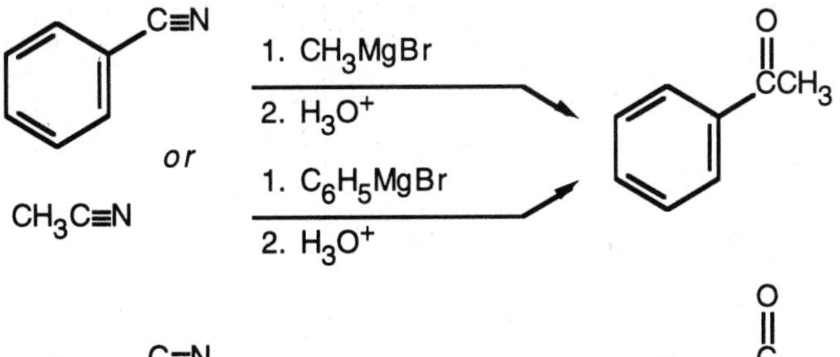 $\xrightarrow[\text{2. } H_3O^+]{\text{1. } CH_3MgBr}$

or

$CH_3C\equiv N$ $\xrightarrow[\text{2. } H_3O^+]{\text{1. } C_6H_5MgBr}$

e) 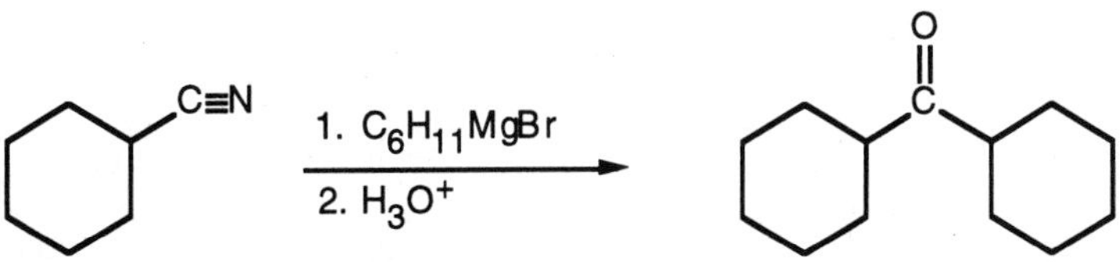 $\xrightarrow[\text{2. } H_3O^+]{\text{1. } C_6H_{11}MgBr}$

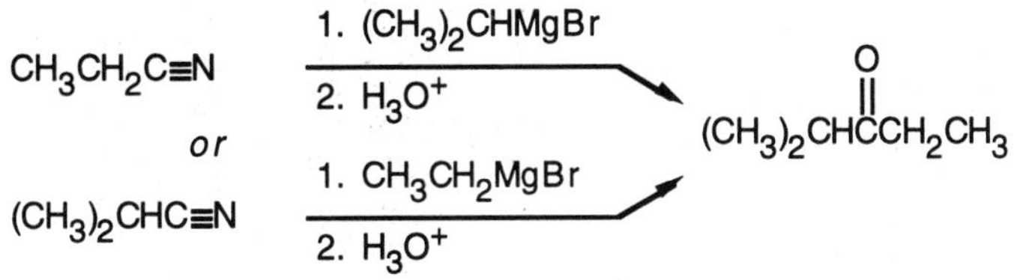

21.26

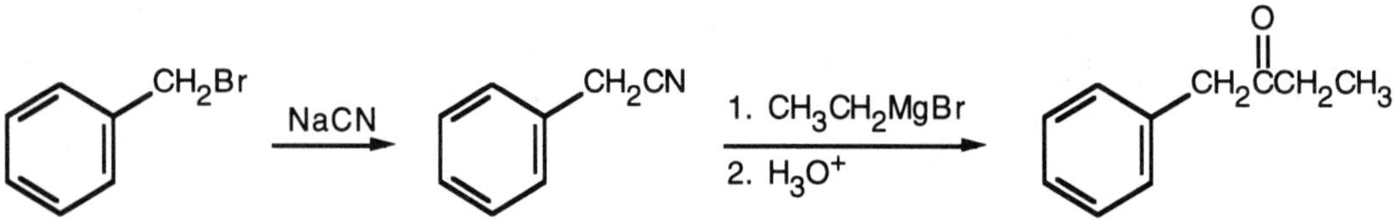

1-Phenyl-2-butanone

21.27

a) $BrCH_2CH_2CH_2Br + HOCH_2CH_2CH_2OH \xrightarrow{Base} -\!\!\{CH_2CH_2CH_2OCH_2CH_2CH_2O\}\!\!-$

b) $HOCH_2CH_2OH + \overset{O}{\overset{\|}{HOC}}(CH_2)_6\overset{O}{\overset{\|}{COH}} \xrightarrow{H_2SO_4}$

$-\!\!\{OCH_2CH_2O\overset{O}{\overset{\|}{C}}(CH_2)_6\overset{O}{\overset{\|}{C}}OCH_2CH_2O\overset{O}{\overset{\|}{C}}(CH_2)_6\overset{O}{\overset{\|}{C}}\}\!\!-$

c) $H_2N(CH_2)_6NH_2 + \overset{O}{\overset{\|}{ClC}}(CH_2)_4\overset{O}{\overset{\|}{CCl}} \longrightarrow$

$-\!\!\{NH(CH_2)_6NH\overset{O}{\overset{\|}{C}}(CH_2)_4\overset{O}{\overset{\|}{C}}NH(CH_2)_6NH\overset{O}{\overset{\|}{C}}(CH_2)_4\overset{O}{\overset{\|}{C}}\}\!\!-$

21.28

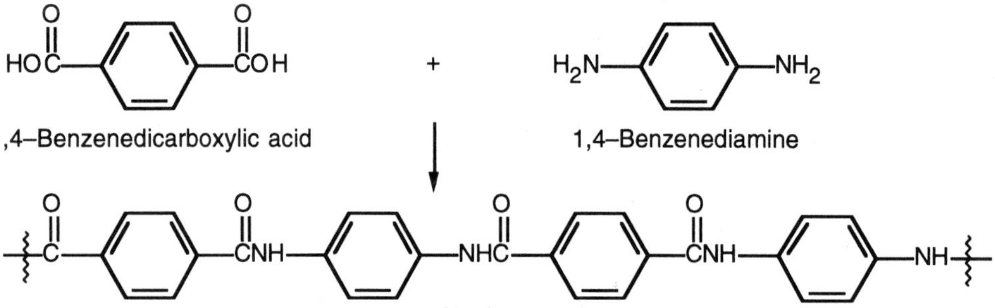

1,4-Benzenedicarboxylic acid   +   1,4-Benzenediamine

Kevlar

21.29

The product of the reaction of dimethyl terephthalate with glycerol has a high degree of cross-linking and is more rigid than Dacron.

21.30

Absorption	Functional group present
a) 1735 cm^{-1}	Aliphatic ester *or* 6-membered ring lactone
b) 1810 cm^{-1}	Aliphatic acid chloride
c) $2500 - 3300 \text{ cm}^{-1}$ and 1710 cm^{-1}	Carboxylic acid
d) 2250 cm^{-1}	Aliphatic nitrile
e) 1715 cm^{-1}	Aliphatic ketone *or* 6-membered ring ketone

21.31 To solve this type of problem:
1. Use the IR absorption to determine the functional group(s) present.
2. Draw the functional group.
3. Use the remaining atoms to complete the structure.

a) 1. IR 2250 cm^{-1} corresponds to a nitrile.
2. $-C \equiv N$
3. CH_3CH_2CN is the structure of the compound.

b) 1. IR 1735 cm^{-1} corresponds to an aliphatic ester.
2.
$$-\overset{\overset{\displaystyle O}{\|}}{C}-O-$$
3. The remaining five carbons and twelve hydrogens can be arranged in a number of ways to produce a satisfactory structure for this compound. For example:

$$CH_3CH_2CH_2\overset{\overset{\displaystyle O}{\|}}{C}OCH_2CH_3 \qquad or \qquad CH_3\overset{\overset{\displaystyle O}{\|}}{C}OCH_2CH_2CH_2CH_3$$

The structural formula indicates that this compound can't be a lactone.

c) $CH_3\overset{\overset{O}{\|}}{C}N(CH_3)_2$

d) $CH_3CH=CH\overset{\overset{O}{\|}}{C}Cl$, $H_2C=C(CH_3)\overset{\overset{O}{\|}}{C}Cl$

21.32 a) *p*–Methylbenzamide
c) Dimethyl succinate *or*
    Dimethyl butanedioate
e) Phenyl benzoate
g) 3,5–Dibromobenzoyl chloride

b) 4–Ethyl–2–hexenenitrile
d) Isopropyl 3–phenylpropanoate
f) *N*–Methyl–3–bromobutanamide
h) 1–Cyclopentenecarbonitrile

21.33

a)

Br—⬡—$CH_2\overset{\overset{O}{\|}}{C}NH_2$

p–Bromophenylacetamide

b)

m–Benzoylbenzonitrile

c)

$CH_3CH_2CH_2CH_2C(CH_3)_2\overset{\overset{O}{\|}}{C}NH_2$

2,2–Dimethylhexanamide

d)

Cyclohexyl cyclohexane-
carboxylate

e)

2–Cyclobutenecarbonitrile

f)

$\overset{\overset{\displaystyle COCl}{|}}{CH_3CH_2CH_2CHCH_2COCl}$

2-Propylbutanedioyl
dichloride

21.34 Many structures can be drawn for each part of this problem.

a)

Cyclopentanecarbonyl
chloride

(*E*)–2–Methyl–2–pentenoyl
chloride

$H_2C=CHCH\overset{\overset{O}{\|}}{C}Cl$
       $|$
     CH_2CH_3

2-Ethyl–3–butenoyl
chloride

b)

1-Cyclohexene-
carboxamide

$CH_3CH_2CH_2C{\equiv}CCH_2\overset{\overset{\displaystyle O}{\|}}{C}NH_2$

3-Heptynamide

$H_2C{=}CHCH{=}CH\overset{\overset{\displaystyle O}{\|}}{C}N(CH_3)_2$

N,N-Dimethyl-2,4-
pentadienamide

c)

Cyclobutanecarbonitrile

$CH_3CH{=}CHCH_2C{\equiv}N$

3-Pentenenitrile

$H_2C{=}\overset{\overset{\displaystyle CH_3}{|}}{C}CH_2C{\equiv}N$

3-Methyl-3-butenenitrile

21.35 The reactivity of esters in saponification reactions is influenced by steric factors. Branching in both the acyl and alkyl portions of an ester hinders attack of the hydroxide nucleophile. This effect is less pronounced in the alkyl portion of the ester than in the acyl portion because alkyl branching is one atom farther away from the site of attack.

Most reactive $\longrightarrow$ Least reactive

$$CH_3\overset{\overset{\displaystyle O}{\|}}{C}OCH_3 > CH_3\overset{\overset{\displaystyle O}{\|}}{C}OCH_2CH_3 > CH_3\overset{\overset{\displaystyle O}{\|}}{C}OCH(CH_3)_2 > CH_3\overset{\overset{\displaystyle O}{\|}}{C}OC(CH_3)_3$$

21.36

2,4,6-Trimethylbenzoic acid

2,4,6-Trimethylbenzoic acid has two methyl groups ortho to the carboxylic acid functional group. These bulky methyl groups block the approach of the alcohol and prevent esterification from occurring under Fischer esterification conditions. Another possible route to the methyl ester:

$\dfrac{1. \ NaOH}{2. \ CH_3I}$

This route succeeds because reaction occurs farther away from the site of steric hindrance.

21.37

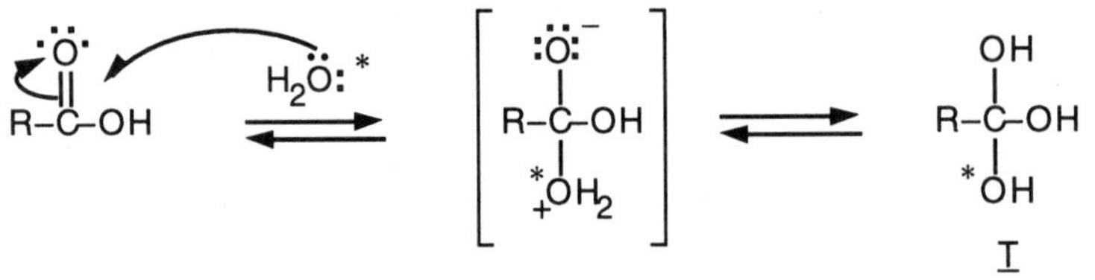

The tetrahedral intermediate T can eliminate any one of the three –OH groups to reform either the original carboxylic acid or labeled carboxylic acid. Further reaction of water with mono-labeled carboxylic acid leads to the doubly labeled product.

21.38

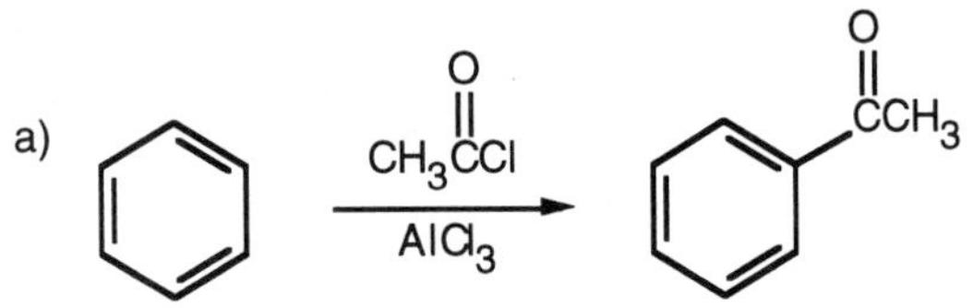

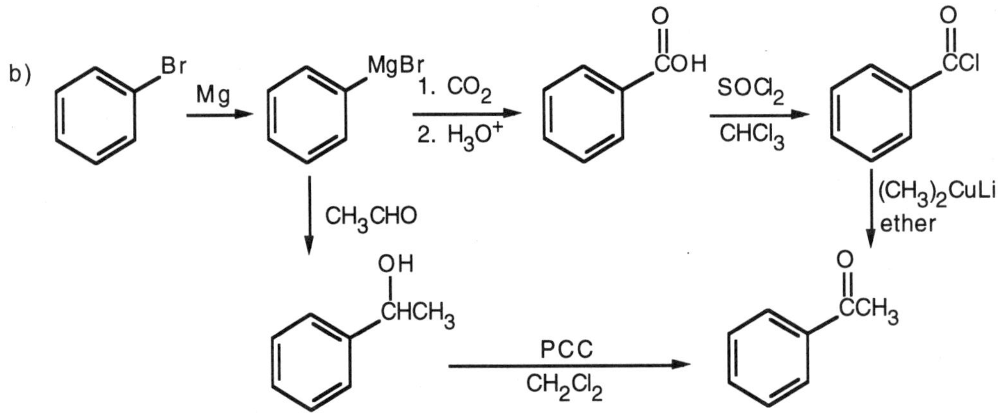

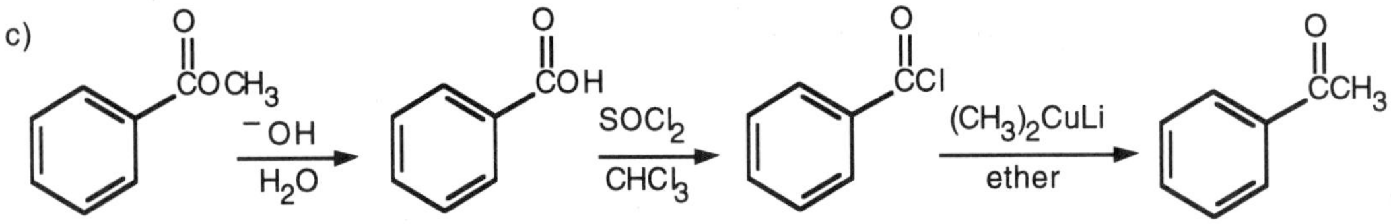

a)

b)

c)

Reaction of an ester with Grignard reagent produces a tertiary alcohol, not a ketone.

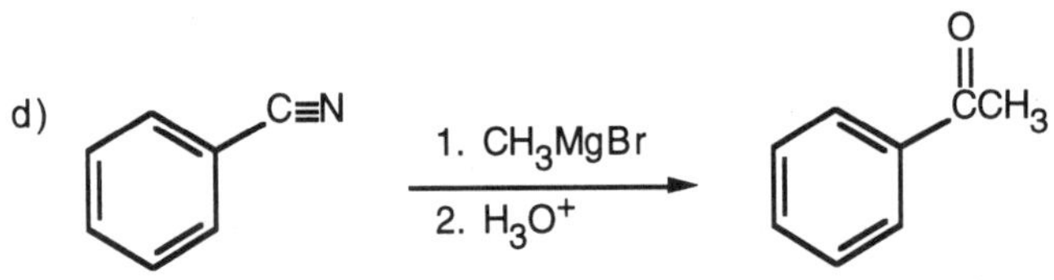

d)

e)

21.39 A negatively charged tetrahedral intermediate is formed when the nucleophile ⁻OH attacks the carbonyl carbon of an ester. An electron-withdrawing substituent can stabilize this negatively charged tetrahedral intermediate and increase the rate of reaction. (Contrast this effect with substituent effects in electrophilic aromatic substitution, in which positive charge developed in the intermediate is stabilized by electron-*donating* substituents.) Substituents that are deactivating in electrophilic aromatic substitution are activating in ester hydrolysis, as the observed reactivity order shows. The substituents –CN and –CHO are electron-withdrawing; –NH₂ is strongly electron-donating.

Most reactive ─────────────────────────────────────► Least reactive

$$X = -NO_2 > -CN > -CHO > -Br > -H > -CH_3 > -OCH_3 > -NH_2$$

21.40

a) $CH_3CH_2CH_2COOH$ $\xrightarrow[\text{2. } H_3O^+]{\text{1. } BH_3}$ $CH_3CH_2CH_2CH_2OH$

b) $CH_3CH_2CH_2CH_2OH$ $\xrightarrow{PCC}$ $CH_3CH_2CH_2CHO$

from (a)   *or*

$CH_3CH_2CH_2COOH$ $\xrightarrow[\text{3. } H_3O^+]{\begin{array}{l}\text{1. } SOCl_2\\\text{2. } LiAlH[OC(CH_3)_3]_3\end{array}}$ $CH_3CH_2CH_2CHO$

c) $CH_3CH_2CH_2CH_2OH$ $\xrightarrow{PBr_3}$ $CH_3CH_2CH_2CH_2Br$

from (a)

d) $CH_3CH_2CH_2CH_2Br$ $\xrightarrow{NaCN}$ $CH_3CH_2CH_2CH_2C\equiv N$

from (c)

e) $CH_3CH_2CH_2CH_2Br$ $\xrightarrow[\text{Ethanol}]{KOH}$ $CH_3CH_2CH=CH_2$

from (c)

f) $CH_3CH_2CH_2CH_2C\equiv N$ $\xrightarrow{H_3O^+}$ $CH_3CH_2CH_2CH_2\overset{\overset{\displaystyle O}{\|}}{C}OH$ $\xrightarrow[\text{2. } CH_3NH_2]{\text{1. } SOCl_2}$

from d)

$CH_3CH_2CH_2CH_2\overset{\overset{\displaystyle O}{\|}}{C}NHCH_3$

g) $CH_3CH_2CH_2CH_2C\equiv N$ $\xrightarrow[\text{2. } H_3O^+]{\text{1. } CH_3MgBr}$ $CH_3CH_2CH_2CH_2\overset{\overset{\displaystyle O}{\|}}{C}CH_3$

from (d)

h)

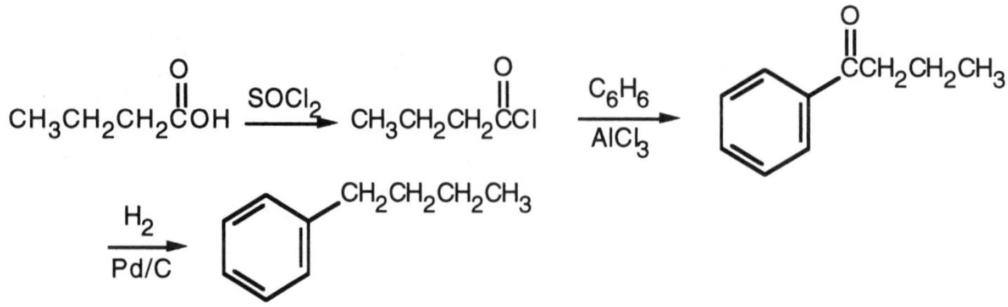

21.41 Dimethyl carbonate is a diester. Use your knowledge of the Grignard reaction to work your way through this problem.

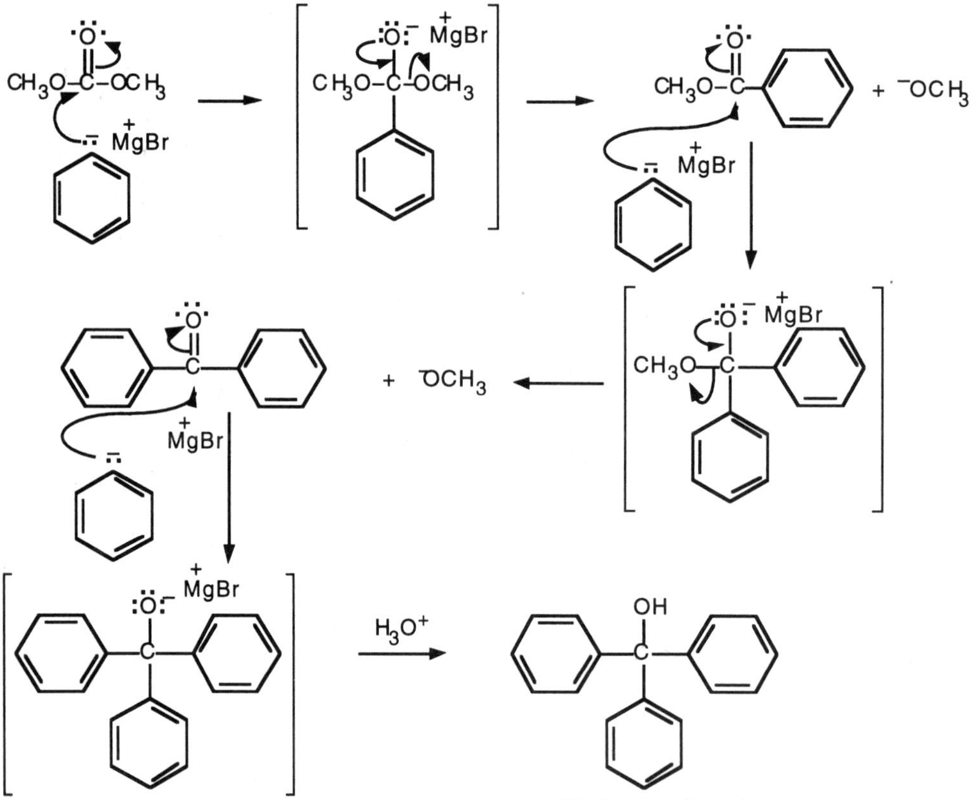

Triphenylmethanol

21.42

$$:O: \quad \quad \overset{+}{O}H \quad \quad OH \quad \quad OH$$

$$R-\overset{\|}{C}-OCH_2CH_3 \; \xrightleftharpoons{H^+} \; \left[R-\overset{\|}{\underset{}{C}}-OCH_2CH_3 \rightleftharpoons R-\overset{|}{\underset{\underset{+}{HOCH_3}}{C}}-OCH_2CH_3 \xrightleftharpoons{-H^+} R-\overset{|}{\underset{OCH_3}{C}}-\ddot{O}CH_2CH_3 \right]$$

$$\ddot{H}\ddot{O}CH_3$$

$$\Big\updownarrow H^+$$

$$O \quad \quad \overset{+}{O}H \quad \quad H\ddot{O}\!\!:\!\!H$$

$$R-\overset{\|}{C}-OCH_3 \; \xrightleftharpoons{-H^+} \; \left[R-\overset{\|}{C}-OCH_3 \rightleftharpoons R-\overset{|}{\underset{OCH_3}{C}}-OCH_2CH_3 \right]$$

$$+ \quad HOCH_2CH_3$$

In acidic methanol, the ethyl ester can react by a nucleophilic acyl substitution reaction to yield a methyl ester. The equilibrium favors the methyl ester because of the large excess of methanol present.

21.43

$$\underset{O \quad \quad O}{HO\overset{\|}{C}(CH_2)_6\overset{\|}{C}OH} + H_2N-\!\!\!\!\bigcirc\!\!\!\!-CH_2-\!\!\!\!\bigcirc\!\!\!\!-NH_2$$

$$\downarrow$$

$$-(\!\!\overset{O}{\overset{\|}{C}}(CH_2)_6\overset{O}{\overset{\|}{C}}-NH-\!\!\!\!\bigcirc\!\!\!\!-CH_2-\!\!\!\!\bigcirc\!\!\!\!-NH\!\!)-$$

Qiana

21.44

$$H\ddot{O}\!\!:\!\!^- \quad \longrightarrow \quad HO\overset{O}{\overset{\|}{C}}CH_2CH_2\ddot{O}\!\!:^- \quad \longrightarrow \quad HO\overset{O}{\overset{\|}{C}}CH_2CH_2O\overset{O}{\overset{\|}{C}}CH_2CH_2\ddot{O}\!\!:^-$$

repeat
many
times

$$\downarrow$$

$$-(\!\!\overset{O}{\overset{\|}{C}}CH_2CH_2O\overset{O}{\overset{\|}{C}}CH_2CH_2O\!\!)-$$

This polymer is a polyester.

21.45

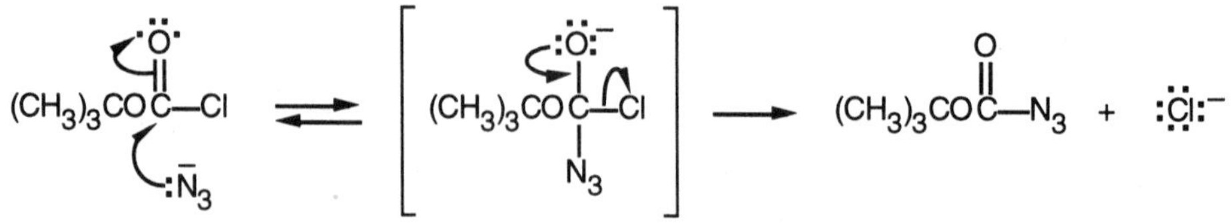

21.46

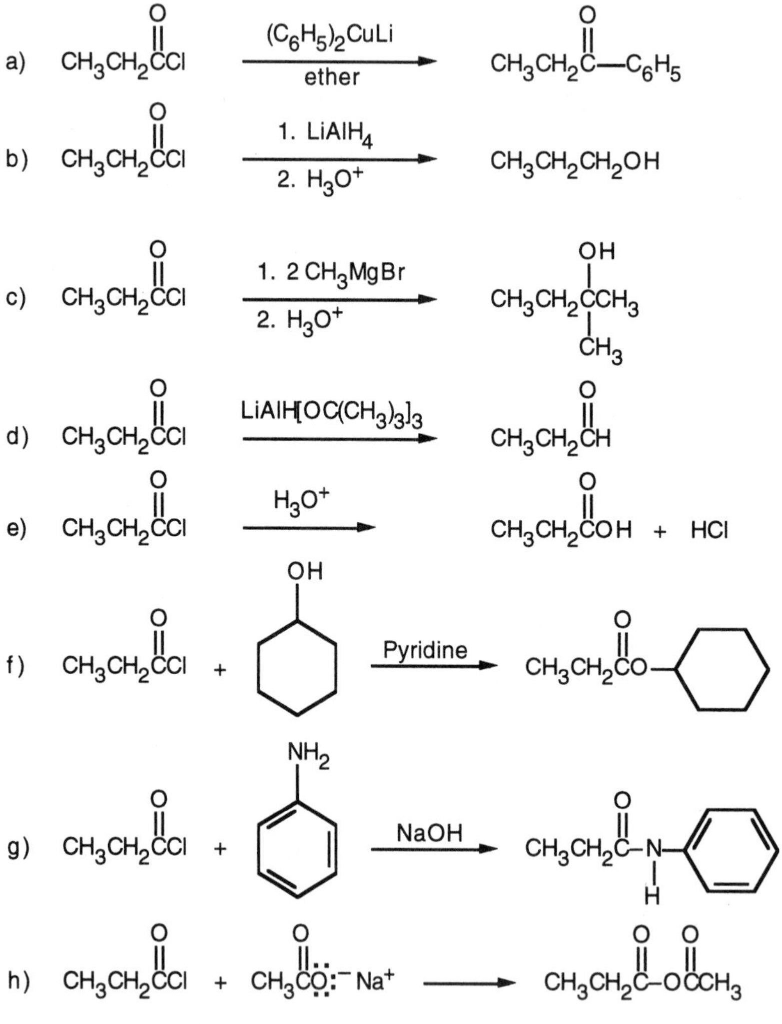

a)

b)

c)

d)

e)

f)

g)

h)

21.47 The reagents in parts a, d, f, and h do not react with methyl propanoate.

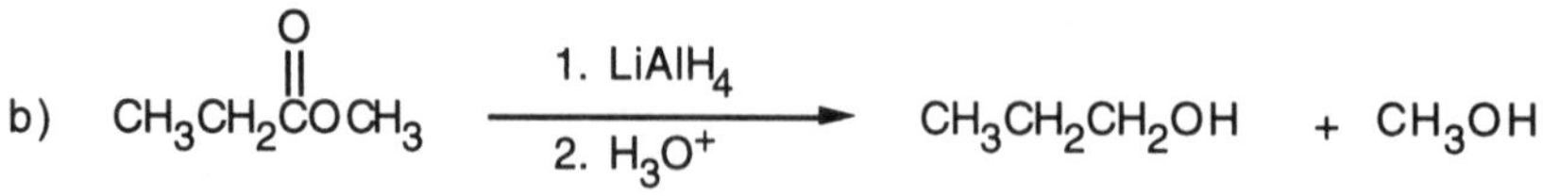

b)

c) $CH_3CH_2\overset{\overset{O}{\|}}{C}OCH_3$ $\xrightarrow[\text{2. } H_3O^+]{\text{1. 2 } CH_3MgBr}$ $CH_3CH_2\overset{\overset{OH}{|}}{\underset{\overset{|}{CH_3}}{C}}CH_3$

e) $CH_3CH_2\overset{\overset{O}{\|}}{C}OCH_3$ $\xrightarrow{H_3O^+}$ $CH_3CH_2\overset{\overset{O}{\|}}{C}OH$ + CH_3OH

g) $CH_3CH_2\overset{\overset{O}{\|}}{C}OCH_3$ + (aniline) $\xrightarrow{NaOH}$ $CH_3CH_2\overset{\overset{O}{\|}}{C}-\underset{\overset{|}{H}}{N}-$(phenyl)

21.48 The reagents in parts a, d, f, g, and h do not react with propanamide.

b) $CH_3CH_2\overset{\overset{O}{\|}}{C}NH_2$ $\xrightarrow[\text{2. } H_2O]{\text{1. } LiAlH_4}$ $CH_3CH_2CH_2NH_2$

c) $CH_3CH_2\overset{\overset{O}{\|}}{C}NH_2$ $\xrightarrow[\text{2. } H_3O^+]{\text{1. } CH_3MgBr}$ $CH_3CH_2\overset{\overset{O}{\|}}{C}\overset{..}{N}HMgBr$ + CH_4

e) $CH_3CH_2\overset{\overset{O}{\|}}{C}NH_2$ $\xrightarrow{H_3O^+}$ $CH_3CH_2\overset{\overset{O}{\|}}{C}OH$ + NH_4^+

The reagent in parts a, d, f, g, and h do not react with propanenitrile.

b) $CH_3CH_2C\equiv N$ $\xrightarrow[\text{2. } H_2O]{\text{1. } LiAlH_4}$ $CH_3CH_2CH_2NH_2$

c) $CH_3CH_2C\equiv N$ $\xrightarrow[\text{2. } H_3O^+]{\text{1. } CH_3MgBr}$ $CH_3CH_2\overset{\overset{O}{\|}}{C}CH_3$

e) $CH_3CH_2C\equiv N$ $\xrightarrow{H_3O^+}$ $CH_3CH_2\overset{\overset{O}{\|}}{C}OH$ + $^+NH_4$

21.49

a)

$CF_3\overset{\overset{O}{\|}}{C}O-\overset{\overset{..\overset{..}{O}..}{\|}}{C}CF_3$ $\rightleftharpoons$ $\left[CF_3\overset{\overset{O}{\|}}{C}O-\overset{\overset{:\overset{..}{O}:^-}{|}}{\underset{\underset{\overset{\|}{O}}{RC-\overset{..}{O}H}}{C}}CF_3 \right.$ $\xrightarrow{-H^+}$ $CF_3\overset{\overset{O}{\|}}{C}O\!\!-\!\!\overset{\overset{:\overset{..}{O}:^-}{|}}{\underset{\underset{\overset{\|}{O}}{RC\overset{..}{O}}}{C}}CF_3 \left. \right]$ $\longrightarrow$ $RCO\overset{\overset{O}{\|}}{C}CF_3$

+ $CF_3\overset{\overset{O}{\|}}{C}OH$

with $RC\overset{..}{O}H$ ($\overset{\|}{\underset{..}{O}}$) attacking.

b) The electron-withdrawing fluorine atoms polarize the nearby carbonyl group and stabilize a tetrahedral intermediate, making the mixed anhydride more reactive than other anhydrides.

c) Because trifluoroacetate is a better leaving group than other carboxylate anions, the reaction proceeds as indicated.

21.50

a) $CH_3CH_2CH_2CH_2C\equiv N$ $\xrightarrow[\text{2. H}_2\text{O}]{\text{1. LiAlH}_4}$ $CH_3CH_2CH_2CH_2CH_2NH_2$

b) $CH_3CH_2CH_2CH_2C\equiv N$ $\xrightarrow{\text{H}_3\text{O}^+}$ $CH_3CH_2CH_2CH_2\overset{\overset{\displaystyle O}{\|}}{C}OH$

$\downarrow$ $\begin{array}{l}\text{SOCl}_2\\\text{CHCl}_3\end{array}$

$CH_3CH_2CH_2CH_2\overset{\overset{\displaystyle O}{\|}}{C}N(CH_3)_2$ $\xleftarrow{\text{2 HN(CH}_3)_2}$ $CH_3CH_2CH_2CH_2\overset{\overset{\displaystyle O}{\|}}{C}Cl$

$\downarrow$ $\begin{array}{l}\text{1. LiAlH}_4\\\text{2. H}_2\text{O}\end{array}$

$CH_3CH_2CH_2CH_2CH_2N(CH_3)_2$

c) $CH_3CH_2CH_2CH_2C\equiv N$ $\xrightarrow[\text{2. H}_3\text{O}^+]{\text{1. CH}_3\text{MgBr}}$ $CH_3CH_2CH_2CH_2\overset{\overset{\displaystyle O}{\|}}{C}CH_3$

$\downarrow$ $\begin{array}{l}\text{1. CH}_3\text{MgBr}\\\text{2. H}_3\text{O}^+\end{array}$

$CH_3CH_2CH_2CH_2\overset{\overset{\displaystyle OH}{|}}{\underset{\underset{\displaystyle CH_3}{|}}{C}}CH_3$

d) $CH_3CH_2CH_2CH_2\overset{\overset{\displaystyle O}{\|}}{C}CH_3$ $\xrightarrow[\text{2. H}_3\text{O}^+]{\text{1. NaBH}_4}$ $CH_3CH_2CH_2CH_2\overset{\overset{\displaystyle OH}{|}}{\underset{\underset{\displaystyle H}{|}}{C}}CH_3$

(from part c)

e) $CH_3CH_2CH_2CH_2C\equiv N$ $\xrightarrow[\text{2. H}_2\text{O}]{\text{1. DIBAH}}$ $CH_3CH_2CH_2CH_2\overset{\overset{\displaystyle O}{\|}}{C}H$

21.51 First, convert cyclohexanol to bromocyclohexane by reaction with PBr₃. The first two
transformations start with bromocyclohexane.

a)

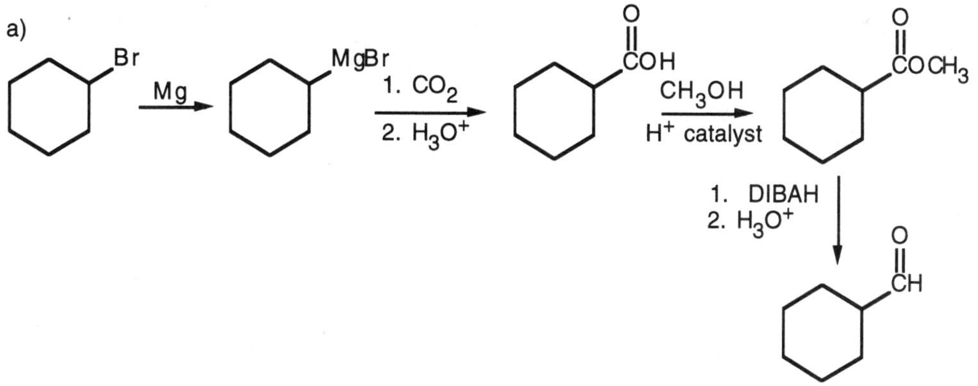

b)

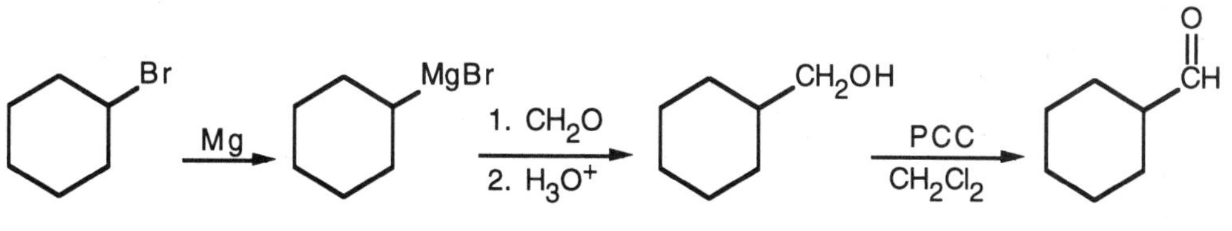

c)

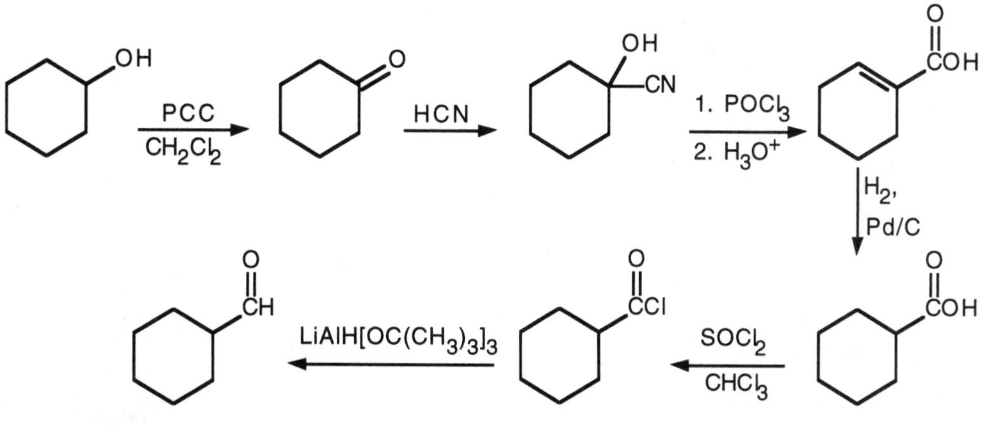

d)

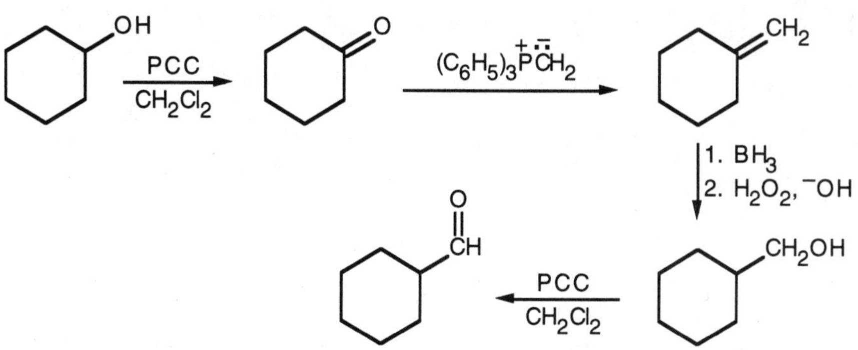

21.52

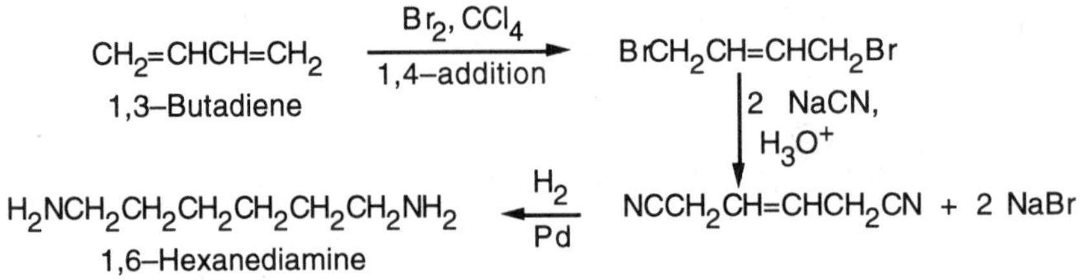

21.53

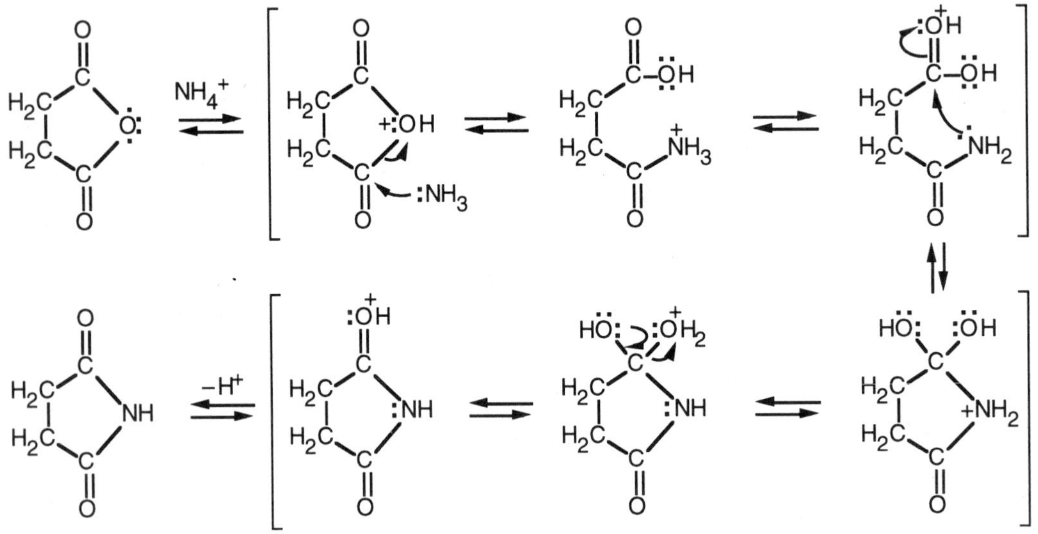

This reaction requires high temperatures because the intermediate amide is a poor nucleophile and the carboxylic acid carbonyl group is unreactive.

21.54 The polyimide pictured is a step-growth polymer of a benzene tetracarboxylic acid and an aromatic diamine.

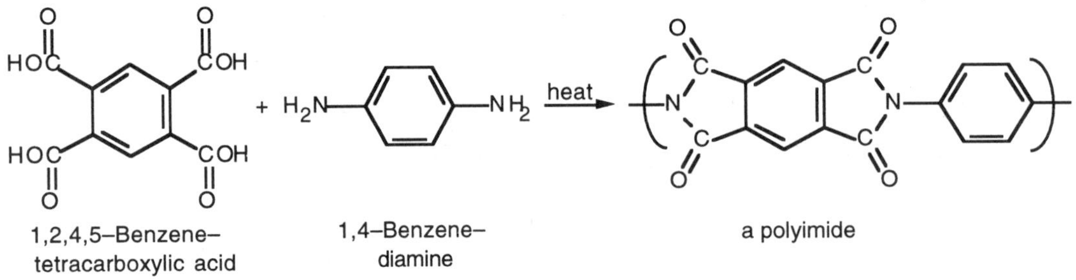

| 1,2,4,5–Benzene–tetracarboxylic acid | 1,4–Benzene–diamine | a polyimide |

21.55 This synthesis requires a nucleophilic aromatic substitution reaction, studied in Section 16.8.

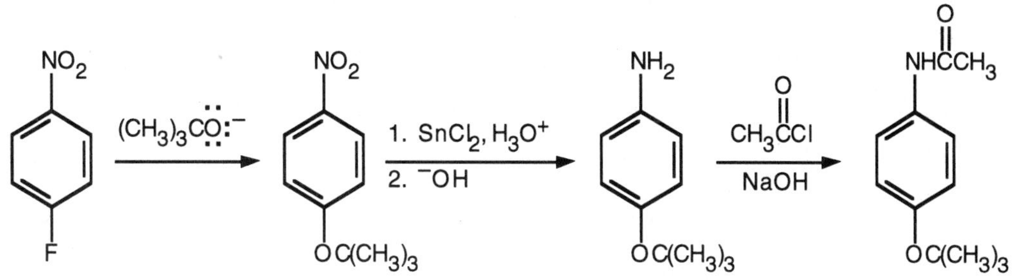

Butacetin

21.56

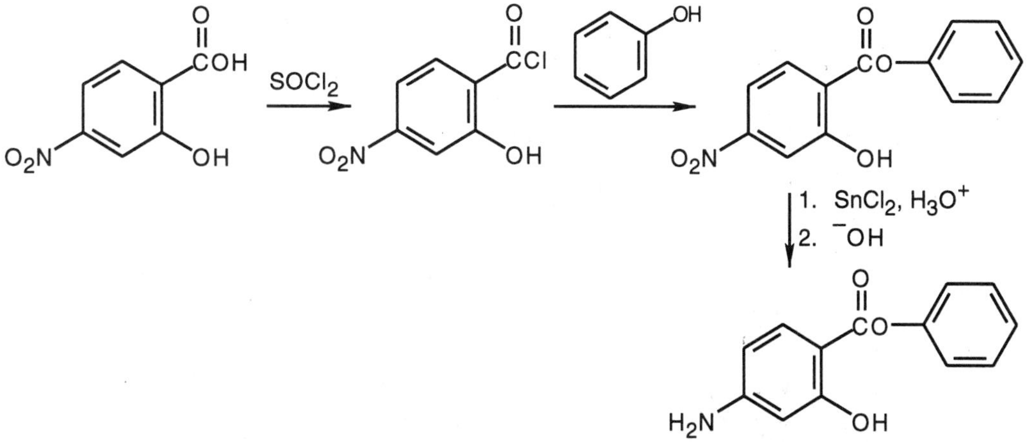

Phenyl 4–aminosalicylate

21.57

a)

	$CH_3CH_2\overset{O}{\overset{\|\|}{C}}NHCH_3$ *N*–Methylpropanamide	$CH_3\overset{O}{\overset{\|\|}{C}}N(CH_3)_2$ *N,N*–Dimethylacetamide
IR	1680 cm^{-1} (*N*–substituted amide)	1650 cm^{-1} (*N,N*–disubstituted amide)
^1H NMR	1 methyl group 1 ethyl group	3 methyl groups

b)

	$HOCH_2CH_2CH_2CH_2C{\equiv}N$ 5–Hydroxypentanenitrile	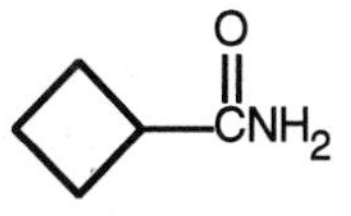 Cyclobutanecarboxamide
IR	3300–3400 cm^{-1} (hydroxyl) 2250 cm^{-1} (nitrile)	1690 cm^{-1} (amide)

c)

$ClCH_2CH_2CH_2\overset{\overset{\displaystyle O}{\|}}{C}OH$

4–Chlorobutanoic acid

$CH_3OCH_2CH_2\overset{\overset{\displaystyle O}{\|}}{C}Cl$

3–Methoxypropanoyl chloride

IR   1710 cm^{-1} (carboxylic acid)   1810 cm^{-1} (acid chloride)

d)

$CH_3CH_2\overset{\overset{\displaystyle O}{\|}}{C}OCH_2CH_3$

Ethyl propanoate

$CH_3\overset{\overset{\displaystyle O}{\|}}{C}OCH_2CH_2CH_3$

Propyl acetate

^{1}H NMR   2 triplets
          2 quartets

          1 singlet
          1 triplet
          1 quartet
          1 multiplet

21.58

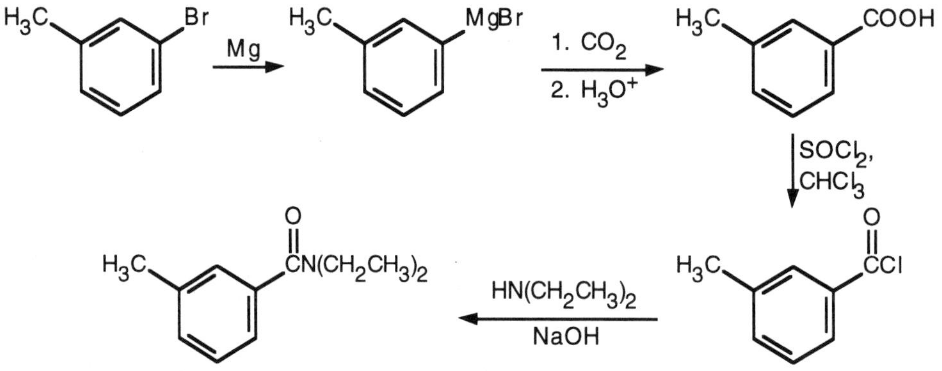

N,N–Dimethyl–m–toluamide

21.59

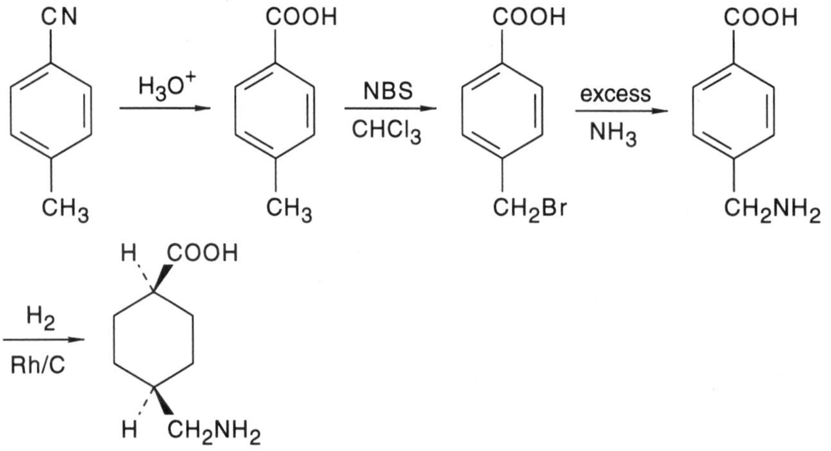

21.60

a)  $H_2\overset{..}{\overset{-}{C}}-\overset{+}{N}\equiv N:$ ⟷ $H_2C=\overset{+}{N}=\overset{..}{\overset{-}{N}}:$

(b) The carboxylic acid first protonates diazomethane:

$$RCOH + H_2\overset{..}{\overset{-}{C}}-\overset{+}{N}\equiv N \longrightarrow CH_3-\overset{+}{N}\equiv N + RCO^-$$

An S$_N$2 reaction then occurs in which the carboxylate ion displaces N$_2$ as leaving group:

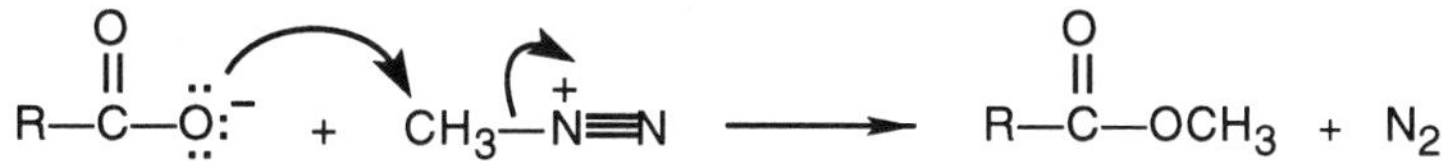

$$R-\overset{O}{\overset{||}{C}}-\overset{..}{\overset{..}{O}}:^- + CH_3-\overset{+}{N}\equiv N \longrightarrow R-\overset{O}{\overset{||}{C}}-OCH_3 + N_2$$

21.61

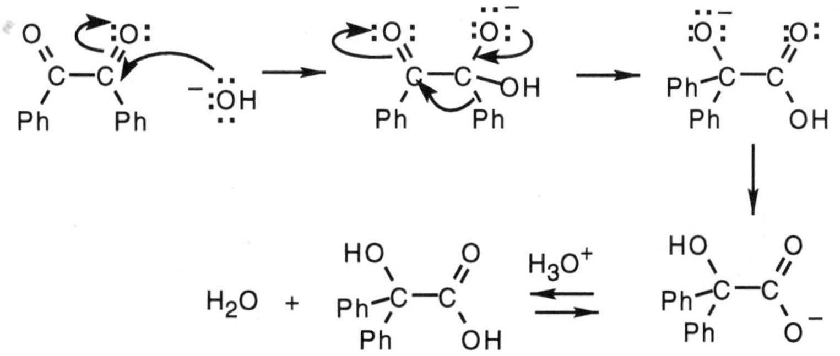

21.62

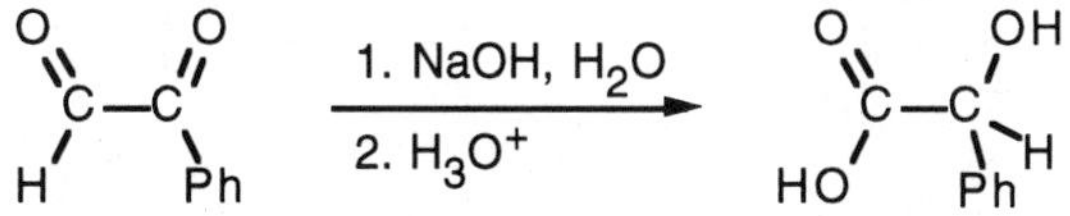

21.63

$$CH_3CHCl\overset{O}{\overset{||}{C}}OCH_3$$

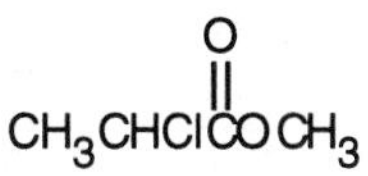

21.64

$$CH_3CH_2CH_2C\equiv N$$

21.65

a) 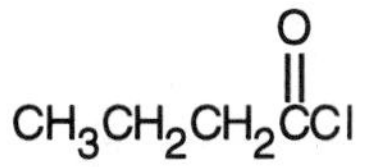
$$CH_3CH_2CH_2\overset{O}{\overset{||}{C}}Cl$$

b)
$$NCCH_2\overset{O}{\overset{||}{C}}OCH_2CH_3$$

c)
$$CH_3\overset{O}{\overset{||}{C}}OCH(CH_3)_2$$

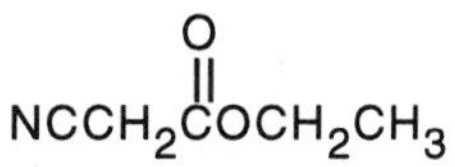

21.66

a)

$$ClCH_2CH_2\overset{O}{\overset{\|}{C}}OCH_2CH_3$$

b)

$$CH_3CH_2O\overset{O}{\overset{\|}{C}}CH_2\overset{O}{\overset{\|}{C}}OCH_2CH_3$$

c)

Ph—CH=CH$\overset{O}{\overset{\|}{C}}OCH_2CH_3$

21.67 Addition of the triamine causes formation of cross-links between prepolymer chains.

$$CH_2CH_2N CH_2\overset{OH}{\overset{|}{C}}HCH_2\left[O-\!\!\!\left\langle\;\right\rangle\!-\overset{CH_3}{\underset{CH_3}{\overset{|}{\underset{|}{C}}}}-\!\left\langle\;\right\rangle\!-OCH_2\overset{OH}{\overset{|}{C}}HCH_2\right]_n\!\!N-$$

with

—N
|
CH$_2$
|
CH$_2$
|
—N

$$CH_2\overset{OH}{\overset{|}{C}}HCH_2\left[O-\!\!\!\left\langle\;\right\rangle\!-\overset{CH_3}{\underset{CH_3}{\overset{|}{\underset{|}{C}}}}-\!\left\langle\;\right\rangle\!-OCH_2\overset{OH}{\overset{|}{C}}HCH_2\right]_n\!\!N-$$

21.68

$$R\!-\!\overset{\overset{\ddot{\cdot}\ddot{O}\cdot}{\|}}{C}\!-\!Cl_3 \;\rightleftharpoons\; \left[R\!-\!\overset{\overset{\ddot{\cdot}\ddot{O}\!:^-}{|}}{\underset{\underset{OH}{|}}{C}}\!-\!Cl_3\right] \;\longrightarrow\; R\!-\!\overset{O}{\overset{\|}{C}}\!-\!OH \;+\; {}^-\!\!:Cl_3 \;\longrightarrow\; R\!-\!\overset{O}{\overset{\|}{C}}\!-\!\ddot{O}\!:^- \;+\; HCl_3$$

with :ÖH$^-$

Study Guide for Chapter 21

After studying this chapter, you should be able to:

(1)   Draw and name carboxylic acid derivatives (21.1, 21.2, 21.32, 21.33, 21.34).

(2)   Explain the relative reactivity of carboxylic acid derivatives (21.3, 21,4, 21.35).

(3)   Prepare carboxylic acid derivatives (21.5, 21.6, 21.7, 21.8, 21.17).

(4)   Predict the products of reactions of:
(a) Acid halides (21.11, 21.12, 21.13, 21.46).
(b) Anhydrides (21.14, 21.15, 21.16).
(c) Esters and lactones (21.19, 21.20, 21.21, 21.47, 21.53).
(d) Amides (21.22, 21.23, 21.48).
(e) Nitriles (21.24, 21.25, 21.26, 21.48, 21.50).

(5)   Formulate nucleophilic acyl substitution mechanisms for carboxylic acid derivatives (21.9, 21.10, 21.18, 21.36, 21.37, 21.39, 21.41, 21.42, 21.45, 21.53, 21.49, 21.60, 21.68).

(6)   Use nucleophilic acyl substitution reactions in synthesis (21.38, 21.40, 21.55, 21.56, 21.58, 21.59).

(7)   Identify carboxylic acid derivatives by spectroscopic techniques (21.30, 21.31, 21.57, 21.63, 21.64, 21.65, 21.66).

(8)   Draw representative segments of step-growth polymers (21.27, 21.28, 21.29, 21.43, 21.44, 21.52, 21.54, 21.67)

22.1–22.2   Acidic hydrogens in the keto form of each of these compounds are underlined.

	Keto Form	Enol Form	Number of Acidic Hydrogens
a)			4
b)	H–C–CCl (H O)	H–C=C–Cl (H OH)	3
c)	H–C–COCH$_2$CH$_3$ (H O)	H–C=COCH$_2$CH$_3$ (H OH)	3
d)	CH$_3$–C–CH (H O)	CH$_3$–C=C–H (H OH)	2
e)	H–C–CO–H (H O)	H–C=C–OH (H OH)	4
f)		or	5
g)			3

22.3

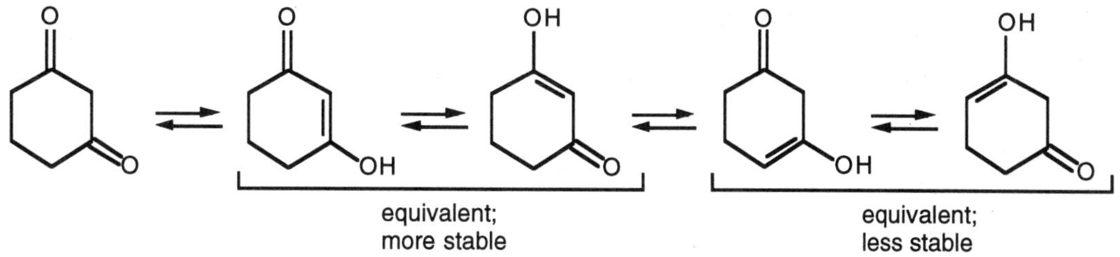

equivalent;
more stable

equivalent;
less stable

The first two mono-enols are more stable because the enol double bond is conjugated with the carbonyl group.

22.4

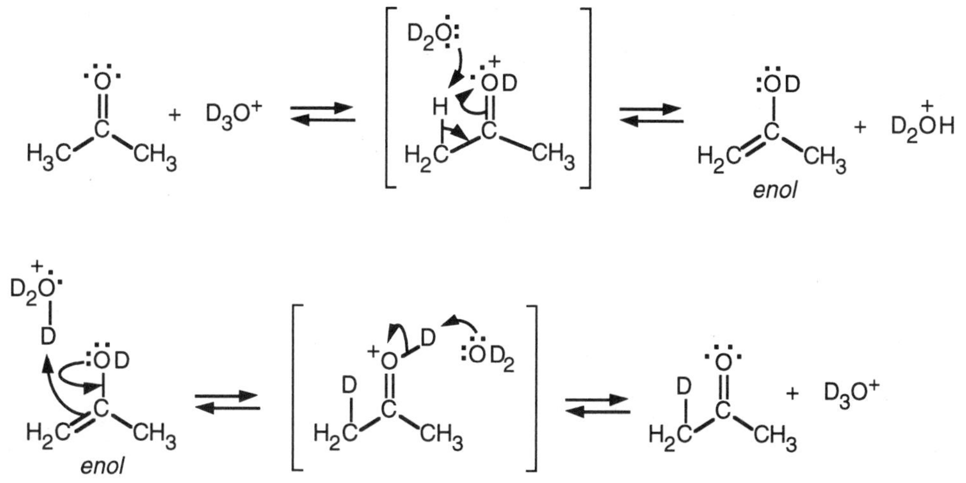

22.5

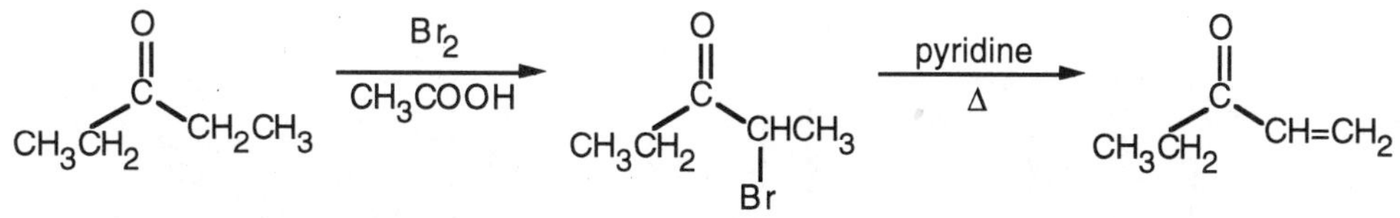

22.6 The Hell-Volhard-Zelinskii reaction involves formation of an intermediate acid bromide enol, so loss of stereochemical configuration occurs at the stereogenic center. Bromination can occur from either face of the enol double bond, producing racemic 2–bromo–2–phenylpropanoic acid.

22.7

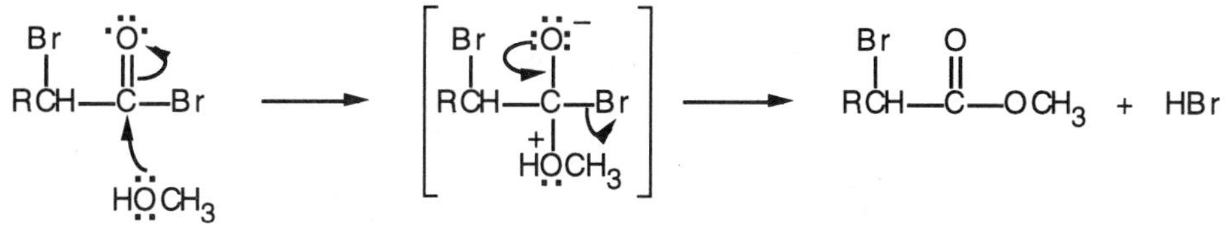

22.8

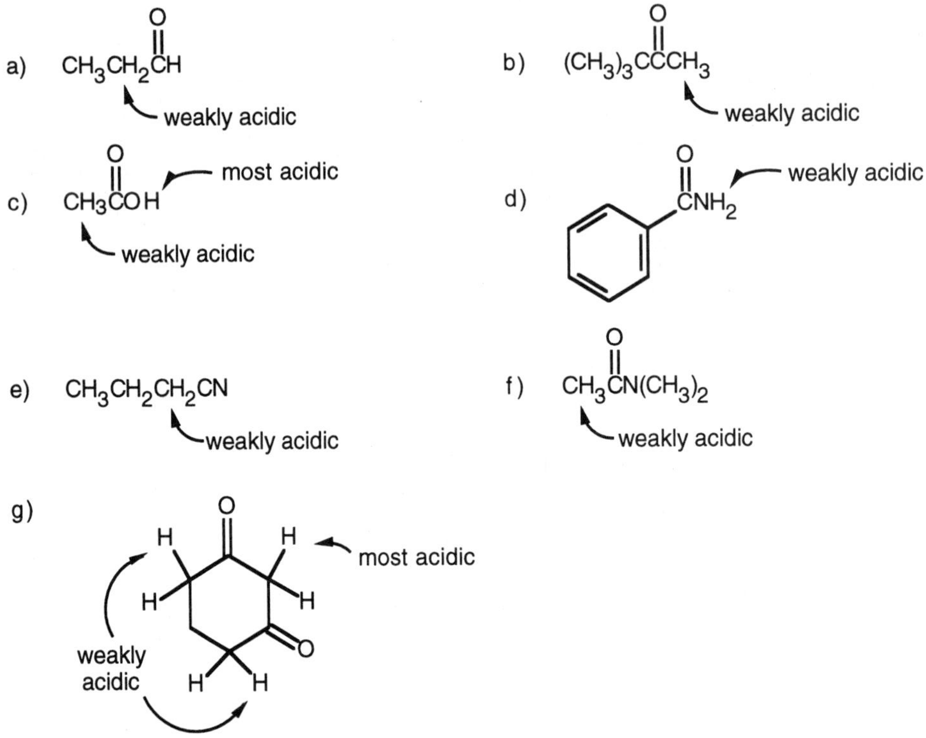

a)

    weakly acidic

b)

    weakly acidic

c)

    most acidic

    weakly acidic

d)

    weakly acidic

e) $CH_3CH_2CH_2CN$

    weakly acidic

f) $CH_3\overset{O}{\overset{\|}{C}}N(CH_3)_2$

    weakly acidic

g)

    most acidic

    weakly acidic

22.9

$$^-:CH_2C\equiv N: \longleftrightarrow H_2C=C=\ddot{N}:^-$$

22.10

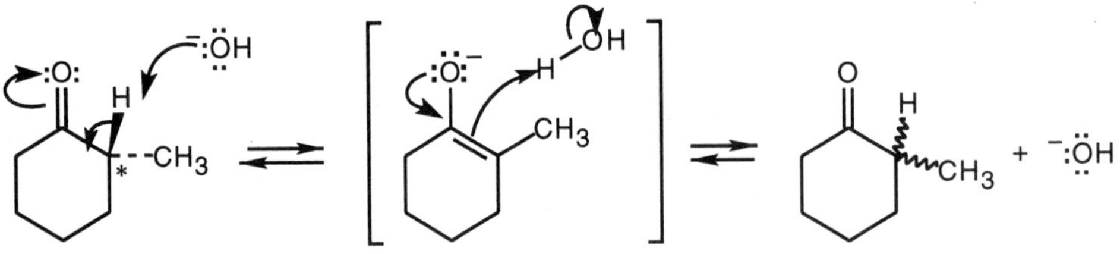

(R)–2–Methylcyclohexanone

Carbon 2 loses its chirality when the enolate ion double bond is formed. Protonation occurs with equal probability from either side of sp^2–hybridized carbon 2, resulting in racemic product.

22.11

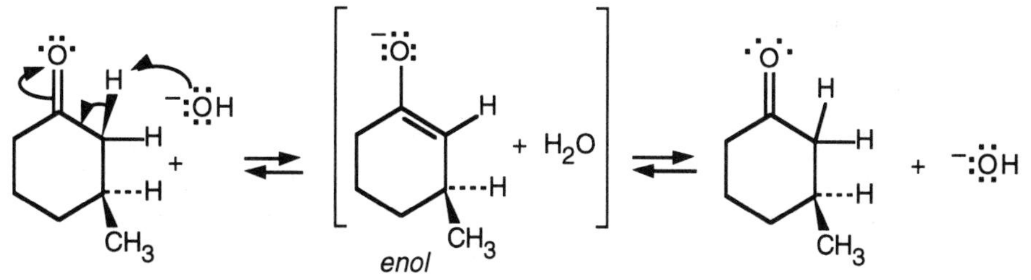

(S)–3–Methylcyclohexanone is not racemized by base because its stereogenic center is not involved in the enolization reaction.

22.12 Since both base-promoted chlorination and base-promoted bromination occur at the same rate, the step involving halogen must come after the slow, or rate-limiting, step. Formation of the enolate ion is the rate-limiting step and is dependent only on the concentrations of ketone and base.

$$\text{rate} = k \, [\text{base}] \, [\text{ketone}]$$

22.13 Halogenation in acid medium is acid-*catalyzed* because hydrogen ions are regenerated:

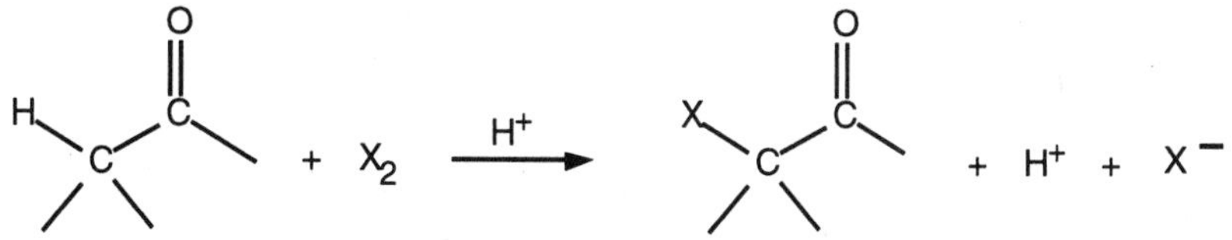

Halogenation in basic medium is base-*promoted* because a stoichiometric amount of base is consumed:

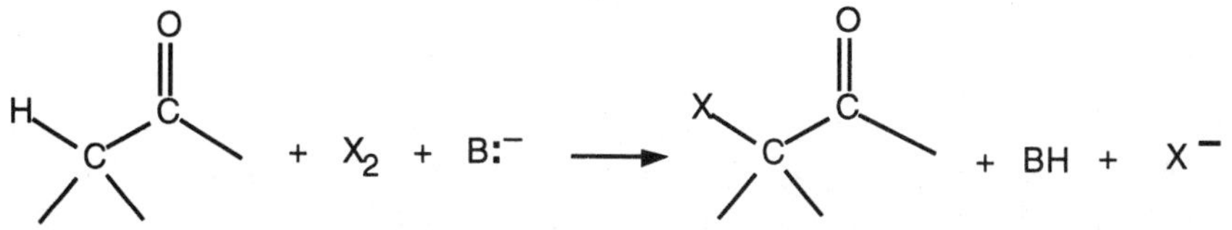

22.14

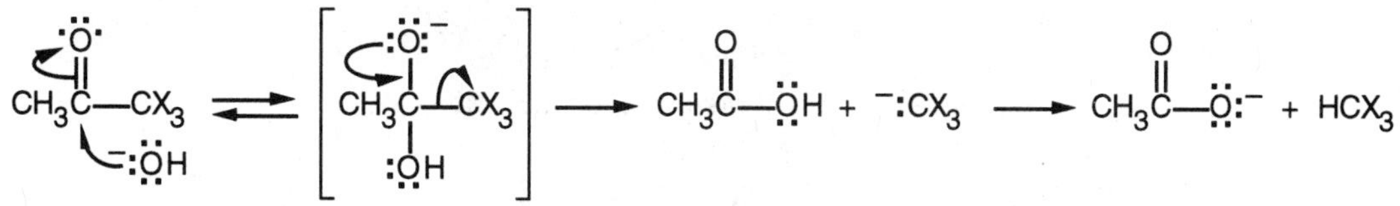

22.15 Since the malonic ester synthesis produces substituted acetic acids, look for the acetic acid component of the compound you want to synthesize. The remainder of the molecule comes from the alkyl halide, which should be primary or methyl.

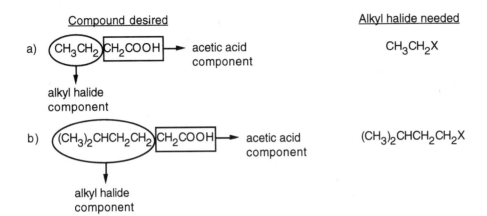

22.16

a) $CH_2(CO_2Et)_2$ $\xrightarrow[\text{2. PhCH}_2\text{Br}]{\text{1. Na}^+ \text{}^-\text{OEt}}$ $PhCH_2CH(CO_2Et)_2$ + $NaBr$

$\downarrow$ $H_3O^+,\ \Delta$

$PhCH_2CH_2CO_2H$ + CO_2 + $2\ C_2H_5OH$

3–Phenylpropanoic acid

b) $CH_2(CO_2Et)_2$ $\xrightarrow[\text{2. CH}_3\text{CH}_2\text{CH}_2\text{Br}]{\text{1. Na}^+ \text{}^-\text{OEt}}$ $CH_3CH_2CH_2CH(CO_2Et)_2$ + $NaBr$

$\downarrow$ 1. Na^+ ^-OEt
2. CH_3Br

$CH_3CH_2CH_2CH(CH_3)CO_2H$ $\xleftarrow{H_3O^+,\ \Delta}$ $CH_3CH_2CH_2C(CH_3)(CO_2Et)_2$ + $NaBr$

2–Methylpentanoic acid
+ CO_2 + $2\ C_2H_5OH$

c) $CH_2(CO_2Et)_2$ $\xrightarrow[\text{2. (CH}_3)_2\text{CHCH}_2\text{Br}]{\text{1. Na}^+ \text{}^-\text{OEt}}$ $(CH_3)_2CHCH_2CH(CO_2Et)_2$ + $NaBr$

$\downarrow$ $H_3O^+,\ \Delta$

$(CH_3)_2CHCH_2CH_2CO_2H$ + CO_2 + $2\ C_2H_5OH$

4–Methylpentanoic acid

d) $CH_2(CO_2Et)_2$ $\xrightarrow[\text{2. BrCH}_2\text{CH}_2\text{CH}_2\text{Br}]{\text{1. 2 Na}^+ \text{}^-\text{OEt}}$ [cyclobutane with CO_2Et and CO_2Et] + $NaBr$

$\downarrow$ $H_3O^+,\ \Delta$

[cyclobutane]—$CO_2CH_2CH_3$ $\xleftarrow[\text{H}^+]{\text{CH}_3\text{CH}_2\text{OH}}$ [cyclobutane]—CO_2H + CO_2 + $2\ C_2H_5OH$

Ethyl cyclobutanecarboxylate

22.17 Since malonic ester has only two acidic hydrogen atoms, it can be alkylated only two times. Formation of trialkylated acetic acids is thus not possible.

22.18 As in the malonic acid synthesis, you should identify the structural fragments of the target compound. The "acetone component" comes from acetoacetic ester; the other component comes from a primary alkyl halide.

a)

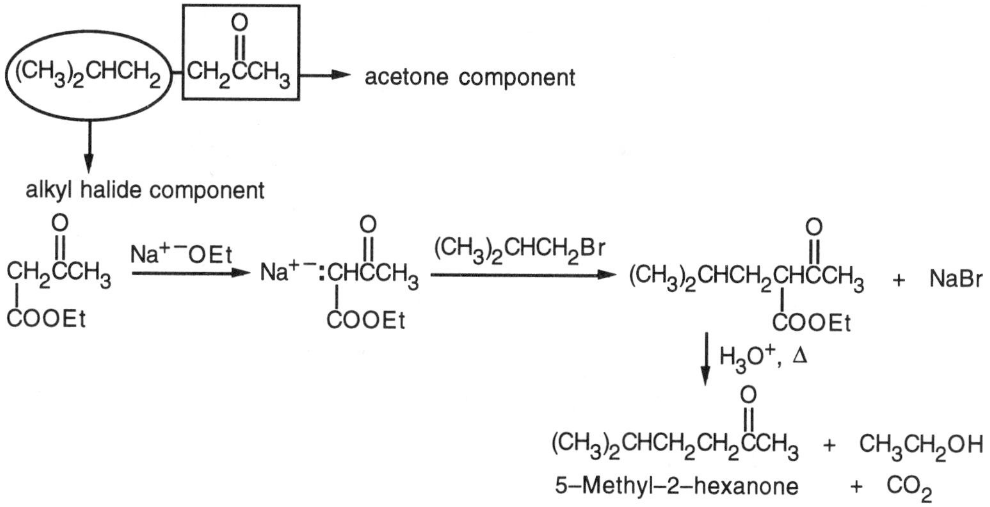

b)

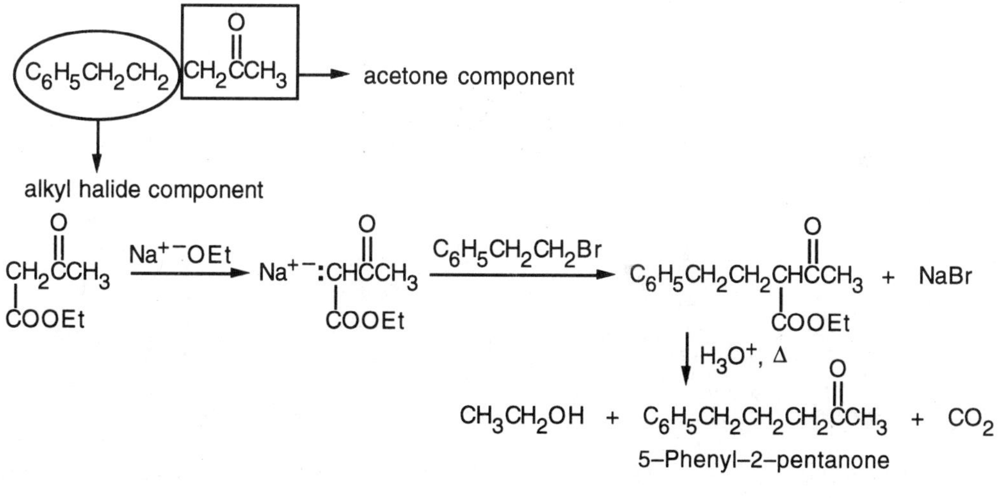

22.19

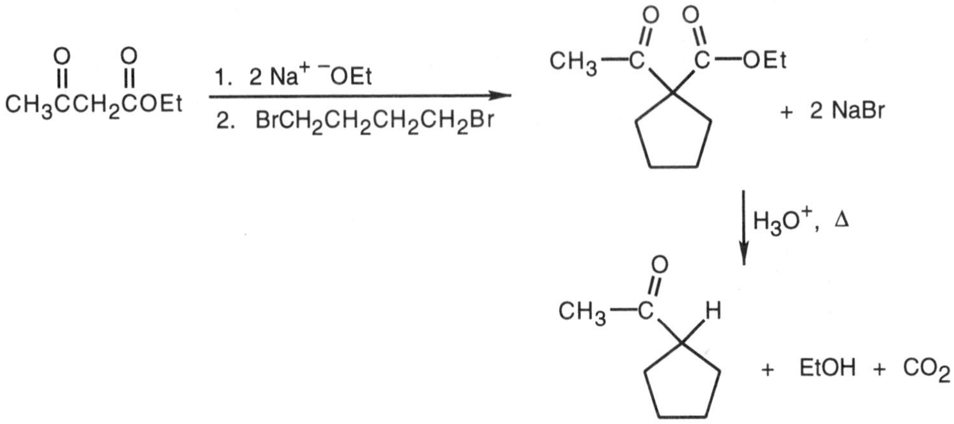

22.20 The acetoacetic ester synthesis can be used only if the desired products meet certain criteria:

(1) Three carbons must originate from acetoacetic ester. In other words, compounds of the type $RCOCH_3$ can't be synthesized by the reaction of RX with acetoacetic ester.
(2) Alkyl halides must be primary or methyl.
(3) The acetoacetic ester synthesis can't be used to prepare compounds that are trisubstituted at the α position.

a)

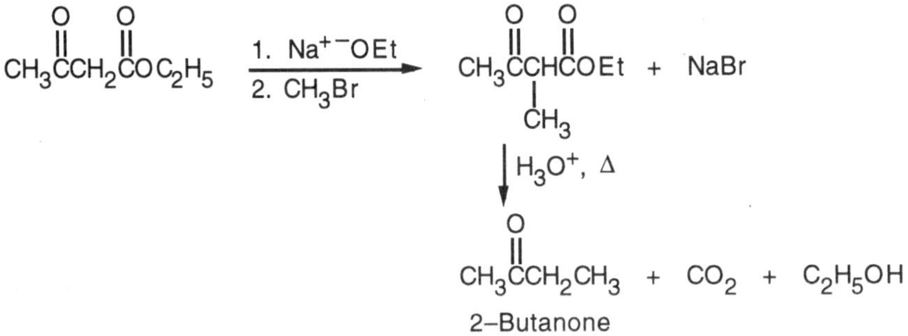

b) Phenylacetone can't be produced by an acetoacetic ester synthesis because bromobenzene, the necessary halide, does not enter into S_N2 reactions. [See (2) above.]

c) Acetophenone can't be produced by an acetoacetic ester synthesis. [See (1) above.]

d) 3,3–Dimethyl–2–butanone can't be prepared because it is trisubstituted at the α position. [See (3) above.]

22.21

a)

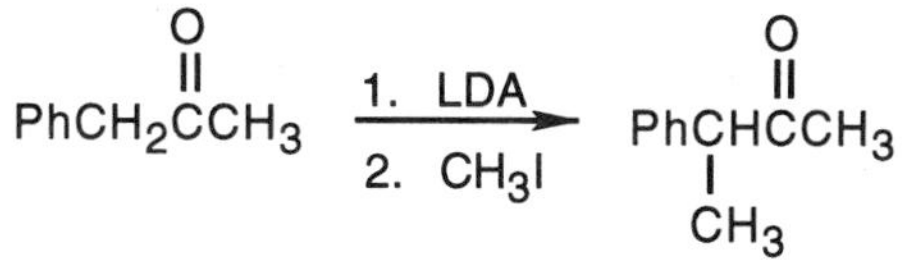

3–Phenyl–2–butanone

Alkylation occurs at the carbon next to the phenyl group because the phenyl group can help stabilize the enolate anion intermediate.

b)

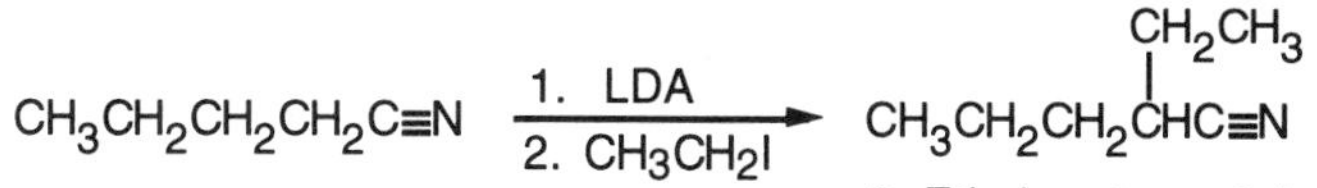

$$CH_3CH_2CH_2CH_2C\equiv N \xrightarrow[\text{2. }CH_3CH_2I]{\text{1. LDA}} CH_3CH_2CH_2\overset{\overset{\displaystyle CH_2CH_3}{|}}{C}HC\equiv N$$

2–Ethylpentanenitrile

c)

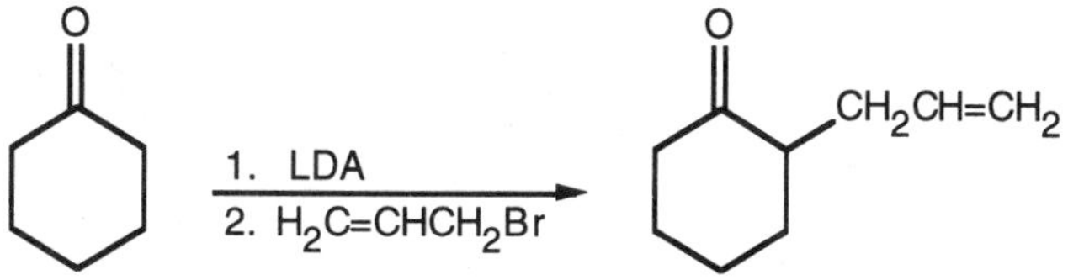

2–Allylcyclohexanone

d)

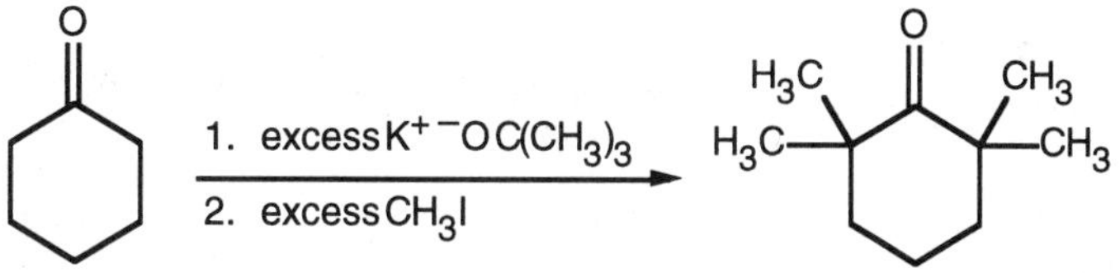

2,2,6,6–Tetramethylcyclohexanone

22.22 The enol form of 2,4–pentanedione is favored over the ketone form for two reasons: (1) Conjugation of the enol double bond with the second carbonyl group stabilizes the enol form. (2) Hydrogen bonding of the enol hydrogen to the other carbonyl group provides further stabilization. Neither of these types of enol stabilization is available to acetone.

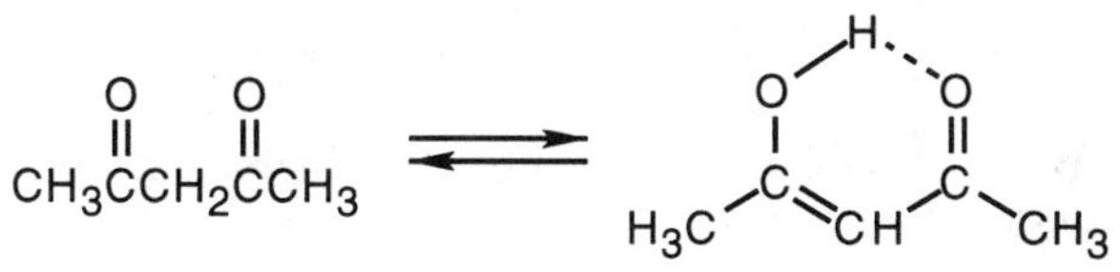

22.23

a)

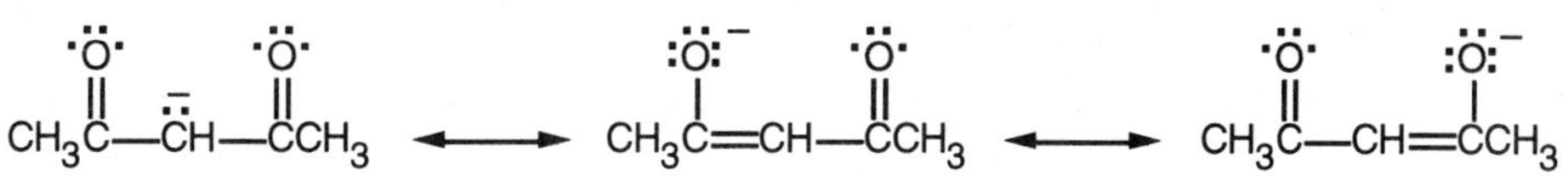

b)

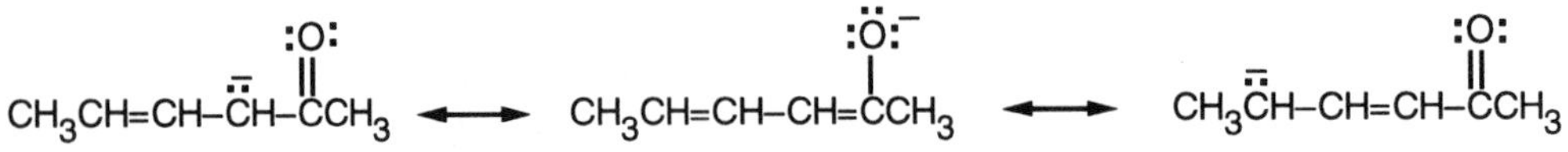

c)

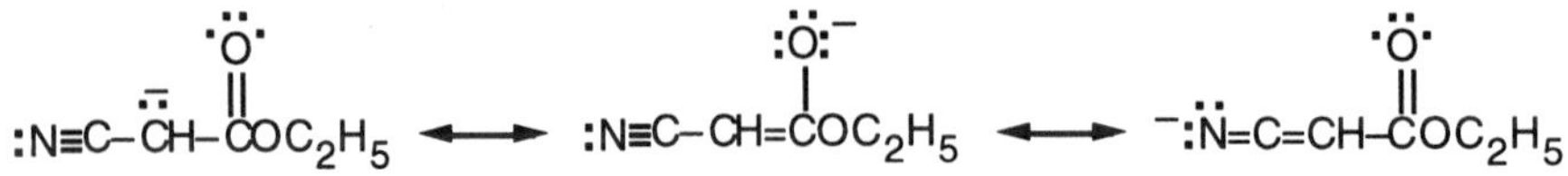

d)

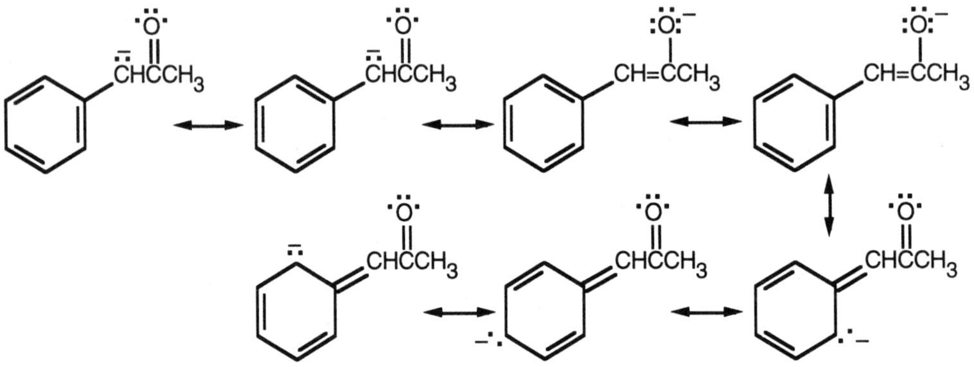

22.24 Acidic hydrogens are underlined.

a)

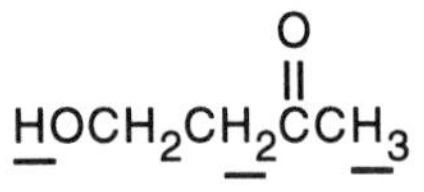

The alcohol hydrogen is more acidic than the α hydrogens

b)

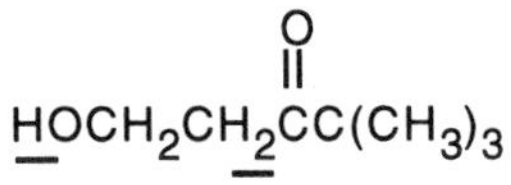

c)

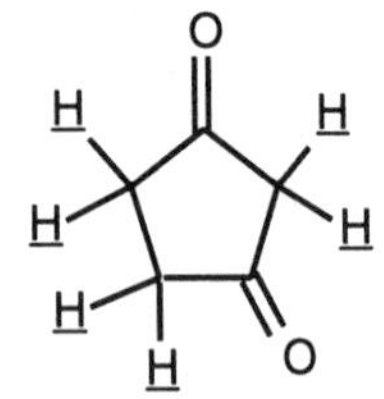

All hydrogens are somewhat acidic, although those between the carbonyls are the most acidic.

22.25 Enolization at the γ position produces an anion that is stabilized by delocalization of the negative charge over the π system of five atoms.

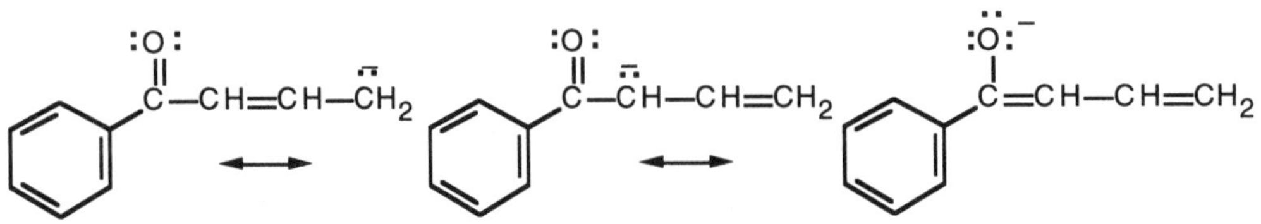

22.26 When a compound containing acidic hydrogen atoms is treated with NaOD in D_2O, all acidic protons are gradually replaced by deuterons. For each proton (atomic weight 1) lost, a deuteron (atomic weight 2) is added. Since the molecular weight of cyclohexanone increases by four after $NaOD/D_2O$ treatment (from 98 to 102), cyclohexanone contains four acidic hydrogen atoms.

22.27

2–Methylcycloheptanone   3–Methylcycloheptanone

2–Methylcycloheptanone has three acidic hydrogen atoms, and 3–methylcycloheptanone has four. After treatment with NaOD in D_2O, the molecular weight of the 2–methyl isomer increases by 3 and that of the 3–methyl isomer increases by 4. A mass spectrum of the deuterated ketones can then distinguish between them.

22.28

Least acidic $\longrightarrow$ Most acidic

$$(CH_3CH_2)_2NH < CH_3\overset{O}{\underset{\|}{C}}CH_3 < CH_3CH_2OH < CH_3\overset{O}{\underset{\|}{C}}CH_2\overset{O}{\underset{\|}{C}}CH_3 < CH_3CH_2\overset{O}{\underset{\|}{C}}OH < Cl_3\overset{O}{\underset{\|}{C}}COH$$

22.29 A nitroso compound is analogous to a carbonyl compound. If there are protons α to the nitroso group, enolization similar to that observed for carbonyl compounds can occur, leading to formation of an oxime. If no protons are adjacent to the nitroso group, enolization to the oxime can't occur, and the nitroso compound is stable.

22.30

a) $CH_2(CO_2Et)_2$ $\xrightarrow[\text{2. } CH_3CH_2CH_2Br]{\text{1. } Na^{+\,-}OEt}$ $CH_3CH_2CH_2CH(CO_2Et)_2$ + NaBr

$\downarrow H_3O^+, \Delta$

$CH_3CH_2CH_2CH_2CO_2C_2H_5$ $\xleftarrow[H^+]{C_2H_5OH}$ $CH_3CH_2CH_2CH_2CO_2H$ + CO_2 + $2\,C_2H_5OH$

Ethyl pentanoate

b)   This ester would be difficult to synthesize by the malonic ester route because the necessary alkyl halide, 2–bromopropane, is a secondary bromide that undergoes elimination as well as substitution.

c) $CH_2(CO_2Et)_2$ $\xrightarrow[\text{2. } CH_3CH_2Br]{\text{1. } Na^+ {}^-OEt}$ $CH_3CH_2CH(CO_2Et)_2$ + NaBr

$\xrightarrow[\text{2. } CH_3Br]{\text{1. } NaOEt}$

$$CH_3CH_2\overset{\overset{\displaystyle CH_3}{|}}{C}HCOOH + 2\, C_2H_5OH + CO_2 \xleftarrow[\Delta]{H_3O^+} CH_3CH_2\overset{\overset{\displaystyle CH_3}{|}}{C}(CO_2Et)_2 + NaBr$$

$\downarrow \begin{array}{l} C_2H_5OH \\ H^+ \end{array}$

$$CH_3CH_2\overset{\overset{\displaystyle CH_3}{|}}{C}HCO_2C_2H_5$$

Ethyl 2–methyl–butanoate

d)   The malonic ester route can't be used to synthesize alpha–trisubstituted carboxylic acids.

22.31

a) $CH_2\overset{\overset{\displaystyle O}{||}}{C}CH_3$ $\underset{\overset{|}{COOEt}}{}$ $\xrightarrow[\text{2. } 2\, CH_3CH_2Br]{\text{1. } 2\, Na^+{}^-OEt}$ $(CH_3CH_2)_2\overset{\overset{\displaystyle O}{||}}{C}CCH_3$ $\underset{\overset{|}{COOEt}}{}$

$\downarrow H_3O^+, \Delta$

$(CH_3CH_2)_2CH\overset{\overset{\displaystyle O}{||}}{C}CH_3 + CH_3CH_2OH + CO_2$

b) $CH_2\overset{\overset{\displaystyle O}{||}}{C}CH_3$ $\underset{\overset{|}{COOEt}}{}$ $\xrightarrow[\text{2. } CH_3CH_2CH_2Br]{\text{1. } Na^+{}^-OEt}$ $CH_3CH_2CH_2\overset{\overset{\displaystyle O}{||}}{C}H\overset{\overset{\displaystyle O}{||}}{C}CH_3$ $\underset{\overset{|}{COOEt}}{}$

$\downarrow \begin{array}{l} \text{1. } Na^+{}^-OEt \\ \text{2. } CH_3Br \end{array}$

$CH_3CH_2CH_2\overset{\overset{\displaystyle H_3C\ \ O}{\overset{|\ \ \ ||}{}}}{C}HCCH_3 \xleftarrow[]{H_3O^+, \Delta} CH_3CH_2CH_2\overset{\overset{\displaystyle H_3C\ \ O}{\overset{|\ \ \ ||}{}}}{C}-CCH_3$
+ CH_3CH_2OH + CO_2 $\underset{\overset{|}{COOEt}}{}$

22.32 Use a malonic ester synthesis if the product you want is an α–substituted carboxylic acid or derivative. Use an acetoacetic acid synthesis if the product you want is an α–substituted methyl ketone.

a) $CH_2(CO_2Et)_2$ $\xrightarrow[\text{2. } 2\, CH_3Br]{\text{1. } 2\, Na^+{}^-OEt}$ $(CH_3)_2C(CO_2C_2H_5)_2$ + 2 NaBr

b)
$$CH_2\overset{\overset{O}{\|}}{C}CH_3 \quad \underset{\text{2. BrCH}_2(CH_2)_4CH_2Br}{\overset{\text{1. 2 Na}^+\text{ }^-OEt}{\longrightarrow}}$$
with COOEt below

cycloheptane ring with $\overset{\overset{O}{\|}}{C}CH_3$ and COOEt substituents

$$\downarrow \overset{\Delta}{H_3O^+}$$

cycloheptane ring with $\overset{\overset{O}{\|}}{C}CH_3$ + CH_3CH_2OH + CO_2

c) $CH_2(COOEt)_2 \quad \underset{\text{2. BrCH}_2CH_2CH_2Br}{\overset{\text{1. 2 Na}^+\text{ }^-OEt}{\longrightarrow}}$

cyclobutane ring with COOEt, COOEt

$$\downarrow \overset{\Delta}{H_3O^+}$$

cyclobutane ring —COOH + CH_3CH_2OH + CO_2

d) $CH_3\overset{\overset{O}{\|}}{C}CH_2\overset{\overset{O}{\|}}{C}OEt \quad \underset{\text{2. H}_2C=CHCH_2Br}{\overset{\text{1. Na}^+\text{ }^-OEt}{\longrightarrow}}$ $H_2C=CHCH_2\overset{}{C}H\overset{\overset{O}{\|}}{C}CH_3$ + NaBr

with CO_2Et below

$$\downarrow H_3O^+, \Delta$$

$$H_2C=CHCH_2CH_2\overset{\overset{O}{\|}}{C}CH_3 \ + \ CO_2 \ + \ C_2H_5OH$$

22.33

a)
cyclohexane ring with COOH, COOH $\overset{\Delta}{\longrightarrow}$ cyclohexane ring with COOH   + CO_2

b)
cyclopentane-1,3-dione $\underset{\text{2. CH}_3I}{\overset{\text{1. Na}^+\text{ }^-OEt}{\longrightarrow}}$ 2-methyl cyclopentane-1,3-dione with CH_3

c) $CH_3CH_2CH_2COOH \quad \underset{PBr_3}{\overset{Br_2}{\longrightarrow}} \quad CH_3CH_2CHBrCOBr \quad \overset{H_2O}{\longrightarrow} \quad CH_3CH_2CHBrCOOH$

$\quad\quad\quad\quad\quad\quad\quad\quad\quad\quad\quad\quad\quad\quad\quad\quad\quad\underline{A} \quad\quad\quad\quad\quad\quad\quad\quad\quad\quad\quad\quad\quad\quad\quad\underline{B}$

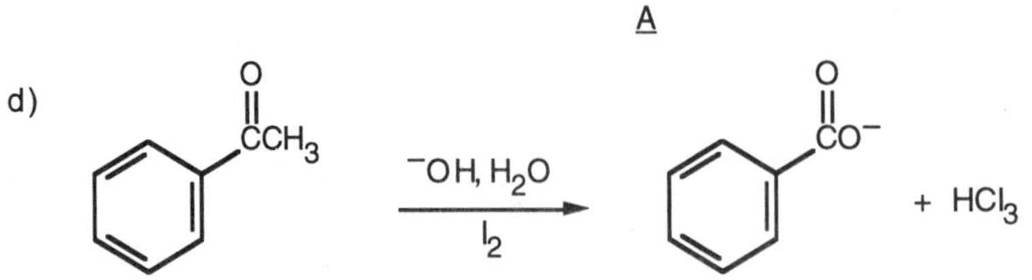

A

d)

^-OH, H_2O

I_2

$+$ HCl_3

22.34

top-side
protonation
H_3O^+

H_3O^+

enol

H_3O^+
bottom-side
protonation

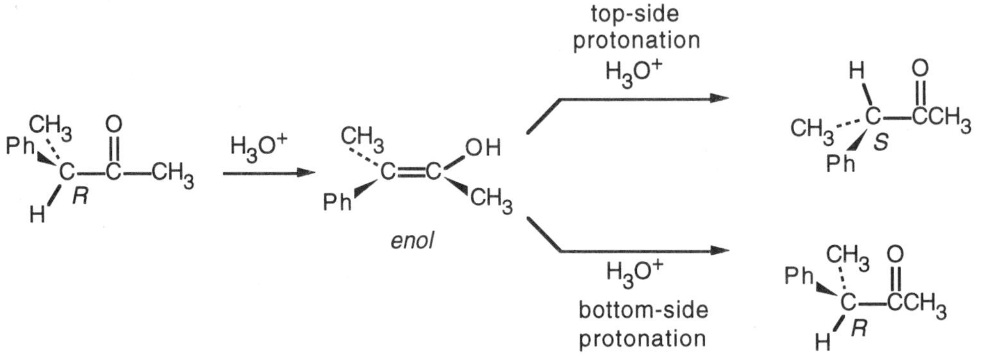

Loss of the proton at C3 during enolization produces a planar, sp^2–hybridized carbon at the original stereogenic center. Protonation occurs from either the top or the bottom face of the double bond to produce racemic 3–phenyl–2–butanone.

22.35

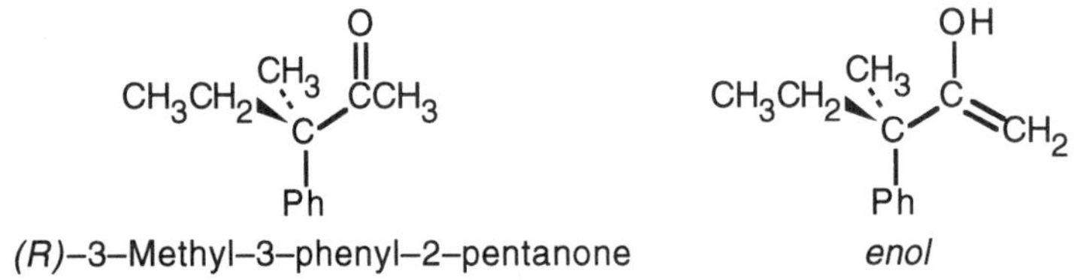

(R)–3–Methyl–3–phenyl–2–pentanone   enol

(R)–3–Methyl–3–phenyl–2–pentanone can enolize only toward the methyl group because there is no proton at carbon 3. Because the stereogenic center is not involved in the enolization process, the ketone is not racemized by acid treatment.

22.36

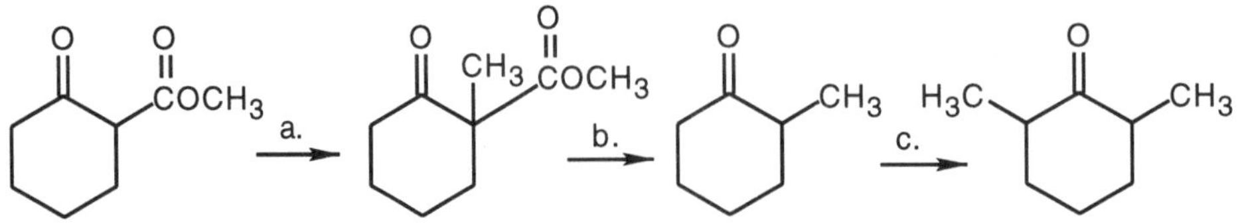

(a) Na^+ ^-OEt, then CH_3I; (b) H_3O^+, heat; (c) LDA, then CH_3I

22.37

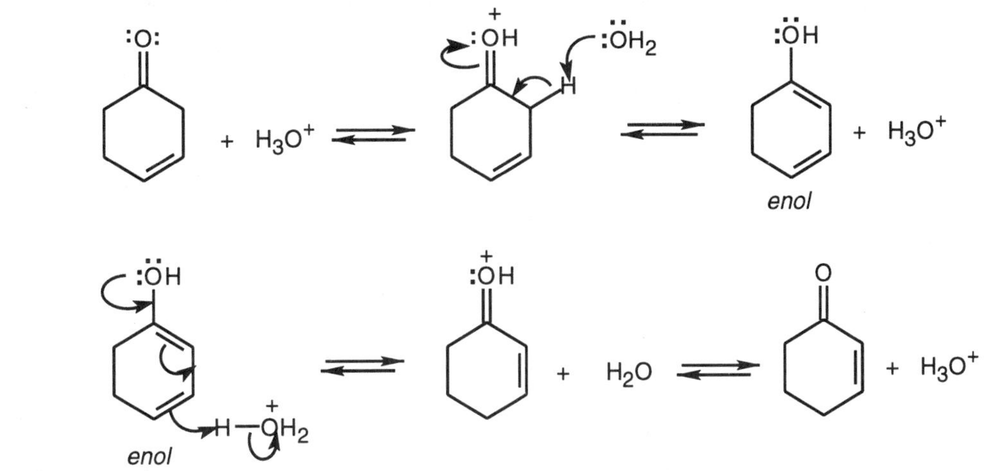

22.38

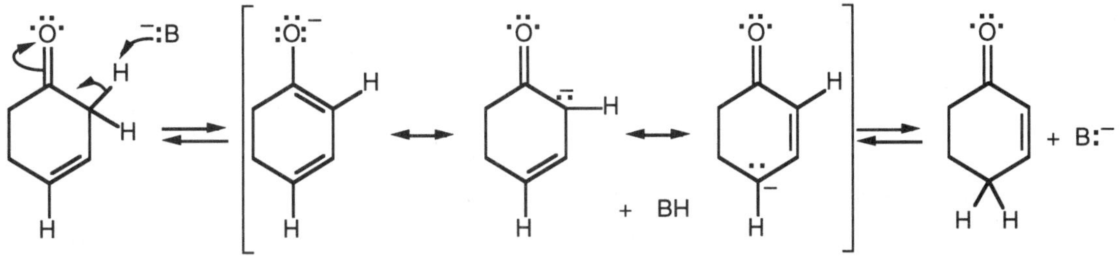

The enolate of 3–cyclohexenone can be protonated at three different positions. Protonation at the γ position yields the α,ß–unsaturated ketone.

22.39

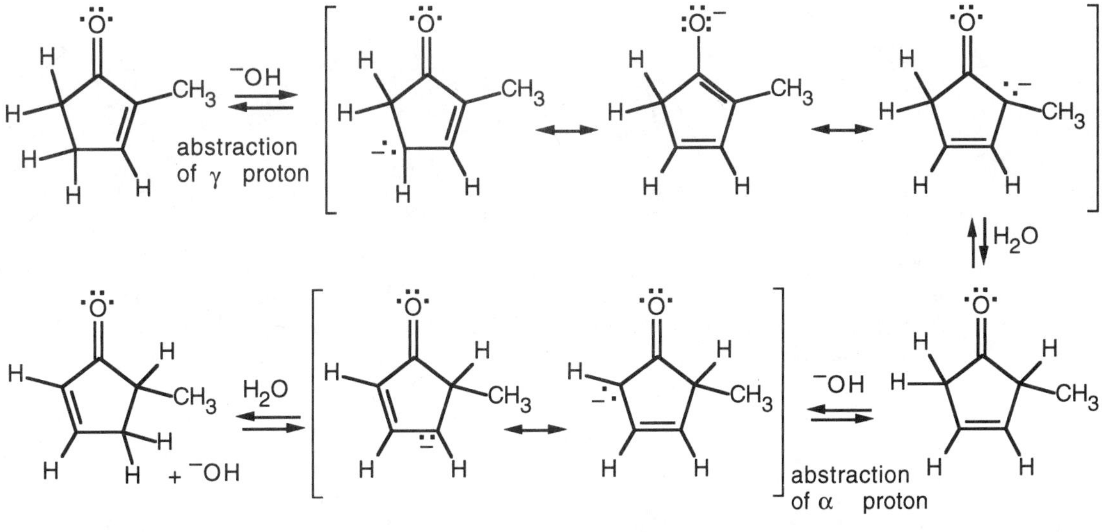

All protons in the five-membered ring can be exchanged by base treatment.

22.40 Protons α to a carbonyl group or γ to an enone carbonyl group are acidic (Problem 22.39). Thus for 2–methyl-2-cyclopentenone, protons at the starred positions are acidic.

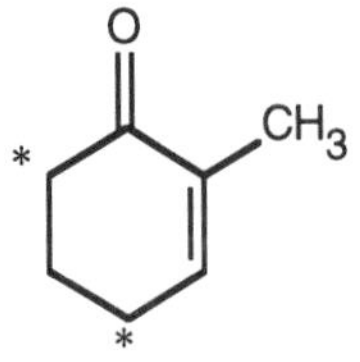

Isomerization of a 2–substituted 2–cyclohexenone to a 6–substituted 2–cyclohexenone requires removal of a proton from the 5–position of the 2–substituted isomer. Since protons in this position are not acidic, double bond isomerization does not occur.

22.41 Start at the end of the sequence of reactions and work backwards. If necessary, cover up pieces of information you are not using at the moment to keep them from distracting you.

a) Because the *keto acid* $C_9H_{13}NO_3$ loses CO_2 on heating, it must be a ß–keto acid. Neglecting stereoisomerism, we can draw the structure of the ß–keto acid as:

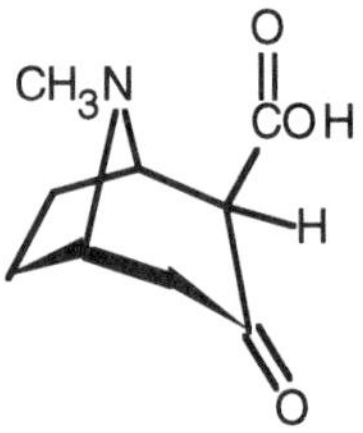

b) When ecgonine ($C_9H_{15}NO_3$) is treated with CrO_3, the keto acid $C_9H_{13}NO_3$ is produced. Since CrO_3 is used for oxidizing alcohols to carbonyl compounds, ecgonine has the following structure. Again, the stereochemistry is unspecified.

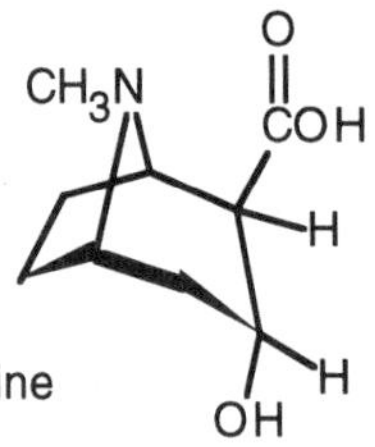

Ecgonine

c) Ecgonine contains carboxylic acid and alcohol functional groups. The other products of hydroxide treatment of cocaine are a carboxylic acid (benzoic acid) and an alcohol (methanol). Cocaine thus contains two ester functional groups, which are saponified on reaction with hydroxide.

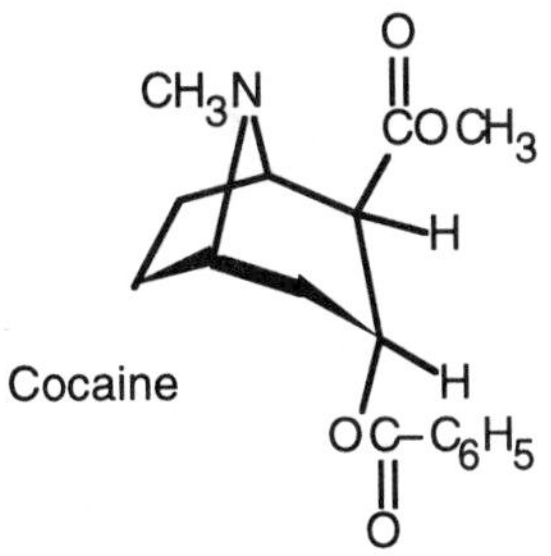

Cocaine

d) The complete reaction sequence:

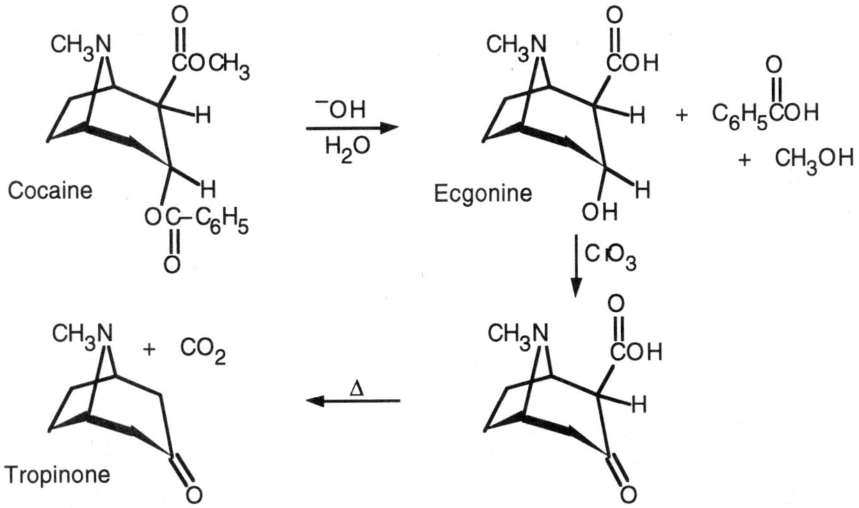

Cocaine

$^-$OH
H_2O

Ecgonine

+ C_6H_5COH

+ CH_3OH

CrO_3

Tropinone + CO_2

Δ

22.42 The haloform reaction occurs only with methyl ketones.

Positive haloform reaction:

CH_3CCH_3, $C_6H_5CCH_3$

Negative haloform reaction:

CH_3CH_2CHO, CH_3COOH, $CH_3C{\equiv}N$

22.43 First treat geraniol with PBr$_3$ to form (CH$_3$)$_2$C=CHCH$_2$CH$_2$C(CH$_3$)=CHCH$_2$Br (geranyl bromide).

a)   CH$_3$CO$_2$Et $\xrightarrow[\substack{\text{2. Geranyl} \\ \text{bromide}}]{\text{1. LDA}}$ (CH$_3$)$_2$C=CHCH$_2$CH$_2$C(CH$_3$)=CHCH$_2$CH$_2$$\overset{\displaystyle O}{\overset{\|}{C}}$OEt

Ethyl geranylacetate

Alternatively:

CH$_2$(CO$_2$Et)$_2$ $\xrightarrow[\substack{\text{2. Geranyl} \\ \text{bromide}}]{\text{1. Na}^+\text{ }^-\text{OEt}}$ (CH$_3$)$_2$C=CHCH$_2$CH$_2$C(CH$_3$)=CHCH$_2$CH(CO$_2$Et)$_2$

$\downarrow$ H$_3$O$^+$, Δ

(CH$_3$)$_2$C=CHCH$_2$CH$_2$C(CH$_3$)=CHCH$_2$CH$_2$$\overset{\displaystyle O}{\overset{\|}{C}}$OH

$\left|\begin{array}{l}\text{1. SOCl}_2 \\ \text{2. CH}_3\text{CH}_2\text{OH,} \\ \quad\text{pyridine}\end{array}\right.$

(CH$_3$)$_2$C=CHCH$_2$CH$_2$C(CH$_3$)=CHCH$_2$CH$_2$$\overset{\displaystyle O}{\overset{\|}{C}}$OEt

Ethyl geranylacetate

b)   CH$_3$$\overset{\displaystyle O}{\overset{\|}{C}}CH_2CO_2$Et $\xrightarrow[\substack{\text{2. Geranyl} \\ \text{bromide}}]{\text{1. Na}^+\text{ }^-\text{OEt}}$ (CH$_3$)$_2$C=CHCH$_2$CH$_2$C(CH$_3$)=CHCH$_2$$\underset{\displaystyle CO_2Et}{CH}$$\overset{\displaystyle O}{\overset{\|}{C}}CH_3$

$\downarrow$ H$_3$O$^+$, Δ

(CH$_3$)$_2$C=CHCH$_2$CH$_2$C(CH$_3$)=CHCH$_2$CH$_2$$\overset{\displaystyle O}{\overset{\|}{C}}CH_3$ + EtOH + CO$_2$

Geranylacetone

22.44

a)

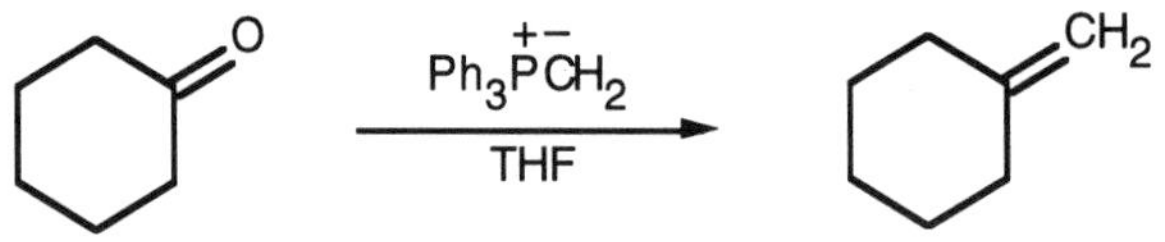

$\xrightarrow[\text{THF}]{\text{Ph}_3\overset{+}{\text{P}}\overset{-}{\text{CH}_2}}$

b)   product of a) $\xrightarrow[\text{peroxides}]{\text{HBr}}$

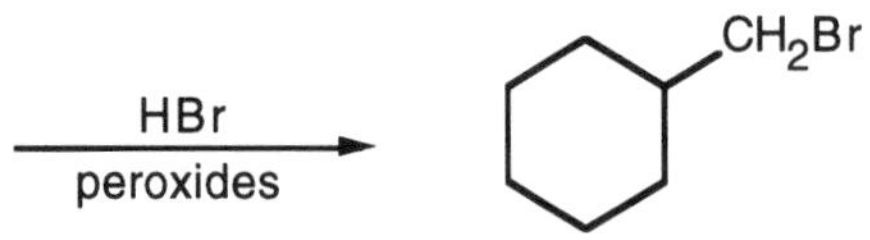

c)

$$\xrightarrow[\text{2. } C_6H_5CH_2Br]{\text{1. LDA}}$$

d)

$$CH_2(CO_2Et)_2 \xrightarrow[\substack{\text{2. product} \\ \text{of b)}}]{\text{1. Na}^+\,^-OEt}$$

CH$_2$CH(CO$_2$Et)$_2$

+ NaBr

$$\xrightarrow[\Delta]{H_3O^+}$$

CH$_2$CH$_2$COOH

+ 2 C$_2$H$_5$OH + CO$_2$

e)

$$\xrightarrow[H_3O^+]{\text{NaCN}}$$

HO   CN

$$\xrightarrow[\text{pyridine}]{\text{POCl}_3}$$

CN

$$\xrightarrow[\Delta]{H_3O^+}$$

COOH

f)

$$\xrightarrow[\text{CH}_3\text{COOH}]{\text{Br}_2}$$

Br

$$\xrightarrow[\Delta]{\text{Pyridine}}$$

g)

product from e)

$$\xrightarrow[\text{Pd/C}]{\text{H}_2}$$

COOH

22.45 Treatment of either the cis or trans isomer with base causes enolization α to the carbonyl group and results in loss of configuration at the α–position. Reprotonation at carbon 2 produces either of the diastereomeric 4–*tert*–butyl–2–methylcyclohexanones. In both diastereomers the *tert*–butyl group of carbon 4 occupies the equatorial position for steric reasons. The methyl group of the cis isomer is also equatorial, but the methyl group of the trans isomer is axial. The trans isomer is less stable because of 1,3 diaxial interactions of the methyl group with the ring protons.

cis   trans

22.46 a) Reaction with Br_2 at the α position occurs only with aldehydes and ketones, not with esters.

b) Aryl halides can't be used in malonic ester syntheses because they don't undergo S_N2 displacement reactions.

c) The product of this reaction sequence, $H_2C{=}CHCH_2CH_2COCH_3$, is a methyl ketone, not a carboxylic acid.

d) Base-promoted bromination of a ketone yields a mixture of mono– to tetra–bromo products. Reaction of the brominated ketones with pyridine and heat produces a mixture of compounds.

22.47

Enolization toward the position between the two carbonyl groups does not occur because the enol product would have a highly strained, nonplanar double bond. Instead, enolization occurs in the opposite direction, and the diketone thus resembles a monoketone rather than a β diketone in its pK_a and degree of enolization.

22.48 Both methylmagnesium bromide and *tert*–butylmagnesium bromide give the expected carbonyl addition products. The yield of the *tert*–butylmagnesium bromide addition product is very low, however, because of the difficulty of approach of the bulky *tert*- butyl Grignard reagent to the carbonyl carbon. More favorable is the acid-base reaction between the Grignard reagent and a carbonyl α proton.

When D_2O is added to the reaction mixture, the deuterated ketone is produced.

22.49 Laurene differs in stereochemical configuration from the starting material at the carbon α to the methylene group. Since this position is α to the carbonyl group in the starting material, enolization and isomerization must have occurred during the reaction.

Isomerization of the starting ketone is brought about by a reversible reaction with the basic Wittig reagent, which yields an equilibrium mixture of two diastereomeric ketones. The isomeric ketone then reacts preferentially with the Wittig reagent to give only the observed product.

22.50 The key step is an intramolecular alkylation reaction of the ketone α carbon, with the tosylate in the adjacent ring serving as the leaving group.

22.51

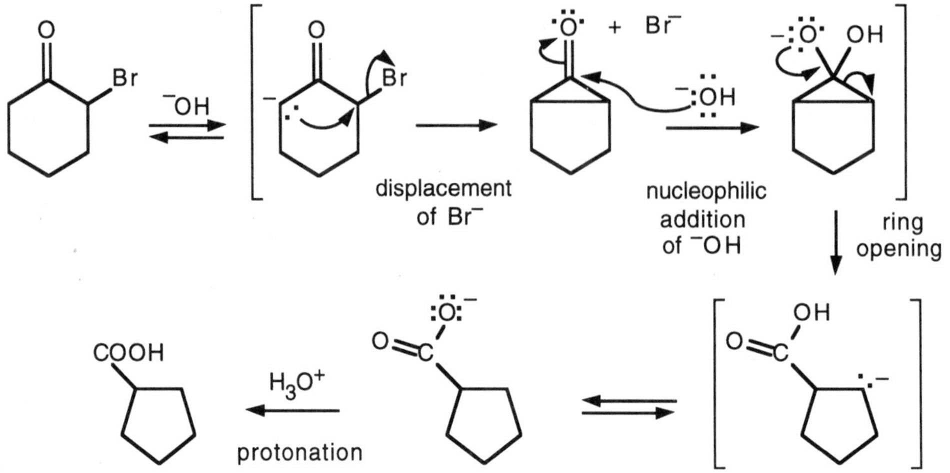

22.52

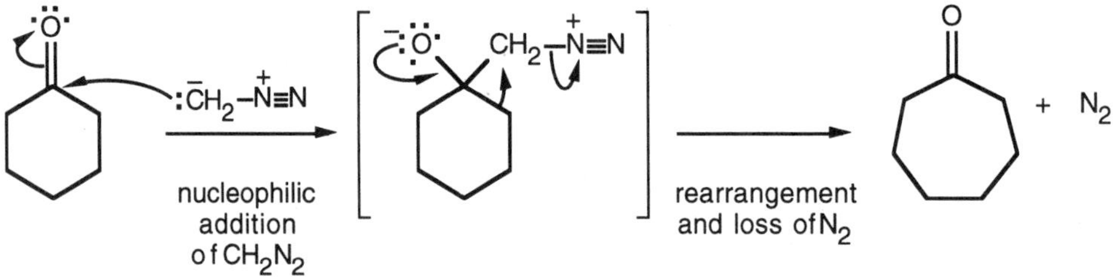

22.53

Acid catalyzes the formation of the enol.

The pi electrons from the enol double bond attack phenylselenenyl chloride.

Loss of a proton from oxygen yields the α–phenylseleno ketone. This reaction is similar to acid-catalyzed α –halogenation of ketones.

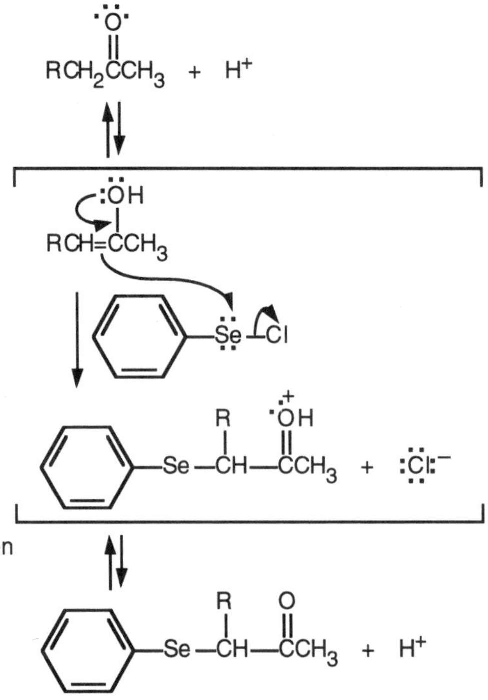

22.54

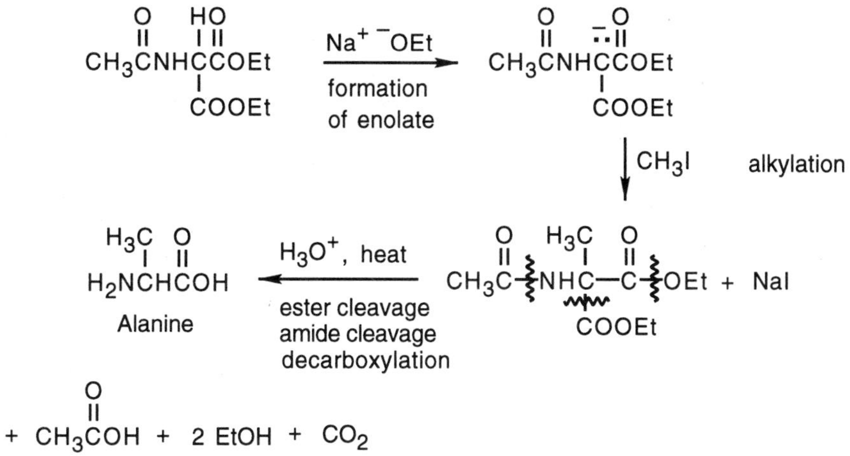

$$\underset{\substack{|\\ \text{COOEt}}}{\overset{\substack{O \quad HO\\ \| \quad | \quad \|}}{CH_3CNHCCOEt}} \xrightarrow[\substack{\text{formation}\\ \text{of enolate}}]{Na^+ \; {}^-OEt} \underset{\substack{|\\ \text{COOEt}}}{\overset{\substack{O \quad {}^-O\\ \| \quad \cdot\cdot\|}}{CH_3CNHCCOEt}}$$

$$\Big\downarrow CH_3I \qquad \text{alkylation}$$

$$\underset{\substack{\text{Alanine}}}{\overset{\substack{H_3C \quad O\\ | \quad \|}}{H_2NCHCOH}} \xleftarrow[\substack{\text{ester cleavage}\\ \text{amide cleavage}\\ \text{decarboxylation}}]{H_3O^+, \; \text{heat}} \underset{\substack{\text{COOEt}}}{\overset{\substack{O \quad H_3C \quad O\\ \| \quad | \quad \|}}{CH_3C\text{---}NHC\text{---}C\text{---}OEt}} + \; NaI$$

$$+ \; \overset{\substack{O\\ \|}}{CH_3COH} \; + \; 2 \; EtOH \; + \; CO_2$$

Study Guide for Chapter 22

After studying this chapter, you should be able to:

(1)   Draw keto and enol tautomers of carbonyl compounds; draw resonance forms of enolates (22.1, 22.3, 22.9, 22.22, 22.23, 22.29).

(2)   Identify acidic hydrogens (22.2, 22.8, 22.24, 22.25, 22.26, 22.27).

(3)   Formulate the mechanisms of acid– and base–catalyzed enolization (22.4, 22.10, 22.11, 22.34, 22.35, 22.37, 22.38, 22.39, 22.40, 22.45, 22.47, 22.48, 22.49).

(4)   Formulate the mechanisms of other alpha–substitution reactions (22.6, 22.7, 22.12, 22.13, 22.14, 22.50, 22.51, 22.52, 22.53).

(5)   Use alpha–substitution reactions of enolates to synthesize the following:
(a) Substituted acetic acid compounds via the malonic ester synthesis (22.15, 22.16, 22.17, 22.30, 22.32).
(b) Substituted methyl ketones via the acetoacetic ester synthesis (22.18, 22.19, 22.20, 22.31, 22.32).
(c) Alkylated carbonyl compounds using LDA (22.21).

(6)   Predict the products of alpha–substitution reactions (22.33, 22.36, 22.43, 22.44, 22.46, 22.54).

Chapter 23 – Carbonyl Condensation Reactions

23.1 When you are first learning the aldol condensation, write all the steps.
(1) Form the enolate of one molecule of the carbonyl compound.

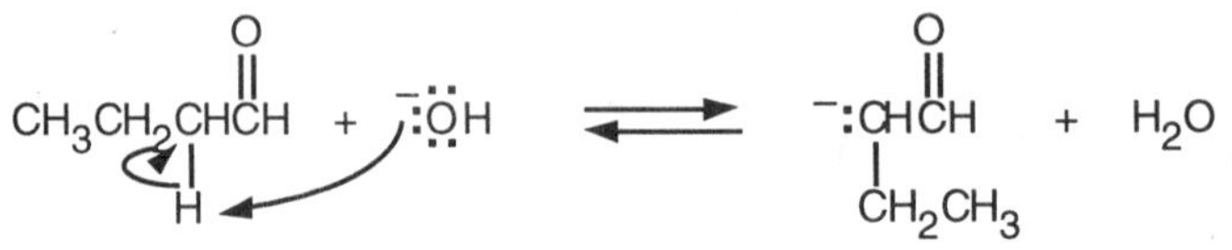

(2) Have the enolate attack the electrophilic carbonyl of the other molecule.

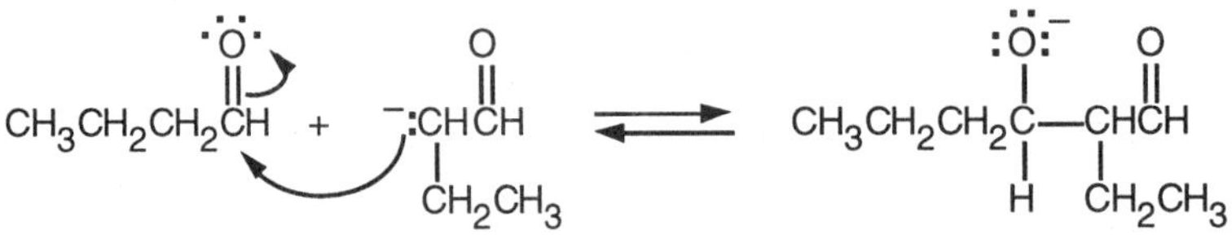

(3) Protonate the anionic oxygen.

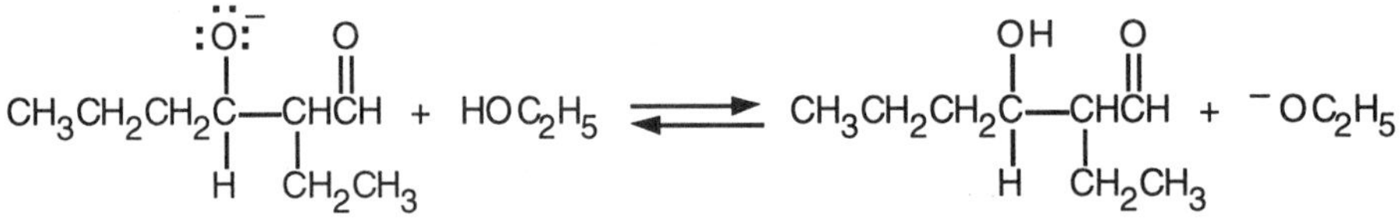

Practice writing out these steps for the other aldol condensations.

b) 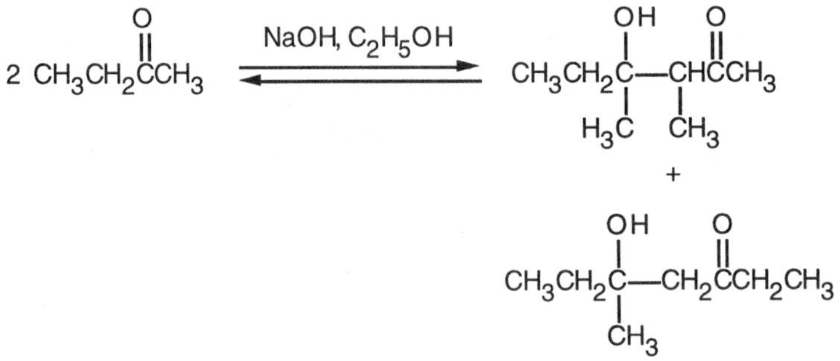

c) 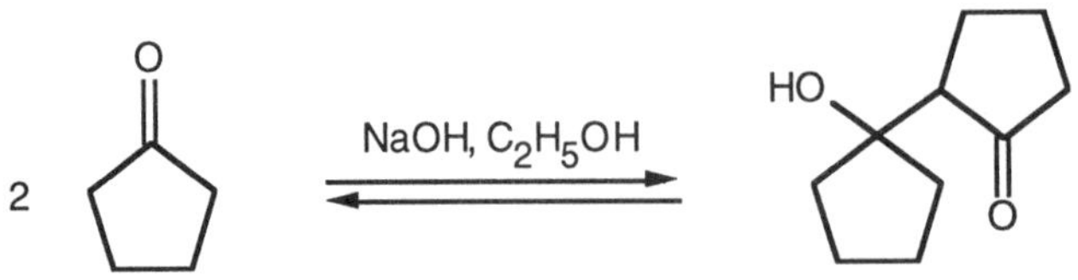

23.2 The reactive nucleophile in the acid-catalyzed aldol condensation is the *enol* of the carbonyl compound. The electrophile is the protonated carbonyl compound.

(1)

$$CH_3CH + H_3O^+ \rightleftharpoons \left[CH_2CH \xrightarrow{H_2O} CH_2=CH + H_3O^+ \right]$$

(2)

$$\left[\begin{array}{c} CH_3CH + CH_2=CH \\ \text{electrophile nucleophile} \end{array} \rightleftharpoons CH_3CCH_2CH \right]$$

(3)

$$\left[CH_3CCH_2CH \right] \xrightarrow{H_2O} CH_3CCH_2CH + H_3O^+$$

23.3

$$\underset{\underset{CH_3}{|}}{\overset{OH}{\underset{|}{CH_3C}}}-CH_2CCH_3$$

4–Methyl–4–hydroxy–2–pentanone

The steps for the reverse aldol are the reverse of those described in Problem 23.1.
(1) Deprotonate the alcohol oxygen.

$$CH_3CCH_2CCH_3 + {}^-{:}\overset{..}{O}H \rightleftharpoons CH_3CCH_2CCH_3 + H_2O$$

(2) Eliminate the enolate anion.

$$CH_3C{-}CH_2CCH_3 \rightleftharpoons CH_3CCH_3 + {}^-{:}CH_2CCH_3$$

(3) Reprotonate the enolate anion.

$${}^-{:}CH_2CCH_3 + H_2O \rightleftharpoons CH_3CCH_3 + {}^-OH$$

23.4

a)

b)

c)

23.5

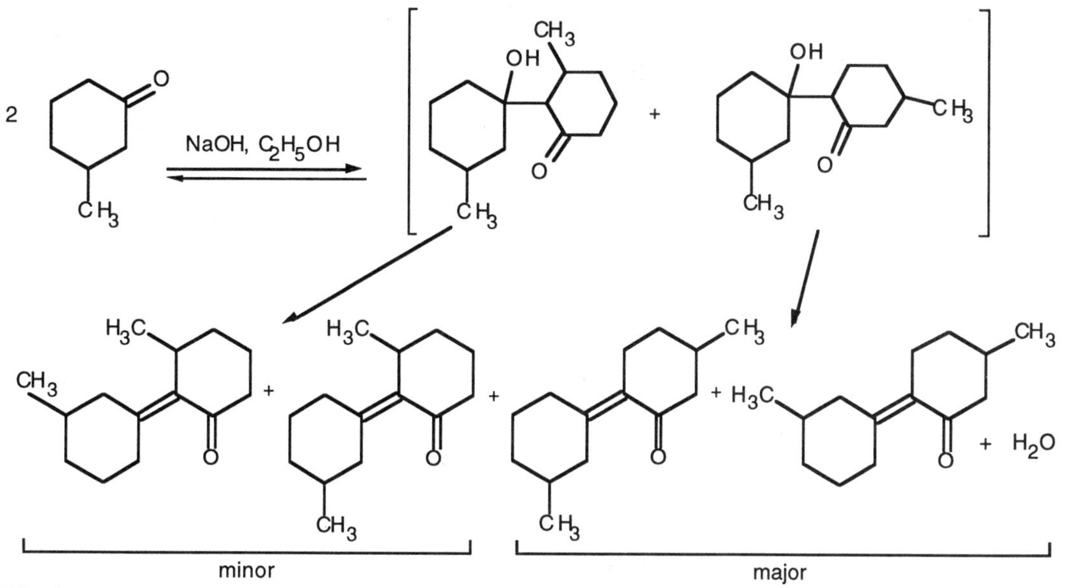

minor

major

23.6

a)

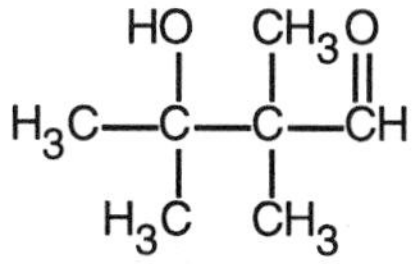

2,2,3–Trimethyl–3–hydroxybutanal is not an aldol self-condensation product. (Note that no aldol *self*-condensation can yield a product with an odd number of carbons.)

b)

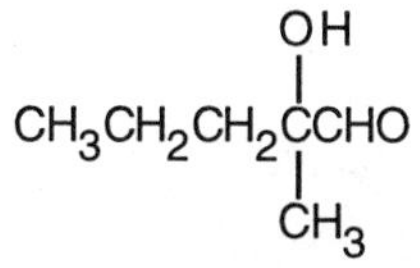

2–Methyl–2–hydroxypentanal

This is not an aldol product. The hydroxyl group in an aldol product must be ß, not α, to the carbonyl group.

c)

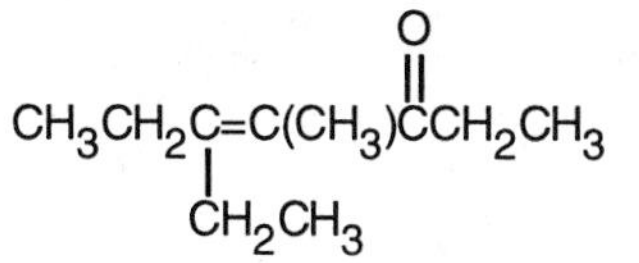

5–Ethyl–4–methyl–4–hepten–3–one

This product results from the aldol self-condensation of 3–pentanone, followed by dehydration.

23.7

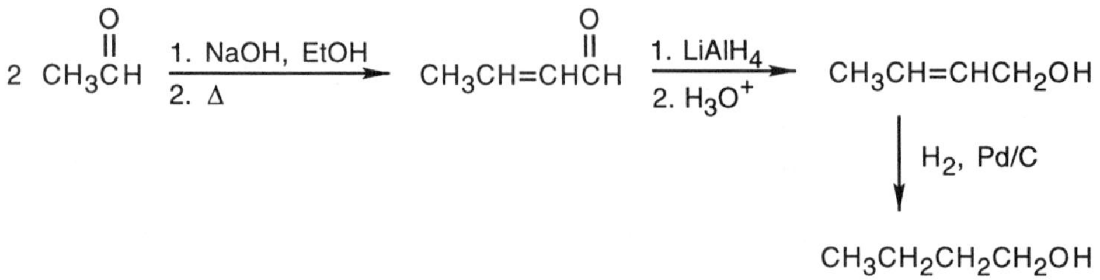

1–Butanol

23.8

a)

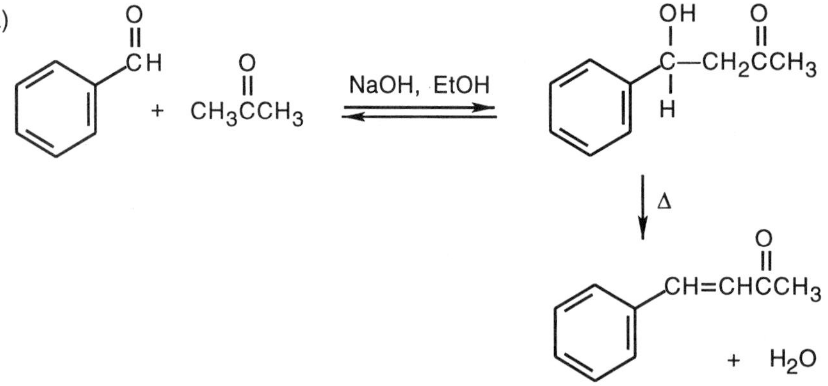

4–Phenyl–3–buten–2–one

This mixed aldol will succeed because one of the components, benzaldehyde,is a good acceptor yet has no α–hydrogen atoms.

b)

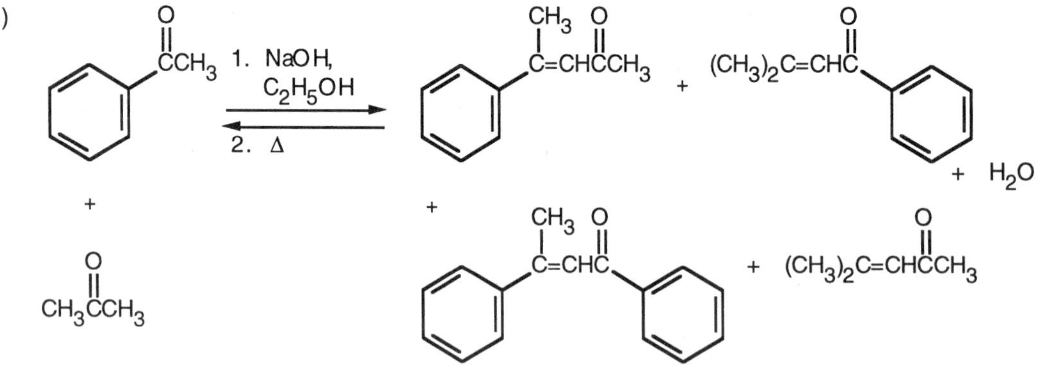

Four products result from the aldol condensation of acetone and acetophenone.

c)

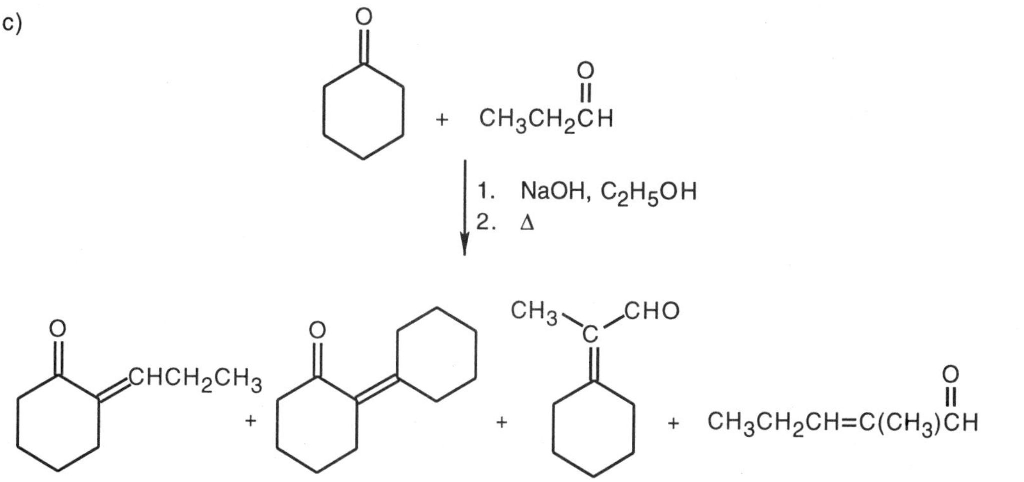

+ CH₃CH₂CH (with C=O)

1. NaOH, C₂H₅OH
2. Δ

CHCH₂CH₃ +   +   + CH₃CH₂CH=C(CH₃)CH (with C=O)

A mixture of products is formed because both carbonyl partners contain α–hydrogen atoms.

d)

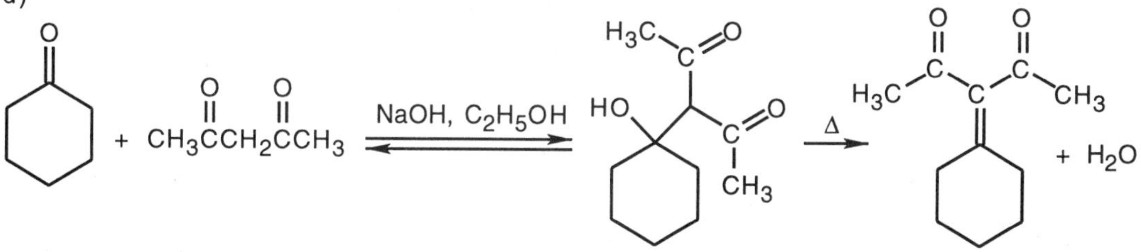

+ CH₃CCH₂CCH₃ (2,4-pentanedione)   NaOH, C₂H₅OH ⇌ Δ → + H₂O

This mixed aldol reaction succeeds because the anion of 2,4–pentanedione is formed more readily than the anion of cyclohexanone.

23.9

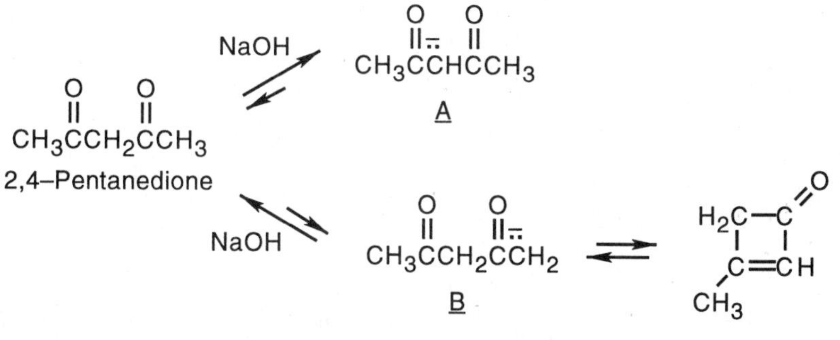

CH₃CCH₂CCH₃
2,4–Pentanedione

NaOH →

O O
‖⁻ ‖
CH₃CCHCCH₃

A

NaOH →

O O
‖ ‖⁻
CH₃CCH₂CCH₂   ⇌

B   CH₃

2,4–Pentanedione is in equilibrium with two enolate ions after treatment with base. Enolate **A** is stable and unreactive, while enolate **B** can undergo internal aldol condensation to form a cyclobutenone product. But because the aldol reaction is reversible and the cyclobutenone product is highly strained, there is little of this product present when equilibrium is reached. At equilibrium, only the stable, diketone enolate ion **A** is present.

23.10

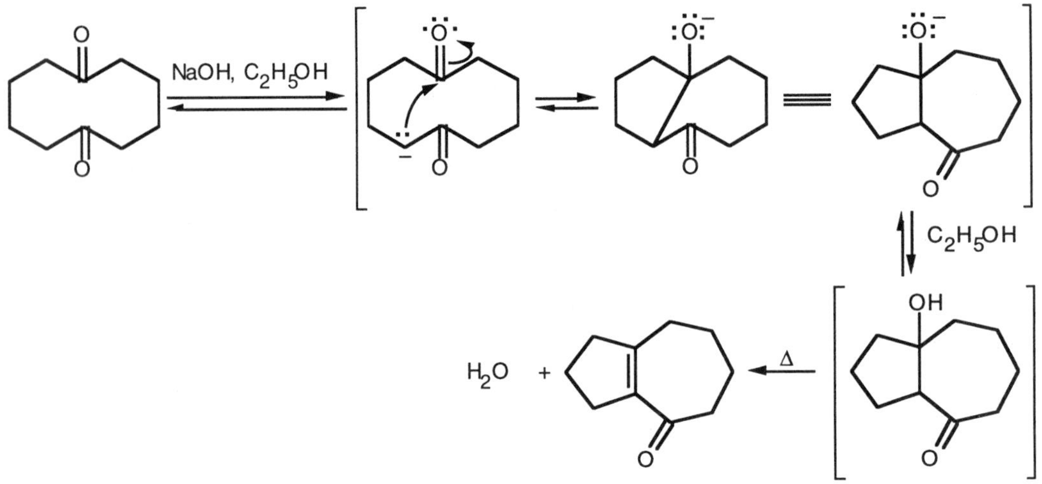

23.11 As in the aldol condensation, writing the two Claisen components in the correct orientation makes it easier to predict the product.

a)

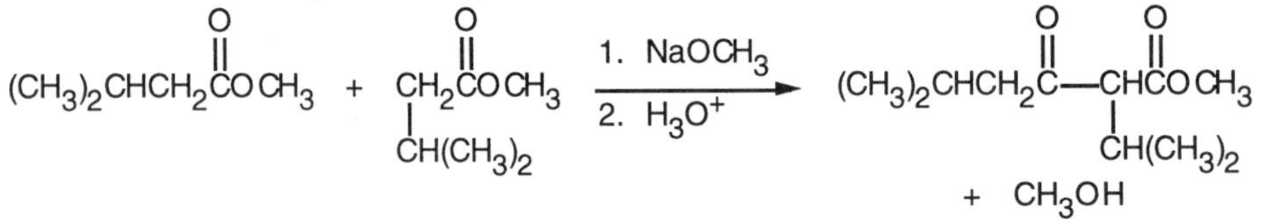

$(CH_3)_2CHCH_2COCH_3$ + CH_2COCH_3 $\xrightarrow[\text{2. } H_3O^+]{\text{1. NaOCH}_3}$ $(CH_3)_2CHCH_2C$—$CHCOCH_3$

with $CH(CH_3)_2$ group; product has $CH(CH_3)_2$ and

+ CH_3OH

b)

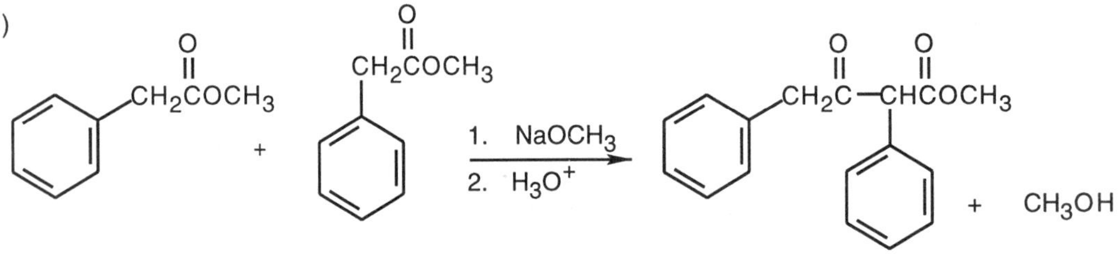

$\xrightarrow[\text{2. } H_3O^+]{\text{1. NaOCH}_3}$

+ CH_3OH

c)

$\xrightarrow[\text{2. } H_3O^+]{\text{1. NaOCH}_3}$

+ CH_3OH

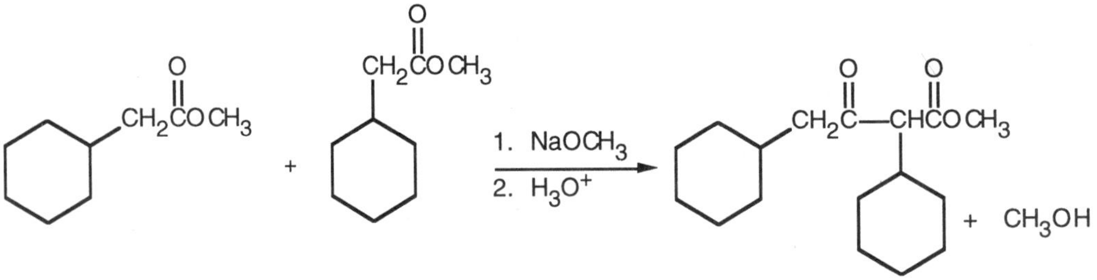

23.12

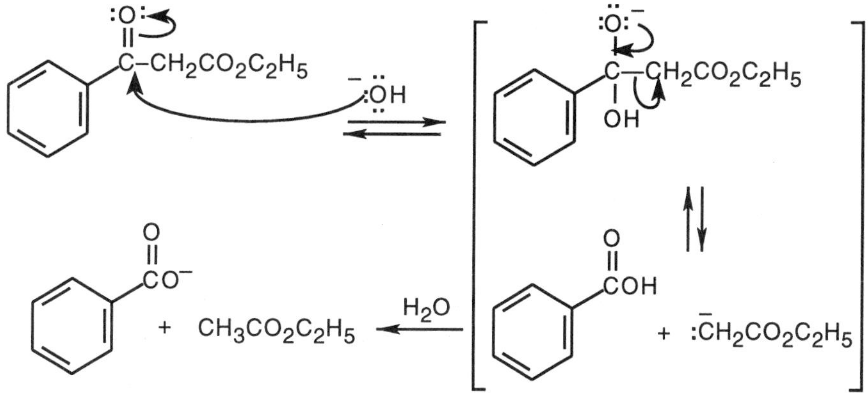

Hydroxide ion can react at two different sites of the ß–keto ester. Abstraction of the acidic α–proton is more favorable but is reversible and does not lead to product. Addition of hydroxide ion to the carbonyl group, followed by irreversible elimination of ethyl acetate, accounts for the observed product.

23.13 Diethyl oxalate gives good yields in mixed Claisen reactions because it has no acidic α hydrogens.

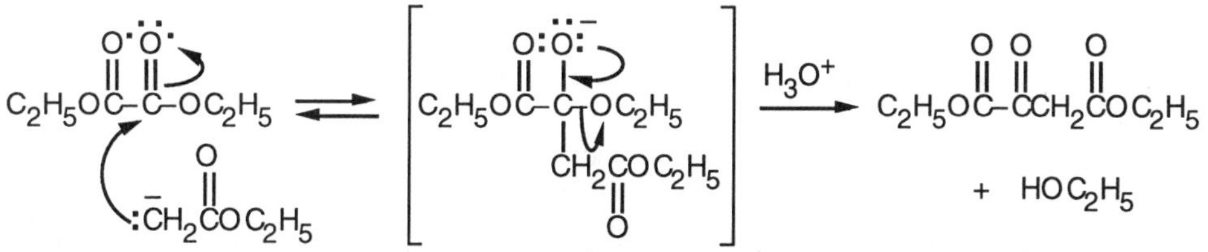

23.14

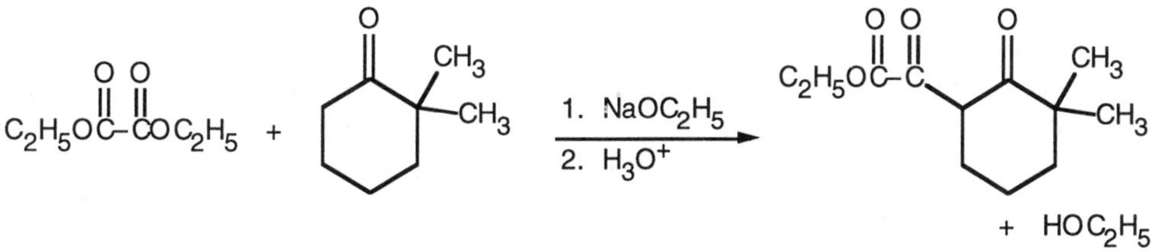

23.15

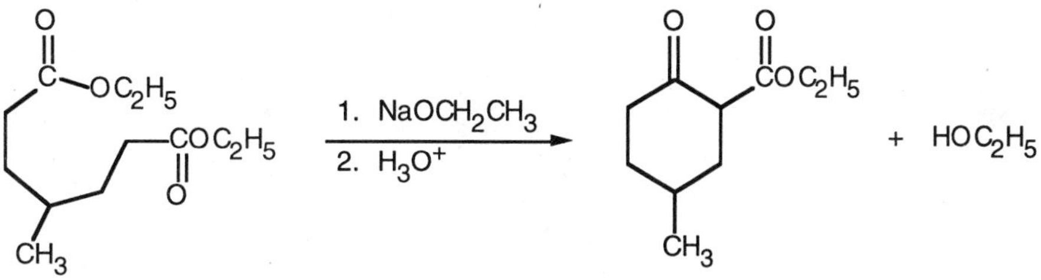

23.16

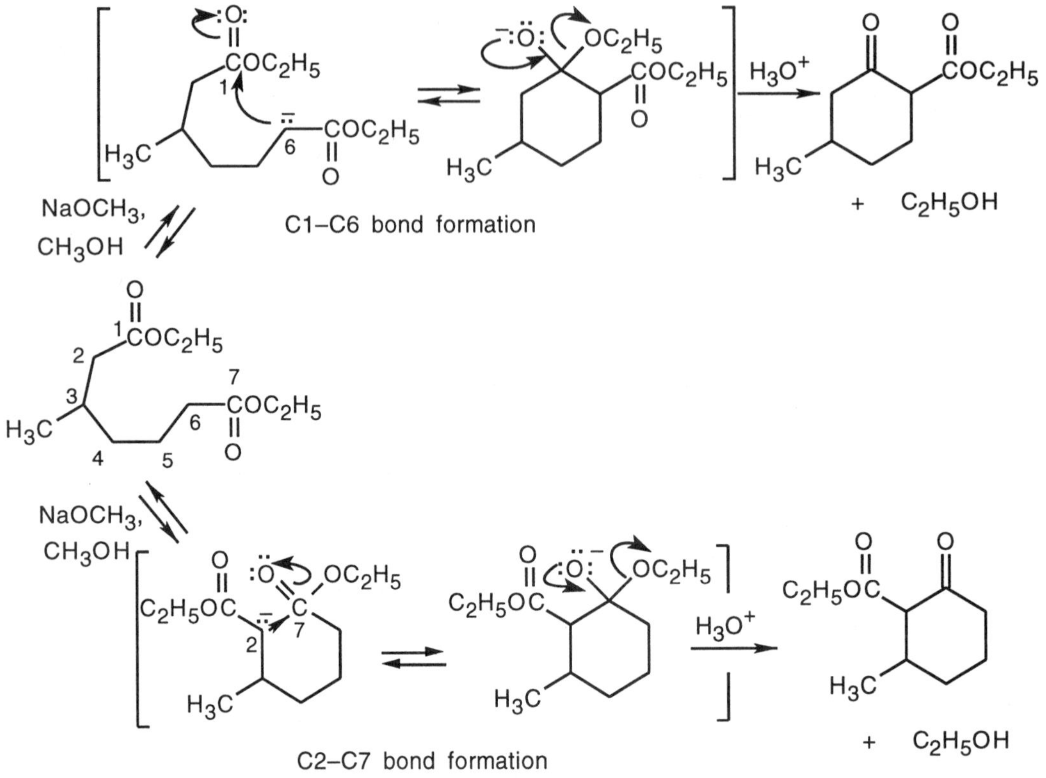

NaOCH₃,
CH₃OH

C1–C6 bond formation

H₃O⁺

+   C₂H₅OH

NaOCH₃,
CH₃OH

C2–C7 bond formation

H₃O⁺

+   C₂H₅OH

Unlike diethyl heptanedioate, diethyl 3–methylheptanedioate is unsymmetrical. Two different enolates can form, and each can cyclize to a different product.

23.17

Michael Donor	Michael Acceptor	Product

a)

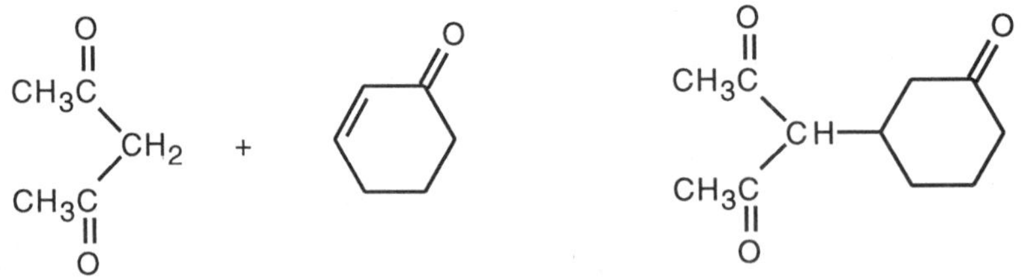

Michael Donor	Michael Acceptor	Product

b)

$$CH_3\overset{\overset{\displaystyle O}{\|}}{C}-CH_2-\overset{\overset{\displaystyle O}{\|}}{C}CH_3 \quad + \quad H_2C=CHC\equiv N$$

$$CH_3\overset{\overset{\displaystyle O}{\|}}{C}-CHCH_2CH_2C\equiv N, \ CH_3\overset{\overset{\displaystyle O}{\|}}{C}$$

c)

$$CH_3\overset{\overset{\displaystyle O}{\|}}{C}-CH_2-\overset{\overset{\displaystyle O}{\|}}{C}CH_3 \quad + \quad \underset{CH_3}{\underset{|}{CH}}=CH\overset{\overset{\displaystyle O}{\|}}{C}OCH_3$$

$$CH_3\overset{\overset{\displaystyle O}{\|}}{C}-\underset{CH_3\overset{\overset{\displaystyle }{\|}}{C}}{CH}-\underset{CH_3}{\underset{|}{CH}}CH_2\overset{\overset{\displaystyle O}{\|}}{C}OCH_3$$

23.18

a)

$$C_2H_5O\overset{\overset{\displaystyle O}{\|}}{C}-CH_2-\overset{\overset{\displaystyle O}{\|}}{C}OC_2H_5 \quad + \quad H_2C=CH\overset{\overset{\displaystyle O}{\|}}{C}CH_3$$

$$C_2H_5O\overset{\overset{\displaystyle O}{\|}}{C}-CHCH_2CH_2\overset{\overset{\displaystyle O}{\|}}{C}CH_3, \ C_2H_5O\overset{\overset{\displaystyle O}{\|}}{C}$$

b)

cyclopentanone-2-COCH₃ $+ \quad H_2C=CH\overset{\overset{\displaystyle O}{\|}}{C}CH_3$

product: cyclopentanone-2,2-disubstituted with CH₂CH₂COCH₃ and COCH₃

c)

$$O_2NCH_3 \quad + \quad H_2C=CH\overset{\overset{\displaystyle O}{\|}}{C}CH_3 \quad \longrightarrow \quad O_2NCH_2CH_2CH_2\overset{\overset{\displaystyle O}{\|}}{C}CH_3$$

23.19

a)

cyclopentanone $+$ pyrrolidine (N–H) $\longrightarrow$ 1-pyrrolidinyl-cyclopentene $+ \quad H_2O$

b)

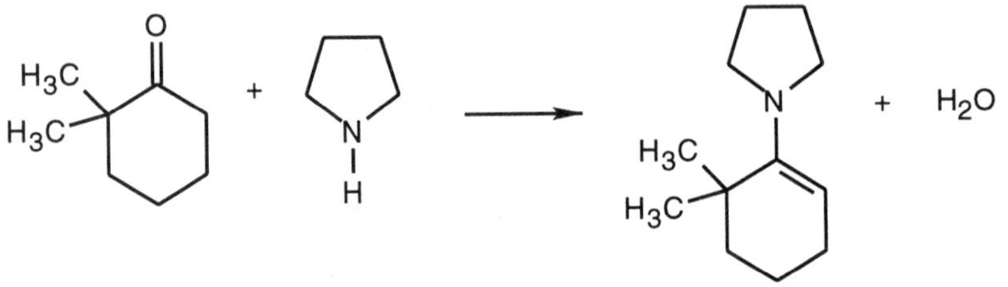

23.20

Enamine	Michael Acceptor	Product (after hydrolysis)

a)

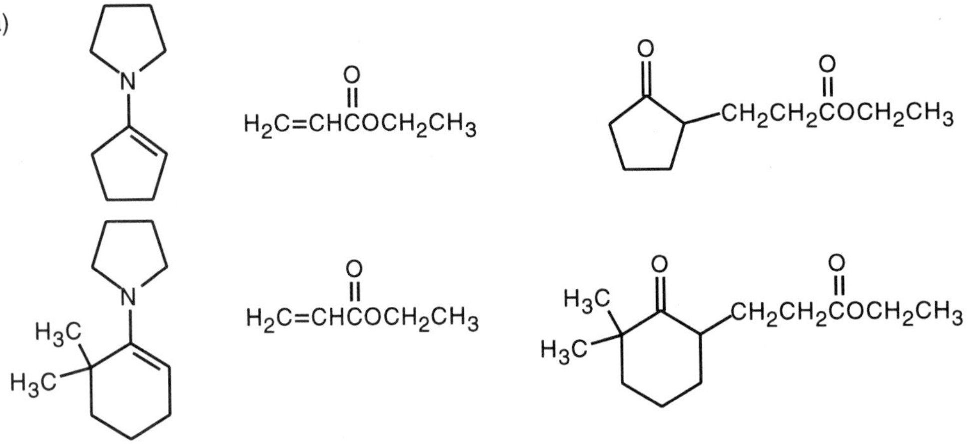

b)

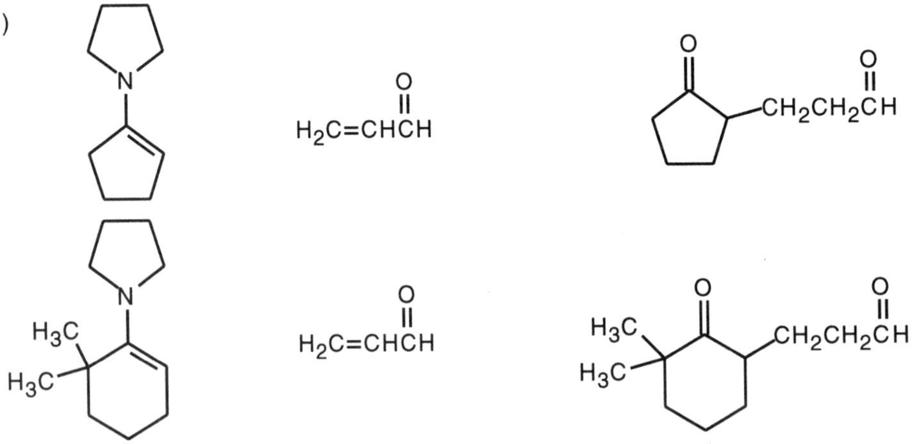

23.21

a)

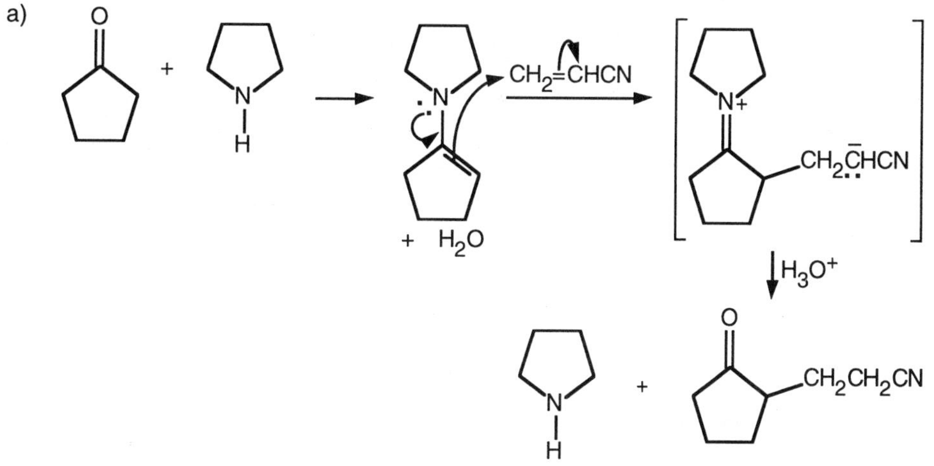

b)

23.22

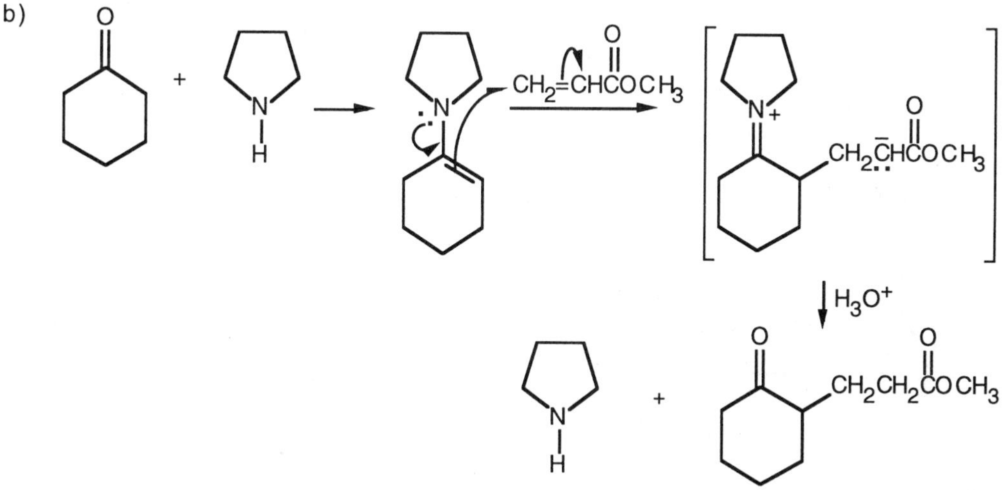

Michael Donor   Michael Acceptor

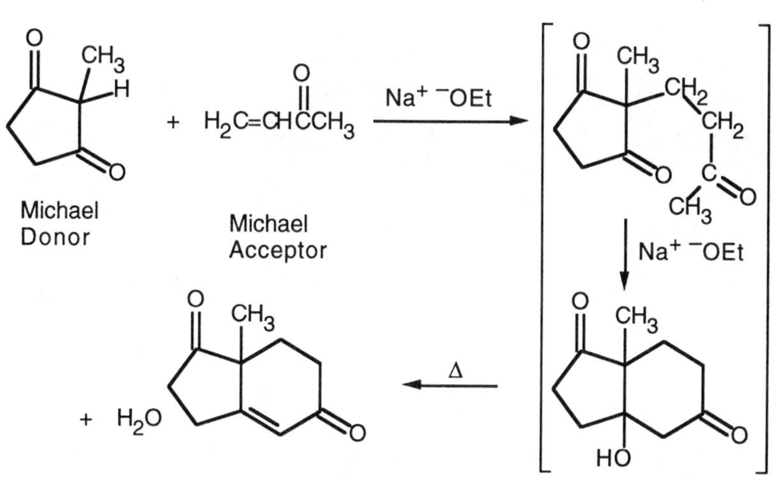

23.23

$CH_3CH_2\overset{O}{\underset{\|}{C}}CH=CH_2$ + [H₃C-methyl dione structure with CH₃, CH₃] $\xrightarrow{Na^+ \ ^-OEt}$ [bicyclic intermediate structure with H₃C, O, CH₃, CH₃, CH₂CH₃]

$\downarrow Na^+ \ ^-OEt$

[product with H₃C, O, CH₃, CH₃, OH]

$\xleftarrow{\Delta}$ [bicyclic dione structure with H₃C, O, CH₃, CH₃, CH₃] + H₂O

23.24–23.25

a) (CH₃)₃CCHO has no α protons and does not undergo aldol self-condensation.

b)

2 [cyclobutanone] $\xrightarrow{NaOH, \ C_2H_5OH}$ [OH, O intermediate structure] $\xrightarrow{\Delta}$ [cyclobutylidene cyclobutanone] + H₂O

c) Benzophenone does not undergo aldol self-condensation.

d)

$2 \ CH_3CH_2\overset{O}{\underset{\underset{CH_2CH_3}{|}}{C}}$ $\xrightarrow{\underset{C_2H_5OH}{NaOH,}}$ $\left[CH_3CH_2\overset{HO}{\underset{\underset{CH_2CH_3}{|}}{C}}-\overset{H_3C}{\underset{|}{CH}}-\overset{O}{\underset{\|}{C}}CH_2CH_3 \right]$

$\downarrow \Delta$

$CH_3CH_2\overset{H_3C}{\underset{\underset{CH_2CH_3}{|}}{C}}=C-\overset{O}{\underset{\|}{C}}CH_2CH_3$ + H₂O

e)

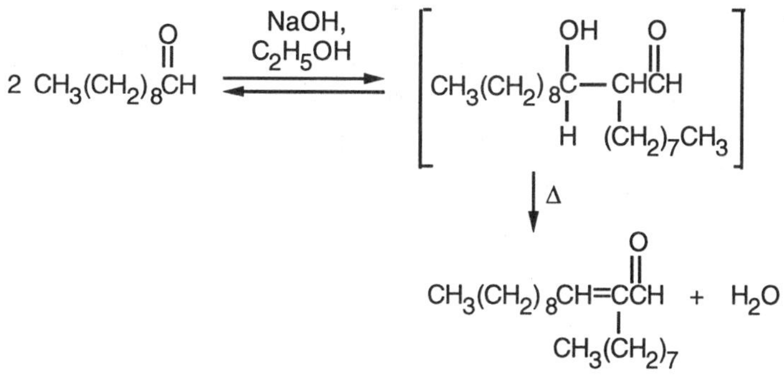

f) C₆H₅CH=CHCHO does not undergo aldol reactions.

23.26

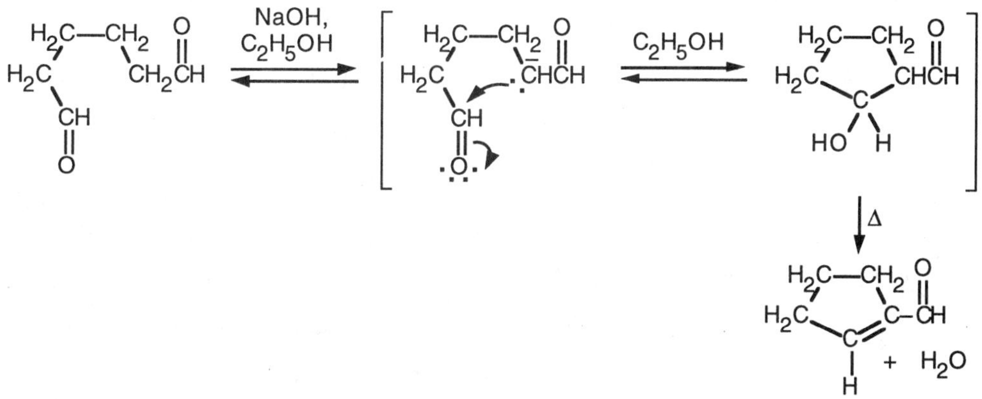

23.27

a)

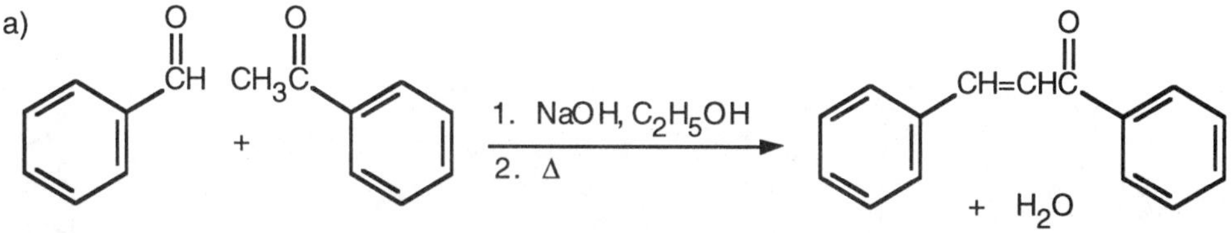

b)

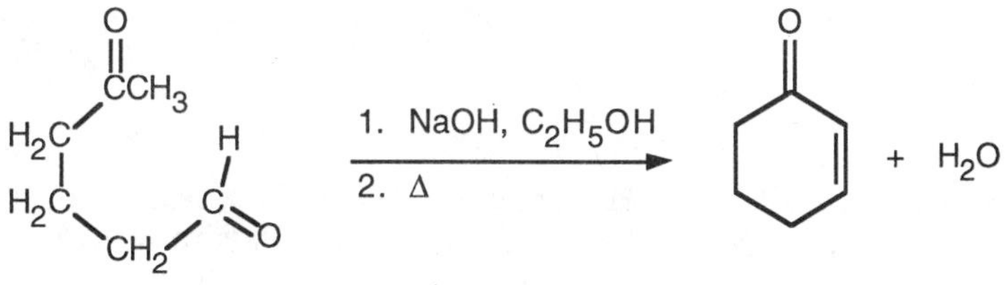

c)

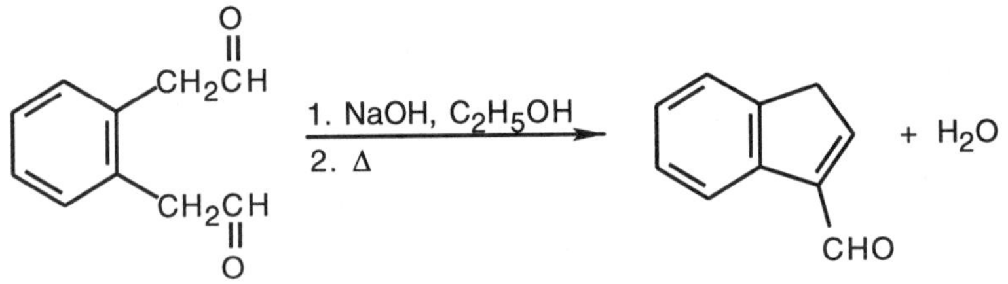

1. NaOH, C_2H_5OH
2. Δ

$+ H_2O$

d)

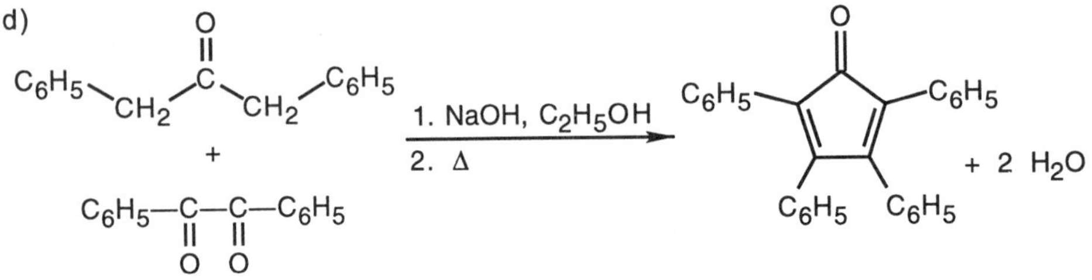

1. NaOH, C_2H_5OH
2. Δ

$+ 2 H_2O$

23.28

$\underline{A}$ $\underline{B}$ $\underline{C}$

$\underline{D}$

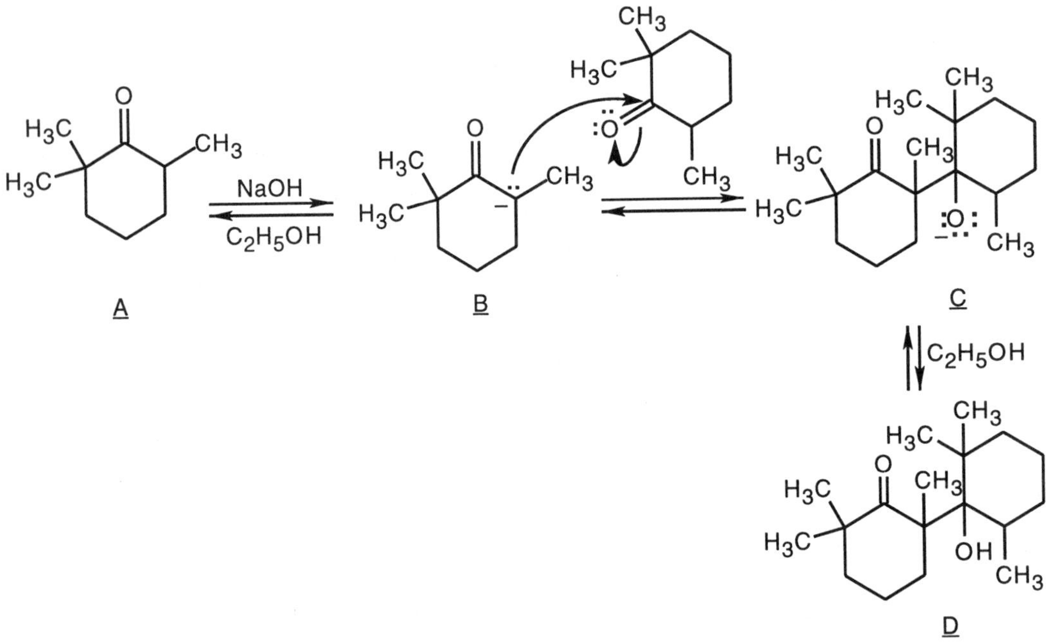

An aldol condensation involves a series of reversible equilibrium steps. In general, formation of product is favored by the dehydration of the ß–hydroxy ketone to form a conjugated enone. Here dehydration to form conjugated product can't occur. In addition, the $\underline{B}$ ⇌ $\underline{C}$ equilibrium favors $\underline{B}$ because of steric hindrance.

23.29

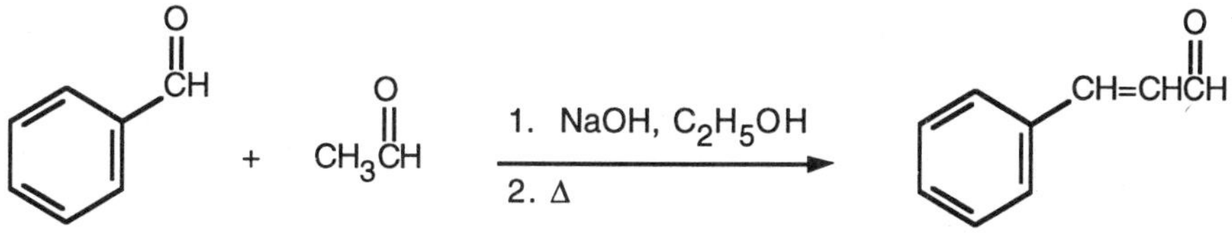

23.30

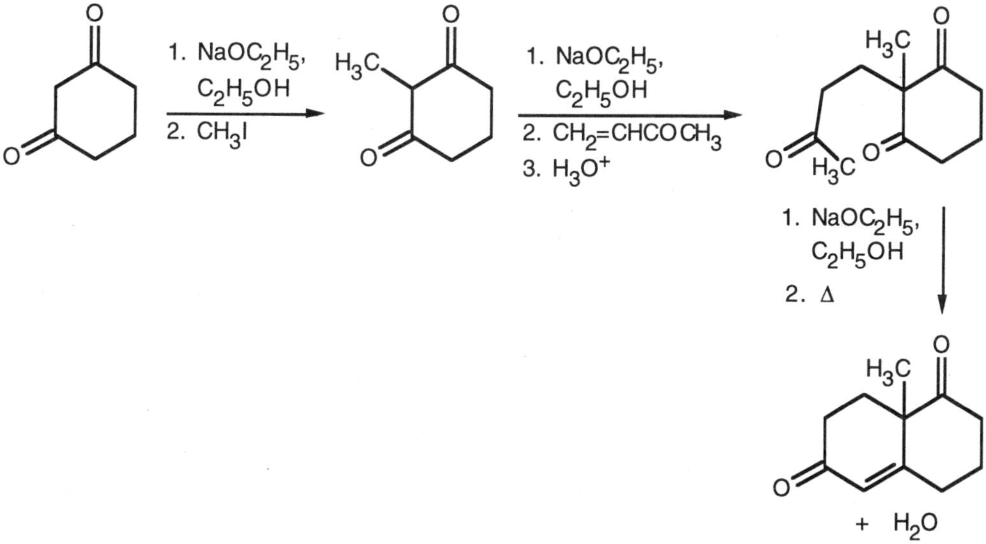

+ H_2O

This sequence of reactions consists of an alkylation of a 1,3-diketone, followed by a Robinson annulation.

23.31 The first step of an aldol condensation is enolate formation. The ketone shown here does not enolize because double bonds at the bridgehead of small bicyclic ring systems are too strained to form. Since the bicyclic ketone does not enolize, it doesn't undergo aldol condensation.

23.32

a)

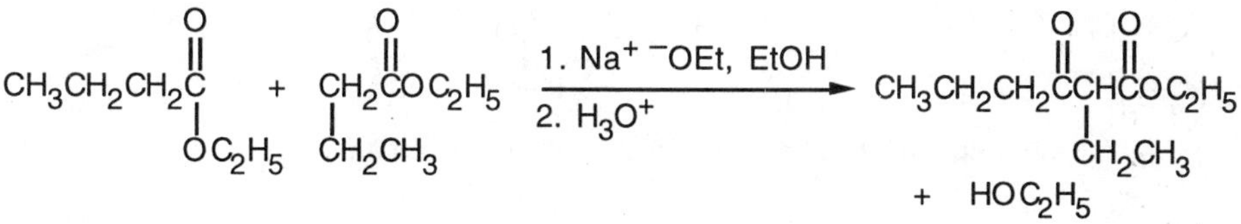

b)

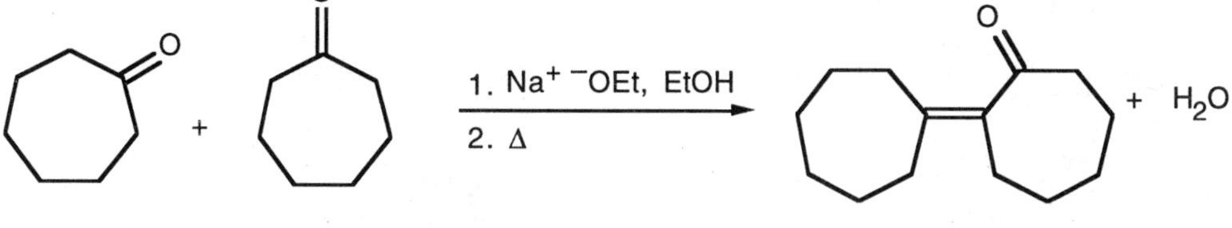

c)

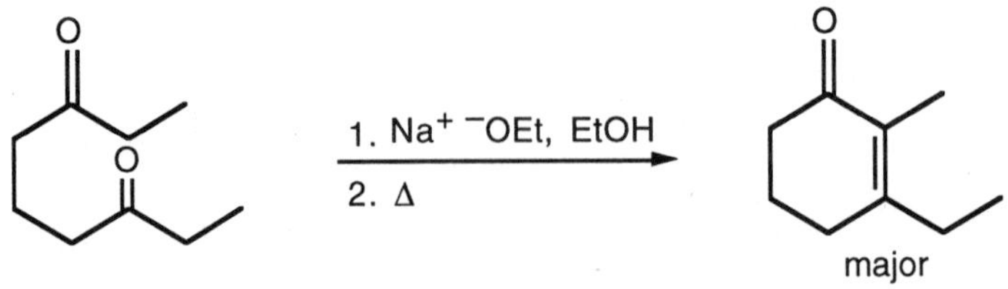

1. Na⁺ ⁻OEt, EtOH

2. Δ

major

d)

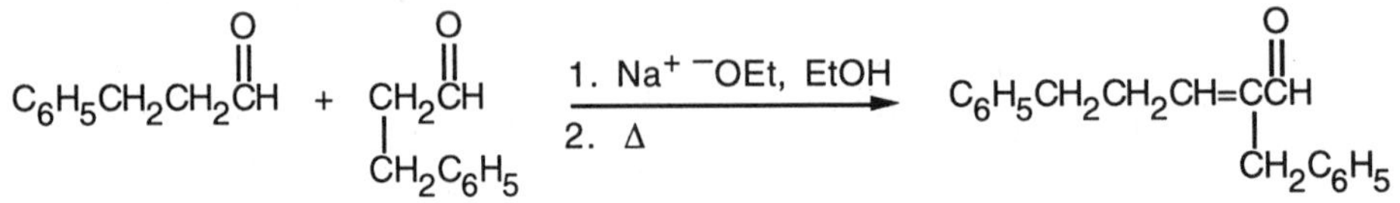

$C_6H_5CH_2CH_2CH$ + CH_2CH

1. Na⁺ ⁻OEt, EtOH

2. Δ

$C_6H_5CH_2CH_2CH=CCH$

23.33

a)

$CH_3CO_2CH_3$ + $CH_3CH_2CO_2CH_3$

Na⁺ ⁻OC₂H₅, C₂H₅OH

self-condensation products   mixed condensation products

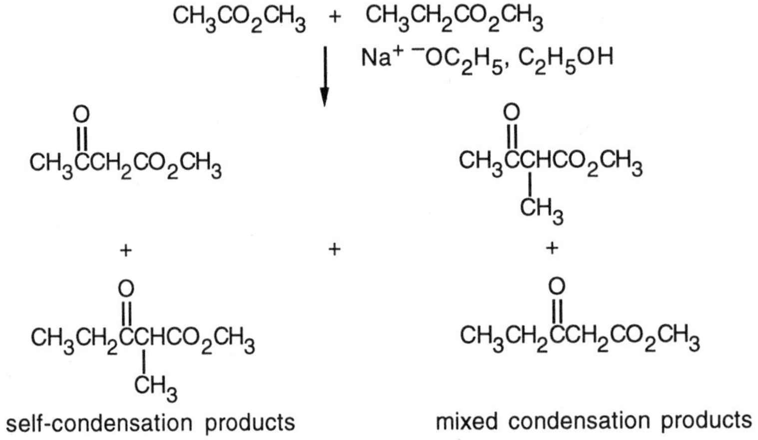

Approximately equal amounts of each product will form.

b)

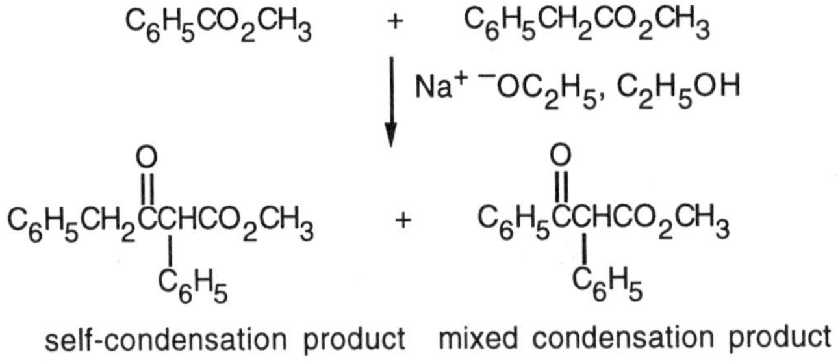

$C_6H_5CO_2CH_3$ + $C_6H_5CH_2CO_2CH_3$

Na⁺ ⁻OC₂H₅, C₂H₅OH

self-condensation product   mixed condensation product

The mixed condensation product predominates.

c)

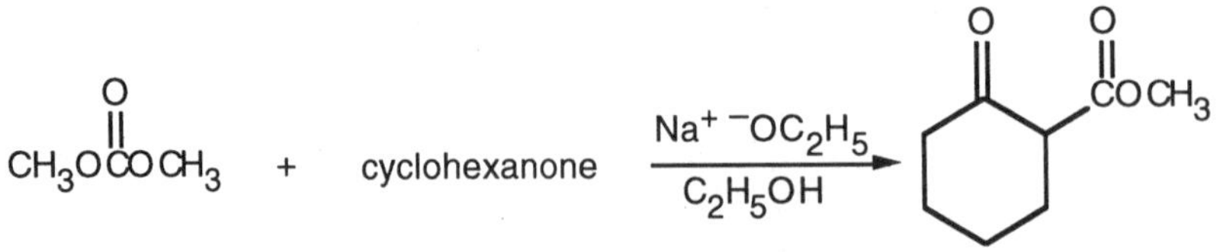

$CH_3O\overset{O}{\overset{||}{C}}OCH_3$ + cyclohexanone

Na⁺ ⁻OC₂H₅
―――――――――
C₂H₅OH

This is the only *Claisen* monocondensation product.

d)

$$C_6H_5\overset{\displaystyle O}{\overset{\|}{C}}H + CH_3CO_2CH_3 \xrightarrow[\text{C}_2\text{H}_5\text{OH}]{\text{Na}^+\ ^-\text{OC}_2\text{H}_5} CH_3\overset{\displaystyle O}{\overset{\|}{C}}CH_2CO_2CH_3 + C_6H_5CH=CHCO_2CH_3$$

self-condensation   mixed condensation
product   product

The mixed Claisen product is the major product.

23.34 a) Several other products are formed in addition to the one pictured. Self-condensation of acetaldehyde and acetone (less likely) can occur, and an additional mixed product is formed.

b) Conjugate addition occurs between a carbon nucleophile and an α,β–unsaturated carbonyl electrophile (the Michael reaction). That is, addition occurs at the double bond, not at the carbonyl group.

$$CH_3\overset{\displaystyle O}{\overset{\|}{C}}CH=CH_2 + CH_2(CO_2C_2H_5)_2 \xrightarrow{\text{base}} CH_3\overset{\displaystyle O}{\overset{\|}{C}}CH_2CH_2CH(CO_2C_2H_5)_2$$

c) There are two problems with this reaction. (1) Michael reactions occur in low yield with mono-ketones. Formation of the enamine, followed by the Michael reaction, gives a higher yield of product. (2) Addition can occur on either side of the ketone to give a mixture of products.

d) Internal aldol condensation of 2,6–heptanedione can product a four-membered ring or a six-membered ring. The six-membered ring is more likely to form because it is less strained.

e) Michael addition occurs at the most acidic position of the Michael donor.

23.35 Michael reactions occur between stabilized enolate anions and α,β–unsaturated carbonyl compounds. Learn to locate these components in possible Michael products. Usually, it is easier to recognize the enolate nucleophile; in a) the nucleophile is the ethyl acetoacetate anion.

$$CH_3\overset{O}{\overset{\|}{C}}CH-CH_2CH_2\overset{O}{\overset{\|}{C}}C_6H_5 \quad \text{comes from} \quad CH_3\overset{O}{\overset{\|}{C}}CH_2$$
$$\underset{\overset{|}{CO_2CH_3}}{} \qquad\qquad\qquad\qquad \underset{\overset{|}{CO_2CH_3}}{}$$

The rest of the compound is the Michael acceptor. Draw a double bond in conjugation with an electron-withdrawing group in this part of the molecule.

$$CH_3\overset{O}{\overset{\|}{C}}CH-CH_2CH_2\overset{O}{\overset{\|}{C}}C_6H_5 \quad \text{comes from} \quad CH_2=CH\overset{O}{\overset{\|}{C}}C_6H_5$$
$$\underset{\overset{|}{CO_2CH_3}}{}$$

$$CH_3\overset{O}{\overset{\|}{C}}CH_2 \;\; + \;\; CH_2=CH\overset{O}{\overset{\|}{C}}C_6H_5 \quad \xrightarrow[\text{2. } H_3O^+]{\text{1. } NaOC_2H_5} \quad CH_3\overset{O}{\overset{\|}{C}}CHCH_2CH_2\overset{O}{\overset{\|}{C}}C_6H_5$$
$$\underset{\overset{|}{CO_2CH_3}}{} \qquad\qquad\qquad\qquad\qquad\qquad\qquad\qquad \underset{\overset{|}{CO_2CH_3}}{}$$

Michael donor Michael acceptor

b) When the Michael product has been decarboxylated after the addition reaction, it is more difficult to recognize the original enolate anion.

$$CH_3\overset{O}{\overset{\|}{C}}CH_2 \;\; + \;\; CH_2=CH\overset{O}{\overset{\|}{C}}CH_3 \quad \xrightarrow[\text{2. } H_3O^+]{\text{1. } NaOC_2H_5} \quad CH_3\overset{O}{\overset{\|}{C}}CHCH_2CH_2\overset{O}{\overset{\|}{C}}CH_3$$
$$\underset{\overset{|}{CO_2CH_3}}{} \qquad\qquad\qquad\qquad\qquad\qquad\qquad\qquad \underset{\overset{|}{CO_2CH_3}}{}$$

Michael donor Michael acceptor

$$CH_3\overset{O}{\overset{\|}{C}}CHCH_2CH_2\overset{O}{\overset{\|}{C}}CH_3 \quad \xrightarrow[\Delta]{H_3O^+} \quad CH_3\overset{O}{\overset{\|}{C}}CH_2CH_2CH_2\overset{O}{\overset{\|}{C}}CH_3$$
$$\underset{\overset{|}{CO_2CH_3}}{}$$

c) $(CH_3O_2C)_2CH_2 \;\; + \;\; CH_2=CHCN \quad \xrightarrow[\text{2. } H_3O^+]{\text{1. } NaOC_2H_5} \quad (CH_3O_2C)_2CHCH_2CH_2CN$

Michael donor Michael acceptor

d) $CH_3CH_2NO_2 \;\; + \;\; CH_2=CH\overset{O}{\overset{\|}{C}}OCH_3 \quad \xrightarrow[\text{2. } H_3O^+]{\text{1. } NaOC_2H_5} \quad CH_3\overset{NO_2}{\overset{|}{C}}HCH_2CH_2\overset{O}{\overset{\|}{C}}OCH_3$

Michael donor Michael acceptor

e) $(CH_3O_2C)_2CH_2 \;\; + \;\; CH_2=CHNO_2 \quad \xrightarrow[\text{2. } H_3O^+]{\text{1. } NaOC_2H_5} \quad (CH_3O_2C)_2CHCH_2CH_2NO_2$

Michael donor Michael acceptor

23.36

(a) LiAlH$_4$, then H$_3$O$^+$; (b) POCl$_3$, pyridine (c) KMnO$_4$, H$_3$O$^+$; (d) CH$_3$OH, H$^+$; (e) Na$^+$ $^-$OEt; (f) H$_3$O$^+$; (g) Na$^+$ $^-$OEt, then CH$_3$Br; (h) H$_3$O$^+$

23.37 If cyclopentanone and base are mixed first, aldol self-condensation of cyclopentanone can occur before ethyl formate is added. If both carbonyl components are mixed together before adding base, the more favorable mixed Claisen condensation occurs with less competition from the aldol self-condensation reaction.

23.38

This is a reverse Claisen reaction.

23.39 Two different reactions are possible when ethyl acetoacetate reacts with ethoxide anion. One possibility involves attack of ethoxide ion on the carbonyl carbon, followed by elimination of the anion of ethyl acetate—a reverse Claisen reaction similar to the one illustrated in 23.38. More likely, however, is the acid-base reaction of ethoxide ion and a doubly activated α–hydrogen of ethyl acetoacetate.

The resonance-stabilized acetoacetate anion is no longer reactive toward nucleophiles, and no further reaction occurs at room temperature. Elevated temperatures are required to make the cleavage reaction proceed.

23.40

$$CH_2(CO_2C_2H_5)_2 \ + \ ^-OC_2H_5 \ \rightleftharpoons \ ^-:CH(CO_2C_2H_5)_2 \ + \ HOC_2H_5$$

23.41 Here is an outline of the Perkin reaction.

a) Acetate ion abstracts an α–proton from acetic anhydride to form the enolate anion.

b) The enolate anion attacks the electrophilic carbon of benzaldehyde.

c)   Protonation of the adduct, followed by heat, forms an unsaturated intermediate.

d) Treatment of the unsaturated intermediate with H_2O cleaves the anhydride and yields cinnamic acid.

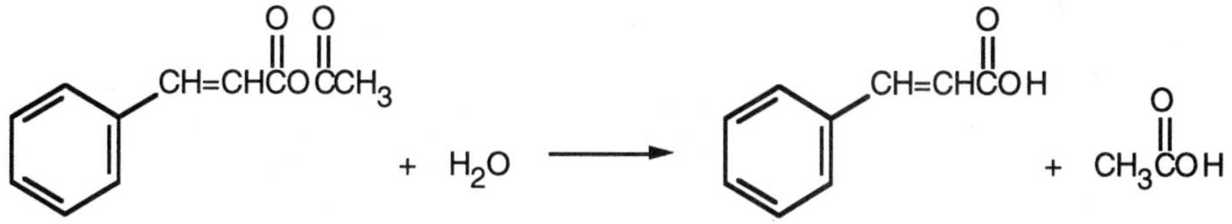

23.42

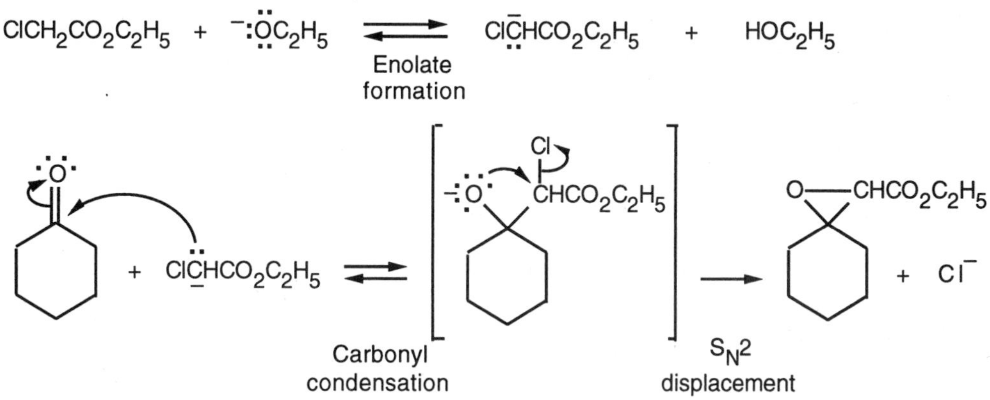

23.43

a)

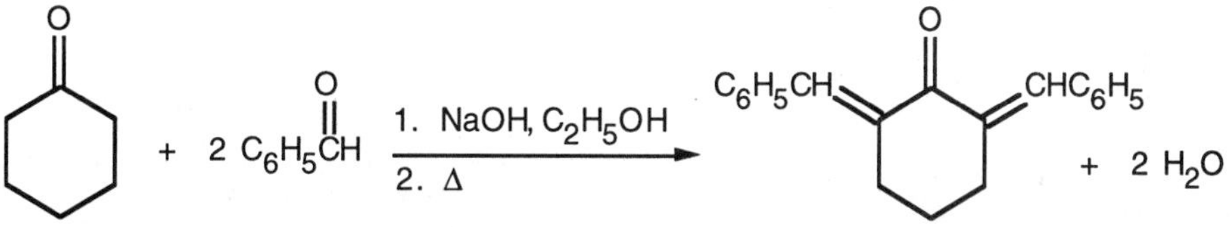

b)

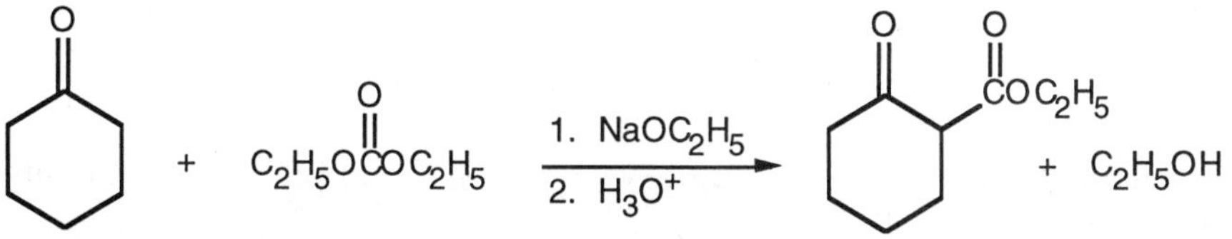

c)

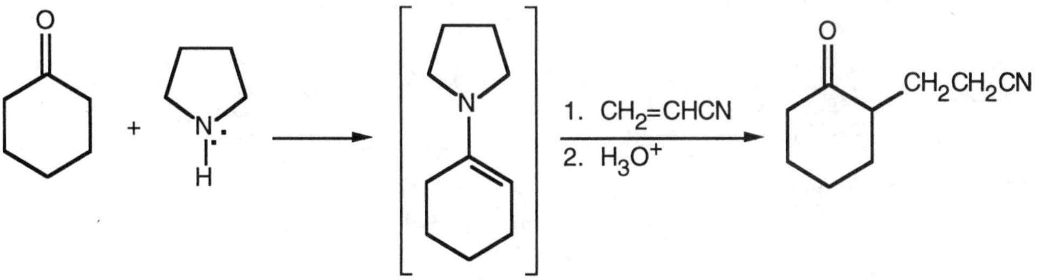

d)

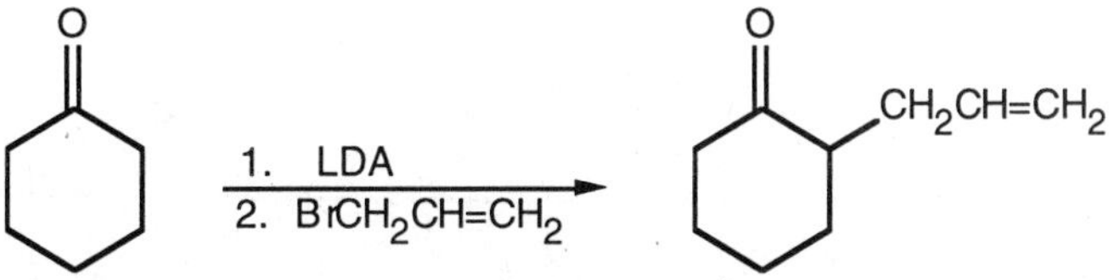

23.44

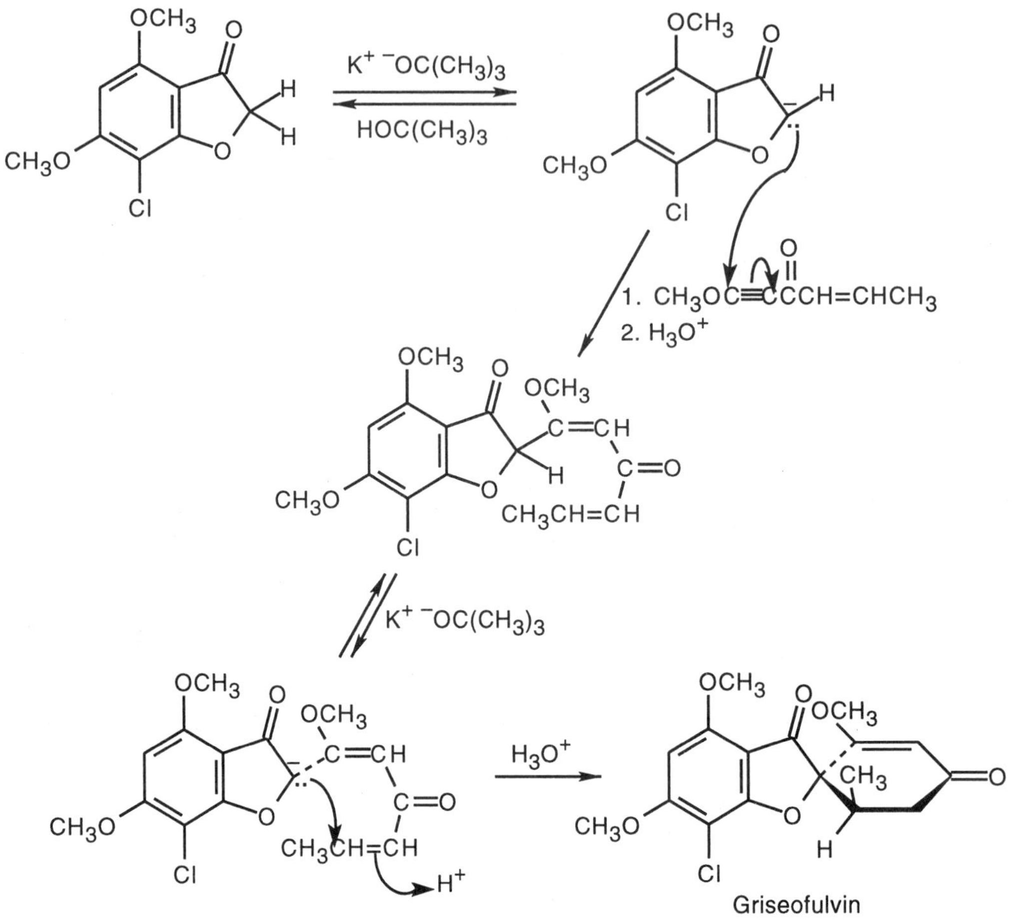

Two Michael reactions are involved in the key step that forms the cyclohexenone ring.

23.45

a)

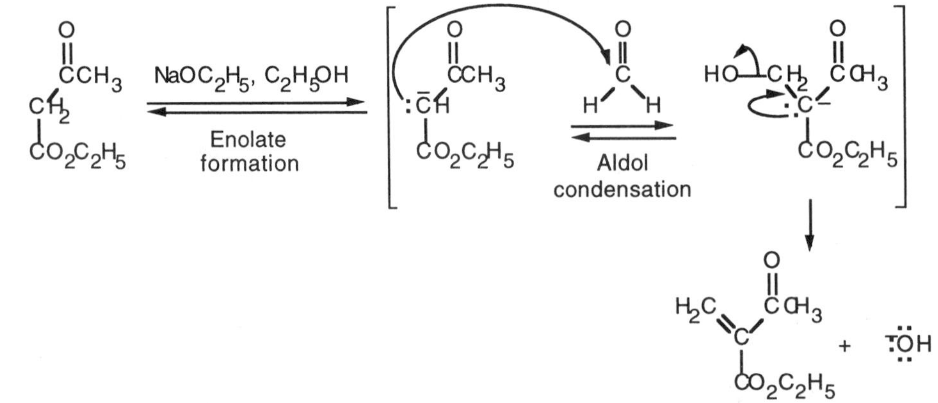

b)

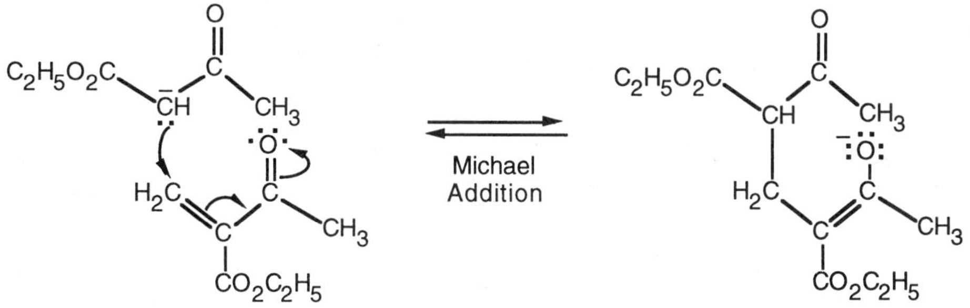

Michael Addition

23.46

3rd)

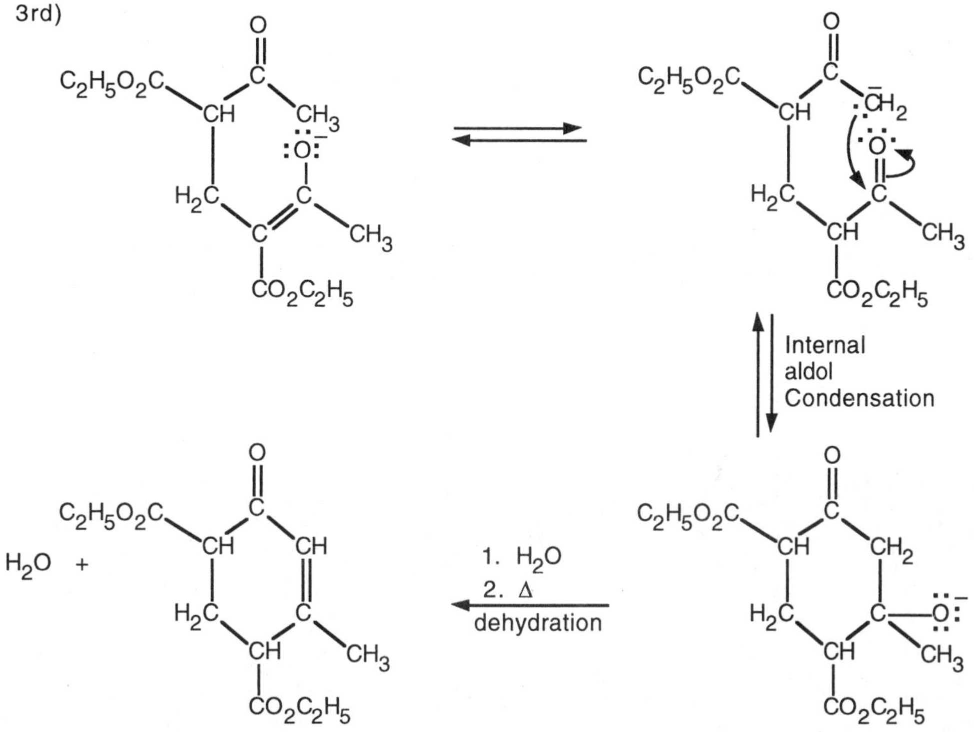

Internal aldol Condensation

1. H_2O
2. Δ
dehydration

H_2O +

4th)

H_3O^+, Δ

Ester cleavage and decarboxylation of a ß–ketoacid

+ CH_3CH_2OH

+ CO_2

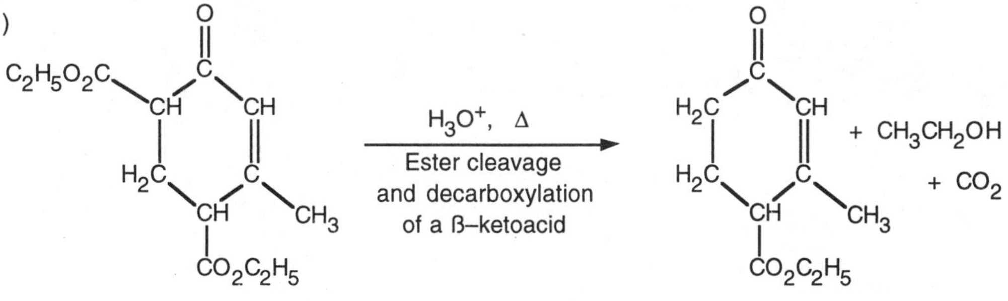

Hagemann's ester

23.47

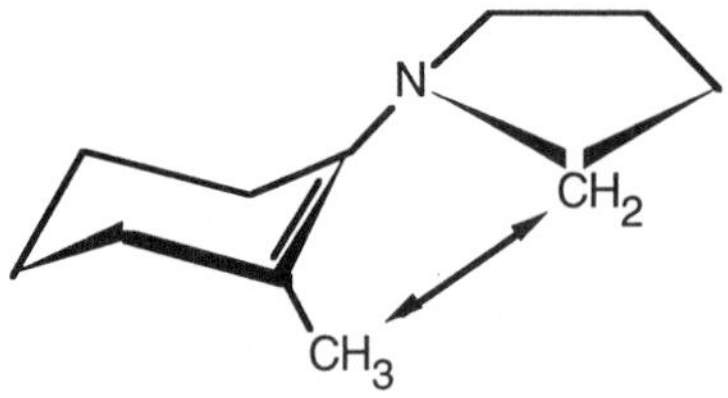

Crowding between the methyl group and the pyrrolidine ring disfavors this enamine.

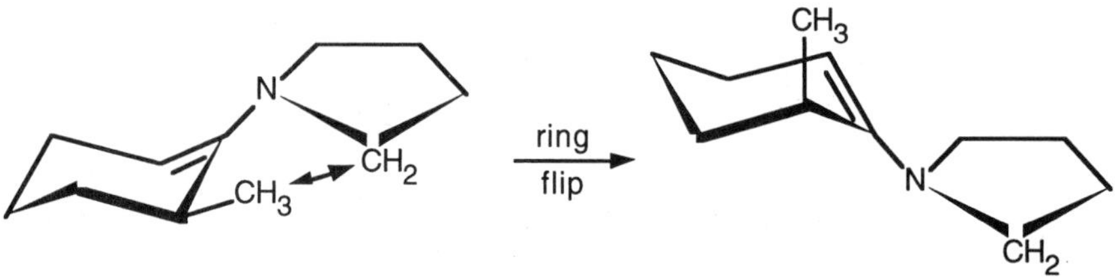

The crowding in this enamine can be relieved by a ring-flip, which puts the methyl group in an axial position. This enamine is the only one formed.

23.48

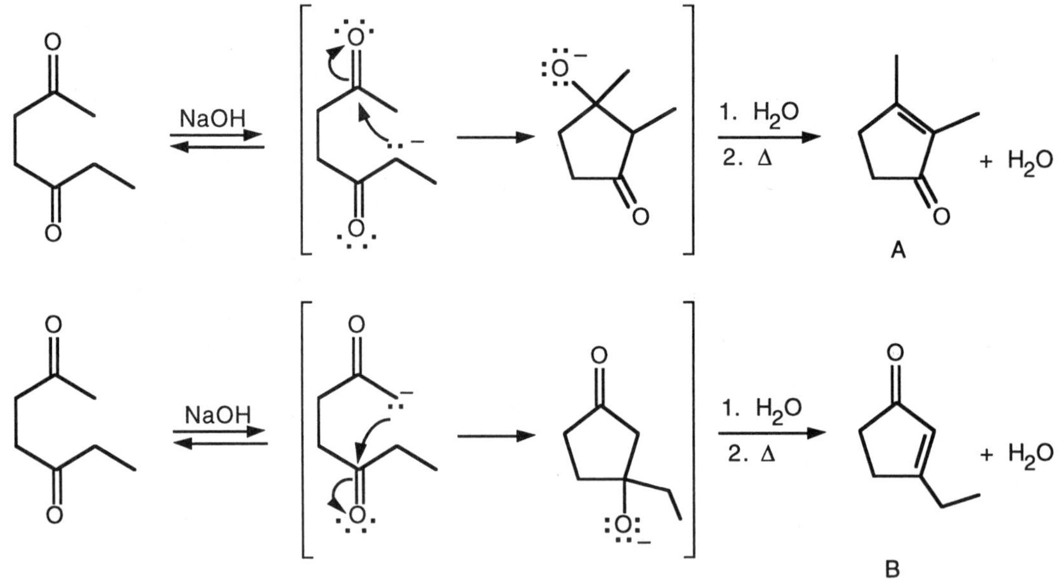

23.49 Product A, which has two singlet methyl groups and no vinylic protons in its ^{1}H NMR, is the major product of the intramolecular cyclization of 2,5–heptanedione.

23.50

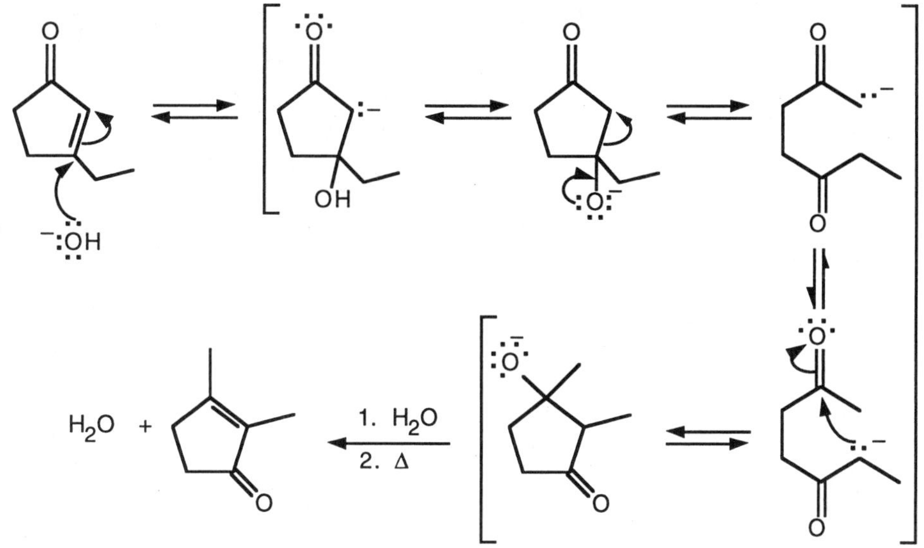

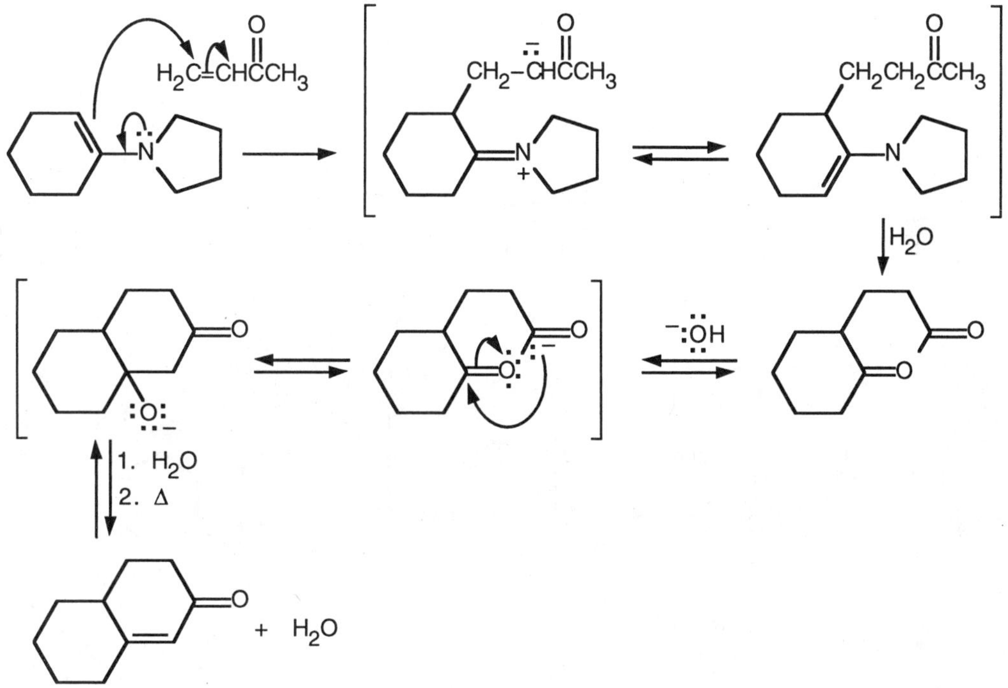

23.51

23.52 a)

b)

23.53 This problem becomes easier if you draw the starting material so that it resembles the product.

23.54

23.55

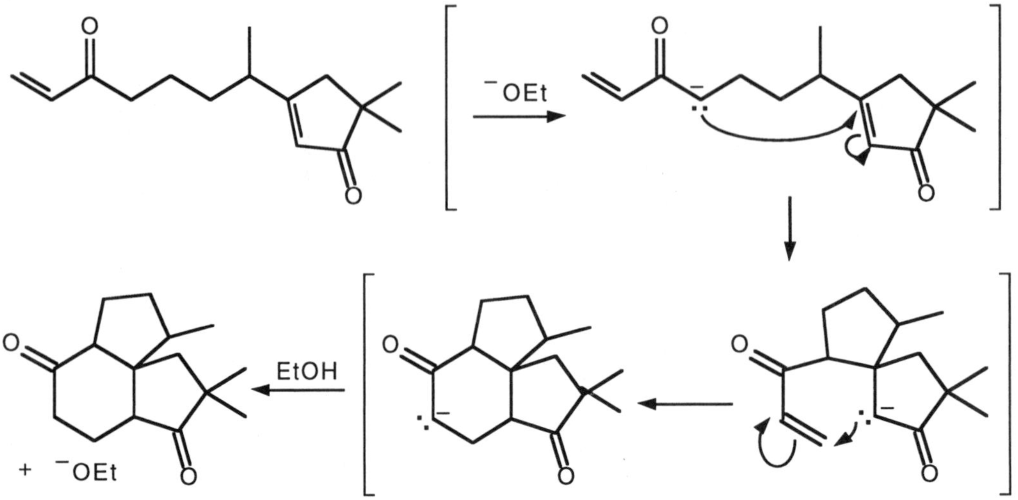

23.56

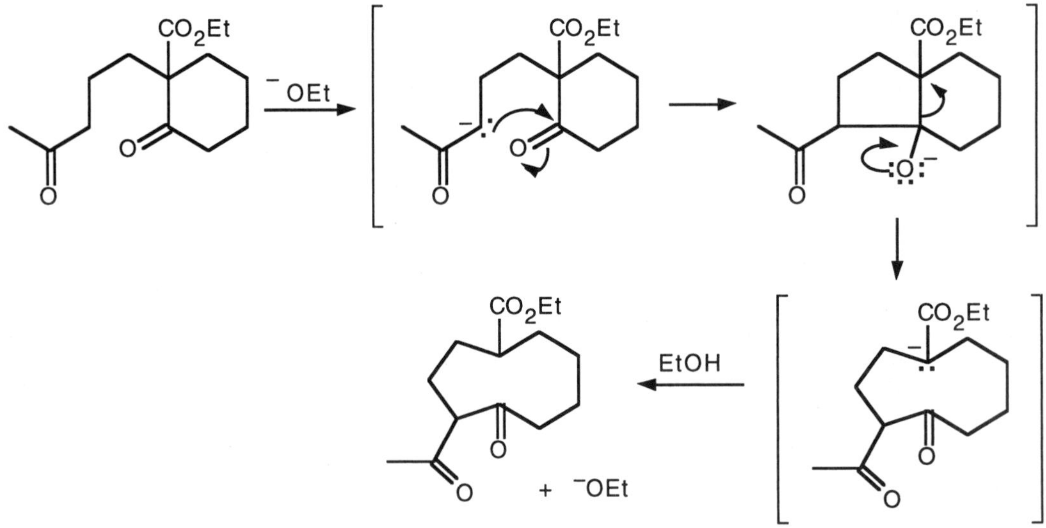

23.57

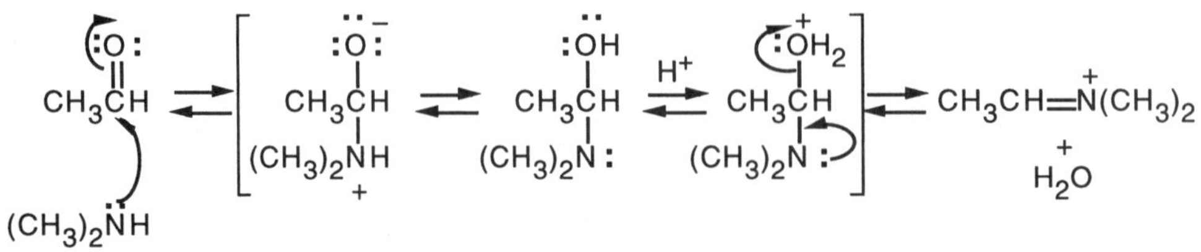

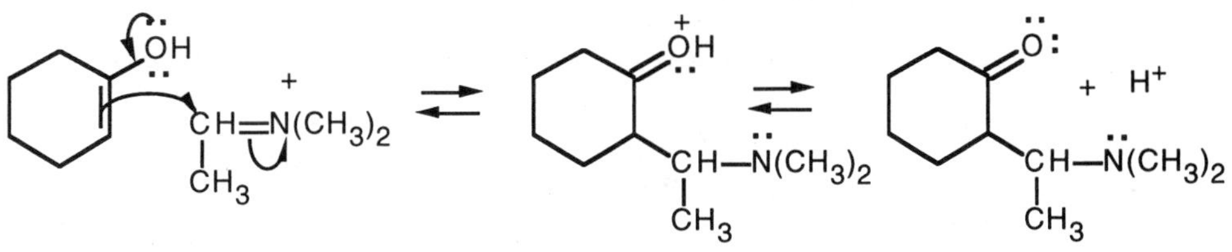

Study Guide for Chapter 23

After studying this chapter, you should be able to:

(1)   Predict the products of:
(a) Aldol condensations (23.1, 23.4, 23.5, 23.7, 23.10, 23.24, 23.25, 23.26);
(b) Claisen condensations (23.11, 23.14, 23.15, 23.16, 23.33, 23.37);
(c) Michael reactions (23.17, 23.18, 23.20, 23.22, 23.23).

(2)   Formulate the mechanism of:
(a) Aldol condensations (23.2, 23.3, 23.9, 23.28, 23.31, 23.48, 23.50);
(b) Claisen condensations (23.12, 23.13, 23.38, 23.39);
(c) Other carbonyl condensations (23.40, 23.41, 23.42, 23.45, 23.46, 23.53, 23.54, 23.55, 23.56, 23.57)).

(3)   Prepare compounds by:
(a) Aldol condensations (23.6, 23.8, 23.27, 23.29);
(b) Michael reactions (23.21, 23.35).

(4)   Use carbonyl condensation reactions in synthesis (23.30, 23.32, 23.43, 23.44, 23.51, 23.52).

24.1

a)

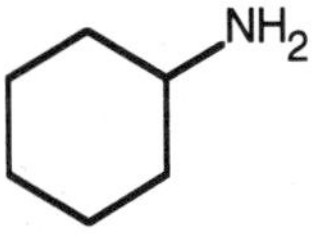

Primary amine

b)

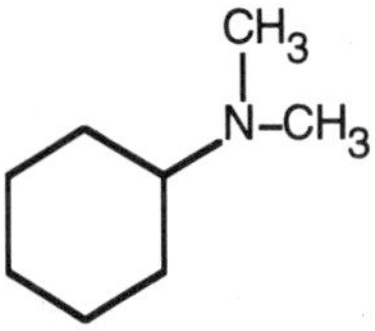

Tertiary amine

c)

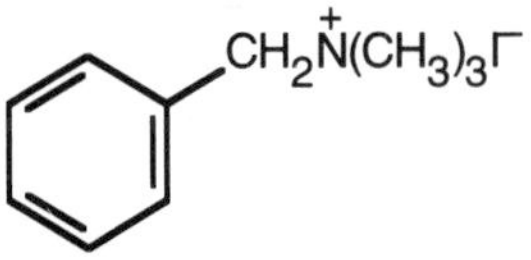

Quaternary ammonium salt

d) $[(CH_3)_2CH]_2NH$

Secondary amine

e)

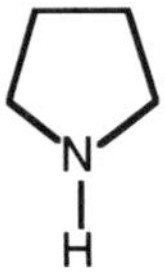

Secondary amine

24.2

a)

$$\underset{\underset{}{\overset{H}{\underset{|}{CH_3NCH_2CH_3}}}}{}$$

N–Methylethylamine

b)

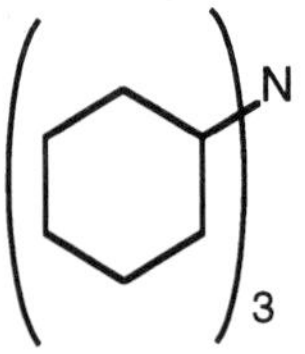

Tricyclohexylamine

c)

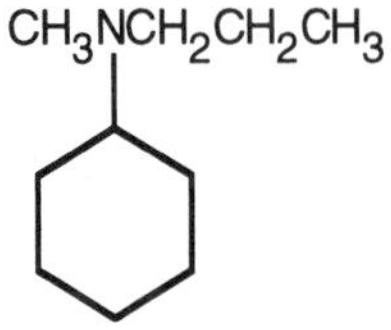

$CH_3NCH_2CH_2CH_3$

N –Methyl–N–propylcyclohexylamine

d)

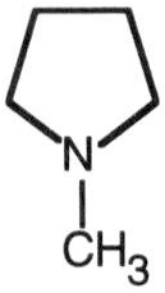

N–Methylpyrrolidine

e) $[(CH_3)_2CH]_2NH$

Diisopropylamine

f)

$$\underset{}{\overset{CH_3}{\underset{|}{H_2NCH_2CH_2CHNH_2}}}$$

1,3–Butanediamine

24.3

a) $(CH_3CH_2)_3N$

Triethylamine

b) $(H_2C{=}CHCH_2)_3N$

Triallylamine

c)

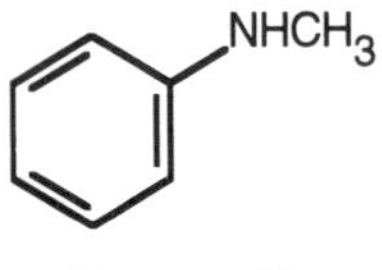

N–Methylaniline

d)

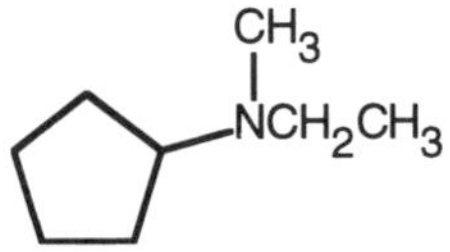

N-Ethyl–N–methylcyclopentylamine

e)

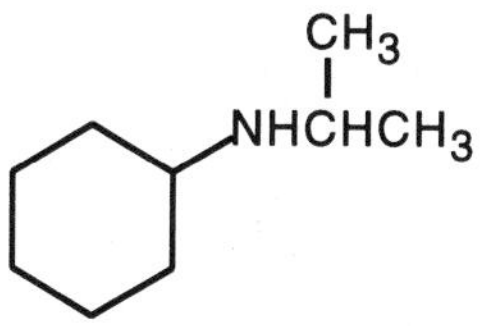

N–Isopropylcyclohexylamine

f)

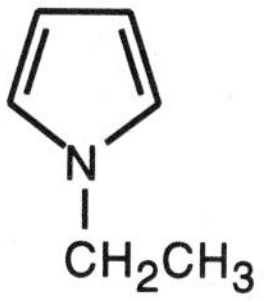

N–Ethylpyrrole

24.4

a)

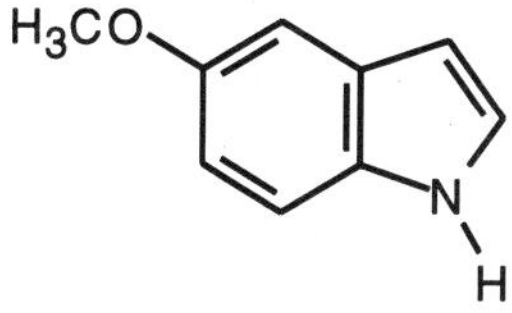

5–Methoxyindole

b)

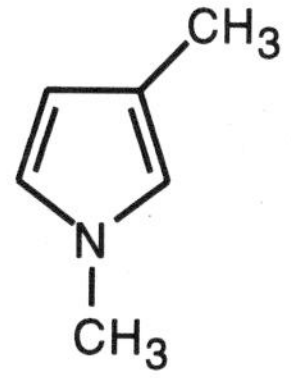

1,3–Dimethylpyrrole

c)

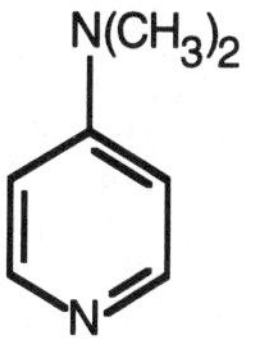

4–(*N*,*N*–Dimethylamino)pyridine

d)

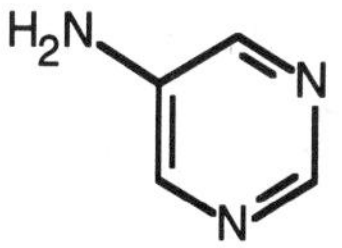

5–Aminopyrimidine

24.5

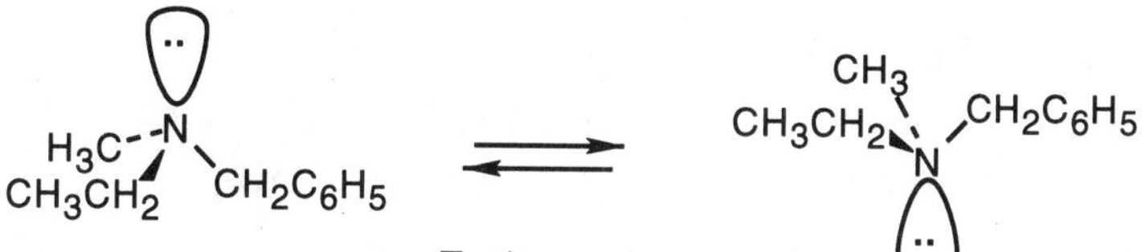

Tertiary amine

Quaternary ammonium salt

Tertiary and quaternary amines such as those pictured above are chiral. Most tertiary amines undergo pyramidal inversion, a relatively low-energy process that interconverts the enantiomers. For quaternary amines, however, interconversion would require breaking and reforming bonds and this does not occur. Each enantiomer is stable and resolvable.

24.6

	More Basic	Less Basic
a)	$CH_3CH_2NH_2$	$CH_3CH_2CONH_2$
	amine	amide
b)	NaOH	CH_3NH_2
	hydroxide	amine
c)	CH_3NHCH_3	CH_3OCH_3
	amine	ether

24.7

$pK_a = 9.33$
stronger acid (smaller pK_a)
weaker base

$CH_3CH_2CH_2\overset{+}{N}H_3$

$pK_a = 10.71$
weaker acid (larger pK_a)
stronger base

The stronger base (propylamine) holds onto a proton more tightly than the weaker base (benzylamine). Thus, the propylammonium ion is less acidic (larger pK_a) than the benzylammonium ion (smaller pK_a).

24.8 Since $K_a \cdot K_b = 10^{-14}$, then $pK_a + pK_b = 14$ and $pK_b = 14 - pK_a$.

$pK_a = 9.33$
Benzylammonium cation

$pK_b = 14 - 9.33 = 4.67$
Benzylamine

$CH_3CH_2CH_2\overset{+}{N}H_3$
$pK_a = 10.71$
Propylammonium cation

$CH_3CH_2CH_2NH_2$
$pK_b = 14 - 10.71 = 3.29$
Propylamine

24.9

N–Protonation
(no resonance stabilization)

O–Protonation
(resonance stabilization)

Protonation occurs at oxygen because an *O*–protonated amide is stabilized by resonance.

24.10

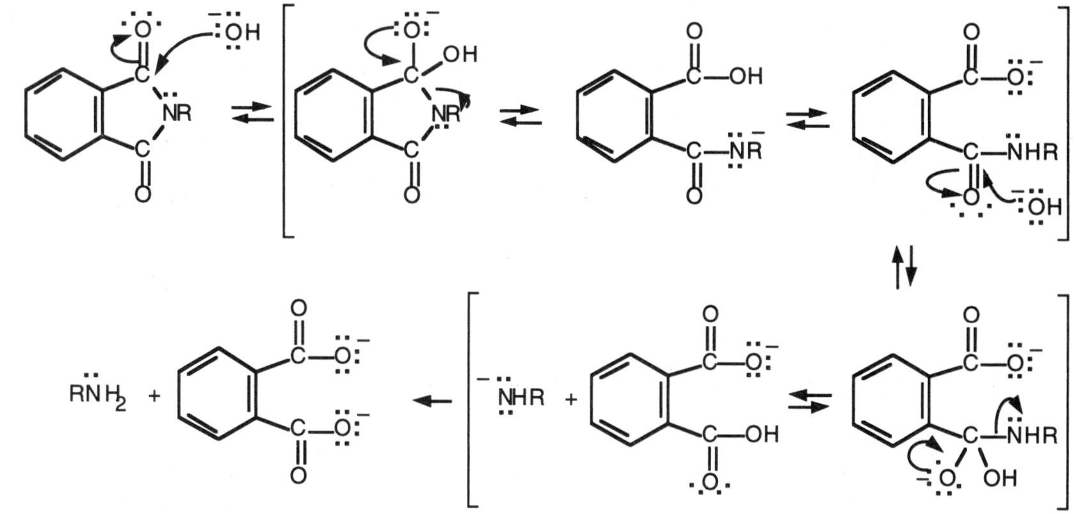

24.11

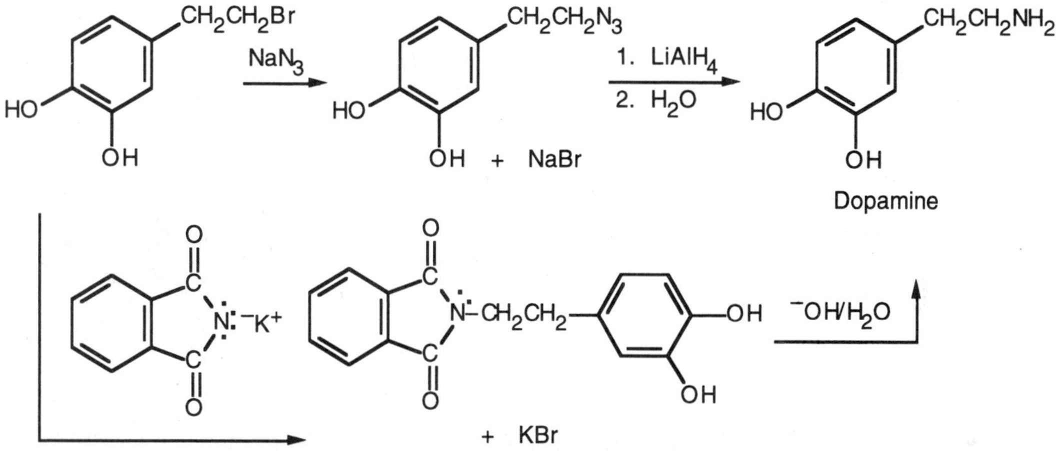

Dopamine

24.12

Amine	Nitrile Precursor	Amide Precursor
a) $CH_3CH_2CH_2NH_2$	CH_3CH_2CN	$CH_3CH_2\overset{O}{\overset{\|}{C}}NH_2$
b) $(CH_3CH_2CH_2)_2NH$		$CH_3CH_2\overset{O}{\overset{\|}{C}}NHCH_2CH_2CH_3$
c) 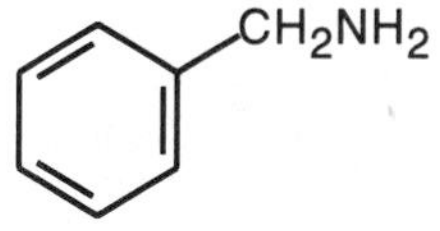 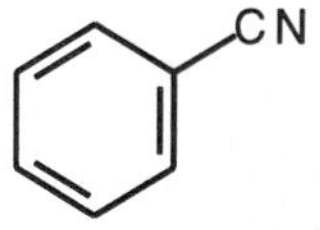 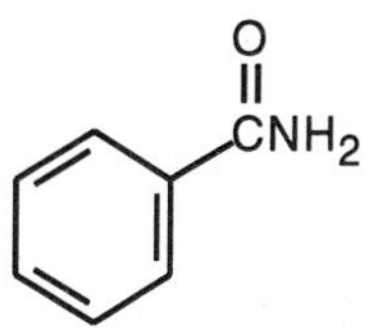		

Amine   Nitrile Precursor   Amide Precursor

d)

The compounds in parts b and d can't be prepared by reduction of a nitrile.

24.13

Amine   Amine Precursor   Carbonyl Precursor

a) $CH_3CH_2NHCH(CH_3)_2$ $CH_3CH_2NH_2$ + $CH_3\overset{O}{\overset{\|}{C}}CH_3$

or

$H_2NCH(CH_3)_2$ + $CH_3\overset{O}{\overset{\|}{C}}H$

b)   + $H\overset{O}{\overset{\|}{C}}CH_3$

c)   + $H\overset{O}{\overset{\|}{C}}H$

or

H_2NCH_3 +

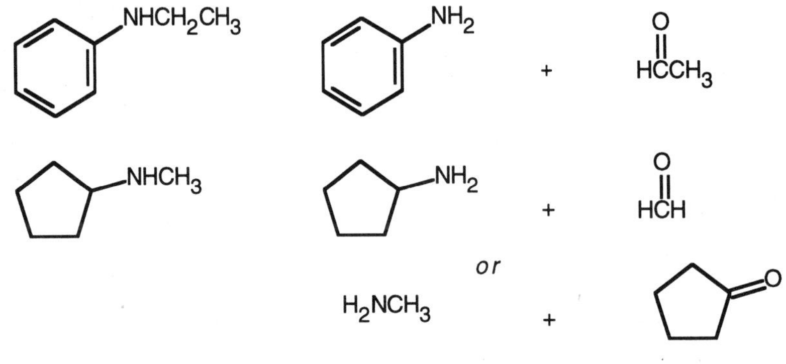

24.14

Tetrahedral   Carbinol–   Iminium
Intermediate   amine   ion

NaBH$_3$CN

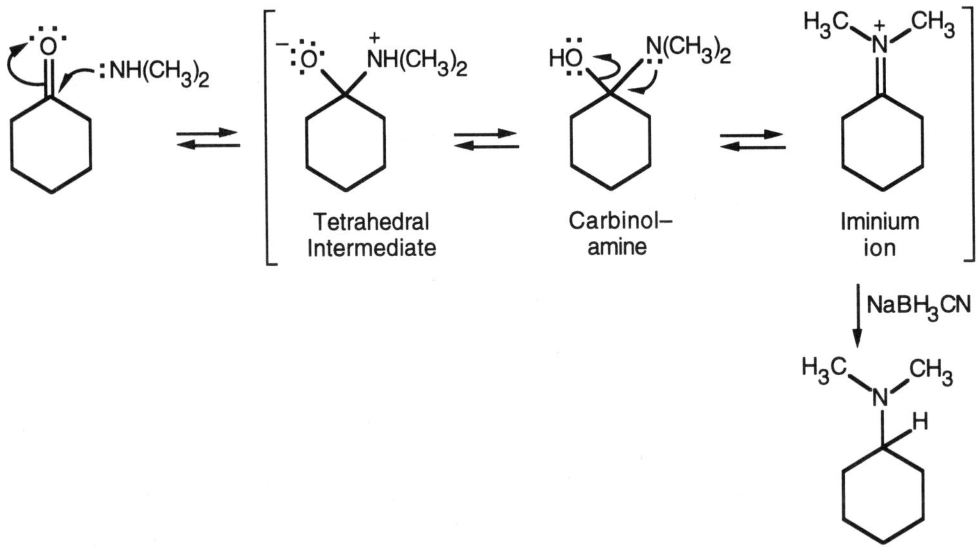

24.15

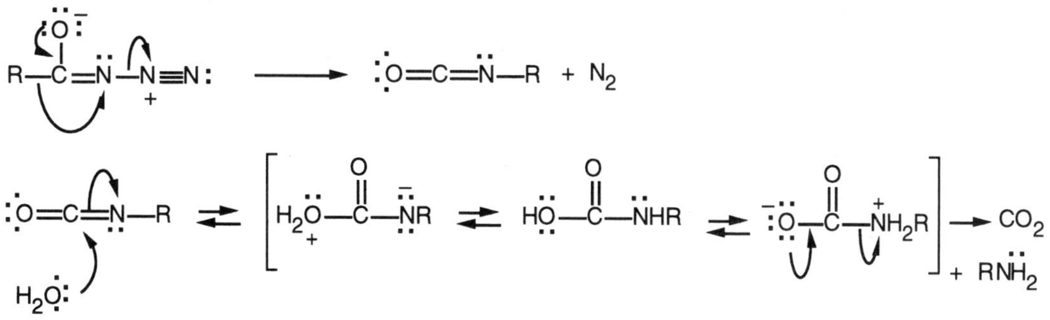

24.16

Amine	Amide Precursor	Acyl Azide Precursor
a) $(CH_3)_3CCH_2CH_2NH_2$	$(CH_3)_3CCH_2CH_2\overset{O}{\overset{\|}{C}}NH_2$	$(CH_3)_3CCH_2CH_2\overset{O}{\overset{\|}{C}}N_3$

b)

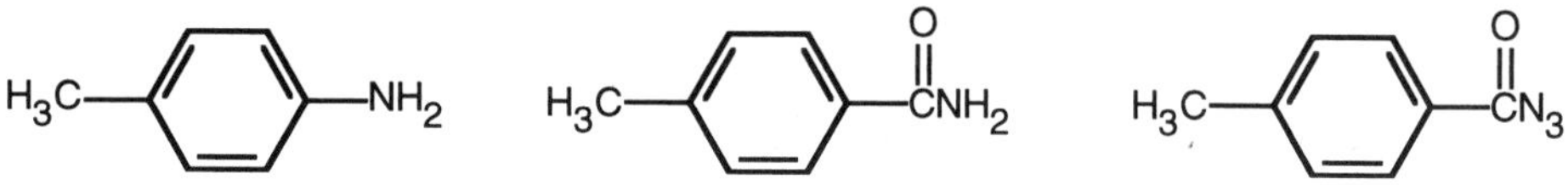

24.17

Amine	Alkene Products	Amine Products
a) $CH_3CH_2CH_2\overset{NH_2}{\overset{\|}{C}}HCH_2CH_2CH_2CH_3$	$CH_3CH_2CH=CHCH_2CH_2CH_2CH_3$	$(CH_3)_3N$
	or	
	$CH_3CH_2CH_2CH=CHCH_2CH_2CH_3$	$(CH_3)_3N$

b)

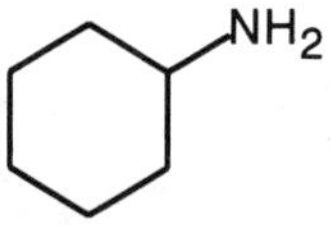

| | | $(CH_3)_3N$ |

c) $CH_3CH_2CH_2\overset{NH_2}{\overset{\|}{C}}HCH_2CH_2CH_3$ $CH_3CH_2CH_2CH=CHCH_2CH_3$ $(CH_3)_3N$

d)

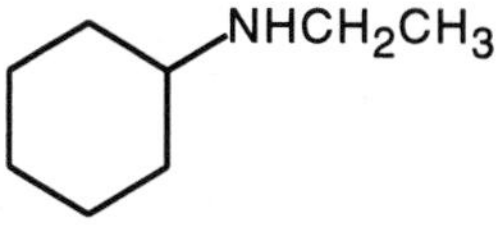

$H_2C=CH_2$ (major)

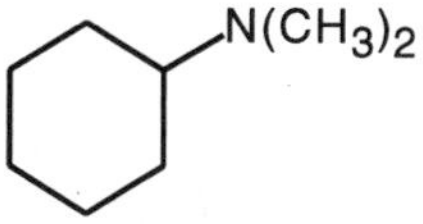

(minor)

or

$N(CH_3)_2$

$(CH_3)_2NCH_2CH_3$

24.18

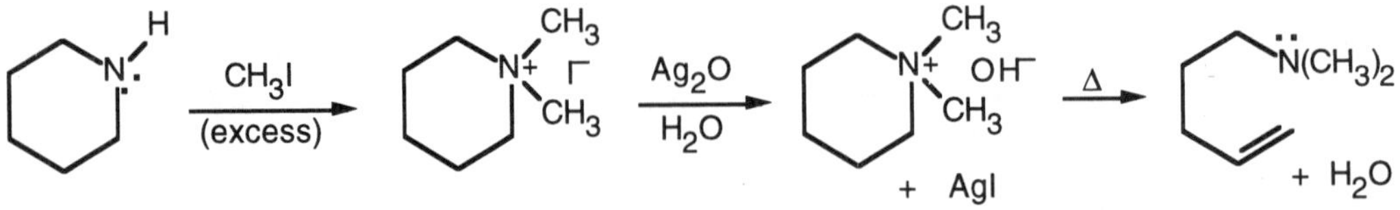

+ AgI

+ H₂O

24.19

$$(CH_3)_3CCCH_3 \xrightarrow[\text{NaBH}_3\text{CN}]{NH_3} (CH_3)_3CCHCH_3$$

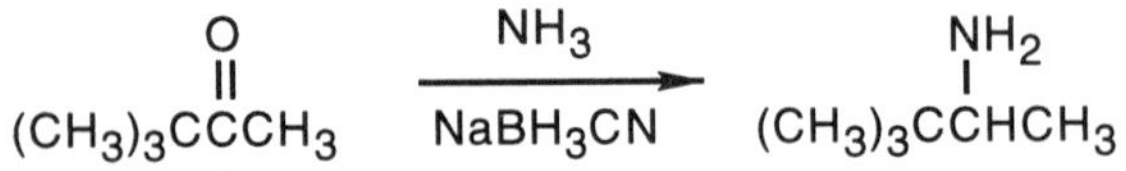

A   B

24.20

a)

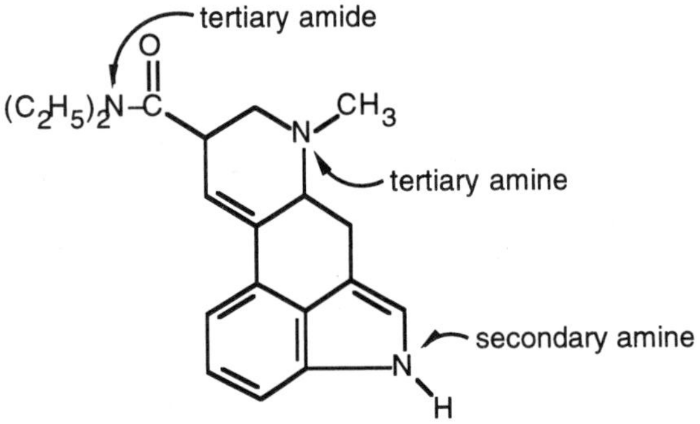

Lysergic acid diethylamide

b)

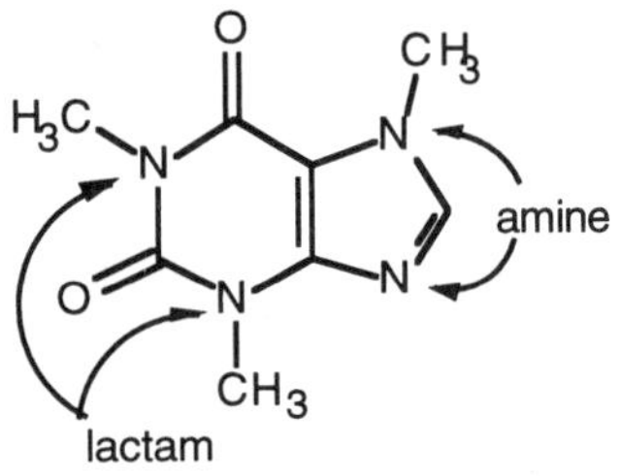

lactam
all nitrogens are tertiary
Caffeine

24.21

a)

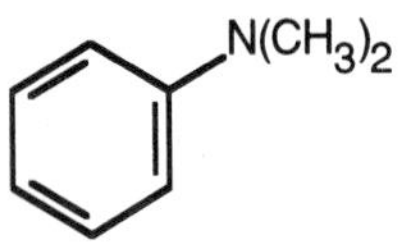

b)

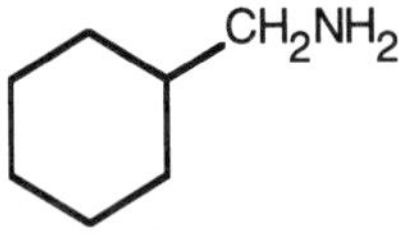

c)

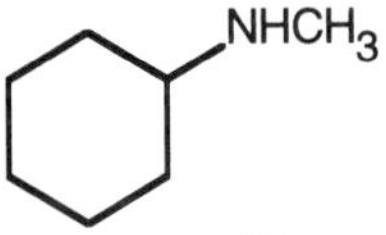

d)

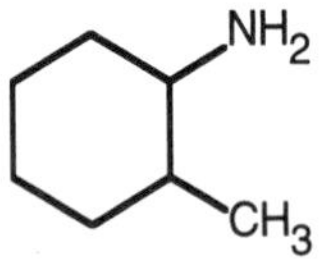

e) $(CH_3)_2NCH_2CH_2COOH$

f)

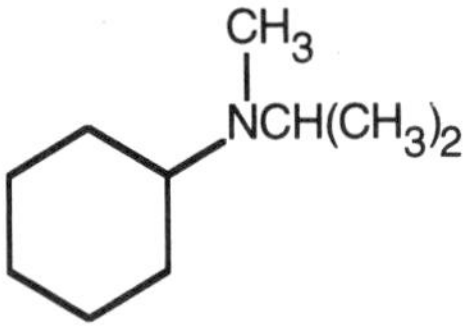

24.22 a) 2,4–Dibromoaniline   b) (2–Cyclopentyl)ethylamine
c) N–Ethylcyclopentylamine   d) N,N–Dimethylcyclopentylamine
e) N–Propylpyrrolidine   f) 4–Aminobutanenitrile

24.23

Dimethylamine

$(CH_3)_3N:$

Trimethylamine

Even though dimethylamine has a lower molecular weight than trimethylamine, it boils at a higher temperature because it forms hydrogen bonds that must be broken in the boiling process. Trimethylamine, by contrast, does not form hydrogen bonds.

24.24

a) $CH_3CH_2CH_2CH_2OH \xrightarrow{PBr_3} CH_3CH_2CH_2CH_2Br \xrightarrow{NaN_3} CH_3CH_2CH_2CH_2N_3$

$\downarrow$ 1. $LiAlH_4$
2. H_2O

$CH_3CH_2CH_2CH_2NH_2$
Butylamine

b) $CH_3CH_2CH_2CH_2OH \xrightarrow{Jones'} CH_3CH_2CH_2\overset{O}{\overset{\|}{C}}OH \xrightarrow[CHCl_3]{SOCl_2} CH_3CH_2CH_2\overset{O}{\overset{\|}{C}}-Cl$

$\downarrow$ $CH_3CH_2CH_2CH_2NH_2$
[from a)]
$NaOH$

$(CH_3CH_2CH_2CH_2)_2NH \xleftarrow[2.\ H_2O]{1.\ LiAlH_4} CH_3CH_2CH_2\overset{O}{\overset{\|}{C}}-NHCH_2CH_2CH_2CH_3$
Dibutylamine

c) $CH_3CH_2CH_2CH_2OH \xrightarrow{Jones'} CH_3CH_2CH_2\overset{O}{\overset{\|}{C}}OH \xrightarrow{SOCl_2} CH_3CH_2CH_2\overset{O}{\overset{\|}{C}}-Cl$

$\downarrow$ 2 NH_3

$CH_3CH_2CH_2NH_2 \xleftarrow[H_2O,\ \Delta]{Br_2,\ ^-OH} CH_3CH_2CH_2\overset{O}{\overset{\|}{C}}-NH_2$
Propylamine

or $CH_3CH_2CH_2\overset{O}{\overset{\|}{C}}-Cl \xrightarrow{NaN_3} CH_3CH_2CH_2\overset{O}{\overset{\|}{C}}N_3 \xrightarrow[2.\ H_2O]{1.\ \Delta} CH_3CH_2CH_2NH_2$

d) $CH_3CH_2CH_2CH_2OH \xrightarrow{PBr_3} CH_3CH_2CH_2CH_2Br \xrightarrow{NaCN} CH_3CH_2CH_2CH_2CN$

$\Big\downarrow \begin{array}{l} 1.\ LiAlH_4 \\ 2.\ H_2O \end{array}$

$CH_3CH_2CH_2CH_2CH_2NH_2$
Pentylamine

e) $CH_3CH_2CH_2CH_2OH \xrightarrow{PCC} CH_3CH_2CH_2\overset{\displaystyle O}{\overset{\|}{C}}H \xrightarrow[NaBH_3CN]{(CH_3)_2NH} CH_3CH_2CH_2CH_2N(CH_3)_2$

N,N–Dimethylbutylamine

f) $CH_3CH_2CH_2NH_2 \xrightarrow[CH_3I]{excess} CH_3CH_2CH_2\overset{+}{N}(CH_3)_3I^- \xrightarrow[H_2O]{Ag_2O} CH_3CH_2CH_2\overset{+}{N}(CH_3)_3{}^-OH$

[from c)]

$\Big\downarrow \Delta$

$CH_3CH=CH_2 + (CH_3)_3N$
Propene

24.25

a) $CH_3CH_2CH_2CH_2\overset{\displaystyle O}{\overset{\|}{C}}OH \xrightarrow{SOCl_2} CH_3CH_2CH_2CH_2\overset{\displaystyle O}{\overset{\|}{C}}-Cl$

$\Big\downarrow 2\ NH_3$

$CH_3CH_2CH_2CH_2\overset{\displaystyle O}{\overset{\|}{C}}NH_2$
Pentanamide

b) $CH_3CH_2CH_2CH_2\overset{\displaystyle O}{\overset{\|}{C}}NH_2 \xrightarrow[H_2O,\ \Delta]{Br_2,\ ^-OH} CH_3CH_2CH_2CH_2NH_2$
[from a)]   Butylamine

c) $CH_3CH_2CH_2CH_2\overset{\displaystyle O}{\overset{\|}{C}}NH_2 \xrightarrow[2.\ H_2O]{1.\ LiAlH_4} CH_3CH_2CH_2CH_2CH_2NH_2$
[from a)]   Pentylamine

d) $CH_3CH_2CH_2CH_2\overset{\displaystyle O}{\overset{\|}{C}}OH \xrightarrow[2.\ H_2O]{1.\ Br_2,\ PBr_3} CH_3CH_2CH_2CH_2\overset{Br}{\underset{}{\overset{|}{C}}}H\overset{\displaystyle O}{\overset{\|}{C}}OH$

2–Bromopentanoic acid

e) $CH_3CH_2CH_2CH_2\overset{\overset{\displaystyle O}{\|}}{C}OH$ $\xrightarrow[\text{2. } H_3O^+]{\text{1. } BH_3}$ $CH_3CH_2CH_2CH_2CH_2OH$

$\downarrow PBr_3$

$CH_3CH_2CH_2CH_2CH_2CN$ $\xleftarrow{\text{NaCN}}$ $CH_3CH_2CH_2CH_2CH_2Br$

Hexanenitrile

f) $CH_3CH_2CH_2CH_2CH_2CN$ $\xrightarrow[\text{2. } H_2O]{\text{1. } LiAlH_4}$ $CH_3CH_2CH_2CH_2CH_2CH_2NH_2$

[from e)] Hexylamine

24.26

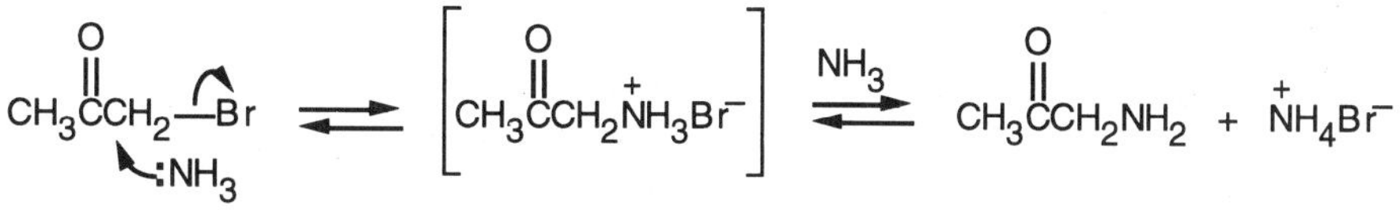

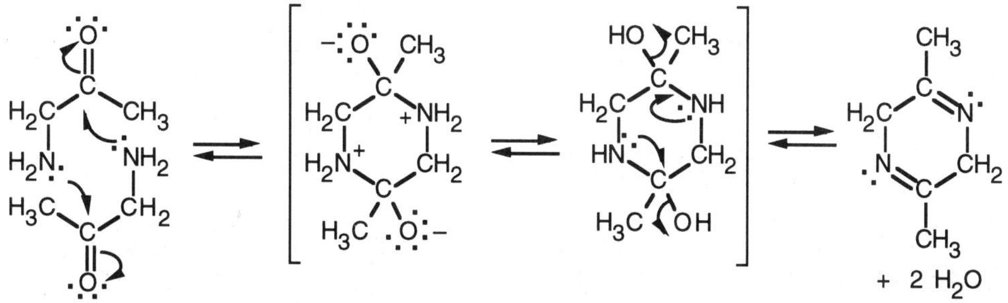

24.27

a)

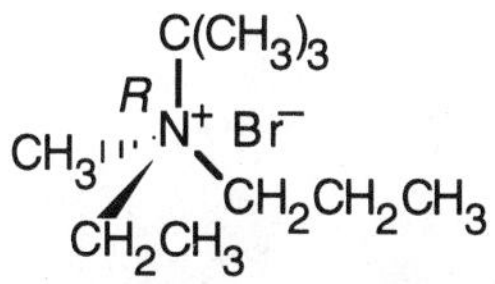

R-$tert$-Butylethylmethyl–
propylammonium bromide

b)

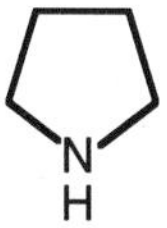

Pyrrolidine

c)

Diallylamine

24.28

a) $CH_3CH_2CH_2CH_2\overset{\overset{\displaystyle O}{\|}}{C}NH_2$ $\xrightarrow[\text{2. } H_2O]{\text{1. } LiAlH_4}$ $CH_3CH_2CH_2CH_2CH_2NH_2$

b) $CH_3CH_2CH_2CH_2CN$ $\xrightarrow[\text{2. } H_2O]{\text{1. } LiAlH_4}$ $CH_3CH_2CH_2CH_2CH_2NH_2$

c) $CH_3CH_2CH=CH_2$ $\xrightarrow[\text{peroxides}]{\text{HBr}}$ $CH_3CH_2CH_2CH_2Br$ $\xrightarrow{\text{NaCN}}$ $CH_3CH_2CH_2CH_2CN$

$\downarrow$ 1. $LiAlH_4$
 2. H_2O

$CH_3CH_2CH_2CH_2CH_2NH_2$

d) $CH_3CH_2CH_2CH_2CH_2\overset{\displaystyle O}{\overset{\displaystyle \|}{C}}NH_2$ $\xrightarrow[\text{H}_2\text{O},\,\Delta]{\text{Br}_2,\,{}^-\text{OH}}$ $CH_3CH_2CH_2CH_2CH_2NH_2$

e) $CH_3CH_2CH_2CH_2OH$ $\xrightarrow{\text{PBr}_3}$ $CH_3CH_2CH_2CH_2Br$ $\xrightarrow{\text{NaCN}}$ $CH_3CH_2CH_2CH_2CN$

$\downarrow$ 1. $LiAlH_4$
 2. H_2O

$CH_3CH_2CH_2CH_2CH_2NH_2$

f) $CH_3CH_2CH_2CH_2CH=CHCH_2CH_2CH_2CH_3$ $\xrightarrow[\text{2. Zn, H}_3\text{O}^+]{\text{1. O}_3}$ $2\ CH_3CH_2CH_2CH_2\overset{\displaystyle O}{\overset{\displaystyle \|}{C}}H$

$CH_3CH_2CH_2CH_2\overset{\displaystyle O}{\overset{\displaystyle \|}{C}}H$ $\xrightarrow[\text{NaBH}_3\text{CN}]{\text{NH}_3}$ $CH_3CH_2CH_2CH_2CH_2NH_2$

g) $CH_3CH_2CH_2CH_2\overset{\displaystyle O}{\overset{\displaystyle \|}{C}}OH$ $\xrightarrow{\text{SOCl}_2}$ $CH_3CH_2CH_2CH_2\overset{\displaystyle O}{\overset{\displaystyle \|}{C}}Cl$

$\downarrow$ 2 NH_3

$CH_3CH_2CH_2CH_2CH_2NH_2$ $\xleftarrow[\text{2. H}_2\text{O}]{\text{1. LiAlH}_4}$ $CH_3CH_2CH_2CH_2\overset{\displaystyle O}{\overset{\displaystyle \|}{C}}NH_2$

24.29

$+\ CH_3NH_2$ $\xrightarrow[\text{CH}_3\text{OH}]{\text{NaBH}_3\text{CN}}$

Ephedrine

24.30

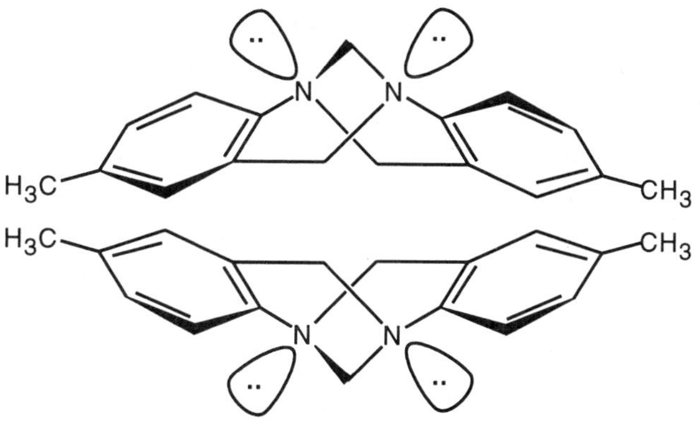

Tröger's base

The enantiomers of Tröger's base do not interconvert. Because of the rigid ring system, the substituents bonded to nitrogen can't be forced into the planar sp^2 geometry necessary for inversion at nitrogen to occur. Since inversion is not possible, the enantiomers are resolvable.

24.31

a)

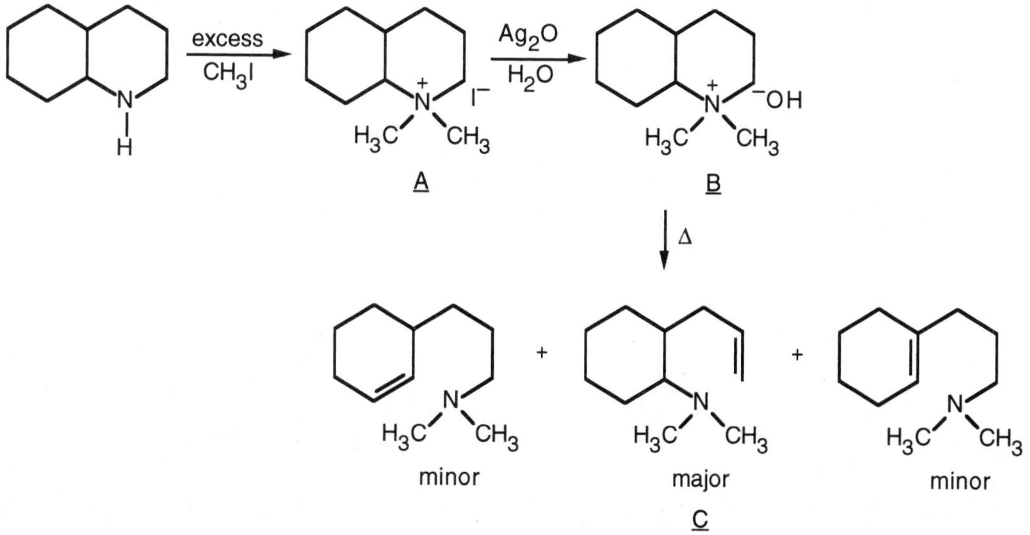

b)

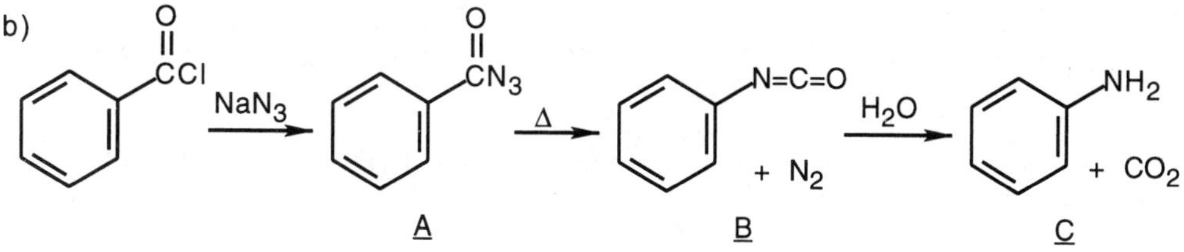

c)

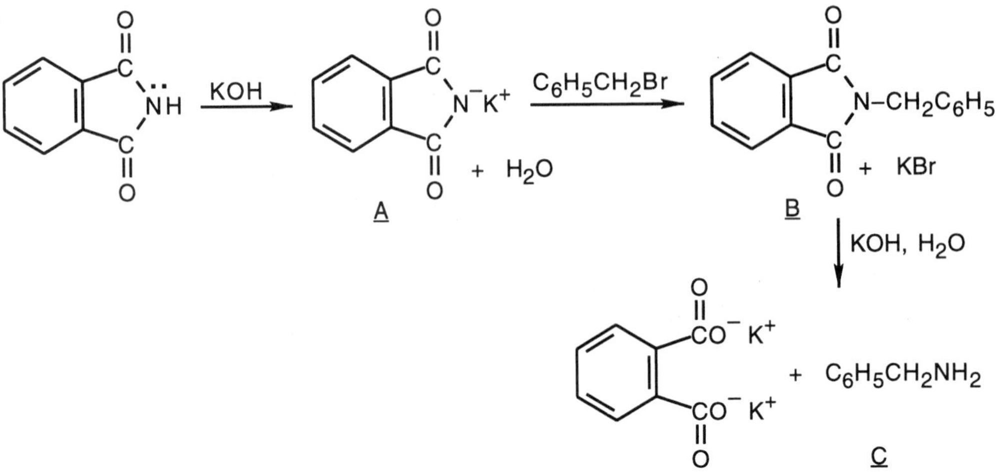

$$+ H_2O$$

$$\underline{A}$$

$$+ KBr$$

$$\underline{B}$$

KOH, H_2O

$$+ C_6H_5CH_2NH_2$$

$$\underline{C}$$

d) $BrCH_2CH_2CH_2CH_2Br$ $\xrightarrow[\text{CH}_3\text{NH}_2]{\text{1 equiv}}$

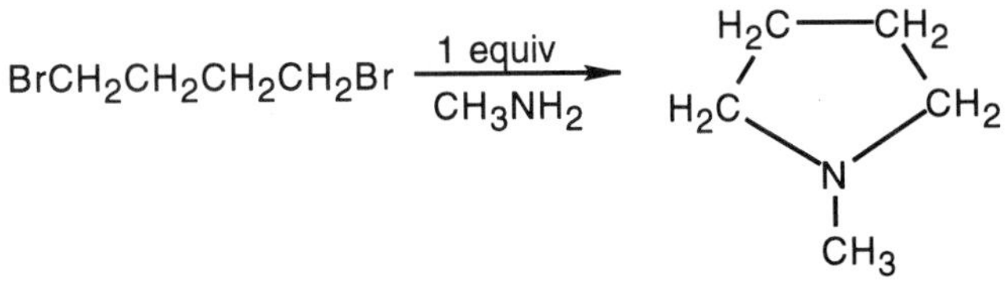

24.32

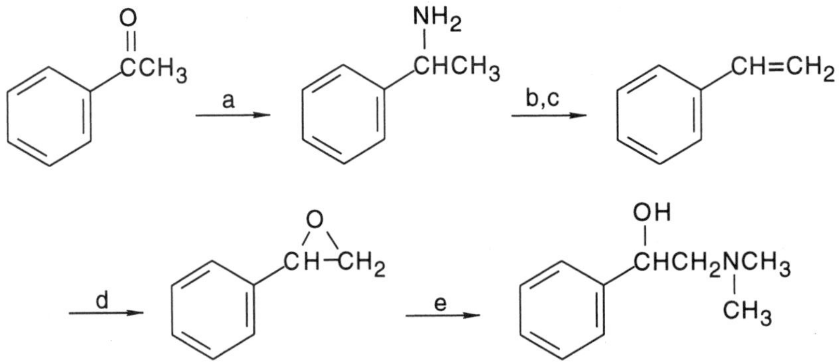

(a) NH_3, $NaBH_3CN$; (b) excess CH_3I; (c) Ag_2O, H_2O, heat; (d) RCO_3H (e) $(CH_3)_2NH$

24.33 a) The Hofmann rearrangement of amides yields an amine containing one less carbon atom than the starting amide. In this case, the product of Hofmann rearrangement is $CH_3CH_2NH_2$, not $CH_3CH_2CH_2NH_2$.

b) No reaction occurs between a tertiary amine and a carbonyl group. To obtain the product shown, use $(CH_3)_2NH$.

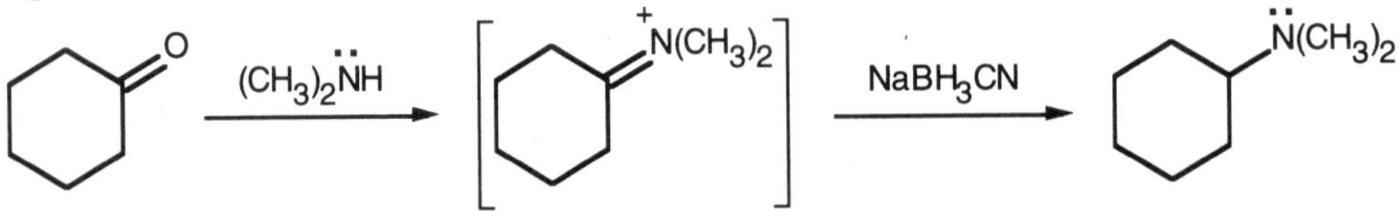

c) An elimination product is obtained when the tertiary bromide $(CH_3)_3CBr$ reacts with ammonia.

$$(CH_3)_3CBr \xrightarrow{\ :NH_3\ } (CH_3)_2C=CH_2$$

d) An isocyanate intermediate in the Hofmann rearrangement results from treatment of an amide with Br_2 and ^-OH. Heat is used to decarboxylate the carbamic acid intermediate.

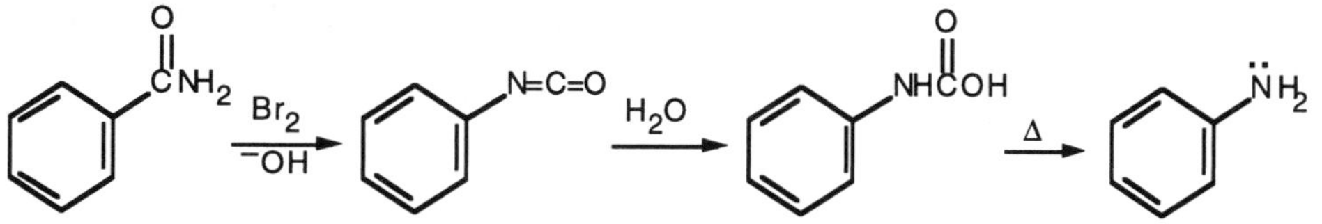

e) The amine in this problem is being subjected to the conditions of Hofmann elimination. The major alkene product, $CH_3CH_2CH_2CH=CH_2$, contains the *less* substituted double bond. The product shown is the minor product.

24.34

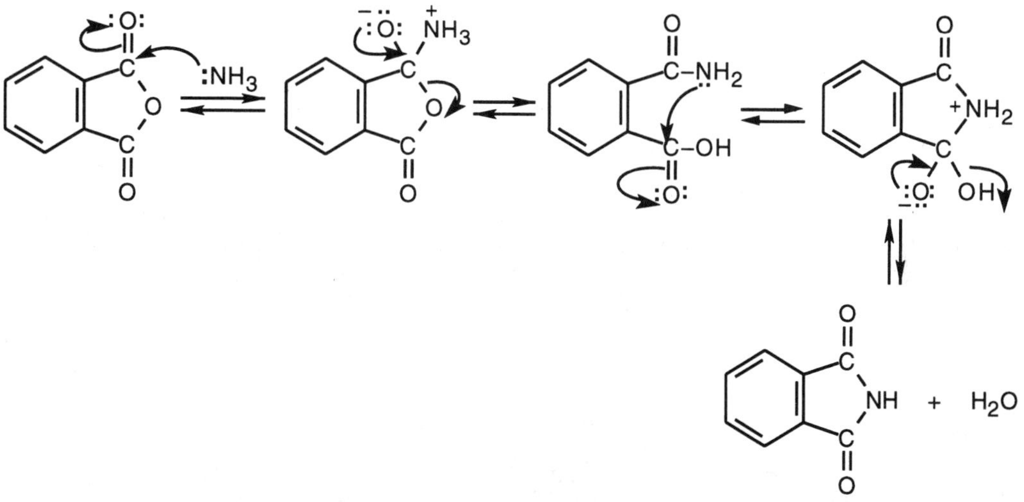

24.35 The structural formula indicates that coniine has one double bond or ring, and the Hofmann elimination product shows that the nitrogen atom is part of a ring.

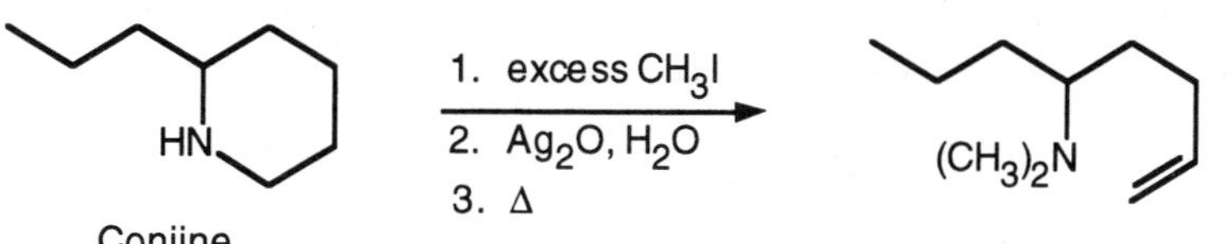

Coniine

1. excess CH_3I
2. Ag_2O, H_2O
3. Δ

$(CH_3)_2N$

5–(*N,N*–Dimethylamino)–1–octene

24.36

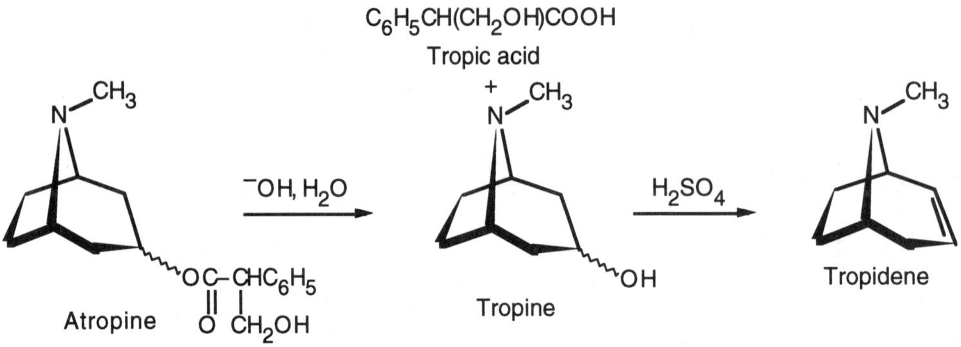

$C_6H_5CH(CH_2OH)COOH$
Tropic acid

Atropine → (⁻OH, H₂O) → Tropine → (H₂SO₄) → Tropidene

24.37

1. CH_3I
2. Ag_2O, H_2O
3. Δ

$N(CH_3)_2$

1. CH_3I
2. Ag_2O, H_2O
3. Δ

+ $(CH_3)_3N\colon$

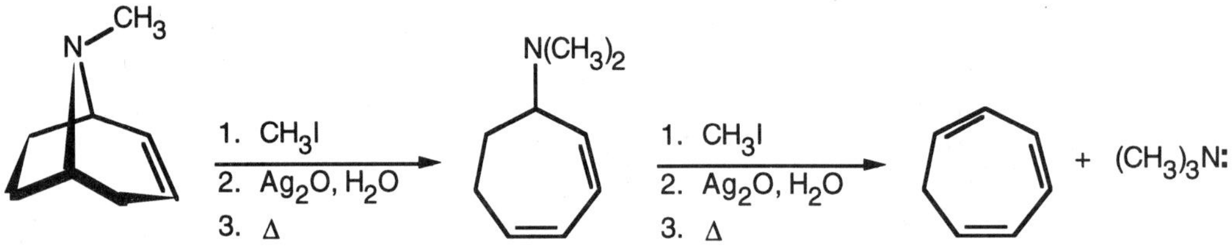

24.38

Benzaldehyde first reacts with methylamine and $NaBH_3CN$ in the usual way to give the reductive amination product *N*-methylbenzylamine. This product then reacts further with benzaldehyde in a second reductive amination to give *N*-methyldibenzylamine.

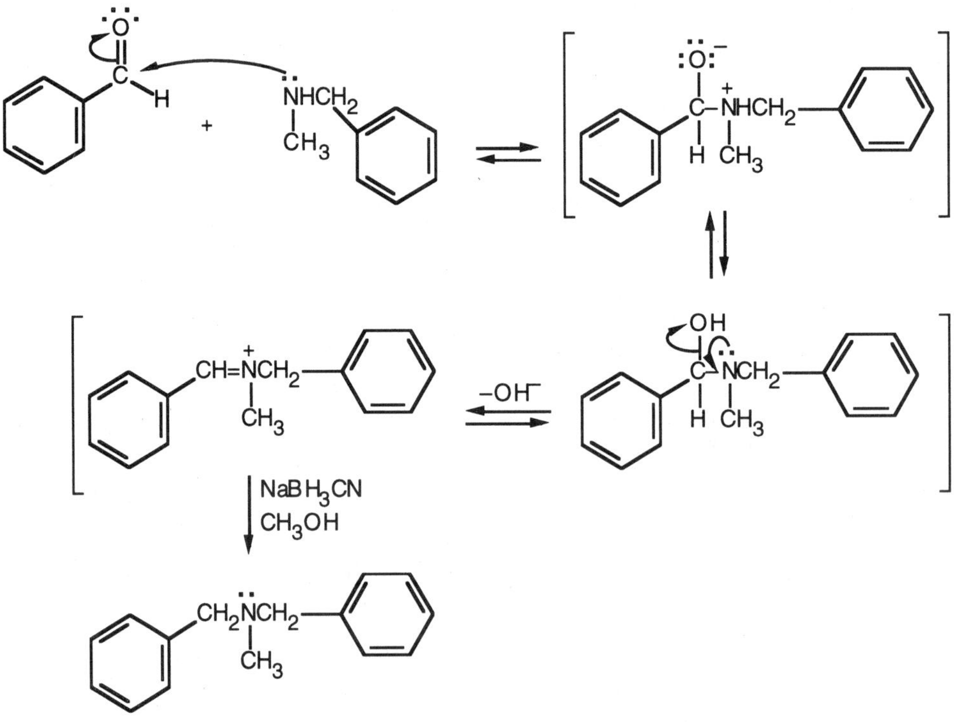

$NaBH_3CN$
CH_3OH

N–Methyldibenzylamine

24.39

$$\text{Amine} \xrightarrow[\begin{array}{l}\text{2. Ag}_2\text{O, H}_2\text{O}\\ \text{3. }\Delta\end{array}]{\text{1. CH}_3\text{I}} \text{Alkene} + \text{Amine}$$

a)

(cyclopentyl-NHCH$_3$) → (cyclopentene) + (CH$_3$)$_3$N:

b)

(CH$_3$)$_2$CHCHCH$_2$CH$_2$CH$_3$ with NH$_2$

(CH$_3$)$_2$CHCH=CHCH$_2$CH$_3$ + (CH$_3$)$_3$N:
(major)

or

(CH$_3$)$_2$C=CHCH$_2$CH$_2$CH$_3$ + (CH$_3$)$_3$N:
(minor)

c)

Ph-NHCHCH$_2$CH$_2$CH$_2$CH$_3$ with CH$_3$

H$_2$C=CHCH$_2$CH$_2$CH$_2$CH$_3$ + Ph-N̈(CH$_3$)$_2$
(major)

or

CH$_3$CH=CHCH$_2$CH$_2$CH$_3$ + Ph-N̈(CH$_3$)$_2$
(minor)

24.40

$$\text{(cyclopentyl)-Br} \xrightarrow[\text{ether}]{\text{Mg}} \text{(cyclopentyl)-MgBr} \xrightarrow[\text{2. H}_3\text{O}^+]{\text{1. CH}_2\text{-CHCH}_3 \text{ (epoxide)}} \text{(cyclopentyl)-CH}_2\text{CHCH}_3 \text{ with OH}$$

↓ PCC
CH$_2$Cl$_2$

(cyclopentyl)-CH$_2$CCH$_3$ with =O

$$\xleftarrow[\text{NaBH}_3\text{CN}]{\text{CH}_3\text{NH}_2}$$

(cyclopentyl)-CH$_2$CHCH$_3$ with NHCH$_3$

Cyclopentamine

24.41

Prolitane

24.42

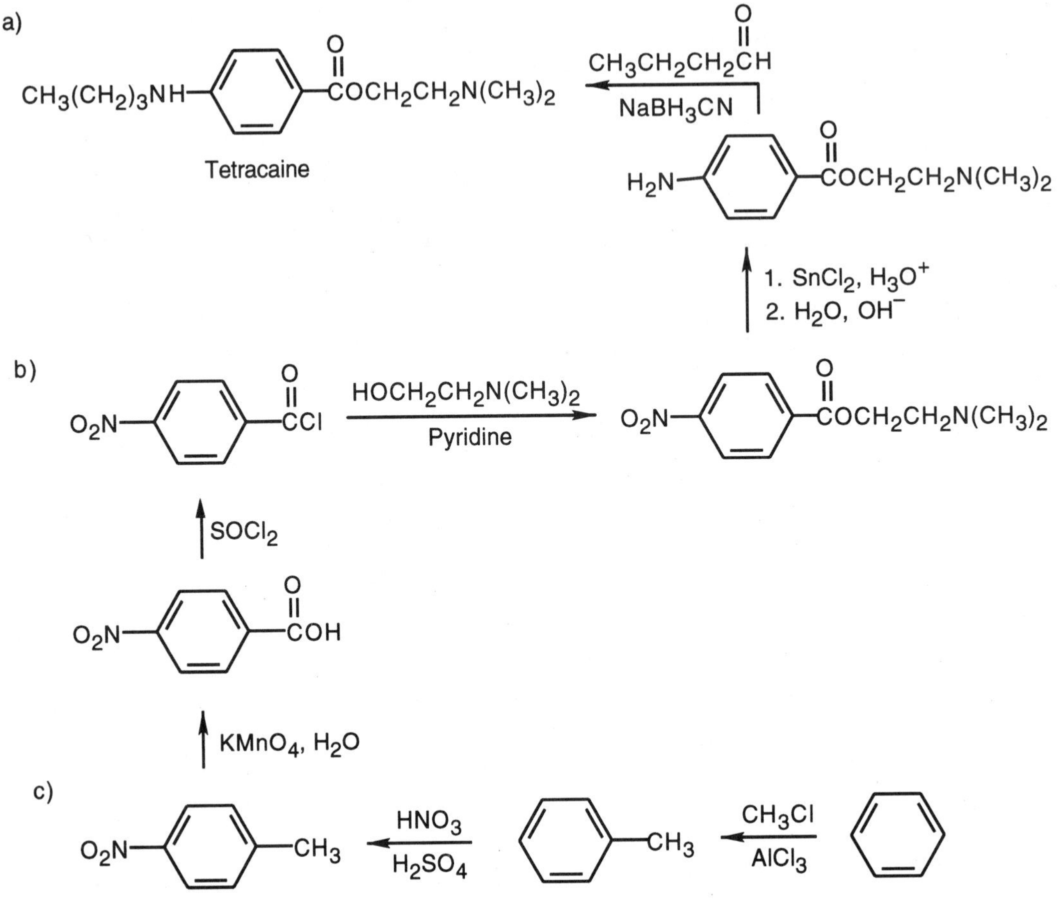

a)

b)

c)

24.43

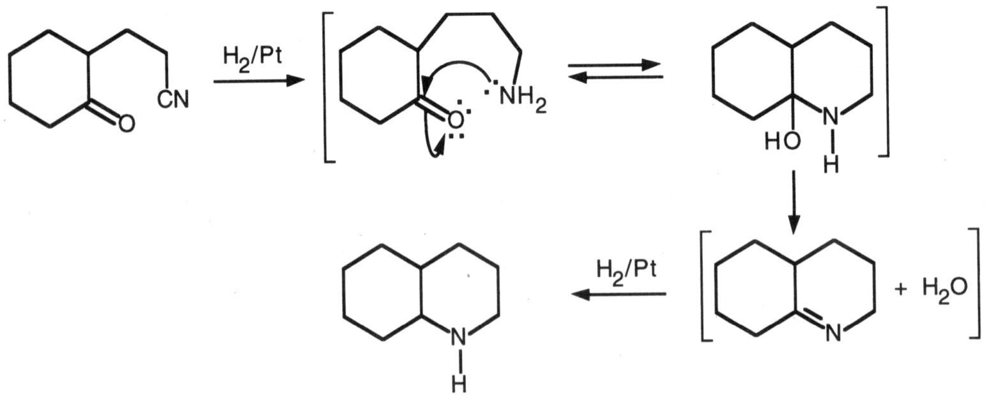

24.44

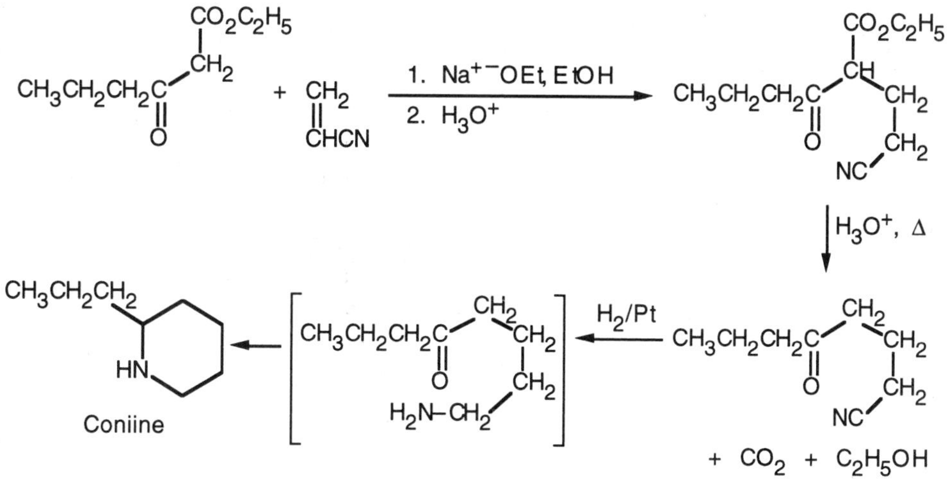

$$CH_3CH_2CH_2\overset{\overset{\displaystyle CO_2C_2H_5}{|}}{\underset{\underset{\displaystyle O}{||}}{C}}\!\!-\!CH_2 \;+\; \underset{\underset{\displaystyle CHCN}{||}}{CH_2} \quad\xrightarrow[\text{2. } H_3O^+]{\text{1. } Na^{+-}OEt,\ EtOH}\quad$$

1. Na⁺⁻OEt, EtOH
2. H₃O⁺

$H_3O^+,\ \Delta$

+ CO₂ + C₂H₅OH

H₂/Pt

CH₃CH₂CH₂ (ring) Coniine

H_2N-CH_2

24.45

OH

1. NaOH

2. CH₂–CHCH₂Cl (epoxide, O)

$OCH_2CH\!-\!CH_2$ (epoxide, O)

(CH₃)₂NH

HO

$OCH_2\overset{}{C}HCH_2N(CH_3)_2$

Propranolol

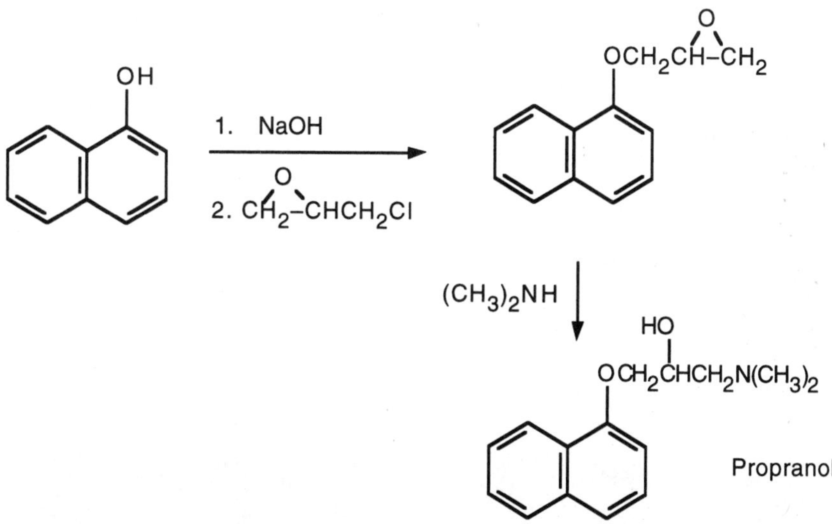

24.46 During pyramidal inversion of nitrogen, the hybridization at nitrogen changes momentarily from sp^3 to sp^2. For an aziridine, this sp^2–hybridized, planar intermediate is of high energy due to severe ring strain in the three-membered aziridine ring. Since the energy barrier for inversion is high, the rate of inversion of (+)–1–chloro–2,2–diphenylaziridine is low.

24.47

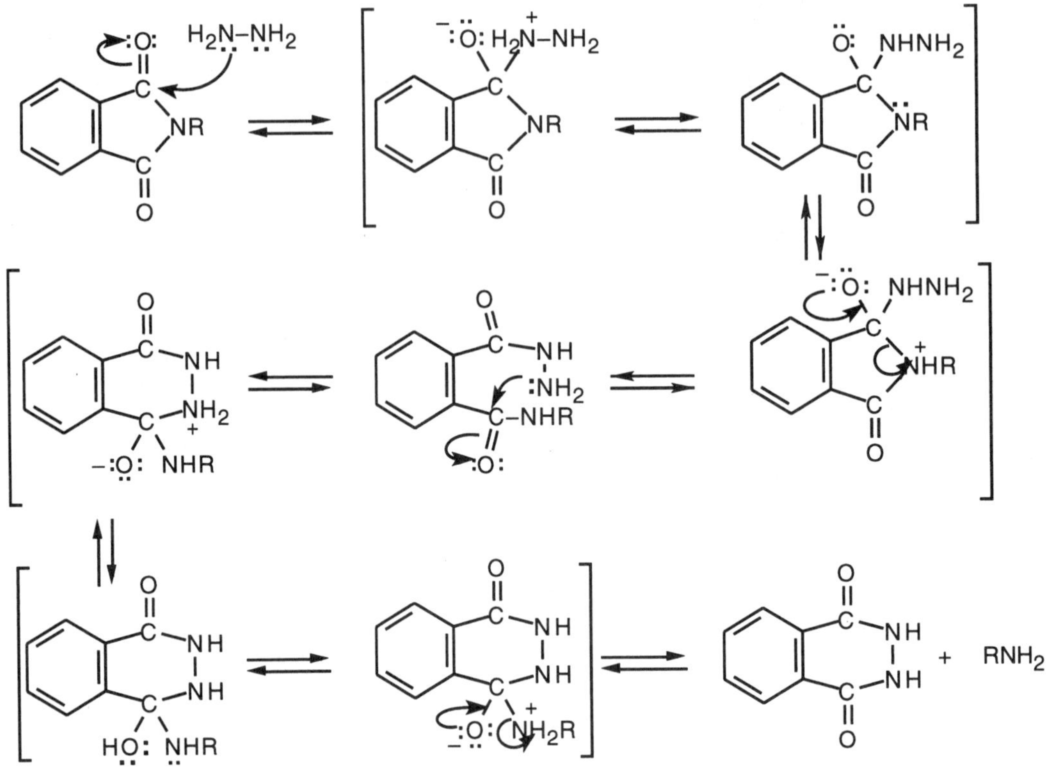

24.48

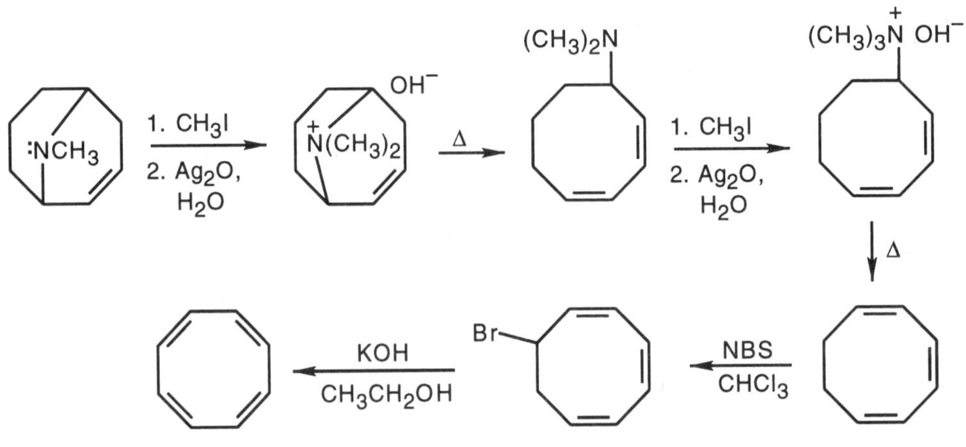

Cyclooctatetraene

24.49

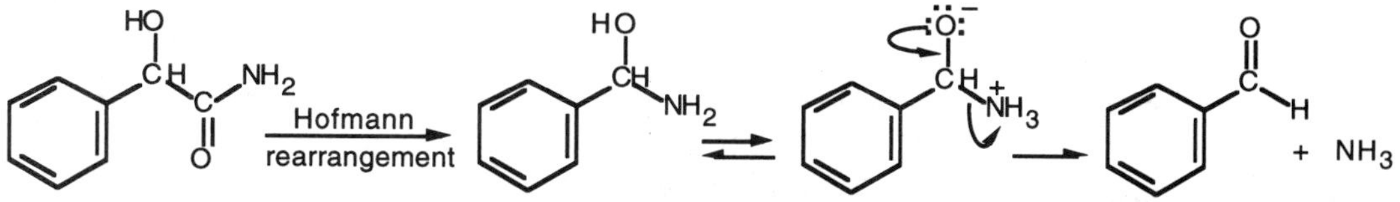

24.50

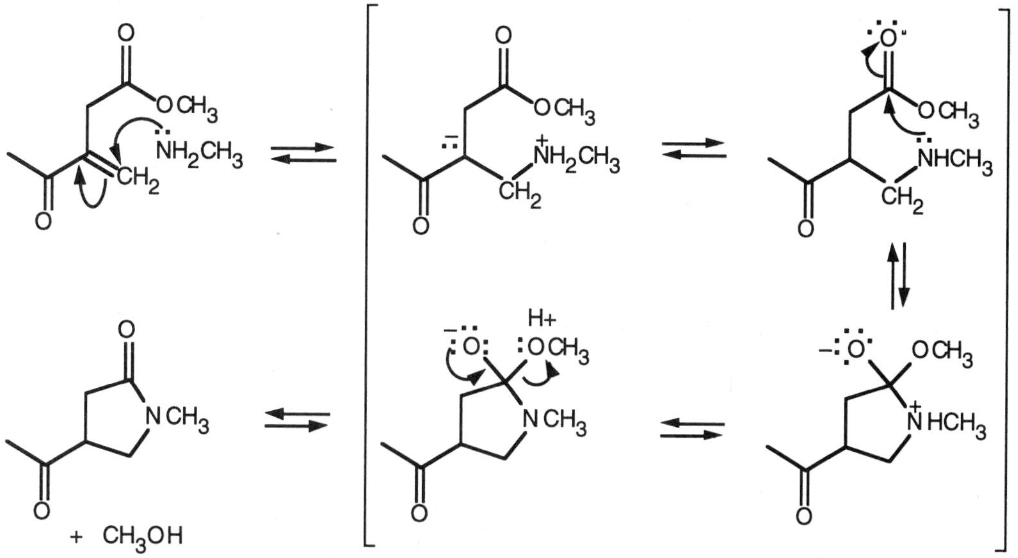

24.51

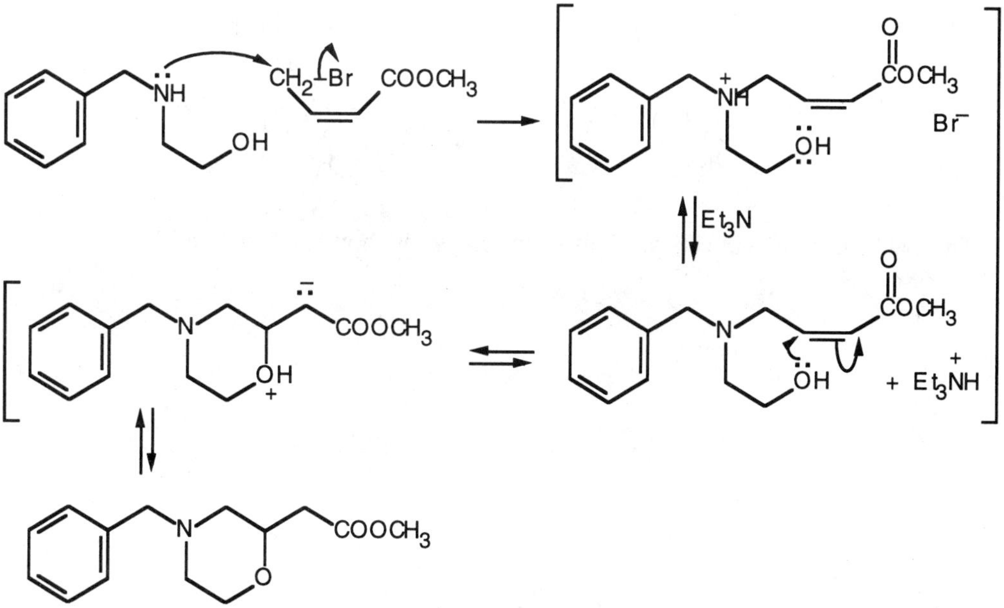

24.52 a) $HOCH_2CH_2CH_2NH_2$ b) $(CH_3O)_2CHCH_2NH_2$

24.53

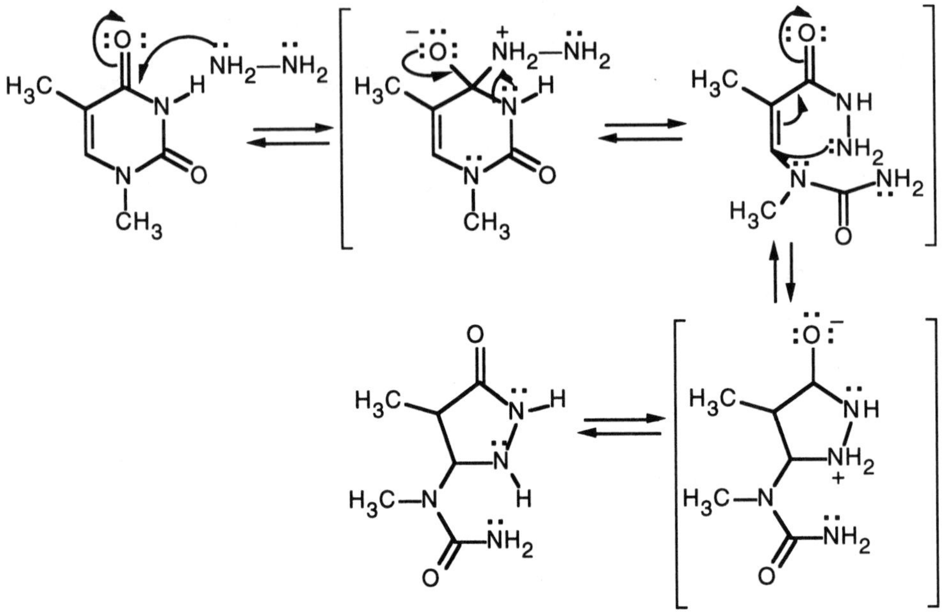

24.54

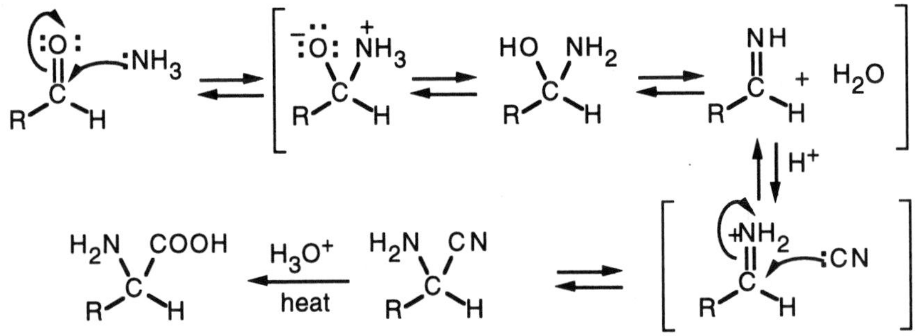

The mechanism of acid-catalyzed nitrile hydrolysis is shown in Problem 21.24.

Study Guide for Chapter 24

After studying this chapter, you should be able to:

(1)   Classify amines as primary, secondary tertiary and quaternary (24.1, 24.20).

(2)   Name and draw amines (24.2, 24.3, 24.4, 24.21, 24.22, 24.27).

(3)   Understand the geometry, stereochemistry and physical properties of amines (24.5, 24.23, 24.30, 24.46).

(4)   Predict the basicity of amines (24.6, 24.7, 24.8, 24.9).

(5)   Synthesize amines by several routes (24.11, 24.12, 24.13, 24.16, 24.24, 24.25, 24.28, 24.29, 24.33, 24.40, 24.41, 24.42, 24.43, 24.44, 24.45).

(6)   Propose mechanisms for reactions involving amines (24.10, 24.14, 24.15, 24.34, 24.47, 24.49, 24.50, 24.51, 24.53, 24.54).

(7)   Predict the products of reactions of amines (24.17, 24.18, 24.26, 24.31, 24.32, 24.37, 24.38, 24.39, 24.48).

(8)   Identify amines by using:
(a) The Hofmann elimination (24.35).
(b) Spectroscopic techniques (24.19, 24.52).

25.1

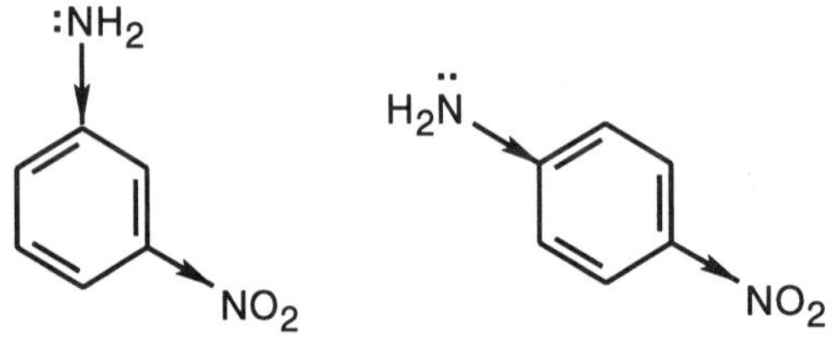

The inductive effect of the electron-withdrawing nitro group makes the amine nitrogens of both *m*-nitroaniline and *p*-nitroaniline less electron-rich and less basic than aniline.

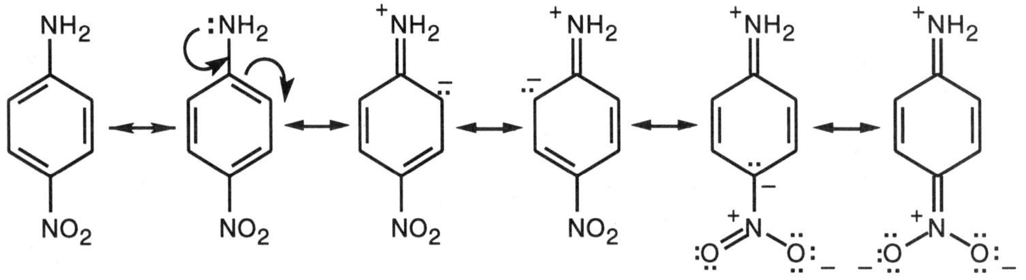

When the nitro group is para to the amino group, conjugation of the amino group with the nitro group can also occur. *p*–Nitroaniline is thus even less basic than *m*–nitroaniline.

25.2

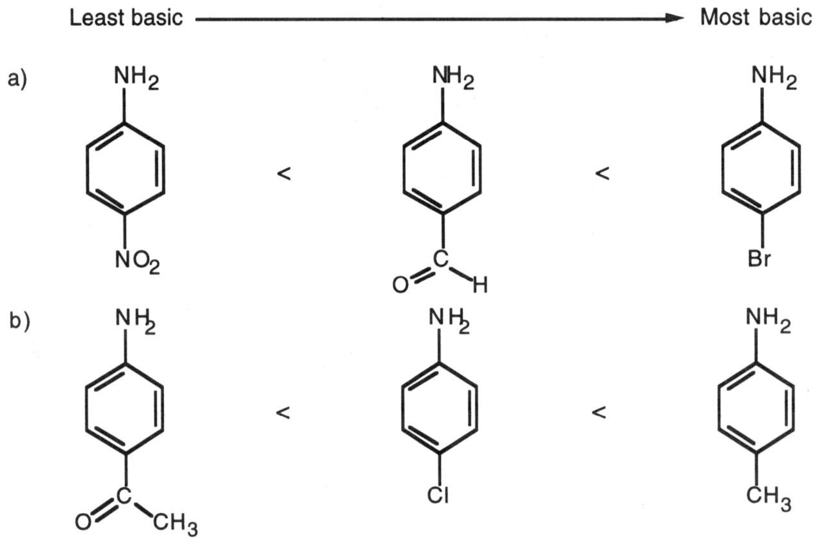

Least basic ⟶ Most basic

a)

b)

c)

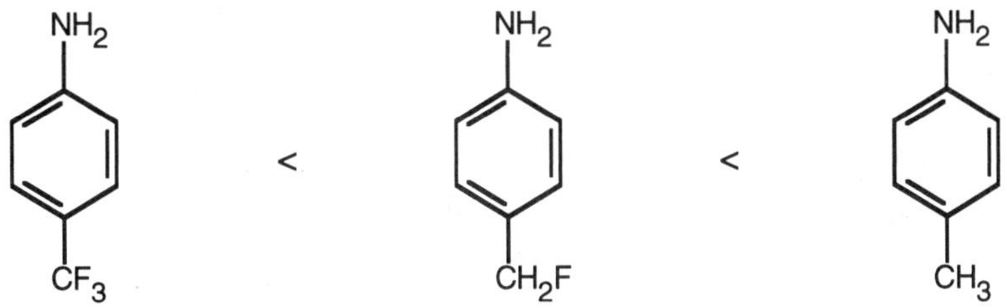

25.3

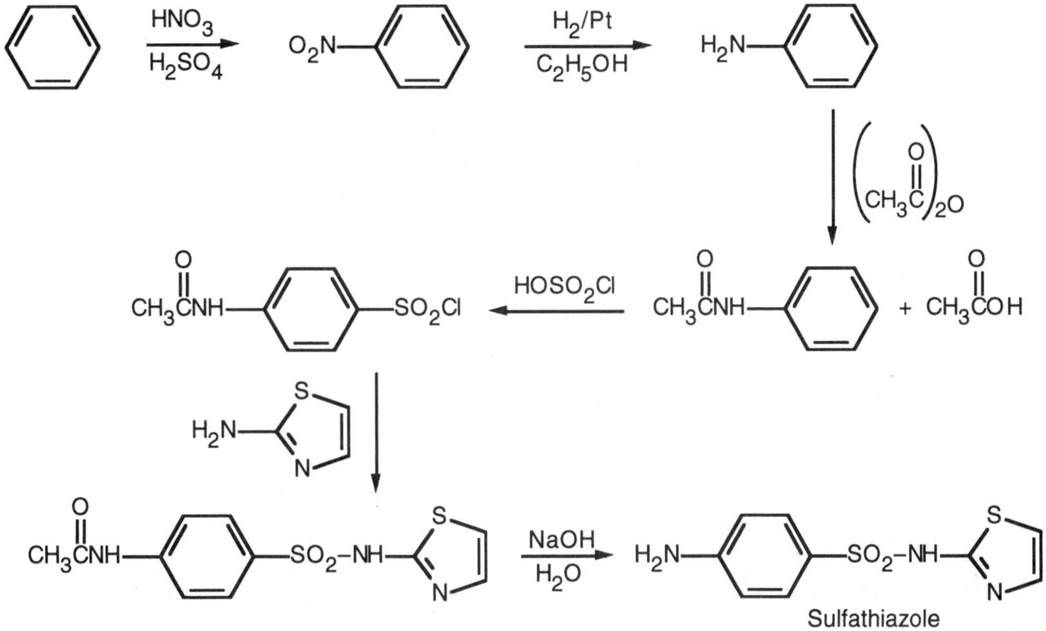

25.4

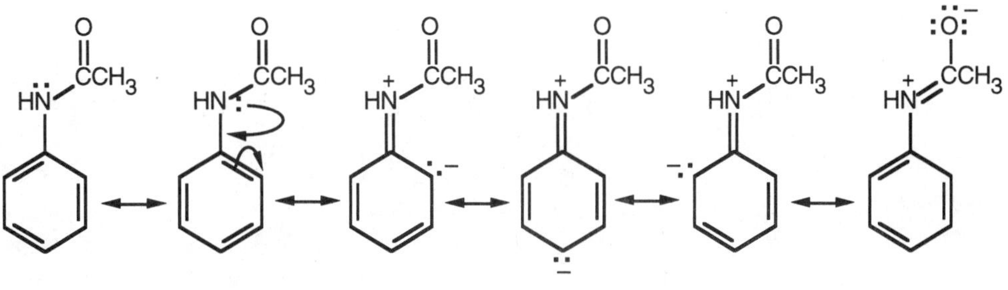

In the last resonance form, the nitrogen lone pair is delocalized onto oxygen, rather than into the aromatic ring. Acetanilide is therefore less activated toward electrophilic aromatic substitution than is aniline.

25.5

a)

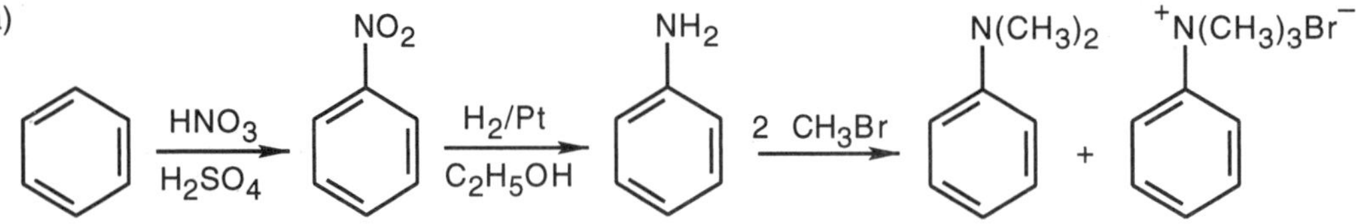
NO2
NH2
N(CH3)2
+N(CH3)3Br−
HNO3
H2SO4
H2/Pt
C2H5OH
2 CH3Br
+

b)

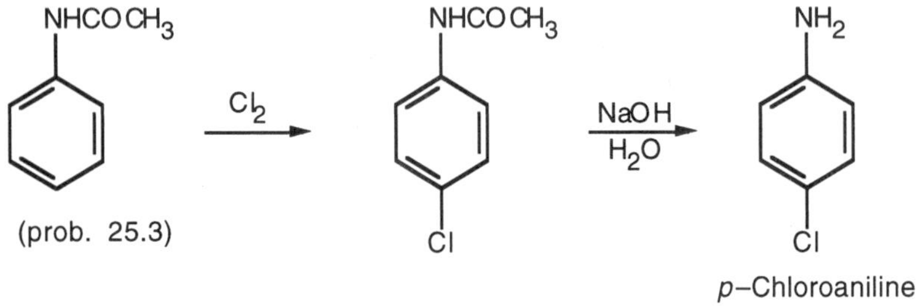
NHCOCH3
(prob. 25.3)
Cl2
NHCOCH3
Cl
NaOH
H2O
NH2
Cl
p–Chloroaniline

c)

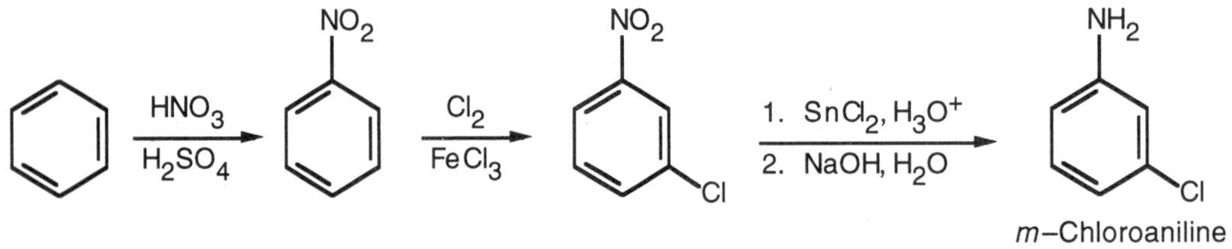
HNO3
H2SO4
NO2
Cl2
FeCl3
NO2
Cl
1. SnCl2, H3O+
2. NaOH, H2O
NH2
Cl
m–Chloroaniline

d)

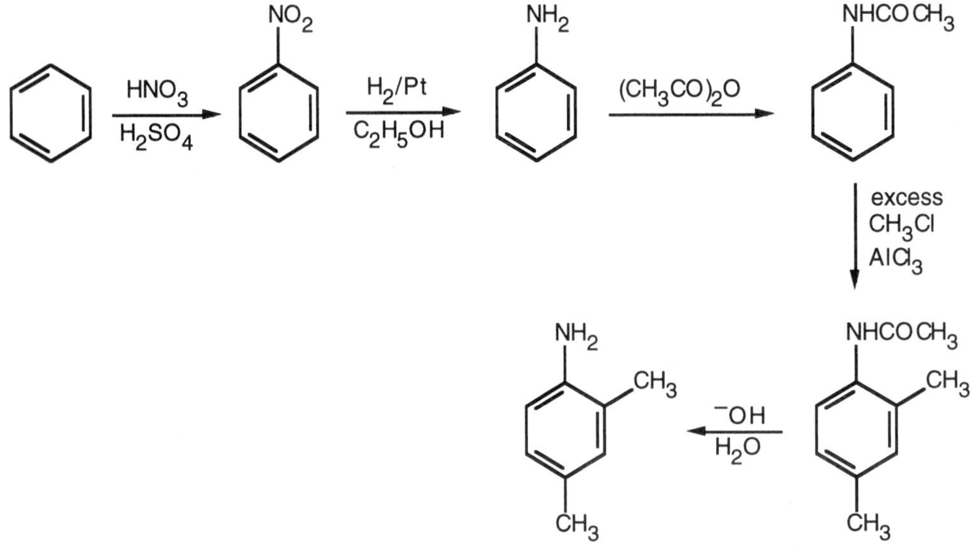
HNO3
H2SO4
NO2
H2/Pt
C2H5OH
NH2
(CH3CO)2O
NHCOCH3
excess
CH3Cl
AlCl3
NHCOCH3
CH3
CH3
−OH
H2O
NH2
CH3
CH3

25.6

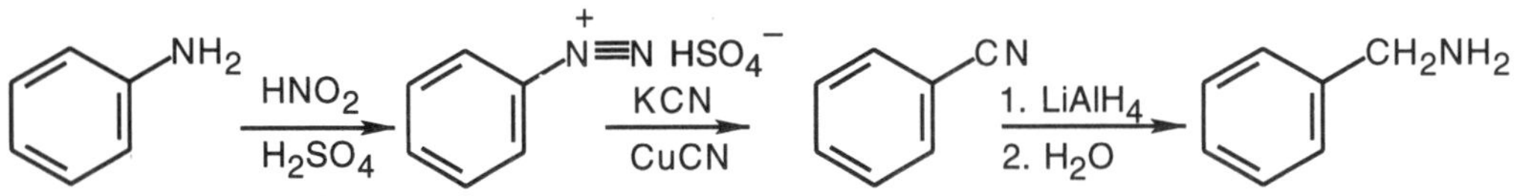
NH2
HNO2
H2SO4
+
N≡N HSO4−
KCN
CuCN
CN
1. LiAlH4
2. H2O
CH2NH2

25.7

a)

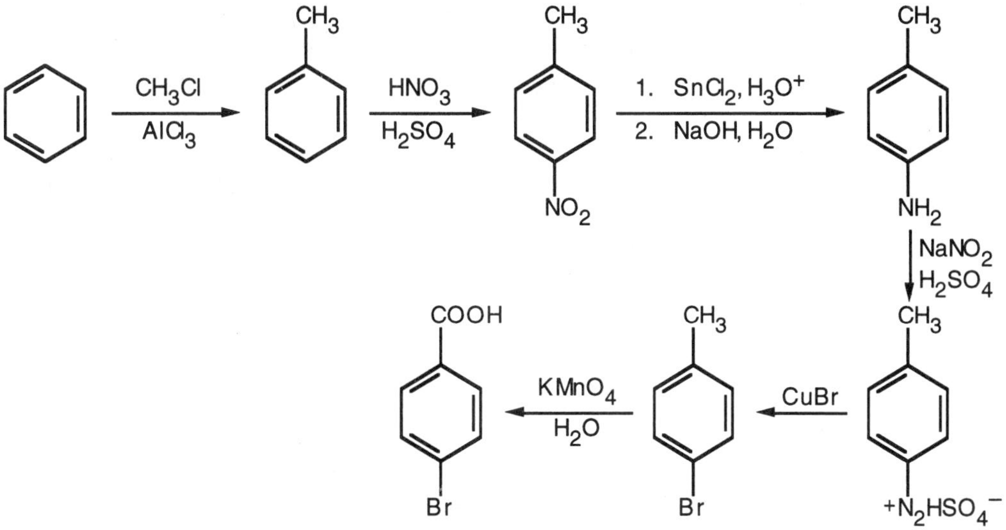

p–Bromobenzoic acid

Other routes to *p*–bromobenzoic acid are also possible.

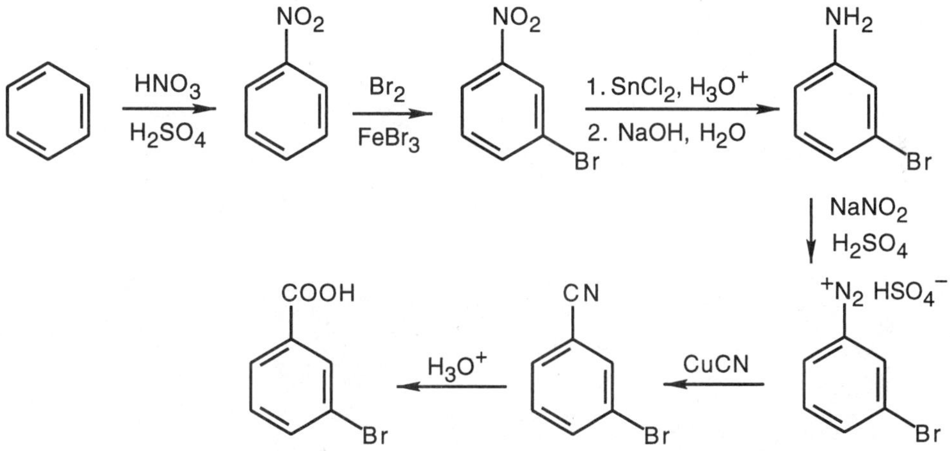

m–Bromobenzoic acid

c)

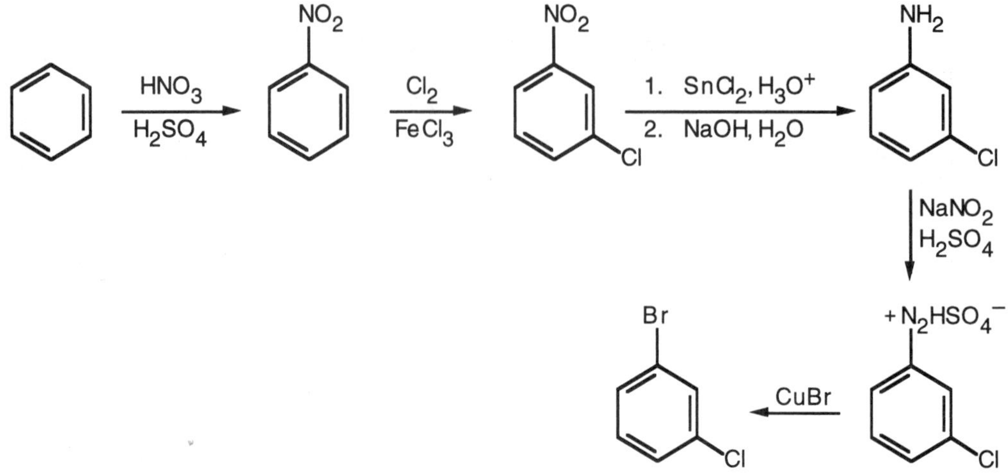

m–Bromochlorobenzene

d)

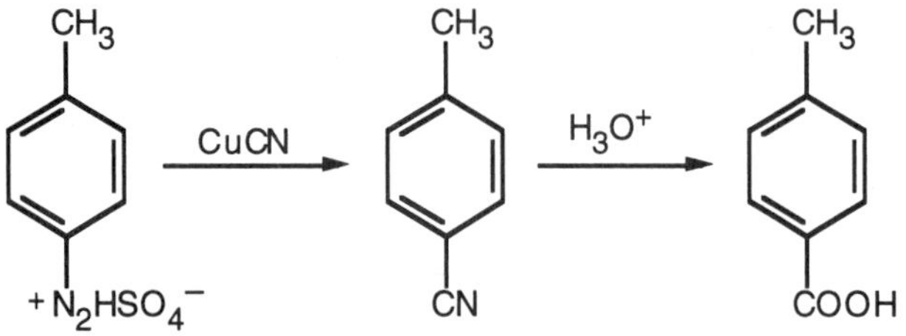

[from 25.7a]   *p*–Methylbenzoic acid

e)

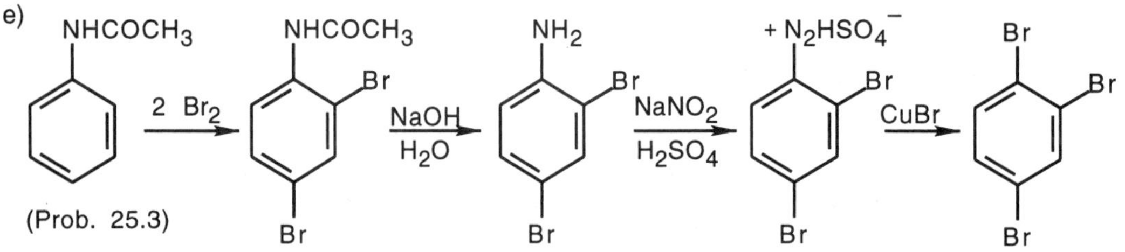

(Prob. 25.3)

1,2,4–Tribromobenzene

25.8

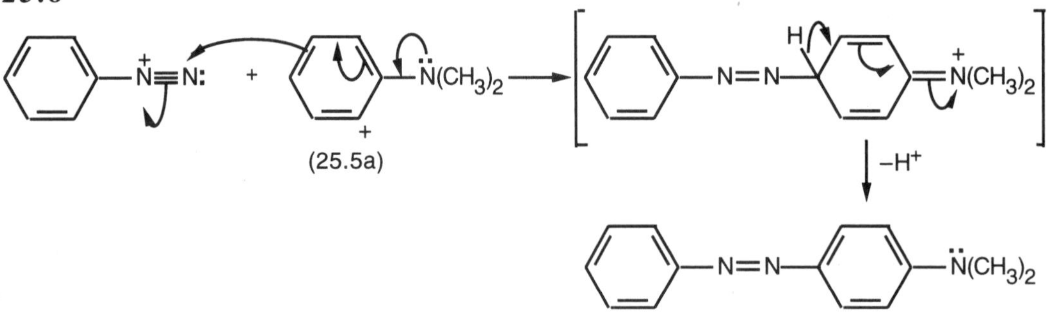

(25.5a)

p–(*N*,*N*–Dimethylamino)azobenzene

25.9

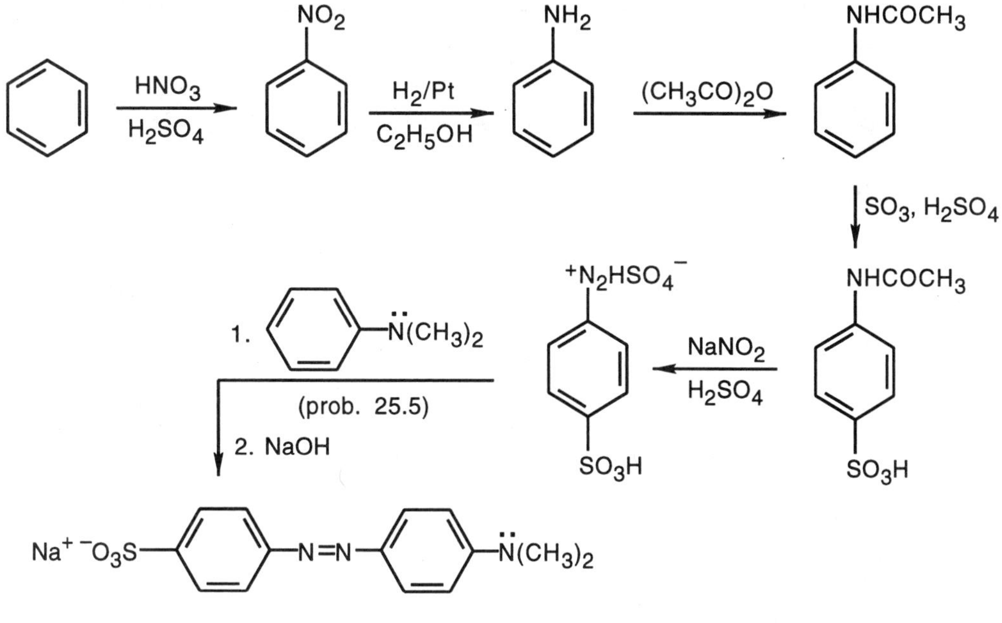

Methyl orange

25.10

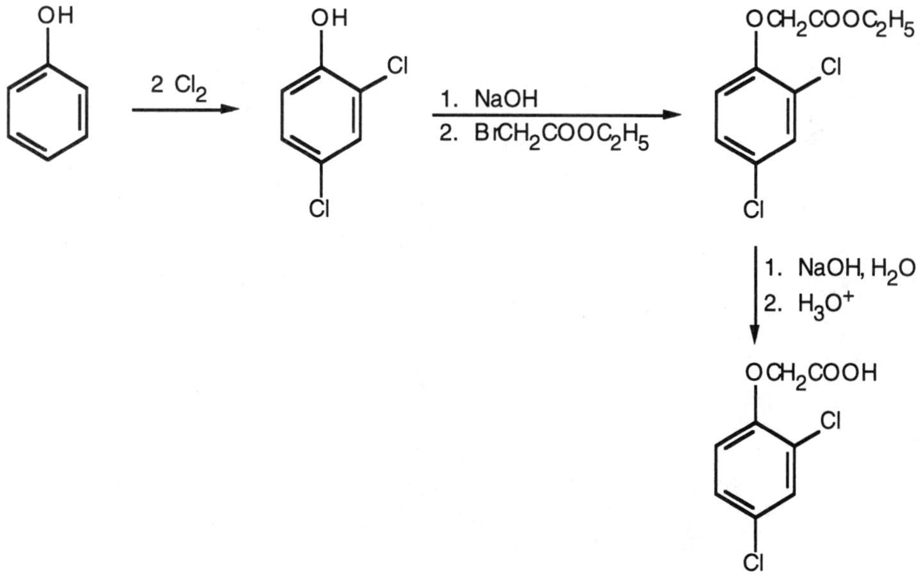

2,4–Dichlorophenoxyacetic acid

25.11

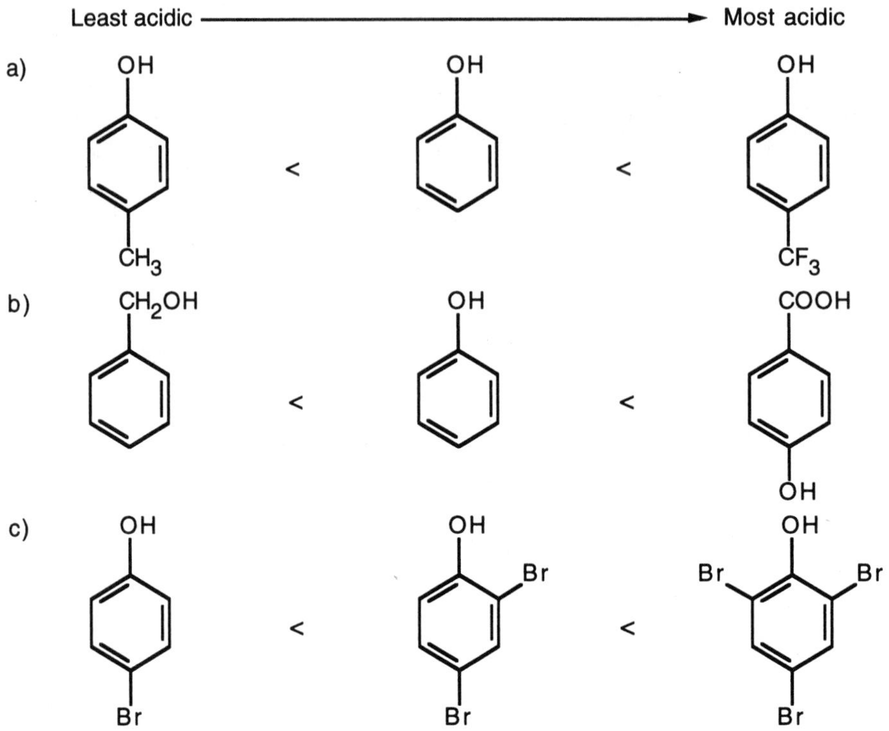

Least acidic ⟶ Most acidic

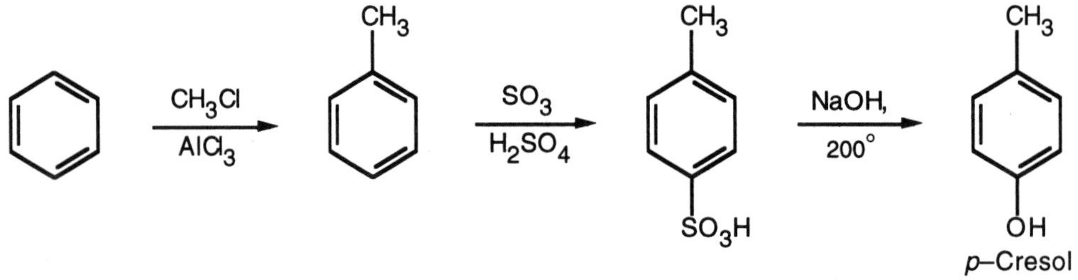

a)

$<$

$<$

b)

$<$

$<$

c)

$<$

$<$

25.12

p–Cresol

25.13

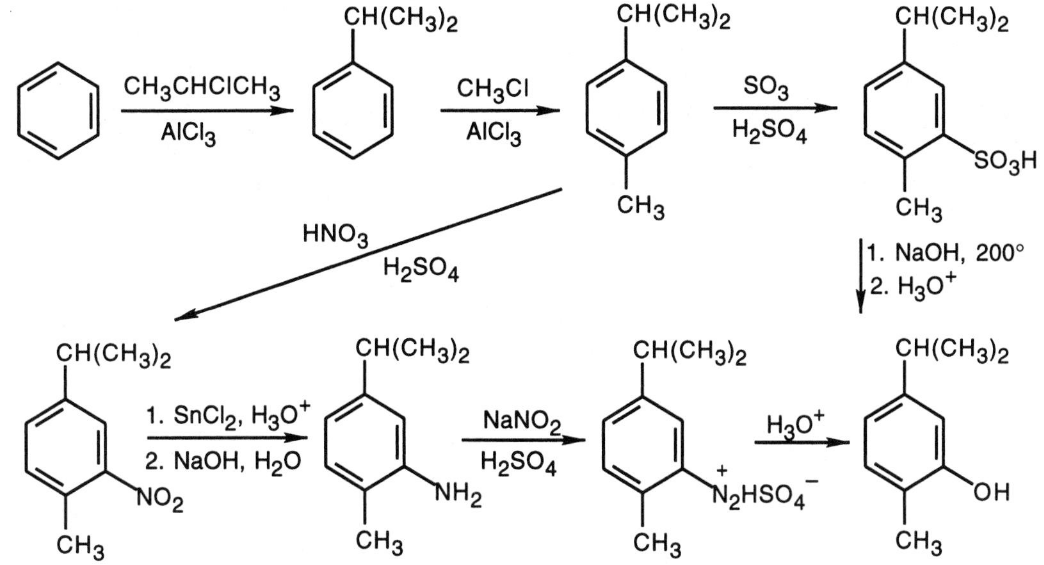

Carvacrol

25.14

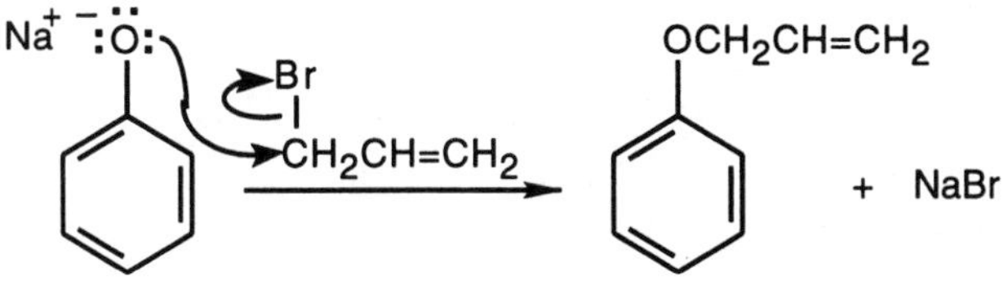

Allyl phenyl ether

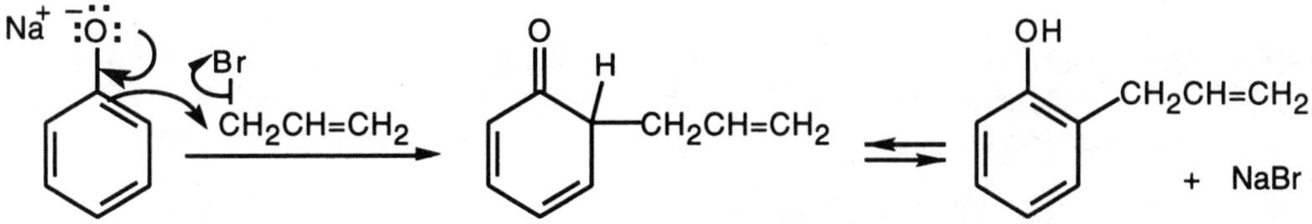

o-Allyphenol

25.15

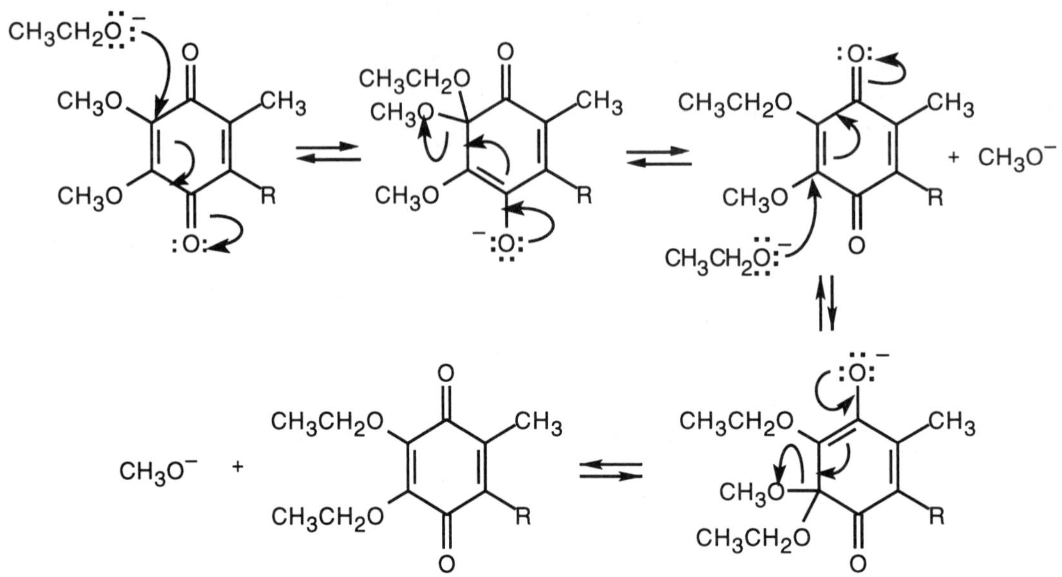

Conjugate addition of ethoxide ion to the double bond produces an intermediate that eliminates methoxide anion. The diethoxy product predominates because of the large excess of ethanol.

25.16

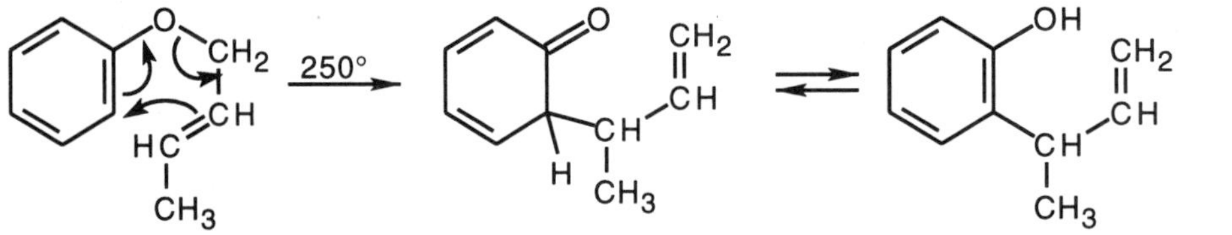

2–Butenyl phenyl ether   o–(1–Methylallyl) phenol

25.17 a) *p*–Bromophenol   b) 2,3–Dichloro–*N*–methylaniline
      c) 2–Methyl–1,4–benzenediamine   d) 3–Methoxy–5–methylphenol
      e) 1,3,5–Benzenetriol   f) *N*,3–Diethyl–*N*,5–dimethylaniline

25.18

a)

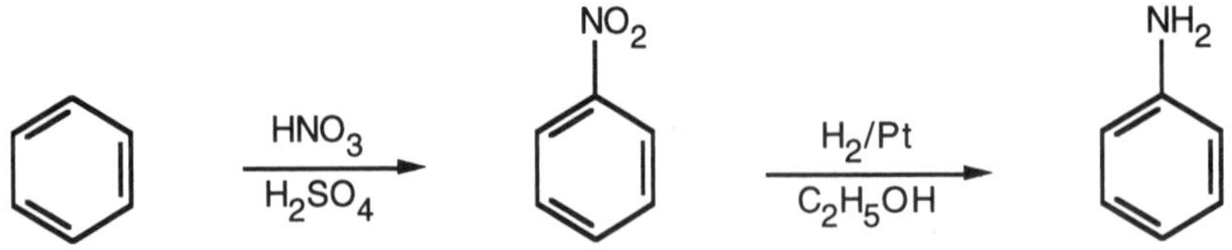

b)

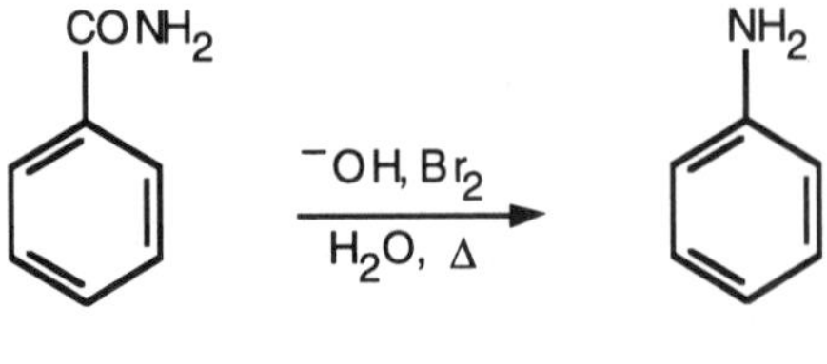

c)

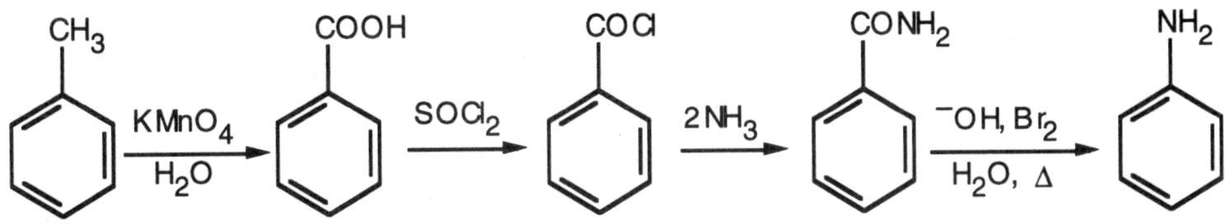

25.19

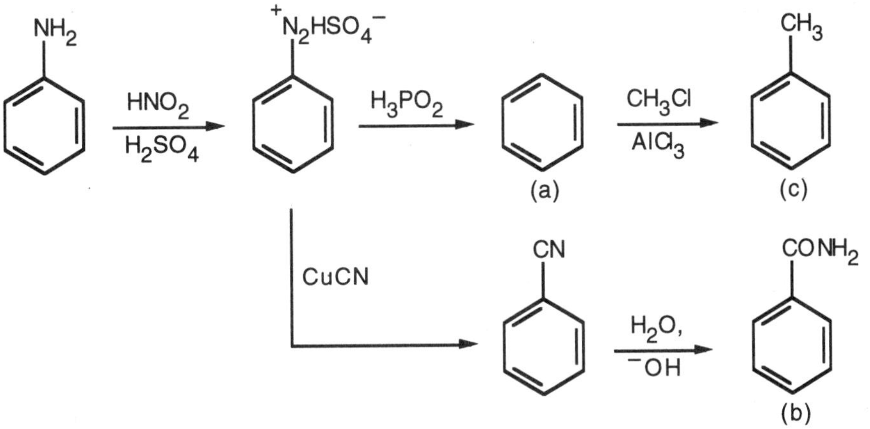

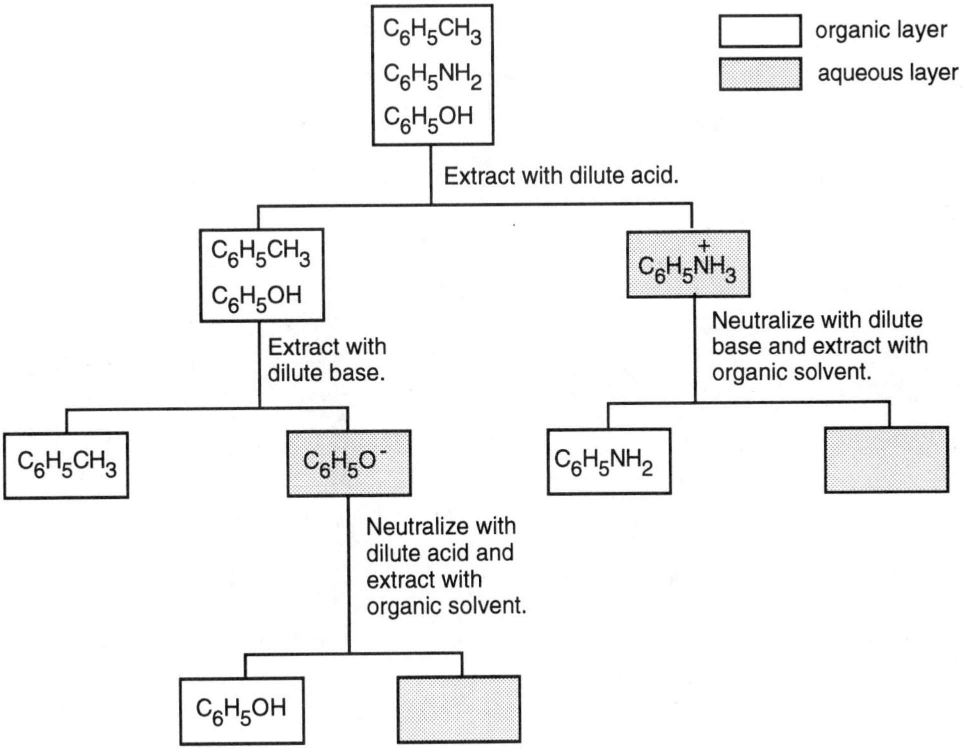

25.20

25.21

a)

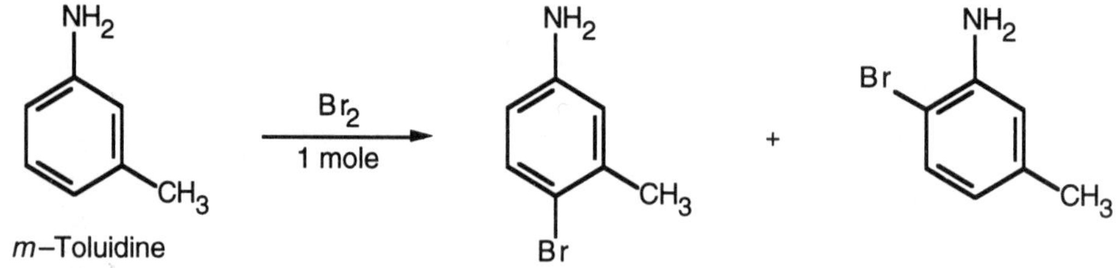

b)

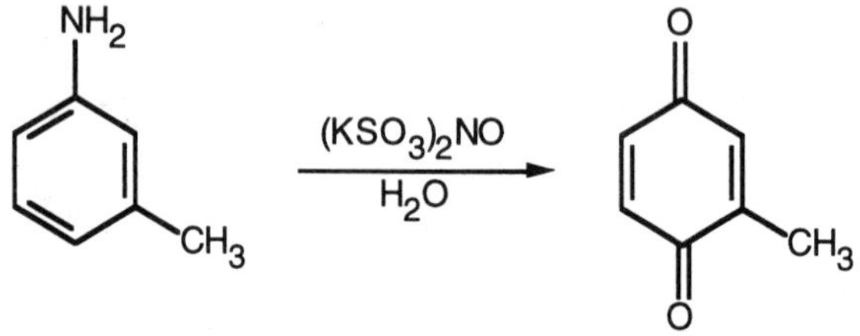

c)

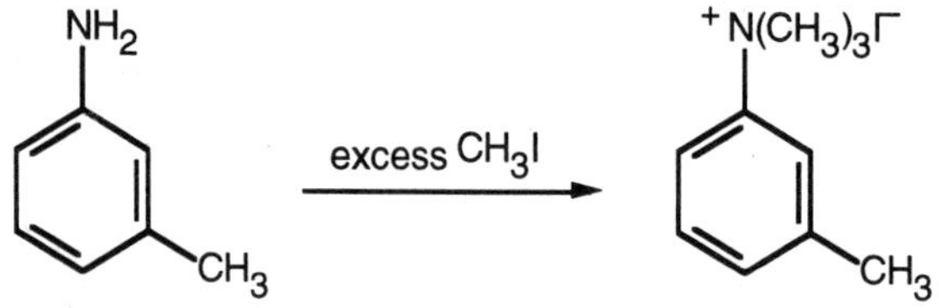

d)

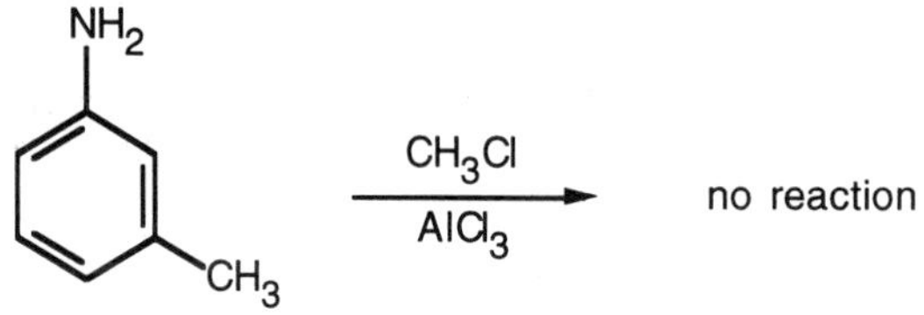

e)

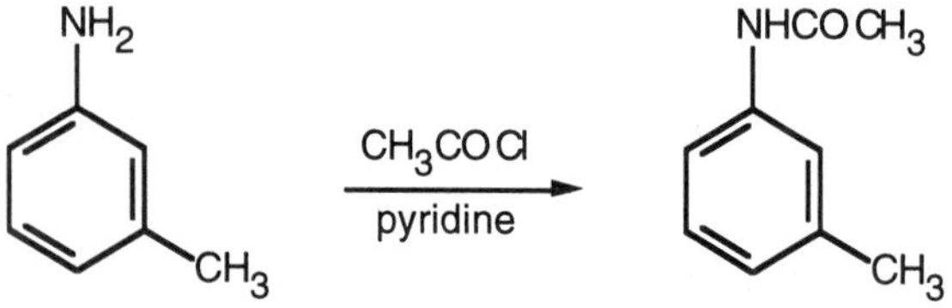

f)

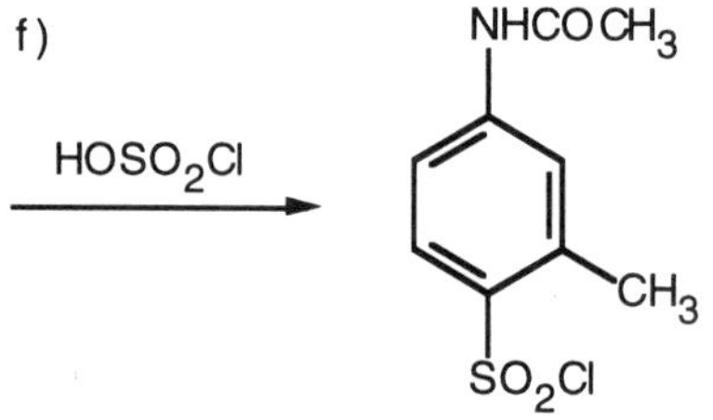

25.22

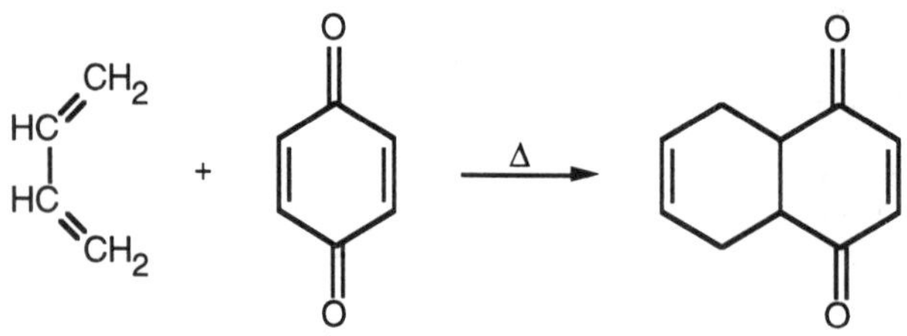

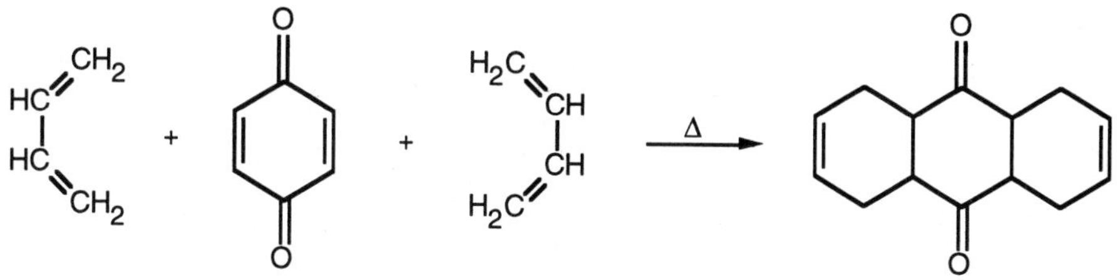

25.23

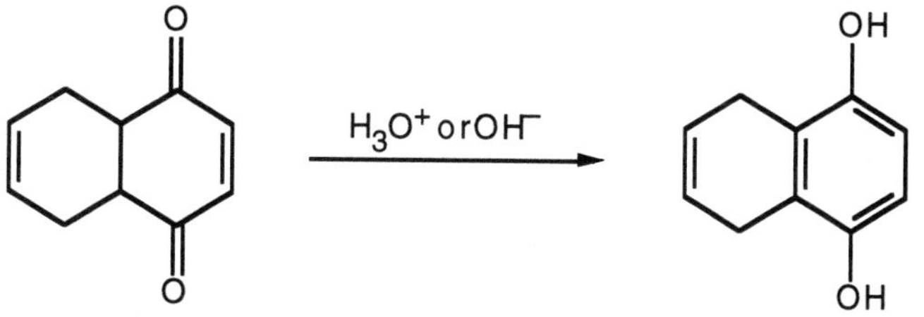

The two-proton singlet at 6.7 δ is due to the aromatic ring protons.

25.24

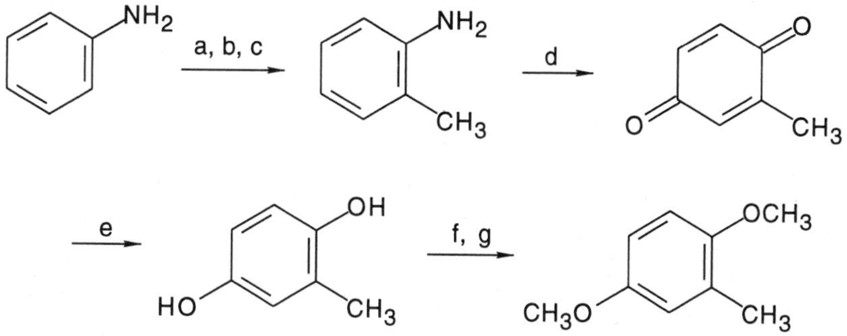

(a) $(CH_3CO)_2O$; (b) CH_3Cl, $AlCl_3$; (c) NaOH, H_2O; (d) $K(SO_3)_2NO$; (e) $SnCl_2$; (f) NaH; (g) CH_3I

25.25

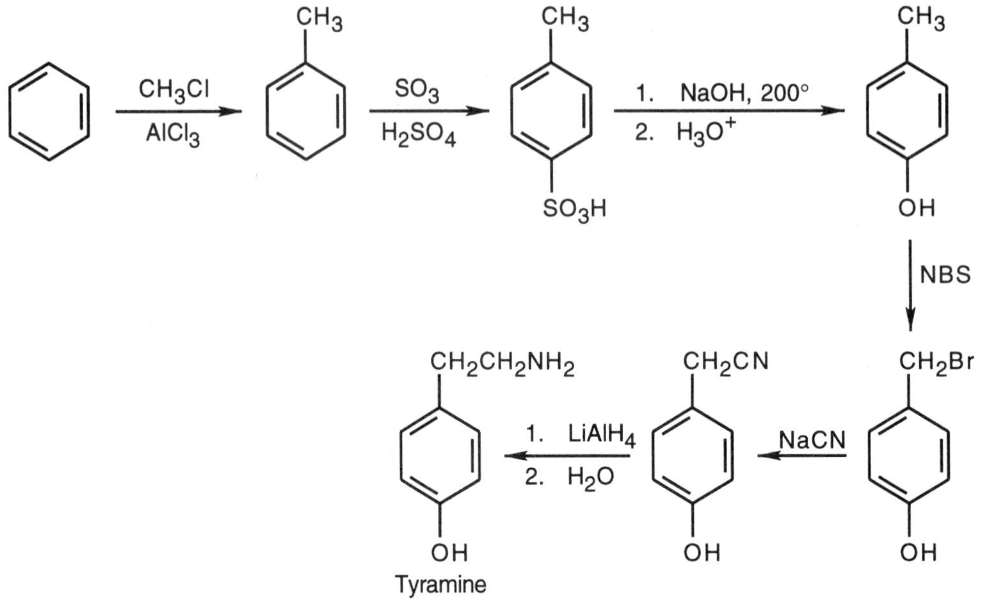

25.26 The nitrogen lone pair of electrons of diphenylamine can overlap the π electron system of either ring. Electron delocalization occurs to an even greater extent for diphenylamine than for aniline. Because the energy difference between non-protonated and protonated amine is much greater for diphenylamine than for aniline, diphenylamine is non-basic.

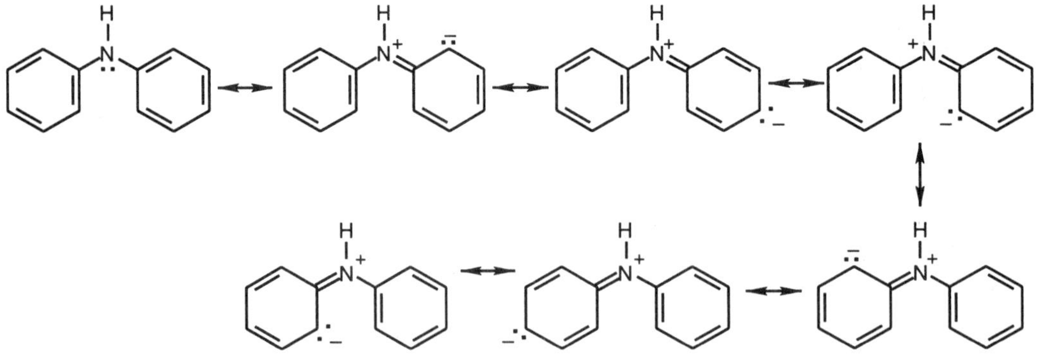

25.27

a)

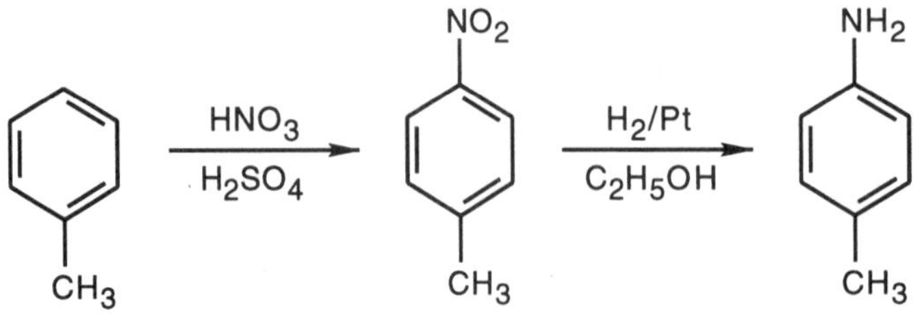

b)

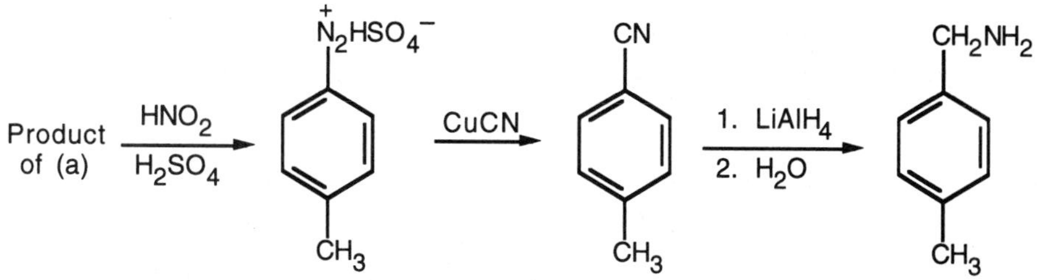

Product of (a) →[HNO₂, H₂SO₄] (N⁺₂HSO₄⁻ on ring, CH₃) →[CuCN] (CN on ring, CH₃) →[1. LiAlH₄, 2. H₂O] (CH₂NH₂ on ring, CH₃)

c)

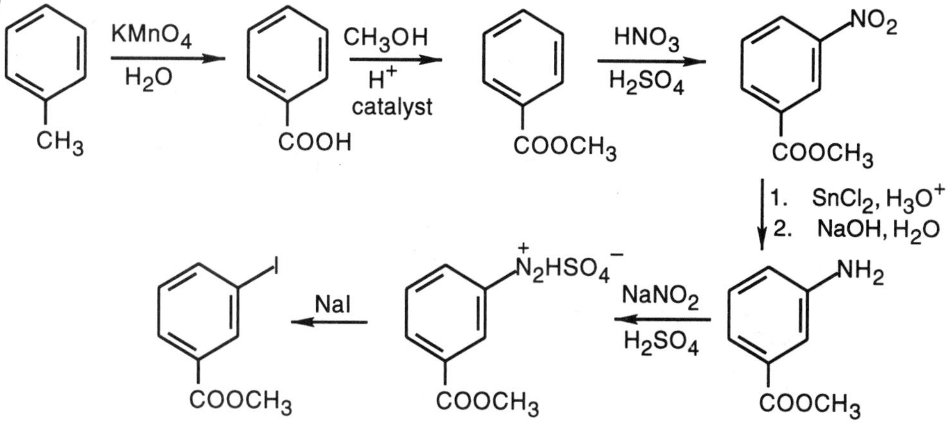

(CH₃ ring) →[KMnO₄, H₂O] (COOH ring) →[CH₃OH, H⁺ catalyst] (COOCH₃ ring) →[HNO₃, H₂SO₄] (NO₂, COOCH₃ ring)

→[1. SnCl₂, H₃O⁺; 2. NaOH, H₂O]

(I, COOCH₃ ring) ←[NaI] (N⁺₂HSO₄⁻, COOCH₃ ring) ←[NaNO₂, H₂SO₄] (NH₂, COOCH₃ ring)

25.28

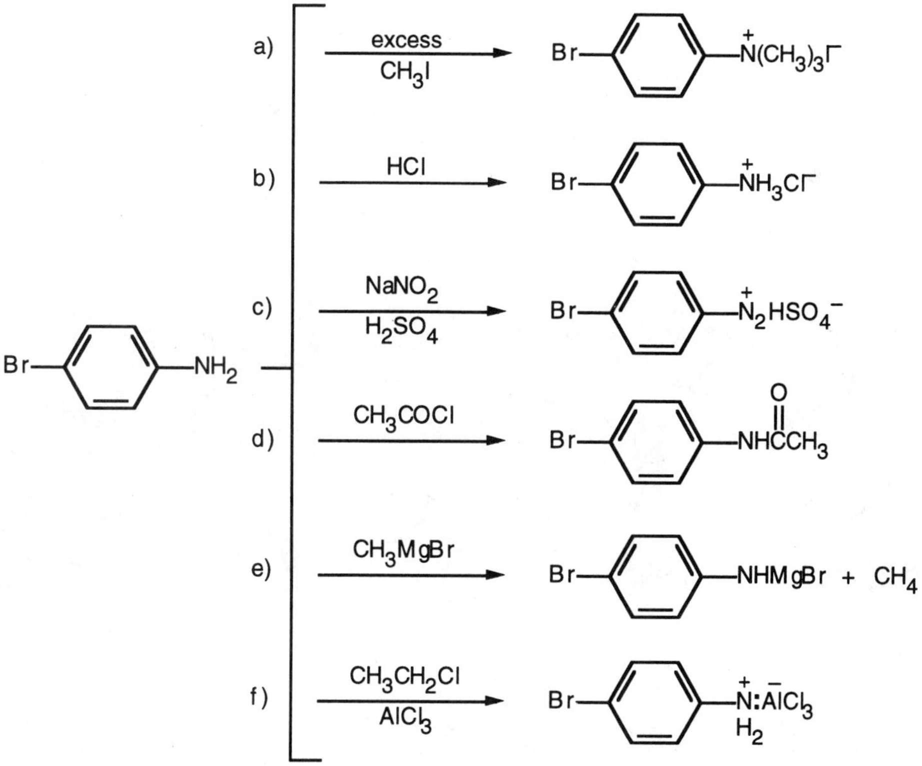

a) →[excess CH₃I] Br—⟨ring⟩—N⁺(CH₃)₃ I⁻

b) →[HCl] Br—⟨ring⟩—N⁺H₃Cl⁻

c) →[NaNO₂, H₂SO₄] Br—⟨ring⟩—N⁺₂HSO₄⁻

d) →[CH₃COCl] Br—⟨ring⟩—NHCCH₃ (with O double bond)

e) →[CH₃MgBr] Br—⟨ring⟩—NHMgBr + CH₄

f) →[CH₃CH₂Cl, AlCl₃] Br—⟨ring⟩—N⁺:AlCl₃⁻ with H₂

g) 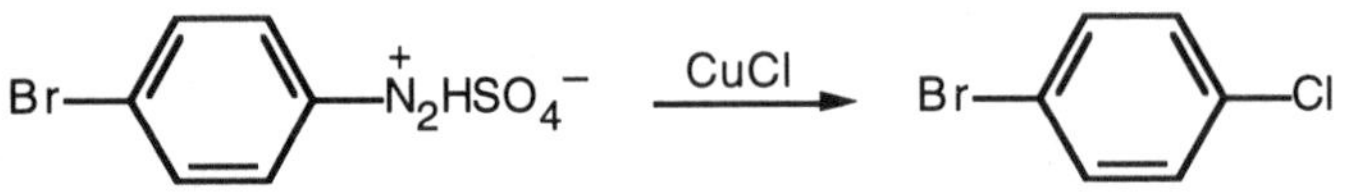
CuCl

h) 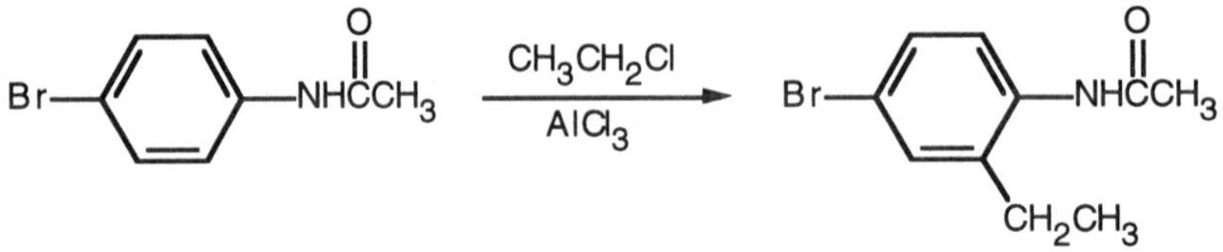
CH_3CH_2Cl
$AlCl_3$

25.29

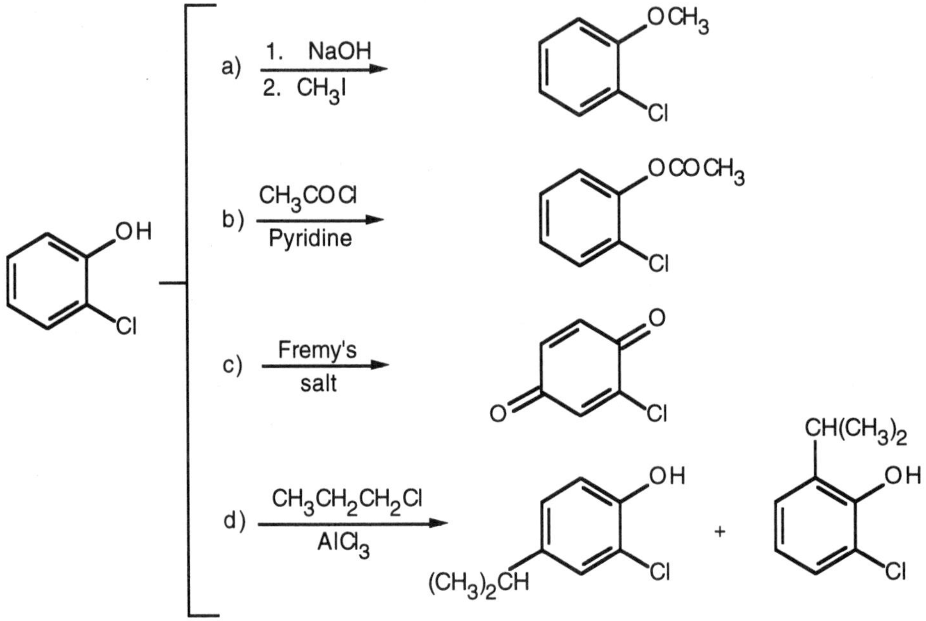
a) 1. NaOH
2. CH_3I
b) CH_3COCl
Pyridine
c) Fremy's
salt
d) $CH_3CH_2CH_2Cl$
$AlCl_3$

25.30

a) 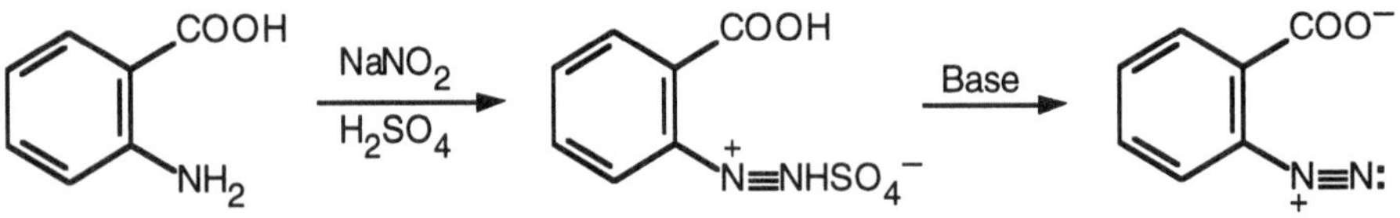
$NaNO_2$
H_2SO_4
Base

b) 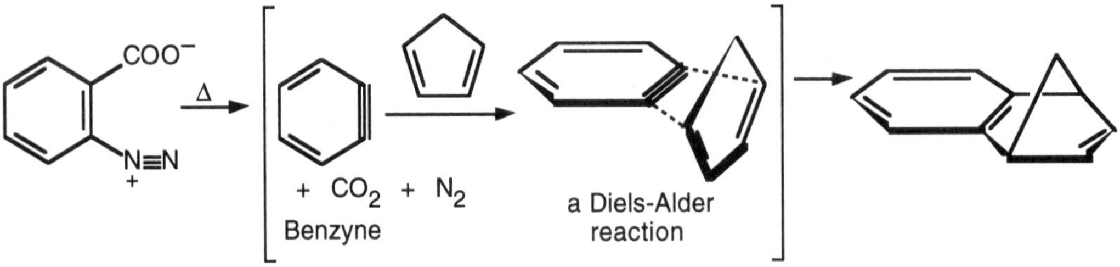
Δ
+ CO_2 + N_2
Benzyne
a Diels-Alder
reaction

25.31

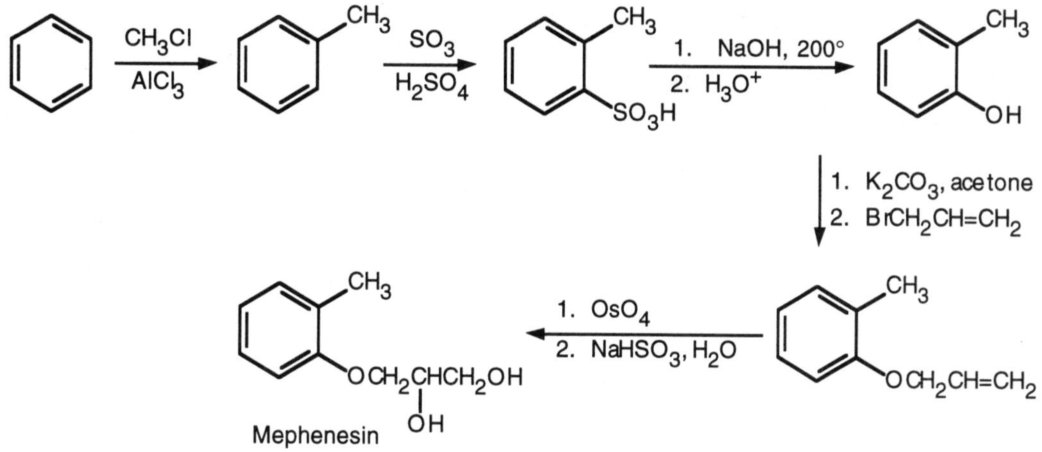

Mephenesin

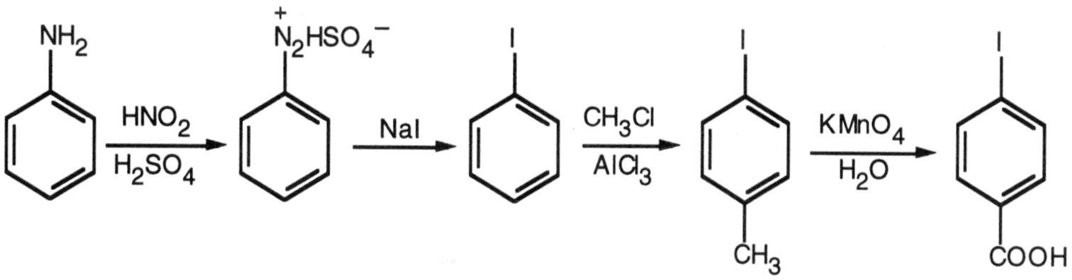

25.32

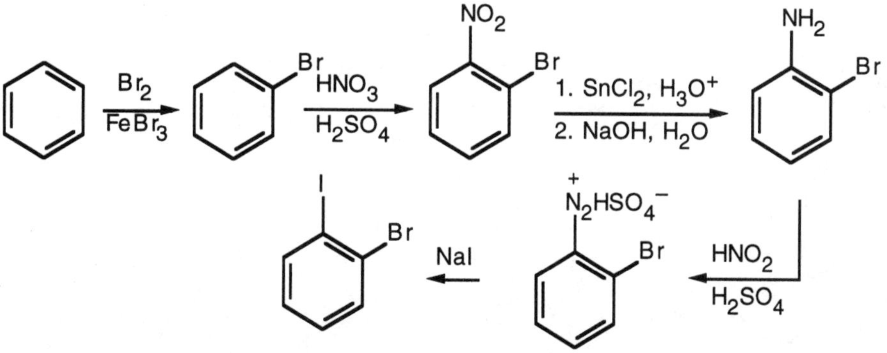

a)

b)

25.33

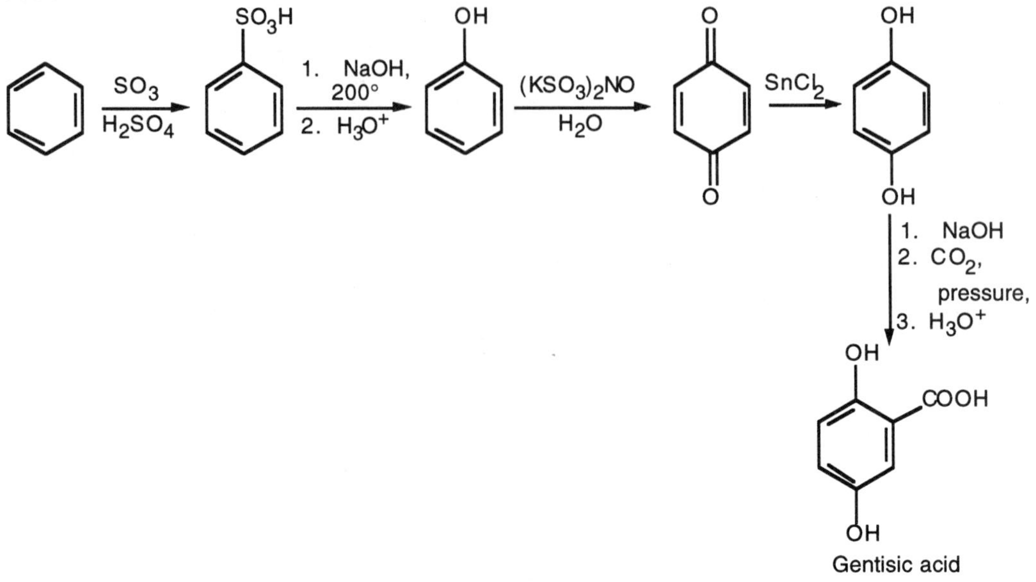

Gentisic acid

25.34

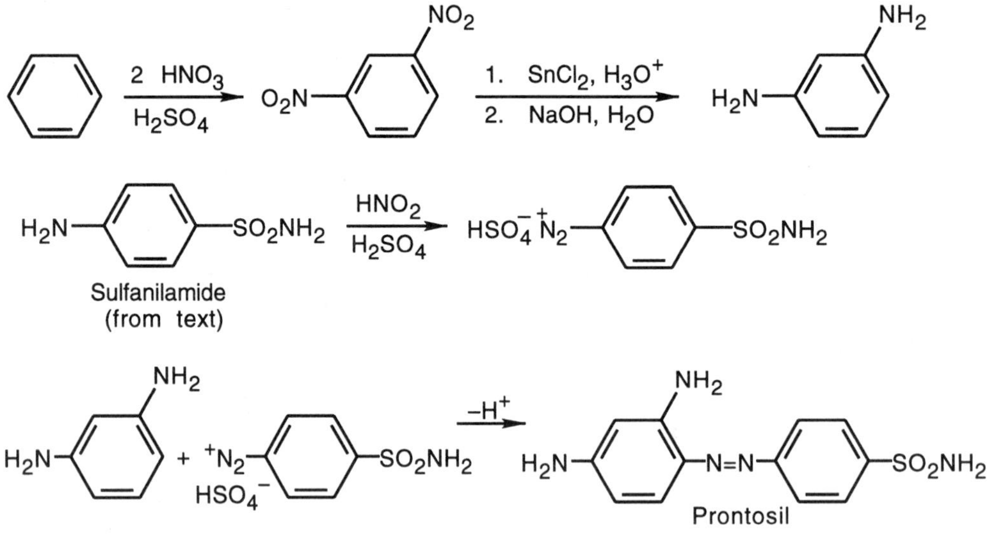

Sulfanilamide
(from text)

Prontosil

25.35

(prob. 25.9)

Orange II

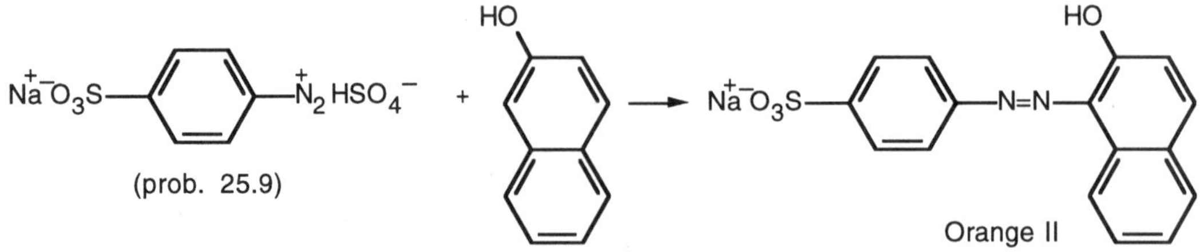

Electrophilic substitution on ß–naphthol occurs between the hydroxyl group and the fused ring.

25.36

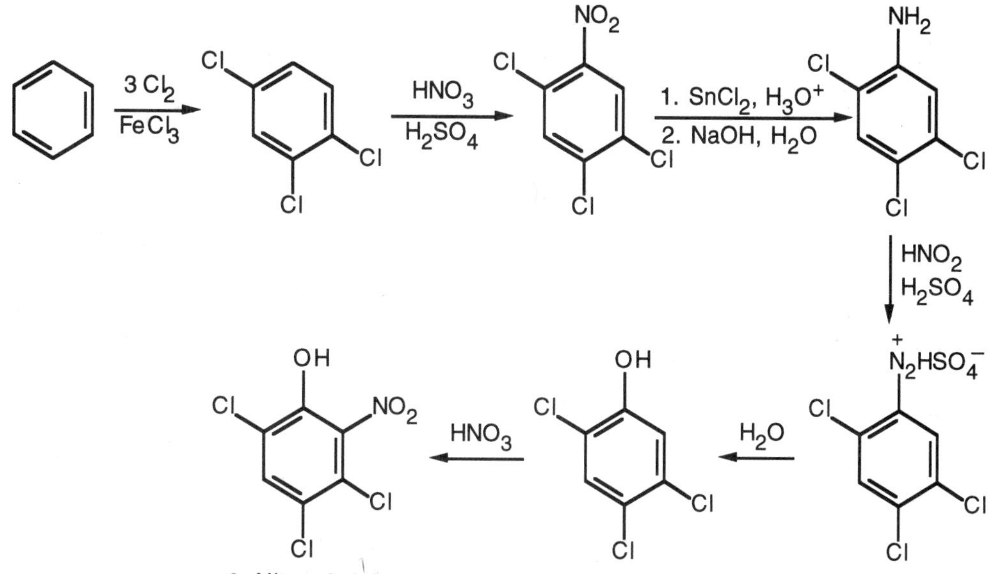

2–Nitro–3,4,6–
trichlorophenol

25.37

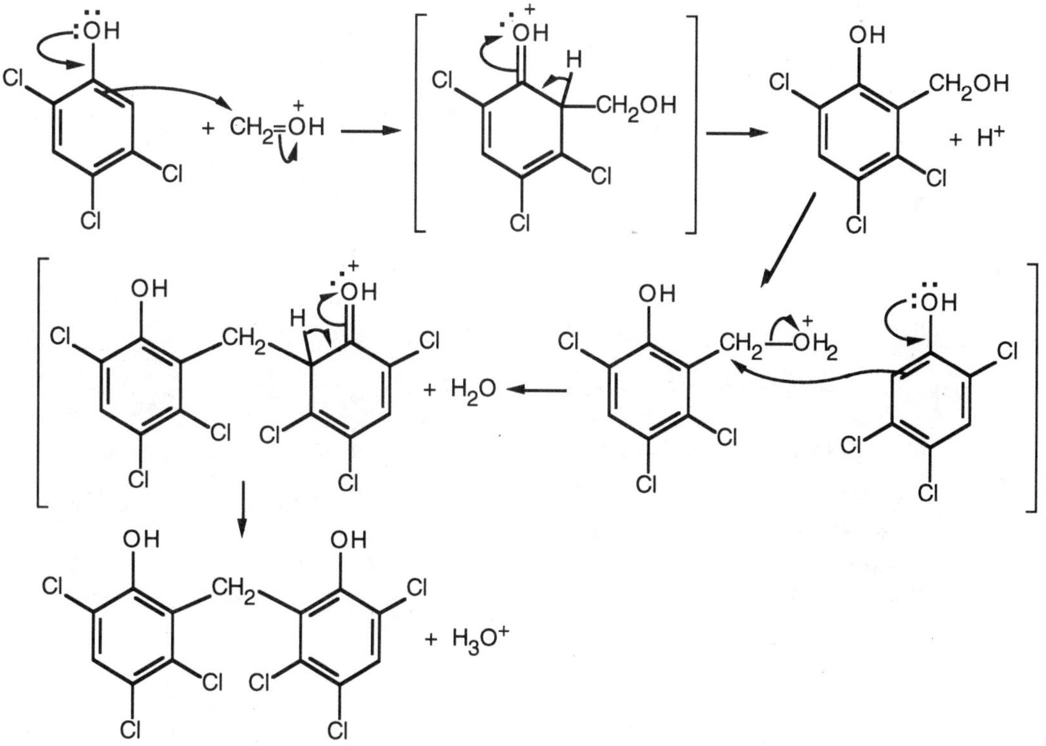

Hexachlorophene

25.38

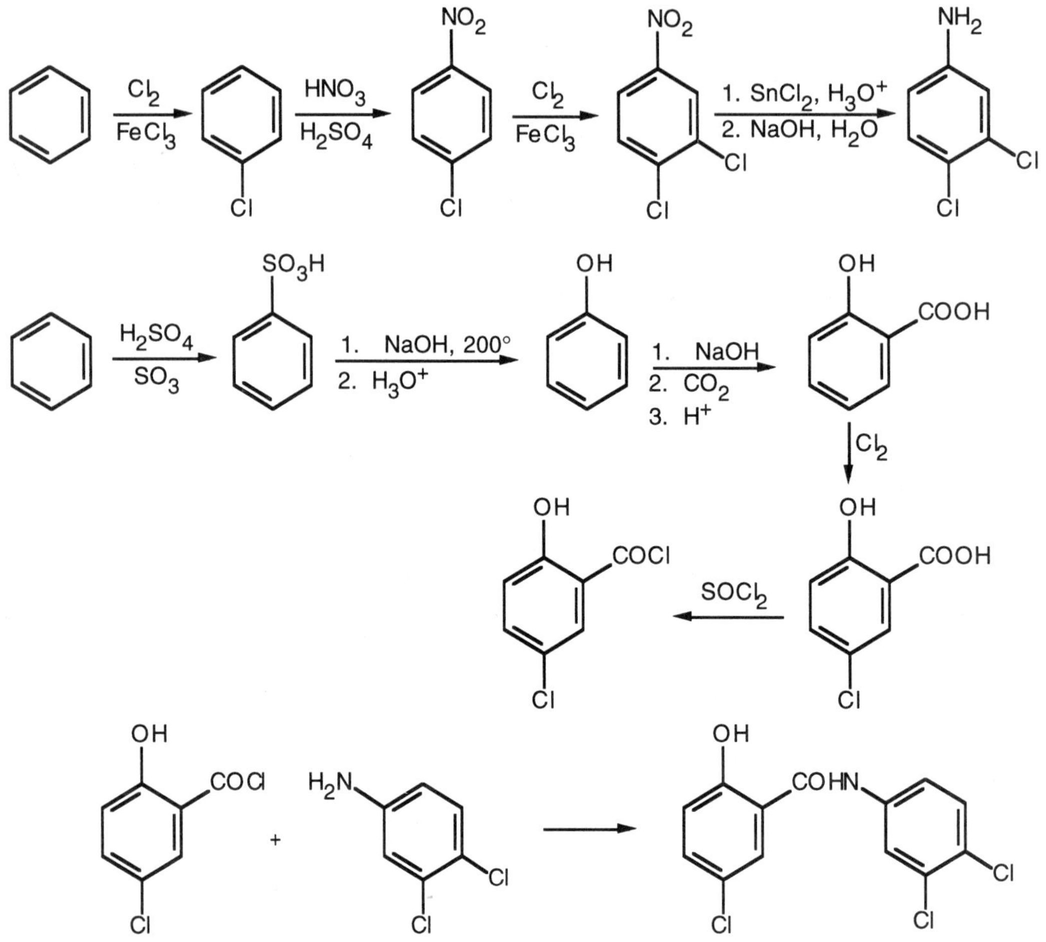

Trichlorosalicylanilide

25.39 <u>Structural formula</u>: $C_8H_{10}O$ contains 4 multiple bonds and/or rings.

<u>Infrared</u>: The broad band at 3500 cm^{-1} indicates a hydroxyl group. The absorptions at 1500 cm^{-1} and 1600 cm^{-1} are due to an aromatic ring. The absorption at 830 cm^{-1} shows that the ring is disubstituted. Compound <u>A</u> is probably a phenol.

<u>^{1}H NMR</u>: The triplet at 1.18 δ (3 H) is coupled with the quartet at 2.56 δ (2 H). These two absorptions are due to an ethyl group.
    The peaks at 6.75 δ-7.05 δ (4 H) are due to an aromatic ring. The symmetrical splitting pattern of these peaks indicate that the aromatic ring is *p*–disubstituted.
    The absorption at 5.50 δ (1 H) is due to an –OH proton.

Compound <u>A</u>

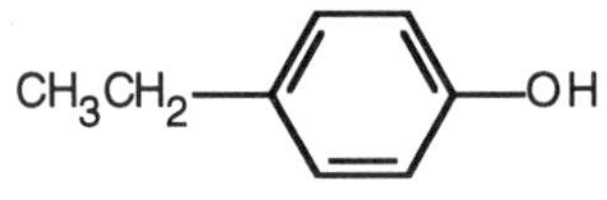

$$CH_3CH_2 - \text{(ring)} - OH$$

p–Ethylphenol

25.40

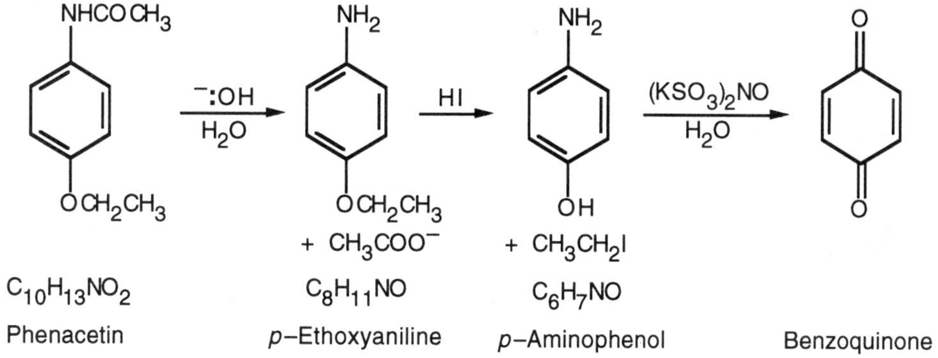
NHCOCH3
⁻:OH
H2O
NH2
HI
NH2
(KSO3)2NO
H2O
O
OCH2CH3
OCH2CH3
+ CH3COO⁻
OH
+ CH3CH2I
O
C10H13NO2
C8H11NO
C6H7NO
Phenacetin
p–Ethoxyaniline
p–Aminophenol
Benzoquinone

25.41

$$CH_3C\equiv N + HCl + ZnCl_2 \rightleftharpoons CH_3C\equiv \overset{+}{N}H\ ZnCl_3^-$$

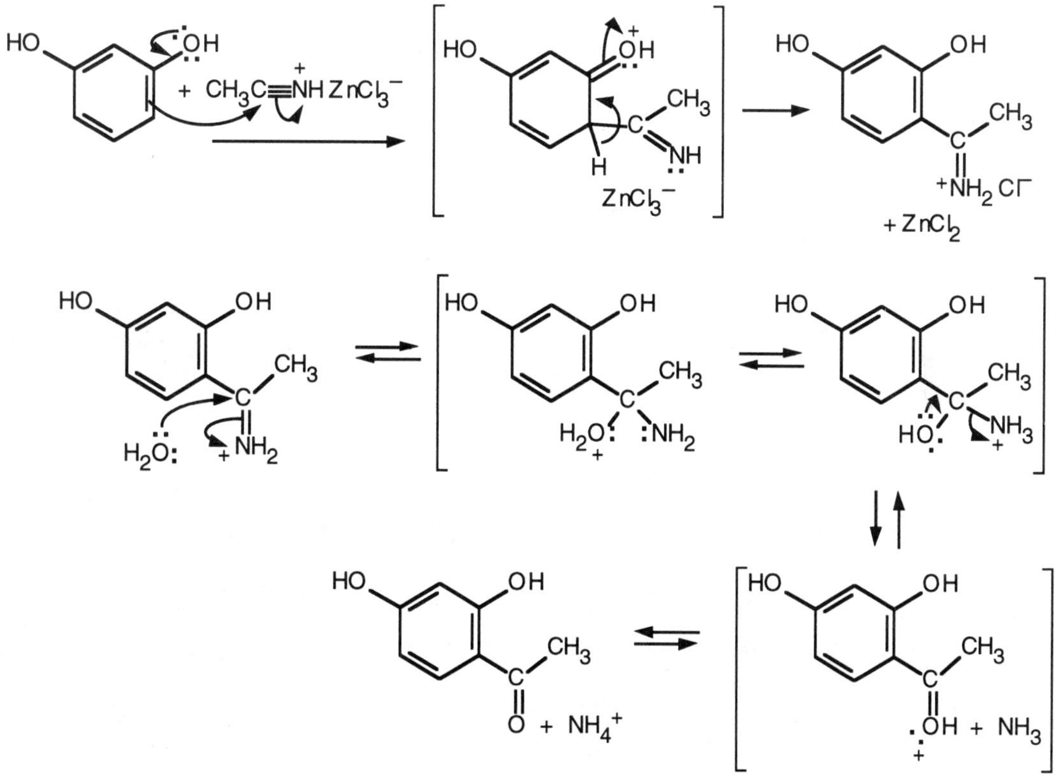

The Hoesch reaction is mechanistically similar to Friedel–Crafts acylation.

25.42

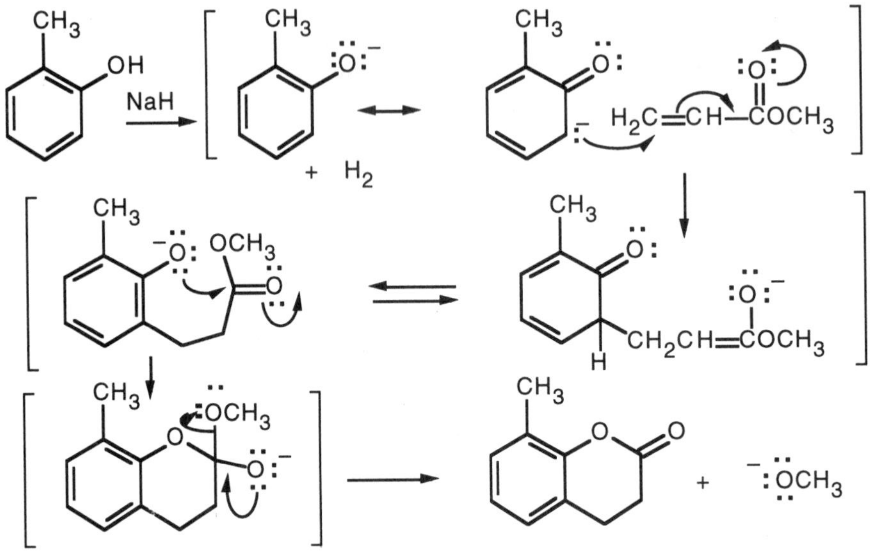

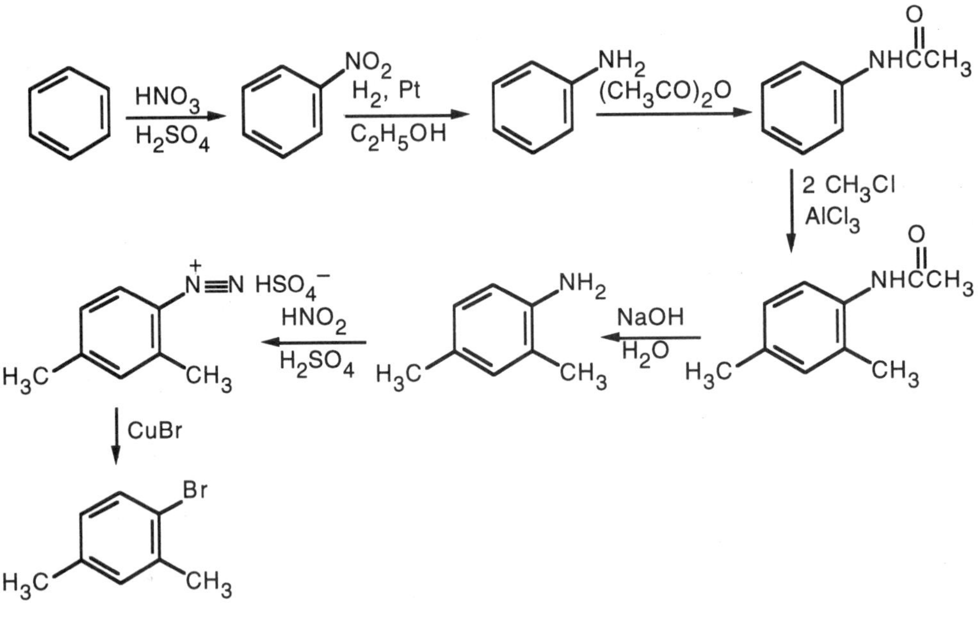

25.43

25.44

a)

(BHT)

b)

25.45

a)

b)

Study Guide for Chapter 25

After studying this chapter, you should be able to:

(1)   Draw and name aromatic amines and phenols (25.17).

(2)   Predict the effects of substituents on the acidity and basicity of
a) Aromatic amines (25.1, 25.2, 25.26).
b) Phenols (25.11).

(3)   Synthesize, by several different methods
a) Aromatic amines (25.3, 25.5, 25.18).
b) Phenols (25.10, 25.12, 25.13, 25.31, 25.33, 25.36, 25.37, 25.38)

(4)   Formulate mechanisms for reactions involving aromatic amines and phenols (25.41, 25.42).

(5)   Use diazonium salts in the synthesis of substituted aromatic compounds (25.6, 25.7, 25.27, 25.30, 25.32, 25.43).

(6)   Synthesize simple dyes using diazo coupling reactions (25.8, 25.9, 25.34, 25.35).

(7)   Predict the products of reactions involving:
a) Aromatic amines (25.19, 25.21, 25.24, 25.25, 25.28).
b) Phenols (25.14, 25.15, 25.16, 25.22, 25.29).

(8)   Identify aromatic amines and phenols spectroscopically (25.23, 25.39, 25.40, 25.44, 25.45).

26.1

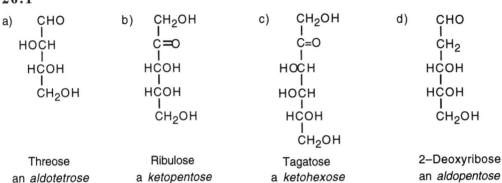

a)
```
    CHO
    |
  HOCH
    |
   HCOH
    |
   CH₂OH
```
Threose

an *aldotetrose*

b)
```
   CH₂OH
    |
   C=O
    |
   HCOH
    |
   HCOH
    |
   CH₂OH
```
Ribulose

a *ketopentose*

c)
```
   CH₂OH
    |
   C=O
    |
  HOCH
    |
  HOCH
    |
   HCOH
    |
   CH₂OH
```
Tagatose

a *ketohexose*

d)
```
    CHO
    |
   CH₂
    |
   HCOH
    |
   HCOH
    |
   CH₂OH
```
2–Deoxyribose

an *aldopentose*

26.2 Projections A, B and C represent the *S* enantiomer of glyceraldehyde; projection D represents the *R* enantiomer.

26.3

a)
```
       COOH
        |
  H₂N┈┈┈┼━━H      =
        |
       CH₃
```
```
       NH₂
        |
    ┈┈┈┼━━ S
   H⁀  |  CH₃
       COOH
```

b)
```
       CHO
        |
   H┈┈┈┼━━OH      =
        |
       CH₃
```
```
       OH
        |
    ┈┈┈┼━━ R
   H⁀  |  CHO
       CH₃
```

c)
```
       CH₃
        |
   H━━━┼┈┈┈CHO    =
        |
       CH₂CH₃
```
```
       CHO
        |
    ┈┈┈┼━━ S
   H⁀  |  CH₃
       CH₂CH₃
```

26.4

a)
```
       CHO
      ┌──┐
  HO──┤S├──H
      ├──┤
  HO──┤S├──H
      └──┘
     CH₂OH
```
L–Erythrose

b)
```
       CHO
      ┌──┐
   H──┤R├──OH
      ├──┤
  HO──┤S├──H
      ├──┤
   H──┤R├──OH
      └──┘
     CH₂OH
```
D–Xylose

c)
```
      CH₂OH
       |
      C=O
      ┌──┐
  HO──┤S├──H
      ├──┤
   H──┤R├──OH
      └──┘
     CH₂OH
```
D–Xylulose

26.5
```
       CHO
      ┌──┐
   H──┤R├──OH
      ├──┤
  HO──┤S├──H
      ├──┤
  HO──┤S├──H
      └──┘
     CH₂OH
```
L–(+)–Arabinose

26.6

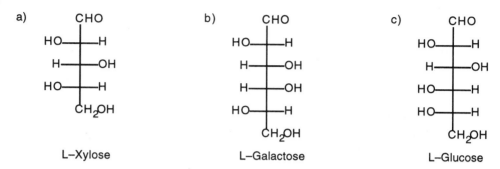

a)
CHO
HO——H
H——OH
HO——H
CH₂OH

L–Xylose

b)
CHO
HO——H
H——OH
H——OH
HO——H
CH₂OH

L–Galactose

c)
CHO
HO——H
H——OH
HO——H
HO——H
CH₂OH

L–Glucose

26.7 An aldoheptose has 5 stereogenic centers. Thus, there are $2^5 = 32$ aldoheptoses — 16 D and 16 L.

26.8

CHO
H——OH
H——OH
HO——H
H——OH
H——OH
CH₂OH

CHO
HO——H
H——OH
HO——H
H——OH
H——OH
CH₂OH

26.9

CHO
H——OH
HO——H
HO——H
H——OH
CH₂OH

D–Galactose

⇌

26.10

CHO
HO——H
H——OH
HO——H
HO——H
CH₂OH

L–Glucose

⇌

CHO
H——OH
H——OH
H——OH
CH₂OH

D–Ribose

26.11

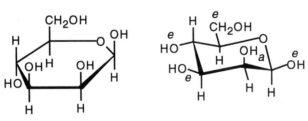

α–D–Fructofuranose   β–D–Fructofuranose

26.12 Let x be the fraction of D–glucose present as the α anomer, and let y be the fraction of D–glucose present as the ß anomer.

Then: $112.2° x + 18.7° y = 52.6°$ $x + y = 1;$ $y = 1 - x$
$112.2°x + 18.7°(1-x) = 52.6°$
$93.5°x = 33.9°$
$x = 0.362$
$y = 0.638$

Thus, 36.2% of glucose is present as the α anomer and 63.8% is present as the ß anomer.

26.13

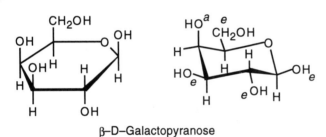

β–D–Galactopyranose

β–D–Mannopyranose

ß–D–Galactopyranose and ß–D–mannopyranose each have one hydroxyl group in the axial position and are therefore of similar stability.

26.14

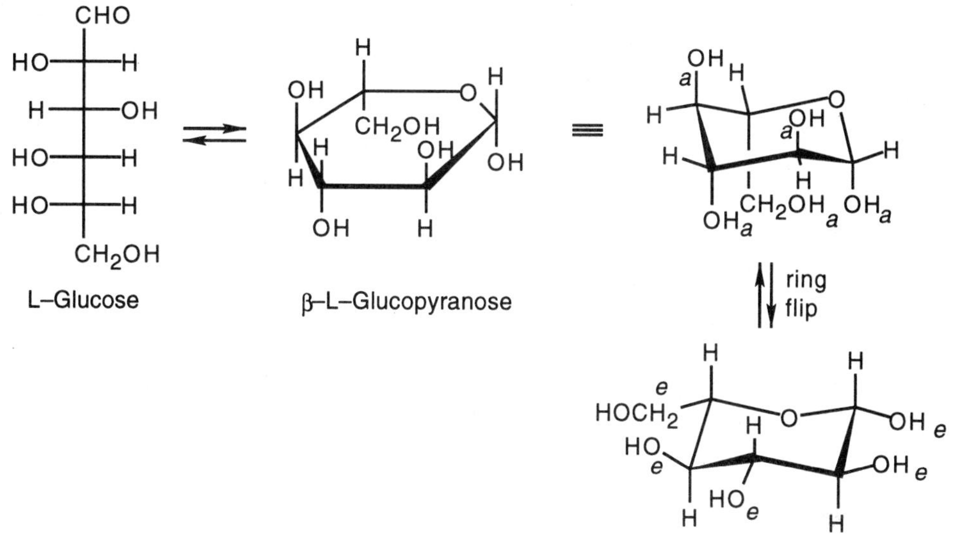

L-Glucose   β-L-Glucopyranose

ring flip

All substituents are equatorial in the more stable conformation of ß–L–glucopyranose.

26.15

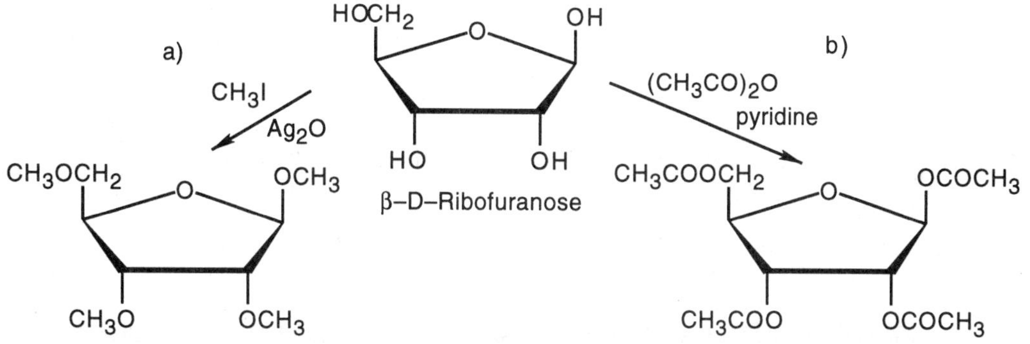

a)   CH₃I / Ag₂O

β-D-Ribofuranose

b)   (CH₃CO)₂O / pyridine

26.16

D–Galactose   Galactitol

1. NaBH₄
2. H₂O

------ Plane of symmetry

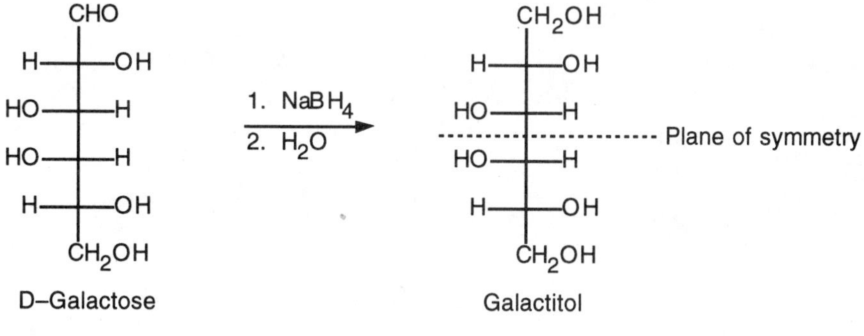

Reaction of D–galactose with NaBH₄ yields an alditol that has a plane of symmetry. Galactitol is a meso compound.

26.17

CHO
H——OH
HO——H
H——OH
H——OH
CH₂OH
D–Glucose

1. NaBH₄
2. H₂O

CH₂OH
H——OH
HO——H
H——OH
H——OH
CH₂OH
D–Glucitol

≡

CH₂OH
HO——H
HO——H
H——OH
HO——H
CH₂OH
L–Gulitol

1. NaBH₄
2. H₂O

CHO
HO——H
HO——H
H——OH
HO——H
CH₂OH
L–Gulose

Reaction of an aldose with NaBH₄ produces a polyol (alditol). Because an alditol has the same functional group at both ends, two different aldoses can yield the same alditol. Here, L–gulose and D–glucose form the same alditol (rotate the Fischer projection of L–gulitol 180° to see the identity).

26.18

CHO
H——OH
HO——H
H——OH
H——OH
CH₂OH
D–Glucose

dil HNO₃
Δ

COOH
H——OH
HO——H
H——OH
H——OH
COOH
D–Glucaric acid

CHO
H——OH
H——OH
H——OH
H——OH
CH₂OH
D–Allose

dil HNO₃
Δ

COOH
H——OH
H——OH
H——OH
H——OH
COOH
Allaric acid

········· plane of symmetry

Allaric acid has a plane of symmetry and is an optically inactive meso compound. Glucaric acid has no symmetry plane.

26.19 D–Allose and D–galactose yield meso aldaric acids. All other D–hexoses produce optically active aldaric acids on oxidation.

26.20

CHO
H——OH
H——OH
H——OH
CH₂OH
D–Ribose

1. HCN
2. H₂, Pd/BaSO₄
3. H₃O⁺

CHO
H——OH
H——OH
H——OH
H——OH
CH₂OH
D–Allose

+

CHO
HO——H
H——OH
H——OH
H——OH
CH₂OH
D–Altrose

26.21

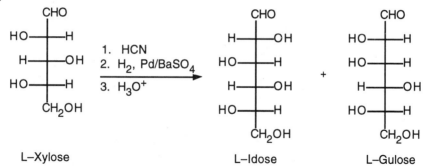

L–Xylose   L–Idose   L–Gulose

26.22

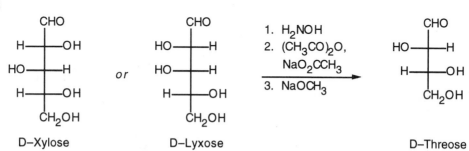

D–Xylose   D–Lyxose   D–Threose

26.23

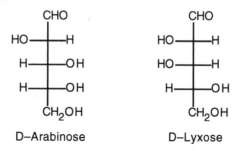

D–Arabinose   D–Lyxose

D–Glucose has the same configuration at C3, C4, and C5 as D–arabinose. D–Lyxose is the only other aldopentose that yields an optically active aldaric acid on nitric acid oxidation.

26.24

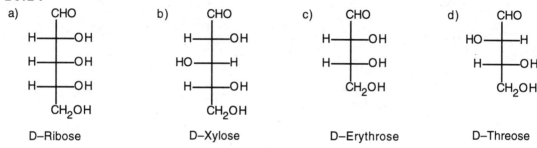

a)   b)   c)   d)

D–Ribose   D–Xylose   D–Erythrose   D–Threose

26.25

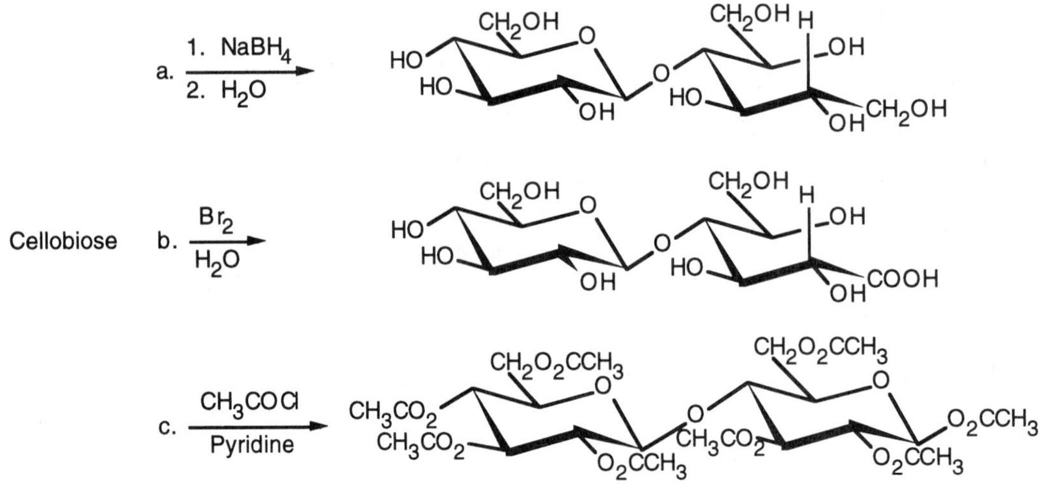

a. $\dfrac{\text{1. NaBH}_4}{\text{2. H}_2\text{O}}$

Cellobiose   b. $\dfrac{\text{Br}_2}{\text{H}_2\text{O}}$

c. $\dfrac{\text{CH}_3\text{COCl}}{\text{Pyridine}}$

26.26

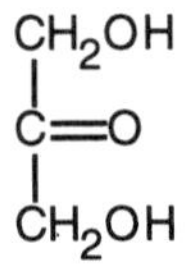

a ketotriose

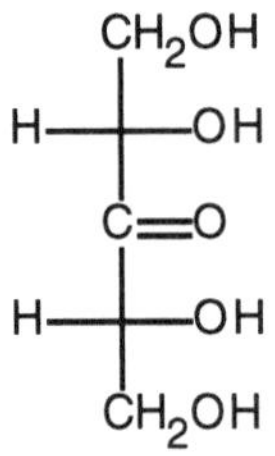

a ketopentose

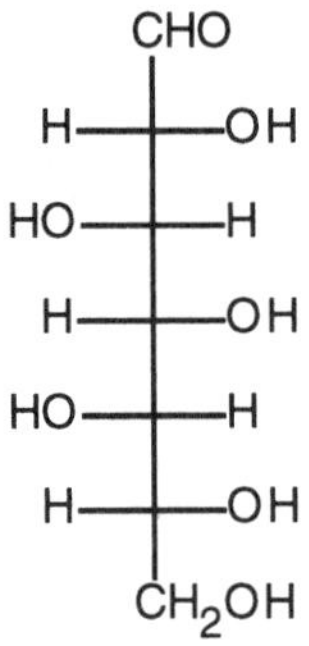

an aldoheptose

26.27

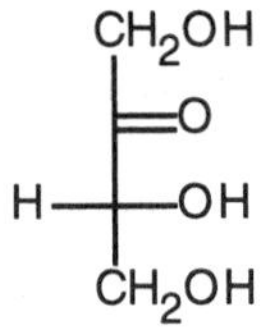

a ketotetrose

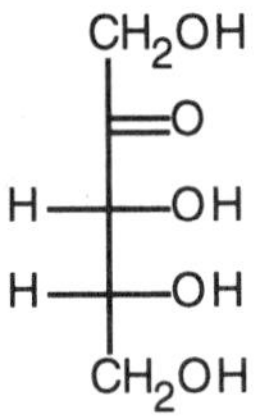

a ketopentose

26.28

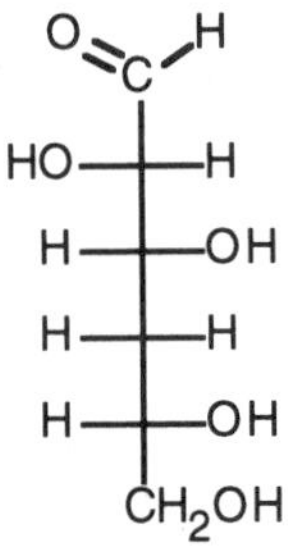

a deoxyaldohexose

26.29

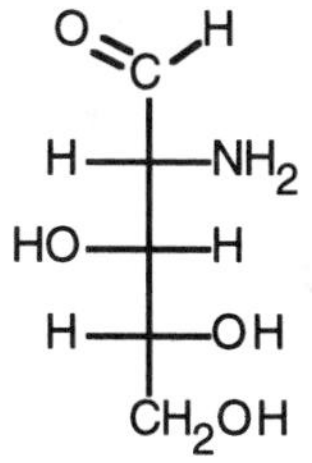

a five-carbon amino sugar

26.30–26.31

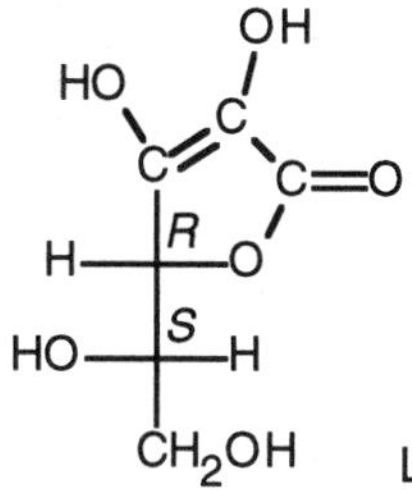

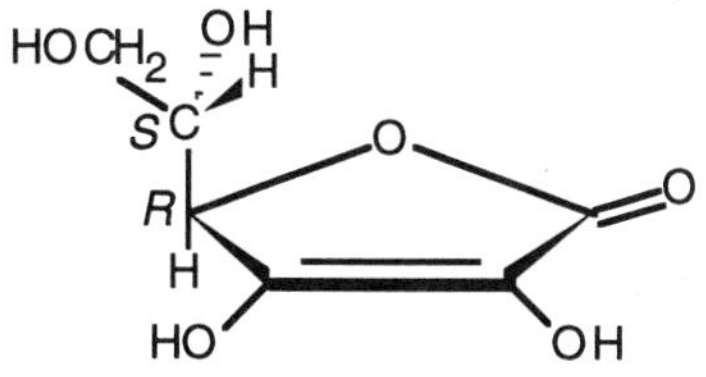

L–Ascorbic Acid

26.32

Definition		Example

a) A *monosaccharide* is a carbohydrate that cannot be hydrolyzed into smaller units.

D–Glucose

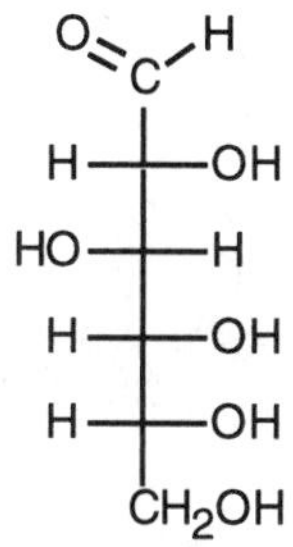

b) An *anomeric center* is a stereogenic center formed when an open chain nonosaccharide cyclizes to a furanose or pyranose ring.

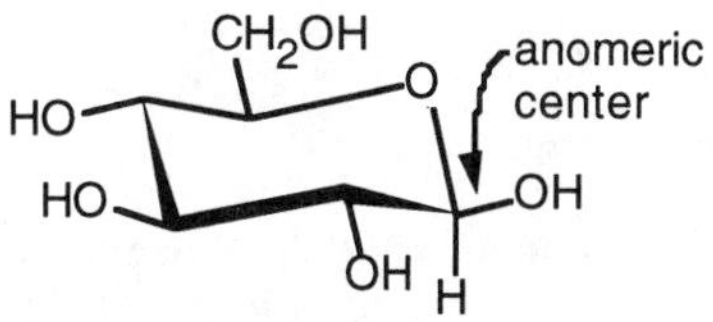

β–D–Glucopyranose

c) A *Haworth projection* is a drawing of a pyranose or furanose in which the ring is drawn as flat. This projection allows the relationship of the ring substituents to be viewed more easily.

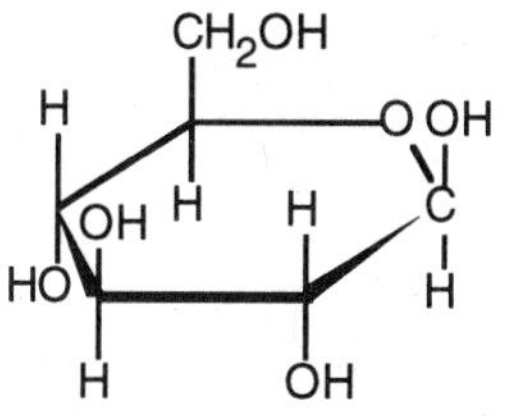

β–D–Glucopyranose

d)   A *Fischer projection* is a drawing of a
     carbohydrate in which each stereogenic center
     is represented as a pair of perpendicular
     lines. Vertical lines represent bonds
     going into the page, and horizontal lines
     represent bonds coming out of the page.

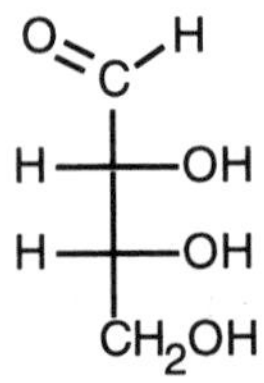

D–Erythrose

e)   A *glycoside* is a acetal of a carbohydrate,
     formed when an anomeric hydroxyl group
     reacts with another compound containing a
     hydroxyl group.

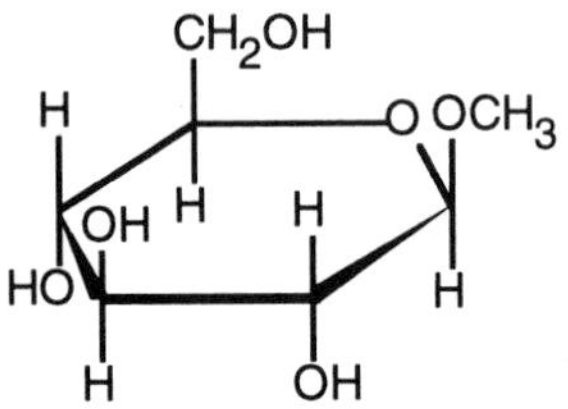

Methyl β –D–glucopyranoside

f)   A *reducing sugar* is a sugar that reacts with
     any of several reagents to yield an oxidized
     sugar plus reduced reagent.

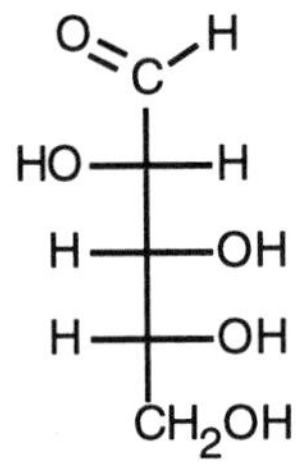

D–Arabinose

g)   A *pyranose* is a six-membered cyclic
     hemiacetal ring form of a monosaccharide.

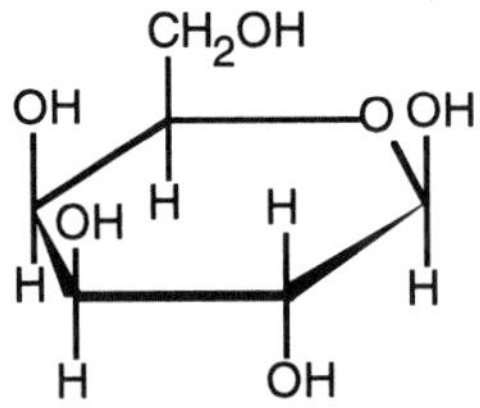

β–D–Galactopyranose

h)   A *1,4' link* occurs when the anomeric hydroxyl
     group (at carbon 1) of a pyranose or furanose
     forms a glycosidic bond with the hydroxyl group
     at carbon 4 of a second nonosaccharide.

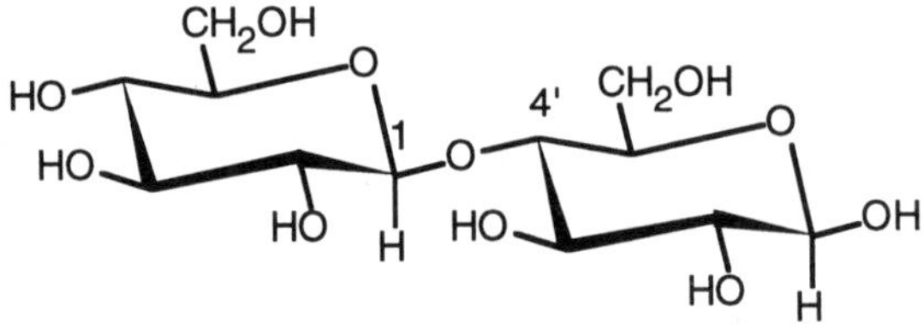

Cellobiose

i) A *D–sugar* is a sugar in which the hydroxyl
    group farthest from the carbonyl group
    points to the right in a Fischer projection.

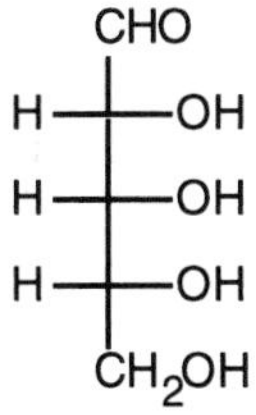

D–Ribose

26.33

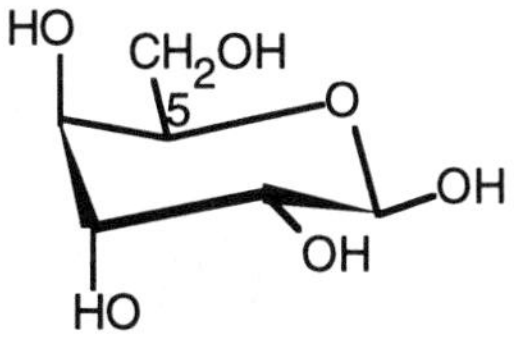

β–D–Gulopyranose

This structure is a pyranose (6-membered
ring) and is a β anomer (the C–1 hydroxyl
group and the –CH$_2$OH groups are *cis*)
It is a D–sugar because the –O– at C5
is on the right in the uncoiled form.

26.34

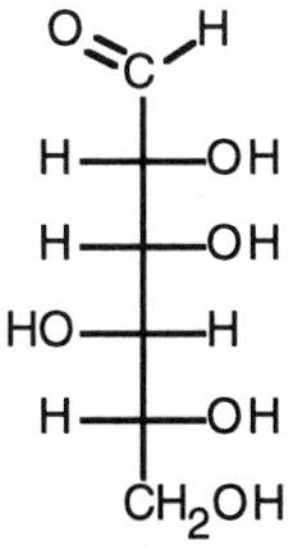

D–Gulose

26.35

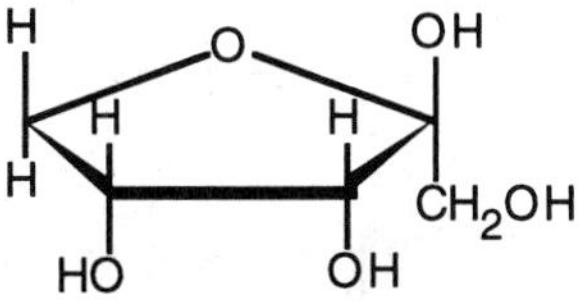

D–Ribulofuranose (β anomer)

26.36–26.37

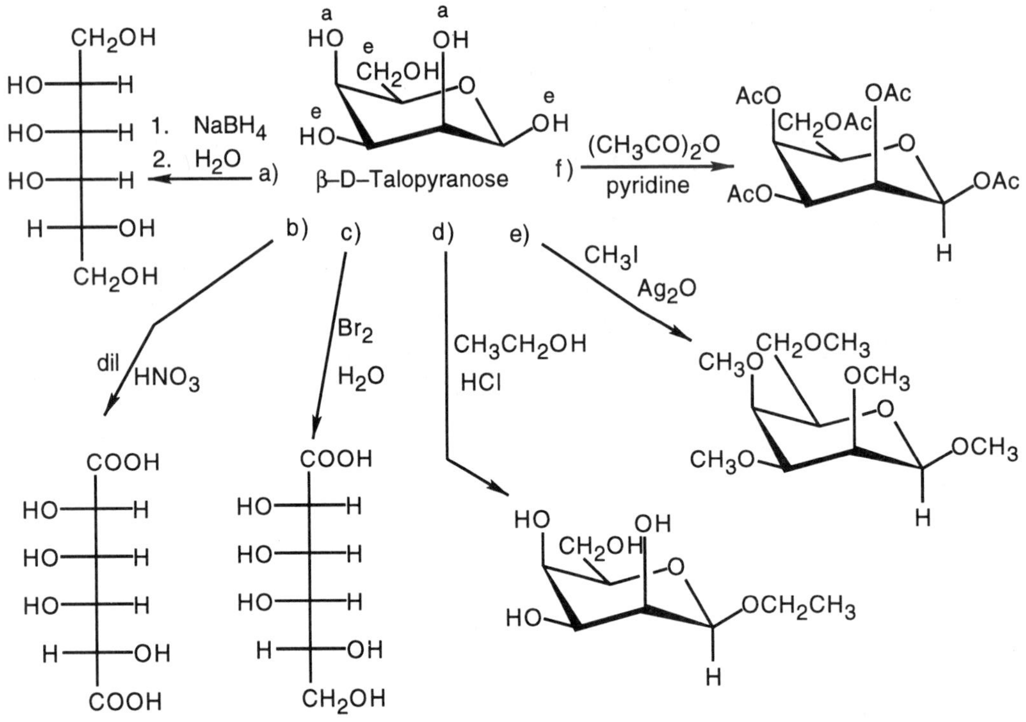

$$\text{a)} \xleftarrow{\begin{array}{l}1.\ \ NaBH_4\\2.\ \ H_2O\end{array}}$$

β–D–Talopyranose

$$\text{f)} \xrightarrow[\text{pyridine}]{(CH_3CO)_2O}$$

b)   c)   d)   e)

c) $\begin{array}{c}Br_2\\H_2O\end{array}$

d) $\begin{array}{c}CH_3CH_2OH\\HCl\end{array}$

e) $\begin{array}{c}CH_3I\\Ag_2O\end{array}$

b) dil HNO₃

26.38

D–Galactose

α–D–Galactopyranose
$[\alpha]_D = +150.7°$

β–D–Galactopyranose
$[\alpha]_D = +52.8°$

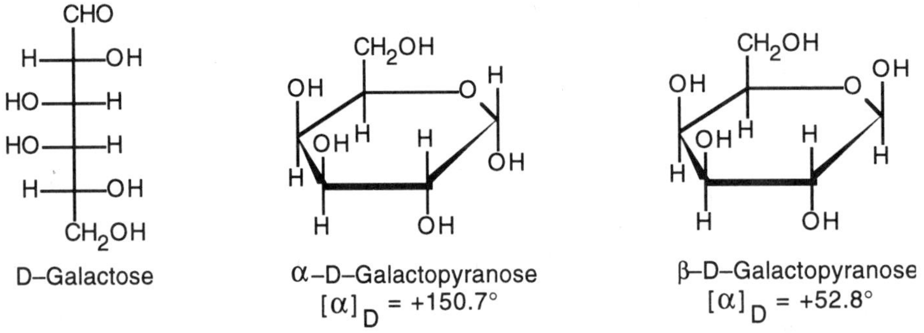

Let x be the fraction of D–galactose present as the α anomer and y be the fraction of D–Galactose present as the ß anomer.

$150.7° x + 52.8° y = 80.2°$ $x + y = 1;\ \ y = 1 - x$
$150.7° x + 52.8°(1-x) = 80.2°$
$97.9° x = 27.4°$
$x = 0.280$
$y = 0.720$

28.0% of D–galactose is present as the α anomer, and 72.0% is present as the ß anomer.

26.39–26.41   Four D–2–ketohexoses are possible.

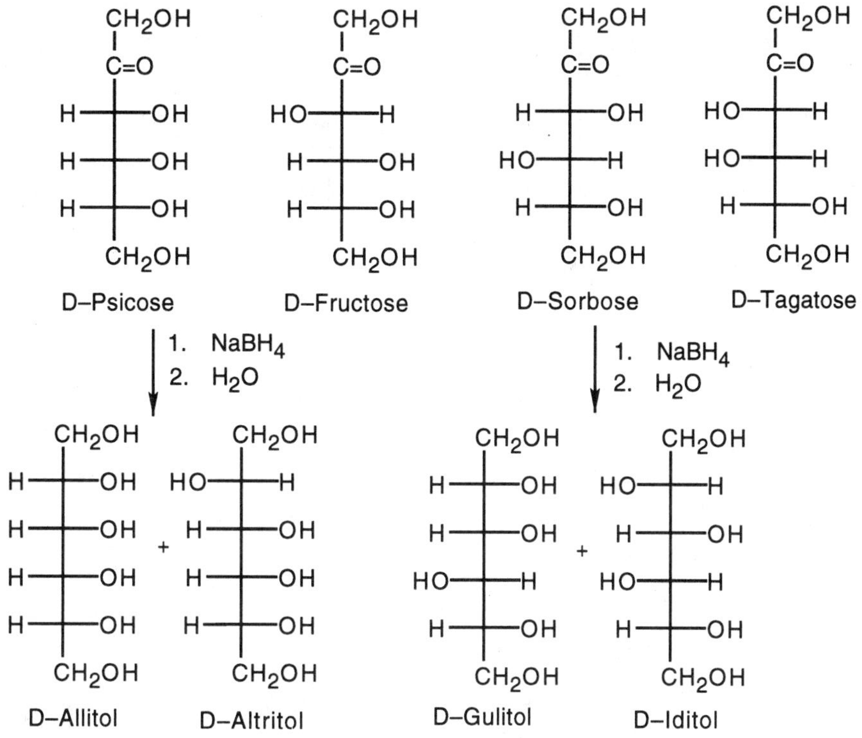

D–Psicose   D–Fructose   D–Sorbose   D–Tagatose

1. NaBH₄
2. H₂O

1. NaBH₄
2. H₂O

D–Allitol   D–Altritol   D–Gulitol   D–Iditol

26.42

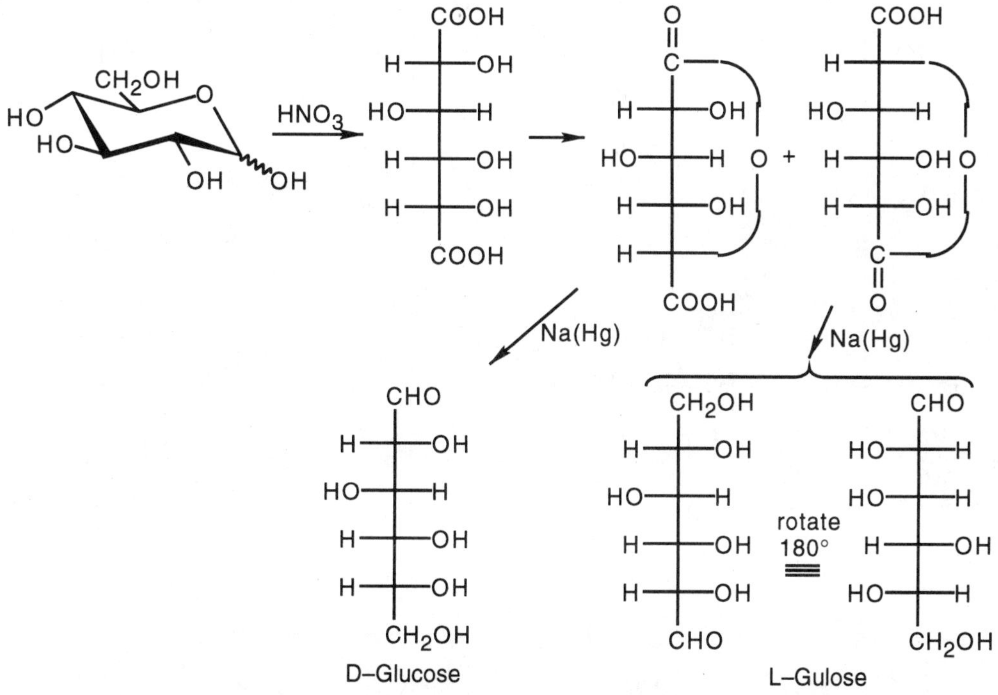

D–Glucose

rotate
180°

L–Gulose

26.43

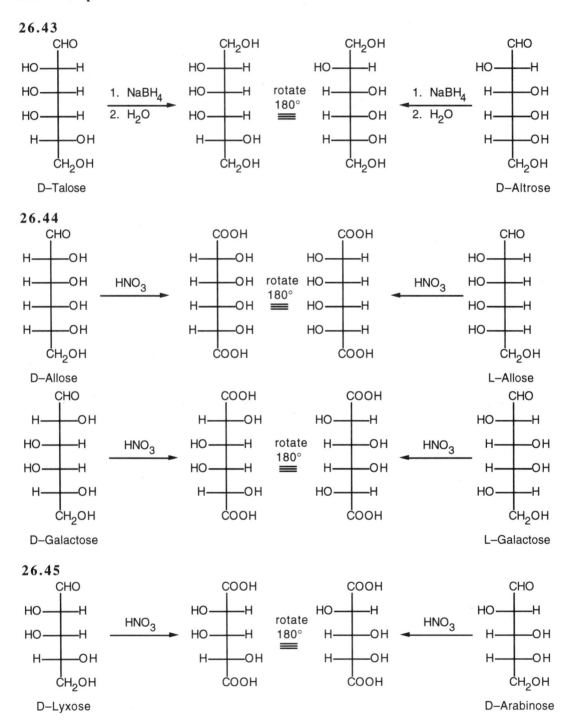

26.44

26.45

26.46

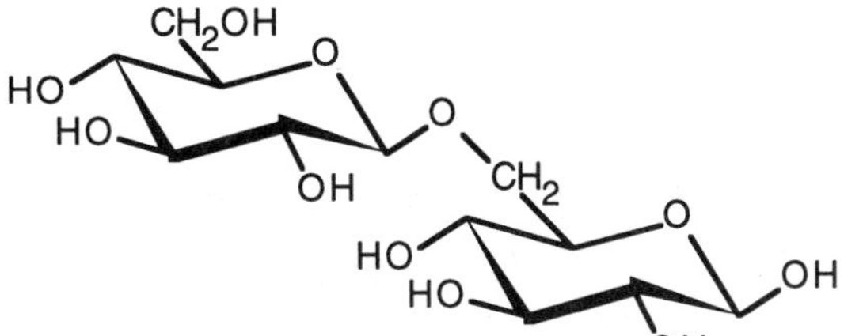

Gentiobiose

6–O–(β–D–Glucopyranosyl)–β–D–glucopyranose

26.47

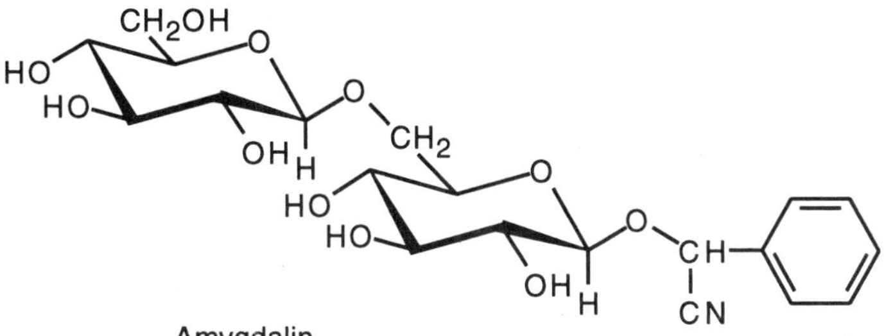

Amygdalin

26.48 Since trehalose is a nonreducing sugar, the two glucose units must be connected through an oxygen atom at the anomeric center of each. There are three possible structures for trehalose. The two glucopyranose rings can be connected (α,α), (ß,ß), or (α,ß).

26.49 Since trehalose is not cleaved by ß–glycosidases, it must have an α,α glycosidic linkage.

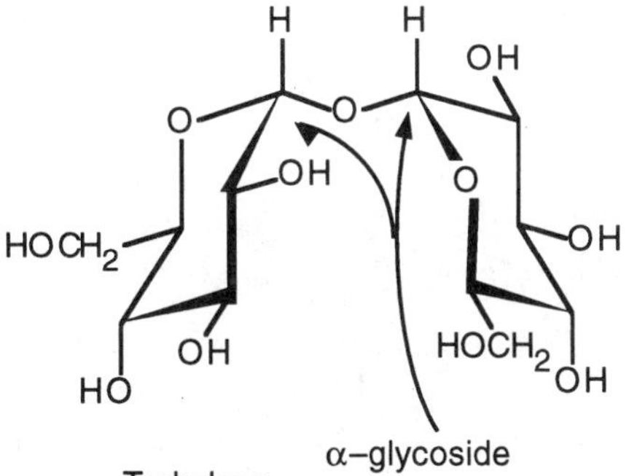

Trehalose

1–O–(α–D–Glucopyranosyl)–α–D–glucopyranose

α–glycoside

26.50

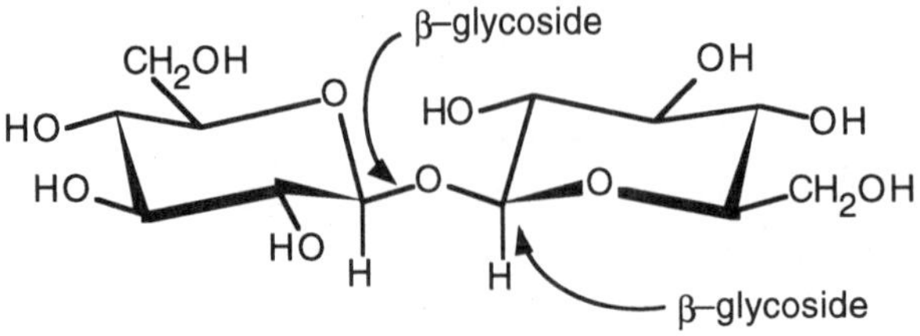

Neotrehalose

1–O–(β–D–Glucopyranosyl)–β–D–glucopyranose

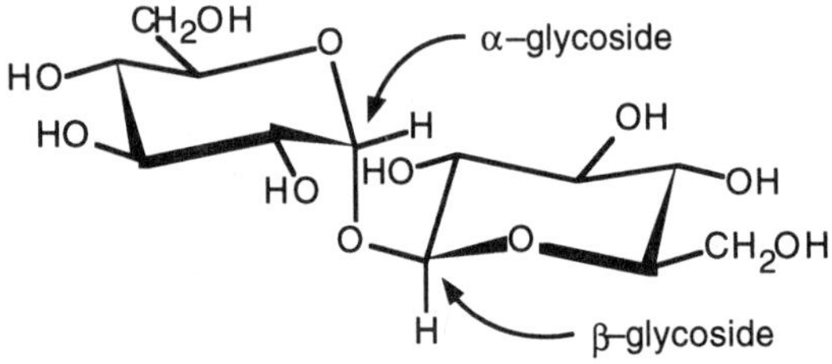

Isotrehalose

1–O–(α–D–Glucopyranosyl)–β–D–glucopyranose

26.51

Glucopyranose is in equilibrium with glucofuranose.

Reaction with two equivalents of acetone occurs *via* the mechanism we saw for acetal formation (Sec. 19.14).

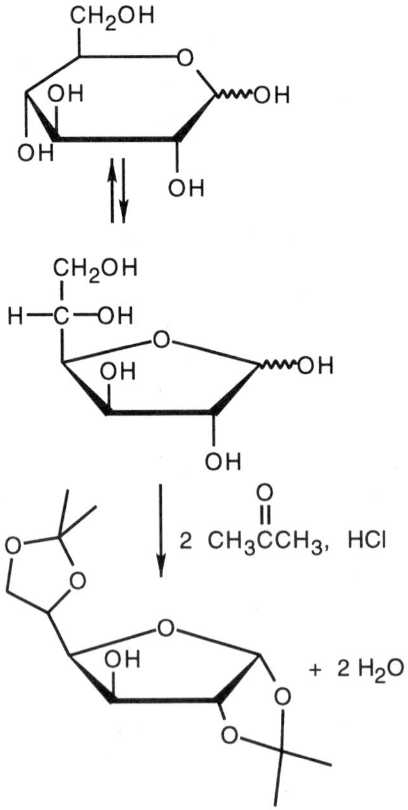

$$2 \ CH_3CCH_3, \ HCl$$

+ 2 H_2O

A five-membered acetal ring forms much more readily when the hydroxyl groups are *cis* to one another. In glucofuranose, the C3 hydroxyl is *trans* to the C2 hydroxyl, and acetal formation occurs between acetone and the C1 and C2 hydroxyls. Since the C1 hydroxyl group is part of the acetone acetal, the furanose is no longer in equilibrium with the free aldehyde, and the diacetone derivative is not a reducing sugar.

26.52

2,3:4,6–Diacetone mannopyranoside

Acetone forms an acetal with the hydroxyl groups at C2 and C3 of D–mannopyranoside because the hydroxyl groups at these positions are *cis* to one another. The pyranoside ring is still a hemiacetal that is in equilibrium with free aldehyde, which is reducing toward Tollens' reagent.

26.53

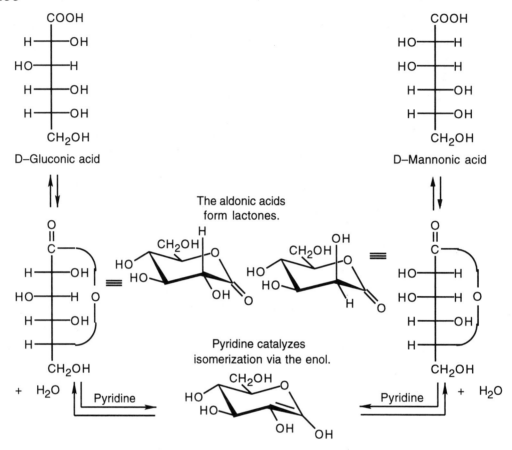

D–Gluconic acid

D–Mannonic acid

The aldonic acids form lactones.

Pyridine catalyzes isomerization via the enol.

Isomerization at C2 occurs because the enediol can be reprotonated on either side of the double bond.

26.54 There are eight cyclitols.

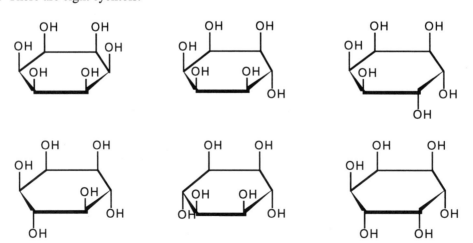

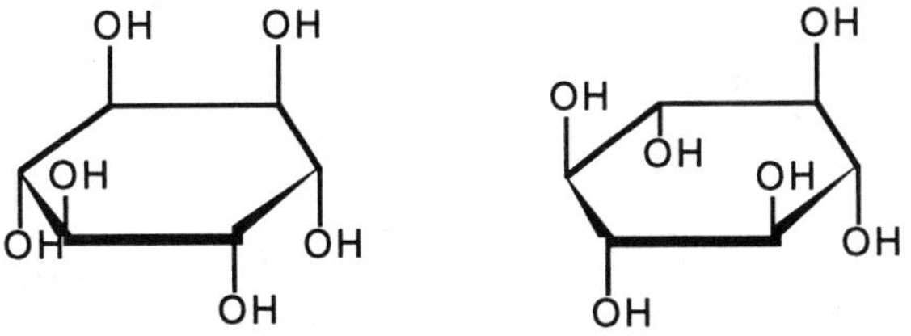

26.55

A ——HNO₃——→ B

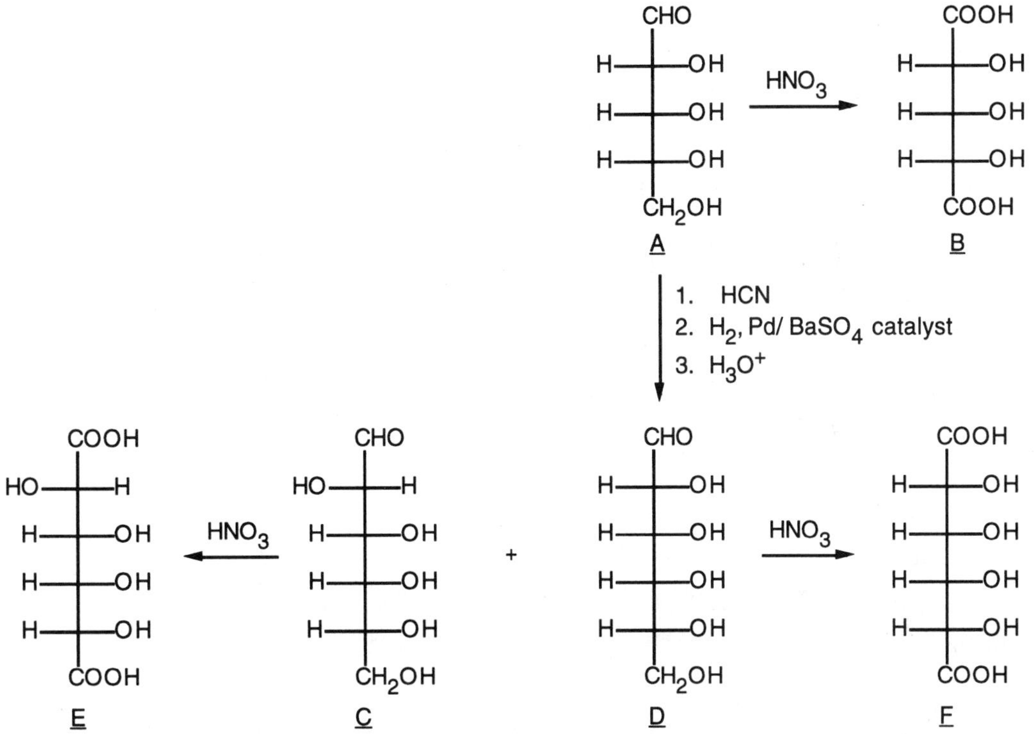

1. HCN
2. H_2, Pd/ $BaSO_4$ catalyst
3. H_3O^+

E ←——HNO₃—— C   +   D ——HNO₃——→ F

26.56

a)

D-Mannose   D-Glucose   D-Fructose   2 H_2NNHPh →   An osazone

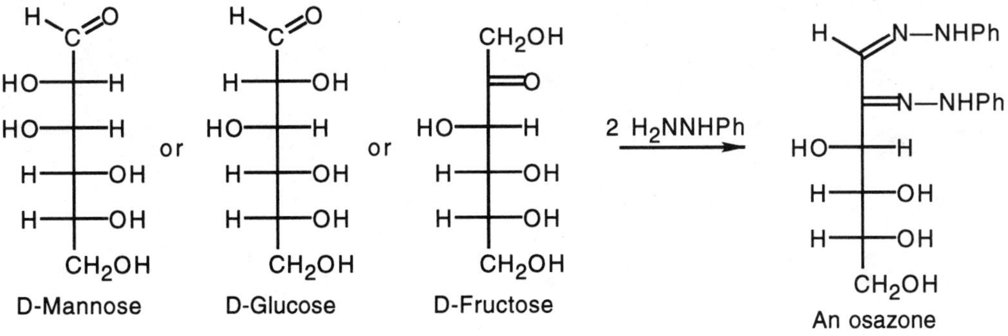

b)

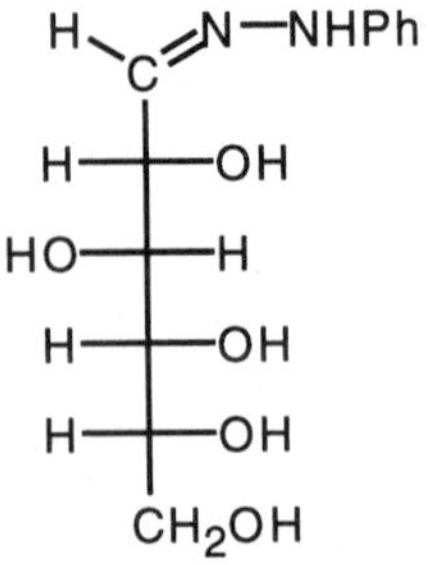

H—C=N—NHPh

H——OH

HO——H

H——OH

H——OH

CH₂OH

phenylhydrazone

c)

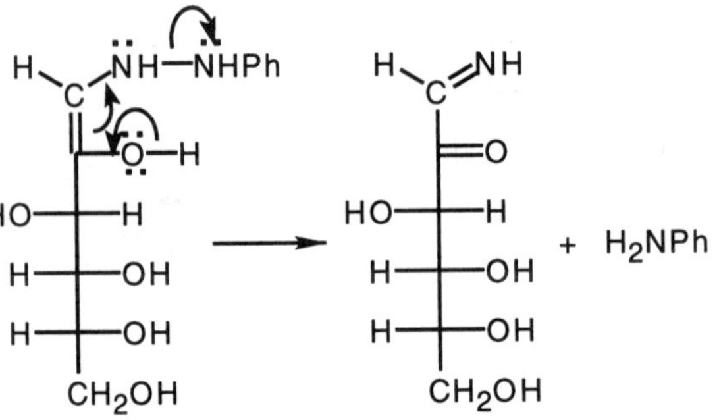

H—C—NH—NHPh ... O—H

HO——H

H——OH

H——OH

CH₂OH

enol

⟶

H—C=NH

C=O

HO——H

H——OH

H——OH

CH₂OH

keto imine

+ H₂NPh

d)

H—C=NH

PhNH—NH₂   C=O

HO——H

H——OH

H——OH

CH₂OH

⇌

H—C=NH

PhNH—NH₂   O⁻

HO——H

H——OH

H——OH

CH₂OH

⇌

H—C=NH

H

PhNH—N——OH

HO——H

H——OH

H——OH

CH₂OH

⇌

H—C=NH

PhNH—N=C

HO——H

H——OH

H——OH

CH₂OH

+ H₂O

H

PhNH—NH₂   C=NH

PhNH—N=

HO——H

H——OH

H——OH

CH₂OH

⇌

H H

PhNH—N—C—NH₂

PhNH—N=

HO——H

H——OH

H——OH

CH₂OH

⇌

H

PhNH—N=C

PhNH—N=

HO——H

H——OH

H——OH

CH₂OH

+ NH₃

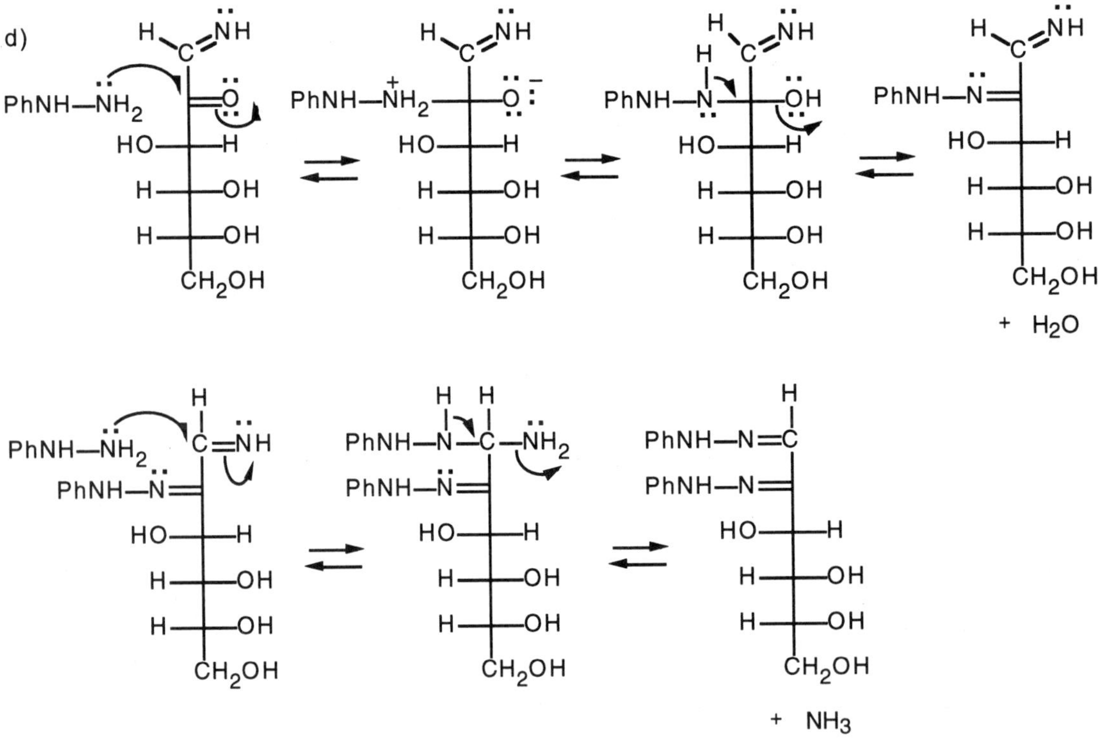

26.57

a) β–D–Idopyranose:

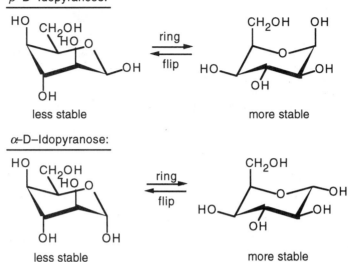

less stable → more stable (ring flip)

α–D–Idopyranose:

less stable ⇌ more stable (ring flip)

b) α-D-Idopyranose is more stable than β-D-idopyranose because it has only one axial group in its more stable chair conformation while β-D-idopyranose has two.

c)

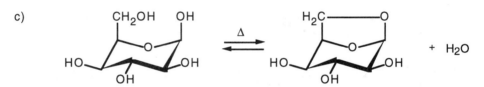

$+ \; H_2O$

1,6-Anhydro-D-idopyranose is formed from the β anomer because the axial hydroxyl groups on carbons 1 and 6 are close enough for the five-membered ring to form.

d) The hydroxyl groups at carbons 1 and 6 of D-glucopyranose are equatorial and are too far apart for a ring to form.

26.58

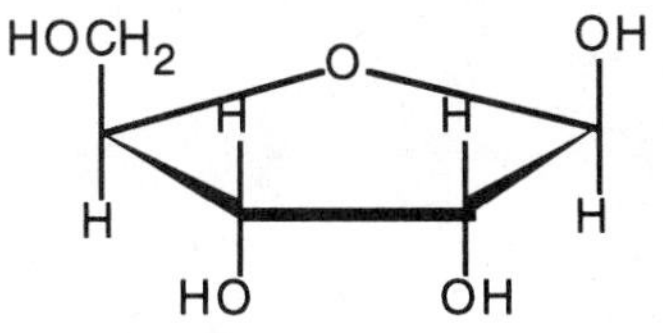

D- Ribofuranose is the sugar present in acetyl CoA

Study Guide for Chapter 26

After studying this chapter, you should be able to:

(1)   Classify carbohydrates as aldoses, ketoses, D or L sugars, monosaccharides or polysaccharides (26.1, 26.2, 26.4, 26.26, 26.27, 26.28, 26.29, 26.33).

(2)   Draw monosaccharides in the following projections:
(a) Fischer projection (26.5, 26.6, 26.8, 26.34).
(b) Haworth projection (26.9, 26.10, 26.11, 26.31, 26.35, 26.54).
(c) Chair conformation (26.13, 26.14, 26.36).

(3)   Calculate the equilibrium percentage of anomers from their specific rotations (26.12, 26.38).

(4)   Predict the products of reactions of monosaccharides (26.15, 26.16, 26.17, 26.18, 26.19, 26.20, 26.21, 26.22, 26.37).

(5)   Deduce the structure of monosaccharides (26.23, 26.24, 26.40, 26.41, 26.42, 26.43, 26.44, 26.45, 26.55).

(6)   Predict the product of reactions of disaccharides (26.25).

(7)   Deduce the structure of disaccharides (26.46, 26.47, 26.48, 26.49, 26.50).

(8)   Formulate mechanisms for reactions involving sugars (26.51, 26.52, 26.53, 26.56).

Chapter 27 – Biomolecules: Amino Acids, Peptides and Proteins

27.1 Amino Acids with aromatic rings: Phe, Tyr, Trp, His.
Amino acids containing sulfur: Cys, Met.
Amino acids that are alcohols: Ser, Thr. (Tyr is a phenol.)
Amino acids having hydrocarbon side chains: Ala, Ile, Leu, Val.

27.2

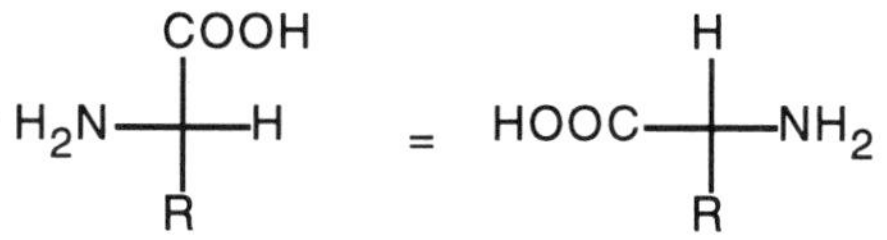

A Fischer projection of the α-carbon of an L–amino acid is pictured above.

For most L–amino acids:

Group	*Priority*
$-NH_2$	1
–COOH	2
–R	3
–H	4

For cysteine:

Group	*Priority*
$-NH_2$	1
$-CH_2SH$	2
–COOH	3
–H	4

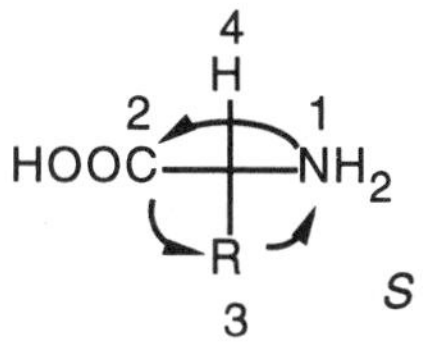

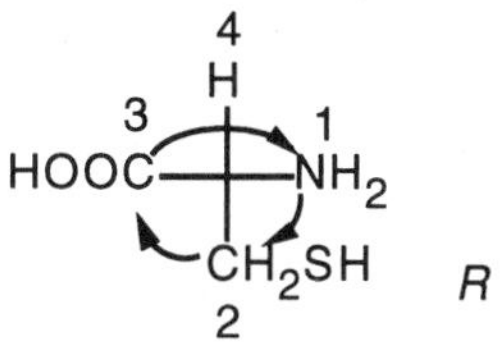

27.3–27.4

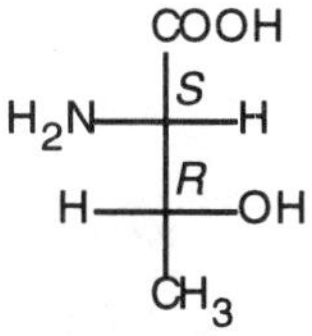

L–Threonine

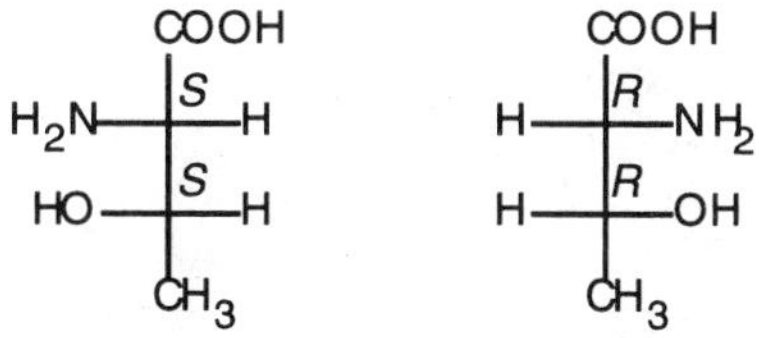

Diastereomers of L–Threonine

27.5

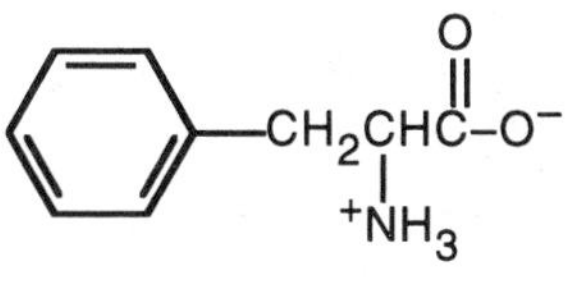

Phenylalanine

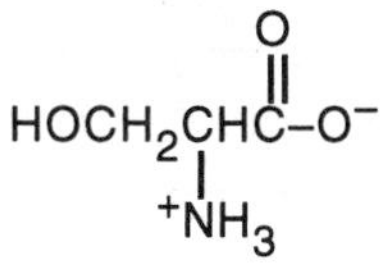

Serine

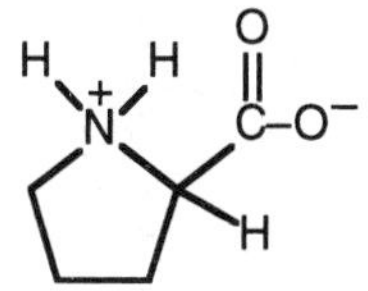

Proline

27.6

a) phenyl–CH$_2$CHCOO$^-$ (with +NH$_3$) + NaOH → phenyl–CH$_2$CHCOO$^-$ (with NH$_2$)

b) phenyl–CH$_2$CHCOO$^-$ (with NH$_2$) + HCl → phenyl–CH$_2$CHCOO$^-$ (with +NH$_3$)

c) phenyl–CH$_2$CHCOO$^-$ (with NH$_2$) + 2 HCl → phenyl–CH$_2$CHCOOH (with +NH$_3$)

27.7

$H_3\overset{+}{N}-CH_2-COOH$ $H_3\overset{+}{N}-CH_2-COO^-$ $H_2N-CH_2-COO^-$

At pH = 2.0   At pH = 6.0   At pH = 10.0

27.8   a) *Amino acid*   *Isoelectric point*

Val   6.0
Glu   3.2
His   7.6

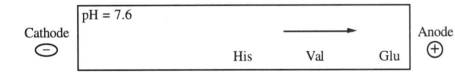

b) *Amino Acid*   *Isoelectric point*

Gly   6.0
Phe   5.5
Ser   5.7

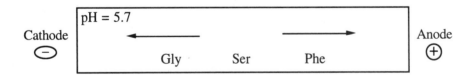

c)

pH = 5.5 ; Cathode (−) ; Anode (+) ; Gly   Ser   Phe

d)

	pH = 6.0	
Cathode $\ominus$	$\longrightarrow$ Gly Ser Phe	Anode $\oplus$

27.9 The Henderson-Hasselbalch equation states:

$$\log \frac{[A^-]}{[HA]} = pH - pK_a$$

For pK_1 of Threonine : $\log \dfrac{[HA]}{[H_2A^+]} = pH - 2.09$

At pH = 1.50   : $\log \dfrac{[HA]}{[H_2A^+]} = 1.50 - 2.09 = -0.59$

or $\log \dfrac{[H_2A^+]}{[HA]} = 0.59$; $\dfrac{[H_2A^+]}{[HA]} = 3.89$

At pH = 1.50 :

$$\overset{+NH_3}{\underset{HO}{CH_3CH-CHCOOH}} \quad 79.6\% \qquad \overset{+NH_3}{\underset{HO}{CH_3CH-CHCOO^-}} \quad 20.4\%$$

For pK_2 of Threonine : $\log \dfrac{[A^-]}{[HA]} = pH - 9.10$

At pH = 10.00   : $\log \dfrac{[A^-]}{[HA]} = 10.00 - 9.10 = 0.90$; $\dfrac{[A^-]}{[HA]} = 7.94$

At pH = 10.00 :

$$\overset{+NH_3}{\underset{HO}{CH_3CH-CHCOO^-}} \quad 11.2\% \qquad \overset{NH_2}{\underset{HO}{CH_3CH-CHCOO^-}} \quad 88.8\%$$

27.10

a) $C_6H_5CH_2CH_2COOH \xrightarrow[\text{2. }H_2O]{\text{1. }Br_2, PBr_3} C_6H_5CH_2\underset{Br}{CHCOOH} \xrightarrow[\text{excess}]{NH_3} C_6H_5CH_2\underset{NH_2}{CHCOOH}$

3–Phenylpropanoic acid

b) $(CH_3)_2CHCH_2COOH \xrightarrow[\text{2. }H_2O]{\text{1. }Br_2, PBr_3} (CH_3)_2CH\underset{Br}{CHCOOH} \xrightarrow[\text{excess}]{NH_3} (CH_3)_2CH\underset{NH_2}{CHCOOH}$

3–Methylbutanoic acid

27.11

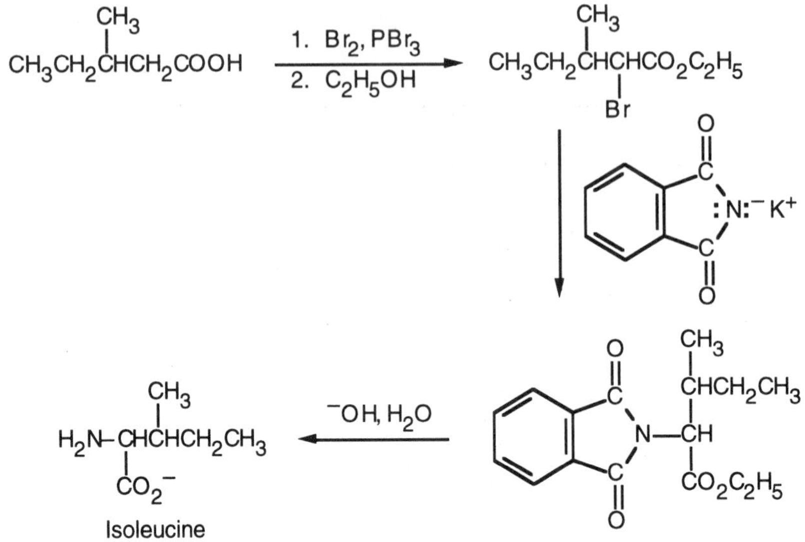

27.12

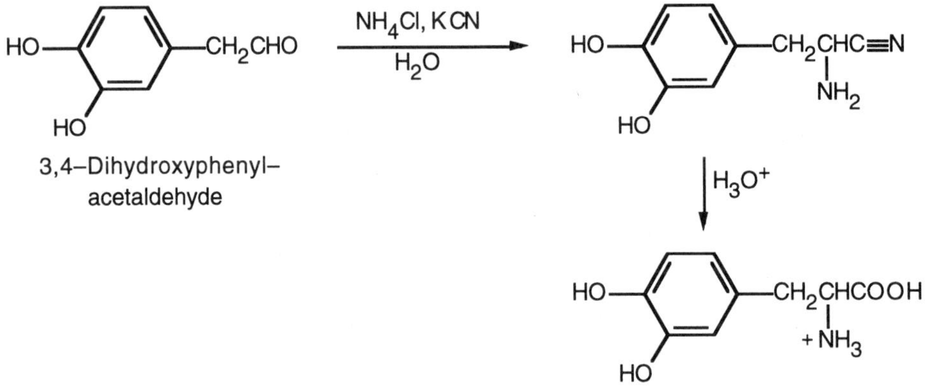

3,4–Dihydroxyphenyl–
acetaldehyde

L–Dopa

27.13

Amino Acid	Halide

a) $(CH_3)_2CHCH_2CHCOOH$
                    $|$
                    NH_2
Leucine

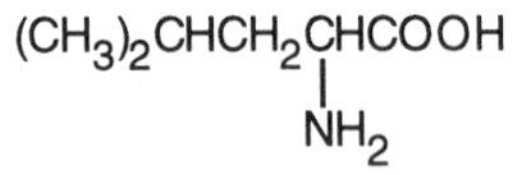

$(CH_3)_2CHCH_2Br$

b)

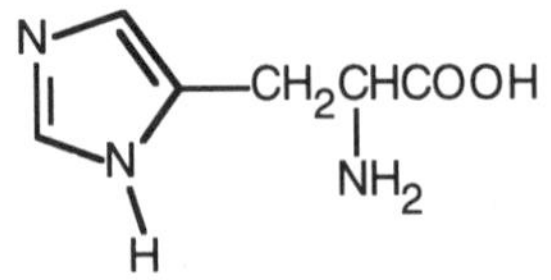

Histidine

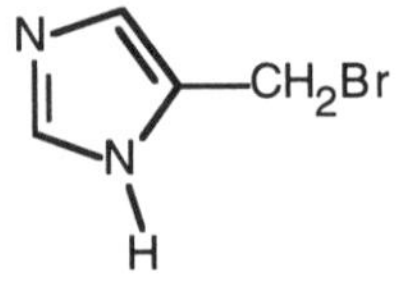

c)

$CH_2CHCOOH$
NH_2

Tryptophan

CH_2Br

d) $CH_3SCH_2CH_2CHCOOH$
NH_2

Methionine

$CH_3SCH_2CH_2Br$

27.14

$$CH_3\overset{O}{\overset{\|}{C}}-NH-\underset{CO_2C_2H_5}{\overset{CO_2C_2H_5}{C:^-}} \quad \overset{H}{\underset{H}{C}}=O \longrightarrow \left[CH_3\overset{O}{\overset{\|}{C}}-NH-\underset{CO_2C_2H_5}{\overset{CO_2C_2H_5}{C}}-CH_2O^- \right]$$

$$\Big\downarrow H_3O^+$$

$$CO_2 + 2\,C_2H_5OH + CH_3COOH + H_3\overset{+}{N}-\underset{\underset{Serine}{}}{\overset{COOH}{\overset{|}{C}}HCH_2OH}$$

27.15 Val–Tyr–Gly   Tyr–Gly–Val   Gly–Val–Tyr
Val–Gly–Tyr   Tyr–Val–Gly   Gly–Tyr–Val

27.16

$CH_3SCH_2CH_2$ CH_2 CH_2 CH_2 $CH(CH_3)_2$

$$H-NHCHC-N-CHC-NHCHC-NHCH_2C-OH$$
$$\underset{O}{\|}\quad\underset{O}{\|}\quad\underset{O}{\|}\quad\underset{O}{\|}$$

Met — Pro — Val — Gly

27.17

Ninhydrin

$$2 \; [\text{Ninhydrin}] + (CH_3)_2CHCHCOOH \xrightarrow{\ ^-OH\ } [\text{product}]$$
$$\underset{NH_2}{}$$

$$+$$
$$(CH_3)_2CHCHO + CO_2 + H_2O$$

27.18 Only primary amines can form the extensively conjugated purple ninhydrin product. A secondary amine such as proline yields a product containing a shorter system of conjugated bonds, which absorbs at a shorter wavelength (440 nm vs. 570 nm).

27.19 Trypsin cleaves peptide bonds at the carboxyl side of lysine and arginine. Chymotrypsin cleaves peptide bonds at the carboxyl side of phenylalanine, tyrosine and tryptophan.

$$\text{Asp–Arg–Val–Tyr–Ile–His–Pro–Phe}$$

Trypsin → Asp–Arg + Val–Tyr–Ile–His–Pro-Phe

Chymotrypsin → Asp–Arg–Val–Tyr + Ile–His–Pro–Phe

27.20 a) Arg–Pro
         Pro–Leu–Gly
             Gly–Ile–Val

The complete sequence:

Arg–Pro–Leu–Gly–Ile–Val

b) Val–Met–Trp
        Trp–Asp–Val
            Val–Leu

The complete sequence:

Val–Met–Trp–Asp–Val–Leu

27.21 Pro–Leu–Gly
         Gly–Pro–Arg
             Arg–Pro

The complete sequence:

Pro–Leu–Gly–Pro–Arg–Pro

27.22 The tripeptide is cyclic.

27.23

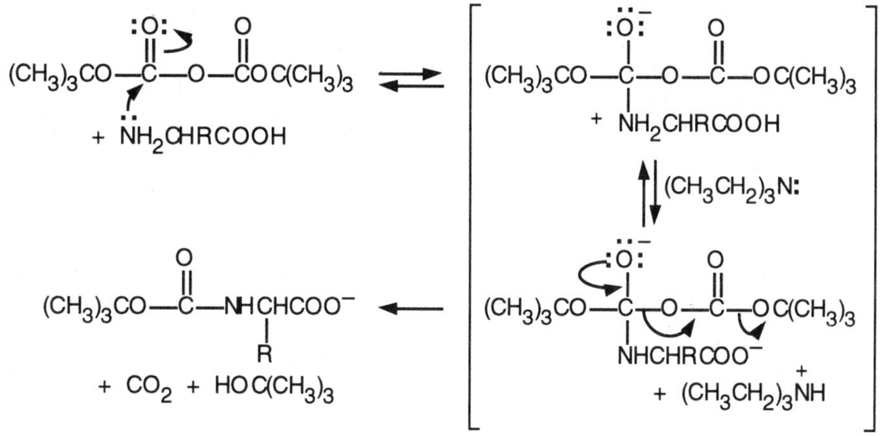

27.24

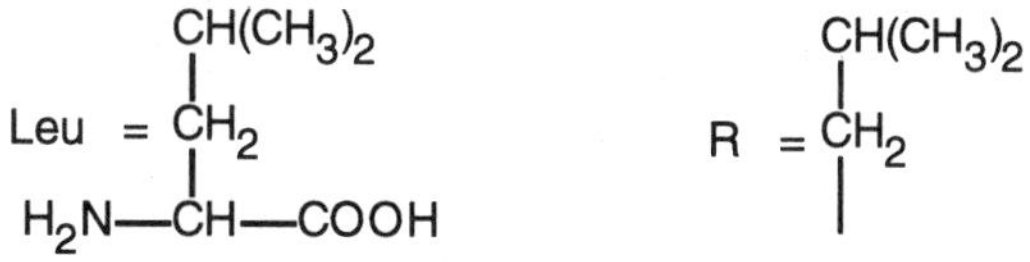

$$\text{Leu} = \begin{array}{c} \text{CH(CH}_3)_2 \\ | \\ \text{CH}_2 \\ | \\ \text{H}_2\text{N}-\text{CH}-\text{COOH} \end{array} \qquad R = \begin{array}{c} \text{CH(CH}_3)_2 \\ | \\ \text{CH}_2 \\ | \end{array}$$

1. Protect the amino group of leucine.

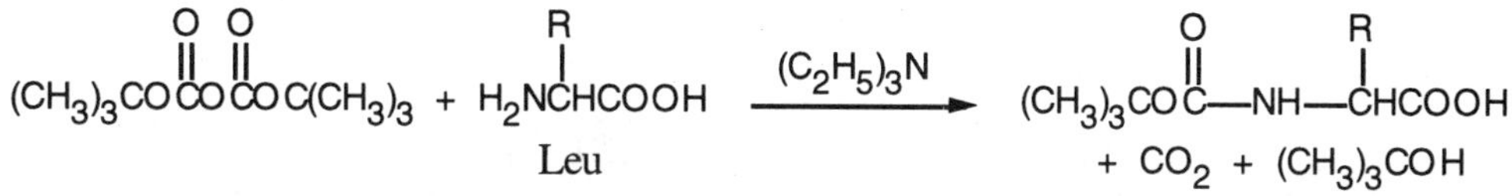

2. Protect the carboxylic acid group of alanine.

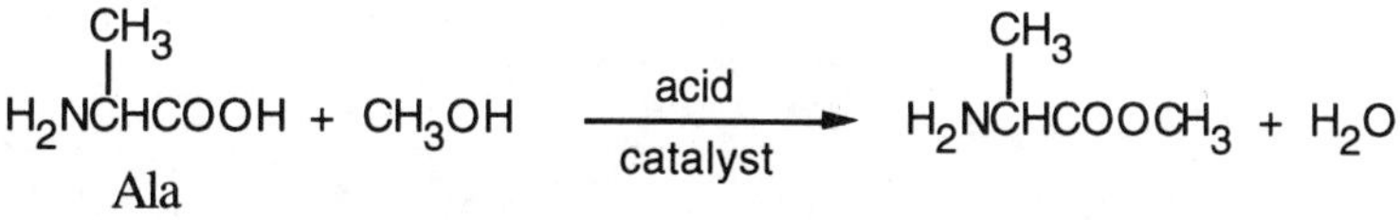

$$\begin{array}{c} \text{CH}_3 \\ | \\ \text{H}_2\text{NCHCOOH} \end{array} + \text{CH}_3\text{OH} \quad \xrightarrow[\text{catalyst}]{\text{acid}} \quad \begin{array}{c} \text{CH}_3 \\ | \\ \text{H}_2\text{NCHCOOCH}_3 \end{array} + \text{H}_2\text{O}$$

Ala

3. Couple the protected acids with DCC.

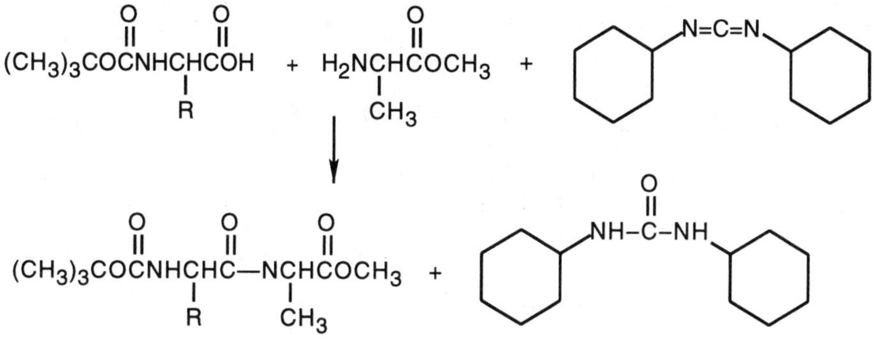

4. Remove the leucine protecting group.

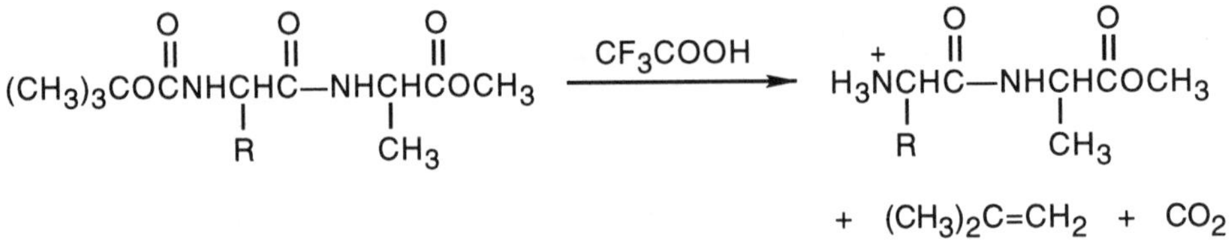

5. Remove the alanine protecting group.

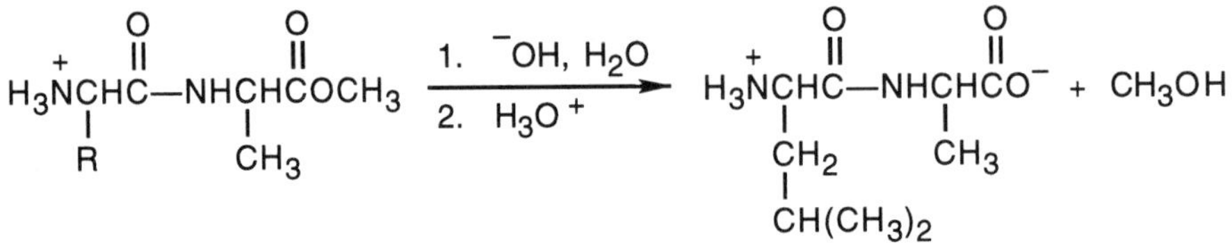

Leu — Ala

27.25

a)

1. Leu–Ala (Prob. 27.24) + $(CH_3)_3CO\overset{O}{\overset{\|}{C}}O\overset{O}{\overset{\|}{C}}OC(CH_3)_3$ $\xrightarrow{(CH_3CH_2)_3N}$ BOC–Leu–Ala

2. Gly + CH_3OH $\xrightarrow{H^+}$ Gly–OCH_3

3. BOC–Leu–Ala + Gly–OCH_3 $\xrightarrow{DCC}$ BOC–Leu–Ala–Gly–OCH_3

4. BOC–Leu–Ala–Gly–OCH_3 $\xrightarrow{CF_3COOH}$ Leu–Ala–Gly–OCH_3

5. Leu–Ala–Gly–OCH_3 $\xrightarrow[2.\ H_3O^+]{1.\ ^-OH,\ H_2O}$ Leu–Ala–Gly

b)

1. Gly + $(CH_3)_3CO\overset{O}{\overset{\|}{C}}O\overset{O}{\overset{\|}{C}}OC(CH_3)_3$ $\xrightarrow{(CH_3CH_2)_3N}$ BOC–Gly

2. BOC–Gly + Leu–Ala–OCH_3 (Prob. 27.24, part 4) $\xrightarrow{DCC}$ BOC–Gly–Leu–Ala–OCH_3

3. BOC–Gly–Leu–Ala–OCH_3 $\xrightarrow{CF_3COOH}$ Gly–Leu–Ala–OCH_3

4. Gly–Leu–Ala–OCH_3 $\xrightarrow[2.\ H_3O^+]{1.\ ^-OH,\ H_2O}$ Gly–Leu–Ala

27.26 A proline residue in a polypeptide chain interrupts α–helix formation because the amide nitrogen of proline has no hydrogen that can contribute to the hydrogen-bonded structure of an α–helix.

27.27 a) Pyruvate decarboxylase is a lyase.
b) Chymotrypsin is a hydrolase.
c) Alcohol dehydrogenase is an oxidoreductase.

27.28

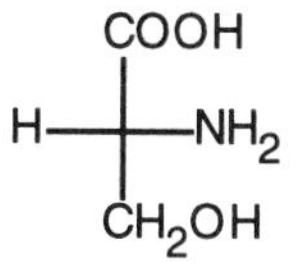

(R)–Serine 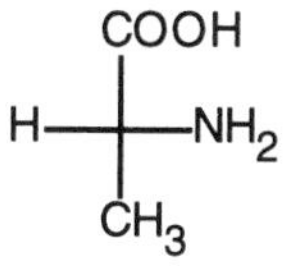   (R)– Alanine

Both acids are D amino acids.

27.29

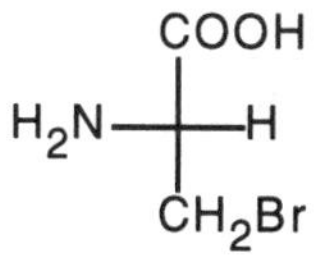

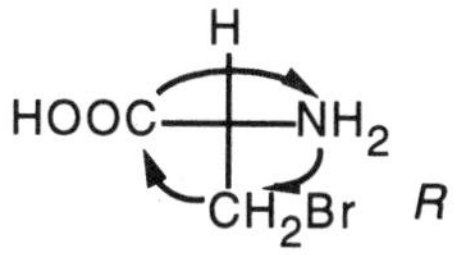

This L "amino acid" also has an R configuration because the –CH_2Br "side chain" is higher in priority than the –COOH group.

27.30

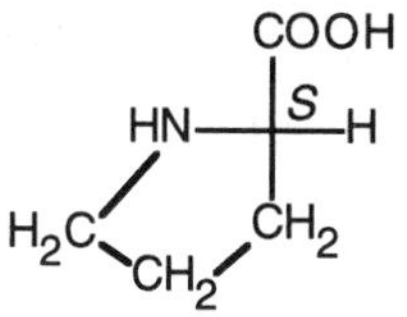

(S)– Proline

27.31

a) 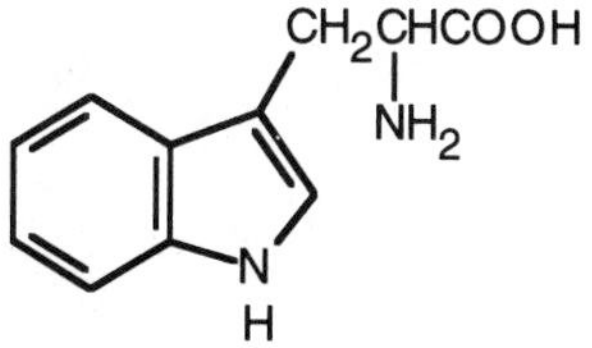
Tryptophan (Trp)

b) 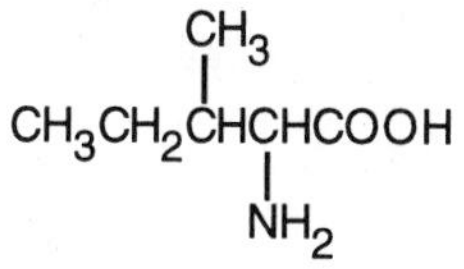
$CH_3CH_2CHCHCOOH$ with CH_3 and NH_2 substituents
Isoleucine (Ile)

c) 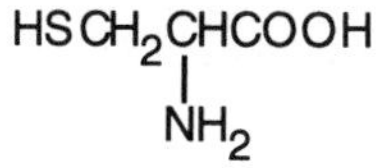 $HSCH_2CHCOOH$ with NH_2
Cysteine (Cys)

d) 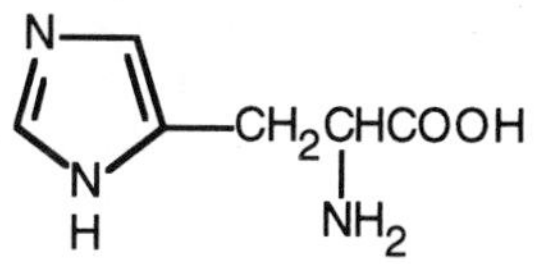
Histidine (His)

27.32

a) HO—⟨benzene ring⟩—CH₂CHCO⁻ with C=O double bond, ⁺NH₃ below

$$\text{HO}\!-\!\!\bigcirc\!\!-\!\text{CH}_2\text{CH}\overset{\text{O}}{\underset{\overset{|}{{}^{+}\text{NH}_3}}{\text{C}}}\text{O}^-$$

Tyrosine

b) $$\text{CH}_3\overset{\text{OH}}{\underset{}{\text{CH}}}\text{CH}\overset{\text{O}}{\underset{\overset{|}{{}^{+}\text{NH}_3}}{\text{C}}}\text{O}^-$$

Threonine

27.33 Water is a polar solvent, and chloroform is nonpolar. Since charged species are less stable in nonpolar solvents than in polar solvents, amino acids exist as the non-ionic amino carboxylic acid form in chloroform.

27.34 *Amino Acid*   *Isoelectric point*

Amino Acid	Isoelectric point
Histidine	7.59
Serine	5.68
Glutamic acid	3.22

The optimum pH for the electrophoresis of three amino acid occurs at the isoelectric point of the amino acid intermediate in acidity. At this pH, one amino acid migrates toward the cathode (the least acidic), one migrates toward the anode (the most acidic), and the amino acid intermediate in acidity does not migrate. In this example, electrophoresis at pH = 5.68 allows the maximum separation of the three amino acids.

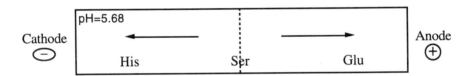

27.35 a) *Amphoteric* compounds can react either as acids or as bases, depending on the circumstances.

  b) The *isoelectric point* is the pH at which a solution of an amino acid or protein is electrically neutral.

  c) A *zwitterion* is a compound that contains both positively charged and negatively charged portions.

27.36

For K_{a1}: $\log \dfrac{[HA]}{[H_2A^+]} = pH - pK_{a1} = 2.50 - 1.99 = 0.51$

$$\dfrac{[HA]}{[H_2A^+]} = 3.24$$

At pH = 2.50, approximately three times as many proline molecules exist in the neutral form as exist in the protonated form.

For K_{a2}: $\log \dfrac{[A^-]}{[HA]} = pH - pK_{a2} = 9.70 - 10.60 = -0.90$

$$\dfrac{[A^-]}{[HA]} = 0.126$$

At pH = 9.70, the ratio of deprotonated proline to neutral proline is approximately 1:8.

27.37 a) Val–Leu–Ser   Ser–Val–Leu
Val–Ser–Leu   Leu–Val–Ser
Ser–Leu–Val   Leu–Ser–Val

b) Ser–Leu–Leu–Pro   Leu–Leu–Ser–Pro
Ser–Leu–Pro–Leu   Leu–Leu–Pro–Ser
Ser–Pro–Leu–Leu   Leu–Ser–Leu–Pro
Pro–Leu–Leu–Ser   Leu–Ser–Pro–Leu
Pro–Leu–Ser–Leu   Leu–Pro–Leu–Ser
Pro–Ser–Leu–Leu   Leu–Pro–Ser–Leu

27.38 100 g of cytochrome C contains 0.43 g iron.

100 g of cytochrome C contains:

$$\dfrac{0.43 \text{ g Fe}}{55.8 \text{ g/mol Fe}} = 0.0077 \text{ moles Fe}$$

$$\dfrac{100 \text{ g Cytochrome C}}{0.0077 \text{ moles Fe}} = \dfrac{X \text{ g Cytochrome C}}{1 \text{ mole Fe}}$$

$$13,000 \text{ g/mol Fe} = X$$

Cytochrome C has a minimum molecular weight of 13,000.

27.39

a) $(CH_3)_2CHCH-COO^-$ $\overset{+}{N}H_3$ L–Valine $\xrightarrow[\text{acid catalyst}]{CH_3CH_2OH}$ $(CH_3)_2CHCHCOOCH_2CH_3$ $\overset{+}{N}H_3$

b) $(CH_3)_2CHCH-COO^-$ $\overset{+}{N}H_3$ $\xrightarrow[(CH_3CH_2)_3N:]{(CH_3)_3OCOOCOC(CH_3)_3}$ $(CH_3)_2CHCHCOO^-$ $NHCOC(CH_3)_3$

c) $(CH_3)_2CHCH-COO^-$ $\overset{+}{N}H_3$ $\xrightarrow{KOH,\ H_2O}$ $(CH_3)_2CHCHCOO^-$ NH_2

d) $(CH_3)_2CHCH-COO^-$ $\overset{+}{N}H_3$ $\xrightarrow[2.\ H_2O]{1.\ CH_3COCl,\ Pyr}$ $(CH_3)_2CHCH-COO^-$ $NHCOCH_3$

27.40

a) $H\overset{O}{\overset{\|}{C}}H$ $\xrightarrow[H_2O]{NH_4Cl\ /\ KCN}$ $H\overset{C\equiv N}{\underset{NH_2}{C}}H$ $\xrightarrow{H_3O^+}$ $H_2\overset{O}{\overset{\|}{C}}-COH$ NH_2 Glycine

b) $CH_3\overset{H_3C}{\overset{|}{C}}H-\overset{O}{\overset{\|}{C}}H$ $\xrightarrow[H_2O]{NH_4Cl\ /\ KCN}$ $CH_3\overset{H_3C}{\overset{|}{C}}H\overset{C\equiv N}{\underset{NH_2}{C}}H$ $\xrightarrow{H_3O^+}$ $CH_3\overset{H_3C}{\overset{|}{C}}H\overset{O}{\overset{\|}{C}}OH$ NH_2 Valine

27.41

a) $\left[\begin{array}{c} CO_2Et \\ ^-:C-CO_2Et \\ | \\ H-N \quad CH_3 \\ \diagdown C \diagup \\ \| \\ O \end{array}\right]$ $\xrightarrow{CH_3CHCH_2Br \text{ (}CH_3\text{)}}$ $CH_3CHCH_2-\overset{CH_3}{\underset{\begin{array}{c}N\\ H \quad C-CH_3\\ \|\\ O\end{array}}{\overset{CO_2Et}{C}}}-CO_2Et$

$\downarrow H_3O^+$

$CH_3\overset{O}{\overset{\|}{C}}OH$ $+\ CO_2$ $+\ 2\ EtOH$ $+$ $CH_3\overset{CH_3}{\overset{|}{C}}HCH_2\overset{O}{\overset{\|}{C}}HCOH$ Leucine NH_2

b)

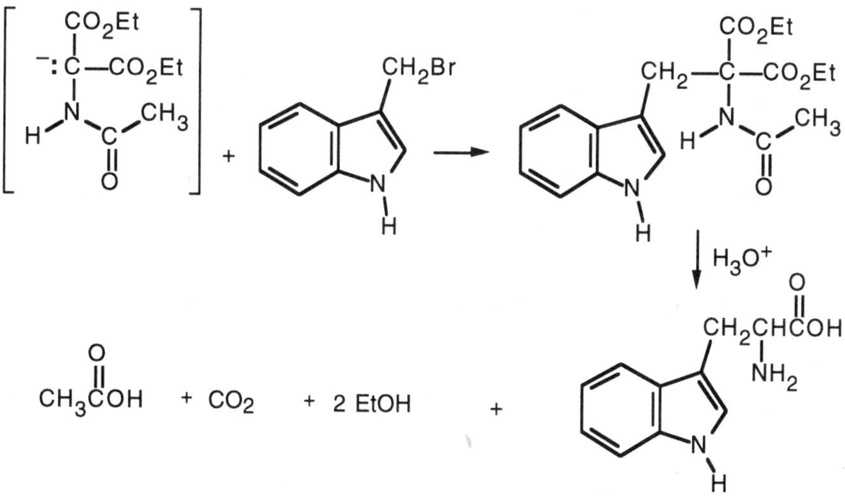

$$CH_3\overset{O}{\overset{\|}{C}}OH \quad + \; CO_2 \quad + \; 2 \; EtOH \quad + $$

Tryptophan

27.42

a)

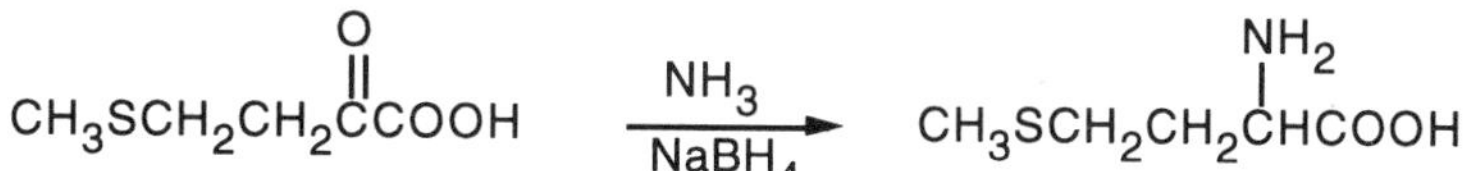

$$CH_3SCH_2CH_2\overset{O}{\overset{\|}{C}}COOH \xrightarrow[NaBH_4]{NH_3} CH_3SCH_2CH_2\overset{NH_2}{\overset{|}{C}}HCOOH$$

Methionine

b)

$$CH_3CH_2\overset{H_3C}{\overset{|}{C}}H\overset{O}{\overset{\|}{C}}COOH \xrightarrow[NaBH_4]{NH_3} CH_3CH_2\overset{H_3C}{\overset{|}{C}}H\overset{NH_2}{\overset{|}{C}}HCOOH$$

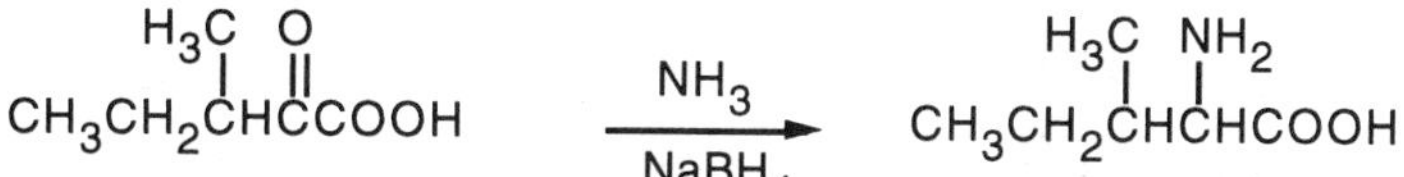

Isoleucine

27.43

a)

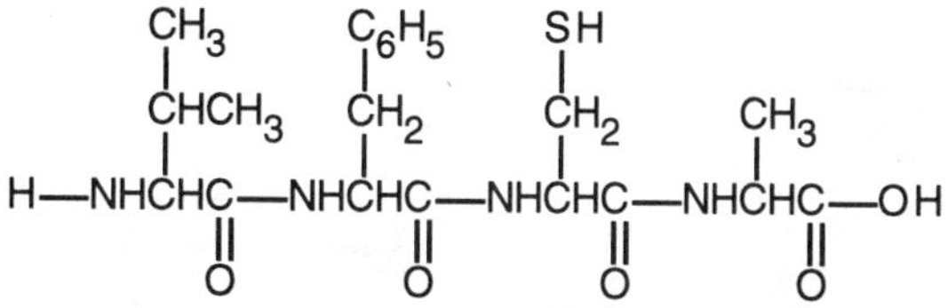

Val — Phe — Cys — Ala

b)

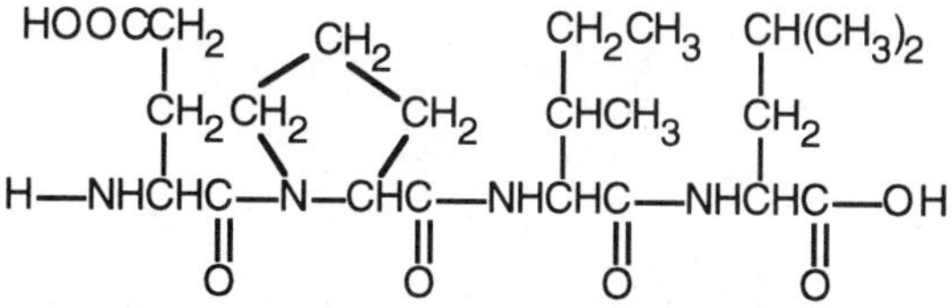

Glu — Pro — Ile — Leu

27.44

1. $(CH_3)_3CO\overset{O}{\overset{\|}{C}}O\overset{O}{\overset{\|}{C}}OC(CH_3)_3$ + Val $\xrightarrow{(CH_3CH_2)_3N:}$ $(CH_3)_3CO\overset{O}{\overset{\|}{C}}$–Val–OH
(BOC–Val–OH)

2. BOC–Val–OH + Cl–CH$_2$–(Polymer) $\longrightarrow$ BOC–Val–OCH$_2$–(Polymer)

3. BOC–Val–OCH$_2$–(Polymer) $\xrightarrow[\text{2. } CF_3COOH]{\text{1. Wash}}$ Val–OCH$_2$–(Polymer)

4. BOC–Ala + Val–OCH$_2$–(Polymer) $\xrightarrow[\text{2. Wash}]{\text{1. DCC}}$ BOC–Ala–Val–OCH$_2$–(Polymer)

5. BOC–Ala–Val–OCH$_2$–(Polymer) $\xrightarrow{CF_3COOH}$ Ala–Val–OCH$_2$–(Polymer)

6. BOC–Phe + Ala–Val–OCH$_2$–(Polymer)
   $\downarrow$ 1. DCC
       2. Wash
   BOC–Phe–Ala–Val–OCH$_2$–(Polymer)

7. BOC–Phe–Ala–Val–OCH$_2$–(Polymer)
   $\downarrow$ HF
   Phe–Ala–Val + HOCH$_2$–(Polymer)

27.45

Peptide $\xrightarrow{\text{PITC}}$ Phenylthiohydantoin   Shortened Peptide

a)   Ile–Leu–Pro–Phe

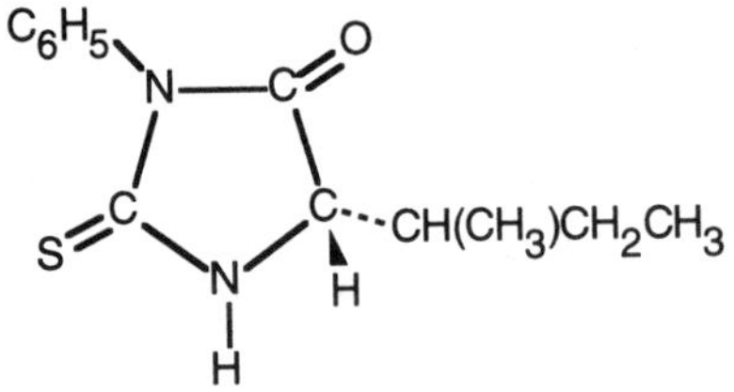

Leu–Pro–Phe

b)   Asp–Thr–Ser–Gly–Ala

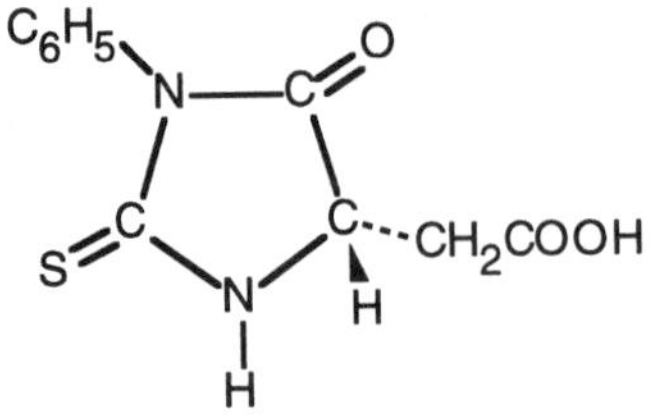

Thr–Ser–Gly–Ala

27.46

$$CH_3OCH_2Cl + SnCl_4 \rightleftharpoons CH_3\overset{+}{O}=CH_2 \ SnCl_5^-$$

27.47

Phe—Leu—Met—Lys—Tyr—Asp—Gly—Gly—Arg—Val—Ile—Pro—Tyr

> Cleaved by Trypsin = - - - - - - ·
> Cleaved by Chymotrypsin = ∿∿∿∿∿

27.48

Tryptophan

Indole ring system

The five-membered ring of indole is like pyrrole in that the lone pair of electrons on nitrogen is part of the aromatic π electron system. Since protonation of the indole nitrogen would disrupt the aromaticity of the indole ring system, tryptophan is not basic, and its isoelectric point is near neutral pH.

Histidine

Imidazole ring

Nitrogen A is "pyrrole-like" and nonbasic because its lone pair is part of aromatic π electron system of the imidazole ring. Nitrogen B is basic because its lone pair electrons lie in the plane of the imidazole ring and are not part of the aromatic π system.. Nitrogen B is less basic than an aliphatic amine nitrogen, however. Thus, histidine is weakly basic, and its isoelectric point is slightly higher than neutral.

27.49

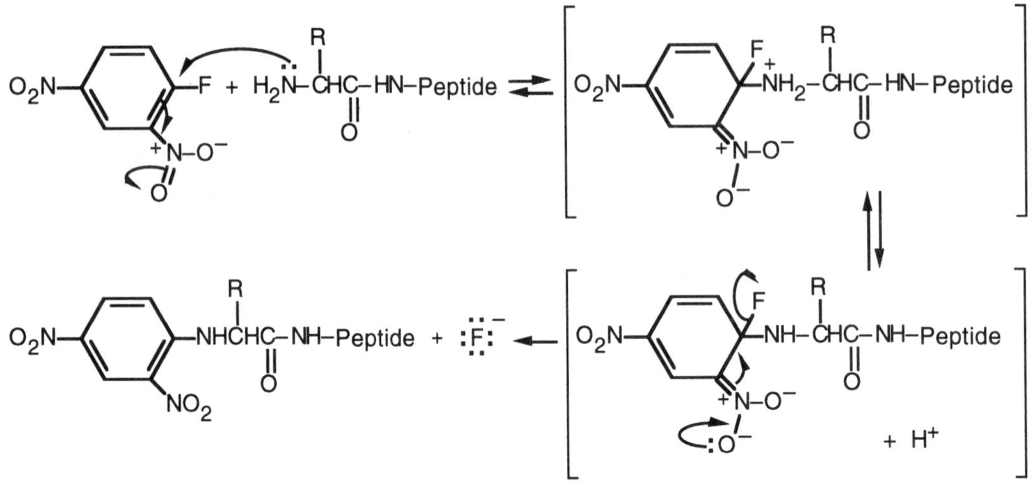

This reaction proceeds via a nucleophilic aromatic substitution mechanism.

27.50 2,4–Dinitrofluorobenzene (Sanger's reagent) reacts not only with the *N*–terminal amino group of a protein, but also with the terminal –NH$_2$ group of amino acids such as lysine or arginine. To use Sanger's reagent for end-group analysis in a protein containing lysine or arginine, these amino acids must first have their basic groups protected.

27.51

a)

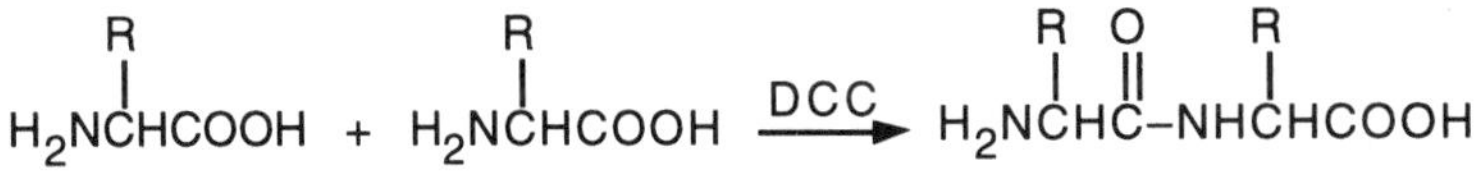

In this sequence of steps, a dipeptide is formed from two equivalents of

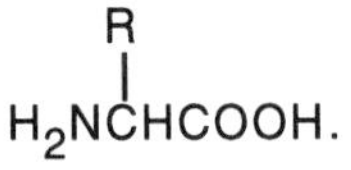

$$\underset{\substack{| \\ \text{H}_2\text{NCHCOOH}}}{\overset{\text{R}}{}}.$$ The mechanism is pictured in Fig. 27.8.

b)

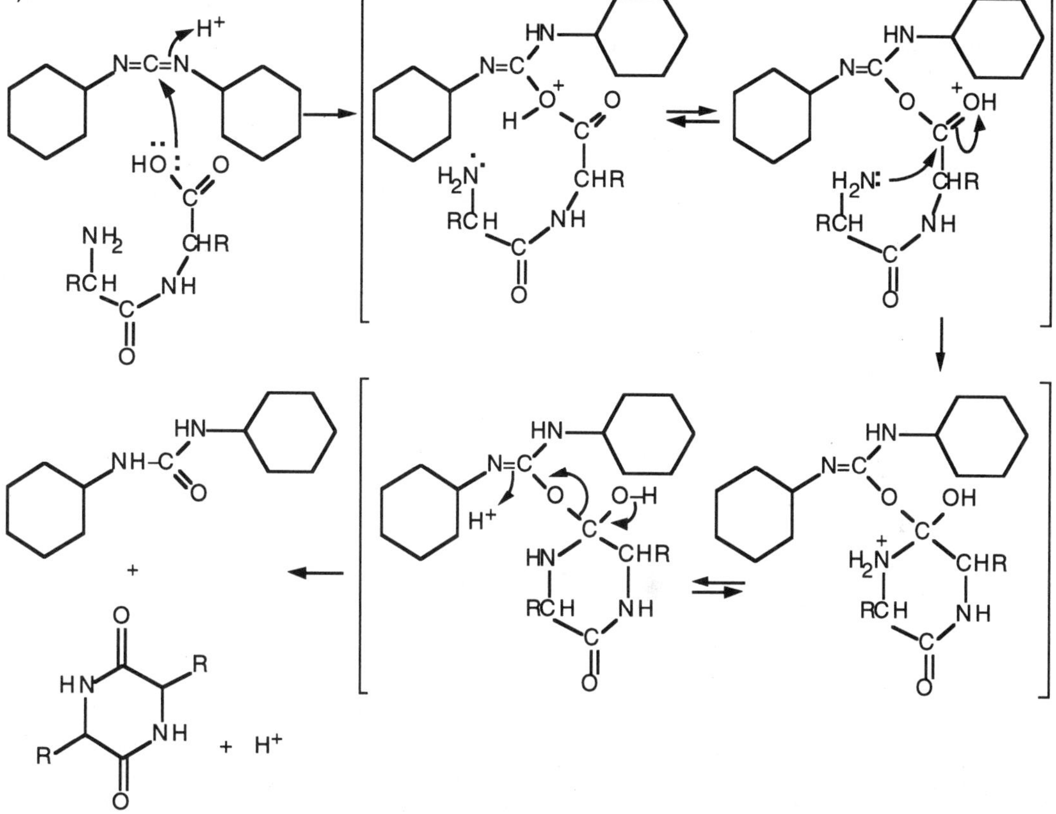

DCC couples the carboxylic acid end of the dipeptide to the amino end to yield the 2,5–diketopiperazine.

27.52

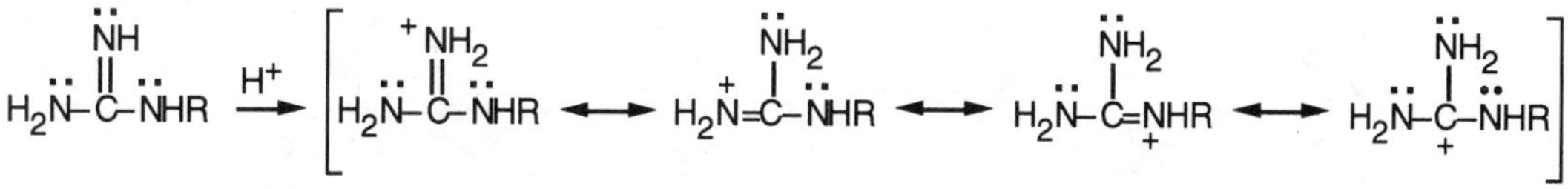

The protonated guanidino group can be stabilized by resonance.

27.53 ^{1}H NMR shows that the two methyl groups of N,N–dimethylformamide are non-equivalent at room temperature. If rotation around the CO–N bond were unrestricted, the methyl groups would be interconvertible, and their ^{1}H NMR absorptions would coalesce into a single signal.

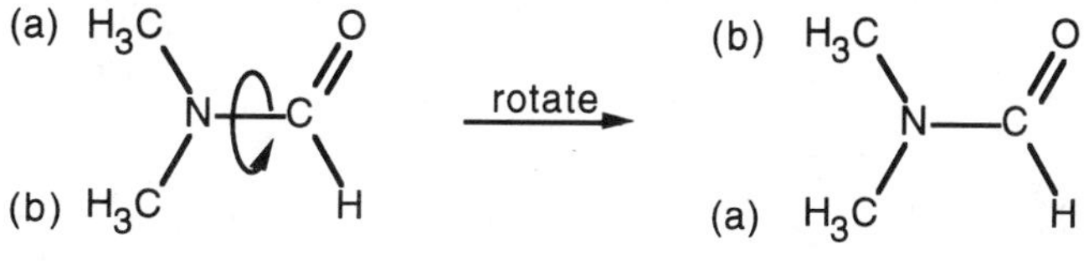

The presence of two absorptions shows that there is a barrier to rotation around the CO–N bond. This barrier is due to the partial double-bond character of the CO–N bond, as indicated by the two resonance forms. Heating to 180° supplies enough energy to allow rapid rotation and to cause the two NMR absorptions to merge.

27.54 Gly–Gly–Asp–Phe–Pro–Val–Pro–Leu

27.55

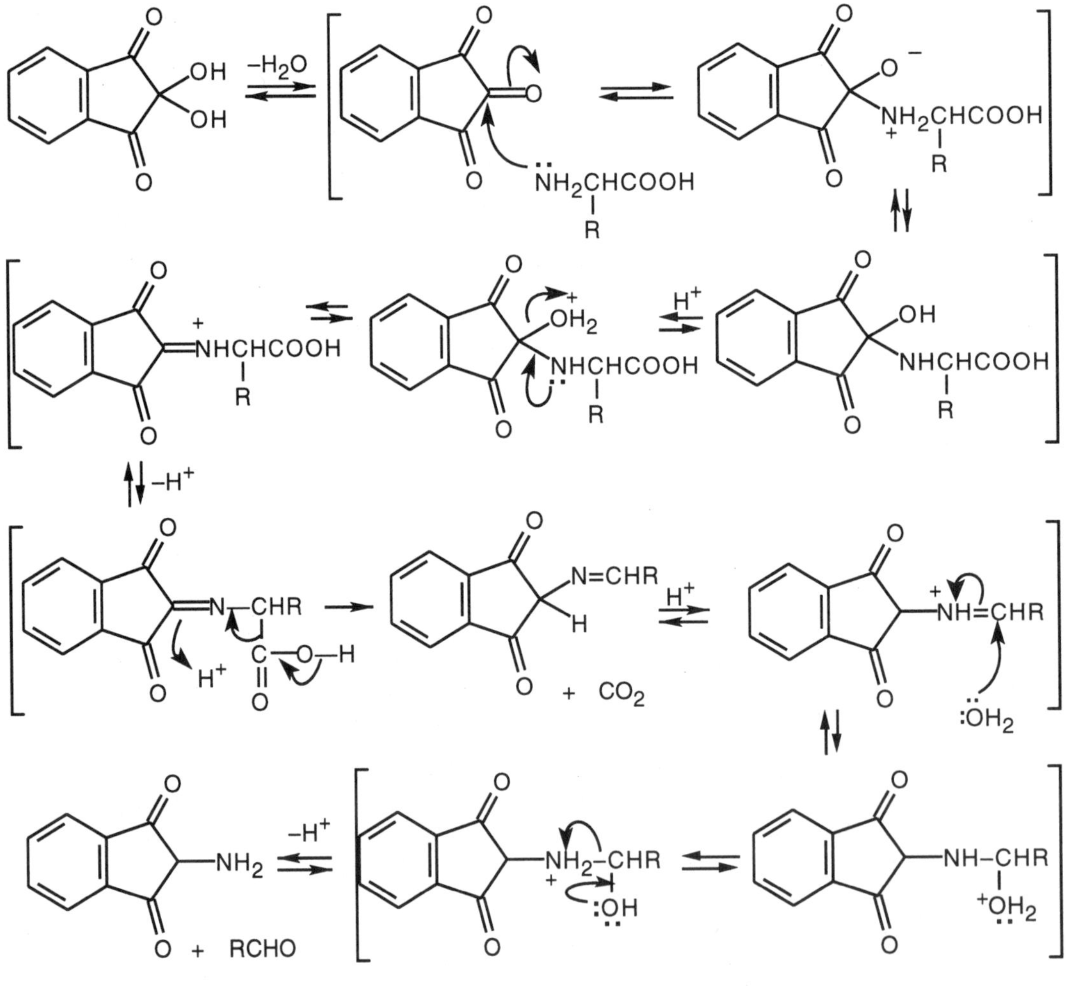

(continued on next page)

27.56

It is also possible to draw many other resonance forms that involve the π electrons of the aromatic 6-membered rings.

27.57

Gly
|
Ile
|
Val
|
Glu
|
Gln–CyS–CyS–Thr–Ser–Ile–CyS–Ser–Leu–Tyr⌇Gln–Leu–Glu–Asn–Tyr⌇CyS–Asn
|
His–Leu–CyS–Gly–Ser–His–Leu–Val–Glu–Ala–Leu–Tyr⌇Leu–Val–CyS
| |
Glu   Gly
| |
Asn   Glu
| |
Val   Arg
〰   ----+----
Phe   Thr┊Lys–Pro–Thr⌇Tyr⌇Phe⌇Phe–Gly

Cleaved by Trypsin = --------
Cleaved by Chymotrypsin = 〰〰〰〰

Thr–Lys–Pro–Thr–Tyr–Phe–Phe–Gly

27.58   Ser–Ile–Arg–Val–Val–Pro–Tyr–Leu–Arg

27.59   Cys–Tyr–Ile–Gln–Asn–Cys–Pro–Leu–Gly–NH$_2$
                    Reduced Oxytocin

Asn–Cys–Pro–Leu–Gly–NH$_2$
|
Gln   S
| |
Ile   S
| |
Tyr–Cys
Oxidized Oxytocin

The C–terminal end of oxytocin is an amide, but this can't be determined from the information given.

27.60

a)

$$H-NHCHC-NHCHC-OCH_3$$

with two C=O groups, CH$_2$ bearing COOH and CH$_2$ bearing phenyl.

b)

$$H-\overset{+}{N}H_2CHC-NHCHC-OCH_3$$

with two C=O groups, CH$_2$ bearing COO$^-$ and CH$_2$ bearing phenyl.

c)

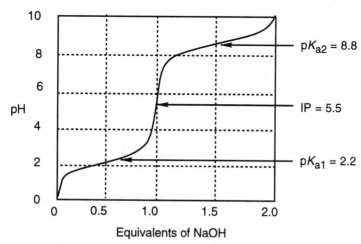

27.61

a) IP = 1/2 (pK_{a1} + pK_{a2}) = 1/2 (2.2 + 8.8) = 5.5

b) This is a neutral amino acid.

27.62

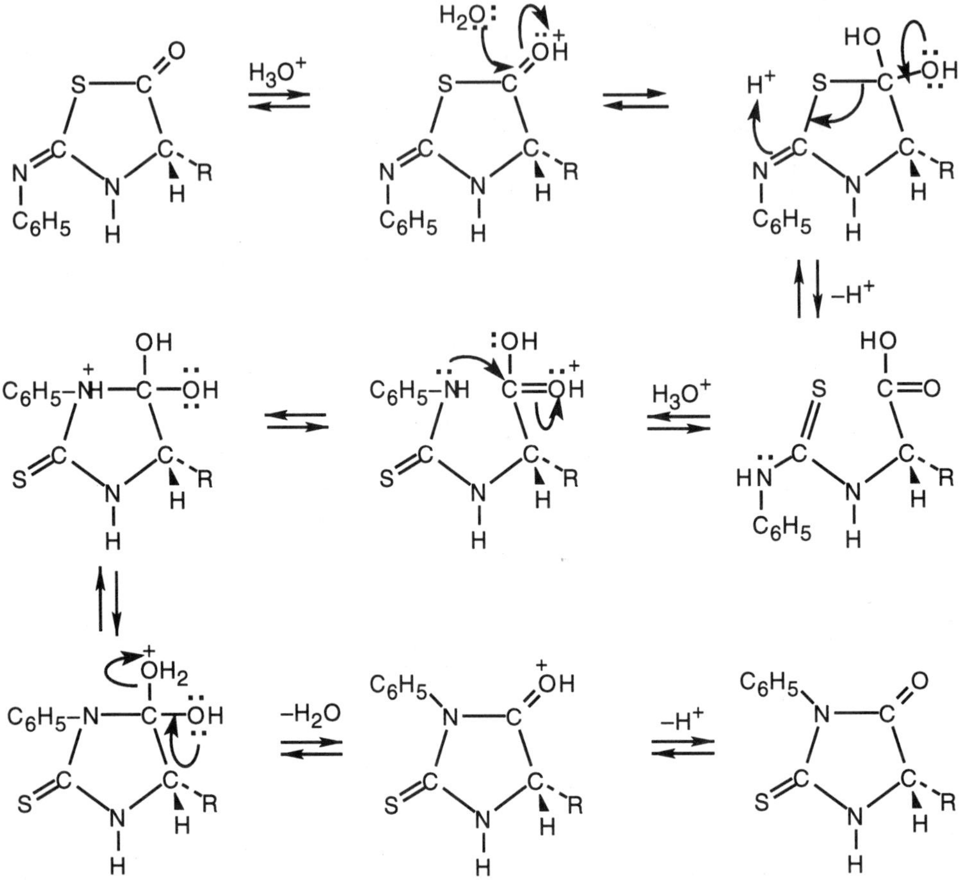

27.63

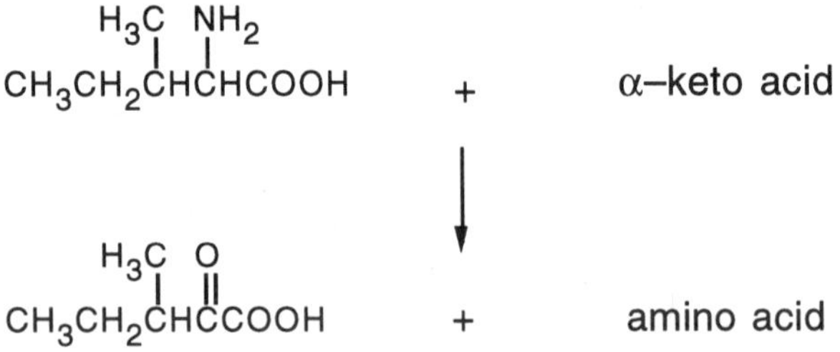

Study Guide for Chapter 27

After studying this chapter, you should be able to:

(1)   Identify the common amino acids and draw them with correct stereochemistry and in dipolar form (27.1, 27.2, 27.3, 27.4, 27.5, 27.6, 27.28, 27.29, 27.30, 27.31, 27.32, 27.52).

(2)   Understand the acid-base behavior of amino acids (27.7, 27.8, 27.9, 27.33, 27.34, 27.35, 27.36, 27.48, 27.52, 27.60, 27.61).

(3)   Synthesize amino acids (27.10, 27.11, 27.12, 27.13, 27.14, 27.40, 27.41, 27.42, 27.63).

(4)   Draw the structure of simple peptides (27.15, 27.16, 27.37, 27.43).

(5)   Determine the structure of peptides and proteins (27.19, 27.20, 27.21, 27.22, 27.47, 27.54, 27.57, 27.58, 27.59).

(6)   Outline the synthesis of peptides (27.24, 27.25, 27.44).

(7)   Understand the classification of proteins and the levels of structure of peptides (27.26, 27.27, 27.48).

(8)   Draw structures of reaction products of amino acids and peptides (27.17, 27.23, 27.39, 27.45, 27.49, 27.50, 27.51, 27.55, 27.56, 27.62).

28.1

$$CH_3(CH_2)_{18}\overset{O}{\overset{\|}{C}}-O(CH_2)_{31}CH_3$$

28.2

a)

$$CH_2O\overset{O}{\overset{\|}{C}}(CH_2)_{14}CH_3$$
$$CHO\overset{O}{\overset{\|}{C}}(CH_2)_{14}CH_3$$
$$CH_2O\overset{O}{\overset{\|}{C}}(CH_2)_{14}CH_3$$

Glyceryl tripalmitate

b)

$$CH_2O\overset{O}{\overset{\|}{C}}(CH_2)_7CH=CH(CH_2)_7CH_3 \quad (cis)$$
$$CHO\overset{O}{\overset{\|}{C}}(CH_2)_7CH=CH(CH_2)_7CH_3 \quad (cis)$$
$$CH_2O\overset{O}{\overset{\|}{C}}(CH_2)_7CH=CH(CH_2)_7CH_3 \quad (cis)$$

Glyceryl trioleate

Glyceryl tripalmitate is higher melting because it is saturated.

28.3

$$CH_3(CH_2)_7C\equiv C(CH_2)_7COOH \xrightarrow[\substack{2. \quad Zn, \\ CH_3COOH}]{1. \ O_3} CH_3(CH_2)_7COOH + HOOC(CH_2)_7COOH$$

Stearolic acid　　　　　　　　　　　　　Nonanoic acid　　　Nonanedioic acid

28.4

$$CH_3(CH_2)_7CH=CH(CH_2)_7\overset{O}{\overset{\|}{C}}O^-Mg^{++}\ ^-O\overset{O}{\overset{\|}{C}}(CH_2)_7CH=CH(CH_2)_7CH_3$$

Magnesium oleate

The double bonds are *cis*.

28.5

a)

Carvone

b)

Camphor

c)

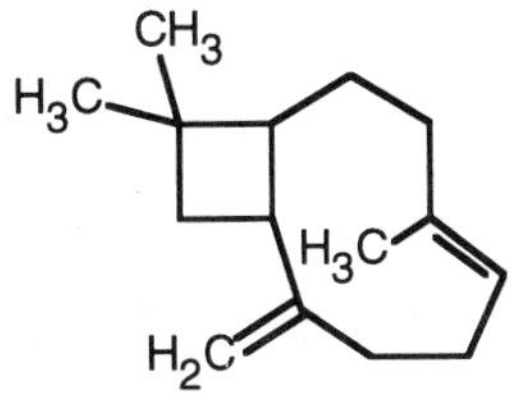

Caryophyllene
(a sesquiterpene)

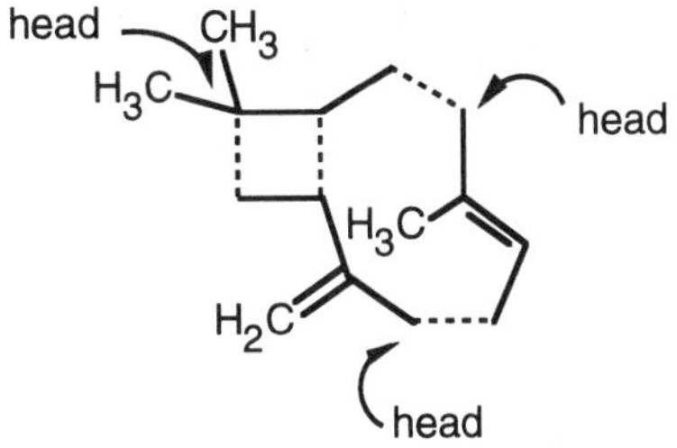

28.6

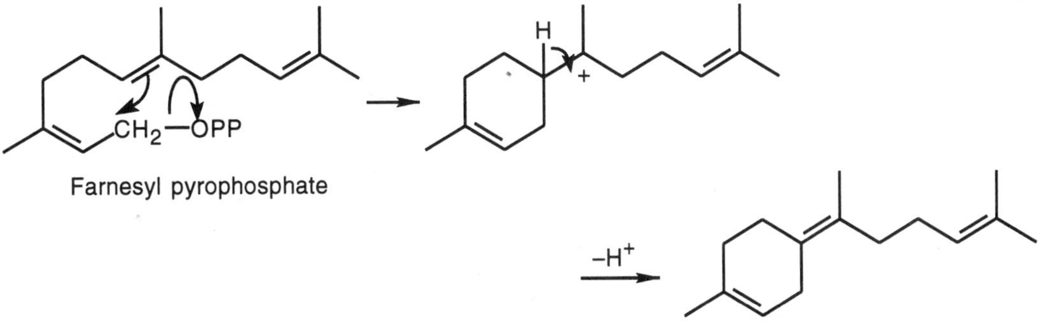

Farnesyl pyrophosphate

$$\xrightarrow{-H^+}$$

γ–Bisabolene

28.7

a)

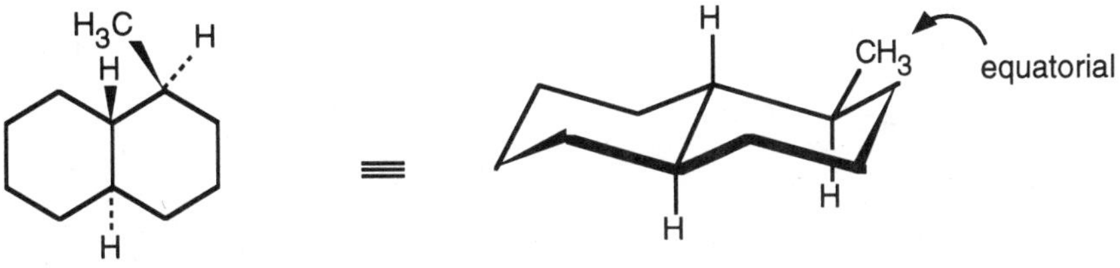

equatorial

b)

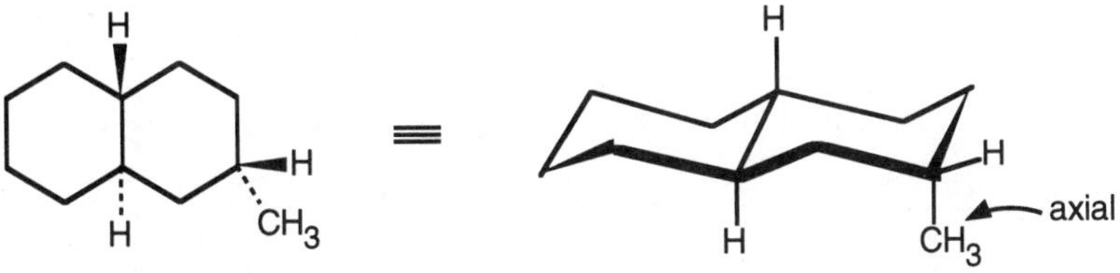

axial

28.8

equatorial

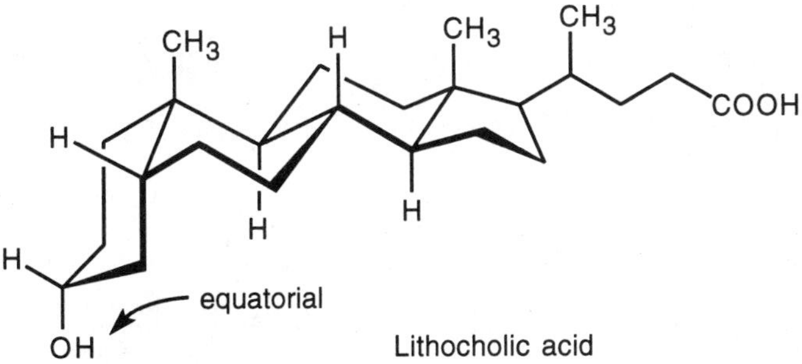

Lithocholic acid

28.9

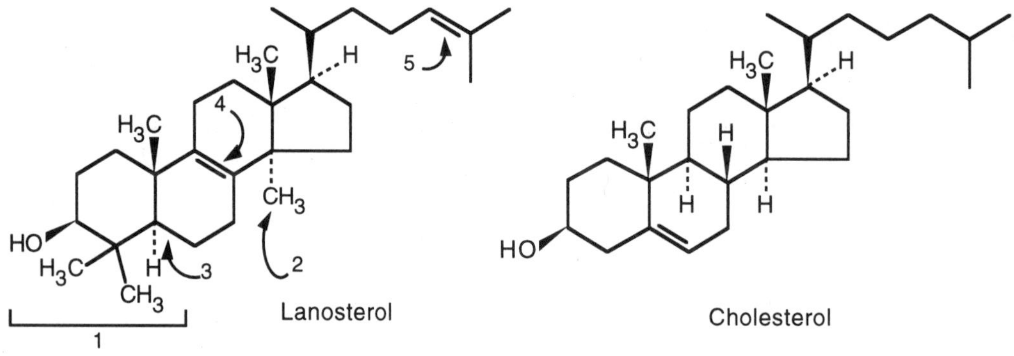

Lanosterol

Cholesterol

Lanosterol

1. Two methyl groups at C4.
2. One methyl group at C14.
3. C5–C6 single bond.
4. C8–C9 double bond.
5. Double bond in side chain

Cholesterol

1. Two hydrogens at C4.
2. One hydrogen at C14.
3. C5–C6 double bond
4. C8–C9 single bond.
5. Saturated side chain.

28.10

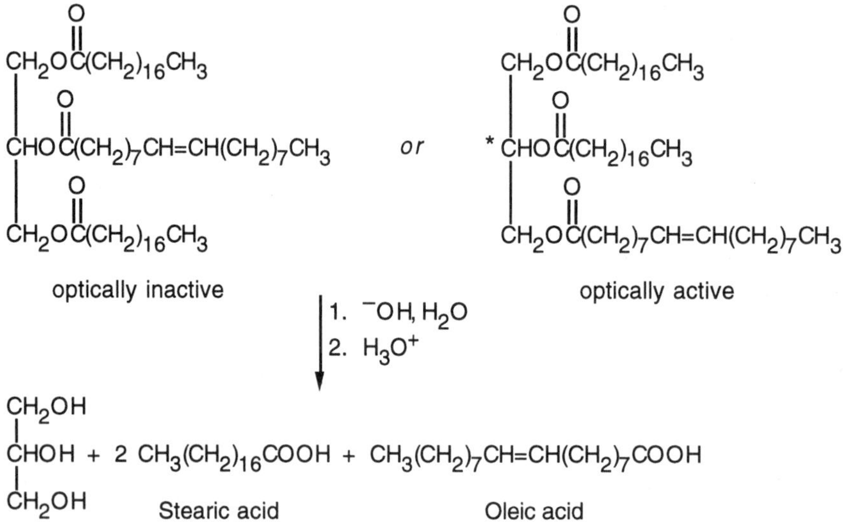

optically inactive   *or*   optically active

1. $^-$OH, H_2O
2. H_3O^+

$$CH_2OH$$
$$CHOH + 2\ CH_3(CH_2)_{16}COOH + CH_3(CH_2)_7CH=CH(CH_2)_7COOH$$
$$CH_2OH$$

Stearic acid   Oleic acid

Four different groups are bonded to the central glycerol carbon atom in the optically active fat.

28.11

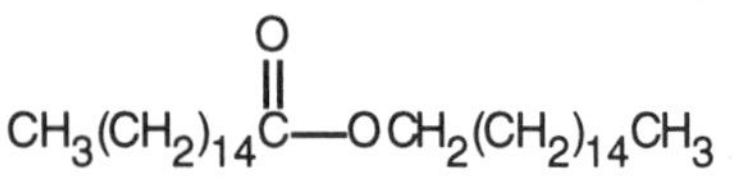

$$CH_3(CH_2)_{14}\overset{O}{\overset{\|}{C}}{-}OCH_2(CH_2)_{14}CH_3$$ Cetyl Palmitate

28.12 Fats, lecithins, cephalins and plasmalogens are all esters of a glycerol molecule that has carboxylic acid ester groups at C1 and C2. The third group bonded to glycerol, however, differs with the type of lipid.

Lipid	Functional group at C3 of glycerol
fat	carboxylic acid ester
cephalin	phosphate ester (also bonded to ethanolamine, an amino alcohol)
lecithin	phosphate ester (also bonded to choline, an amino alcohol)
plasmalogen	vinyl ether

28.13

$$
\begin{array}{ccc}
CH_2OH + HCCH_2R & & CH_2OCH=CHR \\
\overset{O}{\|} & & O \\
CHOH + HOCR' & \xleftarrow{H_3O^+} & CHOCR' \xrightarrow[H_2O]{NaOH} \\
\overset{O}{\|} & & O \\
CH_2OH + HOCR'' & & CH_2OCR''
\end{array}
$$

CH$_2$OCH=CHR
CHOH + Na$^+$ $^-$OCR'
CH$_2$OH + Na$^+$ $^-$OCR''

28.14

RC—OCH$_2$
R'C—OCH
CH$_2$O—P—OCH$_2$CHCH$_2$O—P—OCH$_2$

H$_2$CO—CR'''
HCO—CR''

$\xrightarrow[H_2O]{NaOH}$

3 HOCH$_2$CHCH$_2$OH (OH)
+
2 Na$_3$PO$_4$
+
RCO$^-$ Na$^+$ + R'CO$^-$ Na$^+$ +
R''CO$^-$ Na$^+$ + R'''CO$^-$ Na$^+$

28.15

CH$_2$OC(CH$_2$)$_7$CH=CH(CH$_2$)$_7$CH$_3$
CHOC(CH$_2$)$_7$CH=CH(CH$_2$)$_7$CH$_3$
CH$_2$OC(CH$_2$)$_7$CH=CH(CH$_2$)$_7$CH$_3$

Glyceryl trioleate

a)

Glyceryl trioleate $\xrightarrow[CCl_4]{Br_2}$

CH$_2$OC(CH$_2$)$_7$CHBrCHBr(CH$_2$)$_7$CH$_3$
CHOC(CH$_2$)$_7$CHBrCHBr(CH$_2$)$_7$CH$_3$
CH$_2$OC(CH$_2$)$_7$CHBrCHBr(CH$_2$)$_7$CH$_3$

b)

Glyceryl trioleate $\xrightarrow{\text{H}_2\text{ /Pd}}$

$$\begin{array}{l} \text{CH}_2\text{O}\overset{\text{O}}{\overset{\|}{\text{C}}}(\text{CH}_2)_{16}\text{CH}_3 \\ | \\ \text{CHO}\overset{\text{O}}{\overset{\|}{\text{C}}}(\text{CH}_2)_{16}\text{CH}_3 \\ | \\ \text{CH}_2\text{O}\overset{\text{O}}{\overset{\|}{\text{C}}}(\text{CH}_2)_{16}\text{CH}_3 \end{array}$$

c)

Glyceryl trioleate $\xrightarrow{\text{NaOH, H}_2\text{O}}$

$$\begin{array}{l} \text{CH}_2\text{OH} \\ | \\ \text{CHOH} \quad + \quad 3 \text{ CH}_3(\text{CH}_2)_7\text{CH=CH(CH}_2)_7\text{COO}^-\text{ Na}^+ \\ | \\ \text{CH}_2\text{OH} \end{array}$$

d)

Glyceryl trioleate $\xrightarrow[\text{2. Zn, CH}_3\text{COOH}]{\text{1. O}_3}$

$$\begin{array}{l} \text{CH}_2\text{O}\overset{\text{O}}{\overset{\|}{\text{C}}}(\text{CH}_2)_7\overset{\text{O}}{\overset{\|}{\text{C}}}\text{H} \\ | \\ \text{CHO}\overset{\text{O}}{\overset{\|}{\text{C}}}(\text{CH}_2)_7\overset{\text{O}}{\overset{\|}{\text{C}}}\text{H} \quad + \quad 3 \text{ CH}_3(\text{CH}_2)_7\overset{\text{O}}{\overset{\|}{\text{C}}}\text{H} \\ | \\ \text{CH}_2\text{O}\overset{\text{O}}{\overset{\|}{\text{C}}}(\text{CH}_2)_7\overset{\text{O}}{\overset{\|}{\text{C}}}\text{H} \end{array}$$

e)

Glyceryl trioleate $\xrightarrow[\text{2. H}_3\text{O}^+]{\text{1. LiAlH}_4}$

$$\begin{array}{l} \text{CH}_2\text{OH} \\ | \\ \text{CHOH} \quad + \quad 3 \text{ CH}_3(\text{CH}_2)_7\text{CH=CH(CH}_2)_7\text{CH}_2\text{OH} \\ | \\ \text{CH}_2\text{OH} \end{array}$$

f)

Glyceryl trioleate $\xrightarrow[\text{2. H}_3\text{O}^+]{\text{1. CH}_3\text{MgBr}}$

$$\begin{array}{l} \text{CH}_2\text{OH} \qquad\qquad\qquad\qquad\qquad\qquad\quad \text{OH} \\ | \qquad\qquad\qquad\qquad\qquad\qquad\qquad\qquad\quad | \\ \text{CHOH} \quad + \quad 3 \text{ CH}_3(\text{CH}_2)_7\text{CH=CH(CH}_2)_7\text{C(CH}_3)_2 \\ | \\ \text{CH}_2\text{OH} \end{array}$$

28.16

$(9Z,11E,13E)$–Octadecatrienoic acid
(Eleostearic acid)

$\downarrow$ 1. O$_3$
2. Zn, CH$_3$COOH

$$\text{CH}_3\text{CH}_2\text{CH}_2\text{CH}_2\overset{\text{O}}{\overset{\|}{\text{C}}}\text{H} \;+\; \text{H}\overset{\text{O}}{\overset{\|}{\text{C}}}\text{-}\overset{\text{O}}{\overset{\|}{\text{C}}}\text{H} \;+\; \text{H}\overset{\text{O}}{\overset{\|}{\text{C}}}\text{-}\overset{\text{O}}{\overset{\|}{\text{C}}}\text{H} \;+\; \text{H}\overset{\text{O}}{\overset{\|}{\text{C}}}\text{CH}_2\text{CH}_2\text{CH}_2\text{CH}_2\text{CH}_2\text{CH}_2\text{COOH}$$

The stereochemistry of the double bonds can't be determined from the information given.

28.17

a)

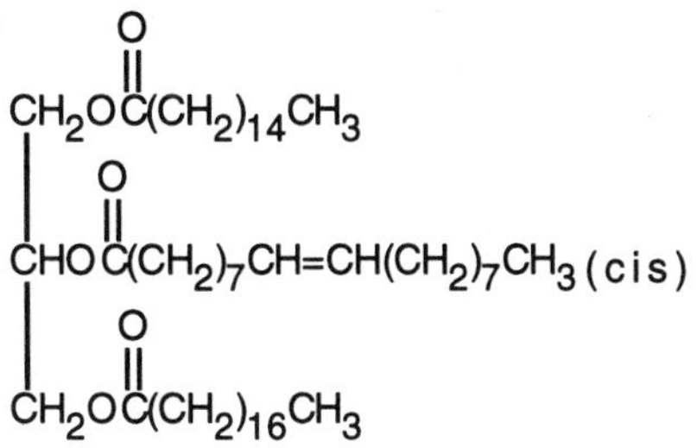

a fat

b)

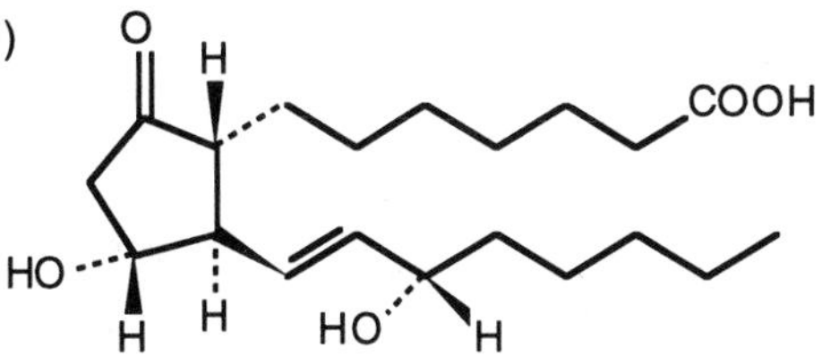

a prostaglandin (prostaglandin E$_1$)

c)

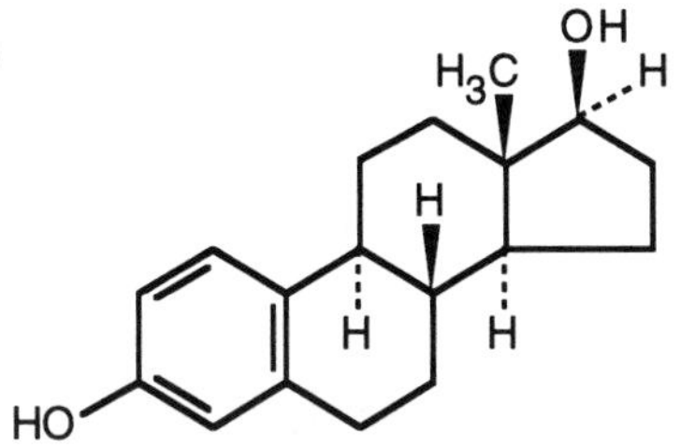

a steroid (Estradiol)

28.18

$$\text{cis–CH}_3(\text{CH}_2)_7\text{CH=CH(CH}_2)_7\text{COOH}$$

Oleic acid

a) Oleic acid $\xrightarrow{\text{CH}_3\text{OH, HCl}}$ CH$_3$(CH$_2$)$_7$CH=CH(CH$_2$)$_7$COOCH$_3$

Methyl oleate

b) Methyl oleate $\xrightarrow[\text{Pd/C}]{\text{H}_2}$ CH$_3$(CH$_2$)$_{16}$COOCH$_3$

(part a)   Methyl stearate

c) Oleic acid $\xrightarrow[\text{2. Zn, CH}_3\text{COOH}]{\text{1. O}_3}$ CH$_3$(CH$_2$)$_7$CHO + OHC(CH$_2$)$_7$COOH

Nonanal   9–Oxononanoic acid

d) 9–Oxononanoic acid (part c) $\xrightarrow[\text{NH}_4\text{OH}]{\text{Ag}_2\text{O}}$ HOOC(CH$_2$)$_7$COOH

Nonanedioic acid

e) Oleic acid $\xrightarrow[\text{CCl}_4]{\text{Br}_2}$ CH$_3$(CH$_2$)$_7$CHBrCHBr(CH$_2$)$_7$COOH

$\downarrow$ 1. 3 NaNH$_2$/NH$_3$
    2. H$_3$O$^+$

CH$_3$(CH$_2$)$_7$C≡C(CH$_2$)$_7$COOH

Stearolic acid

f) Oleic acid $\xrightarrow{\text{H}_2 / \text{Pd/C}}$ CH$_3$(CH$_2$)$_{15}$CH$_2$COOH $\xrightarrow{\substack{\text{1. Br}_2,\ \text{PBr}_3 \\ \text{2. H}_2\text{O}}}$ CH$_3$(CH$_2$)$_{15}$CHBrCOOH

Stearic acid   2–Bromostearic acid

g) 2 CH$_3$(CH$_2$)$_{16}$COOCH$_3$ $\xrightarrow{\substack{\text{1. NaOCH}_3 \\ \text{2. H}_3\text{O}^+}}$ CH$_3$(CH$_2$)$_{16}$CCH(CH$_2$)$_{15}$CH$_3$ + HOCH$_3$

Methyl stearate
(part b)

O
‖
COOCH$_3$

$\downarrow$ H$_3$O$^+$, Δ

CH$_3$(CH$_2$)$_{16}$CCH$_2$(CH$_2$)$_{15}$CH$_3$ + CO$_2$ + CH$_3$OH

This synthesis uses a Claisen condensation, followed by a ß–keto ester decarboxylation.

28.19

CH$_3$(CH$_2$)$_7$C≡CH $\xrightarrow{\text{NaNH}_2}$ CH$_3$(CH$_2$)$_7$C≡C:$^-$ Na$^+$ + NH$_3$

$\downarrow$ I–CH$_2$(CH$_2$)$_5$CH$_2$Cl

CH$_3$(CH$_2$)$_7$C≡C(CH$_2$)$_6$CH$_2$CN $\xleftarrow{\text{NaCN}}$ CH$_3$(CH$_2$)$_7$C≡C(CH$_2$)$_6$CH$_2$Cl + NaI

$\downarrow$ H$_3$O$^+$

CH$_3$(CH$_2$)$_7$C≡C(CH$_2$)$_7$COOH

Stearolic acid

S$_N$2 displacement by acetylide occurs at iodine, rather than at chlorine, because I$^-$ is a better leaving group than Cl$^-$.

28.20

1. O$_3$
2. Zn, CH$_3$COOH

CH$_3$(CH$_2$)$_5$CH + HC(CH$_2$)$_9$COOH
Heptanal   11–Oxoundecanoic acid

CH$_3$(CH$_2$)$_4$CH$_2$ CH$_2$(CH$_2$)$_8$COOH
C=C
H   H
Vaccenic acid

CH$_2$I$_2$ / Zn/Cu

CH$_3$(CH$_2$)$_4$CH$_2$ CH$_2$(CH$_2$)$_8$COOH
H—C—C—H
CH$_2$

Lactobacillic acid

28.21–28.23 Not all the possible stereoisomers of these compounds are found in nature or can be synthesized. Some stereoisomers have highly strained ring fusions; others contain 1,3–diaxial interactions.

a)

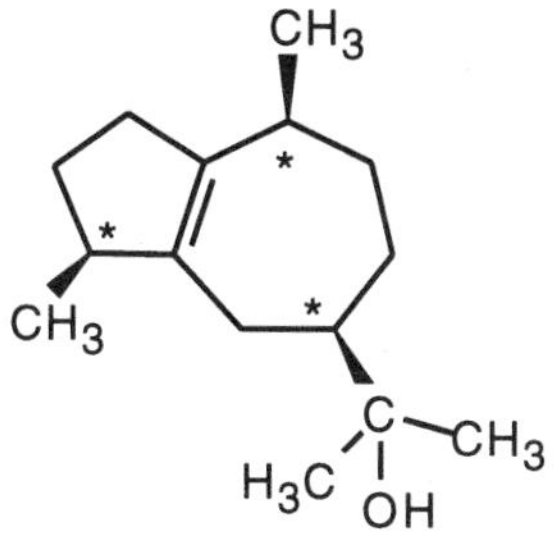

Guaiol
(8 possible stereoisomers)

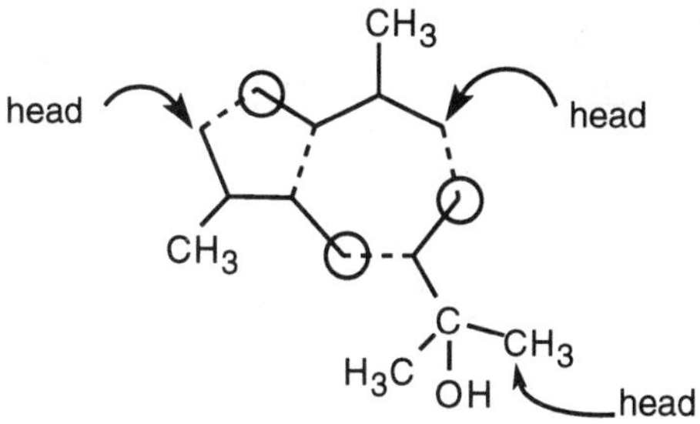

b)

Sabinene
(4 possible stereoisomers)

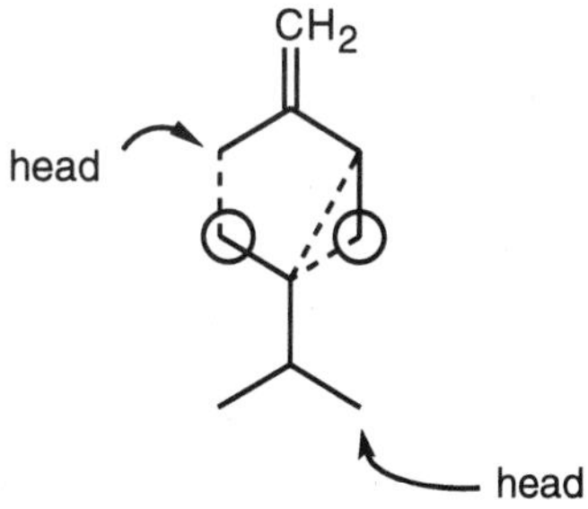

c)

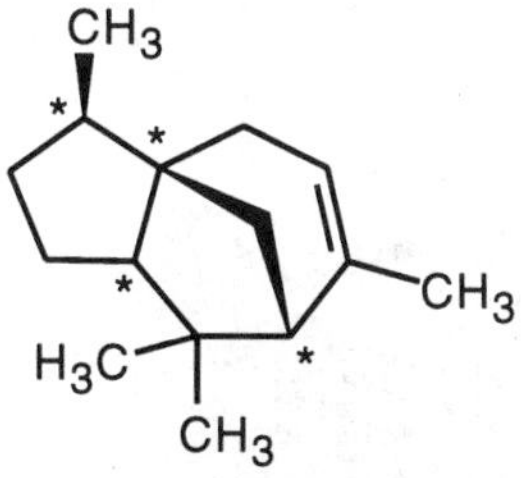

Cedrene
(16 possible stereoisomers)

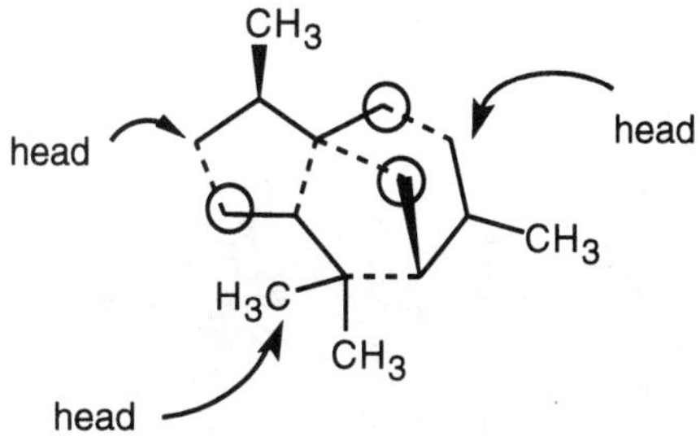

If carbon 1 of each pyrophosphate were isotopically labeled, the labels would appear at the circled positions of the terpenes.

28.24

28.25

ψ–Ionone

β–Ionone

28.26

trans–Decalin

cis–Decalin

Three 1,3–diaxial interactions cause *cis*–decalin to be less stable than *trans*–decalin.

28.27

Dihydrocarvone

28.28

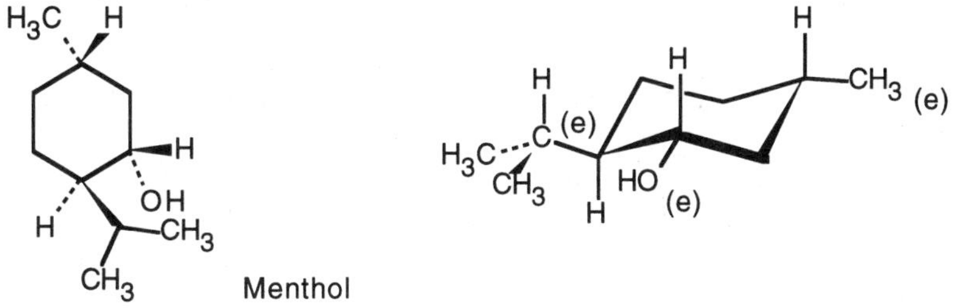

Menthol

28.29–28.30

Cholic acid

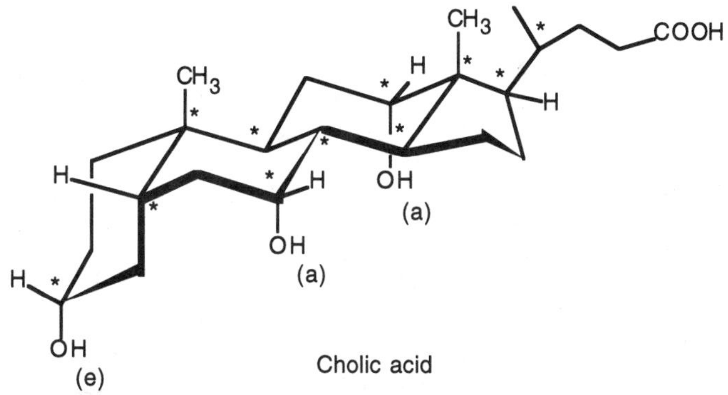

Cholic acid has eleven stereogenic centers and $2^{11} = 2048$ possible stereoisomers (most are very strained).

28.31

a)

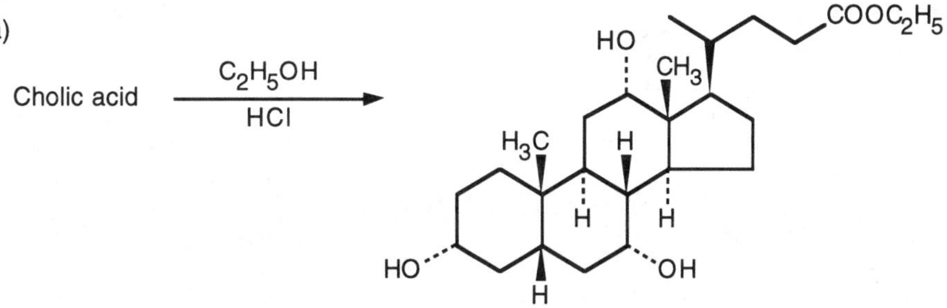

Cholic acid $\xrightarrow[\text{HCl}]{\text{C}_2\text{H}_5\text{OH}}$

b)

Cholic acid $\xrightarrow[\text{CH}_2\text{Cl}_2]{\text{excess PCC}}$

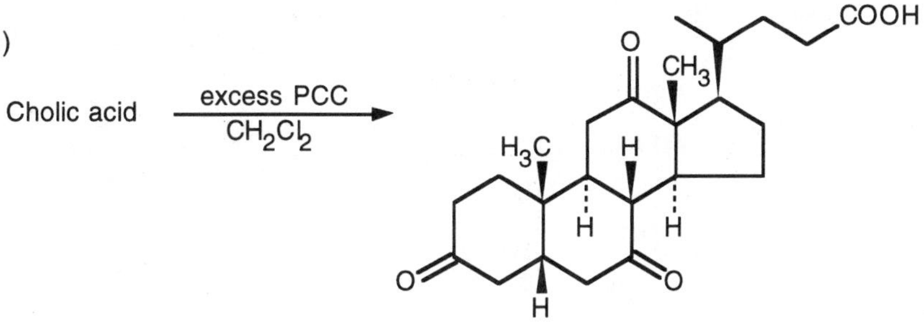

c)

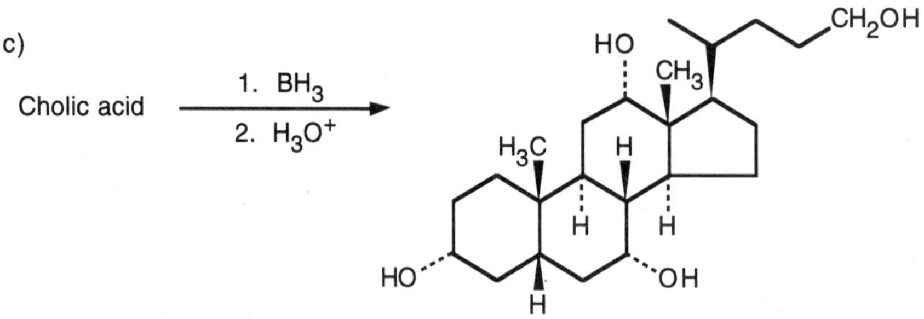

Cholic acid $\xrightarrow{\text{1. BH}_3}{\text{2. H}_3\text{O}^+}$

28.32

a)

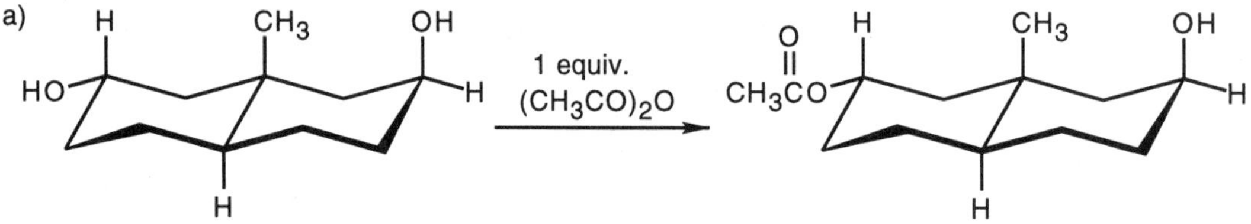

$\xrightarrow{\substack{\text{1 equiv.} \\ (\text{CH}_3\text{CO})_2\text{O}}}$

b)

$\xrightarrow{\substack{\text{1 equiv.} \\ (\text{CH}_3\text{CO})_2\text{O}}}$

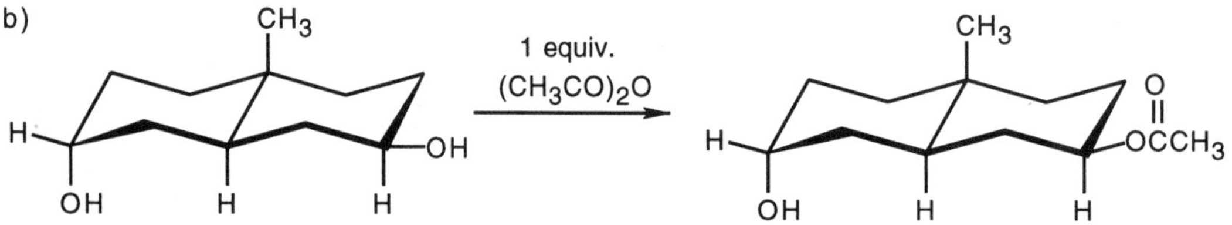

28.33

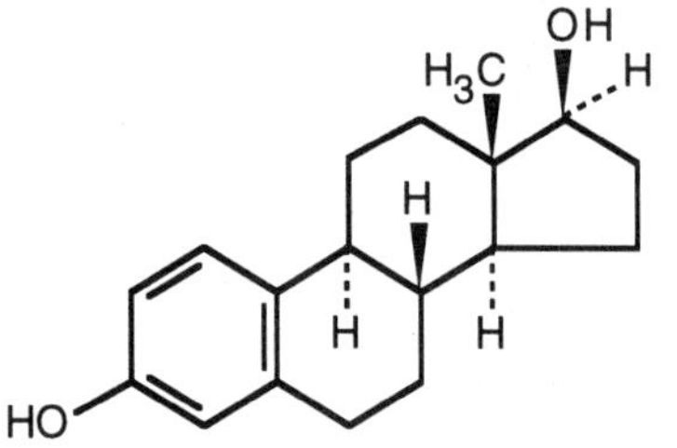

Estradiol

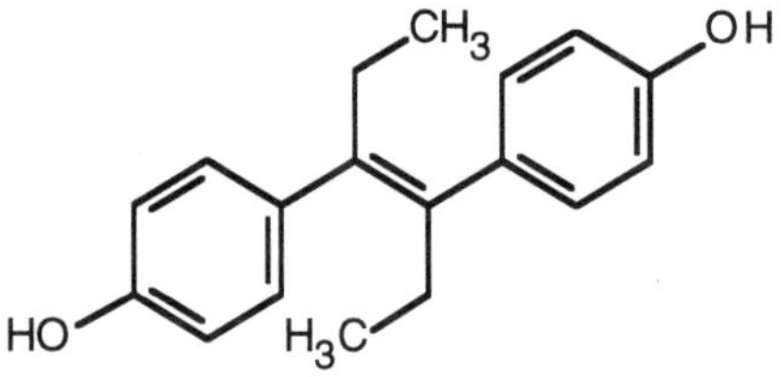

Diethylstilbestrol

28.34

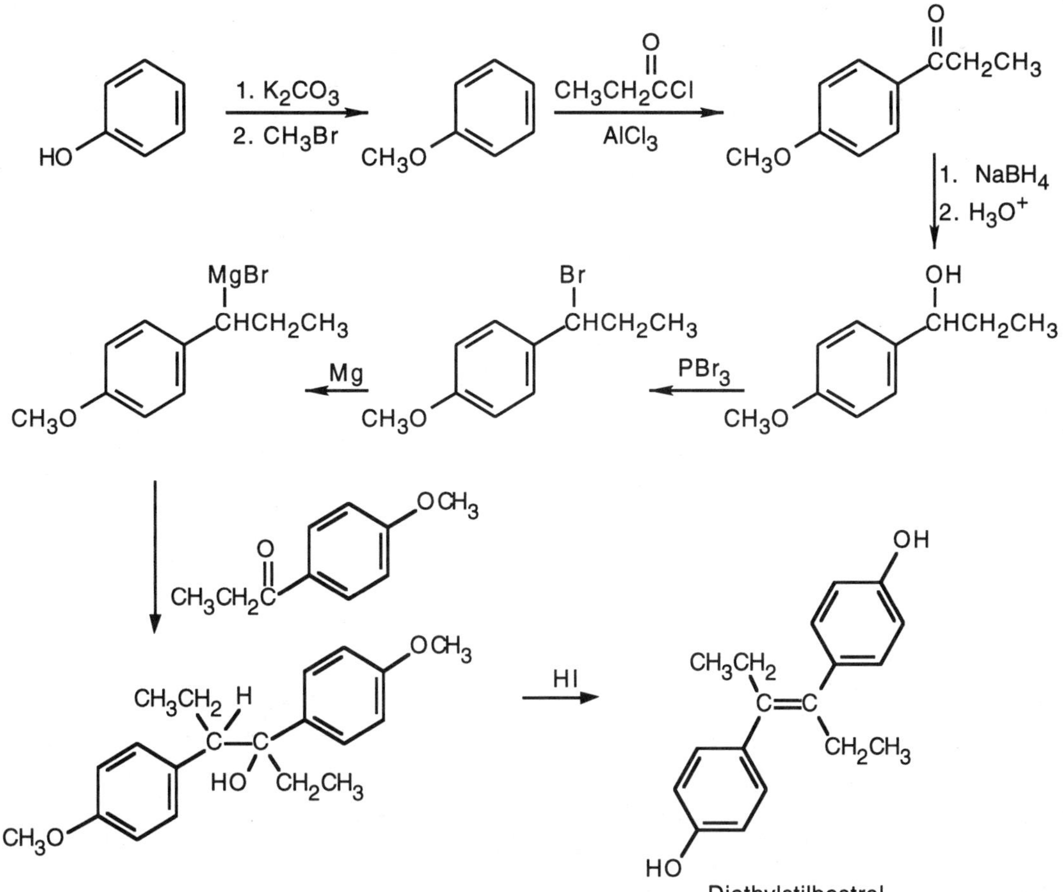

28.35

a) Estradiol $\xrightarrow{\text{1. NaH} \atop \text{2. CH}_3\text{I}}$

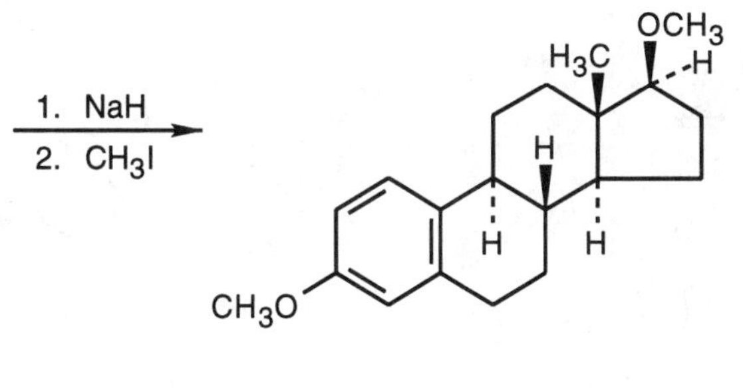

b) 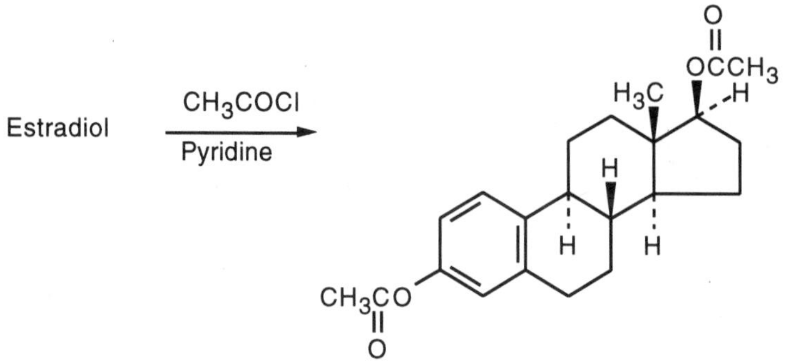
Estradiol
CH3COCl
Pyridine
O
OCCH3
H3C
H
H
H
H
CH3CO
O

c) 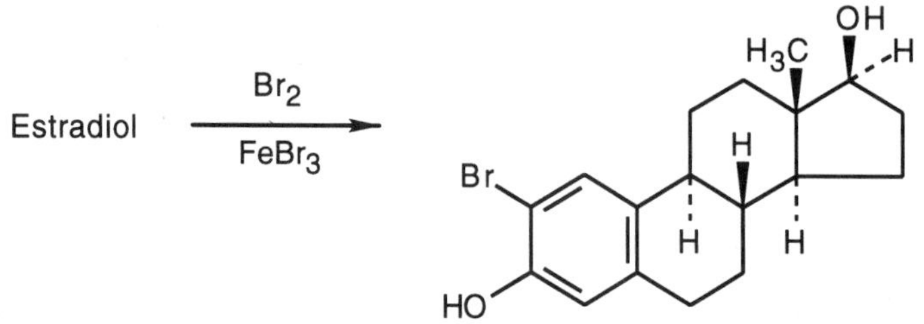
Estradiol
Br2
FeBr3
OH
H3C
H
Br
H
H
H
HO

d) 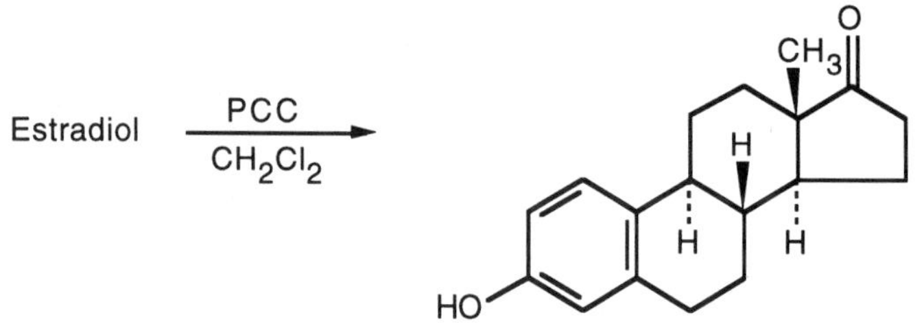
Estradiol
PCC
CH2Cl2
O
CH3
H
H
H
HO

28.36

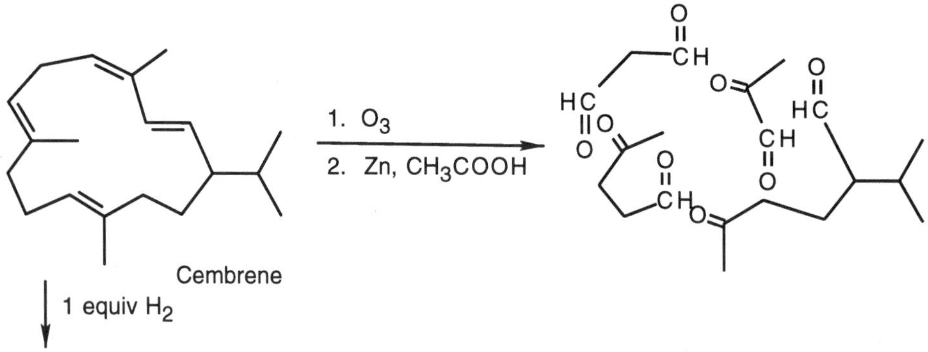
1. O3
2. Zn, CH3COOH
Cembrene
1 equiv H2
O
CH
HC
O
O
O
HC
CH
O
O
CH
O

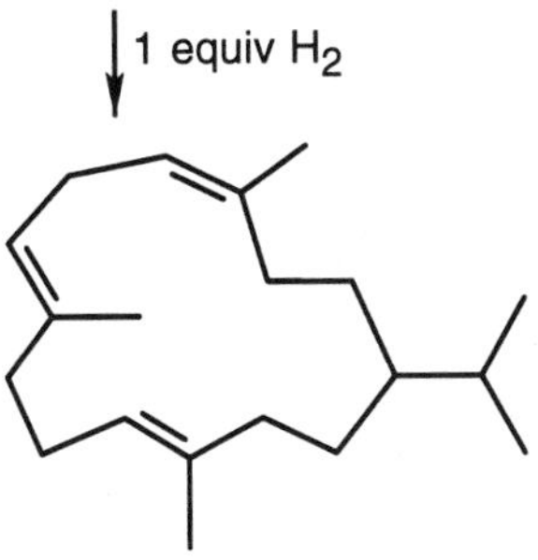

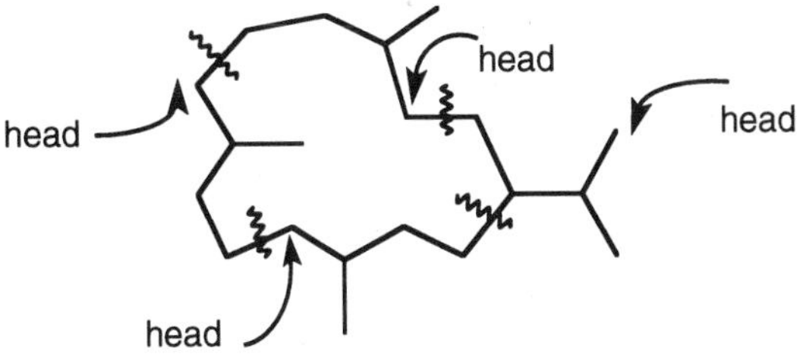

Dihydrocembrene

One equivalent of H_2 hydrogenates the least substituted double bond. Dihydrocembrene has no ultraviolet absorption because it is not conjugated.

28.37

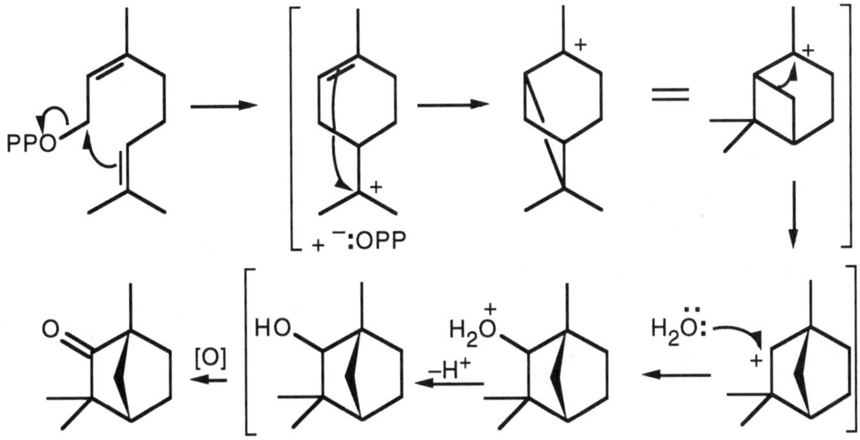

α–Fenchone

28.38

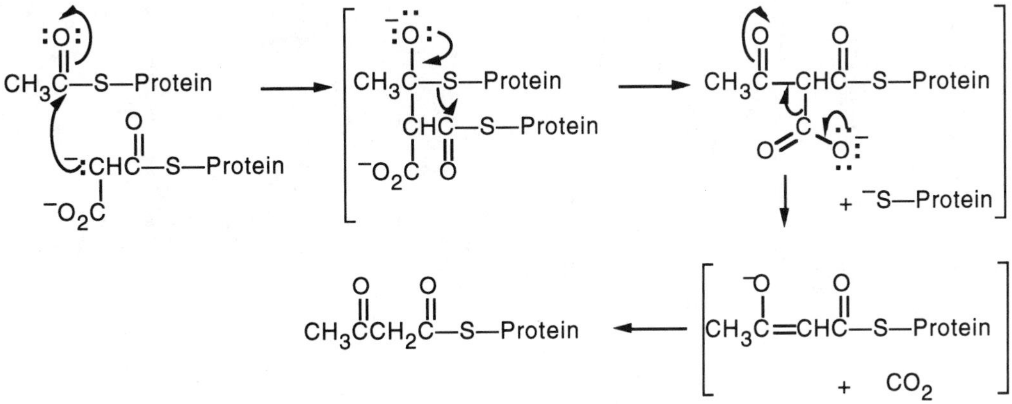

This is a Claisen condensation followed by a decarboxylation.

28.39

This reaction is an aldol condensation.

Study Guide for Chapter 28

After studying this chapter, you should be able to:

(1)   Draw the structures of fats, oils, steroids, and other lipids (28.1, 28.2, 28.4, 28.11, 28.17).

(2)   Determine the structure of a fat (28.3, 28.10, 28.16, 28.20, 28.36).

(3)   Predict the products of reactions of fats and steroids (28.13, 28.14, 28.15, 28.18, 28.19, 28.31, 28.32, 28.35).

(4)   Locate the isoprene units in a terpene (28.5, 28.21, 28.36).

(5)   Understand the mechanism of terpene and steroid biosynthesis (28.6, 28.9, 28.23, 28.24, 28.25, 28.37).

(6)   Draw the structures and conformations of steroids and other fused-ring systems (28.7, 28.8, 28.26, 28.27, 28.28, 28.29, 28.33).

29.1

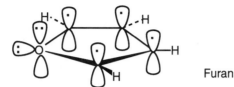

Furan

One oxygen lone pair is in a *p* orbital that is part of the π electron system of furan. The other oxygen lone pair is in an *sp*^{2} orbital that lies in the plane of the furan ring.

29.2

The dipole moment of pyrrole points in the direction indicated because resonance structures make the nitrogen atom electron-poor.

29.3

2–Deuteriopyrrole

29.4

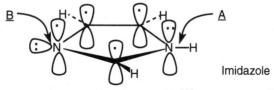

Imidazole

Nitrogen atom **B** is more basic than **A** because its lone pair of electrons lies in an *sp*^{2} orbital and is more available for donation to a Lewis acid than the lone pair of electrons of nitrogen **A**, which is part of the ring π system.

29.5

Attack at C2:

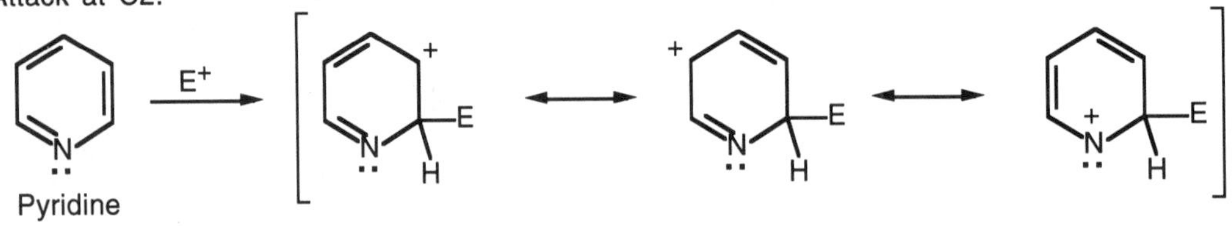

Pyridine

Attack at C3:

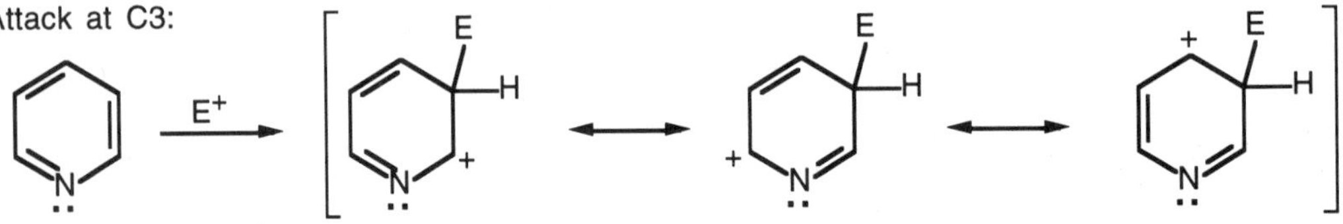

Attack at C4:

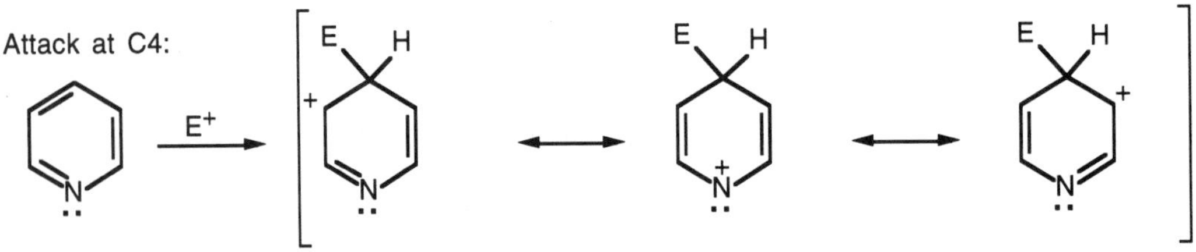

C3 attack is favored over C2 or C4 attack. The positive charge of the cationic intermediate of C3 attack is delocalized onto three carbon atoms, rather than onto two carbons and the electronegative pyridine nitrogen.

29.6

Attack at C3:

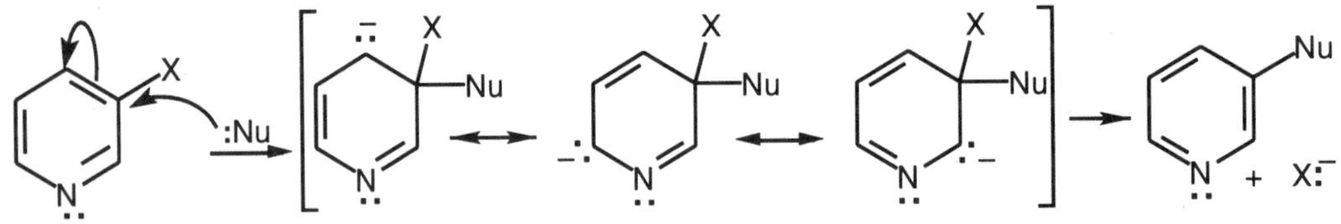

Attack at C4:

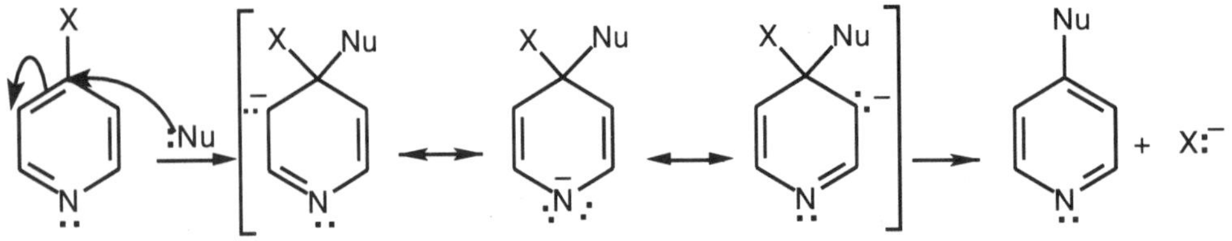

The negative charge resulting from C4 attack can be stabilized by nitrogen. Since no such stabilization is possible for C3 attack, reaction at C3 doesn't occur.

29.7

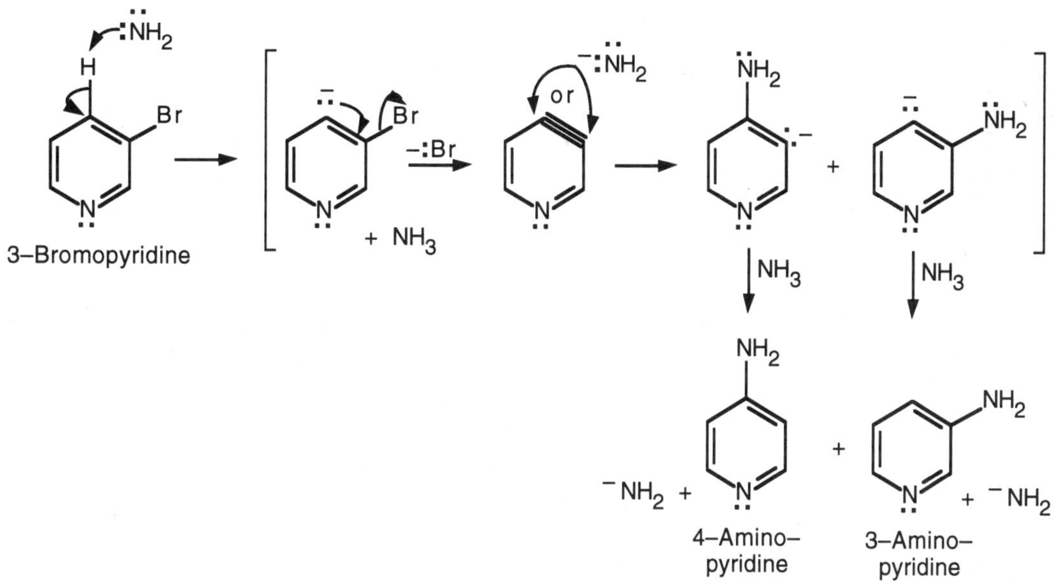

3–Bromopyridine

+ NH$_3$

4–Amino–
pyridine

3–Amino–
pyridine

Reaction of 3–bromopyridine with NaNH$_2$ occurs by a benzyne mechanism. Since $^-$:NH$_2$ can add to either end of the triple bond of the benzyne intermediate, a mixture of products is formed.

29.8

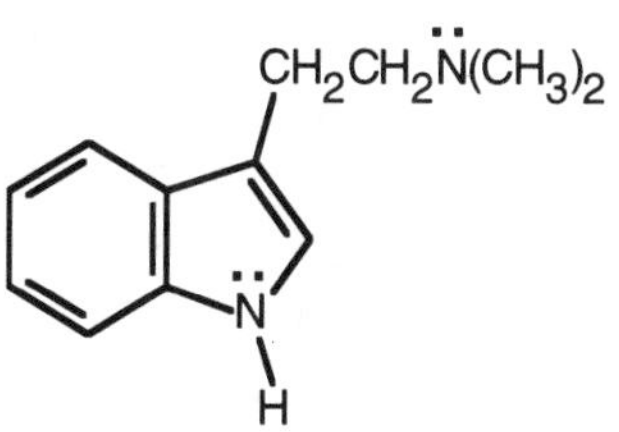

The aliphatic nitrogen atom of *N,N*–dimethyltryptamine is more basic than the ring nitrogen atom because its lone electron pair is more available for donation to a Lewis acid. The aromatic nitrogen electron lone pair is part of the ring π electron system.

29.9

C2 attack:

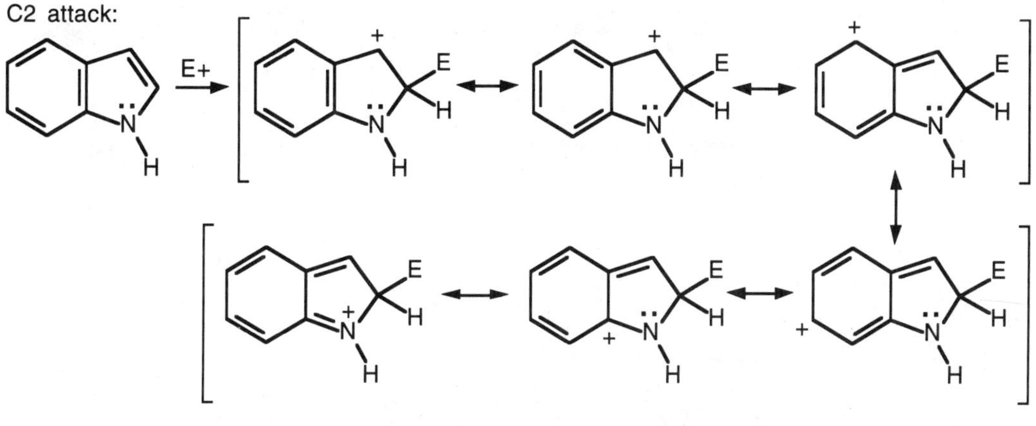

C3 attack:

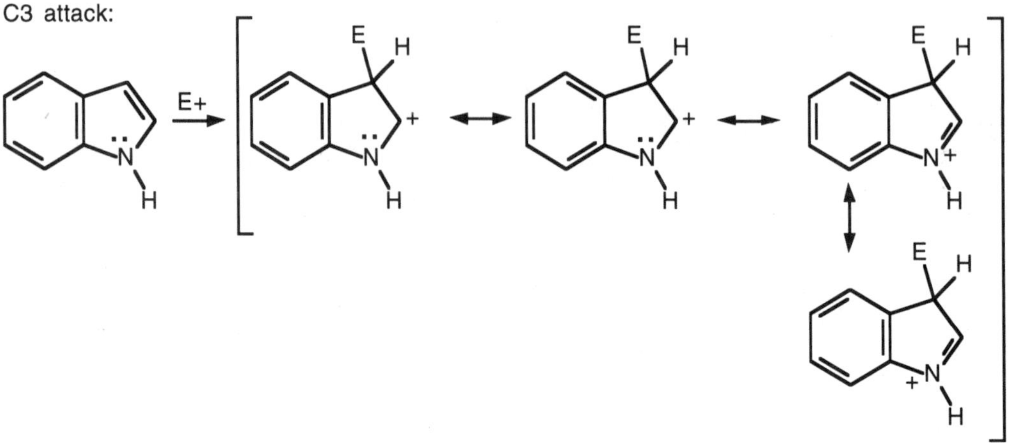

Positive charge can be stabilized by the nitrogen lone pair electrons in both C2 and C3 attack. In C2 attack, however, stabilization by nitrogen destroys the aromaticity of the fused benzene ring. Reaction at C3 is favored, even though the cationic intermediate has fewer resonance forms, because the aromaticity of the six-membered-ring is preserved in the most favored resonance form.

29.10

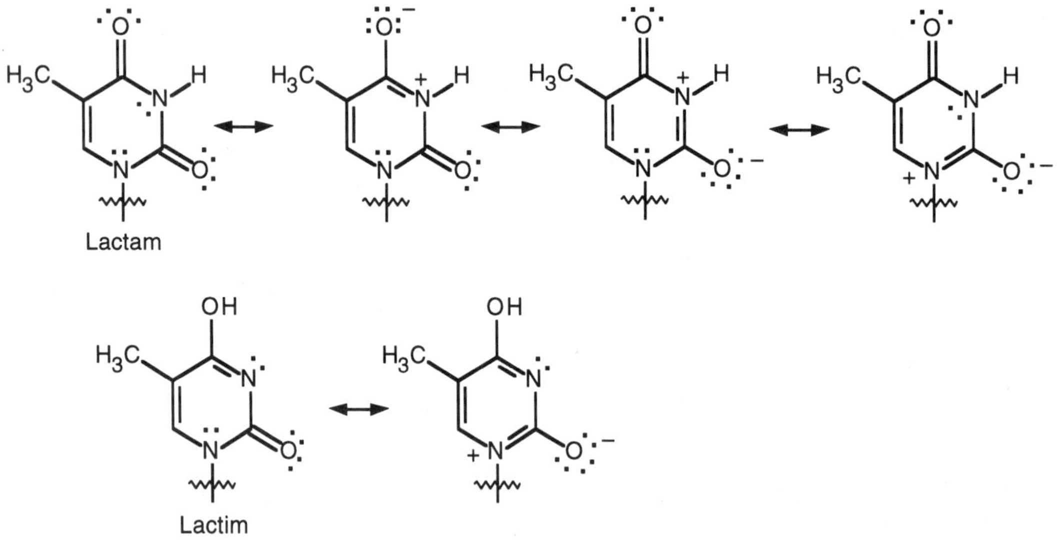

Lactam

Lactim

The lactam form of 2'–deoxythymidine has greater resonance stabilization than the lactim form.

29.11

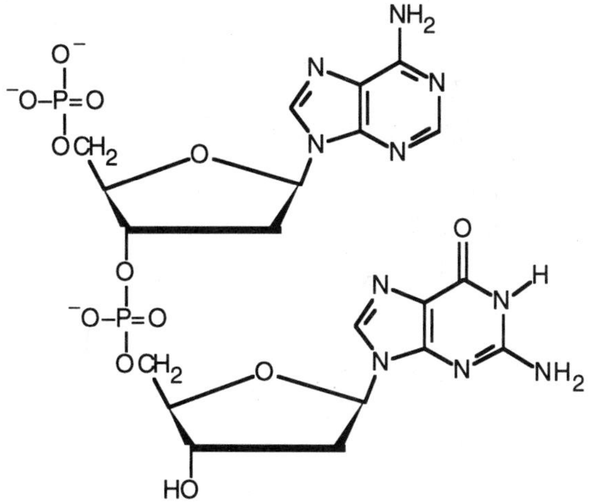

2'–Deoxyadenosine 5'-phosphate (A)

2'–Deoxyguanosine 5'-phosphate (G)

29.12

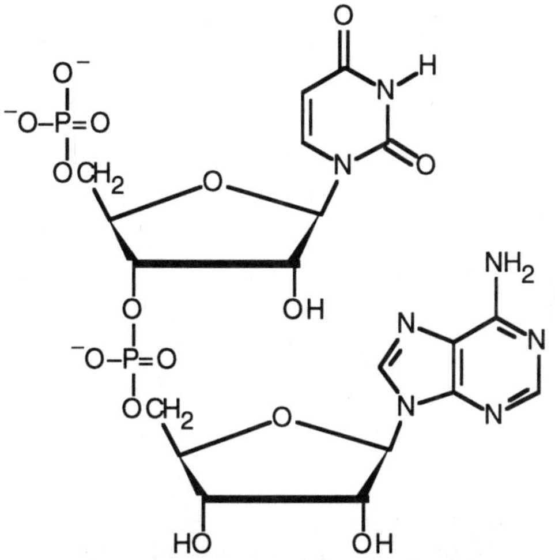

Uridine 5'-phosphate (U)

Adenosine 5'-phosphate (A)

29.13 DNA G–G–C–T–A–A–T–C–C–G–T is complementary to
DNA C–C–G–A–T–T–A–G–G–C–A

29.14

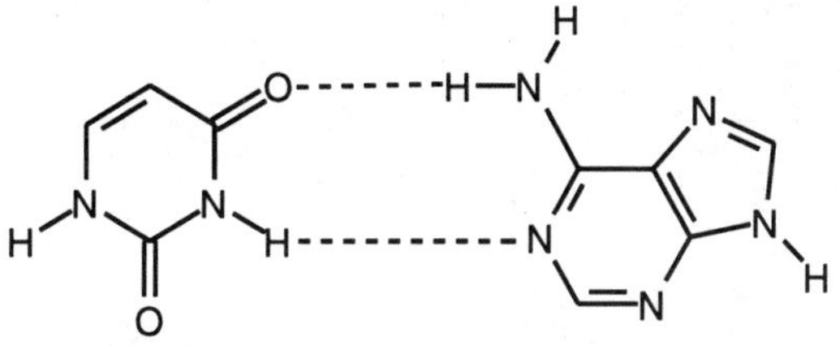

Uracil   Adenine

29.15 DNA G–A–T–T--A–C–C–G–T–A   is complementary to
RNA C–U–A–A–U–G–G–C–A–U

29.16–29.17   Several different codons can code for the same amino acid. The corresponding anticodon follows each codon.

Amino acid:	Ala	Phe	Leu	Tyr
Codon sequence/	GCU/CGA	UUU/AAA	UUA/AAU	UAU/AUA
tRNA anticodon:	GCC/CGG	UUC/AAG	UUG/AAC	UAC/AUG
	GCA/CGU		CUU/GAA	
	GCG/CGC		CUC/GAG	
			CUA/GAU	
			CUG/GAC	

29.18–29.20

The mRNA base sequence:	CUU–AUG–GCU–UGG–CCC–UAA
The amino acid sequence:	Leu—Met—Ala—Trp—Pro–(stop)
The tRNA sequence:	GAA UAC CGA ACC GGG AUU
The DNA sequence:	GAA–TAC–CGA–ACC–GGG–ATT

29.21 Remember:

1. Only a few of the many possible splittings occur in each reaction.

2. Cleavage occurs at both sides of the reacting nucleotide.

$$^{32}P–A–A–C–A–T–G–G–C–G–C–T–T–A–T–G–A–C–G–A$$

Reaction   Fragments

a) A
^{32}P
$^{32}P–A$
$^{32}P–A–A–C$
$^{32}P–A–A–C–A–T–G–G–C–G–C–T–T$
$^{32}P–A–A–C–A–T–G–G–C–G–C–T–T–A–T–G$
$^{32}P–A–A–C–A–T–G–G–C–G–C–T–T–A–T–G–A–C–G$
$^{32}P–A–A–C–A–T–G–G–C–G–C–T–T–A–T–G–A–C–G–A$

b) G
$^{32}P–A–A–C–A–T$
$^{32}P–A–A–C–A–T–G$
$^{32}P–A–A–C–A–T–G–G–C$
$^{32}P–A–A–C–A–T–G–G–C–G–C–T–T–A–T$
$^{32}P–A–A–C–A–T–G–G–C–G–C–T–T–A–T–G–A–C$
$^{32}P–A–A–C–A–T–G–G–C–G–C–T–T–A–T–G–A–C–G–A$

c) C
$^{32}P–A–A$
$^{32}P–A–A–C–A–T–G–G$
$^{32}P–A–A–C–A–T–G–G–C–G$
$^{32}P–A–A–C–A–T–G–G–C–G–C–T–T–A–T–G–A$
$^{32}P–A–A–C–A–T–G–G–C–G–C–T–T–A–T–G–A–C–G–A$

d) C + T ^{32}P–A–A
            ^{32}P–A –A –C –A
            ^{32}P–A–A–C–A–T–G–G
            ^{32}P–A–A–C–A–T–G–G–C–G
            ^{32}P–A–A–C–A–T–G–G–C–G–C
            ^{32}P–A–A–C–A–T–G–G–C–G–C–T
            ^{32}P–A–A–C–A–T–G–G–C–G–C–T–T–A
            ^{32}P–A–A–C–A–T–G–G–C–G–C–T–T–A–T–G–A
            ^{32}P–A–A–C–A–T–G–G–C–G–C–T–T–A–T–G–A–C–G–A

29.22

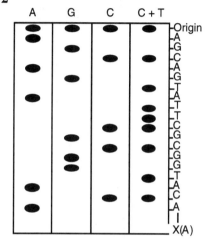

29.23 X–A–T–C–A–G–C–G–A–T–T–C–G–G–T–A–C

29.24

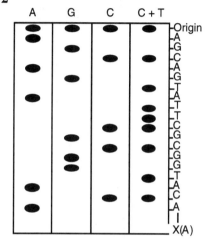

Cleavage of DMT ethers proceeds by an S_N1 mechanism and is rapid because the DMT cation is unusually stable.

29.25

This is an E2 elimination reaction, which proceeds easily because the hydrogen α to the nitrile group is acidic.

29.26 The pyrrole anion, $C_4H_4N:^-$, is a 6 π electron species that has the same electronic structure as the cyclopentadienyl anion. Both of these anions possess the aromatic stability of 6 π electron systems.

29.27

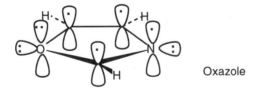

Oxazole

Oxazole is an aromatic 6 π electron heterocycle. Two oxygen electrons and one nitrogen electron are in p orbitals that are part of the π electron system of the ring, along with one electron from each carbon. An oxygen lone pair and a nitrogen lone pair are in sp^2 orbitals that lie in the plane of the ring. Since the nitrogen lone pair is available for donation to acids, oxazole is more basic than pyrrole.

29.28

a)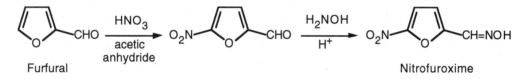

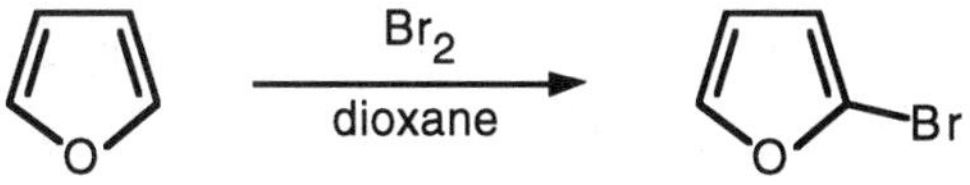

furan $\xrightarrow[\text{dioxane}]{\text{Br}_2}$ 2-bromofuran

b) furan $\xrightarrow[\text{(CH}_3\text{CO}_2)\text{O}]{\text{HNO}_3}$ 2-nitrofuran (NO$_2$)

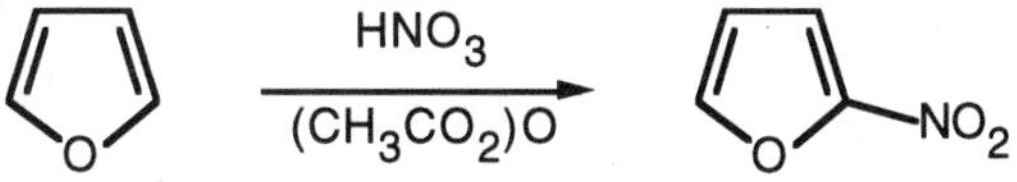

c) furan $\xrightarrow[\text{SnCl}_4]{\text{CH}_3\text{COCl}}$ 2-acetylfuran (CCH$_3$)

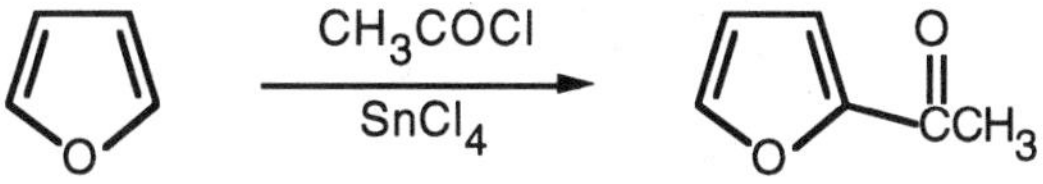

d) furan $\xrightarrow[\text{Pd}]{\text{H}_2}$ tetrahydrofuran

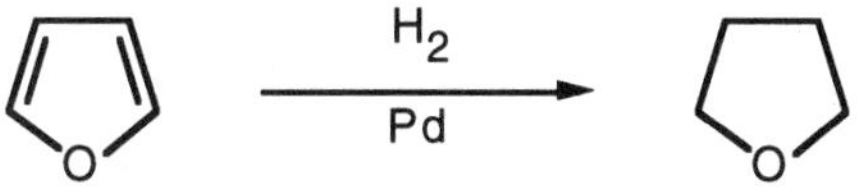

e) furan $\xrightarrow[\text{pyridine}]{\text{SO}_3}$ furan-2-sulfonic acid (SO$_3$H)

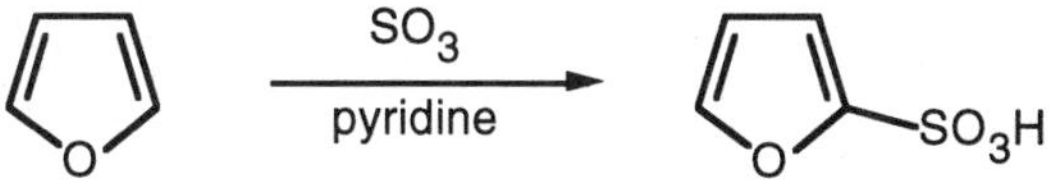

29.29

Furfural —CHO $\xrightarrow[\substack{\text{acetic}\\\text{anhydride}}]{\text{HNO}_3}$ O$_2$N—furan—CHO $\xrightarrow[\text{H}^+]{\text{H}_2\text{NOH}}$ O$_2$N—furan—CH=NOH

Furfural   Nitrofuroxime

29.30

29.31

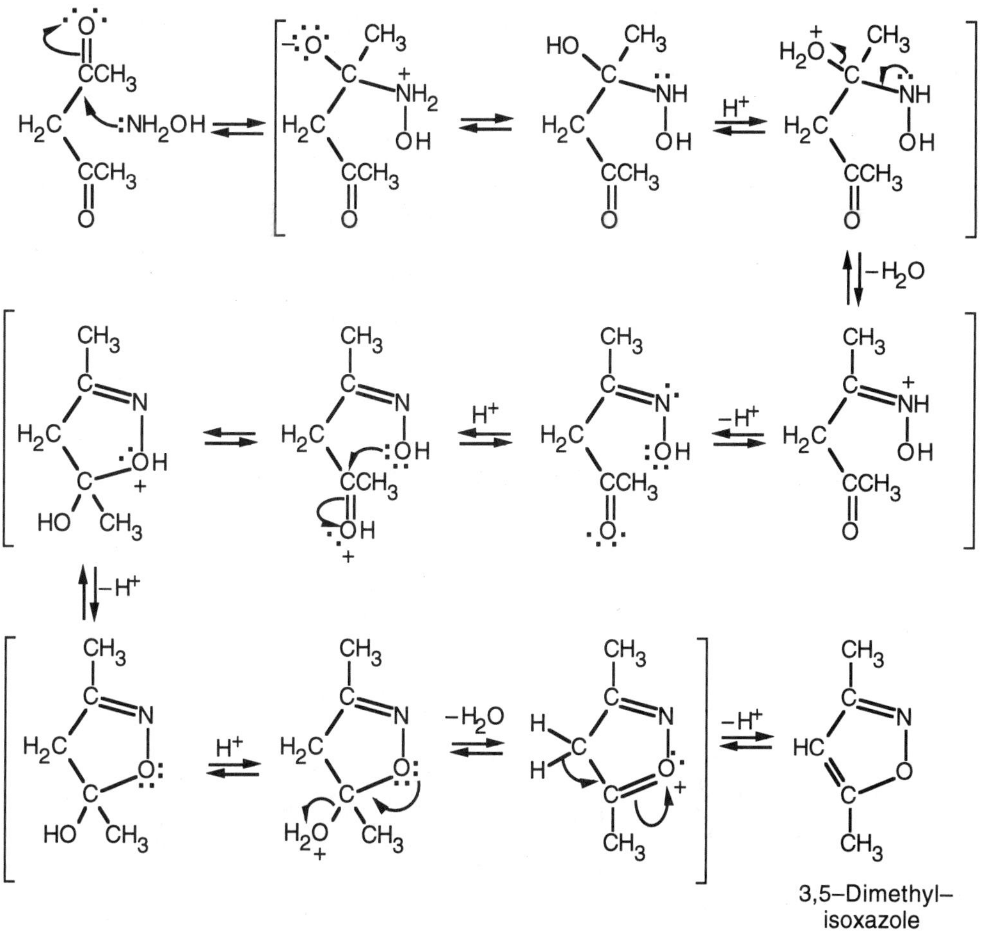

[Reaction scheme showing mechanism of formation of 3,5-Dimethylisoxazole]

3,5–Dimethyl–
isoxazole

29.32 a) A *heterocycle* is a cyclic compound whose ring framework is composed of one or more different elements in addition to carbon.

b) *DNA* is a biological polymer whose monomeric units are nucleotides. A nucleotide is composed of a heterocyclic amine base, deoxyribose, and a phosphate group. DNA is the transmitter of the genetic code of all living organisms.

c) A *base pair* is a specific pair of heterocyclic amine bases that hydrogen-bond to each other in a DNA double helix.

d) *Transcription* is the process by which the genetic message contained in DNA is read by RNA and carried from the nucleus to the ribosomes.

e) *Translation* is the process by which RNA decodes the genetic message and uses the information to synthesize proteins.

f) *Replication* is the enzyme-catalyzed process by which a new molecule of DNA is produced. A DNA double helix separates into two strands, and nucleotide monomers line up and base-pair with each strand. Two new DNA double helices are produced, each one identical to the original DNA helix.

g) A *codon* is a sequence of three mRNA bases that specifies a particular amino acid to be used in protein synthesis.

h) An *anticodon* is a sequence of three tRNA bases that is complementary to the codon sequence. Each tRNA covalently bonds to a specific amino acid and brings that amino acid into correct position for protein synthesis by hydrogen-bonding to its codon.

29.33–29.35

mRNA codon:	a) AAU	b) GAG	c) UCC	d) CAU
Amino acid:	Asn	Glu	Ser	His
DNA sequence:	TTA	CTC	AGG	GTA
tRNA anticodon:	UUA	CUC	AGG	GUA

29.36–29.37   UAC is a codon for tyrosine. It was transcribed from ATG of the DNA chain.

mRNA codon   DNA

29.38 Tyr——Gly——Gly——Phe——Met (stop) is coded by

(UAC)	(GGU)	(GGU)	(UUU)	(AUG)	(UAA)
(UAU)	(GGC)	(GGC)	(UUC)		(UAG)
	(GGA)	(GGA)			(UGA)
	(GGG)	(GGG)			

29.39 Angiotensin II: Asp——Arg——Val——Tyr——Ile——His——Pro——Phe (stop)

mRNA sequence:

GAU	CGU	GUU	UAU	AUU	CAU	CCU	UUU	UAA
GAC	CGC	GUC	UAC	AUC	CAC	CCC	UUC	UAG
	CGA	GUA		AUA		CCA		UGA
	CGG	GUG				CCG		
	AGA							
	AGG							

29.40 mRNA sequence: CUA—GAC—CGU—UCC—AAG—UGA
Amino Acid: Leu——Asp—Arg—Ser—Lys (stop)

29.41

The cyclization is an electrophilic aromatic substitution.

29.42

Conjugate addition of aniline to the α,β-unsaturated aldehyde is followed by an internal electrophilic aromatic substitution.

29.43 1. First, protect the nucleotides.

   a) Bases are protected by amide formation.

Cytosine

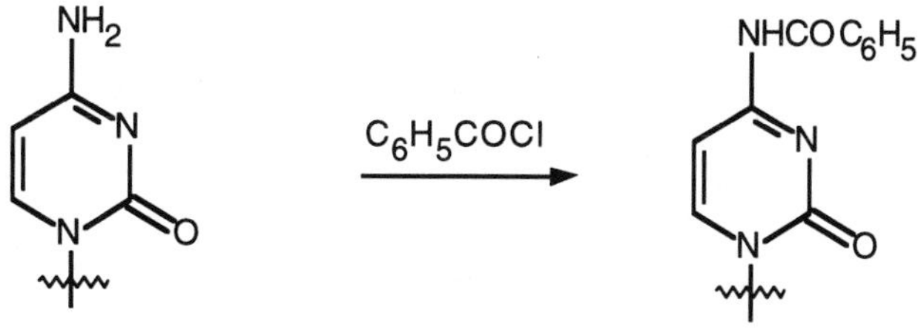
NH2
N
N
O
C6H5COCl
NHCOC6H5
N
N
O

Guanine

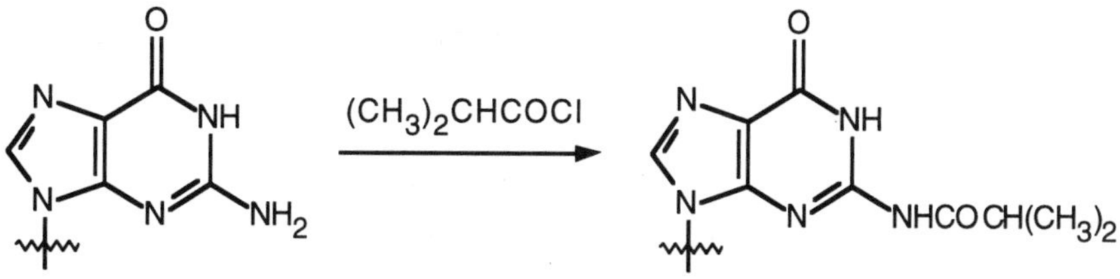
O
N
NH
N
N
NH2
(CH3)2CHCOCl
O
N
NH
N
N
NHCOCH(CH3)2

Thymine does not need to be protected.

b) The 5' hydroxyl group is protected as its *p*–dimethoxytrityl (DMT) ether.

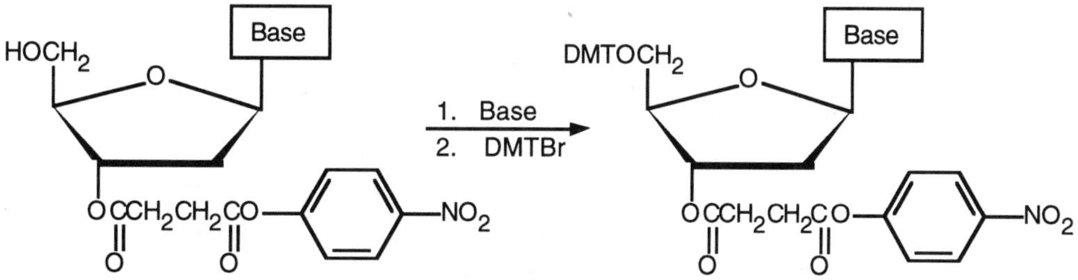
HOCH2
Base
O
OCCH2CH2CO
O
O
NO2
1. Base
2. DMTBr
DMTOCH2
Base
O
OCCH2CH2CO
O
O
NO2

2. Attach a protected 2–deoxycytidine nucleoside to the polymer support.

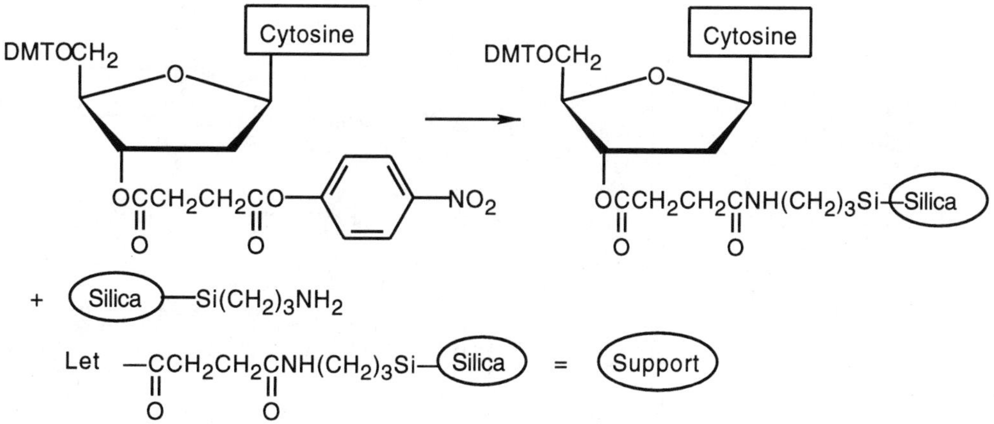
DMTOCH2
Cytosine
O
OCCH2CH2CO
O
O
NO2
DMTOCH2
Cytosine
O
OCCH2CH2CNH(CH2)3Si
Silica
O
O
+
Silica
Si(CH2)3NH2
Let
CCH2CH2CNH(CH2)3Si
Silica
O
O
=
Support

3. Cleave the DMT ether.

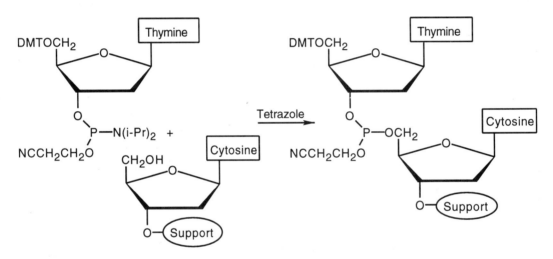

4. Couple protected 2'–deoxythymidine to the polymer–2'–deoxycytidine. (The nucleosides have a phosphoramidite group at the 3' position.)

5. Oxidize the phosphite product to a phosphate triester, using iodine.

6. Repeat steps 3–5 with protected 2'–deoxyadenosine and protected 2'–deoxyguanosine.

7. Cleave all protecting groups with aqueous ammonia to yield the desired sequence.

29.44

Reaction with nitrous acid diazotizes cytosine, and subsequent reaction with water, followed by tautomerization, yields uracil.

29.45

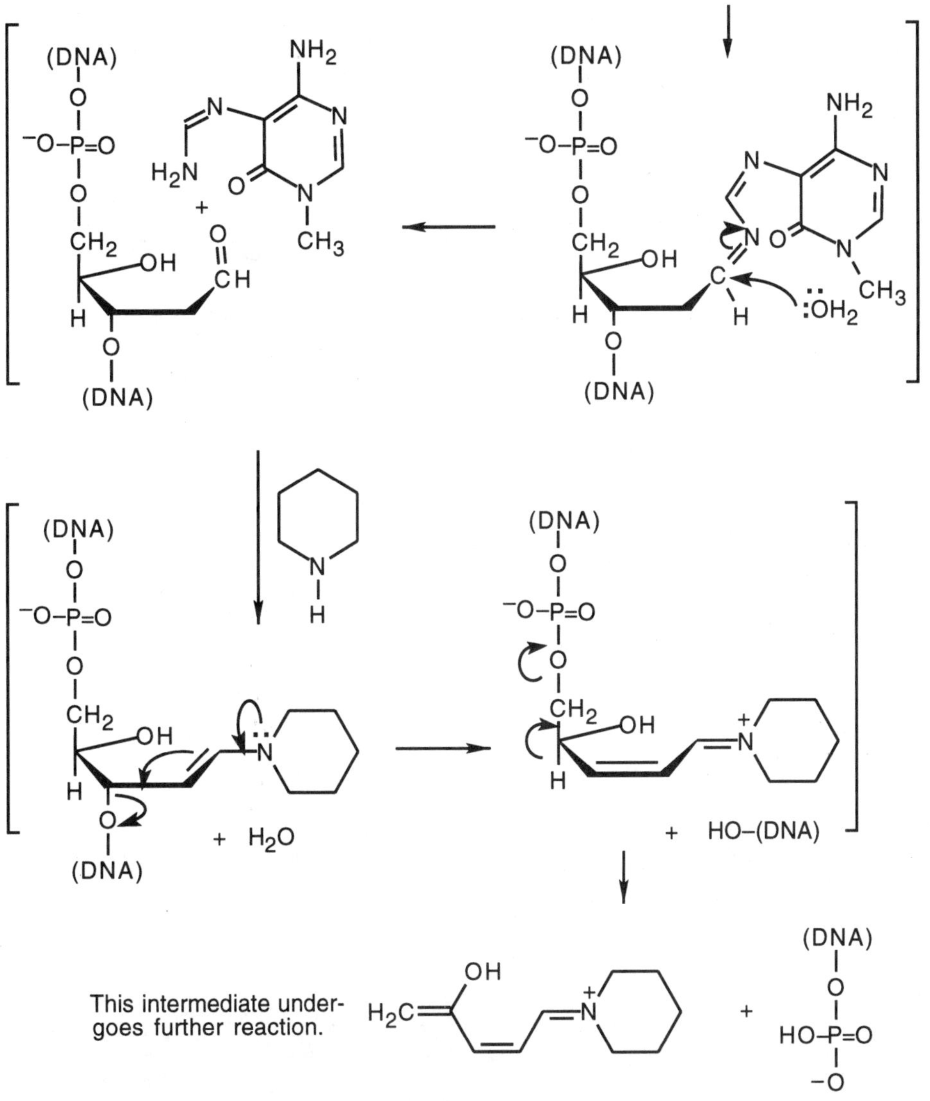

Study Guide for Chapter 29

After studying this chapter, you should be able to:

(1)   Draw orbital pictures of heterocycles and explain their acid-base properties (29.1, 29.2, 29.4, 29.8, 29.26, 29.27).

(2)   Explain orientation and reactivity in aromatic heterocycle reactions (29.5, 29.6, 29.7, 29.9).

(3)   Predict the products of reactions of heterocycles (29.28, 29.29).

(4)   Formulate mechanisms of reactions of heterocycles (29.3, 29.31, 29.41, 29.42, 29.44).

(5)   Draw purines, pyrimidines, nucleosides, nucleotides, and representative segments of DNA (29.10, 29.11, 29.12, 29.36, 29.37).

(6)   Given a DNA or RNA strand, draw its complementary strand (29.13, 29.15, 29.20, 29.34).

(7)   List the codon sequence for a given amino acid or peptide (and *vice versa*) (29.16, 29.33, 29.38)

(8)   Deduce an amino acid sequence from a given mRNA base sequence (and *vice versa*) (29.18, 29.39, 29.40).

(9)   Draw the anticodon sequence of tRNA, given the mRNA sequence (29.17, 29.19, 29.35).

(10)   Outline the process of DNA sequencing, and deduce a DNA sequence from an electrophoresis pattern (29.21, 29.22, 29.23).

(11)   Outline the method of DNA synthesis, and formulate mechanisms of synthetic steps (29.24, 29.25, 29.43, 29.45).

(12)   Define the important terms in this chapter (29.32).

Chapter 30 – The Organic Chemistry of Metabolic Pathways

30.1

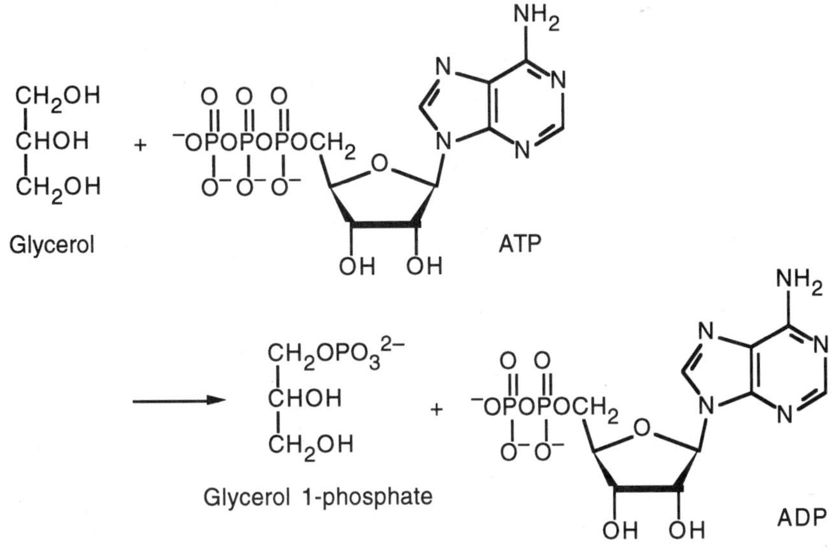

Glycerol + ATP $\longrightarrow$ Glycerol 1-phosphate + ADP

30.2

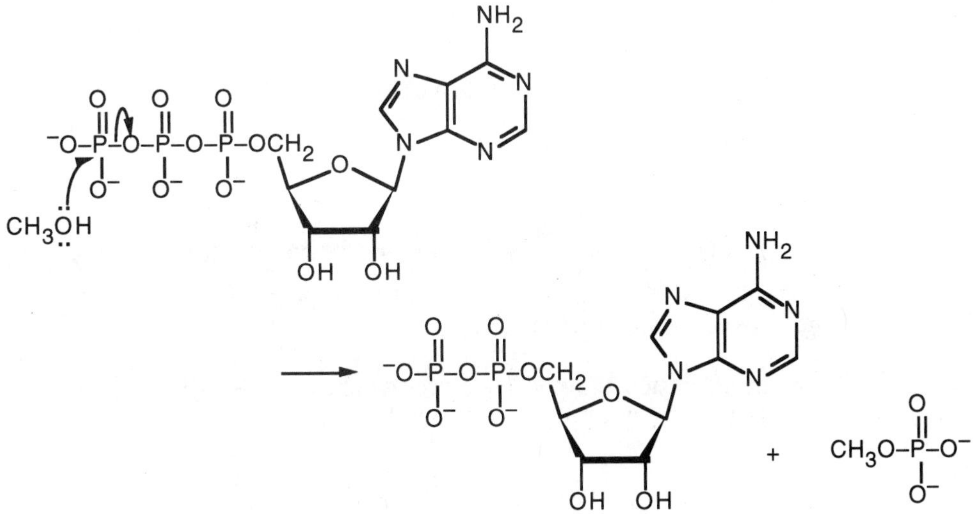

30.3

$$CH_3CH_2-CH_2CH_2-CH_2CH_2-CH_2\overset{\displaystyle O}{\overset{\|}{C}}SCoA$$

Caprylyl CoA

↓ (turn 4)

$$CH_3CH_2-CH_2CH_2-CH_2\overset{\displaystyle O}{\overset{\|}{C}}SCoA \ + \ CH_3\overset{\displaystyle O}{\overset{\|}{C}}SCoA$$

Hexanoyl CoA

↓ (turn 5)

$$CH_3CH_2-CH_2\overset{\displaystyle O}{\overset{\|}{C}}SCoA \ + \ CH_3\overset{\displaystyle O}{\overset{\|}{C}}SCoA$$

Butyryl CoA

↓ (turn 6)

$$CH_3\overset{\displaystyle O}{\overset{\|}{C}}SCoA \ + \ CH_3\overset{\displaystyle O}{\overset{\|}{C}}SCoA$$

30.4

a)

$$CH_3CH_2\overset{\downarrow}{-}CH_2CH_2\overset{\downarrow}{-}CH_2CH_2\overset{\downarrow}{-}CH_2CH_2\overset{\downarrow}{-}CH_2CH_2\overset{\downarrow}{-}CH_2CH_2\overset{\downarrow}{-}CH_2CH_2\overset{\downarrow}{-}CH_2\overset{\displaystyle O}{\overset{\|}{C}}OH$$

↓

$$8 \ CH_3\overset{\displaystyle O}{\overset{\|}{C}}SCoA$$

Seven turns of the spiral are needed.

b)

$$CH_3CH_2-(CH_2CH_2)_8-CH_2\overset{\displaystyle O}{\overset{\|}{C}}OH \longrightarrow 10 \ CH_3\overset{\displaystyle O}{\overset{\|}{C}}SCoA$$

Nine turns of the spiral are needed.

30.5 ATP is produced in step 7 (3-phosphoglyceroyl phosphate —> 3-phosphoglycerate) and in step 10 (phosphoenolpyruvate —> pyruvate).

30.6

$$2^-O_3POCH_2\overset{\displaystyle O}{\overset{\|}{C}}CH_2OH \longrightarrow \left[2^-O_3POCH_2\overset{\displaystyle HO}{\overset{|}{C}}=\overset{\displaystyle OH}{\overset{|}{C}}H \right] \longrightarrow 2^-O_3POCH_2\overset{\displaystyle HO}{\overset{|}{C}}H\overset{\displaystyle O}{\overset{\|}{C}}H$$

Enol

Enzyme-catalyzed enolization is followed by formation of glyceraldehyde 3-phosphate.

30.7

Type of Reaction	Step in Glycolysis
Phosphate transfer from ATP	1, 3
Isomerization	2, 5, 8
Retro-aldol condensation	4
Oxidation	6
Nucleophilic acyl substitution	6
Phosphate transfer to ADP	7,10
E2 Dehydration	9

30.8

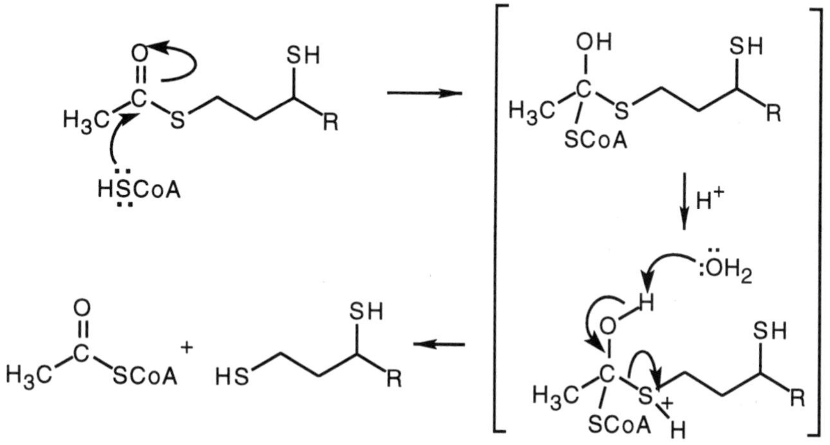

30.9   Citrate and isocitrate are tricarboxylic acids.

30.10

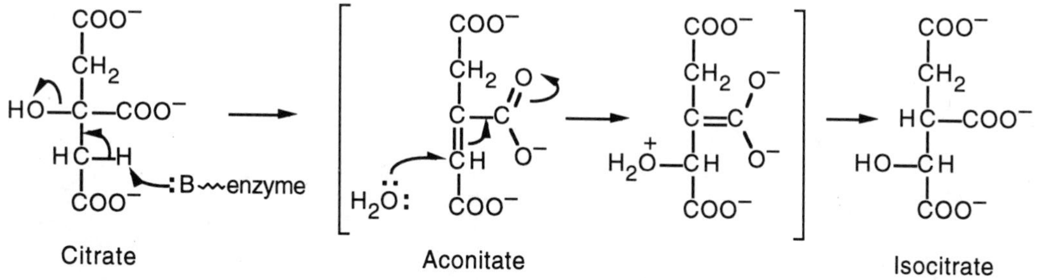

Citrate   Aconitate   Isocitrate

Enzyme-catalyzed E2 elimination of H_2O is followed by nucleophilic conjugate addition of water to produce isocitrate.

30.11 The first step in the conversion of succinyl CoA to succinate is a nucleophilic acyl substitution by phosphate on succinyl CoA to form a mixed anhydride.

Nucleophilic attack by GDP on phosphate results in formation of GTP and succinate.

R = GMP

30.12

30.13

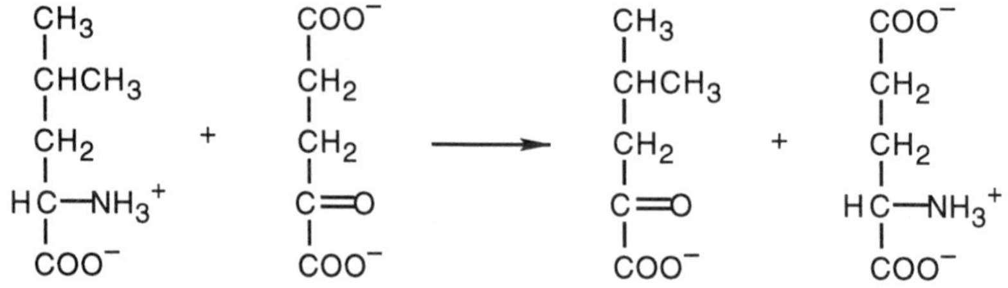

Leucine   α-Ketoglutarate   Glutamate

30.14 You might recognize that β-hydroxybutyryl ACP resembles the β-hydroxy ketones that were described in Chapter 23. These compounds dehydrate readily under both acidic and basic conditions. In this problem, the mechanism will be worked out using base catalysis.

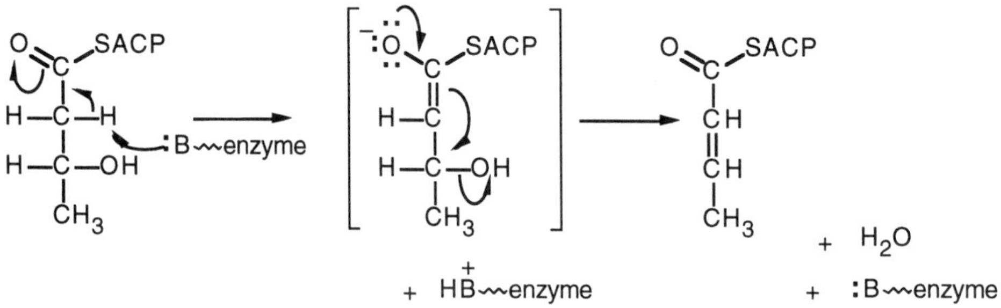

30.15 A fatty acid synthesized from $^{14}CH_3COOH$ has an alternating labeled and unlabeled carbon chain. The carboxylic acid carbon is unlabeled.

$$\overset{*}{C}H_3\overset{}{C}H_2\overset{*}{C}H_2\overset{}{C}H_2\overset{*}{C}H_2\overset{}{C}H_2\overset{*}{C}H_2\overset{}{C}H_2\overset{*}{C}H_2\overset{}{C}H_2\overset{*}{C}H_2\overset{}{C}H_2\overset{*}{C}H_2COOH$$

30.16

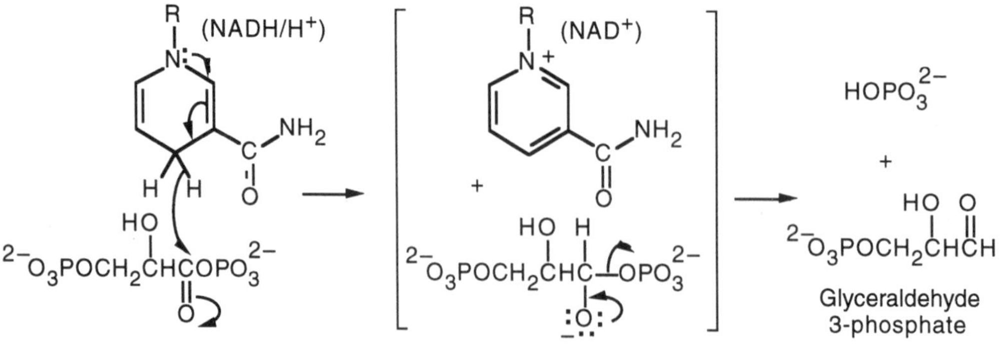

Glyceraldehyde
3-phosphate

30.17 Digestion is the breakdown of bulk food in the stomach and small intestine. Hydrolysis of amide, ester and acetal bonds yields amino acids, fatty acids, and simple sugars.

30.18 Metabolism refers to all reactions that take place inside cells. Digestion is a part of metabolism in which food is broken down into small organic molecules.

30.19 Metabolic processes that break down large molecules are known as catabolism. Metabolic processes that assemble larger biomolecules from smaller ones are known as anabolism.

30.20

AMP

30.21

Cyclic AMP

30.22 ATP transfers a phosphate group to another molecule in anabolic reactions.

30.23 NAD$^+$ is a biochemical oxidizing agent that converts alcohols to aldehydes or ketones, yielding NADH and H$^+$ as byproducts.

30.24 FAD is an oxidizing agent that introduces a conjugated double bond into a biomolecule, yielding FADH$_2$ as a byproduct.

30.25 The exact reverse of an energetically favorable reaction is energetically unfavorable. Since glycolysis is energetically favorable (negative ΔG), its exact reverse has a positive ΔG and is energetically unfavorable. Instead, glucose is synthesized by an alternate pathway that also has a negative ΔG.

30.26

NAD$^+$ is needed to convert lactate to pyruvate.

30.27 a) One mole of glucose is catabolized to two moles of pyruvate, each of which yields one mole of acetyl CoA. Thus,

$$1.0 \text{ mol glucose} \longrightarrow 2.0 \text{ mol acetyl CoA}$$

b) A fatty acid with n carbons yields $n/2$ moles of acetyl CoA per mole of fatty acid. For palmitic acid ($C_{15}H_{31}COOH$),

$$1.0 \text{ mol palmitic acid} \times \frac{8 \text{ mol acetyl CoA}}{1 \text{ mol palmitic acid}} \longrightarrow 8.0 \text{ mol acetyl CoA}$$

c) Maltose is a disaccharide that yields two moles of glucose on hydrolysis. Since each mole of glucose yields two moles of acetyl CoA,

$$1.0 \text{ mol maltose} \longrightarrow 2.0 \text{ mol glucose} \longrightarrow 4.0 \text{ mol acetyl CoA}$$

30.28

	a) Glucose	b) Palmitic acid	c) Maltose
Molecular weight	180.2 amu	256.4 amu	342.3 amu
Moles in 100.0 g	0.5549 mol	0.3900 mol	0.2921 mol
Moles of acetyl CoA produced	2 x 0.5549 mol =1.110 mol	8 x 0.3900 mol = 3.120 mol	4 x 0.2921 mol = 1.168 mol
Grams acetyl CoA produced	898.6 grams	2526 grams	945.6 grams

30.29 Palmitic acid is the most efficient precursor of acetyl CoA on a weight basis.

30.30

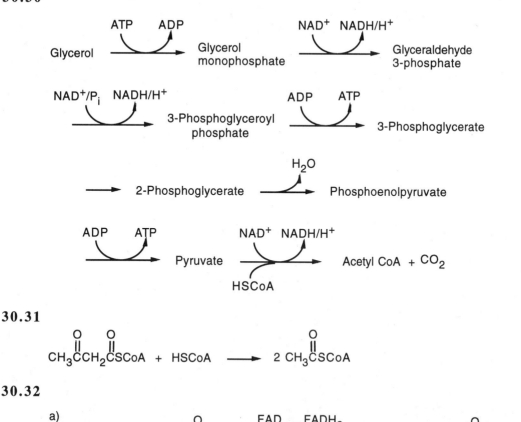

30.31

$$CH_3\overset{O}{\underset{||}{C}}CH_2\overset{O}{\underset{||}{C}}SCoA + HSCoA \longrightarrow 2\ CH_3\overset{O}{\underset{||}{C}}SCoA$$

30.32

a)

$$CH_3CH_2CH_2CH_2CH_2\overset{O}{\underset{||}{C}}SCoA \xrightarrow[\substack{Acetyl\ CoA \\ dehydrogenase}]{FAD \quad FADH_2} CH_3CH_2CH_2CH=CH\overset{O}{\underset{||}{C}}SCoA$$

b)

$$CH_3CH_2CH_2CH=CH\overset{O}{\underset{||}{C}}SCoA + H_2O \xrightarrow[\substack{hydratase}]{Enoyl\ CoA} CH_3CH_2CH_2\overset{OH}{\underset{|}{C}}HCH_2\overset{O}{\underset{||}{C}}SCoA$$

c)

$$CH_3CH_2CH_2\overset{OH}{\underset{|}{C}}HCH_2\overset{O}{\underset{||}{C}}SCoA \xrightarrow[\substack{\beta-Hydroxyacyl\ CoA \\ dehydrogenase}]{NAD^+ \quad NADH/H^+} CH_3CH_2CH_2\overset{O}{\underset{||}{C}}CH_2\overset{O}{\underset{||}{C}}SCoA$$

30.33

Amino acid	α–Keto acid

a)

$$\overset{+NH_3}{CH_3CHCHCOO^-} \quad | \quad CH_3$$

$$\overset{O}{CH_3CHCCOO^-} \quad | \quad CH_3$$

b)

$$\text{(phenyl)}-CH_2\overset{+NH_3}{CHCOO^-}$$

$$\text{(phenyl)}-CH_2\overset{O}{CCOO^-}$$

c)

$$CH_3SCH_2CH_2\overset{+NH_3}{CHCOO^-}$$

$$CH_3SCH_2CH_2\overset{O}{CCOO^-}$$

30.34 (a) Pyridoxal phosphate is the cofactor associated with transamination.
(b) Biotin is the cofactor associated with carboxylation of a ketone.
(c) Thiamine is the cofactor associated with decarboxylation of an α-keto acid.

30.35 As we saw in Section 30.1, formation of glucose 6-phosphate from glucose and ATP is energetically favorable (negative ΔG). The reverse reaction, transfer of a phosphate group to ADP from glucose 6-phosphate, is energetically unfavorable and doesn't occur spontaneously. Phosphate transfers to ADP from either 3-phosphoglyceroyl phosphate or phosphoenolpyruvate have negative ΔG values and are energetically favorable reactions.

In chemical terms, the leaving groups in the reactions of 3-phosphoglyceroyl phosphate and phosphoenolpyruvate are more stable anions than the leaving group in the reaction of glucose, so the reactions are more favorable.

$$RO-\overset{O}{\underset{O^-}{P}}-O^- \longrightarrow RO^- + {}^-O-\overset{O}{\underset{O^-}{P}}-Nu$$

$$:Nu$$

Glucose	3-Phosphoglyceroyl phosphate	Phosphoenolpyruvate

$RO^- =$

CHO	CO_2^-	CHO
HCOH	CHOH	C—O$^-$
HOCH	$CH_2OPO_3^{2-}$	CH$_2$
HCOH		
HCOH		
CH$_2$O$^-$		

30.36

Ribulose 5-phosphate   Enol   Ribose 5-phosphate

The isomerization of ribulose 5-phosphate to ribose 5-phosphate occurs by way of an intermediate enol.

30.37 This is a reverse aldol reaction, similar to step 4 of glycolysis.

30.38 The steps in the conversion α-ketoglutarate —> succinyl CoA are analogous to steps in the conversion pyruvate —> acetyl CoA (shown in Figure 30.5), and the same coenzymes are involved: lipoamide, thiamine, HSCoA and NAD⁺.

30.39

30.40

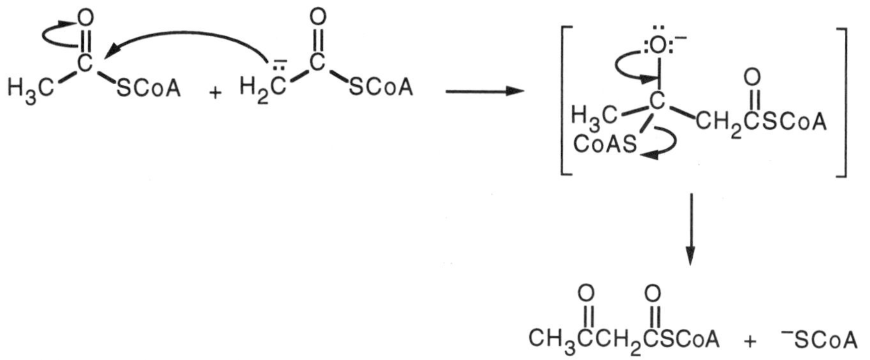

The reaction is a Claisen condensation.

30.41 The first sequence of steps in this mechanism involves formation of the imine (Schiff base) of sedoheptulose 7-phosphate, followed by retro-aldol cleavage to form erythrose 4-phosphate and the Schiff base of dihydroxyacetone.

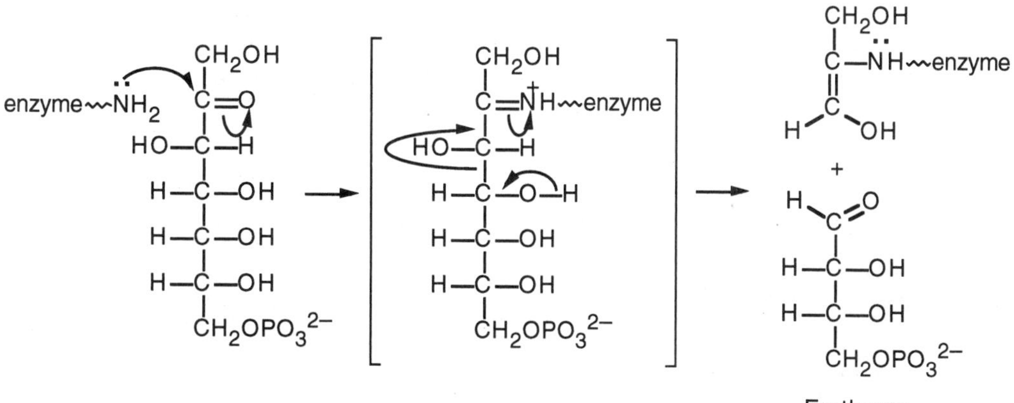

Erythrose
4-phosphate

The Schiff base of dihydroxyacetone undergoes an aldol-like condensation with glyceraldehyde 3-phosphate to yield fructose 6-phosphate. This reaction is almost identical to the reaction pictured for Step 8 of gluconeogenesis in Section 30.8.

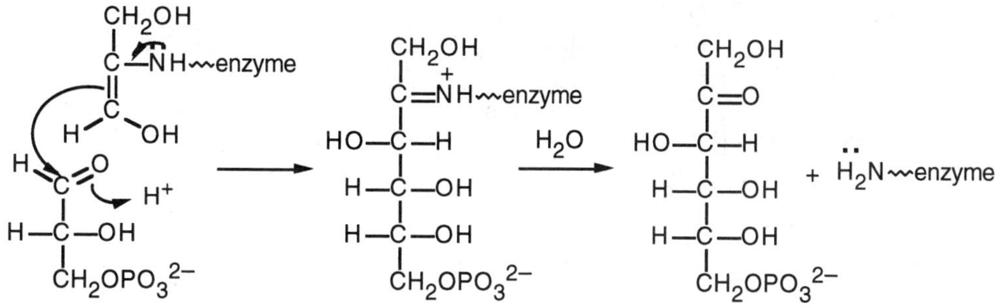

Fructose
6-phosphate

30.42

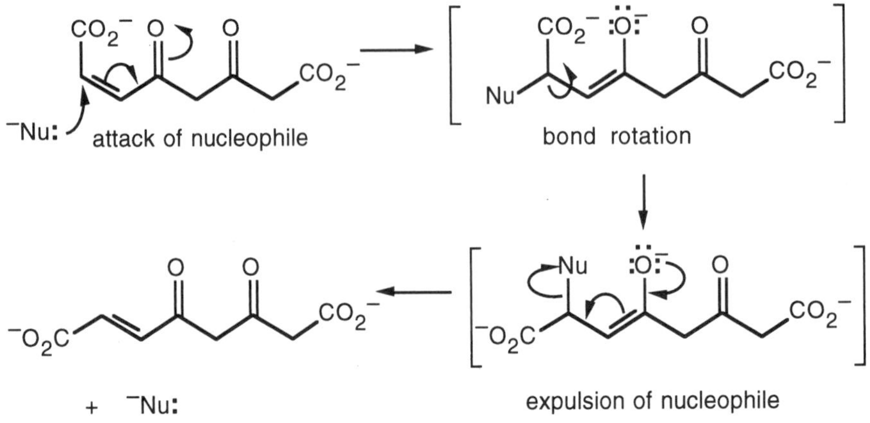

attack of nucleophile

bond rotation

expulsion of nucleophile

+ ⁻Nu:

30.43

nucleophilic addition of enzyme

retro-Claisen condensation

hydrolysis of enzyme-
substrate complex

+   ⁻:B∿enzyme

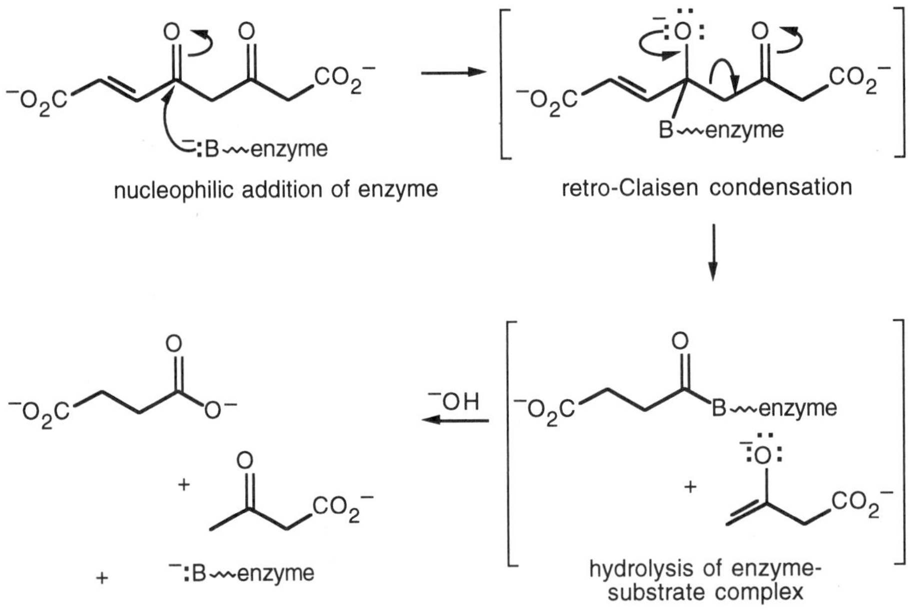

30.44 The first step in the conversion acetoacetate ——> acetyl CoA is the formation of
acetoacetyl CoA. This reaction also occurs as the first step in fatty acid degradation.
Although we haven't studied the mechanism, it involves formation of a mixed carboxylic-
phosphoric anhydride.

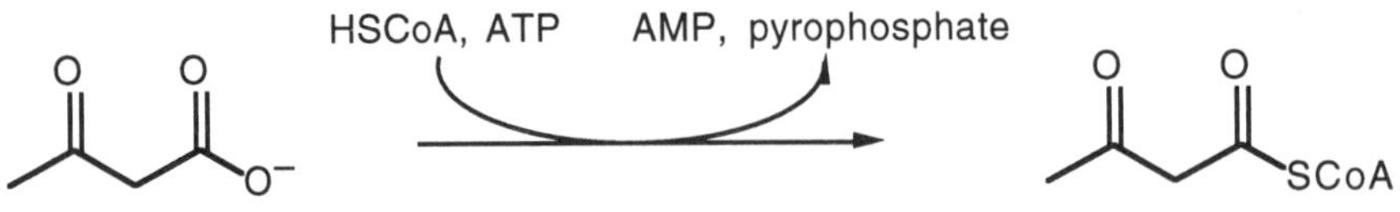

HSCoA, ATP   AMP, pyrophosphate

The final step is a retro-Claisen reaction, whose mechanism is pictured in Section 30.2 as Step 4 of β-oxidation of fatty acids.

30.45 Now is a good time to use retrosynthetic analysis, which you first encountered in Chapter 8. In this degradative pathway, what might be the precursor to acetyl CoA (the final product)? Pyruvate is a good guess, because we learned how to convert pyruvate to acetyl CoA in Section 30.4. How do we get from serine to pyruvate? A transamination reaction is a possibility. However, the immediate transamination precursor to pyruvate is the amino acid alanine, which differs from serine by one hydroxyl group. Thus, we probably have to design a pathway from serine to pyruvate that takes this difference into account.

Many routes are possible, but here's the simplest:

The coenzymes thiamine pyrophosphate and lipoamide are involved in the last step.

30.46

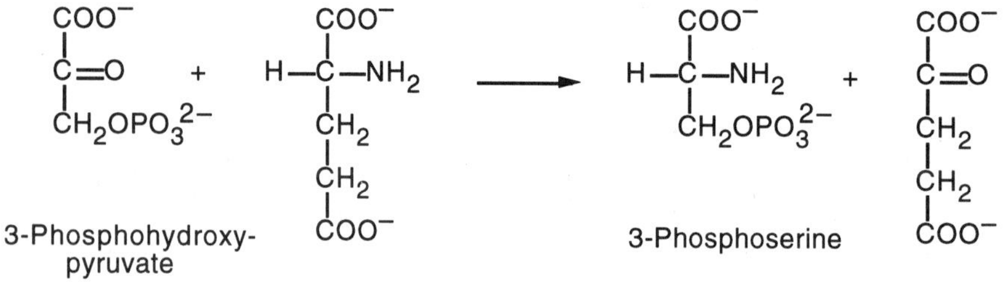

3-Phosphohydroxy-
pyruvate

3-Phosphoserine

This reaction is a transamination that requires the coenzyme pyridoxal phosphate as a cofactor. The mechanism, which is described in Figure 30.7 and Problem 30.12, involves two steps. The first step is the nucleophilic addition of glutamate nitrogen to the aldehyde group of pyridoxal phosphate to yield an imine intermediate, which is hydrolyzed to give α-ketoglutarate plus a nitrogen-containing pyridoxal phosphate byproduct.

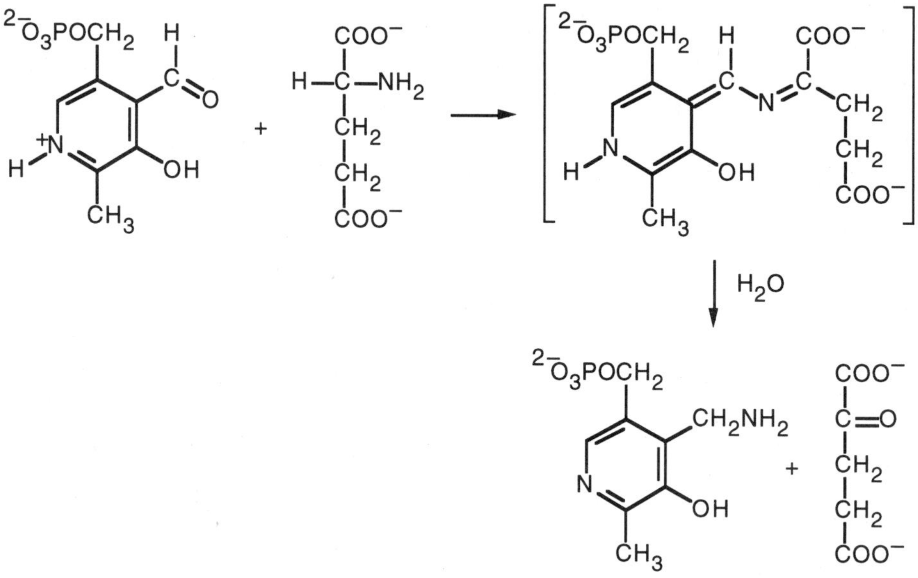

This byproduct reacts with 3-phosphohydroxypyruvate to give 3-phosphoserine plus regenerated pyridoxal phosphate.

30.47

Study Guide for Chapter 30

After studying this chapter, you should be able to:

(1)   Explain the basic concepts of metabolism (30.17, 30.18, 30.19, 30.20, 30.21, 30.22, 30.23, 30.24, 30.34).

(2)   Understand the energy relationships of biochemical reactions (30.25, 30.35).

(3)   Answer questions relating to metabolic pathways of:
(a) Carbohydrates (30.5, 30.7, 30.9, 30.26, 30,27, 30.28, 30.30)
(b) Fatty acids (30.3, 30.4, 30.15, 30.27, 30.28, 30.29, 30.31, 30.32)
(c) Amino acids (30.13, 30.33)

(4)   Formulate reaction mechanisms for:
(a) General biochemical processes (30.1, 30.2, 30.42)
(b) Carbohydrate metabolism (30.6, 30.8, 30.10, 30.11, 30.16, 30.35, 30.36, 30.37, 30.38, 30.41)
(c) Fatty acid metabolism (30.14, 30.39, 30.40, 30.43, 30.44)
(d) Amino acid metabolism (30.12, 30.46)
(e) Combined metabolic pathways (30.43, 30.45, 30.47)

31.1

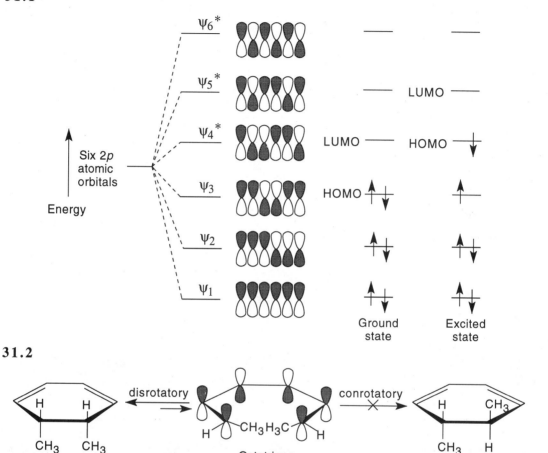

31.2

The symmetry of the octatriene HOMO predicts that ring closure will occur by a disrotatory path and that only *cis* product will be formed.

31.3   Note: *Trans–3,4–dimethylcyclobutene* is chiral; the *S,S* enantiomer will be used for this argument.

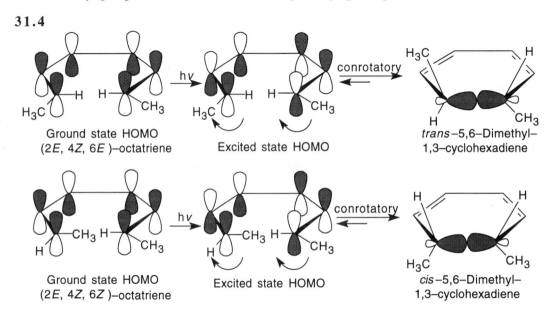

Conrotatory ring opening of *trans*–3,4–dimethylcyclobutene can occur in either a clockwise or a counterclockwise manner. Clockwise opening (path **A**) yields the *E,E* isomer; counterclockwise opening (path **B**) yields the *Z,Z* isomer. Production of (2 *Z*,4 *Z*)–hexadiene is disfavored because of unfavorable steric interactions between the methyl groups in the transition state leading to ring-opened product.

31.4

Ground state HOMO
(*2E*, *4Z*, *6E*)–octatriene

Excited state HOMO

trans–5,6–Dimethyl–
1,3–cyclohexadiene

Ground state HOMO
(*2E*, *4Z*, *6Z*)–octatriene

Excited state HOMO

cis–5,6–Dimethyl–
1,3–cyclohexadiene

Photochemical electrocyclic reactions of 6 π electron systems always occur in a conrotatory manner.

31.5

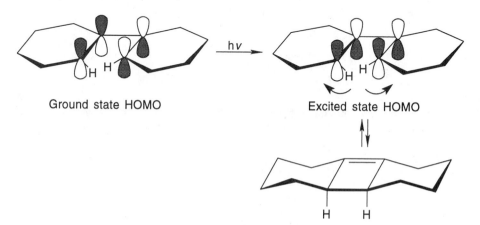

The diene can cyclize by either of two conrotatory paths to form cyclobutenes $\underline{A}$ and $\underline{B}$. Using $\underline{B}$ as an example:

Opening of each cyclobutene ring can occur by either of two conrotatory routes to yield the isomeric dienes.

31.6   A photochemical electrocyclic reaction involving two electron pairs proceeds in a *disrotatory* manner (Table 31.1).

Ground state HOMO   Excited state HOMO

The two hydrogen atoms in the four-membered ring are *cis* to each other in the cyclobutene product.

31.7

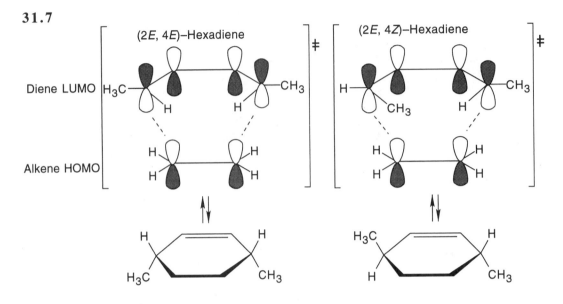

The Diels-Alder reaction is a thermal [4+2] cycloaddition, which occurs with suprafacial geometry. The stereochemistry of the diene is maintained in the product.

31.8

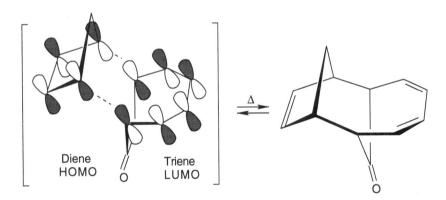

The reaction of cyclopentadiene and cycloheptatrienone is a [6+4] cycloaddition. This thermal cycloaddition proceeds with suprafacial geometry since five electron pairs are involved in the concerted process. The π electrons of the carbonyl group do not take part in the reaction.

31.9

Formation of the bicyclic ring system occurs by a suprafacial [4+2] Diels-Alder cycloaddition process. Only one pair of π electrons from the alkyne is involved in the reaction; the carbonyl π electrons are not involved.

Loss of CO_2 is a reverse Diels-Alder [4+2] cycloaddition reaction.

31.10

The ^{13}C NMR spectrum of homotropilidene would show five peaks if rearrangement were slow. In fact, rearrangement occurs at a rate that is too fast for NMR to detect. The ^{13}C NMR spectrum taken at room temperature is an average of the two equilibrating forms, in which positions 1 and 5 are equivalent, as are positions 2 and 4. Thus, only three distinct types of carbons are visible in the ^{13}C NMR spectrum of homotropilidene.

31.11 Scrambling of the deuterium label of 1–deuterioindene occurs by a series of [1,5] sigmatropic rearrangements. This thermal reaction involves three electron pairs—one pair of π electrons from the six-membered ring, the π electrons from the five-membered ring, and two electrons from a carbon-deuterium (or hydrogen) single bond—and proceeds with suprafacial geometry.

31.12 This [1,7] sigmatropic reaction proceeds with antarafacial geometry because four electron pairs are involved in the arrangement.

31.13

The Claisen arrangement of an unsubstituted allyl phenyl ether is a [3,3] sigmatropic rearrangement in which the allyl group usually ends up in the position *ortho* to oxygen. In this problem both *ortho* positions are occupied by methyl groups. The Claisen intermediate undergoes a second [3,3] rearrangement, and the final product is *p*–allyl phenol.

31.14

Type of reaction	Number of electron pairs	Stereochemistry
a) Thermal electrocyclic	four	conrotatory
b) Photochemical electrocyclic	four	disrotatory
c) Photochemical cycloaddition	four	suprafacial
d) Thermal cycloaddition	four	antarafacial
e) Photochemical sigmatropic rearrangement	four	suprafacial

31.15 a) An *electrocyclic reaction* is a reversible, pericyclic process in which a ring is formed by the reorganization of the π electrons of a conjugated polyene.

b) *Conrotatory* motion in a pericyclic reaction occurs when the lobes of two orbitals both rotate in either a clockwise or a counterclockwise fashion.

c) A *suprafacial* pericyclic reaction occurs between orbital lobes on the same face of one component and orbital lobes on the same face of the other component.

d) An *antarafacial* pericyclic reaction occurs between orbital lobes on the same face of one component and orbital lobes on opposite faces of the other component.

e) *Disrotatory motion* occurs when the lobes of one orbital in an electrocyclic reaction rotate clockwise and the lobes of the other orbital rotate counterclockwise.

f) A *sigmatropic rearrangement* is a pericyclic process in which a σ-bonded group migrates across a π electron system.

31.16

a)

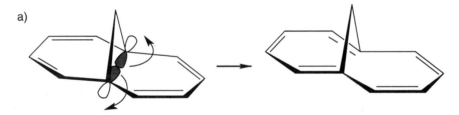

Rotation of the orbitals in the 6 π electron system occurs in a disrotatory fashion. According to the rules in Table 31.1, the reaction should be carried out under thermal conditions.

b)

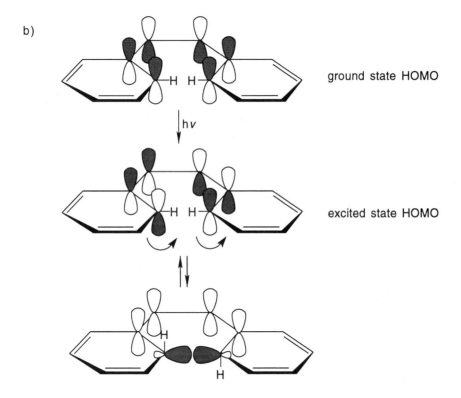

ground state HOMO

hν

excited state HOMO

For the hydrogens to be *trans* in the product, rotation must occur in a conrotatory manner. This can happen only if the HOMO has the symmetry pictured. For a 6 π electron system, this HOMO must arise from photochemical excitation of a π electron. To obtain a product having the correct stereochemistry, the reaction must be carried out under photochemical conditions.

31.17

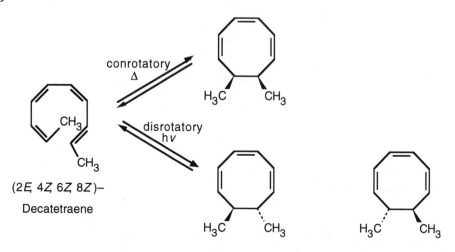

(2E, 4Z, 6Z, 8E)–
Decatetraene

Four electron pairs undergo reorganization in this electrocyclic reaction. The thermal reaction occurs with conrotatory motion to yield a pair of enantiomeric *trans*–7,8-dimethyl–1,3,5–cyclooctatrienes. The photochemical cyclization occurs with disrotatory motion to yield the *cis*–7,8–dimethyl isomer.

31.18

(2E, 4Z, 6Z, 8Z)–
Decatetraene

31.19

Type of reaction	Number of electron pairs	Stereochemistry
a) Photochemical [1,5] sigmatropic rearrangement	3	antarafacial
b) Thermal [4+6] cycloaddition	5	suprafacial
c) Thermal [1,7] sigmatropic rearrangement	4	antarafacial
d) Photochemical [2+6] cycloaddition	4	suprafacial

31.20

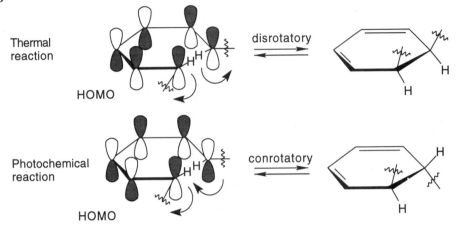

Thermal reaction

HOMO

disrotatory

Photochemical reaction

HOMO

conrotatory

Two electrocyclic reactions, involving three electron pairs each, occur in this isomerization. The thermal reaction is a disrotatory process that yields two *cis*-fused six-membered rings. The photochemical reaction yields the *trans*-fused isomer. The two pairs of π electrons in the eight-membered ring do not take part in the electrocyclic reaction.

31.21

This reaction is a reverse [4+2] cycloaddition. The reacting orbitals have the correct symmetry for the reaction to take place by a favorable suprafacial process.

This [2+2] reverse cycloaddition is not likely to occur as a concerted process because the required antarafacial geometry for the thermal reaction is not possible for a four π-electron system.

31.22

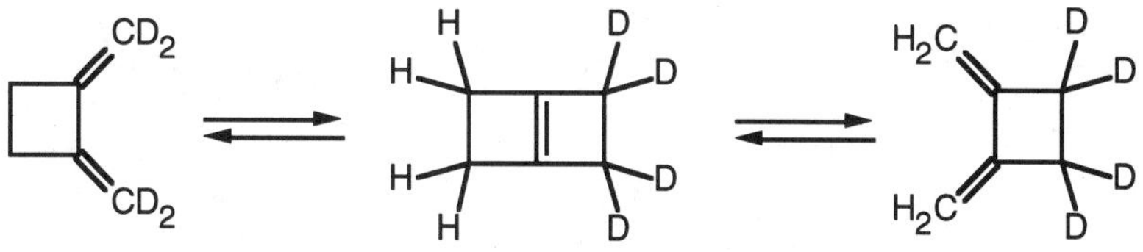

Each electrocyclic reaction involves two pairs of electrons and proceeds in a conrotatory manner.

31.23

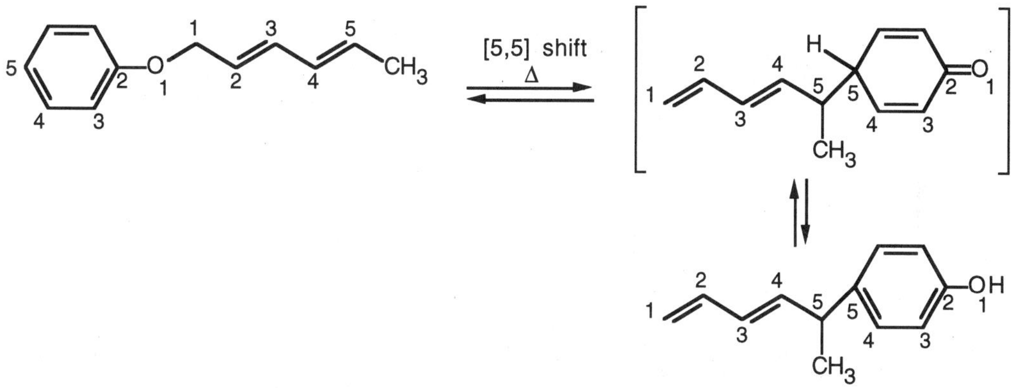

This thermal sigmatropic rearrangement is a suprafacial process since five electron pairs are involved in the reaction.

31.24

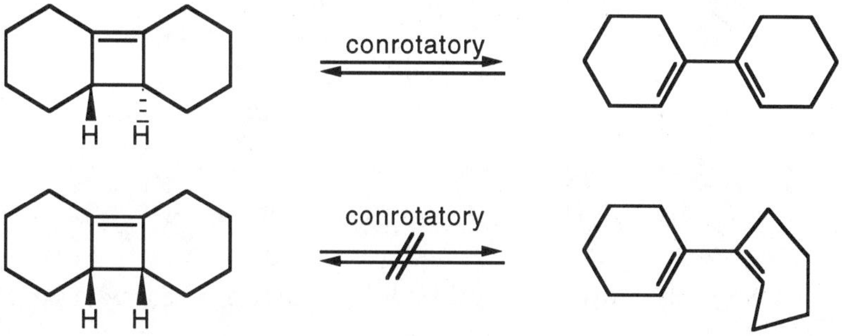

The observed product can be formed by a four-electron *pericyclic* process only if the four-membered ring geometry is *trans*. Ring-opening of the *cis* isomer by a concerted process would form a severely strained six-membered ring containing a *trans* double bond. Reaction of the *cis* isomer to yield the observed product occurs instead by a higher energy, nonconcerted path.

31.25

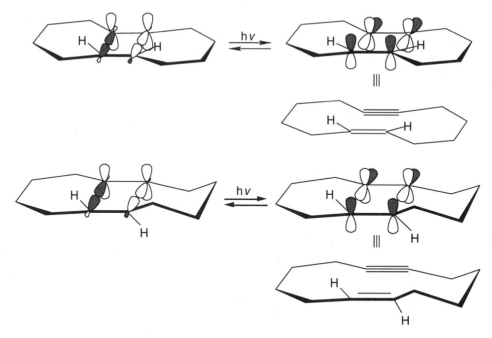

Both reactions are reverse [2+2] photochemical cycloadditions, which occur with suprafacial geometry.

31.26

The first reaction is a Diels-Alder [4+2] cycloaddition, which proceeds with suprafacial geometry.

The second reaction is a reverse Diels-Alder [4+2] cycloaddition.

31.27

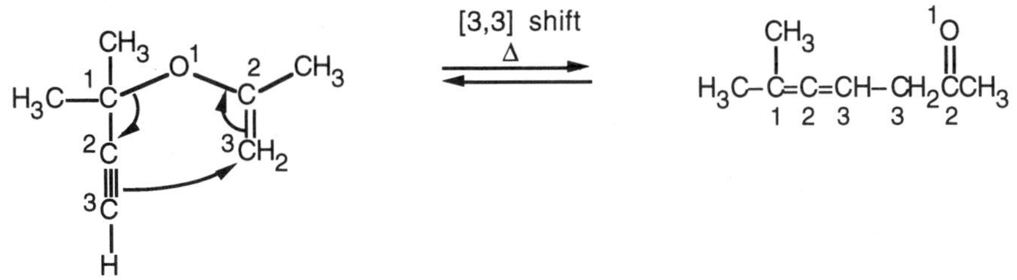

An allene is formed by a [3,3] sigmatropic rearrangement.

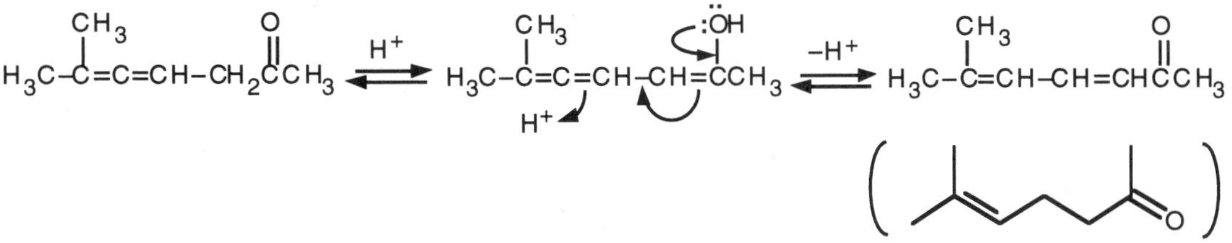

Acid catalyzes isomerization of the allene to a conjugated dienone *via* an intermediate enol.

31.28

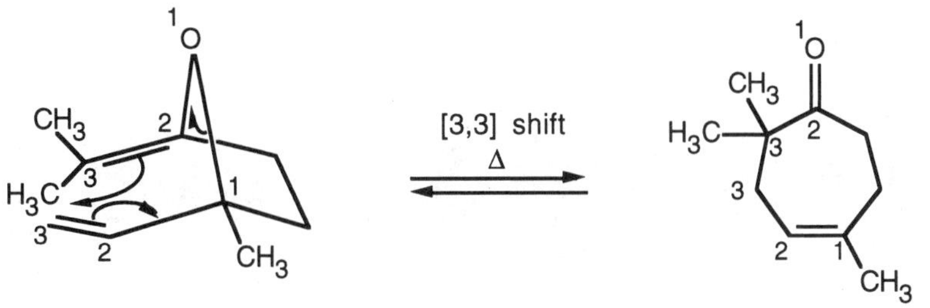

Karahanaenone is formed by a [3,3] sigmatropic rearrangement (Claisen rearrangement).

31.29

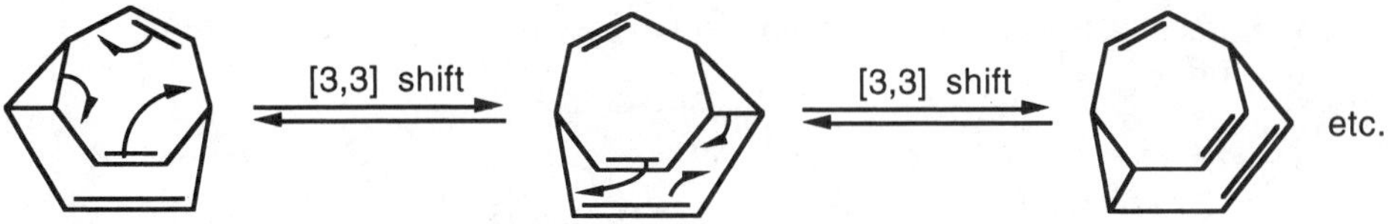

Bullvalene can undergo [3,3] sigmatropic rearrangements in all directions. At 100°, the rate of rearrangement is fast enough to make all hydrogen atoms equivalent, and only one signal is seen in the ^{1}H NMR spectrum.

31.30 Suprafacial shift:

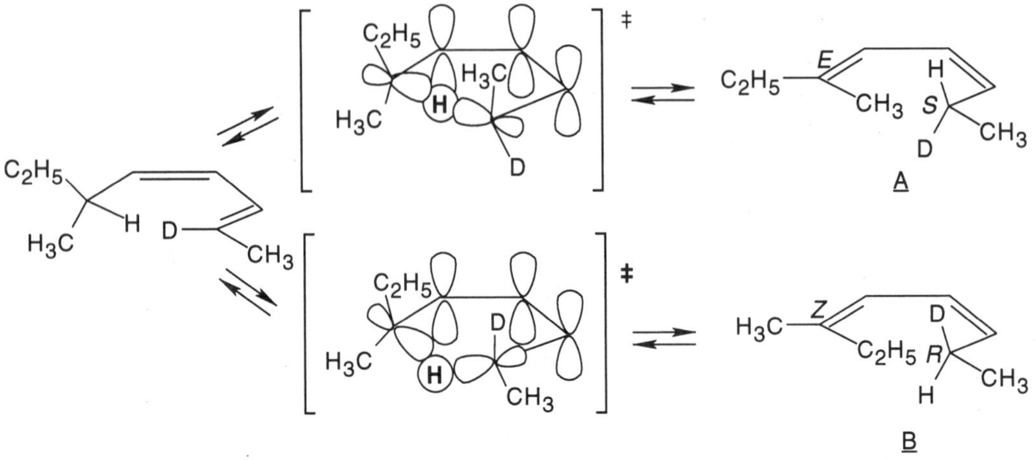

Antarafacial shift:

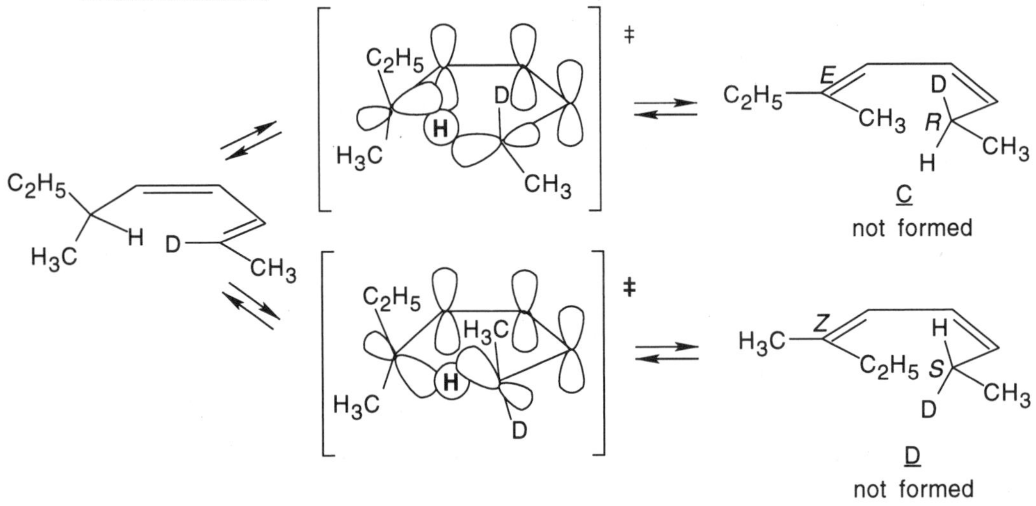

The observed products $\underline{A}$ and $\underline{B}$ result from a [1,5] sigmatropic hydrogen shift with suprafacial geometry, and they confirm the predictions of orbital symmetry. $\underline{C}$ and $\underline{D}$ are not formed.

31.31

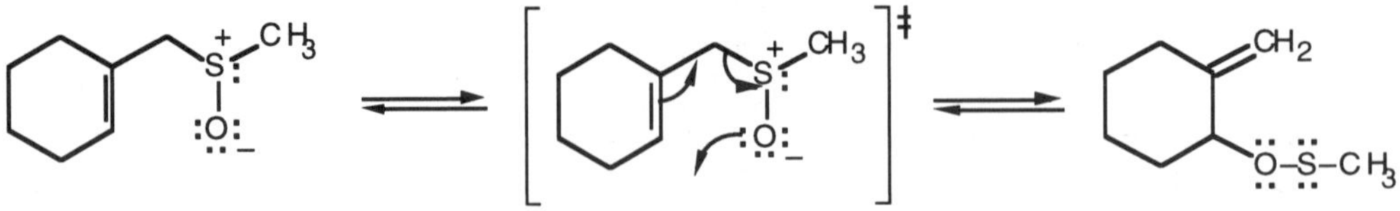

This [2,3] sigmatropic rearrangement involves three electron pairs and should occur with suprafacial geometry.

31.32

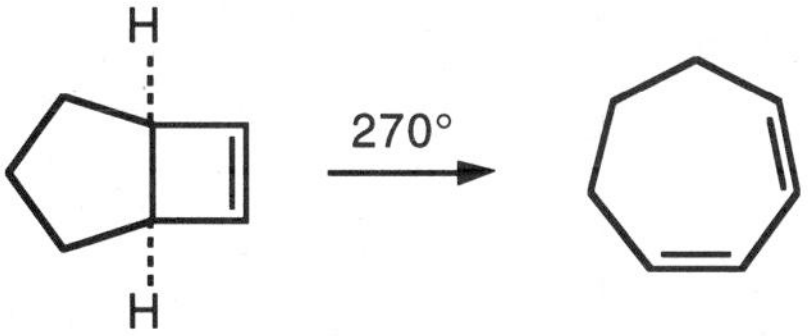

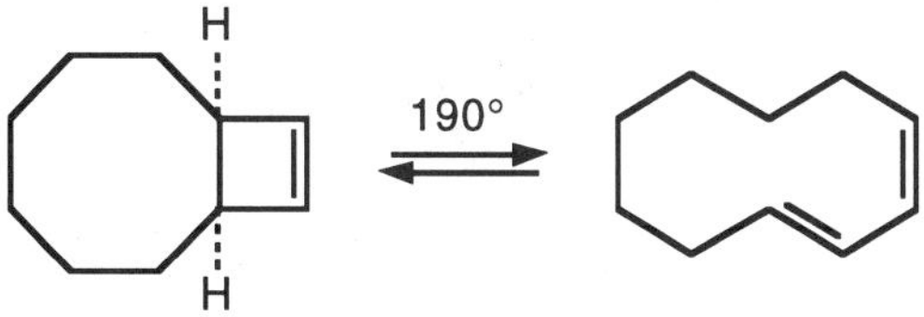

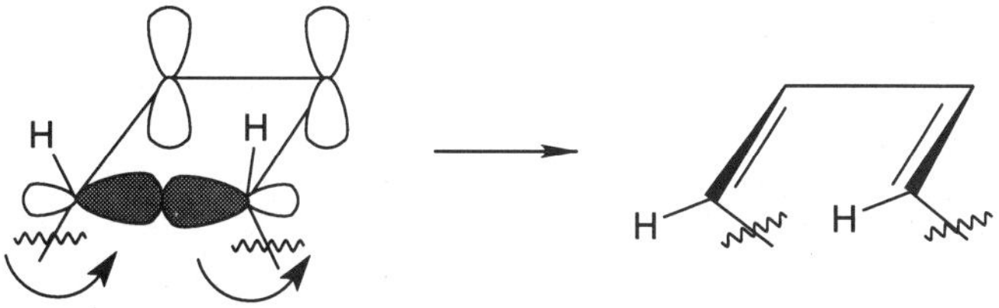

Concerted thermal ring opening of a *cis* fused cyclobutene ring yields a product having one *cis* and one *trans* double bond. The ten-membered ring product of reaction 2 is large enough to accommodate a *trans* double bond, but a seven-membered ring containing a *trans* double bond is highly strained. Opening of the cyclobutene ring in reaction 1 occurs by a higher energy nonpericyclic process to yield a seven-membered ring having two *cis* double bonds.

31.33

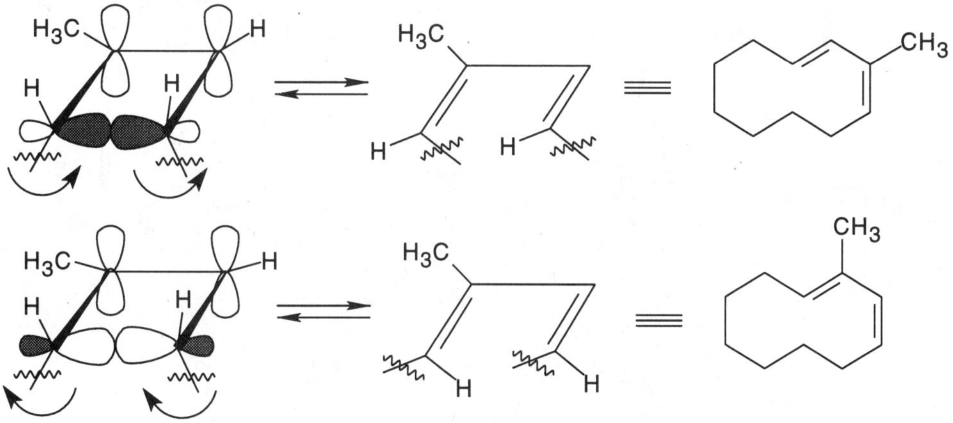

Thermal ring opening of the methylcyclobutene ring can occur by either of two symmetry-allowed conrotatory paths to yield the observed product mixture.

31.34

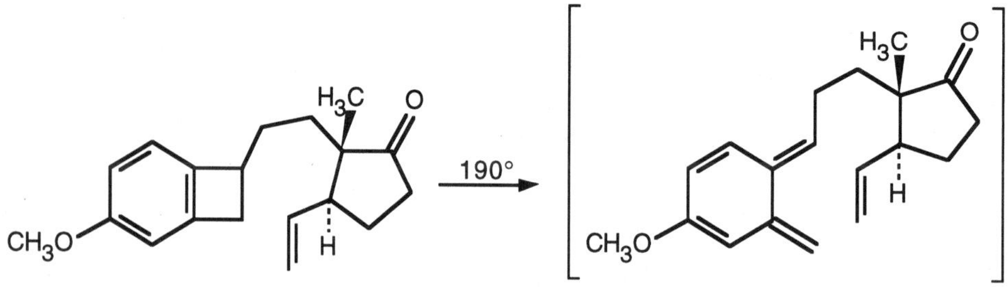

The first reaction is an electrocyclic opening of a cyclobutene ring.

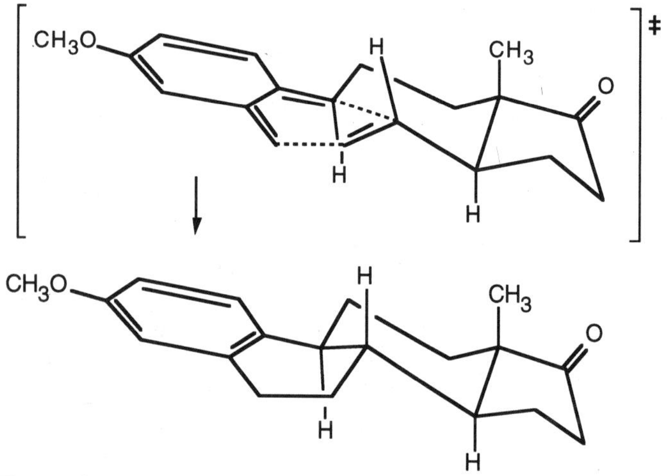

Formation of estrone methyl ether occurs by a Diels-Alder [4+2] cycloaddition.

31.35

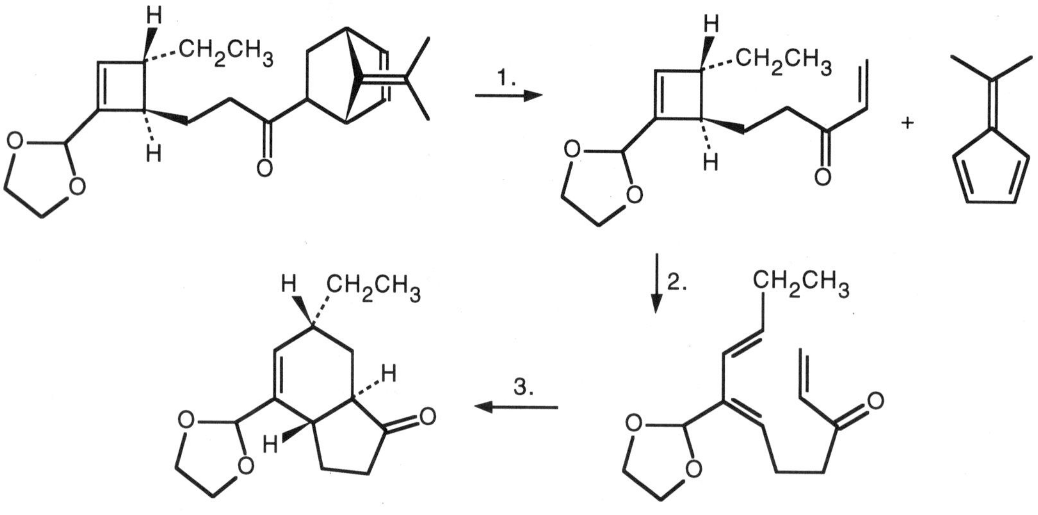

Reaction 1:   Reverse Diels-Alder [4+2] cycloaddition;
Reaction 2:   Conrotatory electrocyclic opening of a cyclobutene ring;
Reaction 3:   Diels-Alder [4+2] cycloaddition.

Treatment with base enolizes the ketone and changes the ring junction from *trans* to *cis*. A *cis* ring fusion is less strained when a six-membered ring is fused to a five-membered ring.

31.36

Study Guide for Chapter 31

After studying this chapter, you should be able to:

(1)   Understand the principle of molecular orbitals.

(2)   Define the terms HOMO, LUMO, conrotatory, disrotatory, suprafacial, antarafacial, electrocyclic reaction, cycloaddition reaction, sigmatropic rearrangement (31.15).

(3)   Locate the HOMO and LUMO of a conjugated π system (31.1).

(4)   Predict the stereochemistry of thermal and photochemical electrocyclic reactions (31.2, 31.3, 31.4, 31.5, 31.6, 31.16, 31.17, 31.18, 31.20, 31.22, 31.24, 31.25, 31.32, 31.33).

(5)   Know the stereochemical requirements for cycloaddition reactions, and predict the products of cycloadditions (31.7, 31.8, 31.9, 31.21, 31.26, 31.34, 31.35).

(6)   Classify sigmatropic rearrangements by order and predict their products (31.10, 31.11, 31.12, 31.13, 31.23, 31.27, 31.28, 31.29, 31.30, 31.31, 31.36).

(7)   Know the selection rules for pericyclic reactions (31.14, 31.19).

Functional-Group Synthesis

The following table summarizes the synthetic methods by which important functional groups can be prepared. The functional groups are listed alphabetically, followed by reference to the appropriate text section and a brief description of each synthetic method.

Acetals, $R_2C(OR')_2$
(Sec. 19.14)   from ketones and aldehydes by acid-catalyzed reaction with alcohols

Acid anhydrides, RCOOCOR'
(Sec. 21.4)   from dicarboxylic acids by heating
(Sec. 21.6)   from acid chlorides by reaction with carboxylate salts

Acid bromides, RCOBr
(Sec. 21.5)   from carboxylic acids by reaction with PBr_3

Acid chlorides, RCOCl
(Sec. 21.4)   from carboxylic acids by reaction with $SOCl_2$

Alcohols, ROH
(Sec. 7.4)   from alkenes by oxymercuration/demercuration
(Sec. 7.5)   from alkenes by hydroboration/oxidation
(Sec. 7.8)   from alkenes by hydroxylation with OsO_4
(Sec. 11.4, 11.5)   from alkyl halides and tosylates by S_N2 reaction with hydroxide ion
(Sec. 18.6)   from ethers by acid-induced cleavage
(Sec. 18.8)   from epoxides by acid-catalyzed ring opening with either H_2O or HX
(Sec. 18.8)   from epoxides by base-induced ring opening
(Sec. 17.6, 19.11) from ketones and aldehydes by reduction with $NaBH_4$ or $LiAlH_4$
(Sec. 17.7, 19.10) from ketones and aldehydes by addition of Grignard reagents
(Sec. 20.8)   from carboxylic acids by reduction with either $LiAlH_4$ or BH_3
(Sec. 21.5)   from acid chlorides by reduction with $LiAlH_4$
(Sec. 21.5)   from acid chlorides by reaction with Grignard reagents
(Sec. 21.6)   from acid anhydrides by reduction with $LiAlH_4$
(Sec. 17.6, 21.7) from esters by reduction with $LiAlH_4$
(Sec. 17.7, 21.7) from esters by reaction with Grignard reagents

Aldehydes, RCHO
(Sec. 7.8)   from disubstituted alkenes by ozonolysis
(Sec. 7.8)   from 1,2-diols by cleavage with sodium periodate
(Sec. 8.5)   from terminal alkynes by hydroboration followed by oxidation
(Sec. 17.9, 19.3) from primary alcohols by oxidation
(Sec. 21.5)   from acid chlorides by partial reduction with $LiAl(O\text{-}t\text{-Bu})_3H$
(Sec. 19.3, 21.7) from esters by reduction with DIBAH [$H\text{-Al}(i\text{-Bu})_2$]
(Sec. 21.9)   from nitriles by partial reduction with DIBAH

Alkanes, RH
(Sec. 7.7)   from alkenes by catalytic hydrogenation
(Sec. 10.8)   from alkyl halides by protonolysis of Grignard reagents
(Sec. 10.9)   from alkyl halides by coupling with Gilman reagents
(Sec. 19.13)   from ketones and aldehydes by Wolff-Kishner reaction
(Sec. 19.13)   from ketones and aldehydes by Clemmensen reduction

Alkenes, $R_2C=CR_2$

(Sec. 7.1, 11.11)	from alkyl halides by treatment with strong base (E2 reaction)
(Sec. 7.1, 17.8)	from alcohols by dehydration
(Sec. 8.6)	from alkynes by catalytic hydrogenation using the Lindlar catalyst
(Sec. 8.6)	from alkynes by reduction with lithium in liquid ammonia
(Sec. 19.16)	from ketones and aldehydes by treatment with alkylidenetriphenylphosphoranes (Wittig reaction)
(Sec. 22.3)	from α-bromo ketones by heating with pyridine
(Sec. 24.7)	from amines by methylation and Hofmann elimination

Alkynes, $RC{\equiv}CR$

(Sec. 8.3)	from dihalides by base-induced double dehydrohalogenation
(Sec. 8.9)	from terminal alkynes by alkylation of acetylide anions

Amides, $RCONH_2$

(Sec. 21.4)	from carboxylic acids by heating with ammonia
(Sec. 21.5, 24.6)	rom acid chlorides by treatment with an amine or ammonia
(Sec. 21.6)	from acid anhydrides by treatment with an amine or ammonia
(Sec. 21.7)	from esters by treatment with an amine or ammonia
(Sec. 21.9)	from nitriles by partial hydrolysis with either acid or base
(Sec. 27.11)	from a carboxylic acid and an amine by treatment with dicyclohexylcarbodiimide (DCC)

Amines, RNH_2

(Sec. 19.17)	from conjugated enones by addition of primary or secondary amines
(Sec. 21.8, 24.6)	from amides by reduction with $LiAlH_4$
(Sec. 21.9, 24.6)	from nitriles by reduction with $LiAlH_4$
(Sec. 24.6)	from primary alkyl halides by treatment with ammonia
(Sec. 24.6)	from primary alkyl halides by Gabriel synthesis
(Sec. 24.6)	from primary alkyl azides by reduction with $LiAlH_4$
(Sec. 24.6)	from acid chlorides by Curtius rearrangement of acyl azides
(Sec. 24.6)	from primary amides by Hofmann rearrangement
(Sec. 24.6)	from ketones and aldehydes by reductive amination with an amine and $NaBH_3CN$

Amino Acids, $RCH(NH_2)COOH$

(Sec. 27.4)	from α-bromo acids by S_N2 reaction with ammonia
(Sec. 27.4)	from aldehydes by reaction with KCN and ammonia (Strecker synthesis)
(Sec. 27.4)	from α-keto acids by reductive amination
(Sec. 27.4)	from primary alkyl halides by alkylation with diethyl acetamidomalonate

Arenes, Ar-R

(Sec. 16.3)	from arenes by Friedel-Crafts alkylation with an alkyl halide
(Sec. 16.11)	from aryl alkyl ketones by catalytic reduction of the keto group
(Sec. 25.3)	from arenediazonium salts by treatment with hypophosphorous acid

Arylamines, $Ar-NH_2$

(Sec. 16.2, 25.2)	from nitroarenes by reduction with either Fe, Sn, or H_2/Pd.

Arenediazonium salts, $Ar-N_2^+ X^-$

(Sec. 25.3)	from arylamines by reaction with nitrous acid

Arenesulfonic acids Ar-SO$_3$H
(Sec. 16.2)   from arenes by electrophilic aromatic substitution with SO$_3$/H$_2$SO$_4$

Azides, R-N$_3$
(Sec. 11.4, 24.6)   from primary alkyl halides by S$_N$2 reaction with azide ion

Carboxylic acids, RCOOH
(Sec. 7.8)   from mono- and 1,2-disubstituted alkenes by ozonolysis
(Sec. 16.10)   from arenes by side-chain oxidation with Na$_2$Cr$_2$O$_7$ or KMnO$_4$
(Sec. 19.5)   from aldehydes by oxidation
(Sec. 20.6)   from alkyl halides by conversion into Grignard reagents followed by
                              reaction with CO$_2$
(Sec. 20.6, 21.9)   from nitriles by vigorous acid or base hydrolysis
(Sec. 21.5)   from acid chlorides by reaction with aqueous base
(Sec. 21.6)   from acid anhydrides by reaction with aqueous base
(Sec. 21.7)   from esters by hydrolysis with aqueous base
(Sec. 21.8)   from amides by hydrolysis with aqueous base
(Sec. 22.7)   from methyl ketones by reaction with halogen and base (haloform
                              reaction)
(Sec. 25.7)   from phenols by treatment with CO$_2$ and base (Kolbe carboxylation)

Cyanohydrins, RCH(OH)CN
(Sec. 19.9)   from aldehydes and ketones by reaction with HCN

Cycloalkanes
(Sec. 7.6)   from alkenes by addition of dichlorocarbene
(Sec. 7.6)   from alkenes by reaction with CH$_2$I$_2$ and Zn/Cu (Simmons-Smith
                              reaction)
(Sec. 16.11)   from arenes by rhodium-catalyzed hydrogenation

Disulfides, R-SS-R'
(Sec. 17.12)   from thiols by oxidation with bromine

Enamines, RCH=CRNR$_2$
(Sec. 19.12)   from ketones or aldehydes by reaction with secondary amines

Epoxides , $R_2C\overset{\displaystyle O}{\overset{\diagup\diagdown}{-}}CR_2$
(Sec. 18.7)   from alkenes by treatment with a peroxyacid
(Sec. 18.7)   from halohydrins by treatment with base

Esters, RCOOR'
(Sec. 21.4)   from carboxylic acid salts by S$_N$2 reaction with primary alkyl halides
(Sec. 21.4)   from carboxylic acids by acid-catalyzed reaction with an alcohol (Fischer
                              esterification)
(Sec. 21.5)   from acid chlorides by base-induced reaction with an alcohol
(Sec. 21.6)   from acid anhydrides by base-induced reaction with an alcohol
(Sec. 22.8)   from alkyl halides by alkylation with diethyl malonate
(Sec. 22.8)   from esters by treatment of their enolate ions with alkyl halides
(Sec. 25.7)   from phenols by base-induced reaction with an acid chloride

Ethers, R-O-R'

(Sec. 16.8) from activated haloarenes by reaction with alkoxide ions

(Sec. 16.9) from unactivated haloarenes by reaction with alkoxide ions via benzyne intermediates

(Sec. 18.4) from primary alkyl halides by S_N2 reaction with alkoxide ions (Williamson ether synthesis)

(Sec. 18.5) from alkenes by alkoxymercuration/demercuration

(Sec. 18.7) from alkenes by epoxidation with peroxyacids

(Sec. 25.8) from phenols by reaction of phenoxide ions with primary alkyl halides

Halides, alkyl, R_3C-X

(Sec. 6.8) from alkenes by electrophilic addition of HX

(Sec. 7.2) from alkenes by addition of halogen

(Sec. 7.3) from alkenes by electrophilic addition of hypohalous acid (HOX) to yield halohydrins

(Sec. 7.10) from alkenes by radical-catalyzed addition of HBr

(Sec. 8.4) from alkynes by addition of halogen

(Sec. 8.4) from alkynes by addition of HX

(Sec. 10.5) from alkenes by allylic bromination with *N*-bromosuccinimide (NBS)

(Sec. 10.7) from alcohols by reaction with HX

(Sec. 10.7) from alcohols by reaction with $SOCl_2$

(Sec. 10.7) from alcohols by reaction with PBr_3

(Sec. 11.4, 11.5) from alkyl tosylates by S_N2 reaction with halide ions

(Sec. 16.10) from arenes by benzylic bromination with *N*-bromosuccinimide (NBS)

(Sec. 18.6) from ethers by cleavage with either HX

(Sec. 22.3) from ketones by alpha-halogenation with bromine

(Sec. 22.4) from carboxylic acids by alpha-halogenation with phosphorus and PBr_3 (Hell-Volhard-Zelinskii reaction)

Halides, aryl, Ar-X

(Sec. 16.1, 16.2) from arenes by electrophilic aromatic substitution with halogen

(Sec. 25.3) from arenediazonium salts by reaction with cuprous halides (Sandmeyer reaction)

(Sec. 29.3) from aromatic heterocycles by electrophilic aromatic substitution with halogen

Halohydrins, $R_2CXC(OH)R_2$

(Sec. 7.3) from alkenes by electrophilic addition of hypohalous acid (HOX)

(Sec. 18.8) from epoxides by acid-induced ring opening with HX

Imines. $R_2C=NR'$

(Sec. 19.12) from ketones or aldehydes by reaction with primary amines

Ketones, $R_2C=O$

(Sec. 7.8) from alkenes by ozonolysis

(Sec. 7.8) from 1,2-diols by cleavage reaction with sodium periodate

(Sec. 8.5) from alkynes by mercuric-ion-catalyzed hydration

(Sec. 8.5) from alkynes by hydroboration/oxidation

(Sec. 16.4) from arenes by Lewis-acid-catalyzed reaction with an acid chloride (Friedel-Crafts acylation)

(Sec. 17.9, 19.4) from secondary alcohols by oxidation

(Sec. 19.4, 21.5) from acid chlorides by reaction with lithium diorganocopper (Gilman) reagents

(Sec. 19.17)   from conjugated enones by addition of lithium diorganocopper reagents
(Sec. 21.9)   from nitriles by reaction with Grignard reagents
(Sec. 22.9)   from primary alkyl halides by alkylation with ethyl acetoacetate
(Sec. 22.9)   from ketones by alkylation of their enolate ions with primary alkyl halides

Nitriles, $R-C \equiv N$

(Sec. 11.5, 21.9)   from primary alkyl halides by S_N2 reaction with cyanide ion
(Sec. 21.9)   from primary amides by dehydration with $SOCl_2$
(Sec. 22.8)   from nitriles by alkylation of their alpha-anions with primary alkyl halides
(Sec. 25.3)   from arenediazonium ions by treatment with CuCN

Nitroarenes, $Ar-NO_2$

(Sec. 16.2)   from arenes by electrophilic aromatic substitution with nitric/sulfuric acids

Organometallics, R-M

(Sec. 10.8)   formation of Grignard reagents from organohalides by treatment with magnesium
(Sec. 10.9)   formation of organolithium reagents from organohalides by treatment with lithium
(Sec. 10.9)   formation of lithium diorganocopper reagents (Gilman reagents) from organolithium reagents by treatment with cuprous halides

Phenols, Ar-OH

(Sec. 16.2, 25.7)   from arenesulfonic acids by fusion with KOH
(Sec. 25.3)   from arenediazonium salts by reaction with aqueous acid
(Sec. 16.9)   from aryl halides by nucleophilic aromatic substitution with hydroxide ion

Quinones,

(Sec. 25.8)   from phenols by oxidation with Fremy's salt [$(KSO_3)_2NO$]
(Sec. 25.8)   from arylamines by oxidation with Fremy's salt

Sulfides, R-S-R'

(Sec. 18.11)   from thiols by S_N2 reaction of thiolate ions with primary alkyl halides

Sulfones, $R-SO_2-R'$

(Sec. 18.11)   from sulfides or sulfoxides by oxidation with peroxyacids

Sulfoxides, R-SO-R'

(Sec. 18.11)   from sulfides by oxidation with H_2O_2

Thiols, R-SH

(Sec. 11.5)   from primary alkyl halides by S_N2 reaction with hydrosulfide anion
(Sec. 17.12)   from primary alkyl halides by S_N2 reaction with thiourea, followed by hydrolysis

```
┌─────────────────────────────────────────────────────────┐
│                                                         │
│            Functional-Group Reactions                   │
│                                                         │
└─────────────────────────────────────────────────────────┘
```

The following table summarizes the reactions that important functional groups undergo. The functional groups are listed alphabetically, followed by a reference to the appropriate text section.

Acetal
1. Hydrolysis to yield a ketone or aldehyde plus alcohol (Sec. 19.14)

Acid anhydride
1. Hydrolysis to yield a carboxylic acid (Sec. 21.6)
2. Alcoholysis to yield an ester (Sec. 21.6)
3. Aminolysis to yield an amide (Sec. 21.6)
4. Reduction to yield a primary alcohol (Sec. 21.6)

Acid chloride
1. Hydrolysis to yield a carboxylic acid (Sec. 21.5)
2. Alcoholysis to yield an ester (Sec. 21.5)
3. Aminolysis to yield an amide (Sec. 21.5)
4. Reduction to yield a primary alcohol (Sec. 21.5)
5. Partial reduction to yield an aldehyde (Sec. 21.5)
6. Grignard reaction to yield a tertiary alcohol (Sec. 21.5)
7. Reaction with lithium diorganocuprate to yield a ketone (Sec. 21.5)

Alcohol
1. Acidity (Sec. 17.4)
2. Oxidation (Sec. 17.9)
   a. Of primary alcohol to yield an aldehyde or acid
   b. Of secondary alcohol to yield a ketone
3. Reaction with carboxylic acid to yield an ester (Sec. 21.4)
4. Reaction with acid chloride to yield an ester (Sec. 21.5)
5. Dehydration to yield an alkene (Sec. 17.8)
6. Reaction with primary alkyl halide to yield an ether (Sec. 18.4)
7. Conversion into an alkyl halide (Sec. 17.8)
   a. Reaction of a tertiary alcohol with HX
   b. Reaction of a primary or secondary alcohol with $SOCl_2$
   c. Reaction of a primary or secondary alcohol with PBr_3

Aldehyde
1. Oxidation to yield a carboxylic acid (Sec. 19.5)
2. Nucleophilic addition reactions
   a. Reduction to yield a primary alcohol (Sec. 17.6)
   b. Reaction with Grignard reagent to yield a secondary alcohol (Sec. 17.7)
   c. Grignard reaction of formaldehyde to yield a primary alcohol (Sec. 17.7)
   d. Reaction with HCN to yield a cyanohydrin (Sec. 19.9)
   e. Wolff-Kishner reaction with hydrazine to yield an alkane (Sec. 19.13)
   f. Clemmensen reduction with acidic zinc amalgam to yield an alkane (Sec. 19.13)
   g. Reaction with alcohol to yield an acetal (Sec. 19.14)
   h. Wittig reaction to yield an alkene (Sec. 19.15)
   i. Reaction with an amine to yield an imine or enamine (Sec. 19.12)
3. Aldol reaction to yield a β-hydroxy aldehyde (Sec. 23.2)
4. Alpha bromination of aldehyde (Sec. 22.3)

Alkane
1. Radical halogenation to yield an alkyl halide (Sec. 10.4)

Alkene
1. Electrophilic addition of HX to yield an alkyl halide (Secs. 6.8 - 6.12)
   Markovnikov regiochemistry is observed. H adds to the less highly substituted carbon, and X adds to the more highly substituted one.
2. Electrophilic addition of halogen to yield a 1,2-dihalide (Sec. 7.2)
   Anti stereochemistry is observed
3. Oxymercuration/demercuration to yield an alcohol (Sec. 7.4)
   Markovnikov regiochemistry is observed, yielding the more highly substituted alcohol.
4. Hydroboration/oxidation to yield an alcohol (Section 7.5)
5. Hydrogenation to yield an alkane (Sec. 7.7)
6. Hydroxylation to yield a 1,2-diol (Sec. 7.8)
7. Radical addition of HBr to yield an alkyl bromide (Sec. 7.10)
8. Oxidative cleavage to yield carbonyl compounds (Sec. 7.8)
9. Reaction with a peroxyacid to yield an epoxide (Sec. 18.7)
10. Simmons-Smith reaction with CH_2I_2 to yield a cyclopropane (Sec. 7.6)

Alkyne
1. Electrophilic addition of HX to yield a vinylic halide (Sec. 8.4)
2. Electrophilic addition of halogen to yield a dihalide (Sec. 8.4)
3. Mercuric-sulfate-catalyzed hydration to yield a methyl ketone (Sec. 8.5)
4. Hydroboration/oxidation to yield an aldehyde (Sec. 8.5)
5. Alkylation of an alkyne anion (Sec. 8.9)
6. Reduction (Sec. 8.6)
   a. Hydrogenation to yield a cis alkene
   b. Reduction to yield a trans alkene

Amide
1. Hydrolysis to yield a carboxylic acid (Sec. 21.8)
2. Reduction with $LiAlH_4$ to yield an amine (Sec. 21.8)
3. Dehydration to yield a nitrile (Section 21.9)

Amine
1. S_N2 alkylation of an alkyl halide to yield an amine (Sec. 24.7)
2. Nucleophilic acyl substitution reactions
   a. Reaction with an acid chloride to yield an amide (Sec. 21.5)
   b. Reaction with an acid anhydride to yield an amide (Sec. 21.6)
   c. Reaction with an ester to yield an amide (Sec. 21.7)
3. Hofmann elimination to yield an alkene (Sec. 24.7)
4. Formation of an arenediazonium salt (Sec. 25.3)

Arene
1. Oxidation of the alkylbenzene side chain to yield a benzoic acid (Sec. 16.10)
2. Catalytic reduction to yield a cyclohexane (Sec. 16.11)
3. Reduction of an aryl alkyl ketone to yield an arene (Sec. 16.11)
4. Electrophilic aromatic substitution (Secs. 16.1 - 16.4)
   a. Halogenation (Secs. 16.1–16.2)
   b. Nitration (Sec. 16.2)
   c. Sulfonation (Sec. 16.2)
   d. Friedel-Crafts alkylation (Sec. 16.3)
      Aromatic ring must be at least as reactive as a halobenzene
   e. Friedel-Crafts acylation (Sec. 16.4)

Arenediazonium salt
1. Conversion into an aryl chloride (Sec. 25.3)
2. Conversion into an aryl bromide (Sec. 25.3)
3. Conversion into an aryl iodide (Sec. 25.3)
4. Conversion into an aryl cyanide (Sec. 25.3)
5. Conversion into a phenol (Sec. 25.3)
6. Conversion into an arene (Sec. 25.3)

Arenesulfonic acid
1. Conversion into a phenol (Secs. 16.2, 25.7)

Carboxylic acid
1. Acidity (Secs. 20.3–20.5)
2. Reduction to yield a primary alcohol (Sec. 17.6)
   a. Reduction with $LiAlH_4$
   b. Reduction with BH_3
3. Nucleophilic acyl substitution reactions (Sec. 21.4)
   a. Conversion into an acid chloride
   b. Conversion into an acid anhydride
   c. Conversion into an ester
      (1) Fischer esterification
      (2) S_N2 reaction with an alkyl halide

Epoxide
1. Acid-catalyzed ring opening with HX to yield a halohydrin (Sec. 18.8)
2. Ring opening with aqueous acid to yield a 1,2-diol (Sec. 18.8)

Ester
1. Hydrolysis to yield a carboxylic acid (Sec. 21.7)
2. Aminolysis to yield an amide (Sec. 21.7)
3. Reduction to yield a primary alcohol (Sec. 17.6)
4. Partial reduction with DIBAH to yield an aldehyde (Sec. 21.7)
5. Grignard reaction to yield a tertiary alcohol (Sec. 17.7)
6. Claisen condensation to yield a β-keto ester (Sec. 23.8)

Ether
1. Acid-induced cleavage to yield an alcohol and an alkyl halide (Sec. 18.6)

Halide, alkyl
1. Reaction with magnesium to form a Grignard reagent (Sec. 10.8)
2. Reduction to yield an alkanes (Sec. 10.8)
3. Nucleophilic substitution (S_N1 or S_N2) (Secs. 11.1 - 11.9)
4. Dehydrohalogenation to yield an alkene (E1 or E2) (Secs. 11.10-11.14)

Halohydrin
1. Conversion into an epoxide (Sec. 18.7)

Ketone
1. Nucleophilic addition reactions
   a. Reduction to yield a secondary alcohol (Secs. 17.6)
   b. Reaction with a Grignard reagent to yield a tertiary alcohol (Sec. 17.7)
   c. Wolff-Kishner reaction with hydrazine to yield an alkane (Sec. 19.13)
   d. Reaction with HCN to yield a cyanohydrin (Sec. 19.9)
   e. Reaction with alcohol to yield an acetal (Sec. 19.14)
   f. Wittig reaction to yield an alkene (Sec. 19.15)
   g. Reaction with an amine to yield an imine or enamine (Sec. 19.12)
2. Aldol reaction to yield a β-hydroxy ketone (Sec. 23.2)
3. Alpha bromination of a ketone (Sec. 22.3)
4. Clemmensen reduction with acidic zinc amalgam to yield an alkane (Sec. 19.13)

Nitrile
1. Hydrolysis to yield a carboxylic acid (Sec. 21.9)
2. Reduction to yield a primary amine (Sec. 21.9)
3. Partial reduction with DIBAH to yield an aldehyde (Sec. 21.9)
4. Reaction with a Grignard reagent to yield a ketone (Sec. 21.9)

Nitroarene
1. Reduction to yield an arylamine (Sec. 25.2)

Organometallic reagent
1. Reduction by treatment with acid to yield an alcohol (Sec. 10.8)
2. Nucleophilic addition to a carbonyl compound to yield an alcohol (Secs. 17.7)
3. Conjugate addition of a lithium diorganocuprate to an α,β-unsaturated ketone (Sec. 19.17)
4. Coupling reaction of a lithium diorganocuprate with an alkyl halide to yield an alkane (Sec. 10.9)
5. Coupling reaction of a lithium diorganocuprate with an acid chloride to yield a ketone (Sec. 21.5)
6. Reaction with carbon dioxide to yield a carboxylic acid (Sec. 20.6)

Phenol
1. Acidity (Sec. 25.6)
2. Reaction with an acid chlorides to yield an ester (Sec. 25.8)
3. Reaction with an alkyl halides to yield an ether (Sec. 25.8)
4. Oxidation to yield a quinone (Sec. 25.8)

Quinone
1. Reduction to yield a hydroquinone (Sec. 25.8)

Sulfide
1. Reaction with an alkyl halide to yield a sulfonium salt (Sec. 18.11)
2. Oxidation to yield a sulfoxide (Sec. 18.11)
3. Oxidation to yield a sulfone (Sec. 18.11)

Thiol
1. Reaction with an alkyl halide to yield a sulfide (Sec. 17.12)
2. Oxidation to yield a disulfide (Sec. 17.12)

Reagents in Organic Chemistry

The following table summarizes the uses of some important reagents in organic chemistry. The reagents are listed alphabetically, followed by a brief description of the uses of each and references to the appropriate text sections.

Acetic acid, CH_3COOH: Used as a solvent for the reduction of ozonides with zinc (Section 7.8) and the α-bromination of ketones and aldehydes with Br_2 (Section 22.3).

Acetic anhydride, $(CH_3CO)_2O$: Reacts with alcohols to yield acetate esters (Sections 21.6 and 26.8) and with amines to yield acetamides (Section 21.6).

Aluminum chloride, $AlCl_3$: Acts as a Lewis acid catalyst in Friedel-Crafts alkylation and acylation reactions of aromatic compounds (Sections 16.3 and 16.4).

Ammonia, NH_3: Used as a solvent for the reduction of alkynes by lithium metal to yield trans alkenes (Section 8.6).
- Reacts with acid chlorides and acid anhydrides to yield amides (Sections 21.5 and 21.6).

Borane, BH_3: Adds to alkenes, giving alkylboranes that can be oxidized with alkaline H_2O_2 to yield alcohols (Section 7.5).
- Adds to alkynes, giving vinylic organoboranes that can be oxidized with alkaline H_2O_2 to yield aldehydes (Section 8.5).
- Reduces carboxylic acids to yield primary alcohols (Section 20.7).

Bromine, Br_2: Adds to alkenes, yielding 1,2-dibromides (Sections 7.2, 14.5).
- Adds to alkynes yielding either 1,2-dibromoalkenes or 1,1,2,2-tetrabromoalkanes (Section 8.4).
- Reacts with arenes in the presence of $FeBr_3$ catalyst to yield bromoarenes (Section 16.1).
- Reacts with ketones in acetic acid solvent to yield α-bromo ketones (Section 22.3).
- Reacts with carboxylic acids in the presence of PBr_3 to yield α-bromo carboxylic acids (Hell-Volhard-Zelinskii reaction; Section 22.4).
- Reacts with methyl ketones in the presence of NaOH to yield carboxylic acids and bromoform (Haloform reaction; Section 22.7).
- Oxidizes aldoses to yield aldonic acids (Section 26.8).

N-Bromosuccinimide (NBS), $(CH_2CO)_2NBr$: Reacts with alkenes in the presence of aqueous dimethylsulfoxide to yield bromohydrins (Section 7.3).
- Reacts with alkenes in the presence of light to yield allylic bromides (Wohl-Ziegler reaction; Section 10.5).
- Reacts with alkylbenzenes in the presence of light to yield benzylic bromides; (Section 16.10).

Di-*tert*-butoxy dicarbonate, $(\underline{t}\text{-BuOCO})_2O$: Reacts with amino acids to give t-BOC protected amino acids suitable for use in peptide synthesis (Section 27.11).

Butyllithium, $CH_3CH_2CH_2CH_2Li$: A strong base; reacts with alkynes to yield acetylide anions, which can be alkylated (Section 8.9).
- Reacts with dialkylamines to yield lithium dialkylamide bases such as LDA [lithium diisopropyl-amide] (Section 22.5).
- Reacts with alkyltriphenylphosphonium salts to yield alkylidenephosphoranes (Wittig reagents (Section 19.15).

Carbon dioxide, CO_2: Reacts with Grignard reagents to yield carboxylic acids (Section 20.6).
– Reacts with phenoxide anions to yield o-hydroxybenzoic acids (Kolbe-Schmitt carboxylation reaction; Section 25.8).

Chlorine, Cl_2: Adds to alkenes to yield 1,2-dichlorides (Sections 7.2 and 14.5).
– Reacts with alkanes in the presence of light to yield chloroalkanes by a radical chain reaction pathway (Section 10.4).
– Reacts with arenes in the presence of $FeCl_3$ catalyst to yield chloroarenes (Section 16.2).

<u>m</u>-Chloroperoxybenzoic acid, m-$ClC_6H_4CO_3H$: Reacts with alkenes to yield epoxides (Section 18.7).

Chlorotrimethylsilane, $(CH_3)_3SiCl$: Reacts with alcohols to add the trimethylsilyl protecting group (Section 17.10).

Chromium trioxide, CrO_3: Oxidizes alcohols in aqueous sulfuric acid (Jones reagent) to yield carbonyl-containing products. Primary alcohols yield carboxylic acids and secondary alcohols yield ketones (Sections 17.9 and 19.5).

Cuprous bromide, $CuBr$: Reacts with arenediazonium salts to yield bromoarenes (Sandmeyer reaction; Section 25.3).

Cuprous chloride, $CuCl$: Reacts with arenediazonium salts to yield chloroarenes (Sandmeyer reaction; Section 25.3).

Cuprous cyanide, $CuCN$: Reacts with arenediazonium salts to yield substituted benzonitriles (Sandmeyer reaction; Section 25.3).

Cuprous iodide, CuI: Reacts with organolithiums to yield lithium diorganocopper reagents (Gilman reagents; Section 10.9).

Dichloroacetic acid, $Cl_2CHCOOH$: Cleaves DMT protecting groups in DNA synthesis (Section 29.16).

Dicyclohexylcarbodiimide (DCC), C_6H_{11}–N=C=N–C_6H_{11}: Couples an amine with a carboxylic acid to yield an amide. DCC is often used in peptide synthesis (Section 27.11).

Diethyl acetamidomalonate, $CH_3CONHCH(CO_2Et)_2$: Reacts with alkyl halides in a common method of α-amino acid synthesis (Section 27.4).

Diiodomethane, CH_2I_2: Reacts with alkenes in the presence of zinc-copper couple to yield cyclopropanes (Simmons-Smith reaction; Section 7.6).

Diisobutylaluminum hydride (DIBAH), $(i\text{-}Bu)_2AlH$: Reduces esters to yield aldehydes (Sections 19.3 and 21.7).
– Reduces nitriles to yield aldehydes (Section 21.9).

2,4-Dinitrophenylhydrazine, $2,4\text{-}(NO_2)_2C_6H_3NHNH_2$: Reacts with ketones and aldehydes to yield 2,4-DNPs that serve as useful crystalline derivatives (Section 19.12).

Ethylene glycol, $HOCH_2CH_2OH$: Reacts with ketones or aldehydes in the presence of a acid catalyst to yield acetals that serve as useful carbonyl protecting groups (Section 19.14).

Ferric bromide, FeBr₃: Acts as a catalyst for the reaction of arenes with Br_2 to yield bromoarenes (Section 16.1).

Ferric chloride, FeCl₃: Acts as a catalyst for the reaction of arenes with Cl_2 to yield chloroarenes (Section 16.2).

Grignard reagent, RMgX: Reacts with acids to yield alkanes (Section 10.8).
- Adds to carbonyl-containing compounds (ketones, aldehydes, esters) to yield alcohols (Section 19.10).
- Adds to nitriles to yield ketones (Section 21.9).

Hydrazine, H₂NNH₂: Reacts with ketones or aldehydes in the presence of KOH to yield the corresponding alkanes (Wolff-Kishner reaction; Section 19.13).

Hydrogen bromide, HBr: Adds to alkenes with Markovnikov regiochemistry to yield alkyl bromides (Sections 6.9 and 14.5).
- Adds to alkenes with non-Markovnikov regiochemistry in the presence of a peroxide catalyst to yield alkyl bromides (Section 7.10).
- Adds to alkynes to yield either bromoalkenes or 1,1-dibromoalkanes (Section 8.4).
- Reacts with alcohols to yield alkyl bromides (Sections 10.7 and 17.8).
- Cleaves ethers to yield alcohols and alkyl bromides (Section 18.6).

Hydrogen chloride, HCl: Adds to alkenes with Markovnikov regiochemistry to yield alkyl chlorides (Sections 6.9 and 14.5).
- Adds to alkynes to yield either chloroalkenes or 1,1-dichloroalkanes (Section 8.4).
- Reacts with alcohols to yield alkyl chlorides (Sections 10.7, 17.8).

Hydrogen cyanide, HCN: Adds to ketones and aldehydes to yield cyanohydrins (Section 19.9).

Hydrogen iodide, HI: Reacts with alcohols to yield alkyl iodides (Section 17.8).
- Cleaves ethers to yield alcohols and alkyl iodides (Section 18.6).

Hydrogen peroxide, H₂O₂: Oxidizes organoboranes to yield alcohols. Used in conjunction with addition of borane to alkenes, the overall transformation effects syn Markovnikov addition of water to an alkene (Section 7.5).
- Oxidizes vinylic boranes to yield aldehydes.(Section 8.5).
- Oxidizes sulfides to yield sulfoxides (Section 18.11).

Hydroxylamine, NH₂OH: Reacts with ketones and aldehydes to yield oximes (Section 19.12).
- Reacts with aldoses to yield oximes as the first step in the Wohl degradation of aldoses (Section 26.8).

Hypophosphorous acid, H₃PO₂: Reacts with arenediazonium salts to yield arenes (Section 25.3).

Iodine, I₂: Reacts with arenes in the presence of CuCl or H_2O_2 to yield iodoarenes (Section 16.2).
- Reacts with methyl ketones in the presence of aqueous NaOH to yield carboxylic acids and iodoform (Section 22.7).

Iodomethane, CH$_3$I: Reacts with alkoxide anions to yield methyl ethers (Section 18.4).
– Reacts with carboxylate anions to yield methyl esters (Section 21.7).
– Reacts with enolate ions to yield α-methylated carbonyl compounds (Section 22.8).
– Reacts with amines to yield methylated amines (Section 25.6).

Iron, Fe: Reacts with nitroarenes in the presence of aqueous acid to yield anilines (Section 25.2).

Lindlar catalyst: Acts as a catalyst for the partial hydrogenation of alkynes to yield cis alkenes (Section 8.6).

Lithium, Li: Reduces alkynes in liquid ammonia solvent to yield trans alkenes (Section 8.6).
– Reacts with organohalides to yield organolithium compounds (Section 10.9).

Lithium aluminum hydride, LiAlH$_4$: Reduces ketones, aldehydes, esters, and carboxylic acids to yield alcohols (Section 17.6).
– Reduces amides to yield amines (Section 21.8).
– Reduces alkyl azides to yield amines (Section 24.6).
– Reduces nitriles to yield amines (Sections 21.9 and 24.6).

Lithium diisopropylamide (LDA), LiN(i-Pr)$_2$: Reacts with carbonyl compounds (aldehydes, ketones, esters) to yield enolate ions (Sections 22.5 and 22.8).

Lithium diorganocopper reagent (Gilman reagent), LiR$_2$Cu: Couples with alkyl halides to yield alkanes (Section 10.9).
– Adds to α,β-unsaturated ketones to give 1,4-addition products (Section 19.17).
– Reacts with acid chlorides to give ketones (Section 21.5).

Lithium tri-$tert$-butoxyaluminum hydride, LiAl(O-t-Bu)$_3$H: Reduces acid chlorides to yield aldehydes (Section 21.5).

Magnesium, Mg: Reacts with organohalides to yield Grignard reagents (Section 10.8).

Mercuric acetate, Hg(OCOCH$_3$)$_2$: Adds to alkenes in the presence of water, giving α-hydroxy organomercury compounds that can be reduced with NaBH$_4$ to yield alcohols. The overall effect is the Markovnikov hydration of an alkene (Section 7.4).

Mercuric sulfate, HgSO$_4$: Acts as a catalyst for the addition of water to alkynes in the presence of aqueous sulfuric acid, yielding ketones (Section 8.5).

Mercuric trifluoroacetate, Hg(OCOCF$_3$)$_2$: Adds to alkenes in the presence of alcohol, giving α-alkoxy organomercury compounds that can be reduced with NaBH$_4$ to yield ethers. The overall reaction effects a net addition of an alcohol to an alkene (Section 18.5).

Methyl sulfate, (CH$_3$O)$_2$SO$_2$: A reagent used to methylate heterocyclic amine bases during Maxam-Gilbert DNA sequencing (Section 29.15).

Nitric acid, HNO$_3$: Reacts with arenes in the presence of sulfuric acid to yield nitroarenes (Section 16.2).
– Oxidizes aldoses to yield aldaric acids (Section 26.8).

Nitrous acid, HNO$_2$: Reacts with amines to yield diazonium salts (Section 25.3).

Osmium tetraoxide, OsO$_4$: Adds to alkenes to yield 1,2-diols (Section 7.8).
- Reacts with alkenes in the presence of periodic acid to cleave the carbon-carbon double bond, yielding ketone or aldehyde fragments (Section 7.8).

Ozone, O$_3$: Adds to alkenes to cleave the carbon-carbon double bond and give ozonides, which can be reduced with zinc in acetic acid to yield carbonyl compounds (Section 7.8).

Palladium on barium sulfate, Pd/BaSO$_4$: Acts as a hydrogenation catalyst for nitriles in the Kiliani-Fischer chain-lengthening reaction of carbohydrates (Section 26.8).

Palladium on carbon, Pd/C: Acts as a hydrogenation catalyst for reducing carbon-carbon multiple bonds. Alkenes and alkynes are reduced to yield alkanes (Sections 7.7 and 8.6).
- Acts as a hydrogenation catalyst for reducing aryl ketones to yield alkylbenzenes (Section 16.11).
- Acts as a hydrogenation catalyst for reducing nitroarenes to yield anilines (Section 25.2).

Periodic acid, HIO$_4$: Reacts with 1,2-diols to yield carbonyl-containing cleavage products (Section 7.8).

Peroxyacetic acid, CH$_3$CO$_3$H: Oxidizes sulfoxides to yield sulfones (Section 18.11)

Phenylisothiocyanate, C$_6$H$_5$–N=C=S: Used in the Edman degradation of peptides to identify N-terminal amino acids (Section 27.9).

Phosphorus oxychloride, POCl$_3$: Reacts with secondary and tertiary and alcohols to yield alkene dehydration products (Section 17.8).

Phosphorus tribromide, PBr$_3$: Reacts with alcohols to yield alkyl bromides (Section 10.7).
- Reacts with carboxylic acids to yield acid bromides (Section 21.5).
- Reacts with carboxylic acids in the presence of bromine to yield α-bromo carboxylic acids (Hell-Volhard-Zelinskii reaction; Section 22.4).

Platinum oxide (Adam's catalyst), PtO$_2$: Acts as a hydrogenation catalyst in the reduction of alkenes and alkynes to yield alkanes (Sections 7.7 and 8.6).

Potassium hydroxide, KOH: Reacts with alkyl halides to yield alkenes by an elimination reaction (Sections 7.1 and 11.11).
- Reacts with 1,1- or 1,2-dihaloalkanes to yield alkynes by a twofold elimination reaction (Section 8.1).

Potassium nitrosodisulfonate (Fremy's salt), K(SO$_3$)$_2$NO: Oxidizes phenols and anilines to yield quinones (Section 25.8).

Potassium permanganate, KMnO$_4$: oxidizes alkenes under neutral or acidic conditions to give carboxylic acid double-bond cleavage products (Sections 7.8).
- Oxidizes alkynes to give carboxylic acid triple-bond cleavage products (Section 8.7).
- Oxidizes arenes to yield benzoic acids (Section 16.10).

Potassium phthalimide, C$_6$H$_4$(CO)$_2$NK: Reacts with alkyl halides to yield N-alkyl-phthalimides, which are hydrolyzed by aqueous sodium hydroxide to yield amines (Gabriel amine synthesis; Section 24.6).

Potassium *tert*-butoxide, KO-*t*-Bu: Reacts with alkyl halides to yield alkenes (Sections 11.10 and 11.11).
- Reacts with allylic halides to yield conjugated dienes (Section 14.1).
- Reacts with chloroform in the presence of an alkene to yield a dichlorocyclopropane (Section 7.6).

Pyridine, C_5H_5N: Acts as a basic catalyst for the reaction of alcohols with acid chlorides to yield esters (Section 21.5).
- Acts as a basic catalyst for the reaction of alcohols with acetic anhydride to yield acetate esters (Section 21.6).
- Reacts with α-bromo ketones to yield α,β-unsaturated ketones (Section 22.3).

Pyridinium chlorochromate (PCC), $C_5H_6NCrO_3Cl$: Oxidizes primary alcohols to yield aldehydes and secondary alcohols to yield ketones (Section 17.9).

Pyrrolidine, C_4H_8N: Reacts with ketones to yield enamines for use in the Stork enamine reaction (Sections 19.12 and 23.12).

Rhodium on carbon, Rh/C: Acts as a hydrogenation catalyst in the reduction of benzene rings to yield cyclohexanes (Section 16.11).

Silver oxide, Ag_2O: Oxidizes primary alcohols in aqueous ammonia solution to yield aldehydes (Tollens oxidation; Section 19.5).
- Catalyzes the reaction of alcohols with alkyl halides to yield ethers (Section 26.8).
- Reacts with tetraalkylammonium salts to yield alkenes (Hofmann elimination; Section 24.6).

Sodium amide, $NaNH_2$: Reacts with terminal alkynes to yield acetylide anions (Section 8.8).
- Reacts with 1,1-or 1,2-dihalides to yield alkynes by a twofold elimination reaction (Section 8.1).
- Reacts with aryl halides to yield anilines by a benzyne aromatic substitution mechanism (Section 16.9).

Sodium azide, NaN_3: Reacts with alkyl halides to yield alkyl azides (Section 24.6).
- Reacts with acid chlorides to yield acyl azides. On heating in the presence of water, acyl azides yield amines and carbon dioxide (Section 24.6).

Sodium bisulfite, $NaHSO_3$: Reduces osmate esters, prepared by treatment of an alkene with osmium tetraoxide, to yield 1,2-diols (Section 7.8).

Sodium borohydride, $NaBH_4$: Reduces organomercury compounds, prepared by oxymercuration of alkenes, to convert the C-Hg bond to C-H (Section 7.4).
- Reduces ketones and aldehydes to yield alcohols (Sections 17.6 and 19.11).
- Reduces quinones to yield hydroquinones (Section 25.8).

Sodium cyanide, NaCN: Reacts with alkyl halides to yield alkanenitriles (Sections 20.6 and 21.9).

Sodium cyanoborohydride, $NaBH_3CN$: Reacts with ketones and aldehydes in the presence of ammonia to yield an amine by a reductive amination process (Section 24.6).

Sodium dichromate, $Na_2Cr_2O_7$: Oxidizes primary alcohols to yield carboxylic acids and secondary alcohols to yield ketones (Sections 17.9 and 19.5).
- Oxidizes alkylbenzenes to yield benzoic acids (Section 16.10).

Sodium hydride, NaH: Reacts with alcohols to yield alkoxide anions (Section 17.4).

Sodium hydroxide, NaOH: Reacts with arenesulfonic acids at high temperature to yield phenols (Section 16.2).
- Reacts with aryl halides to yield phenols by a benzyne aromatic substitution mechanism (Section 16.9).
- Catalyzes the reaction of acid chlorides with alcohols to yield esters – the Schotten-Baumann reaction (Section 21.5).
- Catalyzes the reaction of acid chlorides with amines to yields amides (Section 21.5).
- Reacts with methyl ketones in the presence of iodine to yield carboxylic acids and iodoform (Section 22.7).

Sodium iodide, NaI: Reacts with arenediazonium salts to yield aryl iodides (Section 26.3).

Stannous chloride, SnCl$_2$: Reduces nitroarenes to yield anilines (Sections 16.2 and 25.2).
- Reduces quinones to yield hydroquinones (Section 25.8).

Sulfur trioxide, SO$_3$: Reacts with arenes in sulfuric acid solution to yield arenesulfonic acids (Section 16.2).

Sulfuric acid, H$_2$SO$_4$: Reacts with alcohols and water to yield alkenes (Section 7.4).
- Reacts with alkynes in the presence of water and mercuric acetate to yield ketones (Section 8.5).
- Catalyzes the reaction of nitric acid with aromatic rings to yield nitroarenes (Section 16.2).
- Catalyzes the reaction of SO$_3$ with aromatic rings to yield arenesulfonic acids (Section 16.2).

Tetrazole: Acts as a coupling reagent for use in DNA synthesis (Section 29.16).

Thionyl chloride, SOCl$_2$: Reacts with primary and secondary alcohols to yield alkyl chlorides (Section 10.7).
- Reacts with carboxylic acids to yield acid chlorides (Section 21.4).

Thiourea, H$_2$NCSNH$_2$: Reacts with primary alkyl halides to yield thiols (Section 17.12).

p-Toluenesulfonyl chloride, p-CH$_3$C$_6$H$_4$SO$_2$Cl: Reacts with alcohols to yield tosylates (Sections 11.2 and 17.8).

Trifluoroacetic acid, CF$_3$COOH: Acts as a catalyst for cleaving *tert*-butyl ethers, yielding alcohols and 2-methylpropene (Section 18.6).
- Acts as a catalyst for cleaving the *t*-BOC protecting group from amino acids in peptide synthesis (Section 27.11).

Triphenylphosphine, (C$_6$H$_5$)$_3$P: Reacts with primary alkyl halides to yield the alkyltriphenyl-phosphonium salts used in Wittig reactions (Section 19.15).

Zinc, Zn: Reduces ozonides, produced by addition of ozone to alkenes, to yield ketones and aldehydes (Section 7.8).
- Reduces disulfides to yield thiols (Section 17.12).
- Reduces ketones and aldehydes in the presence of aqueous HCl to yield alkanes (Clemmensen reduction; Section 19.13).

Zinc-copper couple, Zn-Cu: Reacts with diiodomethane in the presence of alkenes to yield cyclopropanes (Simmons-Smith reaction; Section 7.6).

Acetoacetic ester synthesis (Section 22.8): a multi-step reaction sequence for converting a primary alkyl halide into a methyl ketone having three more carbon atoms in the chain.

$$
RCH_2X \ + \ CH_3\text{-}\overset{\overset{O}{\|}}{C}\text{-}\overset{\cdot\cdot}{\underset{}{C}}H\text{-}\overset{\overset{O}{\|}}{C}\text{-}OCH_3 \quad \xrightarrow[\text{2. } H_3O^+,\text{ heat}]{\text{1. heat}} \quad RCH_2\text{-}CH_2\overset{\overset{O}{\|}}{C}CH_3 \ + \ CO_2 \ + \ CH_3OH
$$

Adams' catalyst (Section 7.7): PtO_2, a catalyst used for the hydrogenation of carbon-carbon double bonds.

Aldol condensation reaction (Section 23.2): the nucleophilic addition of an enol or enolate ion to a ketone or aldehyde, yielding a β-hydroxy ketone.

$$
2 \ \ R\text{-}\overset{\overset{O}{\|}}{C}\text{-}\underset{\underset{}{|}}{\overset{\overset{|}{}}{C}}\text{-}H \quad \xrightarrow{\text{NaOH}} \quad R\text{-}\overset{\overset{O}{\|}}{C}\text{-}\underset{\underset{R}{|}}{\overset{\overset{|}{}}{C}}\text{-}\underset{\underset{}{|}}{\overset{\overset{OH}{|}}{C}}\text{-}\underset{\underset{}{|}}{\overset{\overset{|}{}}{C}}\text{-}H
$$

Amidomalonate amino acid synthesis (Section 27.4): a multi-step reaction sequence, similar to the malonic ester synthesis, for converting a primary alkyl halide into an amino acid.

$$
RCH_2X \ + \ {}^-\text{:}CH(NHAc)(CO_2Et)_2 \quad \xrightarrow[\text{2. } H_3O^+]{\text{1. mix}} \quad RCH_2\text{-}\underset{\underset{NH_2}{|}}{CH}\overset{\overset{O}{\|}}{C}OH \ + \ CO_2 \ + \ 2 \ EtOH
$$

Benedict's test (Section 26.8): a chemical test for aldehydes, involving treatment with cupric ion in aqueous sodium citrate.

Cannizzaro reaction (Section 19.16): the disproportionation reaction that occurs when a nonenolizable aldehyde is treated with base.

$$
2 \ R_3C\overset{\overset{O}{\|}}{C}H \quad \xrightarrow[\text{2. } H_3O^+]{\text{1. } HO^-} \quad R_3C\overset{\overset{O}{\|}}{C}OH \ + \ R_3CCH_2OH
$$

Claisen condensation reaction (Section 23.8): a nucleophilic acyl substitution reaction that occurs when an ester enolate ion attacks the carbonyl group of a second ester molecule. The product is a β-keto ester.

$$
2 \ R\text{-}CH_2\text{-}\overset{\overset{O}{\|}}{C}\text{-}OCH_3 \quad \xrightarrow[\text{2. } H_3O^+]{\text{1. } Na^+ \ {}^-OCH_3} \quad R\text{-}CH_2\text{-}\overset{\overset{O}{\|}}{C}\text{-}\underset{\underset{R}{|}}{CH}\text{-}\overset{\overset{O}{\|}}{C}\text{-}OCH_3 \ + \ CH_3OH
$$

Claisen rearrangement (Sections 25.8 and 31.11): the thermal [3.3] sigmatropic rearrangement of an allyl vinyl ether or an allyl phenyl ether.

Clemmensen reduction (Section 19.13): a method for reducing a ketone or aldehyde to an alkane by treatment with amalgamated zinc and aqueous HCl.

Cope rearrangement (Section 31.11): the thermal [3.3] sigmatropic rearrangement of a 1,5-diene to a new 1,5-diene.

Curtius rearrangement (Section 24.6): the thermal rearrangement of an acyl azide to an isocyanate, followed by hydrolysis to yield an amine.

Diazonium coupling reaction (Section 25.3): the coupling reaction between an aromatic diazonium salt and a phenol or aniline.

Dieckmann reaction (Section 23.10): the intramolecular Claisen condensation reaction of a 1,6- or 1,7-diester, yielding a cyclic β-keto ester.

Diels-Alder cycloaddition reaction (Sections 14.8 and 31.7): the thermal reaction between a diene and a dienophile to yield a cyclohexene ring.

Edman degradation (Section 27.9): a method for cleaving the N-terminal amino acid from a peptide by treatment of the peptide with N-phenylisothiocyanate.

Fehling's test (Section 26.8): a chemical test for aldehydes, involving treatment with cupric ion in aqueous sodium tartrate.

Fischer esterification reaction (Section 21.4): the acid-catalyzed reaction between a carboxylic acid and an alcohol, yielding the ester.

Friedel-Crafts reaction (Section 16.3-4): the alkylation or acylation of an aromatic ring by treatment with an alkyl- or acyl chloride in the presence of a Lewis-acid catalyst.

Gabriel amine synthesis (Section 24.7): a multi-step sequence for converting a primary alkyl halide into a primary amine by alkylation with potassium phthalimide, followed by hydrolysis.

Gilman reagent (Section 10.9): a lithium dialkylcopper reagent, R_2CuLi, prepared by treatment of a cuprous salt with an alkyllithium. Gilman reagents undergo a coupling reaction with alkyl halides, and undergo a 1,4-addition reaction with α,β-unsaturated ketones.

Grignard reaction (Section 19.10): the nucleophilic addition reaction of an alkylmagnesium halide to a ketone, aldehyde, or ester carbonyl group.

Grignard reagent (Section 10.8): an organomagnesium halide, RMgX, prepared by reaction between an organohalide and magnesium metal.

Haloform reaction (Section 22.7): the conversion of a methyl ketone to a carboxylic acid and haloform by treatment with halogen and base.

Hell-Volhard-Zelinskii reaction (Section 22.4): the α-bromination of carboxylic acids by treatment with bromine and phosphorus tribromide.

Hofmann elimination (Section 24.7): a method for effecting the elimination reaction of an amine to yield an alkene. The amine is first treated with excess iodomethane, and the resultant quaternary ammonium salt is heated with silver oxide.

Hofmann rearrangement (Section 24.6): the rearrangement of an *N*-bromoamide to a primary amine by treatment with aqueous base.

Jones reagent (Section 17.9): a solution of CrO_3 in acetone/aqueous sulfuric acid. This reagent oxidizes primary and secondary alcohols to carbonyl compounds under mild conditions.

Kiliani-Fischer synthesis (Section 26.8): a multi-step sequence for chain-lengthening an aldose into the next higher homolog.

$$\begin{array}{c} \text{CHO} \\ | \\ \text{R} \end{array} \quad \begin{array}{c} \text{1. HCN} \\ \hline \text{2. H}_2\text{, Pd, BaSO}_4 \\ \text{3. H}_3\text{O}^+ \end{array} \quad \begin{array}{c} \text{CHO} \\ | \\ \text{CH(OH)} \\ | \\ \text{R} \end{array}$$

Kolbe-Schmitt carboxylation reaction (Section 25.8): a method for introducing a carboxyl group in the ortho position of a phenol by treatment of the phenoxide anion with CO_2.

$$\begin{array}{c} \text{1. NaOH} \\ \hline \text{2. CO}_2 \\ \text{3. H}_3\text{O}^+ \end{array}$$

Malonic ester synthesis (Section 22.8): a multi-step sequence for converting an alkyl halide into a carboxylic acid with the addition of two carbon atoms to the chain.

$$\text{R--CH}_2\text{--X} + \text{Na}^+ {}^-\text{:CH--}\overset{\overset{\displaystyle O}{\|}}{\text{C}}\text{--OCH}_3 \quad \begin{array}{c} \text{1. heat} \\ \hline \text{2. H}_3\text{O}^+\text{, heat} \end{array} \quad \text{RCH}_2\text{--CH}_2\overset{\overset{\displaystyle O}{\|}}{\text{C}}\text{OCH}_3 + CO_2 + CH_3OH$$
$$\underset{\displaystyle \text{COOCH}_3}{}$$

Maxam-Gilbert DNA sequencing (Section 29.15): a rapid and efficient method for sequencing long chains of DNA by employing selective cleavage reactions.

McLafferty rearrangement (Section 19.19): a general mass spectral fragmentation pathway for carbonyl compounds having a hydrogen three carbon atoms away from the carbonyl carbon.

Meisenheimer complex (Section 16.8): an intermediate formed in the nucleophilic aryl substitution reaction of a base with a nitro-substituted aromatic ring.

Merrifield solid-phase peptide synthesis (Section 27.12): a rapid and efficient means of peptide synthesis in which the growing peptide chain is attached to an insoluble polymer support.

Michael reaction (Section 23.11): the 1,4-addition reaction of a stabilized enolate anion such as that from a 1,3-diketone to an α,β-unsaturated carbonyl compound.

Robinson annulation reaction (Section 23.13): a multi-step sequence for building a new cyclohexenone ring onto a ketone. The sequence involves an initial Michael reaction of the ketone followed by an internal aldol cyclization.

Sandmeyer reaction (Section 25.3): a method for converting aryldiazonium salts into aryl halides by treatment with cuprous halide.

Schotten-Baumann reaction (Section 21.5): a method for preparing esters by treatment of an acid chloride with an alcohol in the presence of aqueous base.

Simmons-Smith reaction (Section 7.6): a method for preparing cyclopropanes by treating an alkene

Stork enamine reaction (Section 23.12): a multi-step sequence whereby ketones are converted into enamines by treatment with a secondary amine, and the enamines are then used in Michael reactions.

Strecker amino acid synthesis (Section 27.4): a multi-step sequence for converting an aldehyde into an amino acid by initial treatment with ammonium cyanide, followed by hydrolysis.

$$R-\overset{\overset{\displaystyle O}{\|}}{C}-H \xrightarrow{NH_3,\ KCN} R-\overset{\overset{\displaystyle NH_2}{|}}{C}H-CN \xrightarrow{H_3O^+} R-\overset{\overset{\displaystyle NH_2}{|}}{C}H-COOH$$

Tollen's test (Section 19.5): a chemical test for detecting aldehydes by treatment with ammoniacal silver nitrate. A positive test is signaled by formation of a silver mirror on the walls of the reaction vessel.

Walden inversion (Section 11.1): the inversion of stereochemistry at a stereogenic center during S_N2 reactions.

$$Nu:^- + \quad \overset{\diagdown}{\underset{\diagup}{C}}-X \longrightarrow Nu-\overset{\diagup}{\underset{\diagdown}{C}} + :X^-$$

Williamson ether synthesis (Section 18.4): a method for preparing ethers by treatment of a primary alkyl halide with an alkoxide ion.

$$R-O^- Na^+ + R'CH_2Br \longrightarrow R-O-CH_2R' + NaBr$$

Wittig reaction (Section 19 15): a general method of alkene synthesis by treatment of a ketone or aldehyde with an alkylidenetriphenylphosphorane.

$$\underset{R}{\overset{\overset{\displaystyle O}{\|}}{\underset{\diagup}{C}}}{\diagdown}R' + \overset{\diagdown}{\underset{\diagup}{C}}=PPh_3 \longrightarrow \overset{R}{\underset{R'}{\overset{\diagup}{\underset{\diagdown}{C}}=\overset{\diagup}{\underset{\diagdown}{C}}}} + Ph_3P=O$$

Wohl degradation (Section 26.8): a multi-step reaction sequence for degrading an aldose into the next lower homolog.

$$\begin{array}{c} CHO \\ | \\ CH(OH) \\ | \\ R \end{array} \xrightarrow[\substack{\text{2. Ac}_2O \\ \text{3. NaOEt}}]{\text{1. NH}_2OH} \begin{array}{c} CHO \\ | \\ R \end{array}$$

Wohl-Ziegler reaction (Section 10.5): a reaction for effecting allylic bromination by treatment of an alkene with N-bromosuccinimide.

$$\xrightarrow{NBS,\ CCl_4}$$

Wolff-Kishner reaction (Section 19.13): a method for converting a ketone or aldehyde into the corresponding hydrocarbon by treatment with hydrazine and strong base.

$$\underset{R}{\overset{O}{\underset{\displaystyle}{\|}}}\underset{R'}{\overset{\displaystyle C}{}} \quad \xrightarrow{\text{N}_2\text{H}_4, \ \text{KOH}} \quad \text{R-CH}_2\text{-R'}$$

Woodward-Hoffmann orbital symmetry rules (Section 31.12): a series of rules for predicting the stereochemistry of pericyclic reactions. Even-electron species react thermally through either antarafacial or conrotatory pathways, whereas odd-electron species react thermally through either suprafacial or disrotatory pathways (even-antara-con; odd-supra-dis).

Abbreviations

Å   symbol for Angstrom unit (10^{-8} cm = 10^{-10} m)

Ac-   acetyl group, $CH_3\overset{\overset{\displaystyle O}{\|}}{C}-$

Ar-   aryl group

at. no.   atomic number

at. wt.   atomic weight

$[\alpha]_D$ specific rotation

BOC   *tert*-butoxycarbonyl group, $(CH_3)_3\overset{\overset{\displaystyle O}{\|}}{C}O\overset{\overset{\displaystyle O}{\|}}{C}-$

bp   boiling point

n-Bu   *n*-butyl group, $CH_3CH_2CH_2CH_2$-

sec-Bu   *sec*-butyl group, $CH_3CH_2CH(CH_3)$-

t-Bu   *tert*-butyl group, $(CH_3)_3C$-

cm   centimeter

cm^{-1} wavenumber, or reciprocal centimeter

D   stereochemical designation of carbohydrates and amino acids

DCC   dicyclohexylcarbodiimide, C_6H_{11}-N=C=N-C_6H_{11}

δ   chemical shift in ppm downfield from TMS

Δ   symbol for heat; also symbol for change

ΔH heat of reaction

dm   decimeter (0.1 m)

DMF   dimethylformamide, $(CH_3)_2NCHO$

DMSO   dimethyl sulfoxide, $(CH_3)_2SO$

DNA   deoxyribonucleic acid

DNP   dinitrophenyl group, as in 2,4-DNP (2,4-dinitrophenylhydrazone)

(*E*)   entgegen, stereochemical designation of double bond geometry

E_{act} activation energy

E1   unimolecular elimination reaction

E2   bimolecular elimination reaction

Et   ethyl group, CH_3CH_2-

g   gram

hν	symbol for light
Hz	Hertz, or cycles per second (s^{-1})
i-	iso
IR	infrared
J	Joule
J	symbol for coupling constant
K	Kelvin temperature
K_a	acid dissociation constant
KJ	kilojoule
L	stereochemical designation of carbohydrates and amino acids
LAH	lithium aluminum hydride, $LiAlH_4$
Me	methyl group, CH_3-
mg	milligram (0.001 g)
MHz	megahertz (10^6 s^{-1})
mL	milliliter (0.001 L)
mm	millimeter (0.001 m)
mp	melting point
μg	microgram (10^{-6} g)
mμ	millimicron (nanometer, 10^{-9} m)
MW	molecular weight
n-	normal, straight-chain alkane or alkyl group
ng	nanogram (10^{-9} gram)
nm	nanometer (10^{-9} meter)
NMR	nuclear magnetic resonance
-OAc	acetate group, $\overset{\overset{\displaystyle O}{\|}}{-OCCH_3}$
PCC	pyridinium chlorochromate
Ph	phenyl group, $-C_6H_5$
pH	measure of acidity of aqueous solution
pK_a	measure of acid strength ($= -\log K_a$)
pm	picometer (10^{-12} m)
ppm	parts per million
n-Pr	*n*-propyl group, $CH_3CH_2CH_2-$
i-Pr	isopropyl group, $(CH_3)_2CH-$

R–   symbol for a generalized alkyl group

(R)   *rectus*, designation of stereogenic center

RNA   ribonucleic acid

(S)   *sinister*, designation of stereogenic center

sec-   secondary

S_N1 unimolecular substitution reaction

S_N2 bimolecular substitution reaction

tert-   tertiary

THF   tetrahydrofuran

TMS   tetramethylsilane nmr standard, $(CH_3)_4Si$

Tos   Tosylate group

UV   ultraviolet

X–   halogen group (–F, –Cl, –Br, –I)

(Z)   zusammen, stereochemical designation of double bond geometry

⟶   chemical reaction in direction indicated

⇌   reversible chemical reaction

↔   resonance symbol

⤻   curved arrow indicating direction of electron flow

≡   is equivalent to

>   greater than

<   less than

≈   approximately equal to

R⤙   indicates that the organic fragment shown is a part of a larger molecule

◣   single bond coming out of the plane of the paper

- - - - -   single bond receding into the plane of the paper

......   partial bond

δ+, δ–   partial charge

‡   denoting the transition state

Infrared Absorption Frequencies

Functional Group Class		Frequency (cm^{-1})	Text Section
Alcohol	–O–H	3300 - 3600 (s)	17.11
	$-\overset{\displaystyle \backslash}{\underset{\displaystyle /}{C}}-O-$	1050 (s)	17.11
Aldehyde	–CO–H	2720, 2820 (m)	19.19
aliphatic	$\overset{\displaystyle \backslash}{\underset{\displaystyle /}{C}}=O$	1725 (s)	19.19
aromatic	$\overset{\displaystyle \backslash}{\underset{\displaystyle /}{C}}=O$	1705 (s)	19.19
Alkane	$-\overset{\displaystyle \backslash}{\underset{\displaystyle /}{C}}-H$	2850 - 2960 (s)	12.7
	$-\overset{\displaystyle \backslash}{\underset{\displaystyle /}{C}}-\overset{\displaystyle /}{\underset{\displaystyle \backslash}{C}}-$	800 - 1300 (m)	12.7
Alkene	$=\overset{\displaystyle /H}{\underset{\displaystyle \backslash}{C}}$	3020 - 3100 (m)	12.7
	$\overset{\displaystyle \backslash}{\underset{\displaystyle /}{C}}=\overset{\displaystyle /}{\underset{\displaystyle \backslash}{C}}$	1650 - 1670 (m)	12.7
	$RCH=CH_2$	910, 990 (m)	12.7
	$R_2C=CH_2$	890 (m)	12.7
Alkyne	$\equiv C-H$	3300 (s)	12.7
	$-C\equiv C-$	2100 - 2260 (m)	12.7
Alkyl bromide	$-\overset{\displaystyle \backslash}{\underset{\displaystyle /}{C}}-Br$	500 - 600 (s)	12.7
Alkyl chloride	$-\overset{\displaystyle \backslash}{\underset{\displaystyle /}{C}}-Cl$	600 - 800 (s)	12.7

Amine, *primary*	$-N\begin{smallmatrix}H\\\\H\end{smallmatrix}$	3400, 3500 (s)	24.9
secondary	$\backslash N-H$	3350 (s)	24.9
Ammonium salt	$-\overset{+}{N}-H$	2200 - 3000 (broad)	24.9
Aromatic ring	Ar–H	3030 (m)	15.12
monosubstituted	Ar–R	690 - 710 (s)	15.12
		730 - 770 (s)	15.12
o-disubstituted		735 - 770 (s)	15.12
m-disubstituted		690 - 710 (s)	15.12
		810 - 850 (s)	15.12
p-disubstituted		810 - 840 (s)	15.12
Carboxylic acid	–O–H	2500 - 3300 (broad)	20.9
associated	$\backslash C=O$	1710 (s)	20.9
free	$\backslash C=O$	1760 (s)	20.9
Acid anhydride	$\backslash C=O$	1820, 1760 (s)	21.12
Acid chloride			
aliphatic	$\backslash C=O$	1810 (s)	21.12
aromatic	$\backslash C=O$	1770 (s)	21.12

Amide, *aliphatic*	C=O	1690 (s)	21.12
aromatic	C=O	1675 (s)	21.12
N-substituted	C=O	1680 (s)	21.12
N,N-disubstituted	C=O	1650 (s)	21.12
Ester, *aliphatic*	C=O	1735 (s)	21.12
aromatic	C=O	1720 (s)	21.12
Ether	$-\text{O}-\text{C}-$	1050 - 1150 (s)	18.10
Ketone, *aliphatic*	C=O	1715 (s)	19.19
aromatic	C=O	1690 (s)	19.19
6-memb. ring	C=O	1715 (s)	19.19
5-memb. ring	C=O	1750 (s)	19.19
Nitrile, *aliphatic*	$-\text{C}\equiv\text{N}$	2250 (m)	21.12
aromatic	$-\text{C}\equiv\text{N}$	2230 (m)	21.12
Phenol	$-\text{O}-\text{H}$	3500 (s)	25.9

(s) = strong; (m) = medium intensity

Proton NMR Chemical Shifts

Type of Proton		Chemical Shift (δ)	Text Section
Alkyl, primary	$R\text{-}CH_3$	0.7 – 1.3	13.5
Alkyl, secondary	$R\text{-}CH_2\text{-}R$	1.2 – 1.4	13.5
Alkyl, tertiary	$R_3C\text{-}H$	1.4 – 1.7	13.5
Allylic	$-C{=}C\text{-}C\text{-}H$	1.6 – 1.9	13.5
Alpha to carbonyl	$-\overset{\overset{O}{\|}}{C}\text{-}C\text{-}H$	2.0 – 2.3	19.19
Benzylic	$Ar\text{-}C\text{-}H$	2.3 – 3.0	15.12
Acetylenic	$R-C{\equiv}C-H$	2.5 – 2.7	13.5
Alkyl chloride	$Cl\text{-}C\text{-}H$	3.0 – 4.0	13.5
Alkyl bromide	$Br\text{-}C\text{-}H$	2.5 – 4.0	13.5
Alkyl iodide	$I{-}C{-}H$	2.0 – 4.0	13.5
Amine	$N{-}C{-}H$	2.2 – 2.6	24.9
Epoxide	$C{-}C\,H$	2.5 – 3.5	18.10
Alcohol	$HO\text{-}C\text{-}H$	3.5 – 4.5	17.11
Ether	$RO{-}C{-}H$	3.5 – 4.5	18.10
Vinylic	$-C{=}C{-}H$	5.0 – 6.5	13.5
Aromatic	$Ar\text{-}H$	6.5 – 8.0	15.12
Aldehyde	$R-\overset{\overset{O}{\|}}{C}-H$	9.7 – 10.0	19.19
Carboxylic acid	$R-\overset{\overset{O}{\|}}{C}-O{-}H$	11.0 – 12.0	20.9
Alcohol	$R{-}O{-}H$	3.5 – 4.5	17.11
Phenol	$Ar\text{-}O\text{-}H$	2.5 – 6.0	25.9

Ethylene (24,260,000 tons/yr):
prepared by thermal cracking of ethane and propane during petroleum refining; used as starting material for manufacture of polyethylene, ethylene oxide, ethylene glycol, ethylbenzene, 1,2-dichloroethane, and other bulk chemicals.

Propylene (14,420,000 tons/yr):
prepared by steam cracking of light hydrocarbon fractions during petroleum refining; used as starting material for the manufacture of polypropylene, acrylonitrile, propylene oxide, and isopropyl alcohol.

1,2-Dichloroethane (Ethylene dichloride; 9,350,000 tons/yr):
prepared by addition of chlorine to ethylene in the presence of $FeCl_3$ catalyst at 50°C; used as a chlorinated solvent and as starting material for the manufacture of vinyl chloride.

Vinyl chloride (7,405,000 tons/yr):
prepared by addition of chlorine to ethylene followed by elimination of HCl; used as starting material for preparation of poly(vinyl chloride) polymers (hoses, pipes, molded objects).

Benzene (7,330,000 tons/yr):
obtained from petroleum by catalytic reforming of hexane and cyclohexane over a platinum catalyst; used as starting material for the synthesis of ethylbenzene, cumene, cyclohexane, and aniline.

Methyl *tert*-butyl ether (MTBE; 6,835,000 tons/yr):
prepared by acid-catalyzed addition of methanol to isobutylene; used as an octane enhancer in gasoline.

Ethylbenzene (5,935,000 tons/yr):
prepared during catalytic reforming in petroleum refining and by an acid-catalyzed Friedel-Crafts alkylation of benzene with ethylene; used almost exclusively for production of styrene.

Styrene (5,635,000 tons/yr):
prepared by high-temperature catalytic dehydrogenation of ethylbenzene; used in the manufacture of polystyrene polymers (thermoplastics, packaging materials).

Methanol (5,405,000 tons/yr):
prepared by high temperature reaction of a mixture of H_2, CO, and CO_2 ("synthesis gas") over a catalyst at 100 atmospheres pressure; used as a solvent and as starting material for the manufacture of formaldehyde, acetic acid, and methyl *tert*-butyl ether.

Dimethyl terephthalate (4,320,000 tons/yr):
prepared from *p*-xylene by oxidation and esterification; used in the manufacture of polyester polymers (textiles, upholstery, recording tape, and film).

Formaldehyde (3,970,000 tons/yr):
prepared by air oxidation of methanol over a silver or metal oxide catalyst; used in the manufacture of phenolic resins, melamine resins, and plywood adhesives.

Ethylene oxide (3,390,000 tons/yr):
  prepared by high-temperature air oxidation of ethylene over a silver catalyst; used as starting material for the preparation of ethylene glycol and poly(ethylene glycol).

Toluene (3,375,000 tons/yr):
  prepared during catalytic reforming of petroleum; used as a gasoline additive and as a degreasing solvent.

p-**Xylene** (3,115,000 tons/yr):
  prepared by separation from the mixed xylenes that result during catalytic reforming in gasoline refining; used as starting material for manufacture of the dimethyl terephthalate needed for polyester synthesis.

Ethylene glycol (2,775,000 tons/yr):
  prepared by high-temperature reaction between water and ethylene oxide at neutral pH; used as antifreeze and as a starting material for polymers and latex paints.

Cumene (2,580,000 tons/yr):
  prepared by a phosphoric-acid-catalyzed Friedel-Crafts reaction between benzene and propylene; used primarily for conversion into phenol and acetone.

Phenol (2,025,000 tons/yr):
  prepared from cumene by air oxidation to cumene hydroperoxide, followed by acid-catalyzed decomposition; used as starting material for preparing phenolic resins, epoxy resins, and caprolactam.

Acetic acid (1,910,000 tons/yr):
  prepared by metal-catalyzed air oxidation of acetaldehyde under pressure at 80°C and by reaction of methanol with carbon monoxide; used to make vinyl acetate polymers, ethyl acetate solvent, and cellulose acetate polymers.

Propylene oxide (1,850,000 tons/yr):
  prepared from propylene by high-temperature oxidation with air and by formation of propylene chlorohydrin followed by loss of HCl; used in the manufacture of propylene glycol, poly-urethanes, and polyesters.

1,3-Butadiene (1,700,000 tons/yr):
  prepared by steam cracking of gas oil during petroleum refining and by dehydrogenation of butane and butene; used primarily as a monomer component in the manufacture of styrene-butadiene rubber (SBR), polybutadiene rubber, and acrylonitrile-butadiene-styrene (ABS) copolymers.

Acrylonitrile (1,540,000 tons,yr):
  prepared by the Sohio ammoxidation process in which propylene, ammonia, and air are passed over a catalyst at 500°C; used in the preparation of acrylic fibers, nitrile rubber, and acrylo-nitrile-butadiene-styrene (ABS) copolymer.

Vinyl acetate (1,510,000 tons/yr):
  prepared from reaction of acetic acid, ethylene, and oxygen; used for manufacture of poly(vinyl acetate) (paint emulsions, plywood adhesives, textiles).

Acetone (1,385,000 tons/yr):

   prepared by acid-catalyzed decomposition of cumene hydroperoxide and by air oxidation of isopropyl alcohol at 300°C over a metal oxide catalyst; used as a solvent and as starting material for synthesizing bisphenol A and methyl methacrylate.

Cyclohexane (1,550,000 tons/yr):

   prepared by catalytic hydrogenation of benzene; used as starting material for synthesis of the caprolactam and adipic acid needed for nylon.

Caprolactam (840,000 tons/yr):

   prepared from phenol by conversion into cyclohexanone, followed by formation and acid-catalyzed rearrangement of cyclohexanone oxime; used as starting material for the manufacture of nylon-6.

Bisphenol A (740,000 tons/yr):

   prepared by reaction of phenol with acetone; used in the manufacture of epoxy resins and adhesives, polycarbonates, and polysulfones.

1-Butanol (725,000 tons/yr):

   prepared in the oxo process by reaction of propylene with carbon monoxide; used as solvent and as a starting material for synthesis of butyl acetate and dibutyl phthalate.

Isopropyl alcohol (695,000 tons/yr):

   prepared by direct high-temperature addition of water to propylene; used in cosmetics formulations, as a solvent and deicer, and as starting material for manufacture of acetone.

Methyl methacrylate (652,000 tons/yr):

   prepared by acetone cyanohydrin by treatment with sulfuric acid to effect dehydration, followed by esterification with methanol; used for the synthesis of methacrylate polymers such as Lucite.

Aniline (632,000 tons/yr):

   prepared by catalytic reduction of nitrobenzene with hydrogen at 350°C; used as starting material for preparing toluene diisocyanate and for the synthesis of dyes and pharmaceuticals.

Isobutylene (607,000 tons/yr):

   prepared from catalytic cracking of petroleum; used in the manufacture of methyl *tert*-butyl ether, isoprene, and butylated phenols.

Chloromethane (Methyl chloride; 498,000 tons/yr):

   prepared by reaction of methanol with HCl at 0°C and by radical chlorination of methane; used in the manufacture of silicones, synthetic rubber, and methyl cellulose.

Phthalic anhydride (480,000 tons/yr):

   prepared by oxidation of *o*-xylene at 400°C and by oxidation of naphthalene obtained from coal tar; used for the synthesis of polyesters and plasticizers.

Propylene glycol (478,000 tons/yr):

   prepared by high temperature reaction of propylene oxide with water; used in the preparation of polyesters and as an additive in the food industry.

o-**Xylene** (457,000 tons/yr):
  obtained by separation from the mixed xylenes that result during catalytic reforming in petroleum refining; used as starting material for preparation of phthalic acid and phthalic anhydride.

Ethanolamines (377,000 tons/yr):
  prepared by reaction of ammonia with ethylene oxide at 100°C; used in soaps, detergents, cosmetics, and corrosion inhibitors.

2-Ethylhexanol (366,000 tons/yr):
  prepared from butanal by aldol condensation and catalytic hydrogenation (the Oxo Process); used in the manufacture of plasticizers, lubricating-oil additives, and detergents.

Ethanol (357,000 tons/yr):
  prepared by direct vapor phase hydration of ethylene at 300°C over an acidic catalyst; used as a solvent, as a constituent of cleaning preparations, and as starting material for ester synthesis.

2-Butanone (Methyl ethyl ketone; 300,000 tons/yr):
  prepared by oxidation of 2-butanol over a ZnO catalyst at 400°C; used as a solvent for vinyl coatings, lacquers, rubbers, and paint removers.

Trichloromethane (Chloroform; 283,000 tons/yr):
  prepared by radical chlorination of methane or chloromethane with Cl_2; used as a solvent and as a starting material for the manufacture of chlorodifluoromethane refrigerant.

Nobel Prizes in Chemistry

1901 **Jacobus H. van't Hoff** (The Netherlands):
"for the discovery of laws of chemical dynamics and of osmotic pressure"

1902 **Emil Fischer** (Germany):
"for syntheses in the groups of sugars and purines"

1903 **Svante A. Arrhenius** (Sweden):
"for his theory of electrolytic dissociation"

1904 **Sir William Ramsey** (Britain):
"for the discovery of gases in different elements in the air and for the determination of their place in the periodic system"

1905 **Adolf von Baeyer** (Germany):
"for his researches on organic dyestuffs and hydroaromatic compounds"

1906 **Henri Moissan** (France):
"for his research on the isolation of the element fluorine and for placing at the service of science the electric furnace that bears his name"

1907 **Eduard Buchner** (Germany):
"for his biochemical researches and his discovery of cell-less formation"

1908 **Ernest Rutherford** (Britain):
"for his investigation into the disintegration of the elements and the chemistry of radioactive substances"

1909 **Wilhelm Ostwald** (Germany):
"for his work on catalysis and on the conditions of chemical equilibrium and velocities of chemical reactions"

1910 **Otto Wallach** (Germany):
"for his services to organic chemistry and the chemical industry by his pioneer work in the field of alicyclic substances"

1911 **Marie Curie** (France):
"for her services to the advancement of chemistry by the discovery of the elements radium and polonium"

1912 **Victor Grignard** (France):
"for the discovery of the so-called Grignard reagent, which has greatly helped in the development of organic chemistry"

Paul Sabatier (France):
"for his method of hydrogenating organic compounds in the presence of finely divided metals"

1913   **Alfred Werner** (Switzerland):
"for his work on the linkage of atoms in molecules by which he has thrown new light on earlier investigations and opened up new fields of research especially in inorganic chemistry"

1914   **Theodore W. Richards** (U.S.):
"for his accurate determinations of the atomic weights of a great number of chemical elements"

1915   **Richard M. Willstätter** (Germany):
"for his research on plant pigments, principally on chlorophyll"

1916   No award

1917   No award

1918   **Fritz Haber** (Germany):
"for the synthesis of ammonia from its elements, nitrogen and hydrogen"

1919   No award

1920   **Walther H. Nernst** (Germany):
"for his thermochemical work"

1921   **Frederick Soddy** (Britain):
"for his contributions to the chemistry of radioactive substances and his investigations into the origin and nature of isotopes"

1922   **Francis W. Aston** (Britain):
"for his discovery, by means of his mass spectrograph, of the isotopes of a large number of nonradioactive elements, as well as for his discovery of the whole-number rule"

1923   **Fritz Pregl** (Austria):
"for his invention of the method of microanalysis of organic substances"

1924   No award

1925   **Richard A. Zsigmondy** (Germany):
for his demonstration of the heterogeneous nature of colloid solutions, and for the methods he used, which have since become fundamental in modern colloid chemistry"

1926   **Theodor Svedberg** (Sweden):
"for his work on disperse systems"

1927   **Heinrich O. Wieland** (Germany):
"for his research on bile acids and related substances"

1928   **Adolf O. R. Windaus** (Germany):
"for his studies on the constitution of the sterols and their connection with the vitamins"

1929   **Arthur Harden** (Britain):
Hans von Euler-Chelpin (Sweden):
"for their investigation on the fermentation of sugar and of fermentative enzymes"

1930 **Hans Fischer** (Germany):
"for his researches into the constitution of hemin and chlorophyll, and especially for his synthesis of hemin"

1931 **Frederich Bergius** (Germany):
Carl Bosch (Germany):
"for their contributions to the invention and development of chemical high-pressure methods"

1932 **Irving Langmuir** (U.S.):
"for his discoveries and investigations in surface chemistry"

1933 No award

1934 **Harold C. Urey** (U.S.):
"for his discovery of heavy hydrogen"

1935 **Frederic Joliot** (France):
Irene Joliot-Curie (France):
"for their synthesis of new radioactive elements"

1936 **Peter J. W. Debye** (Netherlands/U.S.):
"for his contributions our knowledge of molecular structure through his investigations on dipole moments and on the diffraction of X-rays and electrons in gases"

1937 **Walter N. Haworth** (Britain):
"for his researches into the constitution of carbohydrates and vitamin C"

Paul Karrer (Switzerland):
"for his researches into the constitution of carotenoids, flavins, and vitamins A and B"

1938 **Richard Kuhn** (Germany):
"for his work on carotenoids and vitamins"

1939 **Adolf F. J. Butenandt** (Germany):
"for his work on sex hormones"

Leopold Ruzicka (Switzerland):
"for his work on polymethylenes and higher terpenes"

1940 No award

1941 No award

1942 No award

1943 **Georg de Hevesy** (Hungary):
"for his work on the use of isotopes as tracer elements in researches on chemical processes"

1944 **Otto Hahn** (Germany):
"for his discovery of the fission of heavy nuclei"

1945   **Artturi I. Virtanen** (Finland):
"for his researches and inventions in agricultural and nutritive chemistry, especially for his fodder preservation method"

1946   **James B. Sumner** (U.S.):
"for his discovery that enzymes can be crystallized"

John H. Northrop (U.S.):
Wendell M. Stanley (U.S.):
for their preparation of enzymes and virus proteins in a pure form"

1947   **Sir Robert Robinson** (Britain):
"for his investigations on plant products of biological importance, particularly the alkaloids"

1948   **Arne W. K. Tiselius** (Sweden):
"for his researches on electrophoresis and adsorption analysis, especially for his discoveries concerning the complex nature of the serum proteins"

1949   **William F. Giauque** (U.S.):
"for his contributions in the field of chemical thermodynamics, particularly concerning the behavior of substances at extremely low temperatures"

1950   **Kurt Alder** (Germany):
Otto P. H. Diels (Germany):
"for their discovery and development of the diene synthesis"

1951   **Edwin M. McMillan** (U.S.):
Glenn T. Seaborg (U.S.):
"for their discoveries in the chemistry of the transuranium elements"

1952   **Archer J. P. Martin** (Britain):
Richard L. M. Synge (Britain):
"for their development of partition chromatography"

1953   **Hermann Staudinger** (Germany):
"for his discoveries in the field of macromolecular chemistry"

1954   **Linus C. Pauling** (U.S.):
"for his research into the nature of the chemical bond and its application to the elucidation of the structure of complex substances"

1955   **Vincent du Vigneaud** (U.S.):
"for his work on biochemically important sulfur compounds, especially for the first synthesis of a polypeptide hormone"

1956   **Sir Cyril N. Hinshelwood** (Britain):
Nikolai N. Semenov (U.S.S.R.):
"for their research in clarifying the mechanisms of chemical reactions in gases"

1957   **Sir Alexander R. Todd** (Britain):
"for his work on nucleotides and nucleotide coenzymes"

1958 **Frederick Sanger** (Britain):
"for his work on the structure of proteins, particularly insulin"

1959 **Jaroslav Heyrovsky** (Czechoslovakia):
"for his discovery and development of the polarographic method of analysis"

1960 **Willard F. Libby** (U.S.):
"for his method to use carbon-14 for age determination in archaeology, geology, geophysics, and other branches of science"

1961 **Melvin Calvin** (U.S.):
"for his research on the carbon dioxide assimilation in plants"

1962 **John C. Kendrew** (Britain):
Max F. Perutz (Britain):
"for their studies of the structures of globular proteins"

1963 **Giulio Natta** (Italy):
Karl Ziegler (Germany):
"for their work in the controlled polymerization of hydrocarbons through the use of organometallic catalysts"

1964 **Dorothy C. Hodgkin** (Britain):
"for her determinations by X-ray techniques of the structures of important biochemical substances, particularly vitamin B-12 and penicillin"

1965 **Robert B. Woodward** (U.S.):
"for his outstanding achievements in the 'art' of organic synthesis"

1966 **Robert S. Mulliken** (U.S.):
"for his fundamental work concerning chemical bonds and the electronic structure of molecules by the molecular orbital method"

1967 **Manfred Eigen** (Germany):
Ronald G. W. Norrish (Britain):
George Porter (Britain):
"for their studies of extremely fast chemical reactions, effected by disturbing the equilibrium with very short pulses of energy"

1968 **Lars Onsager** (U.S.):
"for his discovery of the reciprocal relations bearing his name, which are fundamental for the thermodynamics of irreversible processes"

1969 **Sir Derek H. R. Barton** (Britain):
Odd Hassel (Norway):
"for their contributions to the development of the concept of conformation and its application in chemistry"

1970 **Luis F. Leloir** (Argentina):
"for his discovery of sugar nucleotides and their role in the biosynthesis of carbohydrates"

1971 **Gerhard Herzberg** (Canada):
"for his contributions to the knowledge of electronic structure and geometry of molecules, particularly free radicals"

1972 **Christian B. Anfinsen** (U.S.):
"for his work on ribonuclease, especially concerning the connection between the amino acid sequence and the biologically active conformation"

Stanford Moore (U.S.):
William H. Stein (U.S.):
"for their contribution to the understanding of the connection between chemical structure and catalytic activity of the active center of the ribonuclease molecule"

1973 **Ernst Otto Fischer** (Germany):
Geoffrey Wilkinson (Britain):
"for their pioneering work, performed independently, on the chemistry of the organo-metallic sandwich compounds"

1974 **Paul J. Flory** (U.S.):
"for his fundamental achievements, both theoretical and experimental, in the physical chemistry of macromolecules"

1975 **John Cornforth** (Australia/Britain):
"for his work on the stereochemistry of enzyme-catalyzed reactions"

Vladimir Prelog (Yugoslavia/Switzerland):
"for his work on the stereochemistry of organic molecules and reactions"

1976 **William N. Lipscomb** (U.S.):
"for his studies on the structures of boranes illuminating problems of chemical bonding"

1977 **Ilya Pregogine** (Belgium):
"for his contributions to nonequilibrium thermodynamics, particularly the theory of dissipative structures"

1978 **Peter Mitchell** (Britain):
"for his contribution to the understanding of biological energy transfer through the formulation of the chemiosmotic theory"

1979 **Herbert C. Brown** (U.S.):
"for his application of boron compounds to synthetic organic chemistry"

Georg Wittig (Germany):
"for developing phosphorus reagents, presently bearing his name"

1980 **Paul Berg** (U.S.):
"for his fundamental studies of the biochemistry of nucleic acids, with particular regard to recombinant DNA"

Walter Gilbert (U.S.)
Frederick Sanger (Britain):
"for their contributions concerning the determination of base sequences in nucleic acids"

1981 **Kenichi Fukui** (Japan)
Roald Hoffmann (U.S.):
for their theories, developed independently, concerning the course of chemical reactions"

1982 **Aaron Klug** (Britain):
"for his development of crystallographic electron microscopy and his structural elucidation of biologically important nucleic acid - protein complexes"

1983 **Henry Taube** (U.S.):
"for his work on the mechanisms of electron transfer reactions, especially in metal complexes"

1984 **R. Bruce Merrifield** (U.S.):
"for his development of methodology for chemical synthesis on a solid matrix"

1985 **Herbert A. Hauptman** (U.S.):
Jerome Karle (U.S.):
"for their outstanding achievements in the development of direct methods for the determination of crystal structures"

1986 **John C. Polanyi** (Canada):
"for his pioneering work in the use of infrared chemiluminescence in studying the dynamics of chemical reactions"

Dudley R. Herschbach (U.S.):
Yuan T. Lee (U.S.):
"for their contributions concerning the dynamics of chemical elementary processes"

1987 **Donald J. Cram** (U.S.):
Jean-Marie Lehn (France):
Charles J. Pedersen (U.S.):
"for their development and use of molecules with structure-specific interactions of high selectivity"

1988 **Johann Deisenhofer** (Germany):
Robert Huber (Germany):
Hartmut Michel (Germany):
"for their determination of the structure of the photosynthetic reaction center of bacteria"

1989 **Sidney Altman** (U.S.):
Thomas R. Cech (U.S.):
"for their discovery of catalytic properties of RNA"

1990 **Elias J. Corey** (U.S.):
"for his development of the theory and methodology of organic synthesis"

1991 **Richard R. Ernst** (Switzerland):
"for his contributions to the development of the methodology of high resolution NMR spectroscopy"

1992 **Rudolph A. Marcus** (U.S.):
"for his contributions to the theory of electron-transfer reactions in chemical systems"

1993 **Kary B. Mullis** (U.S.):
"for his development of the polymerase chain reaction"

Michael Smith (Canada):
"for his fundamental contributions to the establishment of oligonucleotide-based site-directed mutagenesis and its development for protein studies"

1994 **George A Olah** (U.S.):
"for pioneering research on carbocations and their role in the chemical reactions of hydrocarbons"

1995 **F. Sherwood Rowland** (U.S.)
Mario Molina (U.S.)
Paul Crutzen (Germany)
"for explaining the chemical mechanisms that affect the thickness of the ozone layer."

TO THE OWNER OF THIS BOOK:

We hope that you have found *Study Guide and Solutions Manual for Organic Chemistry*, Fourth Edition, useful. So that this book can be improved in a future edition, would you take the time to complete this sheet and return it? Thank you.

School and address: _____

Department: _____

Instructor's name: _____

1. What I like most about this book is: _____

_____

_____

2. What I like least about this book is: _____

_____

_____

3. My general reaction to this book is: _____

_____

4. The name of the course in which I used this book is: _____

_____

5. Were all of the chapters of the book assigned for you to read? _____

   If not, which ones weren't? _____

6. In the space below, or on a separate sheet of paper, please write specific suggestions for improving this book and anything else you'd care to share about your experience in using the book.

_____

_____

_____

_____

_____

Optional:

Your name: ______________________________ Date: ______________________________

May Brooks/Cole quote you, either in promotion for *Study Guide and Solutions Manual for Organic Chemistry,* Fourth Edition, or in future publishing ventures?

Yes: __________ No: _________

Sincerely,

Susan McMurry

FOLD HERE

NO POSTAGE NECESSARY IF MAILED IN THE UNITED STATES

BUSINESS REPLY MAIL

FIRST CLASS PERMIT NO. 358 PACIFIC GROVE, CA

POSTAGE WILL BE PAID BY ADDRESSEE

ATT: *Susan McMurry*

Brooks/Cole Publishing Company
511 Forest Lodge Road
Pacific Grove, California 93950-9968

FOLD HERE

CUT ALONG DOTTED LINE

FOLD HERE

NO POSTAGE
NECESSARY
IF MAILED
IN THE
UNITED STATES

BUSINESS REPLY MAIL

FIRST CLASS PERMIT NO. 358 PACIFIC GROVE, CA

POSTAGE WILL BE PAID BY ADDRESSEE

ATTN: ___MARKETING_____

**Brooks/Cole Publishing Company
511 Forest Lodge Road
Pacific Grove, California 93950-9968**

FOLD HERE

CUT ALONG DOTTED LINE

FOLD HERE

NO POSTAGE
NECESSARY
IF MAILED
IN THE
UNITED STATES

BUSINESS REPLY MAIL

FIRST CLASS PERMIT NO. 358 PACIFIC GROVE, CA

POSTAGE WILL BE PAID BY ADDRESSEE

ATTN: MARKETING

Brooks/Cole Publishing Company
511 Forest Lodge Road
Pacific Grove, California 93950-9968

FOLD HERE